Among the review chapters are wide-ranging, state-of-the-art exam[ina]-tions of the population dynam[ics] crustacean and molluscan reso[urces] —on scallops and penaeid sh[rimp] fisheries, for example—as well a[s] special topics such as fisheries [fore]casting, the philosophy of fish[ery] regulation, and the principles of gear (lobster trap) design.

Marine Invertebrate Fisheries will be of practical interest to a broad spectrum of readers, ranging from those in the fishing industry to the academic community, as well as members of the public concerned with marine ecology and conservation. It will be particularly valuable to experts from a range of disciplines actively engaged in marine resource management and research who will be able to draw on a diversity of examples of the practical application of marine biology to human affairs.

About the editor

John F. Caddy has been actively involved in fisheries research and its application to managing fisheries since he completed his doctoral research on the ecology of an estuarine shellfish at the University of London twenty years ago. Much of his professional career with the Government of Canada was concerned with the population dynamics of marine invertebrates, and he is the author of numerous scientific papers on this subject. He is currently employed as Senior Fishery Resource Officer in the Food and Agricultural Organization of the United Nations in Rome, and permission from the Organization to undertake this work is gratefully acknowledged. In his present position, he has been able to draw upon his experience in a wide range of national and international fisheries in selected chapters that illustrate the main research approaches currently used in invertebrate fisheries today.

corals, and rare shell harvests. Examples are drawn from tropical and temperate seas, and from arctic and antarctic ecosystems.

MARINE INVERTEBRATE FISHERIES: THEIR ASSESSMENT AND MANAGEMENT

MARINE INVERTEBRATE FISHERIES: THEIR ASSESSMENT AND MANAGEMENT

Edited by
JOHN F. CADDY
FAO
Rome, Italy

WILEY

A WILEY-INTERSCIENCE PUBLICATION

JOHN WILEY & SONS

New York / Chichester / Brisbane / Toronto / Singapore

Library of Congress Cataloging in Publication Data:

Marine invertebrate fisheries: their assessment and management/
  edited by John F. Caddy.

    p. cm.
  "A Wiley-interscience publication."
  Bibliography: p.
  Includes index.
  ISBN 0-471-83237-5
  1. Shellfish fisheries.   I. Caddy, John F.
SH370.M37 1988
333.95'5--dc19   87-32436
                                                          CIP

Printed in the United States of America

10 9 8 7 6 5 4 3 2 1

CONTRIBUTORS

SADAO AOYAMA
Aquaculture Center
Aomori Prefecture
Aomori-Ken, Japan

RICHARD F. J. BAILEY
Department of Fisheries and Oceans
Institut Maurice-Lamontagne
Ste-Flavie, Québec, Canada

DAVID B. BENNETT
Ministry of Agriculture, Fisheries and
   Food
Lowestoft, Suffolk, England

CARL J. BERG, JR.*
Marine Biological Laboratory
Woods Hole, Massachusetts

R. BODDEKE
Netherlands Institute of Fishery
   Investigations
IJmuiden, The Netherlands

B. K. BOWEN
Fisheries Department
Perth Western Australia, Australia

KEITH M. BRANDER
Ministry of Agriculture, Fisheries and
   Food
Lowestoft, Suffolk, England

J. BRAVO DE LAGUNA
Instituto Español de Oceanografia
Madrid, Spain

R. S. BROWN
Western Australian Marine Research
   Laboratories
North Beach, Western Australia,
   Australia

JOHN F. CADDY
Fisheries Resources and Environment
   Division
Food and Agriculture Organization
Rome, Italy

ALAN CAMPBELL
Department of Fisheries and Oceans
St. Andrews, New Brunswick, Canada

Present address: Florida Department of Natural Resources, Bureau of Marine Research, Marathon,
Florida.

N. CAPUTI
Fisheries Department
Western Australian Marine Research
  Laboratories
North Beach, Western Australia,
  Australia

J. STANLEY COBB
Department of Zoology
University of Rhode Island
Kingston, Rhode Island

C. CONAND
Université de Bretagne Occidentale
Brest, France

ERIC EDWARDS
Shellfish Association of Great Britain
London, England

NELSON M. EHRHARDT
Rosenstiel School of Marine and
  Atmospheric Science
University of Miami
Miami, Florida

ROBERT W. ELNER
Department of Fisheries and Oceans
Halifax, Nova Scotia, Canada

MICHAEL J. FOGARTY
National Marine Fisheries Service
Woods Hole, Massachusetts

J. FRÉCHETTE
Sous-ministariat aux Pêches
  Commerciales
Québec, Canada

CARLO FROGLIA
Istituto Ricerche Pesca Marittima
  (CNR)
Ancona, Italy

S. GARCIA
Food and Agriculture Organization
Rome, Italy

RICHARD W. GRIGG
Hawaii Institute of Marine Biology
University of Hawaii at Manoa
Coconut Island, Kaneohe, Hawaii

N. G. HALL
Fisheries Department
Western Australian Marine Research
  Laboratories
North Beach, Western Australia,
  Australia

D. A. HANCOCK
Marine Research Laboratory
Waterman
Perth, Western Australia, Australia

VICTOR S. KENNEDY
University of Maryland
Center for Environmental and Estuarine
  Studies
Cambridge, Maryland

EDWARD F. KLIMA
Southeast Fisheries Center
Galveston, Texas

JAY S. KROUSE
Department of Marine Resources
West Boothbay Harbor, Maine

RICHARD D. METHOT*
Southwest Fisheries Center
La Jolla, California

J. L. MUNRO†
International Center for Living Aquatic
  Resources Management
Townsville, Queensland, Australia

Present address: Northwest & Alaska Fisheries Center, Seattle, Washington.
†*Present address:* International Center for Living Aquatic Resources Management, Coastal Aquaculture Centre, Honiara, Solomon Islands.

MAMORU MURATA
Hokkaido Regional Fisheries Research
  Laboratory
Kushiro, Hokkaido, Japan

STEVEN A. MURAWSKI
National Marine Fisheries Service
Woods Hole Laboratory
Woods Hole, Massachusetts

DAVID A. OLSEN
Fort Lauderdale, Florida

D. G. PARSONS
Science Branch
Department of Fisheries and Oceans
St. John's, Newfoundland, Canada

J. W. PENN
Fisheries Department
Western Australian Marine Research
  Laboratories
North Beach, Western Australia,
  Australia

B. F. PHILLIPS
CSIRO Marine Laboratories
North Beach, Western Australia,
  Australia

VICTOR RICARDO RESTREPO
Rosenstiel School of Marine and
  Atmospheric Science
University of Miami
Miami, Florida

DIETRICH SAHRHAGE
Institut für Seefischerei
Bundesforschungsanstalt für Fischerei
Hamburg, Federal Republic of Germany

FREDRIC M. SERCHUK
National Marine Fisheries Service
Woods Hole Laboratory
Woods Hole, Massachusetts

N. A. SLOAN
Pacific Biological Station
Nanaimo, British Columbia, Canada

MIA J. TEGNER
Scripps Institute of Oceanography
La Jolla, California

S. M. WELLS
IUCN Conservation Monitoring Centre
Cambridge, England

PREFACE

The past few decades have witnessed the rapid development of methods of assessing and managing renewable marine resources, methods that until the 1960s were almost exclusively applied to finfish populations. The management of most invertebrate fisheries has until recently depended heavily on empirical knowledge and procedures, supplemented in the scientific literature by predominantly qualitative studies of these organisms and their rules in marine ecosystems. The 1970s and 1980s, perhaps beginning with the Special ICES Symposium on Population Assessment of Shellfish Stocks held in 1976 (Thomas 1), have seen considerable progress in our knowledge of the dynamics of invertebrate stocks and the application of this knowledge to fishery management. Much of this recent work, however, is widely scattered and difficult for the general reader to access.

Recently there has been a better appreciation of the significance of marine invertebrates in all aspects of human relationship with the sea. They play a key role as components in marine food chains, and this aspect is dealt with in, for example, Chapter 1 on krill; as substrates (e.g., coral reefs) for important ecosystems; and in aquaculture. Their economic importance as food, as ornaments, as items for international trade, and as a source of hard currency for developing countries, is emphasized by a high unit value paid to fishermen at the dockside. Even a superficial inspection of FAO world marine fishery statistics shows that their generally high unit value largely compensates for the smaller landed weights of mollusks and crustaceans, when compared with the combined categories of demersal and small pelagic fish, for example. Thus, although recently shellfish made up only about 15% by landed weight of the world total for these two major categories of seafood combined, they formed at least 35% of the landed value, and their significance in international trade is probably still greater. In fact, exports of crustaceans, mollusks, and other marine invertebrates (excluding corals) account for nearly 40% of the value of world fishery trade. Their share goes up to 55% in the case of developing economies. The number of countries exporting these products, mainly to the markets of developed countries, has increased substantially in the last decade.

This economic importance calls for efficient approaches to management, and the two main sections of the book, Part I dealing with crustacean fisheries, and Part II with a variety of fisheries for mollusks, coelenterates, and echinoderms, summarize a variety of experiences by more than 40 experts in the field, which will be useful to the marine biologist, the fisheries economist and manager, and to others who have an interest in what is commonly referred to nowadays as ocean-ology. The chapters also provide access to a unique combination of relevant publi-cations, some of the most significant of which are in the "gray" literature.

The main objective of this book is to outline, through case studies and reviews, the current practices in management and assessment of "wild" stocks of marine invertebrates, as well as what seem to be the most promising approaches for the future. In the process, it is necessary to assess the extent to which standard methods developed for finfish stock assessment and management can be applied to the wider variety of invertebrate resources. It is beyond our scope here to review the histor-ical development of this subject area, but several landmarks should be mentioned. The theoretical foundation provided by Beverton and Holt (2) continues to be relevant to a study of the dynamics of finfish and invertebrate populations, as does the theory of stock and recruitment proposed by Ricker, and the concept of the surplus production model associated with the names of Graham and Schaefer. All three approaches have been applied to shellfish populations and, with other "global" methodologies, are succinctly described in the fishery manuals of Ricker (3) and Gulland (4). These texts are useful for training workers in numerical methods both for finfish and invertebrates, and would be useful reference material while reading this volume, which concentrates on describing applications rather than on theory.

There can be a few researchers on invertebrate populations, however, who have not experienced difficulties in following procedures that often seem incompatible with existing information on the species in question, or that cannot incorporate types of information that logically seem pertinent to the questions asked by resource managers. One explanation of why this should be so stems from the pervasive influence of the case study in the development of natural sciences. We do not always appreciate what constraints the first well-documented examples in any field of endeavor have on the approach adopted for later problems, which may not prove at all comparable on further study. In the case of the methodologies mentioned above, the species first studied certainly influenced the theoretical structure that evolved. The work of Beverton and Holt (2), for example, was based on infor-mation on North Sea groundfish, particularly plaice and haddock, relatively slow-growing, long-lived species, where ease of sampling and aging favored the devel-opment of age-based methods. Questions of stock and recruitment are of paramount importance to management of salmon resources, and data on the size of adult runs and subsequent smolt production for North American salmon rivers led to the dome-shaped stock–recruit curve. Similarly, the generalized production model was first applied to Pacific tuna; a resource where the difficulty of aging individual fish called for modeling the stock in terms of undifferentiated biomasses and their relationship to yield and fishing effort. For animals such as crustaceans, where individual age cannot be determined from skeletal structures, length as opposed to age-based methods of population analysis have long been in use, and some of these approaches will undoubtedly find a wider application to shellfish populations, for example, to problems of major importance for the farming of marine mollusks.

Several chapters of this book point to comparable advantages, practical and theoretical, that may be expected from a further study of the dynamics of invertebrate populations. The transfer of ideas in the reverse direction, from invertebrate ecology to finfish studies, has already occurred in several occasions, for example, the concept of the community of organisms, developed in the study of marine benthos by Thorson (5). Here, owing to the relative ease of sampling sedentary organisms, information on the pattern and scale of distribution of organisms and, potentially at least, the effects of density on growth and mortality should be easily studied for shellfish populations.

In conclusion, although the current spectrum of stock assessment procedures is probably adequate for coarse-grained management of most types of marine organisms, more refined and cost-effective approaches will require an understanding of the life history, environment, and ecological interactions of the organism in question. It is hoped that this volume will prove a useful step in this ongoing process.

REFERENCES

1. H. J. Thomas, Ed., Population assessment of shellfish stocks. A special meeting held in Charlottenlund, Denmark, 29 Sept–1 Oct 1976. *Rapp. P.-V. Reun., Cons. Int. Explor. Mer*, 175:1–288 (1979).
2. R. J. H. Beverton and S. J. Holt, On the dynamics of exploited fish populations. *Fish. Invest.—Minist. Agric. Fish. Food. (G.B) (Ser. 2)* 19:1–533 (1957).
3. W. E. Ricker, Computation and interpretation of biological statistics of fish populations. *Bull. Fish. Res. Board. Can.* 191:1–382 (1975).
4. J. A. Gulland, *Fish Stock Assessment: A Manual of Basic Methods*. Wiley, New York, 1983, 223 pp.
5. G. Thorson, Bottom communities (sublittoral and shallow shelf). *Mem.—Geol Soc. Am.* 67(1): 461–534 (1957).

JOHN F. CADDY

Rome, Italy
March, 1988

CONTENTS

PART II MOLLUSCAN AND OTHER INVERTEBRATE HARVESTS

PART I
CRUSTACEAN FISHERIES

OVERVIEW OF CRUSTACEAN FISHERIES: ASSESSMENTS AND POPULATION DYNAMICS

The case study approach that forms the basis for this volume follows inevitably from the wide diversity of approaches currently being followed in assessment and management of invertebrate resources. A premature synthesis and extraction of a few "grand principles" for the application of population dynamics theory to highly diverse invertebrate populations therefore seems inappropriate at this early stage. This is not to say that there are no underlying themes in invertebrate assessment, and an attempt will be made to bring out some of these in the editorial overviews preceding the two parts of the book. At the same time there are risks in presenting a superficial summary when the reader is best equipped to form his or her own conclusions from a reading of the case studies and reviews presented here as to what approaches seem most appropriate in each situation.

Although crustacean fisheries can play a role in contributing to human protein needs nowadays, as for many other of the resources described in this book, their prime importance is in the economic context—as high unit-value marine products. In many coastal states in the tropics for example, penaeid shrimp and spiny lobster resources play an important role in contributing to export revenues; this role remains renewable, however, only if fisheries management action is based on an understanding of the resource dynamics of these crustacea and the characteristics of the harvesting sector.

EXPLOITATION OF CRUSTACEAN POPULATIONS IN THE ECOSYSTEM CONTEXT

The first of our case studies deals with what at first sight may seem an atypical invertebrate fishery, that for a pelagic crustacean, the Antarctic krill *Euphausia superba*. This fishery has been carried out in recent years by large distant-water vessels in the frigid waters off Antarctica. Krill form a key component of most

3

food chains leading to Antarctic mammals and birds, and, as such, not only does their rational exploitation have to take into account the sustainable yield concepts discussed in later chapters, but also a societal choice at a global level is implied in setting harvesting levels if populations of other organisms also feeding on it are not to be seriously affected. As a case study therefore, this fishery is also a reminder of the role played by those other small crustaceans (copepods, mysids, amphipods, and others) in marine ecosystems. Here a short life-span and rapid biomass turnover cast these organisms in key roles as pathways from plant material to "higher" organisms such as finfish, which usually depend on crustaceans for food at some stage in their life histories.

The significance of smaller crustaceans in the multispecies context alluded to above extends to the shrimps and prawns discussed in the following chapters. Thus any approach to the elucidation of their population dynamics should ideally integrate information both on the crustacean target species and on the abundance and fisheries for their predators. This consideration is placed in an explicit management framework in Chapter 8, which deals with the species complex fished by European Common Market (EEC) trawler fleets in the Irish Sea. Here the Norway lobster *Nephrops norvegicus* is both the highest unit-value species and a prey to several alternative target species of the trawl fishery. This does not necessarily imply managing the resource to just maximize yield of *Nephrops*, however, although this is one of the first options that comes to mind. Different national fleets have different species and market preferences, and a high degree of cooperation is required in the final critical negotiation of shares. This is possible only in theory, following an assessment of potential yield for the complex of trophically linked species (see below).

Although ecosystem interactions involving crustaceans have been long recognized as important, we are still at an early stage in understanding what appears to be a complex process: the management of multispecies resources that are linked trophically to one another. This implies a balancing of harvesting rates for each component of the food web so as to optimize economic returns to the fullest extent possible while maintaining the integral structure of the food web intact; a possibility that is largely theoretical at this time. Chapter 2 dealing with the small brown shrimp *Crangon crangon* of the North Sea (also a key forage species for fish such as cod and whiting), Chapter 3 on the pink shrimp stocks, and Chapter 12 on spider crab resources of the Northwest Atlantic, all suggest that stock size may in part be a function of cod biomass. These examples all suggest how trade-offs may be implied between the way that different fisheries are prosecuted in the same area. These relationships ideally need to be taken into account in assessing and managing fisheries for most small crustaceans such as shrimp and prawns, as well as for some larger crustaceans whose juvenile stages are prey for commercial fish (e.g., Chapters 9 and 12). Again, societal choices must be made before fisheries scientists can offer advice on the feasibility of attaining a given species mix that may be considered "optimal." Clearly, there must be a broader diffusion of information on the ecological linkages involved and on the potential use of these concepts in management of invertebrate resources. In 1986 we were still in the early stage of this process, and despite the proliferation of ecosystem theory (summarized, for example, in Caddy and Sharp 1), there are a few if any marine fisheries today that can be pointed to as examples where balance has been achieved in practice.

The other related set of problems that are often confounded with species interactions are what may be referred to as "technological interactions." A class of these of particular interest to shrimp fisheries is commonly referred to as the "by-catch problem," and examples are given in Chapters 4 and 13 for penaeid shrimps, the interaction between pandalid shrimp and redfish fisheries is mentioned in Chapter 3, and similar problems in relation to lobster and crab fisheries are mentioned in Chapters 8, 11, 12, and 15. The need to promote the use of gear types that minimize discarding of undersized shrimp (Chapter 2) or reduce the capture of rare or endangered species (such as marine turtles) in the by-catch (Chapter 4) is another question that falls under this heading. A by-catch problem and the rather sophisticated mechanical devices used to reduce its impact are described in Chapter 2; in this situation, strangely enough, the shrimps are less valuable than the juvenile fish by-catch. In cases where shrimps may coexist with nursery areas for other valuable species (or vice versa), season and area closures (e.g., Chapters 4 and 13) are alternatives that if correctly defined, may minimize the capture of juvenile shrimps before they reach their full commercial potential.

BIOLOGICAL CHARACTERISTICS OF POPULATIONS

Although our knowledge of a stock or species is rarely adequate for its proper management before exploitation begins, significant increases in our knowledge of formerly little known species, as described in Chapter 12, can result from a management-oriented research program introduced early on in a fishery. The snow crab fishery in the Gulf of St. Lawrence described in that chapter followed from incidental catches of these crabs being regarded as a nuisance in the trawl fishery for demersals, but snow crabs can now be taken legally only in traps. Recently molted males and all females are discarded at sea, and in fact there is a closure in some areas during the peak molting season. Females are protected and males are sexually mature below the size limit, which would suggest that this resource is not sensitive to overfishing. There is now some debate, however, on the relative importance of the sizes of physiological and functional maturity, and large commercial-sized males may have to be present in adequate numbers for successful reproduction to occur. In a somewhat different context, what is effectively a terminal molt in some crustaceans means that yield calculations should ideally be carried out with discontinuous growth functions. These two points perhaps serve to illustrate the point made many times elsewhere in this volume, that population modeling should follow, rather than precede, a knowledge of species biology.

The problems of modeling fisheries-related processes for organisms showing such special features are obviously not insignificant, and evidently may impose new data requirements; this is touched upon in Chapters 6 and 15. One complication to be faced in modeling shrimp populations is the fact that many species are protandric hermaphrodites: spawning, molt, sex inversion, and migration may all be cross-scheduled, as for *Crangon crangon* (Chapter 2). The ages of sex reversal and of female maturity may also be delayed latitudinally, as for pink shrimps (Chapter 3). Another extreme example of the need to consider particular life-history characteristics in building models is the Florida stone crab fishery (Chapter 10). Here, fishing mortality *sensu strictu* is replaced by a (considerable) discard mortality for the captured crabs, which the fisherman is legally required to throw back at sea after "declawing." A novel yield-per-recruit model, again assuming discrete growth, is proposed as a guide to management of this fishery, where yield comes both from first entrants to the fishery, and from those that have regenerated claws. The importance of proper discard practices is also touched on in Chapters 7 and 11.

IMPACT OF ENVIRONMENTAL FACTORS ON PRODUCTION

The debate between the relative effects of excessive effort and of environmental factors such as rainfall on shrimp recruitment still continues, with the balance perhaps now tipping toward environment as opposed to parent stock size as the main controlling factor for penaeid shrimps except under conditions of extreme stock depletion (Chapter 13); this generalization may also apply for some other heavily fished resources such as homarid lobsters (Chapter 6). Although stock and recruit relationships have proved useful in understanding impacts of fishing on population replacement (e.g., Chapters 5 and 15), caution must be exercised in acting on this tentative conclusion. However, given that under uncontrolled fishing, spawning stock size is usually maintained at very low levels for high unit-value crustaceans, this leaves a very narrow "safety margin" for stock replenishment.

The key role of various aspects of the physical environment and their impacts on crustacean fisheries is also alluded to in a number of cases. Chapter 3, on the pink shrimp *Pandalus borealis*, illustrates this feature well: an arctic species, in the western Atlantic it extends from Greenland to the Gulf of Maine; its life cycle and its annual yield vary latitudinally and through time to reflect year-to-year changes in the environment (principally temperature), in a way that depends to some extent on the position of the stock within the species range.

Chapter 4 discusses another type of environmental impact, namely, that of rainfall and subsequent runoff of fresh water into estuaries on the production of penaeid species in the topics. Here as in later chapters, the relative impacts of fishing effort and environmental fluctuations are discussed, with of course a more immediate and pronounced effect of such fluctuations on yield being evident for short-lived species and for those multiple age group species that are heavily exploited (Chapter 31). The importance of oceanographic conditions in allowing exploitable populations of crustaceans or other marine organisms to develop in a given area is now being fully appreciated; as for the pink shrimp, the presence of a gyre or other oceanic feature leading to retention of water masses in the region may be essential for a stock with pelagic larvae to be self-sustaining. Other resources, such as the Dungeness crab fished off the Pacific coast of North America (Chapter 9), show strong historical oscillations in landings, which may be tied to widespread oceanographic events, such as the El Niño phenomenon. Although the "proximal factor" causing these regular fluctuations is not yet fully elucidated in this case, the key management question is whether it is possible to manage a stock showing such strong oscillations, using steady-state models and management measures. The author of Chapter 9 concludes that the answer is probably no, although obvious management strategies such as pulse fishing have not received universal acceptance. Also essential for sustained fisheries production of species that pass a significant part of their life histories in coastal areas, such as many penaeid shrimp, is the protection of coastal nursery areas in the face of pressure from other human uses of the littoral and subtidal zone; this question is touched on in Chapter 4.

STATISTICS

The need for a continuous series of up-to-date and complete statistical information in managing fisheries is well understood, and Chapters 7 and 16 give some insights here. In addition to logbooks, which are used in the multispecies penaeid shrimp fisheries conducted distant from Australian ports (Chapter 5), other distant-water fleets of factory trawlers such as those fishing shrimps off Labrador are required to carry at-sea observers, as used in monitoring other distant-water fisheries (Chapter 3). For fisheries closer to home, port interview systems and port sampling procedures are widely used, as well as analysis of returns from processors and fish dealers (Chapters 4, 5, and 7). Unfortunately, even in the best of commercial sampling schemes, sources of bias must always be checked for, and research survey cruises are often essential for monitoring stock abundance. Some approaches to surveying crustacean stocks are touched on in Chapters 1 and 3.

The importance of maintaining good statistics and a good exchange of information between managers and fishermen has been a key element in successfully managed shellfish fisheries, and this perspective is well described in Chapters 5 and 11. The thoughtful review of Western Australian fisheries presented in Chapter 16 shows that statistics gathering must also be an intergral part of management as well as research.

TECHNOLOGICAL FACTORS INCLUDING GEAR SELECTION

Many crustaceans are captured in traps, which involves active behavioral responses by the exploited species; these can vary with a variety of physiological factors. Chapter 14 provides a review of some of these considerations. Thus variations in catchability can be a function of temperature, the stage of molting and maturity, availability of food, type of bait, state of the tidal and diurnal cycle, and many other factors. This can pose problems in interpreting catch and effort data, which unfortunately are often the only data available for most crustacean trap fisheries. The use of underwater photography and direct observation techniques is referred to in Chapters 3 and 14, and some of the problems faced in carrying out other field procedures, such as the interpretation of tag and recapture experiments and other special monitoring schemes (e.g., Leslie–Delury methods) for a species that grows by periodic molting, are discussed in Chapters 6 and 12.

Of potential importance for increasing fishery yields are several considerations that, although technologically simple, can have a major impact. Examples touched on are the reduction of damage in discard of small lobsters by better handling practices on board (Chapter 7), the provision of escape gaps in traps to allow small individuals to escape on bottom, or incorporation of biodegradable panels to reduce "ghost fishing" by lost traps (Chapter 14). There is also evidently considerable scope for technologically more sophisticated approaches to minimizing discard damage, such as those used in the Dutch shrimp fishery (Chapter 2), once the importance of these often "invisible" impacts is better appreciated. All these measures are aimed at significantly increasing yield of both crustaceans and by-catch species, or conserving the latter in some cases. In the contrary sense, minor technological improvements or changes in fishing strategy, such as those allowing fishing on deeper "refugia" for larger mature female lobsters in the Bay of Fundy (Chapter 6), could, if uncontrolled, seriously reduce spawning stock size and might be more important then any positive impact that may be hypothesized as a consequence of a law protecting berried females, for example.

STOCK ASSESSMENT

Stock assessment approaches in crustacean fisheries have of course taken and applied the lessons learned in finfish assessment. To the classical dichotomy between yield-per-recruit models or analytical approaches on the one side, and production modeling on the other, have been added several hybrid methods of analysis that take elements from these two formerly separate streams. To these have been added the ability to model simple or complex biological and fishery situations, using the as yet largely unused capabilities of the computer, thus allowing a comparison of model conclusions under various management options with the historical reality embodied in the statistical data series. Examples of more or less continuous changes in assessment methodology are numerous in recent years, at a time when numerical techniques have evolved and continue to evolve rapidly. For instance, in the Western Australian shrimp fishery (Chapter 5), emphasis in research on stock assessment has moved from surplus production models, via concern with size limits (yield-per-recruit analysis), to application of spawner–recruit models. The history of research on pink shrimp stocks has also shown a steady evolution from general to more species-based approaches; yield approximations derived from simple production modeling concepts have preceded use of the analytical catch equation. It is now becoming recognized, however, that uncritical application of yield-per-recruit and production models without taking into account life history characteristics can lead to erroneous results. Techniques developed for finfish analysis, such as cohort analysis, have been used on some crustacean stocks, but face serious problems owing to difficulties in aging the catch.

For many fisheries, resource biologists and managers appear to be marking time, using simple approaches in the interim until more appropriate methods are available. Some new approaches are described in various chapters in this book; the need remains, however, to compare new approaches against old ones in an attempt to see how consistent they are in their predictions and how well they fit into a management framework. As noted above, approaches currently used in crustacean fishery management are generally simple. Experience shows that although better or alternative approaches to population analysis are always welcome, it is the readiness to take management action as a result of threatening changes, rather than because of the sophistication of the model used, that makes for a properly managed fishery.

One of the main difficulties faced in many crustacean fisheries, especially those using passive gear such as traps, is in defining effort units for both commercial gear and research sampling equipment. Problems in effort measurement have also led directly to uncertainties in estimating biomass of stock components, and are also one reason why resource surveys using trapping are uncommon. These problems have led to heavy reliance on yield-per-recruit analysis to assess impacts of size limit and effort changes. In approaching analysis of historical catch (e.g., Chapter 6), past undocumented effects of technological changes have made for difficulties in detecting biologically relevant events. This is the case, for example, with the long series of historical data on fisheries in the maritime provinces of Canada for the lobster *Homarus americanus*, where 100 or more years of landing data have now accumulated. These effects included introduction of trap haulers, depth sounders, and radar and loran navigation and position-finding equipment; but more subtly, new management measures such as limited licensing and trap limits must have also had an impact on the mean catch rate experienced and the effort exerted.

Yield-per-recruit analyses have now begun to take into account molt frequency/ increment data, and together with information on size at maturity and fecundity, have led directly to more relevant formulations that emphasize recruitment overfishing rather than yield overfishing impacts. Parallel to this, and leading to more biologically relevant modeling approaches, are developments of the life table concept mentioned in Chapter 15. Egg-per-recruit models also allow us to take into account the close coordination of molt and maturation schedules in crustaceans (e.g., Chapter 6). New approaches that are beginning to be adopted in several crustacean fisheries result from the recent revival of interest in adapting old methods of analyzing size frequencies to the microcomputer, and inventing new ones.

The definition of unit stocks (management units) still poses a major problem for many crustaceans that often have extended life histories, are difficult to sample at critical stages, and show extensive migrations. This is one reason for the little progress made to date in defining spawner–recruit relationships. For coastally continuous resources such as many crustaceans, stock discrimination is a particular problem (Chapter 6), although as for other marine organisms, many stocks (e.g., rock lobsters, Chapter 7) appear to be associated with oceanic gyres.

FORECASTING

For most heavily exploited crustacean resources, it is apparent that the success of the next year's fishery largely depends on incoming year class size, and as we have noted, this responds to environmental forcing functions. The ability to make even a rough forecast is extremely helpful to all concerned in the fishery, and the Western Australian rock lobster fishery (Chapter 7) is the best example for crustacean fisheries of how time series of catches of larvae on artificial collectors can permit fisheries planning and sensible investment by the industry (see also Chapter 23 for a similar and even longer established system for mollusks). Other indexes of incoming year class strength are commonly used; thus in the Gulf of Mexico (Chapter 4) recruitment prediction is possible from sport fish landings of juvenile shrimps.

MANAGEMENT MEASURES

The strategy of combining a knowledge of the resource with a suite of simple effective measures that are designed to maintain or improve the status quo, and are agreed to by the industry, seems the secret to success in invertebrate resource management, recognizing that for most valuable shellfish resources, "improved" measures for development and management may not be readily accepted if they affect the current allocation of benefits within the industry. This is illustrated by Chapter 11 on British crab fisheries, which also illustrates a common feature of most invertebrate fisheries, namely, that growing market demand is usually a major factor in increasing exploitation rates.

A wide range of measures has been applied or discussed for managing penaeid shrimp resources; for example, mesh-size regulations, seasonal closures, effort limitation, limits on vessel efficiency and on access, monetary measures, and biological measures such as habitat enhancement. Most of these approaches and their application in specific fisheries are discussed in Chapters 4 and 5, and are reviewed in a more general fashion for penaeid shrimps in Chapter 13.

With respect to effort limitation, shellfish fisheries have probably had a longer history of this type of measure than most other types of fishery, and the main features of limited-entry systems are discussed in several chapters (especially Chapter 16). In practice, it seems that this approach has most notably been a success if introduced during the early development of the fishery, hence without causing reduced access to the resource by those already in the fishery; this is the situation when such a measure is introduced in well established, formerly open-access fisheries. Serious problems can result if limited licensing comes after the fishery has already been overfished and overcapitalized, a situation that Chapter 13 indicates is characteristic of most of the world penaeid shrimp fisheries. Vessel replacement should be by boats with comparable power (or it may be required that vessel power be significantly reduced on replacement; see Chapter 5). The generally negative or at best indifferent impact of government subsidies and grants on an overexploited fishery in perpetuating a situation of overcapitalization emerges from several chapters. On the other hand, buyback of surplus effort by the fishing industry is likely to be agreed to only if the latter is convinced of its efficacy and of the industry's own continuing share in any improvements resulting, and if acceptable returns continue to be made. From the socioeconomic perspective, problems of allocation between license holders and those wishing to enter a lucrative fishery can cause social tensions in small coastal communities, which cannot be solved by fisheries assessments alone. This aspect is discussed in Chapter 13, which also touches on the problem of deciding between an overall quota and individual boat quotas within a limited licensing format.

Various measures other than effort or quota control are being successfully used in crustacean fisheries management, in particular, a variety of mechanisms for restricting the fishery by area and/or season. These include closures during periods of molting (e.g., Chapter 12) and area closures of bays to protect sport fishermen (Chapter 4) and to deny access to trawlers. The use of seasonal and permanent closures of nursery areas to protect small shrimp may give economic benefits and the use of "artificial" obstacles to prevent trawling on nursery areas has even been proposed.

SHARING RESOURCES AND OTHER SOCIOECONOMIC CONSIDERATIONS

The problems of fishery management are exacerbated for resources that are "shared" between different jurisdictions, whether these are international (Chapters 1 and 8) or internal, for instance, in the state fisheries context (as in Chapters 4 and 20 for Mollusca). In both cases, problems of deciding on compatible regulations and their enforcement are among the most vexing of all those faced by fisheries managers in their complexity and difficulty of resolution. Usually a hierarchy of decisions and consultations is involved, in a context where other political considerations cannot be excluded. A good example here is the fishery for Norway lobsters in the Irish Sea (Chapter 8), which takes place in a multinational context as one of a series of directed fisheries for different species by member countries of the EEC within community waters. This context explains some of the difficulties in arriving at clear and agreed objectives for different target species that are associated trophically; as described earlier, a particular overall strategy may be optimal economically for EEC fisheries taken as a whole, but face difficulties because it affects the current balance of national shares of the resources concerned.

Simple management measures such as the application of mesh-size regulations can also be complicated in practice, because they also have implications for the allocation of yield between different sectors of society. This occurs, for example, in penaeid shrimp fisheries, where despite the increase in yield predicted if mesh size is increased, a balance usually must be maintained politically between socially disadvantaged coastal fishermen harvesting predominantly small shrimps, inshore or in estuaries with small mesh gear, and large-scale fleets of industrial boats that, although operating closer to the "optimal" size limit, are capital-intensive in operation and more oriented to export markets. Management usually opts for a mixture of strategies in these circumstances, but what are the relevant considerations? Chapter 13 spells out some of these in more detail.

In general, fisheries theorists agree that the point of maximum economic yield (MEY) occurs at a significantly lower level of total effort than maximum sustainable yield (MSY). Maintaining harvesting levels at or around MEY has not apparently met with much success for most crustacean fisheries, however, where high unit values and price elasticity discourage maintenance of the higher unfished biomasses that must be maintained in the sea with this strategy. The difficulties of maintaining a sustainable and economically productive harvesting strategy for open-access resources frequently lead to poor economic returns to developing countries from what seemed initially a promising source of revenue. The problems of overcapitalization and resource sharing noted earlier for penaeid shrimp fisheries are now further complicated by the impact of imports on domestic markets and prices. Shrimp imports now include a growing proportion of cultured shrimps, and on the supply side, conflicts for space and for "seed" shrimps in coastal fisheries are also emerging between artisanal fishermen and shrimp aquaculturists. For most crustacean fisheries, fluctuations in markets and costs of fishing (especially fuel) have had a dominant effect on fisheries where distances and seasonally limiting fishing conditions reduce profit margins. All these considerations call for increased emphasis on bio-economic modeling of invertebrate fisheries once the basic biological criteria for stock conservation are met.

REFERENCE

1. J. F. Caddy and G. D. Sharp, An ecological framework for marine fishery investigations. *FAO Fish. Tech. Pap.* **283:**1–152 (1987).

CASE STUDIES

A / SHRIMPS, PRAWNS, AND KRILL

1 ANTARCTIC KRILL FISHERIES: POTENTIAL RESOURCES AND ECOLOGICAL CONCERNS

Dietrich Sahrhage
Institut für Seefischerei
Bundesforschungsanstalt für Fischerei
Hamburg, Federal Republic of Germany

1. INTRODUCTION

The occurrence of Antarctic krill in the Southern Ocean had already been observed by Captain Cook in 1775, and since the time of Bellingshausen (1820) it was known that baleen whales and penguins feed on these crustaceans. Norwegian whalers used the word krill for the stomach content of whales but the first scientific description of the species *Euphausia superba* was made by Dana in 1850 from a specimen collected in 1840 on board the USS *Porpoise*.

Much more information became available on krill through extensive plankton hauls made during the British *Discovery* investigations between 1925 and 1939 and in the 1950s. A large number of valuable publications emanated from these studies, including the monograph by Marr (1). The interest in and subsequent development of a krill fishery has stimulated an intensification of related research since the early 1960s, particularly in the Soviet Union (Lubimova et al. 2) and in several other countries. Everson (3) reviewed our knowledge on krill and other Antarctic marine living resources in 1977.

Concern over the consequences of the exploitation of these resources on the stocks and on the marine Antarctic ecosystem as a whole led to the establishment of the "Group of Specialists on Living Resources of the Southern Ocean" under the auspices of the Scientific Committee on Antarctic Research (SCAR) and the Scientific Committee on Oceanic Research (SCOR). In 1976 the group developed plans for a large international research program called "Biological Investigations of Marine Antarctic Systems and Stocks" (BIOMASS). The purpose of this program is to gain a deeper understanding of the structure and dynamics of the Antarctic marine ecosystem as a basis for the future management of potential living resources (BIOMASS 4). An intensive study of the biology of krill was identified as a major objective of these international investigations. Besides various national seagoing activities and shore-based studies, two closely coordinated international multiship surveys were carried out under BIOMASS in selected areas of the Atlantic, Indian, and Pacific Oceans, involving 12 research vessels from 10 nations: The First International BIOMASS Experiment (FIBEX) during the 1980–1981 season (Hempel 5) and the Second International BIOMASS Experiment (SIBEX) in two phases during the 1983–1984 and 1984–1985 seasons (Sahrhage 6). These extensive studies resulted in a great increase in scientific knowledge on krill, as can be seen from recent publications, particularly those submitted to the BIOMASS Colloquium in 1982 in Tokyo (Nemoto and Matsuda 7), the International Symposium on Krill at Wilmington, Delaware in 1982 (George 8), the Seminar of the Krill Ecology Group in 1983 in Bremerhaven, Federal Republic of Germany (Schnack 9), and the Fourth SCAR Symposium on Antarctic Biology in Wilderness, South Africa, in 1984 (Siegfried et al. 10).

This improved knowledge is of special importance to the work of the Commission for the Conservation of Antarctic Marine Living Resources (CCAMLR), established under a convention in 1980 (CCAMLR 11, 12).

2. BIOLOGICAL FEATURES

2.1. Distribution

Distribution of krill is almost entirely restricted to the waters south of the Antarctic Polar Front (The Antarctic Convergence) around the southern continent. The species inhabits in particular the seasonal packice zone and its northern fringes characterized by large seasonal differences in ice cover and light conditions at continuously low sea temperatures and high nutrient supplies. Krill have been observed in waters with temperatures up to 4.6°C; however, optimal conditions occur below 2°C. Major concentrations of krill occur in the East Wind Drift, the region along the Antarctic Peninsula, the Bransfield Strait, around South Georgia, near the Kerguelen-Gaussberg Ridge, north of the Ross Sea, and in the Bellingshausen Sea (Fig. 1). Large concentrations are formed regularly during the spring–summer until early autumn season (November until April–May) for feeding and reproduction. Little is known so far on the distribution in winter, but krill have frequently been found under the ice. Feeding and reproduction seem to be especially advantageous in the region between the Antarctic Peninsula and the eastern Scotia Sea. Here a rather regular separation has been observed between large adult krill, living close to the Weddell–Scotia Confluence, and juvenile krill, occurring south and southeast of this zone (Lubimova et al. 2). (The Weddell–Scotia Confluence is the border zone between the Weddell and Scotia Sea water masses.)

Land and island configurations and bottom topography have a considerable influence on the distribution of krill. There is also no doubt that the circulation systems of water rings around the continent, the Antarctic Circumpolar Current and the East Wind Drift further south, are of great importance for the distribution of krill and the variability observed in their abundance (Amos 13). However, it is still not clear how and to what extent the structure and movements of water masses influence the horizontal and vertical distribution of krill. Observations have shown that krill tend to be concentrated in regions of frontal zones, meanders, and eddies, for example, in the Bransfield Strait, the area of the Weddell–Scotia Confluence, off Enderby Land, and in the Prydz Bay region. Such concentrations can be dense and rather stable for some time, whereas in regions with no pronounced frontal interaction, for example south of South Africa, krill appear to be more dispersed, forming widely spaced small swarms. According to Lubimova et al. (2) highest occurrences are observed in cyclonic circulation systems which may retain most of the krill in independent populations. The latter hypothesis has not been proved (see Section 2.6) but the movement of two krill swarms in a poleward direction has been observed (Kanda et al. 14). There are, however, other theories that attempt to account for the distributional control of krill. There could be a continuous drift of krill in a northeasterly direction with circumpolar movements. Another possibility would be that only a part of the population is drifting while the other component remains throughout life more or less in the same location (Marr 1). Furthermore, it is possible that krill undertake seasonal vertical migrations during

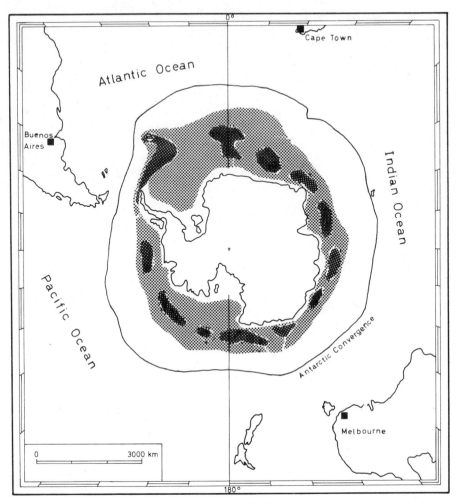

Figure 1. Distribution of Antarctic krill (*Euphausia superba*). Gray, krill moderately abundant; dark, areas of major concentrations.

which they could be carried in a southward direction in the warm deep water in winter, as are other Antarctic plankton.

2.2. Life Cycle

Spawning of krill occurs mainly from mid December until the end of February with annual variations (November–March) related to changing environmental conditions. Females carry spermatophores, and spawning takes place near the surface. Estimates of fecundity vary widely between 630 and 22,000 eggs. The number of eggs depends on krill length and maybe also on environmental conditions. Histological investigations have revealed that only a fraction of the eggs is spawned (Kikuno and Kawamura 15). Egg measurements indicated two groups of egg sizes in the same ovaries. Hence spawning may proceed in two phases or possibly eggs are resorbed, particularly under unfavorable conditions (Denys and McWhinnie 16; Siegel 17). The average number of eggs per female actually laid has been estimated to be 1000–8000.

After release near the surface the eggs sink to considerable depths where hatching occurs. Laboratory experiments indicated that at −1°C hatching takes place after 5–6 days, and that the eggs sink at a speed of 90–320 m/day. Thus it can be calculated that hatching occurs in depths of 1000–1600 m (Marschall 18). This is shallower than assumed by Marr (1) in his hypothesis for the developmental ascent of the young stages of krill. Experiments showed that egg development is inhibited at pressures higher than 200 atm.

The development of krill runs through 12 larval stages (see Table 1 and Fig. 2). Based on the vast material from the *Discovery* investigations, Marr (1) concluded that nauplii and metanauplii rise through the water column to reach the surface layers as first calyptopes larvae, which start to feed and develop through two further calyptopis and six furcilia stages to juvenile krill. Results of recent investigations confirmed Marr's hypothesis (Hempel 19). The overall development from spawning to late furcilias and juvenile krill takes about 1 yr. Calyptopis larvae already appear in January–February, often in large abundance and with a wide oceanic distribution. Guzman (20) observed high densities of furcilias V and VI and small juveniles under the packice in winter. It takes about another year before krill reach maturity at an average length of 33–35 mm for females or at 40 mm for males, which mature later than the females. In both sexes a "rejuvenation" (resorption of gonad products) after spawning and rematuration has been observed, so that it is likely that krill spawn more than once during their lives (Makarov 21; McWhinnie et al. 22).

Large variations in spawning and larval distribution occur because of the influence of environmental conditions, including changes in the ice cover, and variable success in the brood. Siegel (23) concluded that the krill of the Scotia Sea stem mainly from the spawning along the Antarctic Peninsula and in the Bellingshausen Sea. No successful spawning occurs around South Georgia. In the Weddell Sea south of 64°S, *Euphausia superba* was found only in the southeast along the continent as far as about 74°S, where calyptopis larvae also appeared in fair numbers close to the continent near Cape Norvegia. Low growth rates here seem to indicate poor living conditions. No *E. superba* were found in the southwestern and western parts of the Weddell Sea, where *E. crystallorophias* is abundant. Knowledge of egg and larval distribution in the southern Indian and Pacific Oceans is still rather scarce.

2.3. Behavior and Migration

Krill are usually found between the surface and 150 m, often with 90% of the population in water shallower than 65 m, particularly in summer if there is a

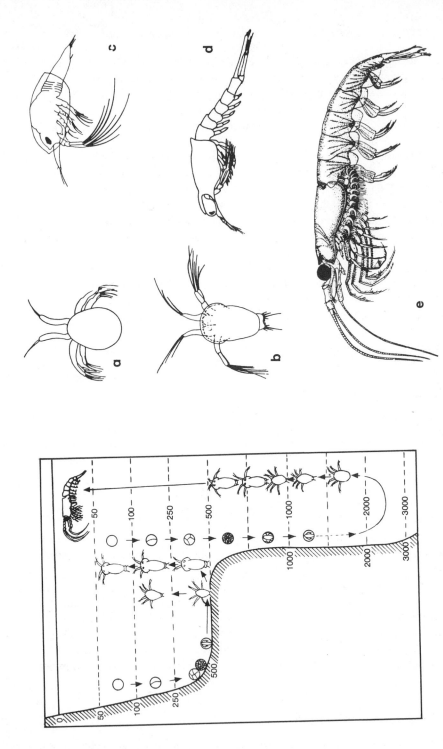

Figure 2. Development of Antarctic krill. (left) Developmental ascent after Marr (1); (right) *a*, nauplius I; *b*, metanauplius; *c*, calyptopis I; *d*, furcilia I; *e*, adult krill.

TABLE 1   Developmental Stages of Krill from Eggs to Adults

Stage	Estimated Age after Spawning	Size Range[a] (mm)	Other Observations
Egg	Until 5–6 days	0.6–0.7	
Nauplius I + II	6–14 days	0.6–0.8	First molting from nauplius II
Metanauplius	Over 13 days	0.7–0.9	to metanauplius
Calyptopis I	Average 23 days	1–2	Start feeding
Calyptopis II	Average 35 days	2–3	
Calyptopis III	Average 70 days	3–5	
Furcilia I–VI	2 months	F_1 3–7	
	Until 1 yr	F_2 4–8	
		F_{3-6} 7–16	
Juveniles	Second year of life	16–40	
Adolescents	Third year of life	30–46	
Adults	Mostly older than 3 yrs	35–63	Females maturing at ~35 mm

[a]Length measured from anterior margin of the eye to tip of the telson.

seasonal warmed surface layer with a thermocline and strong temperature gradients to the winter water layer underneath. Under these conditions, krill tend to be concentrated above the thermocline. However, some krill are also found deeper than 200 m, and are occasionally abundant in near-bottom shelf waters, where fish feed intensively upon them. It has been speculated that krill may descend to below 250 m in autumn after the season of feeding on phytoplankton is ended. However, latest investigations in winter have shown that krill concentrate in larger quantities closely below the pack ice.

Although the larvae are passively dispersed by currents over wide areas, the juveniles, as more active swimmers, may be able to move with some effect against the currents, and adult krill are able to move independently. Krill tend to concentrate in swarms so that the overall distribution is rather patchy but in the waters between the swarms, krill are also present in lower densities.

The formation of swarms is more pronounced in certain areas; however, the influencing factors are not yet well understood. Swarms occur both in oceanic regions and over shelves, often over the shelf break. Oceanographic conditions, for example, local upwellings, other mixing processes, and availability of food are important factors in swarming, in addition to biological determinants such as sex, maturity, and molt stage. The size of swarms varies mostly from a few to several hundred meters across. Macauley et al. (24) described a "super swarm" north of Elephant Island that covered an area of 450 km^2 and may have contained 2.1 million tons of krill. The density of krill in most swarms may be around 100–200 g/m^3 but maximum values of 15 kg/m^3 have been mentioned. Even within the same swarm large differences in density are observed at short distances and behavior is like that in a fog bank without stability. Often swarms contain animals with a definite body orientation, all moving in one direction at a speed of 15–20 cm/sec;

two of these swarms were followed over 46 and 116 miles (Kanda et al. 14). In escaping, krill can move up to 60–100 cm/sec; they have a very high metabolic rate and, as a constantly swimming organism, travel at a relatively high speed through most of their lifetimes, with consequent high energy requirements (Kils 25).

Factors determining the vertical distribution of krill appear complicated, and are poorly understood, but appear to depend on the time of the day and local conditions. Often there are regular diurnal migrations, with krill in deeper layers (50–200 m) during the daytime, and concentrations of mainly small krill at or near the surface at night. Larval krill seem to undertake a somewhat shorter migration than adults, the vertical distance migrated increasing later in development (Nast 26).

2.4. Growth and Mortality

Frequently different sizes of krill are observed separately in closely adjacent localities, which could possibly be due to age-specific differences in swimming activity. Generally, smaller krill are more abundant in shelf areas, whereas larger krill tend to be more numerous offshore.

Smaller krill are also often found under the ice. The length composition of krill in the catches shows considerable variations both within and between seasons, and it is therefore rather difficult to deduce estimates of growth and life-span from length frequency curves. On the basis of bimodal curves early investigators (e.g., Marr 1) assumed a longevity of only 2 yrs, but it was difficult to explain the occasional appearance of another "intermediate" size group (Mackintosh 27). New data analysis indicated that the intermediate group is a regular phenomenon and has to be regarded as one of the age groups. Calculation of growth rates is complicated by the observed "rejuvenation" after spawning (see Section 2.2) and by the finding, in laboratory experiments, that under unfavorable conditions krill undergo a body shrinkage. This has been interpreted as a possible overwintering strategy; however, this phenomenon has to yet be confirmed in nature.

Investigations by Ettershank (28) on the content of the fluorescent age pigment lipofuscin recently allowed the determination of the "physiological age" of krill. They suggested that beyond the larvae and juveniles there may be four to five age classes of adult krill. Three of these classes could be distinguished also by morphometrics, using discriminant analysis. Thus it is likely that krill are a rather long-lived species reaching an age of 6 or even 7 yrs.

Calculations show that growth of krill is rather fast during the first 2 yrs (Table 1), as apparent from the derived value of von Bertalanffy, $K = 0.4018$ (Siegel 23). Maximum length was calculated at $L_\infty = 63.8$ mm, which corresponds well with the maximum size observed at sea, and the average lengths of krill at the end of the second to the sixth year of life (November) may be about 30, 40, 46, 51, and 56 mm, respectively.

Starting with the development from nauplius II to metanauplius, krill undergo regular molts throughout life, including winter. Laboratory experiments revealed intermolt periods of about 20 days at temperatures prevailing in Antarctic waters (1°C). These intervals seem to depend not only on temperature but also on food availability.

Investigations on mortality are still scarce. In the absence of any larger krill fishery, total mortality equals natural mortality. Siegel (23) calculated a value of $Z = 1.0$, corresponding to a survival rate of 36%.

2.5. Food and Feeding

Krill are mainly herbivorous filter feeders. Feeding starts at the calyptopis I stage. There is no apparent difference between the food of the larvae and of the adults. Diatoms seem to play a major role in the food; however, their relative indigestibility is possibly a reason for overestimating their contribution due to the preponderance of diatom remains in the guts. Other food items are algae (particularly ice algae), dinoflagellates, tintinnids, radiolarians, foraminifera, and crustaceans. First investigations in the winter period indicated that detritus is another source of food, and so are bacteria. However, these early winter investigations show that krill is still feeding on low phytoplankton concentrations and on ice algae on the underside of icefloes.

Cannibalism and coprophagy have been observed occasionally but only in aquariums. The omnivorous character of feeding is also reflected in the spectrum of digestive enzymes.

The krill have a filter basket with a rather complex morphology and function (Kils 25) allowing selective feeding through the adaptation of filter sizes and application of different feeding strategies. Because krill need a high-energy diet, food requirements and filtration rates are high and the animals feed almost continuously. Thus food availability may be a dominant factor controlling krill distribution and behavior.

2.6. Stock Separation

In view of the existence of several cyclonic current systems as in the Weddell Sea and off Prydz Bay, the idea has been suggested that there may be a number of independent populations, perhaps with possibilities for some exchange between such stocks (Lubimova et al. 2). Analyses for protein variation using enzyme electrophoresis indicated, however, a single interbreeding population from the waters west of the Antarctic Peninsula throughout the Scotia Sea, eastward to the Prydz Bay region. Furthermore, no significant differences were found in morphometric data for krill from the Bellingshausen and Weddell Seas. Some contradictory results and the lack of data for the eastern Indian Ocean and the Pacific sector of the Antarctic Sea highlight the urgent need for further investigations, especially because this question is of great importance for the management of these resources.

3. POTENTIAL RESOURCES

3.1. Abundance and Biomass Estimations

Estimates of the standing stock of krill vary largely from some tens of millions of tons to 1350 million tons or even more. Because of the patchy distribution it is very difficult to assess the biomass from direct observations. An important question is what proportion of the total krill population is concentrated in swarms and what quantities are dispersed throughout Antarctic waters at low density.

Several attempts have been made to assess the biomass of krill in certain areas, both through quantitative net catches (e.g., with rectangular midwater trawls, RMT) and through hydroacoustic surveys. The largest projects were the FIBEX and SIBEX investigations under BIOMASS (see Section 1). During the echo-integration work of FIBEX in 1980–1981 the total krill biomass at the time of the survey was very tentatively estimated at 200–600 million tons (Hampton 29). A revision of the data on the basis of modified statistical analysis and target strength values led to much lower figures that highlighted methodological deficiencies and stressed the need for further reappraisal. Data from SIBEX are presently being evaluated.

Estimations on the basis of quantitative net sampling indicated large fluctuations between years but also within a season with highest values in summer and low values in spring. The biomass of 650,000 tons of krill calculated for the Bransfield Strait from RMT catches in summer (Siegel 23) is only slightly less than estimates made after hydroacoustic surveys for the same area, namely, ~900,000 tons (Guzman 20) and 1.3 million tons (Hampton 29). First results from SIBEX 1983–1984 and 1984–1985 in the Bransfield Strait from German (F.R.G.) surveys with RMT gave an estimate of 400,000 tons, whereas the Polish estimate of 40,000 tons from echo integration 2 months later (Kalinowski et al. 30) appears remarkably low. Although there are noticeable differences in the various estimates, which may be influenced by differences in the areas covered and in the dates of the surveys, in general there are considerable similarities in the results from echo integration and from net sampling, indicating at least the order of magnitude of krill biomass.

Relatively little is known so far on the abundance and distribution of krill in winter. In August–September 1983 Heywood et al. (31) found an extremely low abundance of krill around South Georgia and in the Scotia Sea but no significant differences between winter and summer catch values near Elephant Island. Guzman (20) described considerable catches of krill in winter in the Bransfield Strait and estimated the biomass at around 520,000 tons. Further investigations are required to quantify the krill biomass in winter as compared to summer and to examine in particular whether there is a large decrease in biomass owing to a massive death rate of adults in late summer after spawning as has been postulated.

The first direct estimate of krill biomass in the Antarctic, based on visual observations from a ship and assuming mean densities, was made by Marr (1) and resulted in figures of 44.5–521 million tons. Marr considered the higher figure as the more reasonable.

3.2. Productivity

Hitherto production of krill has been estimated only indirectly. Based on primary production, that is, an annual carbon fixation assumed to be on the order of 20–50 g C/m^2, and taking a conversion ratio of phytoplankton to krill of between 1 : 10 and 1 : 40, the krill production was calculated at about 100–500 million tons per annum. However, clearly no reliable value of annual primary production is yet available (Heywood and Whitaker 32). This production is extremely variable in different areas of the Antarctic seas, and it is now known that in the open parts of the Southern Ocean it is much smaller than originally assumed (Hempel 33). Another estimate of krill production (500–750 million tons), based on assumptions of the likely total zooplankton biomass in the Antarctic, may be even less reliable. The third method uses the annual consumption of krill by predators, estimated at figures of 400–500 million tons as described in Section 5.1.

A first attempt for a direct assessment of krill production has been made recently by Siegel (23) using data on length and age composition, mortality, and biomass of krill from the Bransfield Strait area. This work resulted in a very tentative estimate of roughly 600,000 tons/yr in the Bransfield strait, 3.4–7.0 million tons in the FIBEX area of the Atlantic sector, and 220 million tons in the whole Antarctic.

In summary, present figures for both the overall biomass and production of krill are still rather uncertain. It appears, however, that the production is relatively low as compared with the total biomass. This limitation must be taken into account in the exploitation and management of the krill resources.

4. THE KRILL FISHERY

4.1. Development of the Fishery

The existence of krill in large quantities in the Antarctic Ocean raised much interest in utilizing this resource as human food and for industrial purposes. Exploratory fishing for krill was started in 1961–1962 by research vessels from the Soviet Union. Japan has carried out fishery expeditions regularly since 1972–1973, and other countries, including Chile, the Federal Republic of Germany, Korea, Poland, and Taiwan, started exploratory fishing around 1975.

Regular large-scale krill fishing began during the 1976–1977 season stimulated by the development of the international law of the sea which resulted in the closure of traditional fishing grounds to distant-water fleets in other parts of the world. Until 1981–1982 annual krill landings rose to a peak of 528,000 metric tons; afterward there was a steep decline to about 220,000 tons in 1982–1983 and only

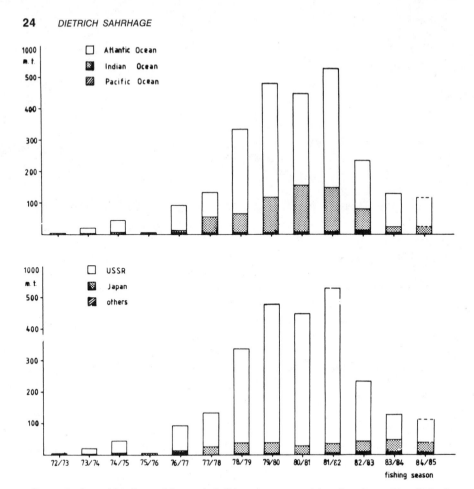

Figure 3. Annual landings of Antarctic krill by ocean areas (above) and countries (below).

128,000 tons were landed in 1983–1984 (Fig. 3). This decline was due entirely to a reduction in the fishery by the Soviet Union because of difficulties in the processing and marketing of krill. In contrast, Japanese landings increased to about 40,000 tons (1984–1985). The share of the Soviet Union in the total annual krill landings decreased from 87% in earlier years to 58% in 1983–1984, whereas the share of Japan increased from about 6 to 39%.* Small quantities were caught by vessels from Poland, the German Democratic Republic, Chile, South Korea, and possibly some other countries. Most landings come from the Atlantic sector of the Antarctic Ocean. Krill fishing in the southern Indian Ocean started slightly later,

---

*More recent statistics show that the krill landings increased again in the 1984–1985 season and during 1985–1986 reached the amount of almost 450,000 tons (85% caught by USSR, 14% by Japan). More than 220,000 tons were taken in a rather limited area near the South Orkneys. It is not yet known whether this had any demonstrable negative effect on local krill-dependent predators.

and catches in the southern Pacific have been minimal so far. Catches consist almost exclusively of krill; only occasionally are there by-catches of small fish.

Major krill fishing grounds are the waters of the Scotia arc from west of the Antarctic Peninsula to South Georgia, particularly the South Orkneys and Elephant Island, in the Atlantic sector, off Enderby Land, near Prydz Bay and on the Kerguelen-Gaussberg Ridge in the southern Indian Ocean, and in the Balleny Islands region and Bellingshausen Sea in the Pacific sector (Fig. 1).

4.2. Vessels, Gear, Catching Techniques, and Strategies

Because the Antarctic fishing grounds are far away from the densely populated areas in the northern hemisphere (cruises from northern Europe and East Asia take about 30 days to reach them), and because of the often unfavorable weather and ice conditions in remote Antarctic seas, fishing is mostly carried out by large factory trawlers. Only this type of vessel is suited also to accommodate the often highly developed machinery necessary for the processing of krill at sea. Some shipyards, for example, in Finland, have designed vessels especially for krill fishing, but so far this fishery is carried out by conventional fish trawlers.

Midwater trawls proved to be the most efficient gear for krill fishing. Other types of gear such as purse seines and frame nets were tried by Soviet and Japanese vessels but were less practical, mainly because of normally rough sea conditions. Compared to other types of pelagic trawls, commercial krill trawls are of smaller size, often having a mouth opening of 400–500 m^2 and a length of about 100 m. Mesh sizes near the cod end are as small as 8–13 mm; some types of gear have large mesh nets with fine mesh liners and security devices against breaking when large catches are made. Nets are constructed of very fine twines in order to reduce drag. Gear is especially adapted to fishing at all depths of krill distribution from near bottom to the surface.

Location of krill concentrations is best achieved by vertical echo sounders of frequencies between 50 and 200 kHz. Good information on krill in the net mouth can be obtained using net sondes. With some experience it is possible to identify clearly krill concentrations on the echograms, and to distinguish them from echo traces caused by other organisms, particularly salps, which can be extremely abundant.

Information on fishing effort is still rather scarce. There is a difference in the fishing strategy employed by fleets from the Soviet Union and Japan. Soviet trawlers are guided by scouting vessels and fisheries research vessels to areas of high krill concentrations. At present, problems with processing the catch limits the level of fishing effort, and few data are available from actual fishing operations. Japanese activities employ single-vessel as well as mother-ship type operations. When large krill are being fished, fishing time is reduced to improve the quality of the catch.

A review of the technical aspects of the harvesting of krill was published by Eddie (34); further information can be found in Nemoto and Nasu (35) and in reports on German (F.R.G.) expeditions (Sahrhage et al. 36; Hempel et al. 37).

4.3. Products and Markets

Krill are a difficult raw material for processing, owing to their small size (about 1 g per animal), their weakness to mechanical stresses, their fast decomposition after capture (high enzyme content), and the difficulties in removing the shell from the meat. Nevertheless krill can be processed in four main ways as whole krill, tail meat, mince, and krill meal (Grantham 38). In Japan, krill are sold as whole frozen product for human consumption but the demand appears to be limited (Suzuki 39). In the Soviet Union frozen whole krill and krill meal have been used in large quantities for feeding fur animals, poultry, and fish (aquaculture). Coagulated krill paste (''Okean'') for human consumption was not well accepted owing to poor quality and storage duration; however, in some other countries good progress has been made in the development of products from minced krill. Automatic peeling machines were developed, particularly in Poland, and peeled canned tail meat is sold in some quantities in the Soviet Union and Chile. Highly valuable chitin–chitosan can be extracted from the shell for industrial uses.

The discovery of rather high concentrations of fluoride in krill by Soevik and Braekkan (40) in 1979 jeopardized the utilization of krill for human consumption (Adelung 41). However, further investigations showed that in living krill fluoride is concentrated in the exoskeleton and that the meat contains very low fluoride quantities if shell and meat are separated quickly after catching (Boone and Manthey 42; Christians and Leinemann 43). Investigations of trace metal content in krill meat confirmed the absence of toxic risks of krill for human food (Stoeppler and Brandt 44).

4.4. Economic Considerations and Perspectives

The krill fishery has to cope with high operating costs. Furthermore, the fishing season, running from about December until May, is rather short. Profitability can be improved if krill fishing can be combined with other fishing activities close to the Antarctic during the remainder of the year and by developing operational bases in countries bordering the Antarctic seas. However, trawling and sophisticated processing of krill are intensive, with high related fuel costs.

Experience has shown that high catch rates of krill of the order of 100–120 tons/day, or even more, are feasible. Considerable variations in the abundance of krill have however been observed from year to year. Further extensions of the market can be expected only if there are regular supplies of krill products with consistent quality and if the prices are competitive with those for similar products on the world market (see detailed study of McElroy 45).

At present (1986) it does not seem very likely that in the near future the krill fishery will develop into a rather large activity with annual landings of several million tons as originally expected. However, the necessary number of suitable vessels and knowhow for locating and catching substantial quantities of krill exist, which could be activated if difficulties in the processing and marketing of krill could be overcome.

4.5. Fisheries Regulations

According to Articles II and IX of the Convention on the Conservation of Antarctic Marine Living Resources (CCAMLR 11), the commission could introduce a wide range of conservation measures. No such measures have been agreed upon so far for krill fisheries, however.

5. ECOLOGICAL CONCERNS

5.1. Role of Krill in the Antarctic Marine Ecosystem

Krill plays a major role in the Antarctic marine ecosystem (Bengtson 46). It forms the major food basis for the baleen whales, most of the seals (particularly the crabeater seals), Antarctic birds (especially penguins), many of the Antarctic fish species, and also squids (Fig. 4). Because the abundance and distribution of these resources directly depend on the availability of krill, so also do the predators at

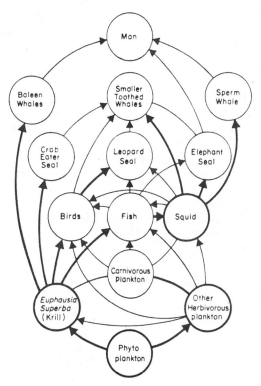

Figure 4. Antarctic marine ecosystem. (Reprinted from G. A. Knox, Antarctic Marine Ecosystem. In R. M. Laws, Ed., *Antarctic Ecology*, Vol. 1, 1970, with kind permission by Academic Press, Inc.)

higher levels of the ecosystem. These, such as elephant seals and sperm whales living on squids; leopard seals living mainly on penguins, seals, krill, fish, and squids; and small-toothed whales living on squids, penguins, and seals, depend indirectly on the availability of the krill resources.

Estimates of stock sizes, biomass, and the annual food consumption of krill and other organisms by predators still vary rather widely, and the analysis of predator–prey interactions is only at an early stage. Nevertheless some rather preliminary figures have been given for the annual consumption of krill (Table 2), which indicate that a total of at least 400–500 million tons of krill may be eaten each year by whales, seals, birds, fishes, and squids. The crabeater seal, consuming around 63 million tons of krill per annum, is the most important predator species.

Through intensive whaling, especially between the 1920s and 1960s, most baleen whale stocks have been reduced to rather low levels. It has been estimated that the pristine stocks of these whales may have consumed about 190 million tons of krill annually as compared to only 40 million tons presently (Laws 47). There are indications that other species have taken advantage of this food "surplus," of the order of 150 million tons not now eaten by the whales. Substantial increases in abundance have been observed for the stocks of several penguin species (Adelie, chin-strap, and gentoo penguins), for some seals, particularly crabeater and fur seals, and possibly also for the minke whales. Even if the ecological relations and biological processes are not yet fully understood, it seems reasonable to conclude that these increases have been due to the increased availability of krill. It is also likely that some whales (sei, fin) and crabeater seals have responded to the more abundant food supply with increased growth and reproductive rates. Following the removal of whales, a change from a whale-dominated system to a seal- and bird-dominated system was recognized, which has recently again shown an increasing intraspecific competition for food.

Besides krill, which mostly dominate in the zooplankton of the seasonal packice zone, there are other herbivorous organisms like other Euphausiacea, particularly

TABLE 2   Rough Estimates of Current Stock Sizes and Biomass of Krill Predators in the Antarctic and Assumed Annual Krill Consumption

	Stock size (1000 individuals)	Antarctic Biomass (million metric tons)	Annual Krill Consumption (million metric tons)
Whales[a]	410	7	40
Seals	33,200	6.5	130
Birds		200[b]	115
Fishes		15(?)	>50(?)
Cephalopods		>35(?)	100(?)

Source: (From Bengtson 46; Laws 47.)

[a]Only blue, fin, sei, humpback, and minke whales, which feed almost entirely on krill.

[b]Of a further 460 million tons of seabirds in the sub-Antarctic some are feeding on krill, for example, off South Georgia.

Thysanoessa sp., amphipods (*Themisto sp.*), and copepods, which sometimes even outnumber the krill and compete to various extents for phytoplankton food. It has been observed at certain times that in the absence of krill, for instance, around South Georgia, predators divert their feeding temporarily to these other herbivores. The main competitor of krill for phytoplankton food may be the salps (*Salpa thompsoni*) and copepods, which compete with larval krill.

5.2. Impact of Krill Fishing on Predators, Competitors, and Other Organisms

The present landings of the commercial krill fishery of around 450,000 metric tons annually are insignificant in comparison to the size of the krill resources. However, the possibility of a sudden substantial increase in the fishery exists (see Section 4.4), and there is no doubt that this could have serious effects on the krill stocks and the marine ecosystem as a whole. Although the standing stock of krill is rather high, it appears that the net production is low (see Section 3), so that there may be relatively little surplus to be taken by outsiders such as humans (Hempel 33). On the other hand, the question is whether we should restrict, or even renounce, krill fishing for the benefit of a further increase in penguin and seal populations that are already larger than originally in the unexploited ecosystem. Refraining from krill fishing does not automatically result in a corresponding recovery of the depleted whale stocks, but this goal could be reached in a reasonable time only with a proper management of all resources, including penguins and seals, taking into account the complex interactions within the whole ecosystem.

5.3. Modeling

Suitable models can be very useful for improving the understanding of the rather complex structure of and interactions in the marine ecosystem (Everson 48). They also provide guidelines for research planning and data evaluation (BIOMASS 4). Experience has shown that it is beneficial first to develop detailed submodels (e.g., krill–whales, krill–whales–seals, krill–cephalopods–sperm whales) prior to modeling the whole ecosystem. Beddington and de la Mare (49) describe theoretical models, strategic simulation models, and estimation models, and such models are expected to form important tools for ecosystem management by CCAMLR.

5.4. Need for Monitoring

Monitoring the resources is a basic requirement for their management. First, the krill resources themselves must be monitored by research and by following closely the activities of the commercial fishery. For the latter it is essential to improve the data basis of CCAMLR as regards fishing effort and catch statistics, with detailed breakdown by small areas.

Second, it is necessary to detect and record significant changes in critical components of the ecosystem by monitoring some key indicator predator and prey species. The ad hoc Working Group on Ecosystem Monitoring of the Scientific Committee of CCAMLR in 1985 identified the crabeater seal, Antarctic fur seal, Adelie, chin-strap, and macaroni penguins, and minke whales as such indicators of changes in prey availability. Furthermore the group identified 13 areas and sites for monitoring programs, giving first priority to the waters of the Prydz Bay, the Bransfield Strait, and around South Georgia (CCAMLR 50). A Working Group for the CCAMLR Ecosystem Monitoring Programme during 1986 elaborated plans by the Commission which will be further developed in 1987. Long-term data sets obtained with standard gear in agreed key areas, such as the German (F.R.G.) observations from 1975 to 1986 in the Elephant Island/Bransfield Strait waters, may be most useful.

5.5. Priorities for Future Management

Management must be based on sound scientific knowledge. Most important are further improvements in the determination of key biological parameters, such as growth, longevity, and mortality of krill; with better stock identification such improvements would allow a better estimation of the productivity of the krill stock(s) in the various regions of Antarctic seas.

The development of an overall management strategy by CCAMLR is urgently needed. In view of the widely differing interests of the members of the commission, this may not be an easy task. Because the net production of krill appears to be rather low, the resources seem vulnerable to overexploitation. Thus it would be desirable for CCAMLR now to establish a conservative interim total allowable catch (TAC), which could be adjusted later on the basis of improved data and scientific knowledge. In doing so, it must be taken into account that even now krill fishing can be rather intensive in local areas of krill concentrations, with possible detrimental effects on local predator populations. In general, however, any influence of the fishery on krill stocks and their consumers could presumably be detected only after the catch of very large quantities of krill.

REFERENCES

1. J. W. S. Marr, The natural history and geography of the Antarctic krill (*Euphausia superba* Dana). *Discovery Rep.* 32:33–464 (1962).

2. T. G. Lubimova et al., *The Ecological Peculiarities, Stocks and Role of E. superba in the Trophic Structure of the Antarctic Ecosystem.* Selected papers presented to the Scientific Committee of CCAMLR 1982-184 (1985), Part II, pp. 391–505.

3. I. Everson, *The Living Resources of the Southern Ocean*, GLO/SO/77/1. FAO, Rome, 1977, 156 pp.

4. BIOMASS, Biological Investigations of Marine Antarctic Systems and Stocks. *Biomass Res. Ser.* 1:1–79 (1977).

5. G. Hempel, FIBEX—An international survey in the Southern Ocean: Review and outlook. *Mem. Nat. Inst. Polar. Res. Spec. Issue (Jpn.)* 27:1–15 (1983).

6. D. Sahrhage, BIOMASS - FIBEX - SIBEX: Internationale Zusammenarbeit zur Erforschung der lebenden Meeresschätze der Antarktis. *Geowiss. Unserer Zeit* 2(4):109–116 (1984).

7. T. Nemoto and T. Matsuda, Eds., Proceedings of the BIOMASS Colloquium in 1982. *Mem. Natl. Inst. Polar Res., Spec. Issue (Jpn.)* 27:1–247 (1983).

8. R. Y. George, Ed., The biology of Antarctic krill *Euphausia superba*. Proceedings of the First International Symposium on Krill held at Wilmington, North Carolina, 16–19 October 1982. *J. Crustacean Biol.* 4, (Spec. No. 1): 1–337 (1984).

9. S. B. Schnack, Ed., On the biology of krill *Euphausia superba*. Proceedings of the Seminar and Report of the Krill Ecology Group, Bremerhaven, 12–16 May 1983. *Ber. Polarforsch., Sonderh.* 4:1–303 (1983).

10. W. E. Siegfried, P. R. Condy, and R. M. Laws, Eds., *Antarctic Nutrient Cycles and Food Webs*. Springer-Verlag, Berlin and New York, 1985, 700 pp.

11. CCAMLR, Commission for the Conservation of Antarctic Marine Living Resources, *Basic Documents*. 1984, 75 pp.

12. CCAMLR, Reports on the meetings of the Commission and the Scientific Committee for the Conservation of Antarctic Marine Living Resources, Hobart, Australia, 1982, 1983, 1984, 1985.

13. A. F. Amos, Distribution of krill (*Euphausia superba*) and the hydrography of the Southern Ocean: Large-scale processes. *J. Crustacean Biol.* 4 (Spec. No. 1):306–329 (1984).

14. K. Kanda, K. Takagi, and Y. Seki, Movement of the larger swarms of Antarctic krill, *Euphausia superba*, populations off Enderby Land during 1976–1977 season. *J. Tokyo Univ. Fish.* 68:25–42 (1982).

15. T. Kikuno and A. Kawamura, Observations of the ovarian eggs and spawning habits in *Euphausia superba* Dana. *Mem. Natl. Inst. Polar Res., Spec. Issue (Jpn.)* 27:104–128 (1983).

16. C. J. Denys and M. A. McWhinnie, Fecundity and ovarian cycles of the Antarctic krill *Euphausia superba* (Crustacea, Euphausiacea). *Can. J. Zool.* 60:2414–2423 (1982).

17. V. Siegel, On the fecundity of Antarctic krill, *Euphausia superba* (Euphausiacea). *Arch. Fischereiwiss.* 36(1/2):185–193 (1985).

18. H. P. Marschall, Sinking speed of krill eggs and timing of early life history stages. *Ber. Polarforsch., Sonderh.* 4: 70–73 (1983).

19. I. Hempel, Studies in eggs and larvae of *Euphausia superba* and *Euphausia crystallorophias* in the Atlantic sector of the Southern Ocean. *Ber. Polarforsch., Sonderh.* 4:30–46 (1983).

20. O. Guzman, Distribution and abundance of Antarctic krill (*Euphausia superba*) in the Bransfield Strait. *Ber. Polarforsch., Sonderh.* 4:169–190 (1983).

21. R. R. Makarov, A study of the second maturation of euphausiid (Eucarida, Euphausiacea) females. *Zool. Zh.* 54(5):670–681 (1975) (in Russian).

22. M. A. McWhinnie, C. J. Denys, R. Parkin, and K. Parkin, Biological investigations of *Euphausia superba* (krill). *Ant. J. U.S.* 14:163–164 (1979).

23. V. Siegel, Untersuchungen zur Biologie des antarktischen Krill, *Euphausia superba*, im Bereich der Bransfield Strasse und angrenzender Gebiete. *Mitt. Inst. Seefisch.* 38:1–237 (1986).

24. M. C. Macauley, T. S. English, and O. A. Mathisen, Acoustic characterization of swarms of Antarctic krill (*Euphausia superba*) from Elephant Island and Bransfield Strait. *J. Crustacean Biol.* 4 (Spec. No. 1): 16–44 (1984).

25. U. Kils, Swimming and feeding of Antarctic krill, *Euphausia superba*—some outstanding energetics and dynamics, some unique morphological details. *Ber. Polarforsch.*, *Sonderh.* 4:130–155 (1983).

26. F. Nast, The vertical distribution of larval and adult krill (*Euphausia superba* Dana) on a time station south of Elephant Island, South Shetlands. *Meeresforschung*, 27 (2):103–118 (1978–1979).

27. N. A. Mackintosh, Life cycle of Antarctic krill in relation to ice and water conditions. *Discovery Rep.* 36:1–94 (1972).

28. G. Ettershank, Age structure and cyclical annual size change in the Antarctic krill *Euphausia superba* Dana. *Polar Biol.* 2:189–193 (1983).

29. I. Hampton, Preliminary report on the FIBEX acoustic work to estimate the abundance of *Euphausia superba*. *Mem. Natl. Inst. Polar Res.*, *Spec. Issue (Jpn.)* 27:165–175 (1983).

30. J. Kalinowski, M. Godlewska, and Z. Klusek, Distribution and stock of krill in the Bransfield Strait and the Drake Passage during December 1983–January 1984, BIOMASS-SIBEX. *Pol. Polar Res.* 8(1–2):151–158 (1985).

31. R. B. Heywood, I. Everson, and J. Priddle, The absence of krill from the South Georgia zone, winter 1983. *Deep-Sea Res.* 32(3):369–378 (1985).

32. R. B. Heywood and T. M. Whitaker, The Antarctic marine flora. In R. M. Laws, Ed., *Antarctic Ecology*, Vol. 2. Academic Press, London, 1984, pp. 373–419.

33. G. Hempel, Antarctic marine food webs. In W. R. Siegfried, P. R. Condy, and R. M. Laws (Eds.), *Antarctic Nutrient Cycles and Food Webs*. Springer-Verlag, Berlin and New York, 1985, pp. 266–270.

34. G. O. Eddie, *The Harvesting of Krill*, GLO/SO/77/2, FAO, Rome, 1977, 76 pp.

35. T. Nemoto and K. Nasu, Present status of exploitation and biology of krill in the Antarctic. *Oceanol. Int.* 75:353–360 (1975).

36. D. Sahrhage, W. Schreiber, R. Steinberg, and G. Hempel, Antarktis-Expedition 1975/ 76 der Bundesrepublik Deutschland. *Arch. Fischereiwiss.* 29(Beih. 1):1–96 (1978).

37. G. Hempel, D. Sahrhage, W. Schreiber, and R. Steinberg, Antarktis-Expedition 1977/ 78 der Bundesrepublik Deutschland. *Arch. Fischereiwiss.* 30(Beih. 1):1–119 (1979).

38. G. J. Grantham, *The Utilization of Krill*, GLO/SO/77/3, FAO, Rome, 1977, 61 pp.

39. T. Suzuki, *Fish and Krill Protein Processing Technology*. Applied Science Publ. London, 1981, 260 pp.

40. T. Soevik and O. R. Braekkan, Fluoride in Antarctic krill (*Euphausia superba*) and Atlantic krill (*Meganyctiphanes norvegica*). *J. Fish. Res. Board Can.* 36:1414–1416 (1979).

41. D. Adelung, Ed., Fluor im antarktischen Ökosystem. *Ber. Polarforsch.* 10:1–74 (1983).

42. R. J. Boone and M. Manthey, The anatomical distribution of fluoride within various body segments and organs of Antarctic krill (*Euphausia superba* Dana). *Arch. Fischereiwiss.* 34(1):81–85 (1983).

43. O. Christians and M. Leinemann, Investigations on the migration of fluoride from the shell into the muscle flesh of Antarctic krill (*Euphausia superba* Dana) in dependence of storage temperature and time. *Arch. Fischereiwiss.* 34(1):87–95 (1983) (in German).

44. M. Stoeppler and K. Brandt, Comparative studies on trace metal levels in marine biota. II. Trace metals in krill, krill products, and fish from the Antarctic Scotia Sea. *Z. Lebensm.-Unters. Forsch.* 169:95–98 (1979).

45. S. McElroy, The potential of krill as a commercial catch. The economics of harvesting krill. In B. Mitchell and R. Sandbrook (Eds.), *The Management of the Southern Ocean.* IIED, London, 1980, 162 pp.

46. J. L. Bengtson, *Review of Antarctic Marine Fauna.* Selected papers presented to the Scientific Committee of CCAMLR 1982–1984 (1985), Part I, pp. 1–226.

47. R. M. Laws, The ecology of the Southern Ocean. *Am. Sci.* 73: 26–40 (1985).

48. I. Everson, Marine interactions. In R. M. Laws, Ed., *Antarctic Ecology*, Vol. 2. Academic Press, London, pp. 783–819, 1984.

49. J. R. Beddington and W. K. de la Mare, *Marine Mammal Fishery Interactions: Modelling and the Southern Ocean.* Selected papers presented to the Scientific Committee of CCAMLR 1982–1984 (1985), Part II, pp. 155–178.

50. CCAMLR, *Report of the Fourth Meeting of the Scientific Committee*, SC-CAMLR IV (1986).

2 MANAGEMENT OF THE BROWN SHRIMP (*CRANGON CRANGON*) STOCK IN DUTCH COASTAL WATERS

R. Boddeke
Netherlands Institute of Fishery Investigations
IJmuiden, The Netherlands

1. INTRODUCTION

The shrimp fisheries in the Netherlands coastal area (Fig. 1), as for those in Belgium, Germany, and along the Danish west coast south of Esbjerg, are aimed at only one species, *Crangon crangon*, the brown shrimp (Fig. 2). With a count of 500–700/kg, this species is one of the smallest shrimps in the world used for

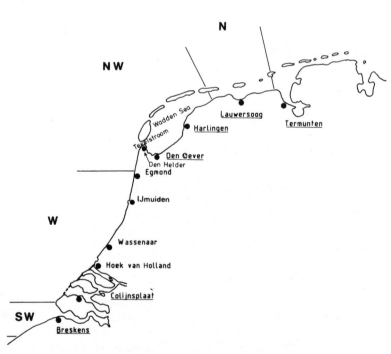

Figure 1. The coastal zone of the Netherlands. Geographic areas and places are indicated; shrimping harbors are underlined.

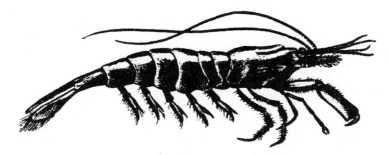

Figure 2. *Crangon crangon*, the brown shrimp.

direct human consumption. The catch is cooked on board in seawater for consumption in peeled form. When salt is added to the seawater in the cooking pot "salted shrimps" are produced, which are served unpeeled.

A fishery for a single species of such a small size (5–8 cm body length) is typical of shrimp fishing in temperate and cold seas, where the benthic ecosystem is dominated by gadoids and flatfish, of which many species reach a large size in comparison to shrimps. The gadoids especially prey not only on juveniles, as do many other fish species, but also heavily on mature brown shrimps.

In tropical seas we can observe a completely different situation. In such areas the shrimp fauna is dominated by the genus *Penaeus*, with a number of species occurring in the same area as a rule; all species reach a maximum length of about 20 cm. Gadoids are virtually absent in the tropics outside upwelling areas, and demersal fish species in tropical areas are generally of a size comparable to penaeid shrimps. This gives the penaeids a position in the benthic community equal to that of fish, whereas the small shrimps in temperate seas occupy a lower place in the food chain.

Another difference arises from the position of the shrimp fishery in the overall picture of demersal fishing. In the tropics, demersal fisheries, thanks to the high price and relative abundance of penaeid shrimps, are mainly shrimp fisheries. In temperate seas, demersal fish are economically much more important than shrimp. The shallow Indonesian archipelago produces only 5% of the world landings of demersal fish, but 30% of the world shrimp catch, whereas the northeastern part of the Atlantic Ocean produces 25% of all demersal fish and only 3% of the world shrimp catch (FAO 1983 data).

Dutch demersal fisheries are even more dominated by fish than the northeast Atlantic as a whole, and the average price per kilogram of *C. crangon* does not differ substantially from the average price of demersal fish species. In 1980, a bad year for shrimp fishing in the Netherlands, brown shrimps accounted for only 2.6% of the landings in weight and for only 2.5% of the value of demersal landings. In 1983, a favorable year for shrimp fishing, these percentages were 3.2% in weight and 4% in value.

2. MANAGEMENT PROBLEMS

Management problems in brown shrimp fisheries result logically from the place of these species in the demersal ecosystem of the North Sea, and the modest place of shrimp fishing in Dutch demersal fisheries as a whole. The following management problems can be distinguished:

1. The fluctuating but always very high natural mortality, especially during the winter half of the year should be taken into account. This high mortality is intrinsic to the population dynamics of the brown shrimp.

2. The landing of huge amounts of undersized shrimps for industrial purposes before 1970 arousing a debate as to whether or not these landings had a negative influence on catches of the more valuable commercial shrimps.

3. The by-catch of large amounts of undersized specimens, discarded dead, in a fishery for commercial shrimps only, resulting from the poor selection properties of shrimp trawls and a too long stay on deck.

4. The necessity of using fine-meshed trawls to catch this tiny shrimp and the incidental catch in these trawls of undersized specimens of commercially important demersal fish species. The modest place of shrimp fishing in the Dutch demersal fishery makes destruction of undersized fish by shrimpers in the Netherlands a very critical issue, socioeconomically speaking.

3. POPULATION DYNAMICS

3.1. Reproduction

The constant battle for survival of the brown shrimp is demonstrated in many aspects by this species' biology, in the first place by the emphasis on egg production. The male sex is not more than a juvenile stage forming a minor part of the population. Sexually mature males are 30–55 mm long. With increasing size, the percentage of spawned males increases from 21% in the category 41–50 mm, to 43% over 50 mm: males of *C. crangon* likely copulate only once. When the maximum amount of sperm is formed in the testis, an elastic sac (spermatophore) is formed around the mass of sperm, which moves into one of the vasa deferentia and is inserted into the oviduct of the female during copulation (Bosschieter 1). In this way the male tissue in the gonad is entirely eliminated, after which rapid development of oocytes begins. The "male" is transformed into a "female" in about 2 months, particularly in March–July.

Over the total area of distribution of the brown shrimp, a remarkably balanced ratio can be observed between numbers of fertile males and sexually mature females (>44 mm) present, irrespective of the density of the stock. To keep this balance, brown shrimps are constantly on the move in a selective way. In very shallow nurseries, no mature males are present among the small shrimps that inhabit these places. The spring migration of mature (mostly unberried) females toward the coast is most likely aimed at finding there younger and smaller mates, while the spawned males stay behind offshore (Boddeke 2).

In August–September no fertile males are present, and females fertilized earlier form a closed group of spawners during these months, thanks to the storage of sperm in their oviducts. The numbers of ripe eggs decrease sharply after September, but ripe eggs can be found the whole year round (Fig. 3).

In October–November a new generation of males appears, resulting in mass spawning and a sharp rise in the percentage of berried females. This spawning coincides with the autumn–winter migration from inshore nurseries to the open sea, which redistributes the large numbers of mature shrimps developed in shallow

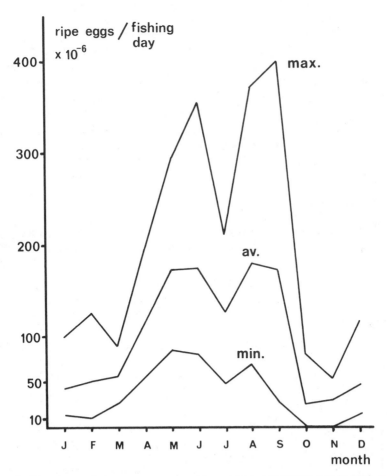

Figure 3. Abundance of ripe eggs expressed in numbers per fishing day in each month. Period 1978–1984; area, Hoek van Holland–Egmond.

nursery areas during summer over the total area of stock distribution. The stimulus for this migration is found in temperature fluctuations of seawater. The patterned character of this migration makes it a miracle of efficiency: unberried females, being close to a molt (and thus in this period of the year close to spawning), are more sensitive to temperature fluctuations than those farther away from the next molt. Berried females with ripe eggs are more sensitive than those with unripe ones. By these criteria, the inshore population is thinned out selectively, and berried females, nonberried females, and fertile males, each at a certain stage of their molting cycle, migrate together to the open sea in consecutive waves, from the end of November until March, depending on the weather (Fig. 4). This patterned migration has a great biological advantage. Even at the low densities that are normal in the open sea, a female that has to be fertilized directly after molting, when her shell is still soft, can still find a male ready to copulate during winter.

Egg production is related to body weight of the female and the time of the year (Boddeke 3). A female of average size (60 mm) in April–September spawns roughly 4500 small "summer" eggs and in November–March 2800 larger "winter" eggs. Eggs are carried by the female for 4–13 weeks, the incubation time depending on water temperature. Thanks to this breeding care, brown shrimp larvae are already in an advanced stage of development at the moment of hatching. This breeding care and long incubation time is also an Achilles heel for this species, because the fate of the eggs is coupled for 4–13 weeks to that of the female. During winter, when natural mortality (also on adult shrimps) is very high, females have to survive for several months to produce larvae.

3.2. Recruitment

In recent years, survival of berried females during winter has been very low, as illustrated by the low densities of females with ripe eggs in spring, lower even than expected from the already low total densities of mature females (Fig. 5).

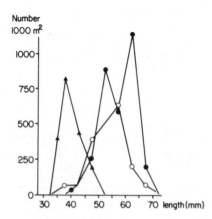

Figure 4. Typical size composition of a wave of *C. crangon* migrating to open sea in winter. Texelstroom, December 12, 1973.

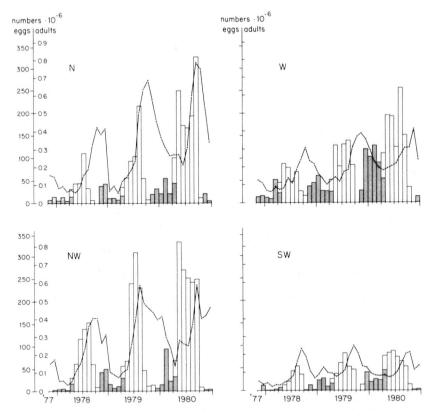

Figure 5. Numbers of adult females (curve) and numbers of ripe winter (hatched histogram) and summer (blank histogram) eggs caught in each month, for the years 1978–1980 per fishing day, in four geographic areas (cf. Fig. 1).

A clear relationship could be observed from 1978 to 1984 between the densities of ripe eggs in March–June (P) and the densities of commercial shrimps four months later in July–October (R). In all geographic zones off the Netherlands coast where the extent of silty inshore nursery areas is the limiting factor to the maximal size of the shrimp population, this relationship can be expressed by the formula $1/R = A + B (1/P)$, in which P is the number of ripe eggs in month x, R is the number of commercial shrimps in month $x + 4$, and A and B are regression coefficients. This relationship is shown for the northern area (N, Fig. 1), where the eastern Wadden Sea functions as the nursery area (Fig. 6).

A special situation in the usual relationship between shallow nurseries and deeper fishing areas was constituted by the sharply defined settlement in May–July of brown shrimp larvae in the sandy coastal zone between Wassenaar and Egmond (Boddeke et al. 4).

Mass settlement of shrimp larvae in relatively deep water in this area coincides

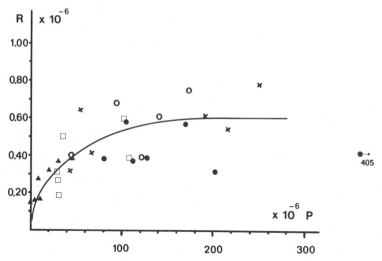

Figure 6. Relation between numbers of ripe eggs caught per fishing day (*P*) in March–July, and catch of consumption shrimps per fishing day (*R*) 4 months later in the northern area (N, Fig. 1), 1978–1984. $1/R = A + B (1/P)$, where $A = 2.38$, $B = 8.86$, and $r^2 = 0.53$. $\triangle$ = March–July, $\square$ = April–August, $\times$ = May–September, $\bigcirc$ = June–October, $\bullet$ = July–November.

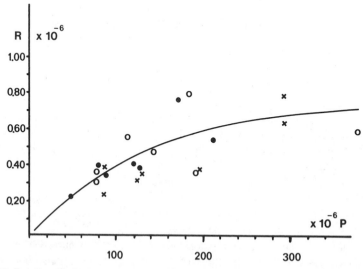

Figure 7. Relationship between number of ripe eggs caught per fishing day (*P*) in May–July and the number of consumption shrimps per fishing day (*R*) 4 months later. Dutch coastal area south of Egmond. $\times$ = May–September, $\bigcirc$ = June–October, $\bullet$ = July–November.

with the annual bloom of calanoid copepods in the Southern Bight. This bloom benefits from the eutrophication of this area by freshwater runoff (rivers Rhine and Meuse), sluiced at Hoek van Holland and carried northward by the residual coastal current, leading to an increased production of flagellates; the preferred food of calanoid copepods (Fransz and Verhagen 5; Klein Breteler and Gonzalez 6).

Calanoid copepods form in their turn the main food of juvenile shrimps. Because the density of (pelagic) copepods is defined by volume instead of area as for the benthic microfauna of shallow inshore nurseries, calanoid copepods form a much less limited food supply for juvenile shrimps than benthic organisms. This seems to be the most likely explanation for the practically linear relationship between parent stock size (P) and recruitment (R) over five times the range of the observed minimum of P in this coastal area (Fig. 7 and also Fig. 5).

Recruitment from ripe summer eggs in the period May–June brings the stock to its annual maximum density, and the maximum densities of adult shrimps reached in autumn decrease from north to south along the Netherlands coast (Fig. 5). In spring of 1979, the density of adult shrimps in the northern area fell to a lower level than in the southwestern area (SW, Figure 1), but increased again by 12-fold in 7 months! (See Table 1.)

3.3. Natural Mortality

In both coastal and inshore nurseries, the annual recovery of the brown shrimp stocks from the minimum level in March–April and September–October, benefits from the low numbers of predators present in April–July. The main predators of shrimps below 30 mm in these months are adult swimming crabs (*Liocarcinus holsatus*) and age-1 sand gobies (*Pomatoschistus minutus*). Owing to high winter mortality the numbers of these are low and rather regular. In August, predation on juvenile shrimps increases greatly under the influence of feeding by 0-group speci-

TABLE 1   Numbers of Brown Shrimp 30–53 mm and over 53 mm per 1000 m² during Biennial Brown Shrimp Surveys off Three Dutch Coastal Areas

	October Numbers			April Numbers			
Shrimp Size (mm)	1978	1979	1980	1978	1979	1980	1981
Netherlands Wadden							
Sea 30–53	1 751	1 504	2 211	183	437	384	556
>53	446	621	532	17	51	59	100
Netherlands coastal area							
30–53	471	823	2 957	74	195	169	333
>53	255	472	617	35	64	119	151
Zeeland estuaries							
30–53	309	896	837	40	76	133	104
>53	164	324	267	12	28	91	38

mens of the swimming crab and a large number of fish species. For 0-groups and gobies, which are very numerous, minimum size for shrimp predation is 25 mm; they mainly eat brown shrimps of 6–18 mm. In spite of the very high numbers of ripe eggs present in July–September, the rate of recruitment to the adult stock cannot continue to cope with adult female mortality after October. By the time of the autumn–winter migration to open sea, mature females leave the relative safety of shallow nursery areas at the same time that the numbers of predators in the Dutch offshore area increases. I and II group whiting (*Merlangius merlangus*) migrate in autumn from the central North Sea to the Belgian coastal area, and are especially numerous along the Dutch coast in the fourth quarter of the year. 0-Group whiting are already present earlier on. Both mature and juvenile cod (*Gadus morhua*) migrate from more northern areas to the Dutch coastal area in autumn and stay there during winter.

The mortality inflicted by these and other fish species on the stock of brown shrimps in Dutch coastal waters is difficult to quantify, owing to fluctuating year class strength and the varying proportion of these year classes present along the Netherlands coast. Prey size preference of cod and whiting of different sizes also plays an important role. Because cod less than 30 cm eat mainly shrimps less than 40 mm long, the effect on commercial shrimp catches of predation by cod in its first winter does not show up before April–May. Predation by age-2 cod, also eating mature (i.e., commercial) shrimps, has a direct effect on catches during winter and also on the subsequent production of larvae. In spite of the extraordinarily high consumption of (mainly small) brown shrimps by juvenile cod of the strong 1969 year class in the winter of 1969–1970, catches of commercial shrimps in January–March 1970 were comparatively high and brown shrimp larvae were numerous along the Dutch coast in January 1970 (Fig. 8a). The effect of predation by the 1969 cod year class on commercial shrimp catches in January–March 1971 was very marked, and *C. crangon* larvae were almost absent in Dutch coastal areas in January 1971 (Fig. 8b, Table 2) (Boddeke 7).

Although the influence of cod predation on stocks of brown shrimp was sometimes very spectacular in the 1970s (the "cod decade"), it is likely that in general the negative influence of whiting is considerably larger, owing to the much higher densities of juvenile whiting in the North Sea compared to juvenile cod since 1970 (Table 3). The estimates of year-class strength shown in Table 3 cannot be used to quantify the predation of whiting on brown shrimps in Dutch coastal areas because the abundance of whiting in different sectors of the North Sea can differ considerably from year to year. The 1983 year class, for example, was extremely numerous in its first summer and autumn in the German coastal areas and had a heavy impact on the local shrimp stock, whereas in the North Sea as a whole, this year class was of moderate strength. The growth rate of whiting can also vary considerably from year to year, making it very difficult to assess the influence of individual year classes. If we follow the stock size of whiting over the last 30 years, however, some clear trends can be observed.

Landings of commercial-size whiting in the North Sea rose from an average of

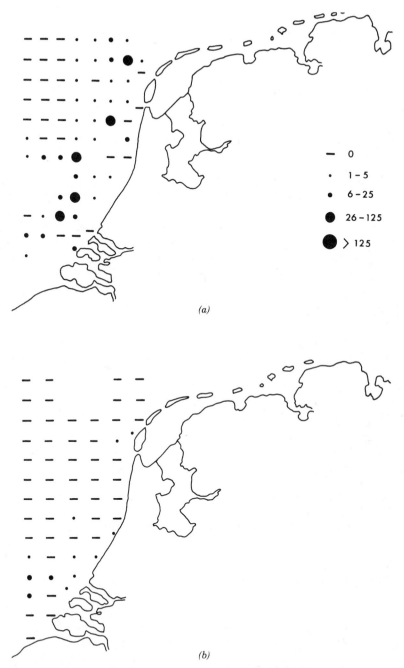

Figure 8. (a) Distribution of shrimp larvae in January 1970. (b) Distribution in January 1971. Surveys carried out with the Encased High Speed Plankton Sampler. Density expressed as numbers under 1 m².

TABLE 2   Consumption by Cod and Commercial Catches of Brown Shrimp

	Consumption (tonnes)				Catches	
Winter	Cod < 30 cm	Cod ≥ 30 cm	Total (tonnes)	First Quarter	Catch landed (tonnes)	Catch/Fishing Day (kg)
1968–1969	216	690	906	1969	306	116
1969–1970	5400	772	6172	1970	982	251
1970–1971	2700	NA	2700	1971	190	92

57,200 tonnes in 1956–1960 to 158,000 tonnes in 1971–1974, and decreased after 1976 (191,000 tonnes) to 99,000 tonnes in 1981–1984 and 74,000 tonnes in 1985. This decreasing trend in the landings of whiting and the below-average year classes after 1979 correspond reasonably well with the improvement in shrimp landings from Dutch coastal areas since 1980. In this connection, when North Sea whiting landings dropped to a very low level in 1985, landings of *C. crangon* from Dutch coastal waters exceeded 6000 tonnes for the first time since 1970.

3.4. Mortality Estimates

Natural mortality of brown shrimp is generally considered to be much higher than fishing mortality (International Council for the Exploration of the Sea 8). Boddeke and Becker (9) estimated average fishing mortality in the fishing area of Den Oever in 1973–1975 at 6.5% per month in July–October, and 11.4% per month in November–March. Tiews and Schumacher (10) estimated the natural mortality of consumption and fodder shrimps (30–53 mm) in the German fishery in 1965–1978 to be on average three times greater than fishing mortality during the fishing season (April–November). Redant (11) calculated the total mortality of postlarval shrimps to be 12.4 times the fishing mortality of commercial shrimps. In Dutch coastal areas numbers of postlarval shrimps longer than 5 mm in October are at a rough estimate 40–80 times higher than the numbers of adult shrimps surviving by the next March–April (see Table 2).

Such high mortality rates explain why monthly densities of adult shrimps (as

TABLE 3   Year-Class Strength at Age 1 of Cod and Whiting in North Sea

	Catches per hour fished in the ICES International Young Fish Survey in February															
	1970	'71	'72	'73	'74	'75	'76	'77	'78	'79	'80	'81	'82	'83	'84	'85
Cod < 25 cm	98	4	38	15	40	8	37	13	10	17	3	9	4	16	4	16
Whiting < 20 cm	274	332	1156	322	893	679	418	513	457	692	227	161	129	397	290	349

expressed in numbers caught per fishing day of commercial shrimps) are largely determined by the recruitment during the same month in Dutch coastal waters (Boddeke and Becker 12).

4. PROTECTION OF UNDERSIZED SHRIMPS

4.1. Landings

The practice of landing undersized shrimps in addition to commercial shrimps has been a common one since 1945 in the Netherlands. In 1959–1964 an average of 9068 tonnes of undersized shrimps and 6144 tonnes of commercial shrimps were landed annually.

Of the undersized shrimps, 77% were landed in July–November (Fig. 9a). The

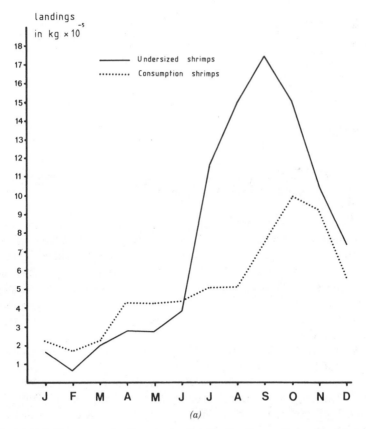

Figure 9. (a) Monthly average landings in the Dutch coastal area of commercial sized and undersized shrimps in 1959–1964. (b) Average monthly landings of commercial shrimps in 1959–1964, 1971–1972, and 1980–1985.

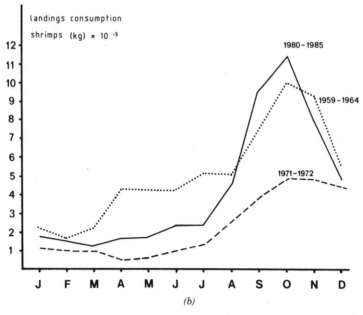

Figure 9. (*Continued*)

financial yield of these landings was very low (Dfl* 0.05/kg) whereas the average price of commercial shrimps in this period was Dfl 1.16/kg.*

The count per kilogram of these undersized shrimps (30–53 mm) was roughly three times that of commercial shrimps (>53 mm); in summer, an average undersized shrimp (42 mm) will exceed the minimum length of commercial shrimps in about 1 month.

Even when taking into account very high natural mortality rates, protecting these undersized shrimps seemed to be a suitable way to increase the yield of Dutch shrimp stocks, on the basis of the considerations mentioned above. A chance to demonstrate the positive effect of protecting undersized shrimps came in 1966, when in the western Wadden Sea the landing of undersized shrimps continued in Harlingen but ceased completely in Den Oever. Although fishing boats from both these harbors fish in adjacent areas, they are separated in the Wadden Sea by extensive tidal flats.

The annual pattern of catches per fishing day of commercial shrimps was rather similar in 1962 when undersized shrimps were landed in both harbors, and the peak in the annual landings in autumn was not very pronounced. In 1966 and 1967, catches per fishing day in Den Oever showed a clear peak in October but the pattern in Harlingen remained unchanged (Fig. 10*a*, *b*). A comparison of shrimp fishing in August–November 1962 and 1966 in these harbors is given in Table 4.

*1 Dfl = 1 Dutch guilder = U.S. $0.55 (1988 value).

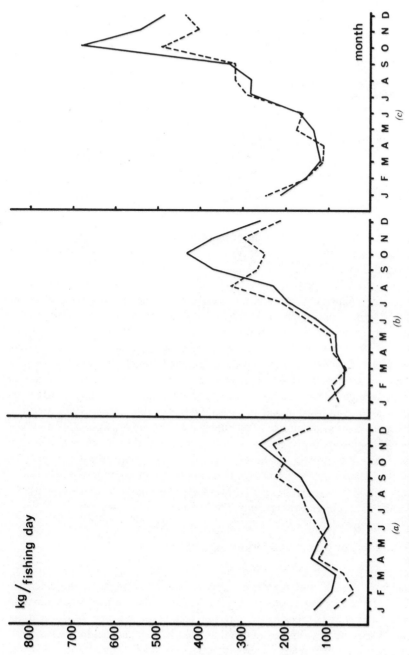

Figure 10. Monthly average catch per fishing day in Harlingen (interrupted line) and Den Oever (solid line). (a) 1962. (b) 1967. (c) 1973.

TABLE 4   **Results of Fishing in Harlingen and Den Oever in 1962 and 1966[a]**

August–November	1962		1966	
	Harlingen	Den Oever	Harlingen	Den Oever
Fishing days (number)	3391	2022	3171	2292
Sold shrimps (tonnes)	646	397	804	764
Unsold shrimps (tonnes)	0	6	96	19
Total landings	646	403	900	783
Average price/kg (Dfl)	1.39	1.45	1.29	1.41
Proceeds (1000 s Dfl)	900	576	1019	1050

[a] In 1962, undersized shrimps were landed at both harbors; in 1966 undersized shrimps were no longer landed at Den Oever.

This comparison showed that fishermen from Den Oever had caught at least an extra 132 tonnes of consumption shrimps by saving undersized shrimps, and grossed Dfl 250,000 extra during August–November 1966.

In financial terms the outcome of this case study was convincing enough, and in 1970, the Ministry of Agriculture and Fisheries bought out the last factory producing shrimp meal, which ended the landings of undersized shrimps in the Netherlands. After 1970, when landings of undersized shrimps in Harlingen had also ceased, the annual pattern of catches per fishing day again became comparable to that in Den Oever (with a clear peak in October; Fig. 10c).

In the years after 1964 when the landings of undersized shrimps gradually declined, average annual landings of commercial shrimps from Dutch coastal areas also decreased from 6144 tonnes in 1959–1964 to 6030 tonnes in 1965–1970.

The sharp drop in the numbers of full-time fishing boats, from 316 in 1964 to 120 in 1971, must have played a role in this modest decline. Since 1971 when this decline ended, the number of fishing days has remained more or less constant (Fig. 11). Nevertheless, monthly landings of commercial shrimps from 1971 to 1972 fell well below the average level of 1959–1964 (Fig. 9b); and annual landings in 1971 and 1972 were on average only 2727 tonnes.

4.2. Survival of Discarded Undersized Shrimps

The protection of undersized shrimps improved gradually since 1970 as a result of the increasing use of rotating sieves and a conveyor system for the catch (see Section 7.2). By the use of these devices, the stay on board became minimal and the catch was stored in basins of seawater before sorting, and kept wet during sorting. In spite of these improved handling practices, annual landings of commercial shrimps after 1970 never exceeded the level of 6000 tonnes until 1985, when 6198 tonnes was landed (see Section 3.4).

A comparison of the monthly pattern of landings in 1980–1985 with that in 1959–1964 shows increased landings of commercial shrimps in September–October but, as a result of the increased natural mortality in winter and spring, landings in

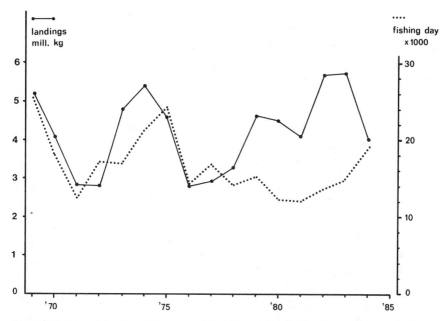

Figure 11. Annual landings of commercial shrimps from the Dutch coastal area (solid line) and number of fishing days for *C. crangon* (dotted line) in this area.

November–August were well below earlier levels of 1959–1964 (Fig. 9*b*). This is the more remarkable because, owing to the modernization of the fleet in the intermediate period, shrimping boats are much more suitable nowadays for fishing in open sea during winter weather conditions than before 1965.

5. THE RECENT SITUATION OF DUTCH SHRIMP FISHERIES

A clear stock–recruitment relationship has been observed in Dutch coastal waters since 1978 that most likely had existed since 1970; (Boddeke and Becker 12). Reduction of fishing during winter and spring will result in a lower mortality among berried females and consequently a higher recruitment in summer and autumn.

Closing of the fishery in winter and spring, however, will have serious social consequences for fishermen who depend totally or largely on shrimping. Economically, it is very hard to predict whether such a measure will result in a financial gain, especially when taking into account the often very high price per kilogram paid for brown shrimps in winter and spring in comparison to summer and autumn.

The positive effect of such a measure will be largely undone biologically by the high level of natural mortality on adult females during winter, and the effect is still more difficult to predict because of the fluctuations in natural mortality. For these reasons combined, a closed season for the shrimp fishery was not considered a viable management alternative. As long as the mortality inflicted by whiting and cod upon the stocks of brown shrimps in Dutch coastal areas remains high, the most suitable way of improving yield seems to be catching as many commercial shrimps as possible in the period August–November, before the gadoids consume them. Owing to a conjunction of circumstances, it is in this direction that the fishing has been evolving in recent years.

By modern techniques of cryogenic freezing, unpeeled commercial shrimps can now be stored for up to 8 months. This makes it possible for the retailers to store in autumn large amounts of shrimps bought at low prices at this time of year. This supply, hanging over the market, depresses prices in winter and spring when fresh shrimps are scarce, often making shrimping unattractive in these seasons. Large, modern coastal fishing boats, also suitable for fishing flatfish, herring, and gadoids have to make high returns, and when prices are low, they can only fish for shrimps when large catches can be made. Thanks to their modern equipment for sorting and cooking of the catch, these vessels are able to process 3 tonnes of commercial shrimps per fishing day, whereas 1 ton per fishing day was about the maximum a crew could handle around 1960.

Also typical of this development is the changed pattern of fishing along the Dutch west coast distant from any shrimping harbor (Fig. 1). In the past, shrimpers came to this area only when shrimps were scarce near their home ports. Nowadays they start fishing here in the second half of August. At this time, shrimps originating from the mass settlement of larvae in May–July in this area begin to reach commercial size (see Section 3.2). This fishery ceases when catches decline in the second half of November, but catches from this area can be extremely large. In September and October of 1982, more than 2000 tonnes of commercial shrimps was landed from the Dutch west coast, whereas the average annual landing in the entire Netherlands in 1976–1981 was only 4000 tonnes!

With the declining cod stock in the North Sea and the unfavorable prospects for sole (which is a badly overfished species), it may be expected that the interest in shrimping will increase further in coming years, especially in September–November, when large catches can be made. If this is the case, annual landings of commercial shrimps are again likely to regularly surpass 8000 tonnes, if natural mortality during winter and spring is reduced as in recent years.

6. FISH BY-CATCH PROBLEMS IN THE DUTCH SHRIMP FISHERIES

6.1. Historical Review

The destruction of the by-catch of undersized flatfish has been a serious problem for the Dutch brown shrimp fisheries for a long time, because the entire coastal area of the Netherlands is a very important nursery area for North Sea stocks of plaice (*Pleuronectes platessa*) and sole (*Solea solea*).

Sole has been the most important target for Dutch fisheries since the 1960s, but the economic relevance of plaice has gradually increased over the years and is practically equal in importance to sole in the 1980s. Both species together now count for more than 50% of the financial output of Dutch fisheries.

Sole, especially juveniles, are often concentrated in areas where consumption shrimps are numerous, and they are consequently intensively fished. Undersized flatfish normally survive capture very well. Towing speed while shrimping is low (3 mph), and the duration of hauls (a very important factor in survival of fish by-catch) is normally not longer than 1 hr.

Traditionally, the main cause of mortality was the sorting of the catch on board by means of a mechanical sieve consisting of two flat sieves on top of each other, making 100–150 beats/min. With this device the catch was divided into three fractions: (1) fish, benthos, and debris; (2) commercial shrimps, and (3) undersized shrimps. The flatfish in fraction 1 especially were fatally damaged, mainly by collisions with the upright borders of the sieve, causing severe brain damage (Kelle 13).

Survival experiments were carried out in the Dutch Wadden Sea in 1964 in which the fish were kept in large tanks with running seawater after sieving. From representative samples of plaice from fraction 1, only 20% was still alive after 24 hr, and only 10% after 48 hr. Of the sole, 33% was still alive after 24 hr, but two-thirds of the survivors were badly damaged.

An additional cause of mortality for these fishes was too often the long stay on deck, owing to the priority given to sorting, cooking, and picking of the commercial catch by the small crew of two or three. Especially during warm sunny weather, a stay on deck quickly becomes fatal for flatfish. Before 1970, the tiny flatfish between the undersized shrimps were landed with these shrimps. After 1970 when landings of undersized shrimps ceased completely in the Netherlands, this fraction was swept overboard but often after a long stay on deck. Smaller flatfish among the commercial shrimps were cooked with the shrimps, and their remains washed out of the batches of shrimps during cooling with seawater after cooking.

The destruction of juvenile sole and plaice by the shrimp fleet reached a peak in 1963–1964 for the following reasons:

1. From 1954 to 1964 the Dutch shrimp fleet was modernized and the average motor power increased from 47 to 100 hp. The number of full-time shrimpers increased from 180 in 1954 to 316 in 1964. (It decreased again sharply after 1964 to 120 in 1971 and the average motor power continued to rise.)

2. Before 1958 very little winter fishing occurred. The introduction of the modern cutter type of boat enabled the fleet to operate during winter in the open sea.

3. The replacement of the otter trawl by a pair of beam trawls increased the effective fishing effort in the 1960s.

4. Landing of undersized shrimps for industrial purposes was still commonplace then.

5. Exceptionally strong year classes of sole and plaice were born in 1963.

Recruitment estimates (at age 1) of these year classes were 1,129 million for plaice and 572 million for sole in the entire North Sea. Also, the much less important dab (*Limanda limanda*) showed a strong year class in 1963.

The losses inflicted on these year classes by the Dutch shrimp fleet were estimated at 155 million sole, 1 billion plaice, and 400 million dabs in the 17-month period from August 1963 to December 1964. These estimates were a shock to both administrators and fishing industry; the more so because high natural and fishing mortality during the severe winter of 1962–1963 had reduced the spawning stock biomass of the North Sea sole stock from 150,000 tonnes in 1962 to 50,000 tonnes in 1963. The negative influence of these losses of juvenile fish on commercial catches could not be exactly quantified because data on natural mortality of juvenile plaice and sole on commercial catches were (and still are) lacking.

However, even conservative estimates demonstrated that a shrimp fishery with such a detrimental effect on other more important fisheries had no right to exist.

7. SOLUTIONS FOR THE BY-CATCH PROBLEM

7.1. Selective Trawl

In 1964, the Netherlands Institute for Fishery Research (RIVO) obtained from France a shrimp net designed (by M. Devismes of le Crotoy) to separate fish and brown shrimps. This net was an otter trawl with wings, divided horizontally in two sections by a "false upper" of large (50-mm) meshes serving as a selective panel. When disturbed by the ground rope, shrimps jump up vertically, pass through the selective panel, and are caught in the fine-mesh (20-mm) cod end. Flatfish when disturbed, flee along the bottom, do not pass the selective panel, and are collected in a wide-meshed fish cod end. Juvenile cod, fleeing to the bottom when approached by a net, also show a behavior favorable for separating. Clupeids and also whiting pass through the selective panel as long as their size permits.

A weak point of this net proved to be the tendency of the false upper to rise in the water, even though provided with lead cords to keep it down. The net was recommended for small, 20-hp boats, but this solution did not work well in the Dutch fishery, with much larger vessels of 100 hp or more.

The Dutch fisherman J. Boersen from Den Oever, in cooperation with RIVO, developed a selective beam trawl suitable for fishing at higher speeds, without the intention of keeping any fish. In this beam trawl net the selective panel is a large-meshed (50-mm) funnel cut off obliquely and fixed at all sides, so that it cannot rise in the water at increased speed or if material becomes entangled in the meshes (Fig. 12) (FAO 14). If constructed of new netting, the selective panel had to be readjusted after some use to correct slack caused by stretching (haul 6, Table 5).

Selection of brown shrimp trawls is normally very poor (Bohl and Koura 15), and the removal from the catch of large amounts of fish, benthos, and debris greatly improves the filtering capacity of such fine-meshed (20-mm cod end) trawls,

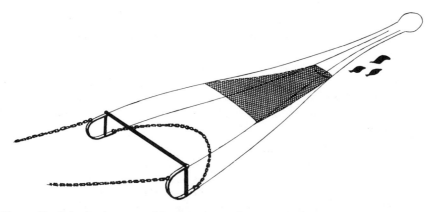

Figure 12. Selective beam trawl for *C. crangon*. Shrimps pass the selective panel and are caught in the cod end. Flatfishes are eliminated from the net through the opening in the belly directly in front of the attachment of the selective panel to the belly of the net.

leading to higher catches of commercial shrimps. Catches of undersized shrimps are normally lower also, because of improved selection in the cod end.

These experimental findings were confirmed in commercial fishing.

7.1.1. Practical Application

Some practical considerations have prevented the general use of this trawl, however:

1. In the southern part of the Netherlands, fishermen do not want to lose an often important by-catch of high-priced eel (*Anguilla anguilla*), a problem not solved by providing the net with a wide-meshed fish cod end.

TABLE 5   Experimental Catches (numbers) with a Selective Shrimp Trawl in Western Wadden Sea, December 17, 1964

		Consumption Shrimps		Undersized Shrimps		Fish By-Catch	
	Haul Time	Normal Net	Selective Net	Normal Net	Selective Net	Normal Net	Selective Net
1 AM	05.58–07.00	20	25	20	15	140	30
2	07.05–08.20	20	25	25	13	135	25
3[a]	08.25–09.20	13	25	20	20	110	15
4	09.40–10.50	15	18	20	18	120	18
5 PM	12.12–13.15	8	12	10	12	60	17
6[b]	13.15–14.15	11	11	10	8	70	5

[a]Haul against the current.
[b]Selective panel stretched.

2. In the Wadden Sea, a side effect of the extensive mussel culture is growth of high amounts of seaweeds, particularly *Ulva lactuca*. These come loose from the substrate when mussel fishing starts in June, hampering shrimping until winter by clogging the nets. The selective panel does not function properly under these conditions. Neither drawback, however, has denied the selective trawl a well-defined place in the Dutch shrimp fishery, and at present it is in common use in Dutch offshore fisheries and also in Germany and Denmark when and where no drifting seaweeds or eels are present.

7.2. Rotating Sieve

In 1967, Verburg's Machine Factory Colijnsplaat (Netherlands), in cooperation with RIVO, started the development of a rotating shrimp sieve to replace the traditional shaking sieve. The final design was introduced into the commercial fishery in 1971, and for several years purchasers received financial support from the government (Fig. 13).

Figure 13. Rotating shrimp sieve, showing the outer sieve selecting the shrimps by size (A), the spiral sieve eliminating the remaining plaice from the commercial shrimps (B), discharge of undersized shrimps (C), discharge of fish and benthos (D), funnel with flexible tube to return fish and undersized shrimps back to sea, together with the water used in the sorting process (E).

7.2.1. Separation of Shrimps and Fish

This rotating sieve has two coaxial cylindrical sieves, both of smooth stainless steel bars. The sieves rotate at a speed of 16 rpm, and the separation of shrimps and fish/benthos/debris takes place in the inner sieve. This consists of spiral-shaped bars of 12-mm diameter, with slots between the bars 12.5 mm wide. Circular bars outside the cylinder divide these slots into openings 36 mm long. A hose over the whole length of the cylinder pumps large amounts of water onto the catch, which greatly improves separation, and the fish fraction coming out of this sieve is automatically swept overboard.

7.2.2. Survival of Undersized Plaice

To investigate the survival of undersized plaice when handled by this rotating sieve in comparison to a shaking sieve, a tagging experiment was carried out in November 1967. In this experiment on the western Wadden Sea, a commercial shrimp boat was chartered and equipped with both a rotating and a shaking sieve (Boddeke and de Veen 16). Three categories of plaice were tagged:

1. Taken directly out of the net (552 specimens).
2. Passed through the rotating sieve (542 specimens).
3. Passed over a shaking sieve (550 specimens).

The results of this experiment are summarized in Table 6. This experiment showed clearly the enormous improvement in the survival chances of undersized plaice with this rotating sieve.

TABLE 6   Survival of Undersized Plaice When Handled by Rotating and Shaking Sieves

Number of Days at Liberty	Directly Out of Net	Tag and Recapture of Shrimps after Various Pretreatments	
		From the Rotating Sieve	From the Shaking Sieve
0–10	40	59	23
11–30	57	39	22
31–60	12	5	0
Over 60	3	2	1
% recaptures	20.3	19.0	8.2
Mean number of days at liberty	17.5	11.7	9.9

7.2.3. Separation of Consumption and Undersized Shrimps

Shrimps falling through the inner sieve cylinder of the rotating sieve are selected by size in the outer sieve cylinder. This is made up of round stainless steel bars with a diameter of 6 mm, and slots 6.5–7.2 mm wide. The sieve has three sections, two with longitudinal and one with spiral bars, to improve selection (Fig. 13, A).

To remove the juvenile plaice 7–8 cm long that otherwise would remain mixed with the commercial shrimps, this sieve cylinder is provided at the end with an extra spiral sieve (Fig. 13, B). Commercial shrimps fall easily through these wide openings and are collected in a basket. Juvenile plaice, however, on entering this sieve headfirst, wrap their lateral fins around two of the bars and are expelled in a few rotations of the sieve and swept overboard together with the fish fraction. Thanks to this extra selection, no significant fraction of juvenile plaice of a specific size remains with the commercial shrimps, as was the case with the shaking sieve.

The undersized shrimps, small plaice (4–8 cm), and a large share of the juvenile sole also fall through the outer sieve cylinder. Thanks to their elasticity, the sole pass more easily through the slots than the more rigid plaice. This fraction, mixed with the large amounts of seawater used in the sorting process, is easily swept overboard together with the fish fraction (Fig. 13, C–E).

7.2.4. Automatic Transporter

The working of the rotating sieve was further improved by an automatic transporter that came into use in 1975. This device has two basins partially filled with running seawater in which the nets are emptied. By an ingenious system of water jets, the catch is brought to a conveyor belt, which transports it at a regular speed into the sieve. The speed can be varied according to the composition of the catch. This automatic transporter has greatly improved the accuracy of the sorting process by taking out the human factor, resulting in considerably increased yields of commercial shrimps. Research carried out in September–November 1975 showed that landings of commercial shrimps by small and large shrimpers increased 37% after the installation of an automatic transporter, in comparison with similar boats equipped with the rotating sieve alone. For the protection of undersized flatfish it is important that fish and shrimps not lie on deck before being sorted, because especially on warm, sunny days this stay is a cause of mortality (see Section 7.2.6).

The combination of rotating sieve and automatic transporter is now widely in use in the Netherlands and in Belgium on boats fishing mainly in the open sea. In the Wadden Sea, however, fishermen with small and old boats continue to use the shaking sieve, which in recent years has also been equipped with a simple automatic transporter. This was partly due to the high costs of a rotating sieve and transporter, and the installation costs of a seawater pump of sufficient capacity (60 m^3/hr) for these small boats. An additional factor was the nuisance caused by drifting seaweeds. Although sorting by a shaking sieve is hindered more by seaweeds than for a rotating sieve, the first can be cleaned more easily during fishing.

7.2.5. Vibrating Sieve

Being aware of the negative effect of the shaking sieve on the survival of under-sized flatfish, Bosker's Machinefabriek Termunterzijl (Netherlands), in coopera-tion with D. Jonk, fisherman from Harlingen, and the RIVO constructed an advanced type of flat sieve (Fig. 14). The differences from the traditional shaking sieve are as follows:

1. A much milder, vibrating action (frequency 160 vibrations/min).
2. An upper sieve with a completely smooth surface and an open end to avoid collisions of the fish with upright parts of the sieve (Fig. 14, A).
3. Devices to transport the catch into the sieve and to transport the undesired fractions of the catch overboard, similar to those for the rotating sieve.

7.2.6. Survival Experiments

Comparative survival experiments were carried out by RIVO in the periods June 13–24 and November 21–December 2, 1983. In these experiments plaice were sorted using the vibrating and rotating sieves, and taken directly out of the net; of

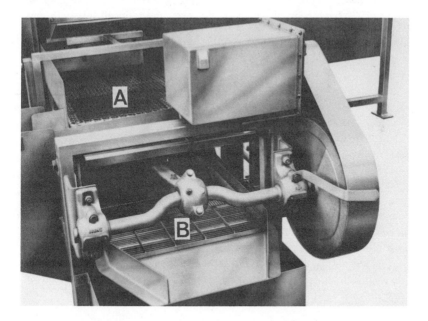

Figure 14. Vibrating sieve. The sieve separating fish–benthos from shrimps (A) is completely flat and has an open end. The sieve separating commercial and undersized shrimps (B) has longitudinal bars followed by a compartment with transverse bars to remove any remaining flatfish.

these three categories representative samples of 250–269 specimens were kept in tanks with running seawater for 96 hr.

Mortality was insignificant with the exception of the period June 20–24, 1983, when 5% of the plaice handled by vibrating and rotating sieves died. This mortality was probably caused by a somewhat longer stay on deck for these fishes in comparison with the fraction taken directly out of the net. In this week of June the weather was exceptionally warm and sunny, with air temperatures exceeding 25°C. No mortality occurred in the week of November 21–25, 1983, when the air temperature was 0°C.

7.3. Effects of Conservation Measures

Recruitment at age 1 of the North Sea stock of plaice has shown a clear upward trend since 1970, when selective trawls and rotating sieves came into common use and the landing of undersized shrimps ceased completely (Fig. 15). Although other factors, namely, the eutrophication of the Southern Bight, may have had a favorable effect on recruitment of plaice as well, a significant contribution to the improved recruitment of plaice owing to the protection of undersized individuals in the Dutch shrimp fishery is likely. For sole a similar trend is quite possible but more difficult to demonstrate, because any positive effects of protecting juvenile sole since 1970 have been obscured by the extra mortality of the 1978 and 1984 year classes in the severe winters of 1978–1979 and 1984–1985.

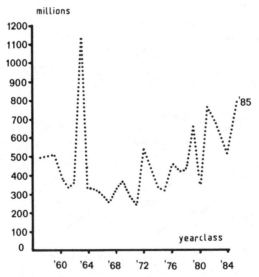

Figure 15. North Sea plaice. Recruitment to commercial fisheries at age 1. Period 1957–1985. Data from ICES North Sea Flatfish Working Group rep. 1986 (ICES, C. M. 1986/Assess.: 13).

The 1985 year class may also be reduced by the severe winter of 1985–1986. An additional complication is the positive effect severe winters can have on the strength of sole year class born in the following spring.

7.4. Management Actions

With the three methods of protecting undersized plaice and sole above, the by-catch problem could be completely solved. On this basis, the Directorate of Fisheries of the Ministry of Agriculture and Fisheries has decided to include in the licensing of shrimp boats, the stipulation that the use of fish-friendly sorting devices on board will be obligatory from January 1, 1987.

REFERENCES

1. J. R. Bosschieter, Onderzoek naar de copulatie bij de garnaal (*Crangon crangon*) RIVO IJmuiden. *Intern. Rep.*—ZE 79-02 (1979).

2. R. Boddeke, The seasonal migration of the brown shrimp *Crangon crangon*. *Neth. J. Sea Res.* 10:103 (1976).

3. R. Boddeke, The occurrence of winter and summer eggs in the brown shrimp (*Crangon crangon*) and the pattern of recruitment. *Neth. J. Sea Res.* 16:151 (1982).

4. R. Boddeke, G. Driessen, W. Doesburg, and G. Ramaekers, Food availability and predator presence in a coastal nursery area of the brown shrimp (*Crangon crangon*). *Ophelia* 26:77 (1986).

5. H. G. Fransz and J. H. G. Verhagen, Modelling research on the production cycle of phytoplankton in the Southern Bight of the North Sea in relation to riverborn nutrient loads. *Neth. J. Sea Res.* 19:241 (1985).

6. W. C. M. Klein Breteler and S. R. Gonzalez, Culture and development of *Temora longicornis* (Copepoda, Calanoidea) at different conditions of temperature and food. *Syllogeus* 58:71 (1986).

7. R. Boddeke, The influence of the strong 1969 and 1970 year-classes of cod on the stock of brown shrimp along the Netherlands coast in 1970 and 1971. *ICES, C. M.* 1971/K:32 (1971) (mimeo).

8. International Council for the Exploration of the Sea (ICES), Report of the working group on Grangonid shrimps. *ICES, C. M.* 1979/K:7 (1979).

9. R. Boddeke and H. B. Becker, Stock size and fishing mortality rates of a brown shrimp (*Crangon crangon*) population along the Dutch coast in the years 1973–1975. *ICES, C. M.* 1977/K:4 (1977) (mimeo).

10. K. Tiews and A. Schumacher, Assessment of brown shrimp stocks (*Crangon crangon* L.) off the German coast for the period 1965–1978. *Arch. Fischereiwiss.* 32:1 (1982).

11. F. Redant, (1978). Consumptie en productie van post- larvale *Crangon crangon* in de Belgische Kustwateren. Vol. 1 en 2. Thesis, Vrije Universiteit, Brussel.

12. R. Boddeke and H. B. Becker, A quantitative study of the fluctuations of the stock of

brown shrimp (*Crangon crangon*) along the coast of the Netherlands. *Rapp. P.-V. Reun., Cons. Int. Explor. Mer* 172:239 (1979).

13. W. Kelle, Verletzungen an untermässigen Plattfischen in der Garnalenfischerei. *Arch. Fischereiwiss.* 28:157 (1977).

14. Food and Agriculture Organization (FAO), Expert consultation on selective shrimp trawls. IJmuiden (Neth.), *FAO Fish. Rep.* No. 139 (1973).

15. H. Bohl and R. Koura, Selektionsversuche mit Garnalenkurren vor der nord-friesischen Küste. *Protok. Fischereitech.* 10:1 (1962).

16. R. Boddeke and J. F. de Veen, Results of a tagging experiment with plaice caught with a Dutch shrimp boat. *ICES, C. M.* 1971/B:14 (1971) (mimeo).

3 FISHERIES FOR NORTHERN SHRIMP (*Pandalus borealis*) IN THE NORTHWEST ATLANTIC FROM GREENLAND TO THE GULF OF MAINE

D. G. Parsons
Science Branch
Department of Fisheries and Oceans
St. John's, Newfoundland, Canada

and

J. Fréchette
Sous-ministariat aux Pêches Commerciales
Quebec, Canada

1. HISTORY OF THE FISHERIES

The northern or pink shrimp, *Pandalus borealis*, occurs in the northwest Atlantic from around 75°N at West Greenland to 42°N at Georges Bank (1). Commercial fisheries have developed in several areas within this range where shrimp are highly concentrated. North to south, these areas include the Davis Strait, deep-water channels off coastal Labrador, the Gulf of St. Lawrence, the Scotian Shelf, and the Gulf of Maine (Fig. 1). Fisheries for the pink shrimp in the northwest Atlantic began in the mid to late 1930s in the extreme northern and southern parts of the range; in 1935, on inshore grounds at West Greenland; and in 1938, in the Gulf of Maine.

The inshore fishery in the Davis Strait, conducted by small side trawlers, expanded rapidly after 1950 with the discovery of new grounds in Disko Bay (2). Inshore catches have averaged 7000–8000 t since the mid-1970s but catches from Disko Bay have been declining (D. M. Carlsson, personal communication). Around 1970, an offshore fishery developed on the slopes of the banks off West Greenland employing large stern trawlers capable of remaining at sea for several weeks. At present, the Davis Strait offshore fishery is the largest for nothern shrimp in the northwest Atlantic, with catches averaging 39,000 t from 1980 to 1986. Countries currently engaged in this fishery are Canada, Denmark, Faroe Islands, France, Greenland, and Norway.

In the mid 1970s, fishing began for shrimp off Labrador primarily in two marginal troughs on the Labrador Shelf—the Hopedale and Cartwright Channels. Large stern trawlers (>45 m) are used in this area, similar to the larger trawlers used in the Davis Strait offshore fishery. Landings increased to more than 4000 t in 1979 and 1980, declined to 1000 t in 1984 then increased again to about 7000 t in 1987. This is a Canadian fishery, in which vessels from the provinces of Quebec, New Brunswick, Nova Scotia, and Newfoundland–Labrador participate.

In the Gulf of St. Lawrence, shrimp fisheries developed off Sept-Iles, Quebec during the mid 1960s, off the west coast of Newfoundland (Port au Choix) in the early 1970s, and north of Anticosti Island in the mid 1970s. In the late 1970s, fisheries also began in an area south of Anticosti Island and in the Gulf of St. Lawrence estuary. Small stern trawlers (about 15–20 m) are used on the west coast of Newfoundland for trips of 1 or 2 days' duration, whereas vessels from Quebec and New Brunswick are generally larger (18–30 m) and can remain at sea up to 7

Figure 1. Areas of commercial concentrations of northern shrimp (*Pandalus borealis*) in the northwest Atlantic: 1, Disko Bay; 2, Davis Strait; 3, Hopedale Channel; 4 Cartwright Channel; 5, Port au Choix; 6, Anticosti Island; 7, Sept-Iles; 8, Gulf of St. Lawrence; 9, Scotian Shelf; 10, Gulf of Maine.

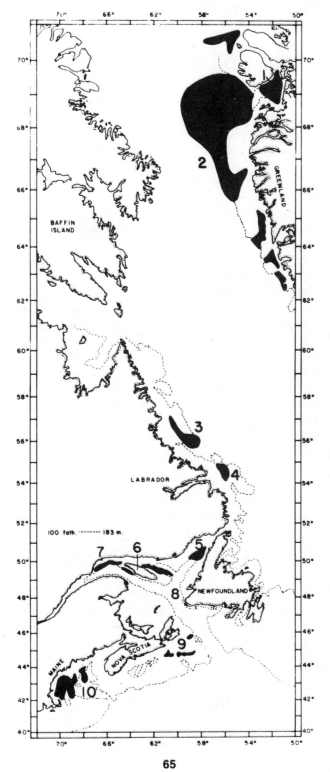

days. Landings from the Gulf of St. Lawrence have increased steadily from less than 100 t in 1965–1966 to about 1200 in 1987.

A small fishery has been conducted in depressions on the Scotian Shelf off Cape Breton Island since 1977 but has not been fully developed mainly because of the limited extent of fishable grounds. During the 1960s, good concentrations of shrimp were known to occur off southwestern Nova Scotia but these no longer exist.

Fishing effort for shrimp in the Gulf of Maine increased during the mid-1940s, but subsequently declined with declining abundance and remained low throughout the 1950s (3). This was followed by a period of rapid expansion in the fishery when landings increased from about 40 t in 1960 to 12,800 t in 1969. High catch levels were maintained from 1969 to 1972, but declined sharply to an average of 400 t from 1979 to 1980. This was followed by an increase to about 3000 t in 1984 and 4100 t in 1985. Small trawlers (less than 50 t) from the states of Maine, New Hampshire, and Massachusetts participate in this fishery (4). Some larger vessels (up to 150 t) operate further offshore.

Small-meshed (~ 45 mm and less) otter trawls, specifically designed for catching shrimp, are used in all northern shrimp fisheries. The smaller vessels of limited power tow relatively small bottom trawls (e.g., 41–5, Western 2A) with horizontal and vertical openings generally less than 15 and 3 m, respectively. In the offshore fisheries, large vessels use much larger trawls (e.g., Sputnik, Kalut) with horizontal and vertical openings in excess of 20 and 7 m, respectively. Pot fishing has occurred in some areas such as the Gulf of Maine and Gulf of St. Lawrence, but has never accounted for a significant proportion of the total catch.

In the Gulf of St. Lawrence, Labrador Sea, and Davis Strait, ice covers the fishing grounds during the winter and spring. Thus fishing in the former area is not extensively carried out from January to April and in the latter areas from January to June. The effects of ice are variable from year to year within each area. Ice is not a limiting factor either on the Scotian Shelf or in the Gulf of Maine. In the latter, the fishery is carried out primarily during the winter, and is directed toward berried (ovigerous) females, which are found in dense concentrations in inshore and shoaler areas during this time of year.

2. BIOLOGICAL AND ECOLOGICAL CONSIDERATIONS

Several aspects of the biology and ecology of the northern shrimp are important in delimiting effective management measures for the stock and in determining the methodology to be used for stock evaluation.

2.1. Biology

Pandalus borealis is a protandric hermaphrodite, a phenomenon well-documented for this species (5) and for several other shrimp of the family Pandalidae (6). The shrimp first develops sexually as a male, functions as a male for one to several years, then undergoes a transitional phase (sex inversion) and spends the rest of its life as a functional female. Sex inversion is facilitated by the degeneration of the androgenic gland, which is responsible for the primary and secondary male characteristics in crustaceans (5, 7). However, some individuals mature directly as females from the juvenile phase and are called primary females as opposed to secondary females that have undergone sex inversion. In most northwest Atlantic shrimp stocks, primary females are rare, indicating that, in these areas, essentially all individuals mature first as males. Haynes and Wigley (8) reported the occurrence of a few primary females in the Gulf of Maine.

Shrimps mature as males generally during the second year of life and may function as males for several years (Fig. 2). Sex inversion takes place between the last functional male period and development and ripening of ovaries; usually a

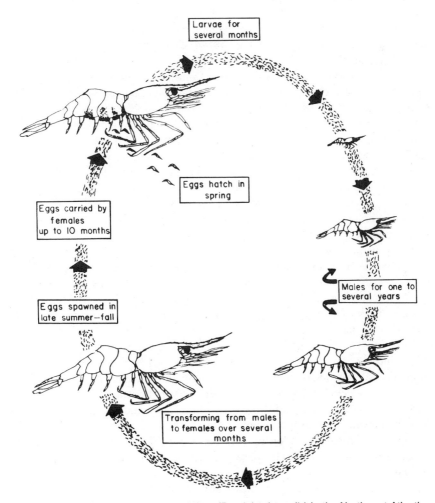

Figure 2. Life cycle of the northern shrimp (*Pandalus borealis*) in the Northwest Atlantic.

period of several months. Allen (9) describes external characteristics of transitionals over the intersex period. The changes that occur in the secondary sex characteristics (e.g., appendix masculina) have been used to detect the phenomenon of sex reversal.

Large variations have been observed throughout the northwest Atlantic in the duration of the different life history stages. Age at sex reversal and age at maturity for females seem to increase from south to north (9), with temperature probably being the most important factor (10). Charnov (11), on the other hand, hypothesized that age at sex reversal could vary due to fishing pressure.

This type of reproductive strategy has important consequences for fishing methods, stock assessment, and management. Females, being the largest individuals in the population, are fully retained in the fishing gears used (12), whereas small male shrimp are only partly retained. Also, the fishing pressure on females can be greatly enhanced when berried females move into shoaler areas in late autumn and winter (8, 13), resulting in dense concentrations in coastal waters just before the extrusion of larvae. In the Gulf of Maine, the winter fishery is directed almost exclusively toward dense concentrations of ovigerous females in inshore areas. In this respect, it is vastly different from fisheries in more northerly areas, where ice cover affords some protection for the ovigerous females in winter and early spring. Reductions in abundance observed for this southern stock have been correlated both with increased fishing effort and high seawater temperature (10), but the impact of very intensive fishing on females during winter, year after year, has not been assessed.

Stock–recruitment relationships have not been demonstrated for northern shrimp stocks and it is difficult, therefore, to determine minimum safe levels of spawning biomass, ensuring future recruitment. If Charnov's (11) theory is correct, in that age at sex reversal can decrease due to increased fishing mortality, assessment of stock status becomes even more complicated.

It is generally recognized that the average size of shrimp increases with depth except when ovigerous females migrate to shallow water. Young males are concentrated in shallow water, whereas females and transitionals are found in greater proportions in deeper water (14). This differential distribution by size prevents heavy fishing pressure on smaller shrimp because fishermen generally search for the larger sizes.

In addition to horizontal movement, northern shrimp also make diel vertical migrations (13, 15), probably related to feeding (16, 17). During the night, a large proportion of individuals move off the bottom, presumably in search of pelagic crustaceans which form part of their diet (1, 16). Recent observations demonstrated the presence of shrimp from 0 to 5 m off the bottom even during the day, with the maximum density occurring from 1 to 4 m off bottom (18).

The diel vertical migration reduces the availability of shrimp to commercial fishing gear. Carlsson et al. (19) observed that mean commercial catch rates during periods of the day could be more than five times the mean catch rate during the night in the Davis Strait. Even during daylight hours, differences in distribution over the bottom can affect the catchability of commercial trawls, especially in inshore fisheries where trawls with small vertical openings have been used.

2.2. Ecology

Shrimp are preyed upon by a variety of finfish but the two most important predators in most areas of the Northwest Atlantic are generally considered to be cod (*Gadus morhua*) and Greenland halibut (*Reinhardtius hippoglossoides*). Increases in shrimp abundance in the Davis Strait were first associated with declining cod stocks in the same area (2). Bowering et al. (20) showed that both species were important predators of shrimp in the Hopedale and Cartwright Channels off Labrador; Greenland halibut because of its very high abundance in these areas and cod because of its intensive predation on shrimp. They concluded that predation by the two species could have a significant effect on the shrimp stocks in these areas. Although redfish (*Sebastes* spp.) frequently occur in the same habitat as northern shrimp, they are not considered to be significant predators but may be significant competitors for food and habitat when abundance levels are high.

The most commonly reported parasites of northern shrimp are bopyrid (phryxid) isopods (10). These, however, are not considered to be a major contributor to shrimp mortality in the northwest Atlantic because of their low incidence and because they are more associated with delayed growth and maturation. Certain protozoan parasites infect muscle tissue (21), requiring culling of the catch in some northern areas, whereas others cause egg mortality in females (22). The latter infection appears to be of some concern as a potential cause of recruitment failure. Apollonio and Dunton (13) suggested that high temperatures in the Gulf of Maine during the period January–March of each year from 1950 to 1953 might have resulted in increases in the proportion of nonviable eggs (due to protozoan infection), thus contributing to poor egg production in those years and subsequently, to a marked reduction of the population.

The habitat of the species has been described for most northwest Atlantic stocks. From the Gulf of Maine to the Davis Strait, the largest shrimp concentrations occur mainly on muddy to sandy–mud bottom (8, 23) within a temperature range of approximately 1.0–6.0°C. These concentrations are found in depressions on the Labrador and Scotian shelves and on the slopes of the Store Hellefiske Banke off West Greenland at depths ranging from 150 to 500 m. In the Gulf of St. Lawrence, shrimp concentrations occur in deep troughs from 200 to 300 m. In the Gulf of Maine, the depth range is broad, extending from less than 10 m to greater than 300 m (8), and is probably related to different seasonal changes in temperature by locality. This stock represents the southern limit of the range of the species in the northwest Atlantic (1).

Horsted and Smidt (24) described populations of shrimp in the fjords of West Greenland distributed in relation to the hydrographic conditions occurring within the fjords. Generally, fjords that are open to ocean currents outside have warm bottom temperatures, resulting in highly productive shrimp populations. Those that have sills cannot receive periodic intrusions of warm water. These "threshold fjords" have very cold temperatures and support shrimp populations with low productivity levels that depend on larval drift from other areas.

There has been no documented discrimination of separate stocks in the northwest Atlantic. However, because concentrations exist in well-defined areas where known physical factors prevail (e.g., temperature and substrate), they have been separated traditionally for assessment purposes and as management units.

Stock maintenance mostly depends on the retention of larvae within an area during the pelagic phase. Ivanov (25) noted the presence of exploitable densities of shrimp around the Pribilof Islands in areas where gyres in the current occur. Several stocks in the northwest Atlantic are also associated with gyres, such as the Sept-Iles stock in the Gulf of St. Lawrence and the offshore stock in the Davis Strait (26). These gyres possibly limit the dispersal of shrimp larvae and may provide the mechanisms that make the stocks self-sustaining.

3. STOCK ASSESSMENT

3.1. Catch Information and Population Structure

Information on catch and fishing effort are recorded routinely through sales slips and/or logbooks. For most stocks, fishermen are required to record estimates of catch weight and the fishing effort in hours fished. All information is provided using either a grid system or by latitude and longitude, thus allowing the separation of catch data by management area. Information is also collected by observers for fisheries off Labrador and in the Davis Strait. These observers collect catch information along with details of catch composition including by-catch and discards.

The reasons for collecting catch and fishing effort data for shrimp can be summarized as follows:

To monitor catch levels and fishing effort in relation to existing regulations.

To estimate mean catch per unit effort as an index of changes in stock abundance over time.

To provide a basis to determine fishing mortality and, subsequently, population size. These data can be used along with data on size structure, which in turn can be used to estimate catch at age.

Information on the size distribution of shrimp catches are collected for all the important stocks in the northwest Atlantic in an attempt to study changes in size distribution and sex ratio during the commercial fishing period, to identify age classes and estimate their abundance through the separation of modal groups (age classes) in size distributions, and to estimate rates of growth and mortality.

Sampling programs should be designed that take into account temporal and spatial changes occurring over the fishing season. In the Gulf of St. Lawrence, where a relatively complete sampling regime has existed since 1977, sampling is conducted over the entire fishing season (April to December) with sampling effort distributed over each stock on a monthly basis. Port sampling is carried out in the Gulf of Maine by individual states over the duration of the fishing season (S. H. Clark, personal communication). In shrimp sampling in the Davis Strait, attempts have been made to account for diel changes in availability related to vertical migration (D. M. Carlsson, personal communication).

Sampling programs are useful only if they provide good representation of the actual size structure of the catch. However, size distributions vary with depth and area (14) and the young male shrimp are not completely available to the commercial trawl, because of selectivity of the mesh size used and because younger (smaller) shrimp are usually found in shallower water. This availability problem becomes more serious when the varying proportions of shrimp in the different age (modal) groups are estimated. Vertical migration activity differs from age group to age group over the diel cycle, which could also affect representation of the catch. Thus combining samples from different depths, locations, and sampling times should be considered with caution.

The sampling scheme used should minimize potential bias due to the above factors. Randomization of samples as well as a high sampling frequency over time can help reduce these biases and correction for selectivity differences can be achieved by applying adjustment factors to the partially selected sizes. Availability and diel variability problems are still difficult to account for, however.

An important objective of shrimp sampling is to separate size distributions into age groups, and a number of graphical and mathematical methods are designed to achieve this. When using these methods it is assumed that the modes occurring in size distributions represent age groups. These modal classes can be readily separated when there is no significant overlap between them. However, in most cases detailed biological observations must be used in conjunction with these methods. Separation of size distributions by sex allows the separation of the younger male and older female age groups. In addition, McCrary (27) developed a method to differentiate females that have not spawned previously (females with sternal spines) from those that have (females without sternal spines). It is assumed that females with spines represent a different (younger) age group from those without.

Figure 3 illustrates the utilization of a statistical method (28) in conjunction with detailed information on stages of sexual development. Good agreement is observed between the two methods in that the separation of male age groups is similar in terms of number of individuals estimated by both. Bimodality, which can be interpreted in the third modal class in the total length frequency, is clearly demonstrated when the biological separation of females with and without sternal spines is considered. These biological observations are therefore very useful in supplementing mathematical methods. However, it is obvious that the success of such an analysis totally depends on the synchrony of maturity events. For several stocks, synchrony is not present; transitionals or first-year females can be present

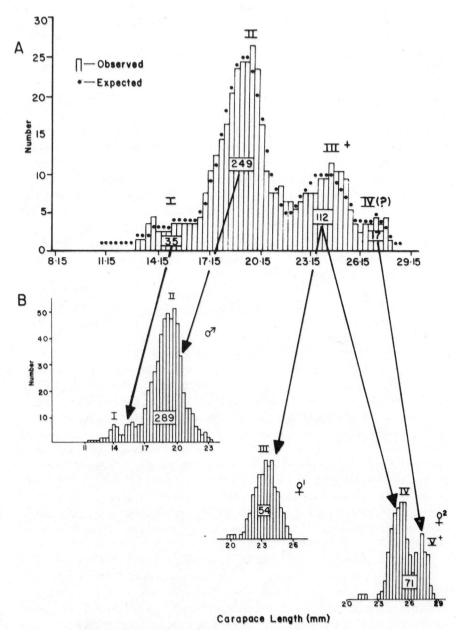

Figure 3. Separation of a shrimp length frequency distribution into age groups by mathematical (*A*) and biological (*B*) methods.

in more than one age group (29). This problem increases the difficulty of age interpretation from modal groups. When aging is possible, the results can be used to estimate growth, mortality, recruitment, and population size.

3.2. Shrimp Abundance Surveys

Most of the shrimp stock size estimates are obtained from research trawl surveys based on the standard methodology of random sampling and stratification by depth. For several of these stocks (e.g., off the east coast of Canada), management strategy depends heavily on direct biomass estimation. However, the utilization of the stratified random survey technique for shrimp is greatly affected by several sources of bias.

Vertical migration of shrimp during the night has been taken into account in some cases by making all fishing sets during daylight or by estimating day/night conversion factors. However, the latter proved to be inappropriate for some stocks off Labrador (30) because even during the day, shrimp occur off the bottom (18) and the pattern varies from day to day. A conversion factor of 1.5 was used in the biomass calculation for the Gulf of St. Lawrence stocks to take into account the difference between the vertical distribution of shrimps and the vertical opening of the trawls used during the surveys.

Also, the problems involved with delineation of stock distribution and the seasonal horizontal migration of shrimp affect the biomass estimation. These factors can influence the interpretation of biomass estimates over a series of years. It also should be noted that for some stocks, large and unexplained within-stratum variability affects the precision of the final estimates (30). This variability could be associated with the effect of tidal currents, which interact with the diel behavior of shrimp.

Despite these limitations, research surveys are still useful in estimating shrimp abundance. For several stocks, age interpretation has not yet been resolved and it is not possible to produce reliable estimates of biological parameters that permit the application of analytical assessment techniques. Development of new techniques that could replace or validate stratified random surveys is greatly needed.

Photographic surveys for shrimp in the West Greenland area have been conducted since 1975. In 1980, Jørgensen and Kanneworff (31) introduced a regression model for the distribution of shrimp from the photographic material. This model incorporates a number of parameters including year, depth, latitude, longitude, and four sets of bottom temperature data. A filtering procedure eliminates all nonsignificant parameters ($p > .10$) each time the model is run (32). The biomass estimates from the photographic analysis have been used along with the catch-per-unit-effort (CPUE) index to determine stock status.

4. ESTIMATION OF SHRIMP GROWTH AND MORTALITY

In some cases, abundance indexes such as CPUE and biomass estimates can be separated by age class. From these, the catch per unit effort per age class for several years as well as biomass per age class from direct shrimp surveys can be estimated. Year-class strength can be estimated using this methodology, but interpretation of the results must be made with caution because younger shrimps are not fully available on the grounds or to the trawls, and it is not always possible to be confident of the age interpretation for older females. It is generally recognized that the last female mode in length distributions corresponds to an accumulation of age groups (14).

The estimation of growth is based on the assumptions that modal classes in length distributions represent different age groups and that the annual cycle of reproduction is synchronous. Displacement of modes in size distributions over time (especially for males) has been observed for several stocks (33). The von Bertalanffy (34) growth model has been routinely used in the absence of a model that accounts for the discontinuous growth of crustaceans due to molting.

The determination of growth is affected, however, by extreme variability within and between age groups. For example, in the Gulf of St. Lawrence and likely in other areas as well, growth is greatly enhanced during sex inversion, although it is essentially absent afterwards, when females become ovigerous. On the other hand, Skúladóttir (35) detected the presence of fast and slow-growing age classes for an Iceland stock. This phenomenon, if present in other stocks, seriously affects shrimp age and growth estimation.

Where time series are available, attempts have been made to estimate rates of mortality. In the Gulf of St. Lawrence, analysis of two independent sets of survey data for two different shrimp stocks resulted in estimates of natural mortality rate (M) ranging from 0.5 to 0.8 for the fully recruited ages (36, 37). Rinaldo (38) estimated total mortality rate (Z) for shrimp in the Gulf of Maine by comparing numbers per tow per age class in successive surveys and showed that in most cases, values were consistently high; ranging from 1.0 to 2.0 and higher. Estimates from catch curves produced similar results using both research and commercial samples, whereas those from a virtual population analysis were generally lower. To obtain Z values from the latter method, an estimate of natural mortality (M) of 0.25 was used based on results of three regression analyses (Z versus total effort).

5. YIELD MODELS

Because of the difficulties in aging shrimp for some stocks, and hence determining mortality rates, yield models for most northwest Atlantic shrimp fisheries are not age-structured and mortality estimates must be assumed or only approximated.

Assessment of the offshore shrimp resource in the Davis Strait was first conducted by the International Commission for the Northwest Atlantic Fisheries (ICNAF) in 1976. Maximum sustainable yield (MSY) was estimated at 26,000 t annually, based on an estimate of stock size and the assumption that yields in the offshore would be similar to those in the Disko Bay, which were assumed to approximate to the MSY level. In later assessments, it was concluded that a safer position would be to maintain a spawning stock at 50% of the virgin level. Ulltang

(39) developed a method for calculating how much the fishery reduces the spawning stock based on the Beverton and Holt (40) yield equations. Critical parameters are mortality after first spawning (M_1) and the time (t) between recruitment to the fishery and first spawning (hatching of eggs). Values of M_1 and t were both estimated at approximately 1.5, and the annual fishing mortality predicted by the model that resulted in a 50% reduction of the spawning stock was $F = 0.4$. Assuming a fishable stock biomass of 100,000 t, a total allowable catch (TAC) of 40,000 t was recommended for 1977. Subsequent assessments, up to 1984, reevaluated catch levels by interpreting changes in fishable biomass. In 1985 it was recognized that previous catches had exceeded advised levels during a period of stability in catch rates, and advice for 1985 was based on the average catch from 1979 to 1984.

Assessment of shrimp resources in Atlantic Canada was first conducted by the Canadian Atlantic Fisheries Scientific Advisory Committee (CAFSAC) in 1978 when a range of options was considered to provide advice on catch levels. These included:

1. Yield $= 0.5MB_0$ where $M = 1.0$ is the natural mortality rate for fully recruited shrimps, and B_0 is the virgin biomass. The validity of this model was questioned owing to uncertainties as to the regularity of recruitment and the value of M.

2. Exploitation was aimed at a 50% reduction in spawning stock size as used by ICNAF/NAFO. The Ulltang (39) model produced an estimate of annual fishing mortality (F) of 0.35, assuming mortality after first spawning to be 1.0 and time between recruitment to the fishery and first spawning as 1.5 yr. These mortality estimates were used in the catch equation:

$$\text{catch} = \frac{B_0 F \,(1 - \exp\,(-Z)}{Z}$$

where B_0 is the biomass at the beginning of the fishing season and F and Z are fishing and total mortality, respectively. The estimates were considered conservative.

3. Maximize yield per recruit accepting Icelandic shrimp stocks as a representative model ($M = 0.35$, $F_{max} = 0.5$). This option recognized that the value of M used in options 1 and 2 above may be too high for the whole period that shrimps are in the fishery. The catch equation was again employed.

4. Exploitation was set at 40% of the available biomass (as determined from swept area methods): a figure interpreted as sustainable in the Sept-Iles fishery.

In subsequent assessments, estimates of natural mortality (M) used in option 1 were 0.5 and 0.7 based on interpretations of the age structure of the standing stock in areas in the Gulf of St. Lawrence. These rates applied to fully recruited sizes (mostly female) and were considerably lower than the value of $M = 1.5$ used by NAFO as the mortality rate after first spawning.

Option 4 became more widely used because it was based on observations from an ongoing fishery where relative stability in abundance had been observed. Also, the method did not require an estimate of M, a parameter to which most models are quite sensitive. The 40% exploitation was revised downward in 1982 to 35% based on new information on trawl efficiency. A review of catch data and biomass estimates in the Sept-Iles fishery indicated that productivity had not been adversely affected by exploitation rates averaging about 35% for the 1975–1984 period. Also, it was generally accepted that such a level of exploitation was appropriate for a species with natural mortality rates of the order of 0.5–0.8/yr.

D. Rivard (personal communication) describes two models used to determine reference catch levels or total allowable catches (TAC) for shrimp in Atlantic Canada.

Model 1: | sampling norm | $\longrightarrow$ | survey | $\longrightarrow$ | survey biomass | $\longrightarrow$ | TAC |

The target catch level for year $t + 1$ is calculated as

$$\text{TAC}_{t+1} = 0.35 \times B_t$$

where 0.35 is the target exploitation rate and B_t is the spring biomass estimate for year t. The variance of the target catch level can be calculated as

$$\text{Var}(\text{TAC}_{t+1}) = 0.35^2 \, \text{Var}(B_t)$$

Model 2: | sampling norm | $\rightarrow$ | survey year 1 <br> survey year 2 <br> survey year n | $\rightarrow$ | biomass year 1 <br> biomass year 2 <br> biomass year n | $\rightarrow$ | TAC |

The target catch level for year $t + 1$ is calculated as

$$\text{TAC}_{t+1} = 0.35 \times \overline{B}$$

where $\overline{B}$ is the average biomass for a number of previous years determined from research vessel surveys. Assuming the annual estimates of B_t are uncorrelated and have equal variance, then

$$\text{Var}\,(\overline{B}) = \frac{\text{Var}\,(B_t)}{k}$$

where k is the number of annual estimates. The TAC variance is calculated as before, substituting $\overline{B}$ for B_t.

Neither general production nor yield-per-recruit modeling have been considered appropriate at this stage for resource assessment of shrimp stock in the Davis Strait or off eastern Canada. Most fisheries are relatively new and relationships between CPUE and effort are poorly defined, often with only a few data points. Attempts to use yield-per-recruit models (with M values ranging from 0.5 to 0.8) have resulted in target fishing mortalities that are too high (about 1.0) to be useful as reference points for exploitation. Yield curves are generally flat-topped, indicating that beyond very low effort values, relatively large increases in effort will result in only small increases in yield per recruit. Changes in mesh size have not been considered because the currently used 40-mm gear is considered optimal for eliminating small, unmarketable shrimp while maintaining good catch rates. An attempt to use yield per recruit for fisheries in the North Sea and off Iceland concluded that such intense fishing as suggested by the model would be directed against females, possibly leading to recruitment failure (12). Therefore, mesh sizes and effort levels were chosen to ensure adequate recruitment, thus compromising yield per recruit.

An assessment conducted for the Gulf of Maine stock in 1982 underscored the limited applicability of yield-per-recruit models in northern shrimp management, but concluded that the technique did provide management insight into the Gulf situation (4). The Ricker (41) model was used over a range of natural mortalities. Assuming M to be constant at 0.25 prior to age 4 followed by an increase to 1.0, it was concluded that a late winter fishery (when age-4 females concentrate in shoaler water) would contribute toward maximizing yield. A later winter fishery could also enhance recruitment by reducing mortality on ovigerous females, because most hatching activity occurs from mid-February to mid-March.

Recent assessments of the Gulf of Maine stock have been based primarily on commercial landings and catch–effort data, biological data from commercial sampling, and research vessel data collected during spring, summer, and fall surveys. The results have indicated that the resource continues to recover after the collapse experienced in the mid 1970s. Length-frequency analysis has indicated that the 1982 year class is the strongest since the early 1970s. Advice forwarded from these assessments has been framed around the appropriateness of the regulated fishing season. In 1985, it was agreed that a season extending from December 1 through May 15 would not adversely affect the stock (S. H. Clark, personal communication).

A cohort analysis was first attempted in 1985 for the Sept-Iles stock in the Gulf of St. Lawrence (42). Sets of data on catch composition and commercial landings for several years were used to produce quarterly catch-at-age matrices from the commercial size distributions. The cohort analysis was calibrated by comparing fishing mortality and fishable biomass for different values of the fishing mortality in the final year (F_t) with standardized fishing effort and catch per effort. The exploitable biomass estimates agreed reasonably well with the CPUE time series. Comparisons with direct biomass estimates were not possible, because only three pairs of data were available.

Cohort analysis is least suitable for species with short life-spans and low exploitation rates. Because the male age groups are not fully available to the fishery, partial recruitment of these ages is low, resulting in a short age span in the fishery. Also, age estimates for females contain considerable error owing to overlapping of several age classes in the length distribution.

Assessment of pandalid shrimp in other areas have considered the traditional methodologies used for finfish as well as new techniques designed specifically for shrimp. Fox (43) determined that yield-per-recruit models are not likely to provide a basis for management in most cases whereas general production models described equilibrium conditions very well. He also developed a simulation model (GXPOPS), which included an age- and sex-specific maturation schedule, a model of copulation in random mating, a provision for specifying the length of the ovigerous period, and a yield computation scheme accounting for stepwise growth. Rinaldo (38) found this model to be adequate for predicting the observed stock decline and approximate yields for the Gulf of Maine shrimp in 1975 and 1976.

Skúladóttir (44) used surplus production models for Icelandic shrimp stocks for which there was good effort data. She concluded that a linear model gave more realistic results when variation in effort is small, but an exponential model is more appropriate when effort varies greatly over the time series.

6. CURRENT FISHERY ISSUES

6.1. Regulations

Northern shrimp fishery regulations have been implemented that attempt to address three primary concerns: maintenance of recruitment, stock protection, and economic viability. Concerns for recruitment have been approached by the implementation of mesh-size regulations and closed areas. Shrimp fisheries in the Davis Strait and Labrador Sea impose a minimum mesh size of 40 mm, and the Gulf of Maine enforces a 44.5-mm mesh regulation. An area adjacent to Disko Bay in the Davis Strait has been designated as an area that possibly contributes to the inshore resource and consequently offshore catches in this area have been controlled. Large offshore vessels fishing at West Greenland are not permitted to fish inshore where there are no regulations. Also, Greenland has recently (1985) introduced a regulation that prohibits the discarding of shrimp over 2 g. This regulation not only minimizes discards, but also assures increased employment in land-based industries, which mainly process smaller shrimp (D. M. Carlsson, personal communication).

TACs have been used as the principal means of stock protection and currently are enforced in the Davis Strait offshore fishery and all Canadian fisheries. The Gulf of Maine has been regulated by seasonal closures since 1975, except in 1978 when there was a complete closure (4). The season is also closed in spring in this area when the shrimp count exceeds 70/lb.

Economic viability has been considered in setting regulations in most fisheries, resulting in some form of limited entry and/or effort control. The number of participants for some Gulf of St. Lawrence fisheries has been regulated to ensure an extended season and acceptable income level. The number of vessels fishing under the Canadian flag off Labrador and for various countries in the Davis Strait have been restricted (with a few exceptions), each operating with individual quotas, thus guaranteeing an acceptable catch level for all license holders.

6.2. Problems

Problems encountered in northern shrimp fisheries are similar to those encountered for finfish, and are related to the product, the environment, and overall economics. Discarding small shrimp is a persistent problem both for the industry and fisheries managers, especially in the Davis Strait and Labrador Sea. Market preference is for the larger shrimp, and often discarding occurs to maintain the desired counts. Discard rates vary between vessels, some being equipped with peelers that can process the smaller shrimp on board whereas others must maintain strict size requirements. Occasionally, off-odors are encountered in the cooked product, possibly related to the feeding behavior of the shrimp, type of substrate, or improper processing. By-catches, especially of cod and Greenland halibut, can be extremely high. Some shrimp vessels are not equipped to handle these species and they are often discarded, whereas in other fisheries, such as in the Gulf of St. Lawrence, the by-catch represents a significant source of income. When abundant, by-catch species can cause damage to the shrimp in the trawl. During the late fall in the Davis Strait, for example, by-catches of Greenland shark are quite high and can cause so much damage that the entire catch has to be discarded. Shrimp also are trawled over muddy substrate, and at times mud is taken up in the trawl, often ruining the catch. In inshore areas at West Greenland, by-catches of a currently unmarketable pelagic shrimp (*Pasiphaea* spp.) are at times quite high, requiring much sorting of the catch (D. M. Carlsson, personal communication).

The discarding of regulated by-catch species has also been a problem in some northwest Atlantic fisheries. Large quantities of small (≤ 25 cm) redfish (*Sebastes* spp.) were routinely discarded in the northeastern Gulf of St. Lawrence shrimp fishery and concerns were raised about deleterious effects on the redfish stocks. Various types of sorting trawls were introduced as a means to reduce the by-catch but gained little acceptance by the fishermen. Subsequent studies indicated that although the discarding practices of the late 1970s were visually alarming (discarded redfish float and virtually all die), the levels attained would not seriously affect the redfish stock. Off Labrador, by-catches of cod (*Gadus morhua*) and Greenland halibut (*Reinhardtius hippoglossoides*) are often high and because these are commercially valuable, a limit of 10% by-catch of regulated species has been applied.

In the offshore areas of Labrador and the Davis Strait, weather conditions can be severe, especially during the fall and winter. Considerable fishing time is lost every year in weathering out the storms. As mentioned previously, ice covers the fishing grounds in northern areas in winter and spring. Residual ice, even in summer, often covers the preferred fishing areas and represents a hazard to the vessels and crews that must work in it. In August of 1984, one of the newest, largest, and most efficient shrimp vessels struck an ice pan in the Davis Strait and sank, fortunately with no casualties.

The economics of fishing are affected by rising production costs, fluctuating markets, variable catch rates, and stock instability. The distance to and from the Davis Strait for east coast Canadian vessels is vast, and the economics of fishing in remote areas is considered to be marginal, at best. Markets for northern shrimp from the northwest Atlantic have been affected recently by increased catches of the same species in the Spitzbergen–Barents Sea. Concentrations of shrimp in most fisheries are high early in the year (late spring–summer) but decline as the season progresses. Prolonged ice conditions, as occurred in the Labrador Sea in 1985, can prevent fishing at optimal times. Stocks also have been noted for their instability as evidenced in the Gulf of Maine where the fishery virtually collapsed in the mid 1970s, as described previously.

Continued demands for new licenses exist in some areas. For fully utilized stocks, additional licenses could result in overexploitation as well as lower returns for participants. For underutilized stocks, there is general apprehension over issuing new licenses for fisheries in which the overall economic picture is uncertain.

6.3. Surveillance and Enforcement

Routine reporting of catch and effort are required as a condition of licensing in all northern shrimps fisheries. Observers are deployed on vessels fishing for Canada in the Labrador Sea and in the Davis Strait. These observers monitor catch and effort, mesh sizes, discards, by-catches, and logbook entries. If violations of fisheries regulations occur, patrol vessels are requested to investigate the situation further. Surveillance vessels also patrol the Canadian Atlantic, and boardings of shrimp vessels are made routinely to ensure regulations are being followed. Surveillance aircraft also patrol the Davis Strait, ensuring that Canadian licensed vessels remain inside the Canadian zone. The Danish Navy conducts similar surveillance in the Davis Strait east of the international boundary and there is surveillance inshore to ensure that large offshore vessels (>80 t) do not fish in these areas.

In the Gulf of Maine, individual states provide enforcement to ensure that the season is adhered to and the 44.5-mm mesh size is being used. State wardens board vessels at dockside to measure the mesh or to ensure that no shrimp are landed during the closed season (S. H. Clark, personal communication).

7. SOCIOECONOMICS

The Canadian fishery for shrimps in the Labrador Sea and Davis Strait presently has two components; Canadian-owned vessels and chartered foreign vessels. The latter have been restricted to the Davis Strait and were allowed as a special consideration, recognizing traditional rights of Labrador fishermen. These charter arrangements have been profitable, and the profits have been used to cross-subsidize other less profitable inshore fisheries. Canadian-owned vessels must travel extensive distances to the fishing grounds and are faced with foreign tariffs associated with Canadian products. The economic situation for these operations is highly variable and depends on good catch rates and markets as well as access to groundfish stocks.

In the Gulf of Maine, an economic study was conducted by Dunham and Mueller (45) in 1976 when the shrimp resource was low. It was concluded that the resource at that time was inadequate for the existing industry but that vessels and processing companies could diversify to other fishery resources. There have been no analyses to reflect the recent recovery of the stock (S. H. Clark, personal communication).

In 1984, some fishermen from the Gulf of St. Lawrence (Quebec) experienced economic losses due to a decrease in the commercial catch rates for the shrimp stock north of Anticosti Island. This situation probably also exists in other fisheries, but may not be entirely explained by decreasing catch rates. Other factors are also important in maintaining profitability in shrimp fishing, especially for vessels

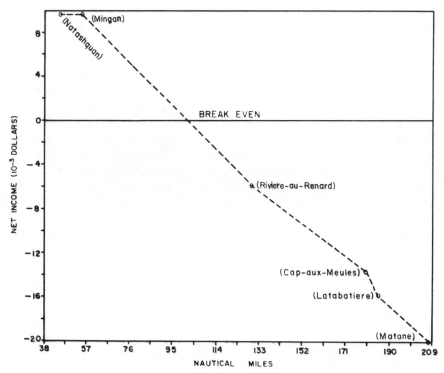

Figure 4. Relationship between net income and distance from fishing grounds for the shrimp fishery north of Anticosti Island in the Gulf of St. Lawrence (from C. Janelle, personal communication).

ranging in size from 15 to 25 m. Figure 4 (C. Janelle, personal communication) illustrates in general terms the expected relationship between net income and the distance from fishing grounds to port of landing, assuming a maximum of 5 days between the first catch taken and the time of landing. It is apparent that the greater the distance to the fishing grounds, the more the effective fishing time will decrease. This is necessary to maintain good quality of the shrimp, which are kept on ice. The break-even point in net income intercepts the distance line at about 100 nautical miles. However, many factors are involved that can affect this relationship, especially yearly catch rate variation, ex-vessel shrimp prices, variation in cost of fishing, and so on. For vessels of this size, which must keep the shrimp catch on ice, the profit margin is expected to be closely related to the proximity of the fishing grounds.

8. PRIORITIES AND APPROACHES

The provision of precise and accurate scientific advice is critical for the effective management of northern shrimp resources. The development of more accurate aging techniques than are presently available would be a major step toward achieving this goal, providing detailed information on mortality, growth, population size and age composition, equilibrium yield, and possible stock–recruitment relationships. In particular, problems related to availability of the smaller shrimp and accumulation of female age groups must be resolved.

Additional research into survey design for shrimp is necessary to explain variability in abundance and distribution in time and space. Advances in this area have implications both for research and commercial fishing. Mechanisms of recruitment also must be investigated to determine whether or not stocks are self-sustaining or dependent on larvae from other areas. These studies require estimates of the reproductive capacity of the parent stock, along with determination of the effects of environmental factors such as currents, gyres, temperature, and ice.

Acoustic methodologies should be developed along with trawl and photographic surveys to determine the distribution and behavior of shrimp. Such studies will also be useful in evaluating the design and performance of various sampling devices and fishing gears. The efficacy of commercial fishing gear should be evaluated in terms of the ability to avoid small shrimp and by-catch species while maintaining economically acceptable catch rates.

Shrimp serve as food for a number of finfish species, some of which also are of commercial importance. These relationships must be fully investigated to determine where dependencies exist and how changes in abundance of either predator or prey interact.

Management should also be concerned with the development of fisheries in unexploited areas. Regulations for developing fisheries should take into account that the stocks may undergo dramatic changes in abundance from year to year. Dense concentrations of a closely related species, *P. montagui*, have been known to exist in the eastern Hudson Strait and Ungava Bay but fisheries have not been extensive in these areas. Even under virgin conditions, the stock in the latter area has shown pronounced changes in abundance. Such stock characteristics preclude the implementation of long-term management plans.

Although conservation of the fishery resource along with maximizing socioeconomic benefits remain important management objectives, the special status of the northern shrimp in terms of its biology and ecological sensitivity must be taken into account in developing effective management regimes.

ACKNOWLEDGMENTS

The authors are indebted to the following people for providing recent, undocumented information on shrimp fisheries and assessment methodology: D. M. Carlsson (Denmark), S. H. Clark (United States), C. Janelle (Canada), and D. Rivard (Canada).

REFERENCES

1. H. J. Squires, Decapod crustaceans of Newfoundland, Labrador and the Eastern Canadian Arctic. *Fish. Res. Board Can. Manuscr. Rep. Ser.* 810:212 (1970).

2. D. M. Carlsson and E. Smidt, Shrimp, *Pandalus borealis* Krøyer, stocks off Greenland: Biology, exploitation and possible protective measures. *ICNAF Sel. Pap.* 4:7–14 (1978).

3. R. L. Dow, Shrimp management in the Gulf of Maine. In T. Frady, Ed., *Proceedings of the International Pandalid Shrimp Symposium*, U.S. Sea Grant Rep. 81-3. University of Alaska, Fairbanks, 1981, pp. 59–61.

4. S. H. Clark, Assessment and management of the Gulf of Maine Northern shrimp (*Pandalus borealis*) fishery. *ICES, C. M. Doc.* K:13:20 (1982).

5. D. B. Carlisle, On the sexual biology of *Pandalus borealis* (Crustacea Decapoda) I. Histological of incretory elements. *J. Mar. Biol. Assoc. U.K.* 38:381–394 (1959).

6. A. A. Berkeley, The post-embryonic development of the common pandalids of British Columbia. *Contrib. Can. Biol. Fish., New Ser.* 6(6):79–163 (1930).

7. J. Berreur-Bonnenfant and H. Charniaux-Cotton, Hermaphrodisme proterandrique et fonctionnement de la zone germinative chez la crevette *Pandalus borealis* Kröyer. *Bull. Soc. Zool. Fr.* 90:243–259 (1965).

8. E. B. Haynes and A. L. Wigley, Biology of the Northern shrimp, *Pandalus borealis* in the Gulf of Maine. *Trans. Am. Fish. Soc.* 98:60–76 (1969).

9. J. A. Allen, On the biology of *Pandalus borealis* Krøyer with reference to a population off the Northumberland coast. *J. Mar. Biol. Assoc. U.K.* 38:189–220 (1959).

10. S. E. Shumway, H. C. Perkins, D. F. Schick, and A. P. Stickney, Synopsis of biological data on the Pink shrimp, *Pandalus borealis* Krøyer, 1838. *NOAA Tech. Rep., NMFS* 30 (FAO Fish. Synop. No. 144):57 (1985).

11. E. L. Charnov, Sex reversal in *Pandalus borealis:* Effects of a shrimp fishery? *Mar. Biol. Lett.* 2:53–57 (1981).

12. International Council for the Exploration of the Sea (ICES), Cooperative Research Report—*Int. Counc. Explor. Sea* 83:60–82 (1978).

13. S. Apollonio and E. E. Dunton, *The Northern Shrimp*, Pandalus borealis, *in the Gulf of Maine*, Completion Report, Proj. 3-12R. Maine (U.S.A.) Department of Marine Research, 1969, p. 81.

14. J. Fréchette and D. G. Parsons, Report of shrimp ageing workshop held at Ste. Foy, Quebec, in May and at Dartmouth, Nova Scotia, in November 1981. *NAFO Sci. Counc. Stud.* 6:79–100 (1983).

15. L. Barr and R. McBride, Surface-to-bottom pot fishing for Pandalid shrimp. *U.S., Fish. Wildl. Serv., Spec. Sci. Rep.-Fish.* 506:7 (1967).

16. L. Barr, Diel vertical migration of *Pandalus borealis* in Kachemak Bay, Alaska. *J. Fish. Res. Board Can.* 27:669–676 (1970).

17. R. W. Wienberg, On the food and feeding habits of *Pandalus borealis* Krøyer 1838. *Arch. Fishchereiwiss.* 31:123–137 (1981).

18. J. Fréchette, S. Pilote, and G. Chabot, Données préliminaires sur la distribution verticale de la crevette, *Pandalus borealis* et ses implications sur les estimations de stocks. *Can. J. Fish. Aquat. Sci., Spec. Publ.* 58:218–226 (1981).

19. D. M. Carlsson, Sv. Aa. Horsted, and P. Kanneworff, Danish trawl surveys on the offshore West Greenland shrimp grounds in 1977 and previous years. *ICNAF Sel. Pap.* 4:67–74 (1978).

20. W. R. Bowering, D. G. Parsons, and G. R. Lilly, Predation on shrimp (*Pandalus borealis*) by Greenland halibut (*Reinhardtius hippoglossoides*) and Atlantic cod (*Gadus morhua*) off Labrador. *ICES, C. M. Doc.* G:54:30 (1984).

21. D. G. Parsons and R. A. Khan, Microsporidiosis in the Northern shrimp, *Pandalus borealis. J. Invertebr. Pathol.* 47:74–81 (1986).

22. A. P. Stickney, A previously unreported peridinian parasite in the eggs of the northern shrimp, *Pandalus borealis. J. Invertebr. Pathol.* 32:212–215 (1978).

23. T. H. Butler, A review of the biology of the pink shrimp *Pandalus borealis*. In *Proceedings-Conference on the Canadian Shrimp Fishery, Saint John, New Brunswick, October 27–29, 1970,* Can. Fish. Rep. No. 17. 1971, pp. 17–24.

24. Sv. Aa. Horsted and E. Smidt, The deep sea prawn (*Pandalus borealis*) in Greenland waters. *Medd. Dan. Fisk.- Havunders.* 1:11–118 (1956).

25. B. G. Ivanov, Distribution patterns of the deep-sea prawn (*Pandalus borealis* Kr.) in the Bering Sea and the Gulf of Alaska. *Oceanology* 7:715–721 (1967).

26. Sv. Aa. Horsted, P. Johansen, and E. Smidt, On the possible drift of shrimp larvae in the Davis Strait. *ICNAF Res. Doc.* 78/XI/93 (Ser. No. 5309):13 (1978).

27. J. A. McCrary, Sternal spines as a characteristic for differentiating between females of some Pandalidae. *J. Fish. Res. Board Can.* 28(1):98–100 (1971).

28. V. Hasselblad, Estimation of parameters for a mixture of normal distributions. *Technometrics* 8:431–444 (1966).

29. B. Rasmussen, Variations in protandric hermaphroditism of *Pandalus borealis*. *FAO Fish. Rep.* 57:1101–1106 (1969).

30. D. G. Parsons and E. J. Sandeman, Groundfish survey techniques as applied to abundance surveys for shrimp. *Can. J. Fish. Aquat. Sci., Spec. Publ.* 58:124–146 (1981).

31. A. G. Jørgensen and P. Kanneworff, Biomass of shrimp (*Pandalus borealis*) in NAFO Subarea 1 in 1977–1980 estimated by means of bottom photography. *NAFO SCR Doc.* 80/XI/169:14 (1980).

32. P. Kanneworff, Biomass of shrimp (*Pandalus borealis*) in NAFO Subarea 1 in 1981–84, estimated by means of bottom photography. *NAFO SCR Doc.* 85/I/8:18 (1985).

33. J. W. Balsiger, A review of pandalid shrimp fisheries in the northern hemisphere. In T. Frady, Ed., *Proceedings of the International Pandalid Shrimp Symposium*, U.S. Sea Grant Rep. No. 81-3. University of Alaska, Fairbanks, 1981, pp. 7–35.

34. L. von Bertalanffy, A quantitative theory of organic growth. *Hum. Biol.* 10:181–213 (1938).

35. U. Skúladóttir, The deviation method: A simple method for detecting year-classes of a population of *Pandalus borealis* from length distributions. In T. Frady, Ed., *Proceedings of the International Pandalid Shrimp Symposium*, U.S. Sea Grant Rep. No. 81-3. University of Alaska, Fairbanks, 1981, pp. 283–306.

36. J. Fréchette and S. S. M. Labonté, Biomass estimate, year-class abundance and mortality rates of *Pandalus borealis* in the northwest Gulf of St. Lawrence. In T. Frady, Ed., *Proceedings of the International Pandalid Shrimp Symposium*, U.S. Sea Grant Rep. No. 81-3. University of Alaska, Fairbanks, 1981, pp. 307–330.

37. J. Fréchette and D. G. Parsons, Preliminary estimation of total mortality for shrimp (*Pandalus borealis*) in the Anticosti Channel. *Can. Atl. Fish. Sci. Advis. Comm. (CAFSAC) Res. Doc.* 81/82:10 (1981).

38. R. G. Rinaldo, Population assessment of the northern shrimp, *Pandalus borealis*, (Krøyer) in the Gulf of Maine, 1965–1975. Ph.D. Thesis, University of Maine, Orono, U.S.A., 1981, 206 pp.

39. Ø. Ulltang, A method for determining the total allowable catch of deep sea shrimp, *Pandalus borealis*, off West Greenland. *ICNAF Sel. Pap.* 4:43–44 (1978).

40. R. J. H. Beverton and S. J. Holt, On the dynamics of exploited fish populations. *Fish. Invest.—Minist. Agric., Fish. Food (G.B.) (Ser. 2)* 19:533 (1957).

41. W. E. Ricker, Computation and interpretation of biological statistics of fish populations. *Bull. Fish. Res. Board Can.* 191:382 (1975).

42. B. Portelance and J. Fréchette, Analyse des cohorts du stock de crevettes de Sept-Iles. *Can. Atl. Fish. Sci. Advis. Comm. (CAFSAC) Res. Doc.* 85/56:49 (1985).

43. W. W. Fox, Jr., Dynamics of exploited pandalid shrimps and evaluation of management models. Ph.D. Thesis, University of Washington, Seattle, U.S.A., 1972, 193 pp.

44. U. Skúladóttir, The experience of the catch per effort versus average effort, the methods of Gulland and Fox, in *Pandalus borealis* fisheries at Iceland. In T. Frady, Ed., *Proceedings of the International Pandalid Shrimp Symposium*, U.S. Sea Grant Rep. No. 81-3. University of Alaska, Fairbanks, 1981, pp. 181–195.

45. W. C. Dunham and J. J. Mueller, *The Economic Impact of a Reduction in Shrimp Landings Under Regulated and Unregulated Programs*. Dept. Res. Agric. Econ., University of Maine, Orono, U.S.A., 1976, p. 49.

4 APPROACHES TO RESEARCH AND MANAGEMENT OF U.S. FISHERIES FOR PENAEID SHRIMP IN THE GULF OF MEXICO

Edward F. Klima
Southeast Fisheries Center
Galveston, Texas

1. INTRODUCTION

Trawl fisheries of the U.S. southeast region harvest two major faunal groups: shrimp and bottom fish. The shrimp fishery is considered to be the most valuable fishery in the United States. In 1984, the Gulf and Atlantic coastal states produced 124,000 metric tons (274 million lb) of whole shrimp valued at more than $449 million. Historically, the Gulf of Mexico has been the major U.S. production area for shrimp and accounts for approximately 80% of the total value of shrimp landed in the United States. Shrimp production in the Gulf has fluctuated between 61,000 metric tons (134 million lb) in 1961 and 121,000 tons (266 million lb) in 1977. The directed shrimp fishery in the Gulf of Mexico harvests brown shrimp, *Penaeus aztecus* (the dominant species, accounting for more than 50% of the total production), white shrimp, *P. setiferus*, and pink shrimp, *P. duorarum* (which account for 25–35% and 15–20%, respectively). This shrimp fishery also incidentally catches and discards up to 900,000 metric tons of bottom fish annually (Klima 1).

Major emphasis on shrimp research was initiated in the late 1950s by the Gulf coastal states and by the Galveston Laboratory of the Bureau of Commercial Fisheries (now the National Marine Fisheries Service, NMFS) to explore shrimp resources and to define basic biological parameters (Temple 2; Caillouet and Baxter 3). Since that time, the Gulf states have continued a strong research role in providing scientific information for management of the shrimp resources. Each state was funded by its legislature and undertook programs that provided appropriate information for management. The federal government provided biological data, catch and fishing effort, and some economic data to the Gulf states. In the early 1960s, intensive surveys by the Bureau of Commercial Fisheries throughout the Gulf of Mexico provided basic information on the biology, spawning cycle, maturation, growth, and life history of the major shrimp stocks. These findings were summarized by Lindner and Cook (4, 5) and Costello and Allen (6). In the late 1960s, the Bureau of Commercial Fisheries significantly decreased its emphasis on shrimp research.

In 1976, the United States Congress enacted The Fishery Conservation and Management Act of 1976 (Magnuson Act), extending U.S. jurisdiction from the edge of state territorial waters to 200 miles offshore and creating regional management councils that have responsibility for developing management plans for all U.S. coastal ocean resources. As a result of this action, emphasis was again placed on providing scientific information for the management of shrimp resources.

This paper summarizes results of current shrimp management, and research programs, including estuarine research programs in the U.S. Gulf of Mexico.

2. BACKGROUND

2.1. The Fishery

The distribution and relative abundance of white, brown, and pink shrimp have been described by Osborn et al. (7) and are depicted in Figures 1–3. The brown shrimp population is found throughout the northern and western Gulf of Mexico, with the center of abundance off Texas. White and pink shrimp are also located throughout the northern Gulf with the center of abundance of white shrimp off Louisiana and that of pink shrimp off southern Florida, with highest concentrations

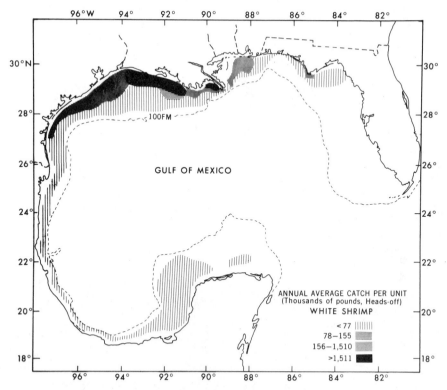

Figure 1. Distribution of catch per unit (thousands of pounds) of white shrimp in the Gulf of Mexico.

around the Dry Tortugas area. White shrimp are believed to have a continuous distribution throughout the northern half of the Gulf and into Mexico. Lindner and Anderson (8) found tagged white shrimp moved across the United States–Mexico border. There appear to be two separate stocks of pink shrimp, one on the Campeche Bank off Mexico and the other off south Florida on the Tortugas and Sanibel grounds. Sheridan et al. (9) and Klima et al. (10) have shown that both brown and pink shrimp stocks freely transit the United States–Mexico border.

Biological information and crude landing statistics have been collected since the turn of the century in the Gulf of Mexico. After World War II, the shrimp fisheries of the Gulf expanded rapidly with the development of the Tortugas fishery off south Florida and the brown shrimp fishery off Texas. Prior to that time, the fishery exclusively pursued white shrimp during daylight hours. Both the pink and brown shrimp fisheries were nocturnal, and production increased rapidly from the early 1950s to the present. The development of more powerful vessels, fishing gear, and electronics allowed the fleet to expand to offshore areas throughout the Gulf of Mexico.

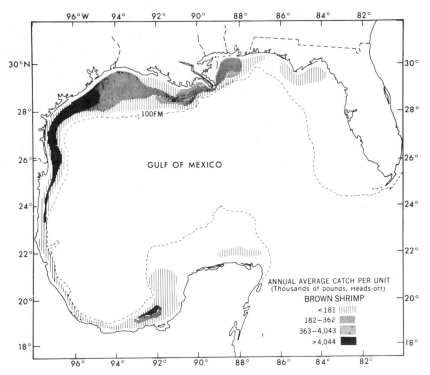

Figure 2. Distribution of catch per unit (thousands of pounds) of brown shrimp in the Gulf of Mexico.

Brown shrimp landings and directed effort have doubled since 1960, despite frequent short-term fluctuations (Fig. 4). Catch-per-unit effort (CPUE) has shown considerable and frequent fluctuations but no increasing or decreasing trend. The average size of captured brown shrimp (Fig. 5) has decreased markedly since 1960 and appears to be inversely related to the increase in effort (Nichols 11). Caillouet et al. (12) likewise have shown a decrease in the average size of brown shrimp in the central Gulf, apparently due to high fishing effort exerted on juveniles in the estuarine areas of Louisiana. In contrast, a relatively constant size off the coast of Texas may be a result of management practices where few immature shrimp have been taken through 1979. This trend is changing as Texas inshore fisheries for brown shrimp began increasing in 1979 (Klima et al. 13).

White shrimp landings have fluctuated appreciably since 1960 (Fig. 6). Peak catches during good years have been relatively constant, but the catches in poor years have been increasing (Nichols 11). The directed effort for white shrimp has more than doubled despite substantial short-term fluctuations. On the other hand, CPUE has shown a long-term decline of 25%, with short-term fluctuations of much

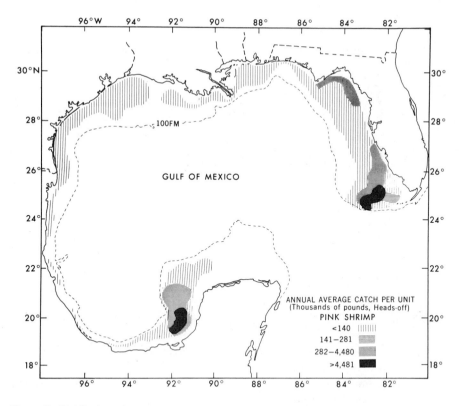

Figure 3. Distribution of catch per unit (thousands of pounds) of pink shrimp in the Gulf of Mexico.

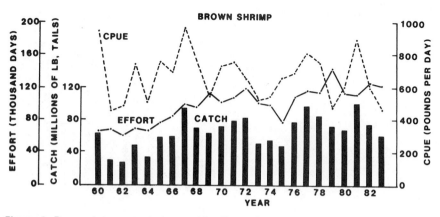

Figure 4. Brown shrimp reported annual landings, directed effort, and average catch per unit effort.

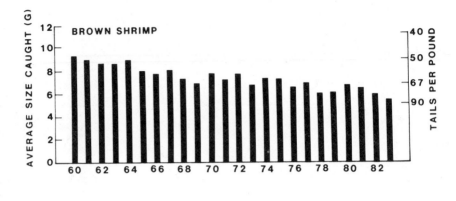

Figure 5. Annual average size of brown shrimp landed.

greater magnitude. The average size of captured white shrimp has declined over the 24-yr period (Fig. 7), but the decline is not as great as the decline observed with brown shrimp (Nichols 11).

Pink shrimp landings have been relatively stable since 1960 with appreciable fluctuations and no apparent trend (Fig. 8) (Klima et al. 14). Fluctuations are appreciably smaller than observed for either the brown or white shrimp stocks. Directed effort appears to fluctuate around two levels with a transition in the early 1970s to appreciably higher levels of effort from 1973 to the present. CPUE appears

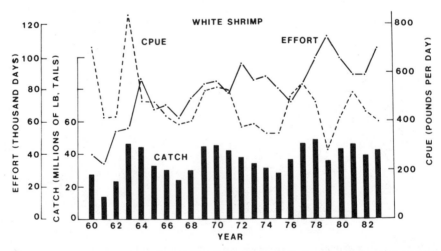

Figure 6. White shrimp reported annual landings, directed effort, and average catch per unit effort.

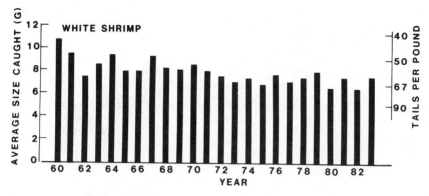

Figure 7. Annual average size of white shrimp landed.

stable throughout the period with no apparent trend. Likewise, the average size of shrimp (Fig. 9) has fluctuated considerably with no continuing trend over the 24-yr time frame (Nichols 11).

2.2. Biological Data Base

The biological data base for shrimp has been accumulating for the past 40 yr. The initial work reported by Lindner and Anderson (8) set the stage for understanding

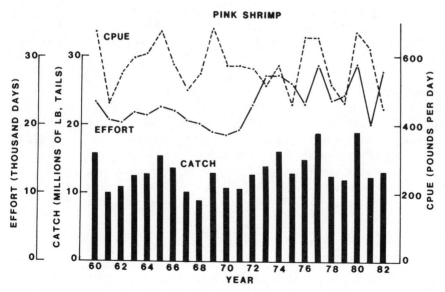

Figure 8. Pink shrimp reported annual landings, directed effort, and average catch per unit effort.

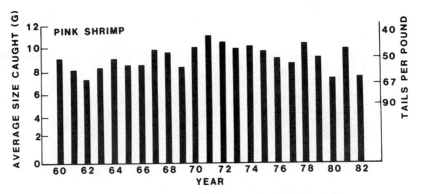

Figure 9. Annual average size of pink shrimp landed.

migration and growth of the white shrimp throughout the northern Gulf and south Atlantic. Recently workers have greatly expanded information on growth, mortality, and other important biological aspects. Some important findings on mortality and growth used to manage shrimp fisheries are reported by Klima et al. (15), Parrack (16), Iversen and Jones (17), Costello and Allen (6), Kutkhun (18), Berry (19, 20), Lindner (21), and Neal (22), and thoroughly reviewed by Christmas and Etzold (23). Growth studies conducted by NMFS, summarized by Klima (24) and reported by Phares (25), show that growth of shrimp is seasonal and is positively related to temperature. Nichols (11) updated the assessment of brown, white, and pink shrimp stocks in the Gulf of Mexico and put into focus information available for management of the shrimp stocks of this area. Nichols (11) not only updates most of the biological information but adds new information on age, growth, and estimates of mortality for these important stocks. Recognizing that estimation of natural mortality (M) may be the single most difficult technical problem in fisheries science, he used two variations of fishing effort to obtain estimates of total mortality. Nichols provided the best estimates for brown shrimp of M as 0.27–0.31/month; best estimates of M for white shrimp were 0.20–0.22/month. He believes that M for adults for both species is between 0.20 and 0.35/month but he has been unable to narrow this range further. Nichols (26) used such data to develop models for the Gulf of Mexico Fishery Management Council related to the Texas Closure (Nichols 27; Poffenberger 28) as well as to provide information on the impact of the Tortugas Sanctuary area on pink shrimp (Nichols 26).

2.3. Management

By the late 1940s, the Gulf coastal states individually expressed concern over the economic welfare of the shrimp fisheries and began implementing state regulations to conserve and manage fishery resources. General objectives of the present state management systems have been to protect the resources and maximize catch among the various user groups. Regulations of the size of harvestable shrimp have been implemented by the various states. Most states regulate the harvestable size of shrimp by opening and closing seasons in state waters and, to some degree, by the restriction of various gears. A summary and comparison of the Gulf states regulations is presented in Table 1. With the implementation of the Magnuson Act in 1976 the Gulf of Mexico Fishery Management Council (GMFMC), one of eight regional management councils, completed a plan in 1981 for managing the shrimp fisheries in the Fishery Conservation Zone (FCZ) (i.e., federal waters that extend from the Territorial Sea out to 200 miles).

Management measures are based on assumptions that no recruitment overfishing occurs, that the stock is a single year class, and that there is a need to protect small shrimp, because growth overfishing is a significant problem in most of the coastal states. Rothschild and Brunenmeister (29) summarized the council's scientific view of the shrimp stocks in the Gulf as follows: (1) there is no demonstrable relationship between stock size and recruitment levels for Gulf of Mexico shrimp stocks; (2) recruitment overfishing of shrimp stocks is impossible; (3) there should be no constraints on the quantity of shrimp taken each year; (4) the environment, especially temperature and salinity and not stock size, controls the success of recruitment; and (5) surplus production models are inadequate for providing guidance on the relationship between stock production and the amount of fishing. These authors discuss the council's views and in general provide information that negates these points. Present management regulations are based more or less on these assumptions.

It is not the intent of this paper to review these concepts except to concur with Rothschild and Brunenmeister (29) that in fact, production models do serve a useful purpose for management, that recruitment overfishing is a distinct possibility with Gulf of Mexico shrimp stocks, and that care should be taken in management of these stocks so that recruitment overfishing does not occur.

The GMFMC identified multiple management problems, and adopted a goal and objectives to resolve these problems:

Goal: To manage the shrimp fishery of the United States waters of the Gulf of Mexico in order to attain the greatest overall benefit to the nation with particular reference to food production and recreational opportunities on the basis of the maximum sustainable yield as modified by relevant economic, social, or ecological factors.

Objectives:

1. Optimize the yield from shrimp recruited to the fishery.
2. Encourage habitat protection measures to prevent undue loss of shrimp habitat.
3. Coordinate the development of shrimp management measures by the

TABLE 1 Comparison of Gulf Shrimp Regulations[a,b,c]

	Florida	Alabama	Mississippi	Louisiana	Texas	Fishery Conservation Zone Plan (Federal Waters)
Shrimp size in outside waters	47 w, 70 h [1,2,3]	68 w [2,3,4]	68 w	100 w (except during spring season) [2]	None	None
Shrimp size in inside waters	47 w, 70 h	68 w [2,3,4]	68 w [1,2,3]	100 w (except during spring season) [2]	50 w, 65 h [1,2,3] Aug. 15 to Oct. 31	None
Closed Gulf seasons	Stone crab areas [4] Tortugas Sanctuary all year	None	May 1–1st Wed. in June in sound and 3 mi out	Jan. 15–March 15 w/15-day flex NTE 60 days [2]	June 1–July 15 15 ×4 fathoms during day for white shrimp [3] To 7 fathoms Dec. 16–Feb. 1 [3] Night all year [3]	June 1–July 15 Texas FCZ [3] Stone crab areas [4] Tortugas Sanctuary all year

Gear restrictions in outside waters | Trawl size limits[1] | None | None | None | NTE 4 trawls 1 Try NTE 16 ft Mesh NLT 5/8 in. Bar ($1\frac{1}{4}$ in. Stretched)[2,3] | NLT 3/4 in. in stretched mesh[2,3] Bay net inside 4 fathoms | None

[a] Provided by T. Leary, Gulf of Mexico Fishery Management Council.

[b] Key:

w = number of whole shrimp/lb.

h = number of headless shrimp/lb.

NTE = not to exceed.

NLT = not less than.

w/15 day flex = flexible opening and closing ± 15 days.

[c] Objective:

[1] To protect small pink shrimp from premature harvest.
[2] To protect small white shrimp from premature harvest.
[3] To protect small brown shrimp from premature harvest.
[4] To prevent gear conflict.

GMFMC with shrimp management programs of the several states, where feasible.

4. Promote consistency with the Endangered Species Act and the Marine Mammal Protection Act.
5. Minimize the incidental capture of finfish by shrimpers when appropriate.
6. Minimize conflicts between shrimp and stone crab fishermen.
7. Minimize adverse effects of underwater obstructions to shrimp trawling.
8. Provide for a statistical reporting system.

The GMFMC adopted, and the NMFS implemented, the following management measures to achieve the desired objectives:

1. Establish the cooperative permanent closure with the state of Florida and the U.S. Department of Commerce off south Florida to protect small pink shrimp until they generally reach a size larger than 68 tails/lb.

2. Establish a cooperative closure of the territorial sea of Texas and the adjacent U.S. FCZ with the state of Texas and the U.S. Department of Commerce during the time when a substantial portion of the brown shrimp in these waters are less than a count of 65 tails/lb.

3. SANCTUARIES AND CLOSURES

3.1. Tortugas Shrimp Sanctuary

The Gulf of Mexico Shrimp Fishery Management Plan established an area commonly known as the Tortugas Sanctuary off south Florida (Fig. 10) and prohibited all trawling activity within that area between May 15, 1981 and April 15, 1983. This regulation was founded on scientific data indicating that the sanctuary is a nursery area for the Tortugas stocks of pink shrimp and that recruitment to the offshore fishery depends on the sanctuary. Lindner (21), Berry (20), and Nichols (26), utilizing growth and mortality data, indicated that the yield of pink shrimp would be greater if harvest was delayed until shrimp were larger than the minimum legal size (69-count*) for landing shrimp in Florida.

The Gulf of Mexico Fishery Management Council's goal in establishing the sanctuary was to protect small, undersized shrimp from fishing. Furthermore, it was assumed that small shrimp were found mainly inside the sanctuary line and that shrimp outside the sanctuary were of legal size or larger. The establishment of a permanent sanctuary was estimated to increase annual yield by about 1 million lb. The "toe" area of the boot-shaped sanctuary was reopened to trawling from April 1983 until August 15, 1984 because of pressure by the shrimp industry, which claimed they could not survive economically without fishing the "toe" of

*The number of shrimp tails that constitutes 1 lb.

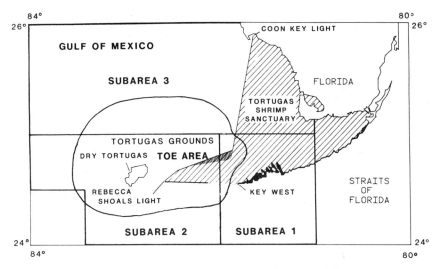

Figure 10. Chart of Dry Tortugas fishery grounds and statistical subareas.

the sanctuary. The prohibition on trawling in the "toe" of the sanctuary was reestablished on August 16, 1984.

The Tortugas fishery has been very stable, with average annual production of about 9.9 million lb that does not fluctuate greatly from year to year. The fishery is bounded by non-trawlable bottoms of loggerhead sponges and coral reefs, where pink shrimp are protected from trawling even though they may be present in high concentrations. The large area of untrawlable bottom surrounding the fishery grounds may explain why this fishery has been so stable since 1960.

Historically, the primary recruitment of small shrimp onto the Tortugas grounds occurs between September and November, although spring recruitment is occasionally strong. In March and April 1981 (prior to the closure of the sanctuary) there was good recruitment of small shrimp onto the Tortugas grounds (Klima and Patella 30; Klima et al. 14). That recruitment continued into the closure period (May 1981) and sustained the fishery through the remainder of 1981. The 1981 catch amounted to 10.2 million lb. However, there was no strong recruitment onto the Tortugas grounds again until March through May 1983. Consequently, the fishery from May 1982 through April 1983 appeared to collapse to an all-time low of about 7 million lb of shrimp.

From April 1983 to August 1984, a portion of the "toe of the boot" area of the sanctuary was opened to fishing and the fleet concentrated its effort on extremely small and abundant shrimp. This was reflected in an increase of the average size count of shrimp landed as well as landings of more than 1 million lb/month (Fig. 11; Klima and Patella 30). Recruitment was poor in the fall of 1983. Above average recruitment was again observed in both April and May 1984. This peak spring 1984 recruitment was again rapidly harvested because the "toe area" was open to fishing, with a catch of more than 1 million lb/month of extremely small shrimp

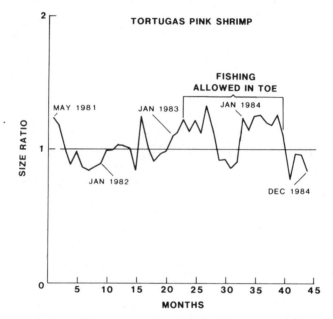

Figure 11. Ratios of monthly mean number of pink shrimp per pound from May 1981 to December 1984 to monthly mean number of pink shrimp per pound from 1960 to 1979.

(Fig. 11). After the closure of the ''toe area'' to fishing in August 1984, the average size landed increased above that of the monthly historical mean.

The fishery from May 1982 through April 1984 produced an all-time low of about 7 million lb for the biological years 1982 and 1983. Production from May to December 1984 was relatively low with a total yield of only 3.0 million lb. The only exceptional month was December 1984, in which 1.94 million lb of large shrimp (33 count) was landed.

The high production of large shrimp during December was probably due to the spring recruitment of shrimp that did not move onto the fishing grounds in August and September, but stayed either in the loggerhead sponge area north of the fishery or in the sanctuary. This stock apparently moved onto the grounds in November–December 1984 (Klima and Patella 30). The authors hypothesize that the shrimp stock was basically protected from the fishery either by the sanctuary or by the loggerhead sponges. They point out that shrimp with a count of 150–200/lb would take about 6 months to reach a size of 20-count and therefore shrimp of this size recruited to the sanctuary area from the Florida Everglades during May would take about 6 months before they reach a size of 20-count shrimp (Berry 19). They further point out the December 1984 fishery was basically concentrated in the extreme northern and northeastern part of the Tortugas fishery.

Klima and Patella (30) state that the production of the fishery from 1981 to 1984 was set by the amount of recruitment and the opening and closing of fishing in the ''toe area.'' With good recruitment in the March through April 1981 period, the stage was set for a good fishery, whereas the lack of recruitment in 1982 and 1983 resulted in devastatingly low production. Good recruitment in the spring of 1983 coincided with the opening of the "toe area" to fishing, which they believed reduced the total yield because of the excessive harvest of small shrimp. These authors conclude that the sanctuary and the "toe area" contain large concentrations of small shrimp and prohibition of trawling in these areas should increase yield to the fishery.

3.2. Texas Closure

The Gulf of Mexico Shrimp Fishery Management Plan (FMP), prepared by the Gulf of Mexico Fishery Management Council and implemented in 1981, seasonally regulates the fishing for brown shrimp in the FCZ off the coast of Texas. This regulation has prohibited shrimp fishing in the FCZ each year since 1981 from late May to mid-July. Concurrently, the state of Texas regulations ban shrimp fishing in its territorial sea, except for the white shrimp fishery inside 4 fathoms. Thus all fishing for brown shrimp was prohibited during these periods in waters along the Texas coast, except for an incidental illegal landing of brown shrimp caught in the white shrimp fishery.

The management objectives of the Texas Closure regulation are designed to increase the yield of shrimp and eliminate the waste caused by discarding undersized shrimp caught during the period of their life cycle when they are growing rapidly. The temporary closure of the offshore fishery from late May to mid-July each year has provided larger shrimp available to the fishery when fishing is again permitted beginning in mid-July. The monetary benefits of this management regulation result from catching larger, higher-priced shrimp, which increased the ex-vessel value of the fishery.

In years prior to the seasonal FCZ closure, discarding of under-sized shrimp commonly occurred because of a Texas law that prohibited fishermen from landing shrimp below a certain size, and lack of a market for small shrimp. The Texas Closure regulation, which was expected to increase the size of shrimp, therefore helped eliminate the need for discarding. The most effective method of eliminating the discarding problem was to delete the application of the Texas law to the Gulf fishery, which the state of Texas did in 1981.

The National Marine Fisheries Service annually evaluates the effectiveness of the Texas Closure by preparing a series of reports for the Gulf of Mexico Fishery Management Council. The first of these annual reports covered abundance and distribution of shrimp off the Texas coast (Matthews 31); a review of the offshore fisheries (Klima et al. 32); impacts on yield (Nichols 27); estimated impacts on the ex-vessel brown shrimp prices and values as a result of the closure (Poffenberger 28); and vessel mobility (Jones and Zweifel 33). Since the implementation of the first closure, the closure has been a success in achieving the objectives of the shrimp fishery management plan. The closure of the FCZ and the territorial sea has increased yield and value of shrimp substantially each year, $59.5 million in 1981, with a low of $31.7 million in 1983 (Tables 2 and 3).

TABLE 2   **Summary of Commercial Catch Statistics and Resource Survey Results for Gulf of Mexico Brown Shrimp Fishery**

	Year				
Statistic	1981	1982	1983	1984	1985

July–August brown shrimp catch[a]

Texas offshore					
Catch	25.0	13.0	9.8	15.3	14.0
Effort	14.8	15.7	10.3	18.6	15.2
CPUE	1,895	922	962	819	918
Louisiana offshore					
Catch	10.5	5.1	4.9	6.6	5.5
Effort	11.9	9.8	11.2	11.2	8.6
CPUE	863	524	439	587	642

May–August brown shrimp catch[b]

Texas					
Inshore	4.2	4.1	5.9	7.1	5.4
Offshore	25.3	13.9	10.5	16.1	14.6
Total	29.5	18.0	16.4	23.5	20.0
Louisiana					
Inshore	15.2	15.1	12.1	14.9	8.8
Offshore	23.1	13.7	8.8	13.6	16.5
Total	38.3	26.8	20.9	28.5	25.3

Source: Klima, Nichols, and Poffenberger, 1986 (62).

[a] Catch in millions of pounds, fishing effort in thousands of days, and catch per trip.

[b] Catch in millions of pounds.

Not only did the economic yield increase, but CPUE off Texas is always substantially greater during July–August than CPUE off Louisiana (Tables 2 and 3). Small emigrating brown shrimp are protected and allowed to grow to a larger size. Discarding was a problem only in 1985 when unusual biological conditions prevailed resulting in smaller shrimp at the opening of the season, and approximately 1.1 million lb was discarded immediately after the area was open to fishing.

Without the prohibition on trawling during the period of brown shrimp emigration, it is believed that large quantities of small brown shrimp would have been caught resulting in wastage and lower yield to the fishery both biologically and economically (Klima et al. 13). That is not to say there are no problems associated with the Texas Closure; quite the contrary, there are several problems that cause concern not only for Texas shrimpers, but shrimpers from other Gulf coastal states. These are as follows:

1. Loss of migrating shrimp to Mexico occurs during the closure period.

TABLE 3 Summary of Analytical Results of Texas Closure Shrimp Fishery Management Measure, 1981–1985[a]

Statistic	Year				
	1981	1982	1983	1984	1985
FCZ Closure					
CPUE ratio Texas:elsewhere[b]					
July	2.26	2.06	2.34	1.86	1.69
August	1.56	1.35	1.40	1.34	0.95
Increase in Y/R at $F = 1.0$ (M = 0.15–0.28)	+14 to +37%	−10 to +10%	+12 to +33%	+15 to +33%	+14 to +33%
Change in Gulfwide yield (million pounds)					
May–August	+4.0 (5%)	+0.7 (1%)	−0.5 (1%)	−0.6 (1%)	−0.8 (1%)
May–April	+4.2 (4%)	+1.4 (2%)	+0.4 (1%)	+1.4 (2%)	[c]
Change in Gulfwide value (million dollars)					
May–August	+10.4 (7%)	+5.3 (3%)	+2.1 (2%)	+8.5 (6%)	+0.8 (1%)
May–April	+9.7 (4%)	+6.0 (3%)	+6.7 (3%)	18.7 (9%)	[c]
Combined closures (FCZ and territorial sea)					
Change in Gulfwide yield					
May–April (million pounds)	+9.8 (10%)	+4.9 (7%)	+3.5 (6%)	+5.1 (6%)	[c]
Change in Gulfwide value					
May–April (million dollars)	+59.5 (25%)	+43.2 (19%)	+31.7 (16%)	+37.4 (18%)	[c]

[a]Values shown are the statistics used to measure the effects of the closure for the Fishery Conservation Zone (FCZ) alone and for the territorial sea and FCZ combined.

[b]Long-term average CPUE ratios (Texas:elsewhere) for 1960–1980 are July, 1.27; August, 1.06.

[c]Data required for estimate not yet available.

Y/R = Yield per recruit.

Source: Klima, Nichols and Poffenberger, 1986 (62).

2. Too many vessels fishing off Texas during the open season leave few shrimp for the resident vessels during the remainder of the fishing season and cause a concentration of vessels at the opening.

3. Increasing inshore shrimp fisheries harvest small juvenile shrimp, thereby reducing the potential yield to the offshore fishery.

4. Closure is not consistently applied throughout the northern Gulf.

5. Lower prices have been paid to the fishermen during July and August since 1984.

6. Tie up of vessels occurs during the closure and illegal fishing in Mexico, and high cost and long distance to fish off Louisiana.

The loss of migrating shrimp to Mexican waters, where U.S. fishermen are prohibited from fishing, does cause concern to U.S. fishermen. A major study is underway to define this loss rate, but data from 1978 to 1980 do not show a significant loss of shrimp from the United States into Mexico (Klima et al. 10).

Local Texas fishermen have complained seriously about the concentration of vessels off the Texas coast at the opening of the Texas shrimp season. Many of these vessels are from other parts of the Gulf with the resulting concentration of vessels causing problems at the opening of the fishing season. If catch rates are high, discarding increases because of the crews' inability to handle and sort the shrimp before the next tow. Poorer-quality shrimp may also be associated with high catch rates and inferior handling at the opening. Pulse fishing and an increased fishing effort at the beginning of the season opening are postulated to have an impact on individual fishermen by decreasing the yield per individual fisherman (Nichols 33a). If fishing effort is increased by 12–25%, decreases in yield per individual fisherman are noted; however, no decrease in value would be observed until fishing effort increased by 45%.

The total number of vessels fishing in the Gulf of Mexico has not drastically increased since 1976, when approximately 4177 vessels operated. In 1983, a total of 4999 vessels was operating in the Gulf of Mexico; an increase of only 16%. The increase in gear efficiency is more devastating. In 1976 less than 5% of the offshore vessels were quad-rigged, whereas in 1985 probably more than 60% were quad-rigged (four nets). A quad-rigged vessel sweeps twice the bottom area of a double-rigged vessel and at least four times the bottom area of a single-rig vessel. This indicates a substantial increase in the efficiency or fishing power of the Gulf fleet of offshore vessels. Although there is only a small increase in the number of vessels (i.e., 16%), the dragging efficiency is almost 75% greater than in 1976, and a total increase in fishing potential of 91%. Because total catch has remained the same there is more capacity than is needed to harvest the shrimp resources of the Gulf of Mexico. This overcapacity is one of the root problems of the shrimp fishery.

Nichols (34) has also shown that an increase in inshore fishing can have a marked impact on the offshore fleet. A decrease of 50% in inshore fishing effort increases offshore poundage yields by approximately 56%. Contrarily, a doubling of inshore effort reduces offshore yields by 59%. He further showed that offshore yields are less responsive to an increase in fishing mortality. A doubling of offshore fishing would decrease offshore yields by 28%. Nichols (34) has also shown that offshore yield-per-recruit in dollars is more responsive to changes in inshore fishing, but not to changes in offshore fishing. A doubling of inshore fishing mortality would decrease offshore dollar yields per recruit by 59%, and total yield-per-recruit in dollars would decrease 31% with a 100% increase in inshore fishing at present levels. A decrease of 50% in inshore fishing would increase total yield per recruit in dollars by 25%. A doubling of offshore fishing mortality would increase total dollar yield only by 4%. Klima et al. (13) pointed out that inshore landings were very similar in 1981, 1982, 1983, and 1984, despite evidence of considerably lower recruitment in some years, and that the increasing inshore fisheries in Texas may soon begin to have an impact on offshore yields.

The economic condition of the shrimp fishery has changed since the late 1970's. Vessel numbers have increased slightly, fishing power has doubled, access to Mexican fishing grounds has been stopped, insurance premiums have increased 300%, and the ex-vessel price paid for shrimp is lower than indexes of other food products. The overcapacity and excess fishing power of the fleet in the Gulf of Mexico, the increasing landings of small shrimp by the inshore fisheries in the northern Gulf, and the changing economic condition of the fishery are major problems and appear to be the main factors causing economic hardship on U.S. fishermen. Poffenberger (35) has shown that the price structure paid for shrimp during July and August is not directly impacted by opening of the season, but by other factors such as the imports of shrimp into the United States. The combination of poor recruitment during 1982–1985, no closure of other areas in the northern Gulf, overcapacity of the fleet, increased inshore fisheries, and perceptions by fishermen of lack of enforcement and loss of migrating shrimp to Mexico have caused serious concern related to the Texas Closure. As a result, in January 1986 at the GMFMC meeting the Council voted to suspend the Texas Closure for 1 year.

A lack of a Gulfwide closure similar to the one off Texas is perceived by many fishermen to hold potential for some financial gains. Nichols (36) has shown that a delay until June 1 in the harvesting of the extremely small shrimp (> 100 count) found in the bays and estuaries of Louisiana and Texas would significantly increase shrimp yield. Klima et al. (13) have pointed out that shrimp of at least 100 count or larger are harvested in May and June throughout Louisiana and Texas inshore areas. Optimum size of harvest for brown shrimp is somewhere between 40 and 50 count (Nichols 11). Therefore, any management measure that restricts harvesting of small shrimp should substantially increase the gain in both pounds and dollars to the fishermen.

A Gulfwide closure would decrease the possibility of pulse fishing and increase the potential for additional revenues from the brown shrimp stock. Such a closure is unlikely in the immediate future because Louisiana, Mississippi, and Alabama regulations are geared for social and economic goals and therefore the need for small shrimp is prevalent in these areas in order to maintain high employment levels in both fishing and processing sectors. Therefore, the likelihood of anticipating a closure in other areas of the Gulf of Mexico is low.

3.3. Research and Management

A current major cooperative program in the Gulf of Mexico is SEAMAP, in which the various coastal states participate with the federal government in sampling both offshore and nearshore areas throughout the Gulf of Mexico (Bane 37). Sampling is concentrated in three areas: demersal finfish and shrimp, ichthyoplankton, and environmental information. Surveys are conducted at various times of the year utilizing both federal and state research vessels in a cooperative effort to minimize duplication and cut costs. Data are shared between the states and the federal agencies. Plans for these surveys are formulated by SEAMAP working groups under the auspices of the Gulf States Marine Fisheries Commission. Data from these surveys have been used to evaluate the Texas Closure (Nichols 27; and Matthews 31) and are useful in assessements of king mackerel and Spanish mackerel, also under federal regulation. The information obtained can be used by any of the partners involved in the SEAMAP investigation.

4. ESTUARIES

The Gulf of Mexico estuaries serve as nurseries, providing protection and food for shrimp, menhaden, flounder, spot, croaker, spotted seatrout, redfish, oysters, and a host of other important organisms. Many of these species spawn in offshore waters and their young stages are swept into the bays, bayous, and tributaries by currents. They utilize shallow portions of salt marshes, mangrove, and seagrass areas to grow and be protected from predators (Zimmerman and Minello 38). As juveniles, they utilize the open bays where they are first exposed to fishing pressures and then migrate to offshore waters.

Human population growth in the southeast coastal regions of the United States is substantially greater than the national average and is accompanied by industrial and real estate development. Alteration or destruction of 1% of estuarine habitats required by commercial and recreational fishery species occurs each year. Louisiana marshes are vanishing at the rate of approximately 50 square miles annually (Gaglian et al. 39; Hatton et al. 40; Baumann et al. 41). Sea level is expected to rise substantially (Titus and Barth 42), compounding the loss of critical estuarine areas in the Gulf of Mexico. The freshwater demand for industrial and urban uses in Texas has increased from 2 million acre-ft in 1930 to about 17.9 million acre-ft in 1980 and is expected to increase with the projected expanding population growth. The increased use of freshwater will limit inflows into Texas bays and impact their productivity.

Highly productive estuarine areas are presently being dramatically altered by humans. The economic value and demand for nearshore property has increased rapidly in recent years. Sustained production of marine fisheries is strongly linked to maintenance of nearshore habitats, and the protection, management, and restoration of these habitats is a critical fisheries issue. The maintenance of important fishery habitats is the responsibility of both state and federal governments, and NOAA/NMFS is taking a leading role because fish do not recognize political boundaries.

As part of its responsibility, NMFS makes recommendations to the U.S. Army Corps of Engineers on permit applications that modify estuarine habitats. The National Marine Fisheries Service began quantifying the cumulative acreage or habitat in the Corps of Engineers permit program in 1981 in the southeast region. Habitat preserved as a result of these recommendations in 1981 alone was estimated to be worth more than $33 million to the commercial and recreational fishing industry (Lindall and Thayer 43). However, these recommendations are based heavily on available information concerning relationships between habitat and fishery production. All too often the information base is not sufficient to allow informed decisions on the impacts of chemical and physical alterations. As a result, the agency and others are forced into a holding action to slow changes until the necessary information can be developed. In the meantime, changes do occur and habitat is eroded or lost.

The shrimp fishery of the Gulf of Mexico has been studied intensively, and the available information may be adequate for maintenance of the fishable stocks if current conditions remain the same. But conditions are changing and information is not adequate for understanding functional relationships and impacts of various habitat alterations on the survival, growth, and distribution of the juveniles in the estuaries or the adults offshore (Fig. 12).

	ESTUARIES			OFFSHORE		
	VEGETATED BOTTOMS	RIVER DELTAS	PASSES/ CHANNELS	OPEN BAYS	ROUGH BOTTOM	SMOOTH BOTTOM
SPAWNING AREAS	PARTIAL	INADEQUATE	INADEQUATE	INADEQUATE	PARTIAL	INADEQUATE
FEEDING GROUNDS	PARTIAL	PARTIAL	INADEQUATE	INADEQUATE	PARTIAL	INADEQUATE
MIGRATORY ROUTES	INADEQUATE	INADEQUATE	INADEQUATE	INADEQUATE	INADEQUATE	INADEQUATE
FUNCTIONS	PARTIAL	INADEQUATE	INADEQUATE	INADEQUATE	PARTIAL	INADEQUATE
VALUE	PARTIAL	PARTIAL	PARTIAL	INADEQUATE	PARTIAL	INADEQUATE

ADEQUATE PARTIAL INADEQUATE

Figure 12. Status of habitat information on shrimp.

5. RESEARCH

Shrimp research has been directed at monitoring landings and evaluating management measures designed to increase yield by allowing more time for shrimp to grow before they are caught. Unfortunately, limited research has been conducted on factors that affect recruitment. Some understanding exists on general environmental conditions that influence survival of young in estuarine nurseries, but this understanding is inadequate either for predicting yield or for understanding environmental impacts such as alteration of freshwater inflow and destruction of estuarine habitat. The information is inadequate to quantify the value of habitats within estuarine systems for shrimp production. Such information will be extremely important in terms of offsetting damage or alterations to the environment.

Temperature and salinity are the two most important factors controlling survival and growth of young shrimp (Zein-Eldin and Aldrich 44; Aldrich et al. 45). However, there are obviously other factors that are important as well. Renaud (46) has shown hypoxia (less than 2.0 ppm dissolved oxygen) off the coast of Louisiana affects the distribution of penaeid shrimp. Zimmerman et al. (47) and Minello and Zimmerman (48–50) have shown that vegetated habitat and predation affect the abundance of juvenile penaeids. Cause–effect relationships and interactions between variables such as temperature, salinity, substrate availability, predation, precipitation, river discharge, and habitat type are not known for any of the penaeids at this time. Zein-Eldin and Renaud (51) noted a lack of information necessary to protect and even enhance the habitat for penaeid shrimp. Temperature and salinity have been shown to be correlated with brown shrimp harvest in some areas. Matylewich and Mundy (52), Barrett and Gillespie (53) and Barrett and Ralph (54) have shown that poor catches of brown shrimp in the spring are associated with heavy precipitation and low water temperatures in Louisiana. The use of environmental variables to predict future harvest varies from region to region and from state to state. Sutter and Christmas (55) used multiple regression analysis including water temperature, salinity, and the number of postlarval brown shrimp in nursery areas to predict the June–July commercial harvest in Mississippi waters. Klima et al. (32) used catch of shrimp by the bait shrimp fishery of Galveston Bay to predict annual offshore harvest of brown shrimp from Texas waters. Standing stock of juvenile shrimp in estuaries and postlarval abundance indexes have also been used to predict annual Texas offshore catches of brown shrimp (Sullivan et al. 56; and Baxter 57).

Forecasting shrimp production is in its infancy. Generally, some reasonable forecasts can be made for brown shrimp; however, this is not true for white or pink shrimp stocks. Environmental parameters are useful in forecasting if extreme conditions are observed but forecasts are not accurate under average environmental conditions. In contrast, bait shrimp indexes of juvenile abundances based on line bait catches are better for forecasting in average years than in extremely poor or excellent years.

Direct exploitation of intertidal habitats by fishery organisms in the Gulf of

Mexico may be more important than previously thought. Juveniles of commercial species such as brown shrimp, spotted seatrout (*Cynoscion nebulosus*), and southern flounder (*Paralichthys lethostigma*) have been shown to invade intertidal marsh habitat in preference to subtidal open water (Zimmerman and Minello 38). Because direct exploitation of intertidal areas depends on inundation, tidal dynamics are important in controlling accessibility for these organisms to preferred habitat. In the Gulf of Mexico, seasonal tides dominate daily tides (Provost 58), resulting in higher and longer inundation events on the Gulf coast during the spring and fall (Hicks et al. 59). Historical postlarval recruitment of brown shrimp coincides with these spring and fall high tides, suggesting a relationship and perhaps dependency on access to marsh habitats (Zimmerman et al. 47). Brown shrimp may benefit additionally where marshes are lower owing to sea level rises and local land subsidence. The northwestern Gulf of Mexico has one of the highest rates of apparent sea-level rise of all U.S. coastal regions (Hicks et al. 59), and one consequence is increased flooding of cordgrass (*Spartina alterniflora* and *S. patens*) marshes. This flooding increases the area and duration of accessibility to optimal habitat for brown shrimp, and may result in higher production.

Such production appears to be related to increased food supply. Growth rates of brown shrimp are higher in *S. alterniflora* habitat than in open water (Zimmerman and Minello 60). Higher growth rates are apparently caused by greater abundances of food organisms (pericarid crustaceans and polychaete worms) in *Spartina* habitat.

Zimmerman and Minello (61) suggest that food abundance and temperature are primary factors controlling growth, but neither affect mortality except under extreme conditions. Survival of postlarval and juvenile shrimp in salt marsh nurseries appears to be largely regulated by predation. Estimates of juvenile brown shrimp (11–40 mm total length) mortality in a Galveston Bay salt marsh have been made from an examination of size-frequency distributions, and actual mortality during spring months ranged from 43 to 70% over a 2-week period. In field experiments, where fish predators were excluded from experimental enclosures, brown shrimp mortality over a similar length of time was only 3–11%. Studies of stomach contents indicate that the dominant fish predator on shrimp in the marsh during the spring months is the southern flounder, (*Paralichthys lethostigma*). During the late summer and fall, red drum, (*Scianenops ocellatus*), and spotted seatrout, (*Cynoscion nebulosus*), feed on juvenile shrimp.

The presence of emergent marsh vegetation reduces fish predation on juvenile brown shrimp (Minello and Zimmerman 48; Zimmerman and Minello 60), but does not appear to protect white shrimp, (*Penaeus setiferus*), in a similar manner (Minello and Zimmerman 50). Other characteristics of nursery habitats that protect brown shrimp from at least some predators include turbid water and a suitable substrate for burrowing by shrimp (Minello and Zimmerman 49). Continued research in these areas will provide information needed to model mortality of shrimp. Shrimp mortality can be affected by predator and prey density, predator and prey size, density of alternative prey, the presence of various habitat characteristics, physical and chemical variables, and interactions among many factors. The model could be useful for predicting survival of shrimp under various conditions and for determining the most important protective habitats for juvenile shrimp.

REFERENCES

1. E. F. Klima, A Review of the Fishery Resources in the Western Central Atlantic. *WECAF Stud.* 3:1–77 (1976).

2. R. F. Temple, Shrimp research at the Galveston Laboratory of the Gulf Coastal Fisheries Center. *Mar. Fish. Rev.* 25(3):16–20 (1973).

3. C. W. Caillouet and K. N. Baxter, Gulf of Mexico shrimp resource research. *Mar. Fish. Rev.* 35(3/4):21–24 (1973).

4. M. J. Lindner and H. L. Cook, Synopsis of biological data on the white shrimp, *Penaeus setiferus* (Linn.), 1767. *FAO Fish. Rep.* 57:1439–1468 (1970).

5. H. L. Cook and M. J. Lindner, Synopsis of biological data on the brown shrimp, *Penaeus aztecus* Ives 1981. *FAO Fish. Rep.* 57:1471–1497 (1970).

6. T. J. Costello and D. M. Allen, Synopsis of biological data on the pink shrimp, *Penaeus duorarum duorarum* Burkenroad 1939. *FAO Fish. Rep.* 57:1499–1537 (1970).

7. K. W. Osborn, B. W. Maghan, and S. B. Drummund, Gulf of Mexico shrimp atlas. *U.S. Fish and Wildl. Serv. Circ.* 312:1–20 (1969).

8. M. L. Lindner and W. W. Anderson, Growth, migration, spawning and size distribution of shrimp, *Penaeus setiferus. Fish. Bull.* 56(106):554–645 (1956).

9. P. F. Sheridan, F. J. Patella, Jr., K. N. Baxter, and D. A. Emiliani, Movement of brown shrimp, *Penaeus aztecus*, and pink shrimp, *P. duorarum*, across the Texas-Mexico Border. *Fish. Bull.* (1986) (submitted for publication).

10. E. F. Klima, R. G. M. Castro, K. N. Baxter, F. J. Patella, and S. Brunenmeister, Summary of cooperative Mexico–United States shrimp research program, 1978–1982. *Mar. Fish. Rev.* (1987) (to be published).

11. S. Nichols, *Updated Assessments of Brown, White and Pink Shrimp in the U.S. Gulf of Mexico.* Southeast Fisheries Center, NMFS, Miami, FL, 1984.

12. C. W. Caillouet, F. J. Patella, and W. B. Jackson, Trends toward decreasing size of brown shrimp, *Penaeus aztecus*, and white shrimp, *P. setiferus*, in reported annual catches from Texas and Louisiana. *Fish. Bull.* 774:985–989 (1980).

13. E. F. Klima, K. N. Baxter, and F. J. Patella, Review of the 1984 Texas Closure for the Shrimp Fishery off Texas and Louisiana. *NOAA Tech. Memo.* NMFS-SEFC-156, 1–33 (1985).

14. E. F. Klima, G. A. Matthews, and F. J. Patella, A Synopsis of the Tortugas Pink Shrimp Fishery, 1960–83. *North Am. J. Fish. Manage.* 6:301–310 (1986).

15. E. F. Klima, A white shrimp mark-recapture study. *Trans. Am. Fish. Soc.* 103(1):107–113 (1974).

16. M. L. Parrack, Aspects of brown shrimp, *Penaeus aztecus*, growth in the northern Gulf of Mexico. *Fish. Bull.* 76(4):827–836 (1979).

17. E. S. Iversen and A. C. Jones, Growth and migrations of the Tortugas pink shrimp,

Penaeus duorarum, and changes in the catch per unit of effort of the fishery. *Fl. Geol. Surv. Tech. Ser.* 34: 1–22 (1961).

18. J. H. Kutkuhn, Dynamics of a penaeid shrimp population and management implications. *Fish. Bull.* 65:313–333 (1966).

19. R. J. Berry, Dynamics of the Tortugas (Florida) pink shrimp population. Ph.D. Thesis, University of Rhode Island, Kingston, 1967, 160 pp. (also available from University Microfilms, Ann Arbor, MI, 1968, 177 pp.).

20. R. J. Berry, Shrimp mortality rates derived from fishery statistics. *Proc. Annu. Gulf Caribb. Fish. Inst.* 22:66–78 (1969).

21. M. J. Lindner, What we know about shrimp size and the Tortugas fishery. *Proc. Annu. Gulf Caribb. Fish. Inst.* 18:18–25 (1965).

22. R. A. Neal, Methods of marking shrimp. *FAO Fish. Rep.* 57:1149–1165 (1969).

23. J. Y. Christmas and D. J. Etzold, *The Shrimp Fishery of the Gulf of Mexico United States: A Regional Management Plan*, Tech. Rep. Ser. No. 2. Gulf Coast Res. Lab., Ocean Springs, MS, 1977, 128 pp.

24. E. F. Klima, Proceedings of the International Shrimp Releasing Marking and Recruitment Workshop, 25–29 November 1978, Salmiya, State of Kuwait. *Kuwait Bull. Mar. Sci.* 2:185–207 (1981).

25. P. L. Phares, Temperature-associated growth of white shrimp in Louisiana. Southeast Fisheries Center, U. S. National Marine Fisheries Service. *NOAA Tech. Memo.* NMFS-SEFC-56 (1980).

26. S. Nichols, Updated yield per recruit information about the Tortugas pink shrimp fishery. *North Am. J. Fish. Manage.* 6:339–343 (1986).

27. S. Nichols, Impacts on shrimp yields of the 1981 Fishery Conservation Zone Closure off Texas. *Mar. Fish. Rev.* 44(9–10):31–37 (1982).

28. J. R. Poffenberger, Estimated impacts on ex-vessel brown shrimp prices and values as a result of the Texas Closure regulation. *Mar. Fish. Rev.* 44(9–10):38–43 (1982).

29. B. J. Rothschild and S. L. Brunenmeister, The dynamics and management of shrimp in the northern Gulf of Mexico. In J. A. Gulland and B. J. Rothschild, Eds., *Penaeid Shrimps: Their Biology and Management*. Fishing News Books, Farnham, Surrey, England, 1984, pp. 145–172.

30. E. F. Klima and F. J. Patella, A synopsis of the Tortugas pink shrimp, *Penaeus duorarum*, fishery, 1981–84, and the impact of the Tortugas Sanctuary. *Mar. Fish. Rev.* 47(4) (1986).

31. G. A. Matthews, Relative abundance and size distributions of commercially important shrimp during the 1981 Texas closure. *Mar. Fish. Rev.* 44(9–10):5–15 (1982).

32. E. F. Klima, K. N. Baxter, and F. J. Patella, Jr., A review of the offshore shrimp fishery and the 1981 Texas closure. *Mar. Fish. Rev.* 44(9–10):16–30 (1982).

33. A. C. Jones and J. R. Zweifel, Shrimp fleet mobility in relation to the 1981 Texas closure. *Mar. Fish. Rev.* 44(9–10):50–54 (1982).

33a. S. Nichols, personal communication, Southeast Fisheries Center, Miami Laboratory, Miami, FL 33149.

34. S. Nichols, Impacts of the combined closures of the Texas territorial sea and FCZ on brown shrimp yields. *NOAA Tech. Memo.* NMFS-SEFC-141 (1984).

35. J. R. Poffenberger, *Estimated Impacts of Texas Closure Regulation on Ex-vessel Prices and Value, 1982 and 1983*. Southeast Fisheries Center, NMFS, Miami, FL, 1984.

36. S. Nichols, *Analysis of Alternative Closures for Improving Brown Shrimp Yield in the Gulf of Mexico*. Report to Gulf of Mexico Council, 1986.

37. N. Bane, Annual Report of the Southeast Area Monitoring and Assessment Program (SEAMAP) October 1, 1984–September 30, 1985, 10 pp.

38. R. J. Zimmerman and T. J. Minello, Densities of *Penaeus aztecus, Penaeus setiferus*, and other natant macrofauna in a Texas salt marsh. *Estuaries* 7:421–433 (1984).

39. S. M. Gaglian, K. J. Meye-Arendt, and K. M. Wicker, Land loss in the Mississippi River deltaic plain. *Trans. Gulf Coast Assoc. Geol. Soc.* 31:295–300 (1981).

40. R. S. Hatton, R. D. DeLanne, and W. H. Patrick, Jr., Sedimentation, accretion, and subsidence in marshes of Barataria Basin, Louisiana. *Limnol. Oceanogr.* 28:393–502 (1983).

41. R. H. Baumann, J. W. Dag, Jr., and C. A. Miller, Mississippi deltaic wetland survival: Sedimentation versus coastal submergence. *Science* 224:1093–1095 (1984).

42. J. G. Titus and M. C. Barth, An overview of the causes and effects of sea level rise. In M. Barth and J. Titus, Eds., *Greenhouse Sea Level Rise: A Challenge for this Generation*. Van Nostrand-Reinhold, New York, 1984, pp. 1–56.

43. W. N. Lindall, Jr. and G. W. Thayer, Quantification of National Marine Fisheries Service. Habitat conservation efforts in the southeast region of the United States. *Mar. Fish. Rev.* 44(12):18–22 (1982).

44. Z. Zein-Eldin and D. Aldrich, Growth and survival of postlarval *Penaeus aztecus* under controlled conditions of temperature and salinity. *Biol. Bull. (Woods Hole, Mass.)* 129:199–216 (1965).

45. D. Aldrich, C. Wood, and K. Baxter, An ecological interpretation of low temperature responses in *Penaeus aztecus* and *Penaeus setiferus* post-larvae. *Bull. Mar. Sci.* 18:61–71 (1968).

46. M. L. Renaud, Hypoxia in Louisiana coastal waters during 1983: Implications for fisheries. *Fish. Bull.* 84(1):19–26 (1986).

47. R. J. Zimmerman, T. J. Minello, and G. Zamora, Jr., Selection of vegetated habitat by brown shrimp, *Penaeus aztecus*, in a Galveston Bay salt marsh. *Fish. Bull.* 82:325–336 (1984).

48. T. J. Minello and R. J. Zimmerman, Fish predation on juvenile brown shrimp, *Penaeus aztecus* Ives: The effect of simulated *Spartina* structure on predation rates. *J. Exp. Mar. Biol. Ecol.* 72:211–231 (1983).

49. T. J. Minello and R. J. Zimmerman, Selection for brown shrimp, *Penaeus aztecus*, as prey by the spotted seatrout, *Cynoscion nebulosus*. *Contrib. Mar. Sci.* 27:159–167 (1984).

50. T. J. Minello and R. J. Zimmerman, Differential selection for vegetative structure between juvenile brown (*Penaeus aztecus*) and white (*P. setiferus*) shrimp, and implications in predator-prey relationships. *Estuarine Coastal Shelf Sci.* 20:707–716 (1985).

51. Z. P. Zein-Eldin and M. L. Renaud, Inshore environmental effects on brown shrimp, *Penaeus aztecus*, and white shrimp, *P. setiferus*, populations in coastal waters, particularly in Texas. *Mar. Fish. Rev.* 48(3):9–19.

52. M. Matylewich and P. Mundy, Evaluation of the relevance of some environmental factors to the estimation of migratory timing and yield for the brown shrimp of Pamlico Sound, North Carolina. *North Am. J. Fish. Manage.* 5:197–209 (1985).

53. B. Barrett and M. Gillespie, 1975 environmental conditions relative to shrimp production in coastal Louisiana. *La. Dept. Wildl. Fish., Tech. Bull.* 15:1–22 (1975).

54. B. Barrett and E. Ralph, 1977 environmental conditions relative to shrimp production in coastal Louisiana along with shrimp catch data for the Gulf of Mexico. *La. Dept. Wildl. Fish., Tech. Bull.* 26:1–16 (1977).

55. F. C. Sutter, III and J. Y. Christmas, Multilinear models for the prediction of brown shrimp harvest in Mississippi waters. *Gulf Res. Rep.* 7(3) (1983).

56. L. Sullivan, D. Emiliani, and N. Baxter, Standing stock of juvenile brown shrimp, *Penaeus aztecus*, in Texas coastal ponds. *Fish Bull.* (1985).

57. K. N. Baxter, Abundance of postlarval shrimp—an index of future shrimping success. *Proc. Annu. Gulf Caribb. Fish. Inst.* 15:79–87 (1983).

58. M. W. Provost, Tidal datum planes circumscribing salt marshes, *Bull. Mar. Sci.* 26:558–563 (1983).

59. S. D. Hicks, H. A. Debaugh, Jr., and L. E. Hickman, Jr., *Sea Level Variations for the United States 1855–1980*, NOAA/NOS Rep. National Ocean Service, Tides and Water Levels Branch, Rockville, MD, 1983, 170 pp.

60. R. J. Zimmerman and T. J. Minello, Fishery habitat requirements: utilization of nursery habitats by juvenile penaeid shrimp in a Gulf of Mexico salt marsh. In B. J. Copeland, K. Hart, N. Davis, and S. Friday, Eds., *Research for Managing the Nation's Estuaries*, UNC Sea Grant Publ. 84-08. University of North Carolina, Raleigh, NC, 1984, pp. 371–380.

61. R. J. Zimmerman and T. J. Minello, personal communication, Southeast Fisheries Center, Galveston Laboratory, Galveston, TX 77550.

62. E. Klima, S. Nichols, and J. Poffenberger, Executive Summary, Texas Closure, *NOAA Tech. Memo.* NMFS-SEFC-172, 1–10 (1986).

5 RESOURCE ASSESSMENT AND MANAGEMENT PERSPECTIVES OF THE PENAEID PRAWN FISHERIES OF WESTERN AUSTRALIA

J. W. Penn, N. G. Hall, and N. Caputi
Fisheries Department
Western Australian Marine Research Laboratories
North Beach, Western Australia, Australia

1. INTRODUCTION

Penaeid prawn stocks in Australian waters provide the largest component in value of the Australian fishing industry, with catches up to 26,000 tonnes having been recorded (1). The fisheries for prawns fall into two broad groups, with the long-standing fisheries of the eastern seaboard, where fishing occurs in both estuarine and offshore zones, falling into one group, and the more recently developed, exclusively offshore industrial fisheries of the northern, western, and southern coastlines forming a second group (2). In addition to the different development histories, the stocks exploited by these fisheries also cover a wide range of penaeid habitat types. On the east coast the stocks generally follow the most typical situation (3) of depending on freshwater-dominated estuaries for nursery areas (4), whereas the western stocks [and to a lesser extent northern stocks (5)] represent the alternative end of the environmental spectrum by utilizing marine to hypersaline (up to 60‰) estuaries and embayments as nursery habitats (6).

The latter arid-zone stocks, although occupying atypical penaeid habitats, are similar to those occurring in parts of northwest Africa (3), the Arabian Gulf (7), and Mexico (8), also in similar habitats. Together these fisheries form a distinctive but small segment of the world's penaeid resources. In addition to belonging to a relatively unusual group of fisheries, Australia's arid-zone penaeid fisheries also have a number of features that make them unique. They have all developed and remained purely as offshore export-based fisheries. The west and south coast fisheries have also been developed relatively slowly since their discovery in the 1960s, under a system of limited-entry management. This controlled development has extended the usual short development period to full exploitation of less than 5 to 15–20 yr, during which a detailed set of fisheries data has been maintained (9).

Although both the geographic and management characteristics of these fisheries, and particularly the Western Australian (W.A.) fishery, may not be typical of world penaeid fisheries, their relative simplicity and reliability in terms of fishing statistics have allowed a number of basic attributes of exploited penaeid stocks to be investigated and documented. In this chapter, the majority of the information provided has been derived from the research work and practical management experiences in the two major W.A. fisheries of Shark Bay and Exmouth Gulf.

2. THE WESTERN AUSTRALIAN PRAWN FISHERIES

The major W.A. prawn trawl fisheries operate within the large embayments of Shark Bay (24–26°S) (Fig. 1) and Exmouth Gulf (21–22°S) (Fig. 2) on the central

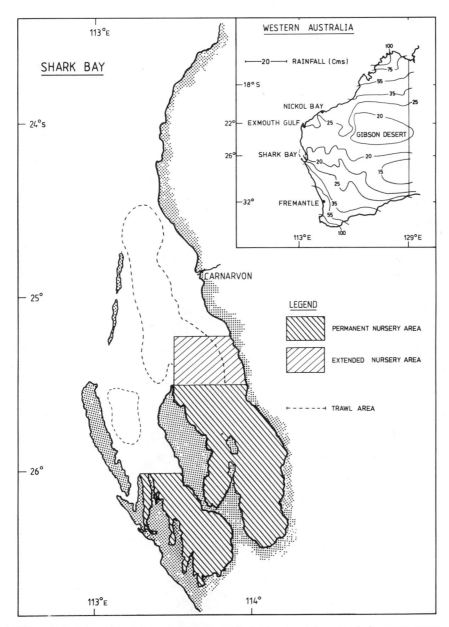

Figure 1. A chart of Shark Bay showing the main fishing grounds, extended nursery areas, and permanently closed nursery areas. Inset shows the coastline and rainfall contours of Western Australia.

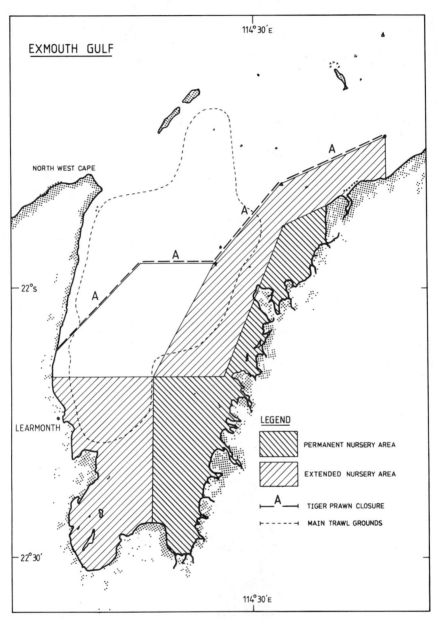

Figure 2. A chart of Exmouth Gulf showing the main trawl grounds, the tiger prawn stock closure line A, the extended nursery area, and the permanently closed nursery areas.

west coast where the desert approaches the ocean. In these areas evaporation (2.2 m/yr) exceeds rainfall by a factor of 10, with the result that both embayments range from marine at the ocean entrance to hypersaline (up to 60‰) in the bay heads (10, 11). In both locations the offshore movement of the prawn stocks, and fishing, extends only to the marine waters of the outer embayment (12, 13). Detailed reviews of the biological studies (14) and management regimes (9) of these and other Australian fisheries have been presented elsewhere.

2.1. Shark Bay

Commercial trawling for prawns in Shark Bay began in 1962, following the discovery of western king (*Penaeus latisulcatus*) and tiger (*P. esculentus*) prawns during a series of research vessel surveys from 1951 to 1961 (15, 16). As a result of the success of the first year's operations and under the threat of a large influx of vessels, limitations on entry to the fishery were introduced in 1963. The historical record of catches and fishing effort for the fishery to 1985 is presented in Figure 3.

The original fleet was composed of mostly small (15–18 m, engines less than 120 kW), single-rig trawlers, many of which were rock lobster vessels operating

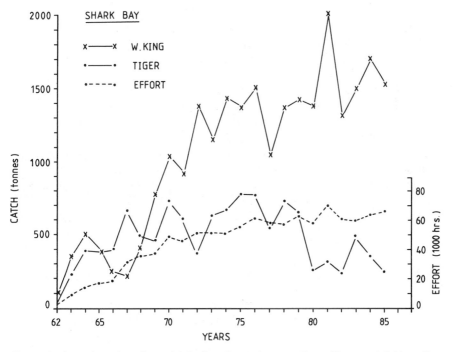

Figure 3. Annual catches (by weight) of each species, together with annual fishing effort (actual hours trawled) for the Shark Bay prawn fishery (1962–1985).

Figure 4. A typical W.A. design, 18 m prawn trawler, which has been developed from the original Gulf of Mexico style vessels built during the 1960s.

on a part-time basis (17). However, by 1972, twin-rig Gulf of Mexico style trawlers, 18–20 m long with 200 kW engines (Fig. 4), had become standard for the fleet and all part-time vessels had been phased out. Since 1976 larger replacement vessels (23–25 m, 260–375 kW engine power) have entered the fishery as a result of a variety of economic factors unrelated to fishing, including federal government shipbuilding subsidies on vessels larger than 21 m waterline length (9).

Trawling in Shark Bay is carried out almost exclusively at night. The gear used consists of conventional rectangular wooden otter boards and flat, low-opening nets (18) constructed of 50-mm mesh (cod ends 45 mm). The catch is snap frozen (whole) on board and unloaded at 2- to 4-week intervals for processing in shore-based factories. Prawns are exported to the Japanese market as either whole prawns or prawn tails (12). A change in processing that has significantly affected fishing was the introduction of mechanical prawn-peeling machines to the Shark Bay factory in 1976, which for the first time allowed small (>66/kg count) prawns to be taken in quantity and economically processed.

2.2. Exmouth Gulf

The Exmouth Gulf fishery began in 1963 when vessels from the newly discovered Shark Bay fishery explored the area (200 km farther north) at the end of the 1963 season (19). Initially the fishery was based on daylight fishing for schooling banana prawns (*P. merguiensis*), but changed entirely to a night fishery for tiger (*P. esculentus*), western king (*P. latisulcatus*), and endeavor (*Metapenaeus endeavouri*) prawns as the banana prawn stock declined (Fig. 5) (20).

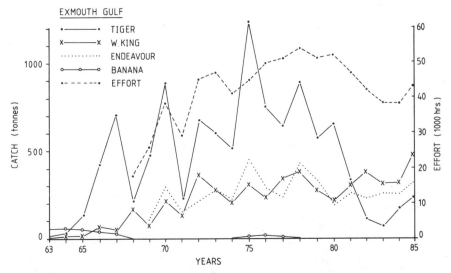

Figure 5. Annual catches (by weight) of each species together with fishing effort (actual hours trawled) for the Exmouth Gulf prawn fishery (1963–1985).

The fishery has been subject to controlled expansion under a system of limited-entry management introduced in 1965, when 15 licenses were issued. Although vessel numbers have been controlled, vessel replacements in the early 1970s resulted in the original single-rigged fleet of largely part-time converted rock lobster boats being replaced with specifically designed 180–225-kW-engine-power, 18–20-m twin-rig trawlers. Since 1973, all trawlers have operated in the fishery on a full-time basis.

Since 1980, larger, 23-m (260–375-kW-engine-power) replacement trawlers were built for the fishery in response to federal government shipbuilding subsidies on larger vessels (9). The trawling gear used by the fleet has been similar to that in Shark Bay (Section 2.1), with the exception of the early years (1963–1967), when high-opening balloon nets (18) were used to take schooling banana prawns.

Because of the small size of Exmouth Gulf, most of the fleet has always operated on a short trip duration of 1–3 days, with the catch being stored in refrigerated brine. The remainder of the fleet are freezer boats that unload at 10–14 day intervals. Essentially all of the catch is landed to the two shore-based factories for processing. Mechanical prawn-peeling machines were also introduced to the Exmouth fishery in 1976 and had a significant effect on fishing practice. As in Shark Bay, the ability to process small prawns allowed changes in the timing of fishing, but also added significantly to profitability by allowing the economical processing of lower-value endeavor prawns that were often previously discarded. After processing, the processed catch is transported 1000 km by road to Fremantle (Fig. 1) for export largely to the Japanese market.

3. FISHERIES BIOLOGY

3.1. Life Histories

Taxonomic and distributional information for the exploited Australian penaeid prawns has been previously summarized (21). The basic life histories of *P. latisulcatus* and *P. esculentus* in Western Australia have also been extensively reported (6, 12, 15, 19, 22).

In both fisheries, western king prawns are in the center of their latidudinal range, and spawning occurs throughout the year. However, population fecundity indexes have been used to show that the peak period of egg production occurs from late autumn to spring (23). After settling inshore, the postlarvae/juveniles undergo a period of rapid growth during spring–summer, before migration offshore in autumn (6). Age at recruitment to the adult stock (fishery) varies between 6 and 12 months, with the spring-spawned modal group (6–8 months) being dominant in the catch. The timing of arrival of recruits in the fishery each year shows minor variations owing to the timing of lunar cycles (15).

Tiger prawns in Shark Bay and Exmouth Gulf are close to the southern end of their range and have a more restricted spawning season. Spawning occurs from late winter through to summer with peak egg production occurring in spring (20), following the pattern typical of *Penaeus* species (12). The postlarvae grow rapidly over summer and migrate offshore to join the adult stock in autumn at approximately 6–8 months of age. Investigations of the timing and location of recruitment for both species have provided the basis for the position and duration of the extended nursery closures to regulate the size at first capture in the W.A. fisheries (9). In each fishery spawning stocks for both species essentially remain within each embayment (20, 23), allowing each to be treated as a unit stock for assessment purposes.

3.2. Habitat Descriptions

The major stocks of western king and tiger prawns on the W.A. coastline are located on the arid northwest coastline in large marine–hypersaline embayments. Nursery areas are within the hypersaline (35–55‰) inshore zone, with western king prawn juveniles occurring on the bare sand areas of the shallow sand flats, and tiger prawns on adjacent sea grass areas (6, 12).

Juveniles in these areas are subject to relatively little hydrological variation, with the exception of infrequent disturbances due to summer cyclone activity. In Shark Bay, these events are particularly rare; however, in Exmouth Gulf where severe cyclones occur on average about every 5 yr, their impact has been associated with both positive and negative effects on recruitment (13). Cyclonic rainfall has also been found to improve recruitment in banana prawn fisheries both in Western Australia and in the Gulf of Carpentaria where rainfall has been successfully used to predict catches (24). The adult stocks, upon which most fishing takes place, occur in the outer marine sections of the embayments; here western king prawns are associated with hard sand bottoms, whereas the tiger prawns show preference for areas with softer mud sediments (13, 25).

3.3. Prawn Behavior and Catchability

Western king and tiger prawns in the W.A. fisheries are taken almost exclusively at night. Western king prawns are strongly nocturnal and have behavioral patterns (26) that result in considerable variations in catchability related to moonlight and temperature (27). Tiger prawns, although generally nocturnal, appear to have a less variable behavior (26), although recent aquarium observations (28) have shown that temperature may also affect activity. The species generally appears more catchable than the king prawn however, and is often taken during daylight on occasions when the water becomes turbid (19).

An understanding of these differences in behavioral characteristics and their impact on catchability has been found to be an essential prerequisite to the use of catch and effort data as measures of abundance from the W.A. fisheries (26). Catchability changes related to factors such as lunar cycle, for example, have been found to be particularly important when making year-to-year comparisons of detailed catch rate data, and in the timing of surveys to estimate abundance of recruits and spawning stocks.

4. DATA ACQUISITION

4.1. Commercial Logbook Systems

Because of the relatively small fleet sizes in the two major W.A. fisheries and the need for detailed current data with limited-entry management (9), a maximum coverage for all fishing activity has been attempted. Logbooks have been maintained in both the Shark Bay and Exmouth Gulf fisheries since their inception in the early 1960s (29), although reliable data on catch rates in Exmouth Gulf are available only since 1970 (20).

Information recorded in the logbooks is detailed at the shot level, which is then accumulated to provide daily records of catch and effort within each fishing ground. Although location of fishing activity was initially recorded with reference to an arbitrary 10 × 10 min grid system based on latitude and longitude, it is now recorded in terms of the fishing ground on which the shot was made. The arbitrary divisions of the geographical grid created artificial boundaries that bore no relationship to the distribution of prawns or to the fishing grounds, which are well identified and recognized by vessel skippers. Accurate records of landings by species for each vessel are also obtained for both fisheries from local shore-based factories, and these are used to adjust the estimates of catch provided in logbooks. Effort data are standardized both within and between years using relative fishing powers calculated for each vessel (25). If required, weights are converted to numbers of prawns using information obtained from factory sampling of commercial catches.

Annual effective effort is calculated by dividing an estimate of the average catch rate for a species into the total annual catch of that species for each year (25).

As additional years of data are added to the logbook data base, the data set has become increasingly valuable. This is largely due to the comprehensive nature of the system; virtually every skipper maintains a logbook. Accuracy and completeness of information is maintained by continued liaison between research staff and fishermen, with considerable feedback to fishermen being required to ensure that the quality of recording is maintained.

In addition to the logbook data system general information on the operation of the fishery is also recorded for each year. This includes data on product prices, crew commissions, and landings of by-catch, which directly influence the application of fishing effort within the fishery, and data on individual vessel characteristics and replacements, which affect fishing power.

4.2. Biological Sampling

Biological data on the prawns caught in Shark Bay and Exmouth Gulf have been obtained from a routine monthly sampling program in the local processing factories and from research vessel surveys (9). Factory samples are obtained from the major grounds being fished during 5–7 day sampling periods each month of the season. Samples consist of approximately 200 individuals, which are separated by sex and weighed as a group before carapace length and spawning condition (females only) are recorded. Research vessel catches are treated similarly but include additional data on specific location sampled (13).

4.3. Research Vessel Surveys

Stock surveys have been carried out using a 21-m twin-rig research vessel *Flinders* and chartered commercial trawlers. A series of annual (10–20 day) research vessel trawl surveys on an approximate grid of 8 km covering all trawlable areas within each fishery have been carried out in Shark Bay (1974–1980) and Exmouth Gulf (1975–1977). These surveys provide unbiased distributional data on size and abundance, which have been used to validate and assist in the interpretation of the data from the commercial logbook and factory sampling systems. Surveys have also been utilized to obtain data on recruitment and spawning patterns for use in development of independent stock–recruitment indexes (13).

4.4. Mark–Recapture Experiments

Tag recapture experiments are used extensively in fisheries resource assessment, and were first applied to the W.A. prawn fisheries at Shark Bay in 1966 (15). A review of these and other Australian penaeid tagging studies (1953–1978) has been presented previously (6). That review suggested that penaeid prawns as a group have physical characteristics that tend to cause significant bias in the recapture results almost irrespective of the tag used. The major problem identified (30) was related to the highly variable initial tagging mortality generated largely by capture and handling. Significant mortality was also found to occur at the first post-tagging molt, but those prawns that survived this event were then found to be largely free of the effects of the tag.

These factors were considered to be the cause of the similar pattern of recapture observed in many penaeid tagging studies (6). Typically a high rate of return (of prawns with reduced growth) occurs during the first 4–6 weeks, followed by a steep decline, after which a low rate of return continues for a longer than expected period. This in general causes natural mortality (M) to be overestimated and growth rate (K) and fishing mortality (F) to be underestimated. For this reason and because of the local difficulties associated with obtaining fully documented recaptures from the high-catch-rate, freezer boat fisheries, these methods for estimating M and K were not pursued in the W.A. fisheries.

5. RESOURCE ASSESSMENT

As a result of the early implementation of limited-entry management to the W.A. prawn fisheries, initial resource assessments employed surplus production models to predict likely consequences of changes in effort levels (17). Once effort had begun to approach the predicted optimal levels, the research emphasis changed to regulation of the size at first capture. More recently, assessment of stock-recruitment relationships for tiger prawns has been given priority (20).

5.1. Surplus Production Model Assessments

The results of fitting surplus production models to the catch and effective effort data for Shark Bay have been reported previously (9). For Exmouth Gulf these techniques have been less satisfactory owing to the wider variability in the catch data (Fig. 5).

Reanalysis of the Shark Bay data from 1962 to 1985, with effective effort calculated using weight rather than numbers of prawns, has resulted in the curves shown in Figure 6. The results for king prawns from this current assessment are similar to those reported for the period from 1962 to 1980 (31). In contrast, reduced catches for tiger prawns associated with high effective effort since 1980 have caused the tail of the generalized production model to swing downward in comparison with the earlier analysis. This has led to a much more conservative estimate of the optimum effort for this species.

From this analysis of the Shark Bay data it has become apparent that surplus production models fitted to catch and effort data are relatively insensitive to reduced catches associated with declining recruitment at high effort levels. The models continue to provide optimistic estimates of the optimum effort until sufficient data points are obtained at the reduced catch levels. Furthermore, as seen by the tiger prawn data up to 1979, the models are unable to provide any indication that recruitment is likely to be greatly affected by further increases in effort.

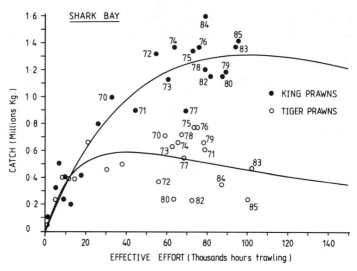

Figure 6. Relationships between catch (by weight) and effective effort (1962–1985) for the western king and tiger prawn stocks in Shark Bay fitted using the generalized production model. Note: logbook data not available for 1981.

5.2. Yield-Per-Recruit Assessments

Yield-per-recruit (Y/R) assessment techniques have been widely used for penaeid stocks (32), including the Exmouth Gulf tiger prawn stock (19). They have not, however, been used as the basis for management of the industrial W.A. fisheries, because of a lack of confidence in the estimates of natural mortality and growth parameters from tagging data (see Section 4.4). An empirical approach to the problem of growth overfishing has therefore been adopted for management.

Soon after the prawn fishery was established in Shark Bay, nursery areas where very small prawns consistently occurred were closed to trawling. In any case growth overfishing was not considered a problem in the 1960s and early 1970s, because the factories in both Shark Bay and Exmouth Gulf would not accept small prawns (>66/kg whole). However, with generally higher prices for prawns in the late 1970s, competition between vessels led to a general increase in effort, particularly earlier in the season (February/March) and near the closed nursery areas. This trend was exacerbated by the introduction of mechanical prawn-peeling machines in both fisheries in 1976, which allowed whole small prawns to be reduced to prawn meat with minimal labor costs. The greatest impact of this technology and effort redistribution was in Shark Bay, where up to 40% of the catch was below export sizes by 1980–1981 compared with less than 5% before 1976. A similar redistribution of effort into February/March also occurred in the Exmouth Gulf fishery, resulting in significant increases in the catch of both tiger and king prawns under export size.

A preliminary analysis of the Exmouth Gulf tiger prawn data, relating the recruitment indexes of (13) to the resulting catch (a direct measure of Y/R), suggested that the catch for the years 1976–1980 was generally less than obtained during 1970–1975 for equivalent recruitment levels. This occurred despite the higher effort during the latter years.

To counter these problems with size at first capture, closed seasons and an experimental system of extended nursery areas were implemented to provide short-term protection for new recruits at the start of each year. These measures have effectively reversed the trend of increasing fishing pressure on small prawns and changed the timing and distribution of effort closer to that occurring in the early 1970s. For the Shark Bay king prawn stock the catches of 1982–1985 under this system have been above average (Fig. 3) and the proportion of nonexport prawns has again fallen to 5–10% of the total catch. These results have contrasted with the widely accepted view (33) that fishing of small prawns at or below the size at first maturity would produce the highest catch, because the high values of natural mortality (M) outweighed the increase in biomass from growth rate (K). However, with improvement in tagging techniques it is to be expected that the sources of bias in tag-based estimates of M and K (27, 30) will be reduced, so that true estimates of M will be lower and K higher. In the banana prawn (*P. merguiensis*) fishery of the Gulf of Carpentaria (34), early Y/R analysis suggested that the start of fishing should be in mid-March. However, a subsequent analysis with a more detailed data set suggested that a mid-April or later starting date would be more appropriate, particularly at present effort levels (35). This study also showed that the optimal starting date can vary in the short term owing to the timing of recruitment, and that preseason sampling is a necessary prerequisite to proper management of prawn stocks. Preseason sampling has been adopted as a research/management tool in setting opening dates for extended nursery areas in the W.A. fisheries (Shark Bay 1978–1980 and Exmouth Gulf 1983 onward) and occurs regularly in the South Australian fisheries (31). This pragmatic approach to growth overfishing in Australian penaeid fisheries appears to have been successful at this time in maintaining or enhancing the yield both in weight and in value.

5.3. Assessment of Stock–Recruitment Relationships

Although stock–recruitment relationships (SRRs) are fundamental to the management of fisheries, they have "received remarkably little attention" in the assessment of penaeid shrimp stocks (36). Concern was expressed by a number of countries in 1981 that effort levels might be affecting recruitment (37), but a subsequent critical review (38) concluded that the data available at that time did not present convincing evidence of SRRs or recruitment overfishing. However, since 1980 there have been significant declines in the levels of catch from the two major stocks of tiger prawns in Shark Bay and Exmouth Gulf. These sustained decreases in catch were inconsistent with the past high degree of variation shown in the catch record (Figs. 3 and 5), suggesting that recruitment overfishing may be implicated.

5.3.1. Exmouth Gulf

Analysis of the Exmouth Gulf fishery data has enabled both a SRR and the complementary replacement relationship between recruitment and surviving spawning stock (modified by fishing) to be documented for the tiger prawn stock (13, 20). These papers also contain details of the indexes of stock and recruitment utilized, the environmental factors (namely cyclones) likely to influence recruit survival, and the measurement of fishing effort affecting survival to the spawning stock. An evaluation of the robustness of the SRR and of the recruitment to spawning stock relationship (RSR) with respect to assumptions made and potential sources of bias has also been presented.

Using multiple regression techniques, a Ricker SRR, adjusted by rainfall occurring in January (RJ) and February (RF) of the form

$$RI_t = 3.71SI_{t-1} \exp(-0.053SI_{t-1} - 0.0029RJ_t + 0.0030RF_t)$$

has been developed from the historical data. The index of recruitment (RI_t) shown was based on March–May catch rates of year t in the recruitment area and SI_{t-1} represents the index of spawning stock based on August to October catch rates of the preceding year in the spawning area. The multiple correlation for this relationship was 0.97; that is, 94% of the variation in recruitment was explained by the equation (13). The negative and positive parameters for January and February rainfall (cyclones), respectively, suggest a hypothesis that cyclones have the potential to increase or decrease recruit survival depending upon their timing (13).

To illustrate the combined effects of rainfall and spawning stock on recruitment, predicted SRRs under three different environmental situations are presented in Fig. 7 together with the historical data used in the analysis. This shows that 13 of the 15 year classes occur near curve *B*, which represents no rainfall (i.e., no cyclones) in January–February, which is close to the usual situation from historical rainfall records. The two year classes, 1971 and 1975, that appear as outliers from this line were both associated with particularly severe cyclones. The predicted SRRs corresponding to these rainfall situations are shown as curves *A* and *C* representing recruitment with January/February rainfalls of 0 mm/200 mm and 300 mm/0 mm, respectively.

The complementary relationship that determines how effective effort applied to autumn recruits (ER_t) influences survival to spring of the spawning stock was also examined using multiple regression. The equation of the form

$$SI_t = 1.27RI_t \exp(-0.0232ER_t)$$

resulted in a multiple correlation of 0.94 (88% of variation in spawning stock explained), with the parameter associated with effective effort being significant at the 0.01 level. This relationship, illustrated in Fig. 8, showed that increasing effort from the moderate level (about 40 units) experienced until the mid-1970s to the high level (about 60 units) experienced in the 1978–1980 period would result in a

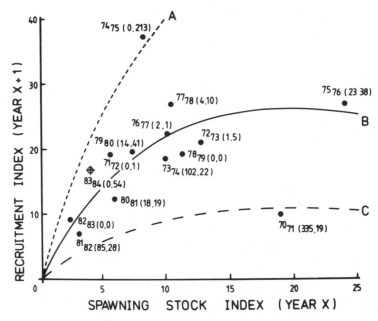

Figure 7. The relationship between spring spawning stock, autumn recruitment, and January/February rainfall for the Exmouth Gulf tiger prawn stock. Spawning/recruit years and January/February rainfall (mm) are presented as follows: 7576$_{(23, 38)}$ (1983/1984 data point included but not used in regression) (reproduced from Ref. 20).

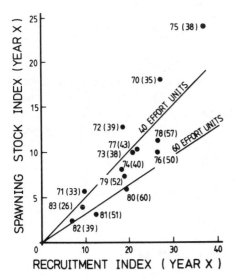

Figure 8. The relationship between recruitment, effective fishing effort (recruitment to spawning season), and resultant spawning stock for the Exmouth Gulf tiger prawn stock from 1970 to 1983. Year and effective effort units are given as 70(35), for example (reproduced from Ref. 20).

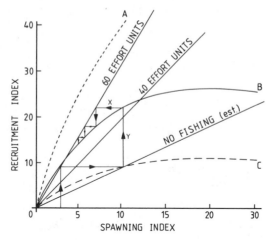

Figure 9. Interaction between SRR and RSR for the Exmouth Gulf tiger prawn stock under a variety of environmental and effort conditions (reproduced from Ref. 13).

reduction of the expected spawning stock by about 37% for any given level of recruitment.

For management purposes an understanding of the interaction between the various SRRs and the RSR (Fig. 9) has been essential to determine the equilibrium spawning stock and recruit levels that are likely to be generated by given levels of effort. For example, under the typical environmental conditions of low January and February rainfall (curve *B*), the equilibrium stock and recruitment level resulting from the high fishing effort levels (about 60 units) that built up during 1978–1980 corresponds to the intersection of the two lines at *B*60. This low stock and recruitment level, although at an equilibrium point, corresponds to a low yield (100–200 t) and may not remain stable. That is, a small increase in effort could result in no intersection of the two lines and theoretically cause a decline to extinction if effort levels were maintained.

On the basis of this model, the recent high effort levels have been accepted as the most likely cause of the reduction in catches of *P. esculentus* in Exmouth Gulf (see stepped line *x* on Fig. 9). This contrasts with the historical situation where lower effort levels of about 40 units occurred and higher equilibrium stock and recruitment levels (i.e., at *B*40) with yields of the order of 600 t annually were maintained. However, effort levels lower than 40 units are not warranted, for although this would result in a higher spawning stock, it would only provide a similar recruitment level and so produce an overall reduction in catch.

The zero effort line (i.e., no fishing), which has been extrapolated from the data, indicates that the recovery from low stock and recruitment levels could be quite rapid, say, 1–2 yr, if fishing ceased (see stepped line *y* on Fig. 9). This suggests that with unrestricted fishing, where low catch rates would lead to a cessation or severe reduction of fishing effort as vessels left the fishery, a rapid buildup of stocks would occur, thus encouraging fishing effort to increase again. High fishing effort would then lead to a reduction in abundance, followed again by a reduction in effort. This sequence of events would tend to repeat itself and produce cycles in abundance, which would be difficult to explain unless one had knowledge of the underlying SRR and RSR and of any important environmental factors affecting recruitment. Similar difficulties in interpretation of SRRs can also occur in the more usual situation, where environment is the dominant influence on recruitment (38). Even in the Exmouth Gulf situation, where the environment was relatively stable, identification of environmental effects played an important part in being able to identify the SRR within the scatter of points. These problems highlight the need always to consider spawning stock and environmental effects simultaneously when assessing SRRs.

As a consequence of the environmental effects on recruitment in the Exmouth tiger prawn fishery it has become obvious that in future the management must be flexible enough to make allowance for both high or low recruitment whenever it occurs. Should a year with poor environmental conditions occur (curve C in Fig. 7) strong consideration would have to be given to an immediate reduction in effort to below 40 units so as to avoid a severe reduction in the spawning stock. However, if a high recruitment level (curve A in Fig. 7) occurred because of advantageous environmental conditions, then effort could be allowed to expand above 40 units for that year without adversely affecting the long-term equilibrium stock and recruitment level ($B40$ in Fig. 9).

Since the sustained high levels of effort in the late 1970s and the resulting decrease in the spawning stock to critical levels in 1982, the necessary rebuilding of the spawning stock has been achieved primarily by greatly reduced fishing on the recruits as well as a complete ban on fishing for tiger prawns during the main period of spawning. Because of the intense nature of the effort on recruits, it would have been pointless banning fishing on the spawning stock unless effort on the recruits was reduced to allow sufficient escapement to the spawning stock. The impact of these controls on effort has been to decrease the catch during 1983, but produce a doubling of the spawning index. In 1984, recruitment and spawning stock levels both improved and a significant increase in catch to 167 tonnes occurred (Fig. 5), while in 1985 a catch of 225 tonnes was achieved and similar spawning stock levels were maintained.

5.3.2. Shark Bay

Although yet to be completed, a preliminary evaluation of the data for the Shark Bay tiger prawn stock has indicated that a SRR shaped similarly to that of the Exmouth Gulf stock, but with little environmental variation, will be produced. Of particular interest is the apparently new lower plateau of catches (Fig. 3), which has been evident since 1979 and which is now showing about the same proportional variation that occurred during the preceding high-catch years. This suggests that a new equilibrium has been established between spawning stock, recruitment, and effort, and that the annual variability in recruitment now showing is still due to underlying environmental factors.

Because of the multispecies nature of the Shark Bay fishery, with tiger prawns being the minor species in the catch, management measures to reduce effort (Section 6.2) on tiger prawns directly have yet to be developed. As a consequence, high effort levels are continuing to be applied to the stock (Fig. 6), with the result that a considerable number of low-recruitment years have occurred, supporting the hypothesis that recruitment overfishing is occurring.

6. MANAGEMENT

The history and effectiveness of management on the W.A. prawn fisheries prior to 1981 has been previously reported (9, 17). Administration of these fisheries since their inception has been under the effective control of a single authority, the Western Australian Fisheries Department, which has provided the legislative, research, and enforcement support for management policies.

Initially, the main considerations for management were given as ''the economic viability of fishing units and processing establishments, and the effects of exploitation on the prawn resources'' (39). These considerations led to the early implementation of a system of limiting entry to each fishery, so that development could occur at a controlled pace towards a point ''somewhere between the maximum economic yield and maximum sustainable yield'' (17).

This limited-entry management regime was successful during the 1960s and 1970s in controlling the expansion in total catch, while keeping catch rate at economically viable levels (9). However, this apparently successful strategy was based on the widely accepted view (40) that recruitment overfishing does not occur with penaeid fisheries. As a consequence management felt no need to respond rapidly to the continuing escalation of effective effort per licensed vessel until some clear evidence of recruitment failure was produced. Changes in fishery regulations and management strategies in response to the documentation of recruitment overfishing (Section 5.3) have now occurred.

6.1. Fishery Regulations

The regulations are summarized in Table 1.

Limited-Entry. The philosophy of limiting the number of vessels with access to specific fisheries was first introduced as a management method for Australian prawn fisheries when 25 trawlers were licensed to operate in Shark Bay for 1963. This number was increased to 30 for the 1965–1971 period, then to 32 (1972–1974), and to 35 from 1975 to the present. Preceding each increase a review of catch rates and economic performance of the fleet was undertaken (41).

In Exmouth Gulf, limited-entry was introduced in 1965, two years after Shark Bay when 15 vessels were licensed to operate. This number was increased to 17

TABLE 1   Summary of Current (1986) Fishery Management Regulations Applied to Shark Bay and Exmouth Gulf Prawn Fisheries of Western Australia

Regulation	Shark Bay Fishery	Exmouth Gulf Fishery
Limited entry		
Number of licensed trawlers	35	19
Vessel replacement limit (max. size)	375 vessel units (See Section 6.2)	375 vessel units
Fishing gear controls		
Trawl nets (number and max. head-rope length)	2 × 14.6 m	2 × 13.7 m
Otter boards (max. dimensions)	2.4 × 0.9 m	2.25 × 0.9 m
Ground chains (max. number and size)	2 chains/net, 10 mm link diameter	2 chains/net, 10 mm link diameter
Closed seasons	Nov. 1 to March 1	Nov. 15 to March 1
Closed areas	See Fig. 1	See Fig. 2
Permanent nursery areas	Closed at all times	Closed at all times
Temporary nursery areas	Closed Aug. 1 to April 15	Closed Nov. 15 to April 15
Spawning stock closure (for tiger prawns)	Nil	See Line *A*, Fig. 2 (closure time varies with recruit abundance)

in 1967, 20 in 1970, 22 in 1972, and 23 in 1979. Subsequently, in 1985 a reduction in fleet size to 19, was effected by a buyback scheme.

Closed Areas. Permanently closed nursery areas in the south of Shark Bay (Fig. 1) were introduced in 1963 (15). In addition, a temporary closure of an adjacent area was introduced experimentally in 1978, and with modifications became a permanent feature of the fishery in 1982. This extended nursery area is closed to fishing from August 1 to April 15 annually.

In the Exmouth Gulf fishery a similar system of permanent closed and temporarily extended nursery areas (Fig. 2) has been developed since 1975 to protect small prawns. In addition, an extensive closure line (line A, Fig. 2) was introduced in 1982 to control fishing on the tiger prawn stock that generally occurs in the southeastern Gulf, as part of a short-term management strategy to increase the tiger prawn spawning stock levels.

Closed Seasons. A summer closed season from November 1 to March 1 was introduced for the Shark Bay fishery in 1982. Summer closures have also been introduced experimentally in Exmouth Gulf since 1983. Presently the closure extends from November 15 to March 1.

Gear Controls. Since 1976 a limitation on the maximum number and size (head-rope length) of nets, at 2 × 14.63 m (8 fathoms) has been enforced in both fisheries.

Mesh sizes are not regulated in either fishery; however, standard mesh sizes (50 mm) have been used throughout the history of the fishery.

6.2. Current Management Strategies

As a result of the documentation of recruitment overfishing of tiger prawns in Exmouth Gulf and similar implications for the Shark Bay tiger prawn stock (Section 5.3), recent management has been focused clearly on the issues of increases in effective effort and its impact on survival of recruits to spawning. To address this problem in the most affected Exmouth Gulf fishery, the revised management strategy has endeavored to:

1. Reduce the total effort being applied by the fleet.
2. Segregate the effort applied to each stock, so that specific controls could be applied to effort directed toward the affected tiger prawn stock (closure line A, Fig. 2).
3. Delay the application of effort on tiger prawns until later in the life cycle so that growth overfishing does not occur and survival to spawning is enhanced.

Specific measures to achieve these objectives have been to encourage vessels with alternative fisheries licenses to leave the fishery temporarily (1983–1985) and to shorten the overall fishing season for all species to $9\frac{1}{2}$ months. In addition, effort within the tiger prawn area (within closure line A) has not been permitted prior to April and has been restricted to a third of the fleet by a roster fishing system (1983–1985). The closing date for this tiger prawn fishing has been varied depending on observed catch rates during each season and the relationship between recruitment catch rates and subsequent spawning stock levels. Fishing has, as a consequence, ceased between May and July with a complete closure being enforced during the spring spawning season from August onward (1982–1985). These measures, which severely restricted fishing on tiger prawns, have produced a significant improvement in the spawning stock levels since 1982, which has corresponded to a gradual increase in yield each year since 1983 (Fig. 5).

With the intention of permanently restructuring the Exmouth fishery, an industry-funded buyback scheme has been implemented (1985) which has removed four licensed vessels (17%) from the fishery. A reduction in maximum net sizes to 2×13.7 m ($7\frac{1}{2}$ fathoms) and restrictions on otter board sizes and ground chains have also been introduced for 1986, to avoid vessels escalating trawling speed to negate the controls on net size (Table 1). Vessel replacement rules limiting the maximum size/power of new vessels to 375 units have also been introduced for the 1986 season. The vessel units involved are the sum of the installed kilowatt output of the main engine and the under deck volume in cubic meters.

In the Shark Bay fishery, where tiger prawns are the minor species in the catch, and the habitats of the two species are more interspersed (25), the radical management measures applied in Exmouth Gulf were not feasible because they could impact on the western king prawn yields. Here measures to reduce total effort have been restricted to the conventional seasonal closures noted previously and stricter controls on gear. The same vessel replacement policy as introduced for Exmouth Gulf has been added to the Shark Bay regulations and maximum otter board size regulations are to be introduced in 1987.

Because of the more complex nature of this fishery, more novel approaches to segregating and controlling effort being applied to the tiger prawn stock are under consideration. Firstly an effort reduction system based on short-term (7-day) total fishery closures over each full moon period is being evaluated for impact on both species. This could reduce the catch of tiger prawns during the recruitment period (March–May), thus allowing greater survival to spawning while increasing the average catch rate for king prawns, which have low catchability during full moon periods (27). A second research proposal being considered is to segregate effort applied to the tiger prawn recruitment by placing "obstacles" to trawling on known areas of tiger prawn habitat to produce a self-policing area closure system specifically for tiger prawns. Such a measure applied to the recruitment areas would theoretically delay exploitation until closer to the spawning season, when most of the stock would have migrated farther offshore onto open trawl grounds. Should these measures not prove efficient or practical, the most likely alternative to provide a cost-effective effort reduction mechanism will be an industry-funded buyback scheme to remove vessels, as was applied in the Exmouth Gulf fishery.

In contrast to the difficult issues of effort control the management policies regarding size at first capture have been more successful in Shark Bay. Here the extended nursery closure system has succeeded in reversing the trend toward smaller size at first capture so that few ($<10\%$) below-export size (66/kg) king or tiger prawns are now taken compared with up to 40% in the late 1970s. In Exmouth Gulf, similar closures for small tiger prawns have also been successfully implemented; however, smaller than optimal export size western king prawns are presently fished at the start of each season. To address this problem the summer closed season has been lengthened progressively, and may be extended further when the tiger prawn stock regains its former abundance.

6.3. Enforcement

The application of limited-entry management to developing fisheries, rather than fully exploited fisheries, has brought significant benefits in the area of enforcement and cooperation with industry members. The demonstrated profitability of these fisheries (41) has brought with it a value attached to a license usually equivalent to one year's gross catch, currently (1985) about A$500,000 in both fisheries. (This "value" is the additional amount paid to a vessel owner by the purchaser when a licensed vessel is sold.) Under these circumstances, the possibility of license suspension or cancellation for an infringement of the regulations is a significant deterrent, but more importantly the relationship between license transfer value and the state of the stock gives all license holders a vested interest in supporting management rules for the fishery. For these reasons day-to-day enforcement requirements tend to be minimal, with most enforcement actions being based on industry reporting of infringements. However, this situation would certainly not have been achieved had the fishery been unrestricted with a large number of marginally profitable vessels, as in the more usual situation in penaeid fisheries. This high degree of cooperation between industry and the managing authority has been a notable feature of the W.A. prawn fisheries and has been attributed to the implementation of limited entry at a time when fishing was profitable (9). The buildup of this high degree of cooperation has been best illustrated in the Exmouth Gulf fishery, where the various closure lines and roster fishing arrangements operated successfully without an on-site enforcement presence throughout the economically difficult low-recruitment years (1982–1985). This cooperation, however, has brought with it a special responsibility for precise data collection, analysis, interpretation, and implementation to which the fishing industry has responded by providing a high level of information.

7. GENERAL DISCUSSION AND CONCLUSIONS

Assessments using surplus production model techniques provided an adequate basis for management to expand fishing effort on Western Australia's penaeid fisheries during the early stages of development (9). However, these methods could not foresee the twin problems of growth and recruitment overfishing (20) that have appeared as effort levels have continued to increase beyond those that are now recognized to be above the MSY levels for individual stocks.

Although not a major segment of the W. A. research programs, experience with optimizing yield per recruit, utilizing empirical methods, has suggested that growth overfishing is not a particularly critical issue for purely offshore penaeid fisheries. Relatively simple management changes to delay the start of fishing have proved effective in significantly changing size composition and appear to have increased yield in both weight and value terms. These practical results have been in contrast to the traditional suggestions, based on tagging data, that yield would be enhanced by smaller sizes at first capture. However, these results do have considerable implications for management of fisheries where both inshore (artisanal) and offshore fishing occurs on stocks of similar ["brown shrimp" species (32)] penaeids.

In contrast to the growth overfishing problem, the recent documentation of an underlying spawning stock–recruitment relationship and recruitment overfishing in the W.A. fisheries, has proved a more difficult issue to address in management terms. Although the combination of geographic, environmental, and effort factors affecting the tiger prawn stocks may have been unusual in allowing the stock–recruitment relationship to become evident (13, 20) the results are considered to have wider implications for management of penaeid stocks in general. That is, with the use of technology resulting in escalating effort levels in most penaeid fisheries, whether limited-entry or not, there is now a clear need to include spawning stock with environmental parameters when assessing recruitment variations. For this purpose, detailed historical catch effort statistics have been found to be essential for the production of reliable indexes of stock and recruitment, as well as effort expended on recruits. Without such data it would not be possible to separate the effects of spawning stock levels on recruitment from those of environment [which usually dominates penaeid recruitment (38)] and give advanced warning of recruitment overfishing should it arise. In this regard it may be useful to look for features of the affected W.A. stocks that may help identify other stocks at risk (13, 20). Such stocks appear to include those confined to geographically discrete areas, species with high catchability, particularly through aggregation effects, and stocks in multispecies fisheries situations where the alternative species can provide an effective subsidy supporting fishing on very low catch rates for individual species. It was also noted that the past history of limited-entry management was also indirectly implicated as a factor contributing to the maintenance of excessive levels of effort after the stock had collapsed. That is, the past history of profitability enabled the fleet to continue to operate when catch rates became uneconomic.

These issues also highlighted the need for flexibility in the control of effort levels. Within Western Australia, the application of limited-entry management to each developing prawn fishery has forced vessels to rely on single fisheries because access to alternative stocks is no longer available. This lack of fleet mobility, in addition to the biological consequences of increasing effort within each limited-entry fishery, suggests that the fishing industry will have to accept increased management controls both to reduce effort and to regulate vessel or gear changes that can increase effective effort, if yields are to be maintained.

REFERENCES

1. N. M. Haysom, Review of the penaeid prawn fisheries of Australia. In P. C. Rothlisberg, B. J. Hill, and D. J. Staples, Eds., *Proceedings of the Second Australian National Prawn Seminar.* Simpson Halligan, Brisbane, Australia, 1985, pp. 195–203.

2. N. V. Ruello, An historical review and annotated bibliography of prawns and the prawning industry. In P. C. Young, Ed., *Proceedings of the First Australian National Prawn Seminar.* Aust. Govt. Publ. Serv., Canberra, 1975, pp. 305–334.

3. S. Garcia and L. Le Reste, Life cycles, dynamics, exploitation and management of coastal penaeid shrimp stocks. *FAO Fish. Tech. Pap.* **203:**1–215 (1981).

4. A. A. Racek, Prawn investigations in eastern Australia. *Res. Bull. State Fish. N.S.W.* **6:**1–57 (1959).

5. I. S. R. Munro, Biology of the banana prawn (*Penaeus merguiensis*) in the south-east corner of the Gulf of Carpentaria. In P. C. Young, Ed., *Proceedings of the First Australian National Prawn Seminar.* Aust. Govt. Publ. Serv., Canberra, 1975, pp. 60–78.

6. J. W. Penn, A review of mark recapture and recruitment studies on Australian panaeid shrimp. *Kuwait Bull. Mar. Sci.* **2:**227–248 (1981).

7. K. H. Mohamed, M. El-Musa, and A. R. Abdul-Ghaffar, Observations on the biology

of an exploited species of shrimp, *Penaeus semisulcatus* de Haan, in Kuwait. *Kuwait Bull. Mar. Sci.* **2**:33–52 (1981).

8. R. R. C. Edwards, The fishery and fisheries biology of penaeid shrimp on the Pacific coast of Mexico. *Oceanogr. Mar. Biol.* **16**:145–180 (1978).

9. B. K. Bowen and D. A. Hancock, The limited entry prawn fisheries of Western Australia: research and management. In J. A. Gulland and B. J. Rothschild, Eds., *Penaeid Shrimps: Their Biology and Management.* Fishing News Books, Farnham, Surrey, England, 1984, pp. 272–289.

10. B. W. Logan, R. G. Brown, and P. G. Quilty, *Quarternary Sediments, Sedimentary Processes and Environments, Exmouth Gulf. Carbonate Sediments of the West Coast of Western Australia*, Part 2. Dept. of Geology, University of Western Australia, Nedlands, 1976, pp. 61–73.

11. B. W. Logan and D. E. Cebulski, Sedimentary environments of Shark Bay, Western Australia. *Mem. Am. Assoc. Pet. Geol.* **13**:1–37 (1970).

12. J. W. Penn and R. W. Stalker, The Shark Bay prawn fishery (1970–76). *West. Aust. Dep. Fish. Wildl. Rep.* **38**:1–38 (1979).

13. J. W. Penn and N. Caputi, Spawning stock-recruitment relationships and environmental influences on the tiger prawn (*Penaeus esculentus*) fishery in Exmouth Gulf, Western Australia. *Aust. J. Mar. Freshwater Res.* **37**:491–505 (1986).

14. W. Dall, A review of penaeid prawn biological research in Australia. In P. C. Rothlisberg, B. J. Hill, and D. J. Staples, Eds., *Proceedings of the Second Australian National Prawn Seminar.* Simpson Halligan, Brisbane, Australia, 1985, pp. 11–21.

15. R. J. Slack-Smith, The prawn fishery of Shark Bay, Western Australia. *FAO Fish. Rep.* **57(3)**:717–734 (1969).

16. R. J. Slack-Smith, Early history of the Shark Bay prawn fishery, Western Australia. *Fish. Bull. West. Aust.* **20**:1–44 (1978).

17. D. A. Hancock, The basis for the management of West Australian prawn fisheries. In P. C. Young, Ed., *Proceedings of the First Australian National Prawn Seminar*, Aust. Govt. Publ. Serv., Canberra, 1975, pp. 252–269.

18. J. W. Watson, I. K. Workman, C. W. Taylor, and A. F. Serra, Configurations and relative efficiencies of shrimp—trawls employed in south eastern United States waters. *NOAA Tech. Rep.*, NMFS **3**:1–12 (1984).

19. T. F. C. White, Population dynamics of the tiger prawn *Penaeus esculentus* in the Exmouth Gulf prawn fishery, and implications for the management of the fishery. Ph.D. Thesis, University of Western Australia, Nedlands, 1975.

20. J. W. Penn and N. Caputi, Stock-recruitment relationships for the tiger prawn (*Penaeus esculentus*) fishery in Exmouth Gulf, Western Australia and their implications for management. In P. C. Rothlisberg, B. J. Hill, and D. J. Staples, Eds., *Proceedings of the Second Australian National Prawn Seminar.* Simpson Halligan, Brisbane, Australia, 1985, pp. 165–173.

21. D. L. Grey, W. Dall, and A. G. Baker, *A Guide to the Australian Penaeid Prawns.* Northern Territory Department of Primary Production, Darwin, Australia, 1983, 140 pp.

22. J. W. Penn, The influence of tidal cycles on the distributional pathway of *Penaeus latisulcatus* Kishinouye in Shark Bay, Western Australia. *Aust. J. Mar. Freshwater Res.* **26**:93–102 (1975).

23. J. W. Penn, Spawning and fecundity of the western king prawn, *Penaeus latisulcatus* Kishinouye, in Western Australian waters. *Aust. J. Mar. Freshwater Res.* **31:**21–35 (1980).

24. D. J. Staples, W. Dall, and D. J. Vance, Catch prediction of the banana prawn, *Penaeus merguiensis*, in the south-eastern Gulf of Carpentaria. In J. A. Gulland and B. J. Rothschild, Eds., *Penaeid Shrimps: Their Biology and Management.* Fishing News Books, Farnham, Surrey, England, 1984, pp. 259–267.

25. N. G. Hall and J. W. Penn, Preliminary assessment of effective effort in a two species trawl fishery for Penaeid prawns in Shark Bay, Western Australia. *Rapp. P.-v. Réun. Cons. int. Explor. Mer* **175:**147–154 (1979).

26. J. W. Penn, The behaviour and catchability of some commercially exploited penaeids and their relationship to stock and recruitment. In J. A. Gulland and B. J. Rothschild, Eds., *Penaeid Shrimps: Their Biology and Management.* Fishing News Books, Farnham, Surrey, England, 1984, 173–186.

27. J. W. Penn, Tagging experiments with the western king prawn *Penaeus latisulcatus* Kishinouye. II. Estimation of population parameters. *Aust. J. Mar. Freshwater Res.* **27:**239–250 (1976).

28. B. J. Hill, Effect of temperature on duration of emergence, speed of movement, and catchability of the prawn *Penaeus esculentus.* In P. C. Rothlisberg, B. J. Hill, and D. J. Staples, Eds., *Proceedings of the Second Australian National Prawn Seminar.* Simpson Halligan, Brisbane, Australia, 1985, pp. 77–83.

29. R. J. Slack-Smith and A. E. Stark, Automatic processing of Western Australian prawn fishermen's log books. *FAO Fish. Rep.* **57(2):**309–319 (1968).

30. J. W. Penn, Tagging experiments with the western king prawn, *Penaeus latisulcatus* Kishinouye. I. Survival, growth and reproduction of tagged prawns. *Aust. J. Mar. Freshwater Res.* **26:**197–211 (1975).

31. B. K. Bowen and D. A. Hancock, Review of penaeid prawn fishery management regimes in Australia. In P. C. Rothlisberg, B. J. Hill, and D. J. Staples, Eds., *Proceedings of the Second Australian National Prawn Seminar.* Simpson Halligan, Brisbane, Australia, 1985, pp. 247–265.

32. S. Garcia, Reproduction, stock assessment models and population parameters in exploited penaeid shrimp populations. In P. C. Rothlisberg, B. J. Hill, and D. J. Staples, Eds., *Proceedings of the Second Australian National Prawn Seminar.* Simpson Halligan, Brisbane, Australia, 1985, pp. 139–158.

33. J. H. Kutkuhn, Dynamics of a penaeid shrimp population and management implications. *Fish. Bull.* **65(2):**313–338 (1966).

34. C. Lucas, G. Kirkwood, and I. Somers, Assessment of the stocks of the banana prawn *Penaeus merguiensis* in the Gulf of Carpentaria . *Aust. J. Mar. Freshwater Res.* **30(5):**639–652 (1979).

35. I. F. Somers, Maximising value per recruit in the fishery for banana prawns, *Penaeus merguiensis*, in the Gulf of Carpentaria. In P. C. Rothlisberg, B. J. Hill, and D. J. Staples, Eds., *Proceedings of the Second Australian National Prawn Seminar.* Simpson Halligan, Brisbane, Australia, 1985, pp. 185–191.

36. J. A. Gulland, Introductory guidelines to shrimp management: Some further thoughts. In J. A. Gulland and B. J. Rothschild, Eds., *Penaeid Shrimp: Their Biology and Management.* Fishing News Books, Farnham, Surrey, England, 1984, pp. 290–298.

37. J. A. Gulland and B. J. Rothschild, Eds., *Penaeid Shrimp: Their Biology and Management.* Fishing News Books, Farnham, Surrey, England, 1984.

38. S. Garcia, The stock-recruitment relationship in shrimps: Reality or artifacts and misinterpretations? *Oceanogr. Trop.* **18:**25–48 (1983).

39. B. K. Bowen, The economic and sociological consequences of licence limitation. In P. C. Young, Ed., *Proceedings of the First Australian National Prawn Seminar.* Aust. Govt. Publ. Serv., Canberra, 1975, pp. 270–275.

40. R. A. Neal, The Gulf of Mexico research and fishery on penaeid prawns. In P. C. Young, Ed., *Proceedings of the First Australian National Prawn Seminar.* Aust. Govt. Publ. Serv., Canberra, 1975, pp. 60–78.

41. T. F. Meany, Limited energy in the Western Australian rock lobster and prawn fisheries—an economic calculation. *J. Fish. Res. Board Can.* **36:**789–798 (1979).

CASE STUDIES

B / LOBSTERS AND CRABS

6 THE LOBSTER FISHERY OF SOUTHWESTERN NOVA SCOTIA AND THE BAY OF FUNDY

Alan Campbell
Department of Fisheries and Oceans
St. Andrews, New Brunswick, Canada

1. INTRODUCTION

The American lobster, *Homarus americanus*, is distributed throughout the coastal waters of the northwest Atlantic from Labrador to North Carolina (Squires 1). The largest yields of lobsters are from the Gulf of Maine, southwestern Nova Scotia, and the southern Gulf of St. Lawrence (Fig. 1). During 1984, lobster landings from the Bay of Fundy and southwestern Nova Scotia totaled 6,991 t with a landed value of about Canadian $54.2 million. The landings were 24 and 36%, respectively, of the Canadian total: (28,671 t and about Canadian $152.5 million).

Lobsters from the Bay of Fundy and off southwestern Nova Scotia probably are from a single stock (Campbell and Mohn 2; Ennis 3). Fishing for lobsters occurs mostly at depths from a few meters to 50 m in the inshore area, which is partitioned into four lobster fishing districts (LD) or management units (Fig. 2). Since 1971, a lobster fishery has developed offshore beyond 93 km from the coast of southwestern Nova Scotia (Fig. 2) (Pezzack and Duggan 4, 5). Only a few boats have fished the area between the inshore and offshore areas (Sharp and Duggan 6), although the number is increasing. This chapter reviews the biology of and fishery for lobsters in this general area from a management perspective.

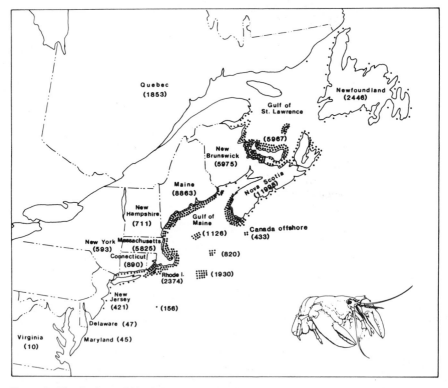

Figure 1. Distributions of North American lobster landings during 1984. One dot equals 100 tonnes.

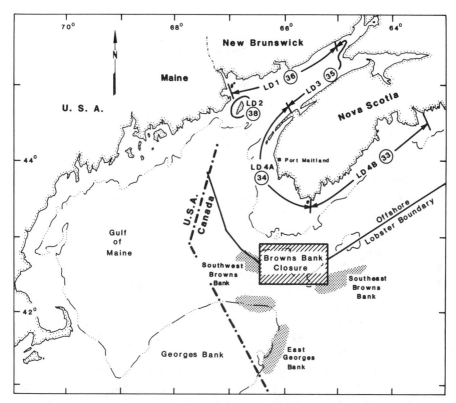

Figure 2. Inshore lobster districts and offshore lobster fishing areas (stippled) off southwestern Nova Scotia and the Bay of Fundy. Numbers in circles are newly designated lobster district numbers to replace the old numbers (e.g. 1, 2, 3 and 4A).

2. HISTORY OF FISHERY

2.1. Fishing Gear

During the eighteenth century, lobsters were caught in Canada by hook, spear, and hoop net (netting stretched across a hoop with bait in the center; e.g., see McLeese and Wilder 7, p. 54) but by the late 1860s Nova Scotia fishermen began to use wooden traps (DeWolf 8). Now traps are the only legal lobster fishing gear allowed in Canada. In recent years, wooden traps have been replaced by lighter, more durable plastic-coated-wire mesh traps. Although in shallow water (<50 m) traps usually are set one per buoy, in deeper water, fishermen attach 15–20 traps to a single rope or "string" with a buoy at each end. The bait used in traps varies, but normally consists of salted, fresh, or frozen herring, fish heads, or fresh or frozen mackerel.

For over 100 years, the lobster fishery in the Bay of Fundy and southwestern Nova Scotia has used relatively small boats. In the early 1900s, sailboats and rowboats 4–5 m long and traps hauled by hand were replaced by gasoline-powered motor boats and mechanical trap haulers (Prince 9; Halkett 10). The average size of boat has increased gradually over the years. Now the inshore fishing boats are generally of the Cape Island style and are 10–14 m long, constructed of wood or fiberglass. Only eight lobster boats fish in the offshore areas of Georges and Browns Banks and they are large (20–35 m), constructed of wood or steel. Most fishermen fish for lobsters within 40 km of their home port. However, the addition of hydraulic trap haulers, depth sounders, radar, and loran-C equipment have allowed fishermen to travel further and faster and to fish deeper and more efficiently than their predecessors (Pringle et al. 11). For example, the average number of trap hauls per boat per day in the Port Maitland area (Fig. 2) ranged from 84 to 117 during the 1944–1962 period, but has steadily increased to an average of 233 for the 1985–1986 fishing season (A. Campbell, unpublished data).

2.2. Regulations

The lobster fishery is a common property resource under the jurisdiction of the federal government of Canada in accordance with the British North America Act of 1867. This original fisheries act gave the Minister of Fisheries the responsibility to protect and enhance fish stocks, and since 1878, egg-bearing females have been protected and size limits and fishing season restrictions imposed. Starting in the late 1960s, there have been limits on licenses and the maximum number of traps per license (Table 1). The present minimum legal size at recruitment is 81-mm carapace length. The fall–spring fishing season has been largely maintained since the turn of the century, with a few modifications. The fishing seasons were

TABLE 1   Summary of Landings, Value, Licenses, and Median Carapace Lengths of Legal-Sized Recruited Lobsters for the Inshore and Offshore Lobster Fishery in the Bay of Fundy and Southwestern Nova Scotia during 1984[a]

| | Inshore Lobster District[a] | | | | | |
	1	2	3	4A	Offshore	Total
Landings (t)	228	365	195	5770	433	6991
	(3.3)	(5.2)	(2.8)	(82.5)	(6.2)	
Landed value (Can. $1000)	1834	3016	1646	45054	2621	54171
	(3.4)	(5.6)	(3.0)	(83.2)	(4.8)	
Number of licenses	185	109	91	962	8	1355
	(13.7)	(8.0)	(6.7)	(71.0)	(0.6)	
Trap limit/license	300	375	300	375	1000	—
Mean landings (t/license)	1.2	3.3	2.1	6.0	54.1	5.2
Mean value (Can. $1000/ license)	9.9	27.7	18.1	46.8	327.6	40.0
Median carapace length (mm)[b]	88	89	94	88	120	89

[a] Values in parentheses are percentages of the total in the row above.
[b] For 1983–1984, after Campbell and Pezzack (12).

intended, in part, to protect lobsters while molting, mating, and extruding eggs, to reduce exploitation rates, and to adjust for seasonal market demands, especially to take advantage of seasonal high prices in the fall–spring period. Presently, the open fishing seasons are slightly different for each of the four lobster districts: LD1 = second Wednesday in November to January 14 and April 1 to June 29; LD2 = second Wednesday in November to fourth Thursday of June; LD3 = October 15 to December 31 and March 1 to July 31; LD4 = last Tuesday in November to May 31.

For the Canadian offshore lobster fishery, regulations include a limit of eight licensed boats, 1000 traps per boat, and an annual quota of 90 t per boat. There is a closed area for brood stock protection on Browns Bank (Fig. 2) (Pezzack and Duggan 4, 5).

2.3. Yields

Lobster landings peaked at the turn of the century, declined from 1903 to the 1930s, recovered in the early 1940s, and remained relatively constant from the late 1940s to the early 1980s at around 4600 t (Fig. 3). During 1984, inshore south-western Nova Scotia produced the bulk (82.5%) of lobsters compared with the Bay of Fundy (11.3%) and the offshore areas (6.2%) (Table 1). The bulk of the inshore landings is made up of juveniles (subadults) newly recruited into the fishery; 60–

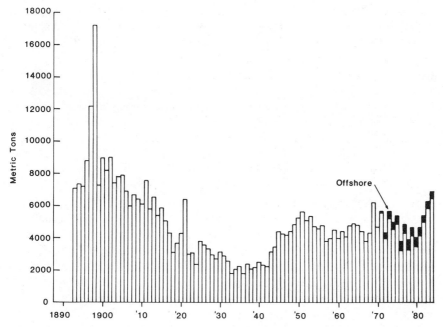

Figure 3. Annual lobster landings for the Bay of Fundy and southwestern Nova Scotia 1897–1984.

85% of lobsters are within the first molt of legal size (81–94-mm carapace length). The median size of recruited lobsters for the inshore areas is 89 mm carapace length (Table 1). In contrast, for most deep-water areas and offshore, most lobsters are large, with a median size of 120 mm (Table 1) (Campbell and Pezzack 12).

2.4. Markets

During the period 1840–1900 most caught lobsters were canned as lobster meat (Venning 13; Found 14). Beginning in 1881, live lobsters from the Bay of Fundy and southwestern Nova Scotia were exported to the United States, and only about half of the catch was still canned during 1925 (Found 15). A few years later, as a result of improved handling, shipping, and storage (McLeese and Wilder 7), canneries began to disappear from the area (DeWolf 8; Robinson 16). Now lobsters caught in this area are sold live, some locally, but most are exported to Japan, the United States, and various European countries (McLeese and Wilder 7; Shellfish Market Bulletin, Marketing Directorate, Department of Fisheries and Oceans, Ottawa).

3. LIFE HISTORY

3.1. General Biology

The eggs of *H. americanus* are incubated on the pleopods of the female for 9–12 months, and development of the eggs is closely related to temperature (Templeman 17; Perkins 18). The larvae hatch during summer (July–September) (Campbell 19) and are free-swimming for 1–2 months, the time depending on water temperature (Templeman 20). Stage IV larvae are stronger swimmers than the smaller larval stages and are capable of swimming at up to 9 cm/s (Ennis 21). Stage IV larvae settle to the bottom in late summer with the early postlarvae probably remaining cryptic in burrows, preferring gravel to silt–clay sediments (Cobb 22, 23; Pottle and Elner 24). As the individuals become bigger than about 30 mm carapace length, their foraging areas generally become larger (Stewart 25; P. Lawton, personal communication). Juvenile and adult lobsters occupy a wide variety of substrates (e.g., mud, sand, rock, bedrock), although the most common habitat is sand-gravel bottom with overlying rocks and boulders (Cooper and Uzmann 26).

Lobsters reaching the minimum legal size of 81-mm carapace length are estimated to be 6–8 yr old (Wilder 27). At this size, they molt on average once a year, but as they become larger, molting becomes less frequent (Campbell 28). The majority of lobsters are immature at their first molt into legal size (81–94-mm carapace length). Fifty percent of females reach physiological maturity (ovaries) at about 95-mm carapace length, and reach functional maturity (external eggs present) at about 108-mm carapace length, respectively (Campbell 28; Campbell and Robinson 29). The smallest and largest egg-bearing females recorded in the area were 82- and 200-mm carapace length (Campbell and Pezzack 12).

Sexual selection prior to mating may involve chemical communication (Atema 30), and although mating usually occurs after the female molts while the shell is soft (Templeman 31), sperm transfer also may occur in the intermolt hard-shell stage (Dunham and Skinner-Jacobs 32). Mature females are capable of extruding many egg broods during their life (Waddy and Aiken 33), and fecundity increases logarithmically with size (Campbell and Robinson 29), with as many as 97,440 eggs per brood recorded (Herrick 34).

3.2. Adult Migration

Tagging studies (Campbell et al. 35; Campbell and Stasko 36, 37; Campbell 19) indicate that mature lobsters move considerably further than immature ones and that long-distance movement (>100 km) allows some interchange of lobsters between the Bay of Fundy, the Gulf of Maine, and the adjoining continental shelf. Many mature lobsters also make seasonal migrations into shallow water during summer–fall and into deep water during winter–spring. Many mature lobsters return to the location where they were tagged initially. This can involve a round-trip movement of 10–400 km within 1 yr, depending on the bottom topography and temperature (Campbell and Stasko 37; Campbell 19; Pezzack and Duggan 38). Such supposedly temperature-dependent seasonal migrations can explain some of the long- and short-distance movements of lobsters that have been recorded (Campbell and Stasko 37). These migrations may not be well coordinated between lobsters because some lobsters may move laterally rather than up and down the slopes of the continental shelf (Cooper and Uzmann 26; Pezzack and Duggan 38). Although some lobsters apparently return to the same location year after year, about 10–20% of tagged mature lobsters move to other areas. Over many years, there is probably a general mixing of mature lobsters in the area, which means that lobsters lost to one area may be replaced by others from another area.

For *H. americanus*, seasonal depth migrations appear to be associated with an attempt to maintain the highest local ambient temperatures to maximize degree-days needed for molting, growth, gonad development, egg extrusion (Cooper and Uzmann 26), and egg development (Campbell and Stasko 37; Campbell 19). Berried females hatching eggs in relatively warm shallow water, whether on the shoals of Georges or Browns Banks or in coastal areas, may confer a survival advantage on the pelagic larvae by decreasing development time to the benthic stage (Huntsman 39; Caddy 40).

Seasonal migrations by mature lobsters have several implications for lobster fishery management in this area. The movements complicate estimation of brood stock size in a particular location because of the vast areas that must be sampled over a short time period. The movements also make stock differentiation difficult. Indeed, a jurisdictional problem arises in stock assessment and management of offshore lobsters that is complicated by the movement of lobsters across the U.S.–Canada international border. By coincidence, the movement of mature lobsters to shallow water during summer, when the fishing season is closed, protects these lobsters from exploitation. However, as mature females tend to move to deeper water earlier in the fall than mature males, the males are more vulnerable to the early fall fishery in the Bay of Fundy (Campbell 41; Campbell and Stasko 37).

4. RECRUITMENT

4.1. Sources

The relationships between seasonal movement of adult lobsters, distribution of brood stock, larval recruitment processes, and oceanographic features still are unclear. Concentrations of berried females (with eggs about to hatch) are found on Browns Bank, Georges Bank, and the New Brunswick side of the Bay of Fundy during summer months (Campbell and Pezzack 12). To a lesser extent, berried females are found in the coastal waters of southwestern Nova Scotia. Lobsters from coastal waters of southwestern Nova Scotia and the Bay of Fundy were estimated by Campbell and Pezzack (12) to produce between 23 and 53% of the total egg production in the area and those from the offshore Browns and eastern Georges Bank to produce 47–77%. These results suggest that most mature females of both the coastal and offshore areas are capable of producing many lobster eggs.

Where the planktonic lobster larvae settle as benthic juveniles is unknown. The processes that determine where larvae hatch, how they are distributed in the plankton, and where they eventually settle are still debated. There are two main theories on the larval recruitment mechanisms (for review, see Ennis 3). The first assumes that larvae are transported passively by residual currents. The second is based on larvae showing well developed behavioral responses to environmental stimuli and being able to maintain their general position in an area by vertical migrations. Recent studies indicate that the general surface currents in the Gulf of Maine and adjoining continental shelf are more complicated than previously thought (cf. Perry and Hurley 42, for review). There may be more local gyres (e.g., on Browns and Georges Banks) in this area (Perry and Hurley 42) (Fig. 4) than just the large circular current system in the Gulf of Maine suggested by Bumpus and Lauzier (43; Stasko and Campbell 44, p. 214). Whether larvae are retained in these gyres is unknown, although the question is under investigation (G. C. Harding and J. D. Pringle, personal communication).

Consequently, larvae that recruit to coastal southwestern Nova Scotia are believed either to be from local sources in some areas or to drift from Browns Bank, Georges Bank, or the eastern shore of Nova Scotia (Stasko and Campbell 44; Stasko 45; Dadswell 46; Harding et al. 47; Stasko and Gordon 48). Whether or not larval recruitment comes primarily from inside or outside the Bay of Fundy is equally uncertain. Residual surface currents tend to flow into the Bay of Fundy on the Nova Scotia side and out of the bay on the New Brunswick side (Fig. 4) (Bumpus and Lauzier 43). Therefore, larvae released by females on the offshore banks or in coastal areas of southwestern Nova Scotia may enter and settle in the Bay of Fundy. Larvae from the New Brunswick side of the Bay of Fundy may be transported by surface currents along the Gulf of Maine coastline, or may be caught

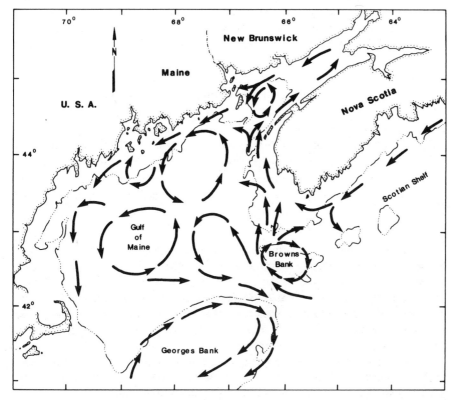

Figure 4. Composite representation of the spring–summer water currents in the upper water layers of the Bay of Fundy and Georges and Browns Banks. Compiled from Bumpus and Lauzier (43), Butman et al. (43a), Greenberg (43b), Smith (43c), Brooks (43d), and Perry and Hurley (42).

in eddies within bays along the mainland coast or the southern coast of Grand Manan or in some years remain in a gyre in the Bay of Fundy (Dickie 49).

4.2. Variability

In most inshore lobster fisheries, exploitation rates are sufficiently high (60–95%) to remove many of the legal-sized lobsters each year (Anthony 50; Campbell 51). Consequently, most of the inshore lobster landings are made up of new recruits, suggesting that annual landings are reasonable indicators of recruitment into the fishery (Campbell and Mohn 2; Ennis 52).

Suggested causes that contribute to fluctuations in recruitment or lobster landings in an area include the following:

Temperature variation (Flowers and Saila 53; Dow 54; Orach-Meza and Saila

55; Fogarty 56), which acts mainly to regulate activity and catchability (McLeese and Wilder 56a) and growth rates of larvae, juveniles, and adults and consequently, the numbers of lobsters recruiting as a year class (Caddy 57).

River freshwater discharge (Sutcliffe 58), which affects the survival of lobsters because they are sensitive to low salinity (McLeese 59; Phillips and Sastry 60, p. 22).

High exploitation rates, which can affect recruitment by reducing eggs per recruit (Campbell and Robinson 29), resulting in possible recruitment failure and decline in landings in some areas (e.g., eastern Nova Scotia: Robinson 61).

Other factors such as man-made interruptions of larval supply (Dadswell 46) and ecosystem changes (Wharton and Mann 62; Miller 63), which also may affect recruitment in some areas (for review, see Ennis 3).

The shape of the stock–recruitment relationship, whether it is a dome-shaped Ricker (64) or asymptotic Beverton–Holt (65) or other type of curve, has not been resolved for *H. americanus* in the Bay of Fundy and southwestern Nova Scotia areas, because the brood-stock size and levels of recruitment are difficult to measure accurately. Fogarty and Idoine (66) calculated the relationship between late stage IV larval abundance (from 1949 to 1963, time series from Scarratt 67, 68) and stock size 5–7 yr later in the Northumberland Strait region (southern Gulf of St. Lawrence). The relationship was asymptotic, which suggested a density-dependent regulatory mechanism between larval settlement and subsequent recruitment. Such a mechanism might have been a result of competition for resources, such as shelter, for example, and one that could provide for a generally stable population that is relatively resilient to exploitation (Fogarty and Idoine 66). However, other factors may cause variation in the stock–recruitment curve, theoretically making populations less stable. For example, the impact of the nemertean egg predator (*Pseudocarcinonemertes homari*) on lobster egg loss (Campbell and Brattey 69) or gaffkemia disease (Stewart 70) at high host densities could conceivably lead to dome-shaped spawning–recruitment curves and host–parasite oscillations (e.g., Botsford and Wickham 71; Fogarty and Idoine 66). There are no studies to date on lobster populations resolving the density-dependent effects of predation, parasitism, starvation, or cannibalism on the survival of eggs or larvae as a result of high berried female densities.

The relatively stable lobster landings from southwestern Nova Scotia and the Bay of Fundy during the last 40 yr (Fig. 3) may be due, in part, to a large continental shelf area providing sufficient refugia to protect reproductive lobsters from high exploitation (Anthony and Caddy 72; Campbell and Robinson 29). With low survival to maturity of inshore lobsters, the mature lobsters moving to deeper water may constitute a refugium in space. Also, the seasonal movement of mature lobsters into shallow water during summer and into deeper water in winter (Campbell and Stasko 37; Campbell 19) may allow the brood stock to escape much of the fall–spring fishing season, and would thus constitute a refugium in time (Anthony and

Caddy 72). Increased exploitation of large lobsters in deep water in coastal areas (e.g., Campbell and Duggan 73) and offshore areas (Pezzack and Duggan 4) could therefore reduce the level and stability of recruitment.

5. FISHERY SAMPLING

Lobster landings were recorded as early as 1869, although segregation of landings by county did not start until 1892 and by statistical district until 1947. The latter approach allowed yield trends of separate small areas to be monitored (annual reports of the Department of Marine and Fisheries). Average weights of lobsters and the percentages of size classes in the annual catch are available for the turn of the century (Pringle et al. 11; Venning 13; Robinson 61). From 1944 to 1969, D. G. Wilder maintained a commercial sampling program on lobsters at one study port (Port Maitland) by collecting data on catch, effort, size distribution, and tag recapture data to obtain information on growth (e.g., Campbell 74), stock abundance and exploitation rates (Paloheimo 75), and movements (e.g., Campbell 76). Since 1978, sampling was resumed at Port Maitland and was started at several other similar representative study ports. Details of commercial sampling data available from statistical systems, interviews, and in-port and at-sea sampling are as follows (see also Rowell 77; Pringle and Duggan 78). For inshore boats, information is obtained from purchase slips by boat name on lobster landings, days fished, and port landed. For offshore boats, vessel logs provide vessel name, daily catch, soak time of traps, depth and location fished, bait used, and location landed. Interviews with fishermen in port provide weight of catch, depth, number of traps set and hauled, soak days, and number of crew by boat. Sampling of catch at sea provides the location and depth where traps are fished, total catch by numbers and weight, number of traps set and hauled, soak time, trap escape spacing and entrance hoop size, surface and bottom temperatures, length frequency, sex ratio, molt stage, number of berried females with egg stage, and by-catch such as crab species.

6. YIELD MODELS

Surplus yield models (e.g., Schaefer 79), based on the concept that stock yield is a function of stock size and fishing effort, have been modified in various ways and applied to *H. americanus* fisheries of Rhode Island and Maine (Marchesseault et al. 80; Jensen 81; see reviews by Saila and Marchesseault 82; Cobb and Wang 83). For the whole of the Bay of Fundy and southwestern Nova Scotia lobster fishery, the lack of sufficiently long time series of accurate fishing effort data has restricted the use of such models.

Adaptations of the Beverton–Holt (65) dynamic pool model for predicting yield per recruit have been used for various lobster fisheries (Ennis 84; Fogarty 85). Recently, a yield and egg per recruit model based on size (instead of age) was developed by Caddy (86, 87) and modified by Campbell and Robinson (29) and

Campbell (88, 89) for lobsters of the Bay of Fundy and southwestern Nova Scotia. The analyses indicate that increasing the recruited (minimum legal) size from 81 to as much as 94 mm carapace length would be beneficial in terms of a substantial increase in yield and in eggs per recruit. Reductions in fishing mortality would increase the number of eggs per recruit more than it would the yield per recruit. Alternatively, application of a maximum size regulation (e.g., 127 mm carapace length) to protect larger, highly fecund lobsters concurrent with the present minimum legal size would reduce yield per recruit slightly but increase eggs per recruit substantially (Campbell 88).

Although the yield and egg per recruit models are useful tools for fishery analysis, caution is required in their interpretation (Bannister and Addison 90; Botsford and Hobbs 91) because several assumptions are used and the models therefore do not mimic reality completely. The models assume average growth rates and do not take into account natural variability of growth in each size group of lobsters. Fishing and natural mortality rates are assumed constant for all size groups and density-dependent effects on mortality, growth, and reproductive rates are considered negligible. The lack of a realistic stock–recruitment function limits these models to providing only yield per recruit instead of forecasting the effect of fishing on total yield. Also, the inclusion of more detailed functions for ovarian-molt cycles in these models could fine-tune these predictions (Caddy 57). Although the egg-per-recruit model (e.g., Campbell 88) takes multiple egg extrusions between molts into account, the frequency of multiple extrusions (Waddy and Aiken 33) in lobsters of various sizes and intermolt periods under different natural environmental conditions still has to be determined.

7. MANAGEMENT

The process of lobster fisheries management in this area involves the interaction between biologists, economists, fishermen, managers, and regulation enforcement personnel. Assessment-oriented studies and biological advice to fisheries managers are vetted through peer review by the Canadian Atlantic Fisheries Scientific Advisory Committee (CAFSAC). Consultation with industry on lobster management issues, such as regulation changes, is maintained through annual or semi-annual meetings with an advisory committee for each lobster district. The committees are made up of fishermen, federal and provincial government managers, a biologist, an economist, and fisheries officers. These advisory committees were implemented a few years ago and should work well as individuals gain the confidence of each other (Pringle 92). Any change in lobster regulations, however, must be approved at a higher level, ultimately by the Minister of the Federal Department of Fisheries and Oceans.

Lobsters have been the mainstay of a valuable inshore multispecies fishery for more than 100 yr in southwestern Nova Scotia and the Bay of Fundy. For the last 40 yr, lobster landings were relatively stable despite increasing fishing effort and worldwide consumer demand for a high unit value commodity (Pringle et al. 11).

The present management regulations involving limiting the number of boats, traps, and fishing seasons and protecting egg-bearing females and lobsters below minimum legal size may have had some benefit in protecting the resource. However, without adequate empirical stock-recruitment information, it is difficult to determine exactly at what level these regulations are effective conservation tools. The increasing exploitation of large mature lobsters in recent years concurrent with high exploitation of the newly recruited lobsters in the inshore fishery has caused concern among biologists and managers that fishery-caused recruitment failure could occur. A number of management initiatives have been proposed recently to counteract this potential problem, for example, reducing effort, increasing recruit size, and institution of a maximum size (Campbell 89). However, the current socioeconomic climate is such that there is resistance to any changes in the management regulations that are designed to promote conservation and/or economic performance (Pringle et al. 11). Important biological questions on recruitment still must be answered, such as whether the brood stock in the Browns Bank closure area is an important source of recruitment for inshore southwestern Nova Scotia lobsters, and whether a reliable juvenile index as used for palinurid lobsters (e.g., Caputi and Brown 93) can also be developed to predict recruitment and fishery yields for the American lobster.

ACKNOWLEDGMENTS

I thank D. E. Graham and A. M. Williamson for technical assistance; F. Cunningham and W. McMullon for drawing the figures; J. Hurley and J. Thomas for typing; and D. W. McLeese, R. J. Miller, and R. I. Perry for reviewing earlier drafts of the manuscript.

REFERENCES

1. H. J. Squires, Distribution of decapod Crustacea in the northwest Atlantic. *Ser. Atlas Mar. Environ.*, *Am. Geogr. Soc.*, Folio No. 12 (1966).

2. A. Campbell and R. K. Mohn, Definition of American lobsters stocks for the Canadian maritimes by analysis of fishery-landing trends. *Trans. Am. Fish. Soc.* **112**:744–759 (1983).

3. G. P. Ennis, Stock definition, recruitment variability, and larval recruitment processes in the American lobster, *Homarus americanus*: A review. *Can. J. Fish. Aquat. Sci.* **43**:2072–2084 (1986).

4. D. S. Pezzack and D. R. Duggan, The Canadian offshore lobster (*Homarus americanus*) fishery 1971-1982. *Shellfish Comm.*, *ICES, C. M.* (1983)/K (1983).

5. D. S. Pezzack and D. R. Duggan, The Canadian offshore lobster fishery 1971-1984, catch history, stock condition and mangement options. *Can. Atl. Fish. Sci. Advis. Comm. (CFSAC), Res. Doc.* **85/89**:26 (1985).

6. G. J. Sharp and R. E. Duggan, An aerial survey of near-shore and mid-shore lobster fishing distribution off southwestern Nova Scotia, spring and fall 1983-84. *Can. J. Fish. Aquat. Sci.*, *Manuscr.* Rep. **1847**:37 (1985).

7. D. W. McLeese and D. G. Wilder, Lobster storage and shipment. *Bull. Fish. Res. Board Can.* **147**:69 (1964).

8. A. G. DeWolf, The lobster fishery of the maritime provinces: Economic effects of regulations. *Bull. Fish. Res. Board Can.* **187**:59 (1974).

9. E. E. Prince, Report of the Canadian Lobster Commission, 1898. In *Thirty-First Annual Report of the Department of Marine Fisheries, 1899*, Suppl. No. 1. Can. Lobster Comm., 1899, p.41.

10. A. Halkett, in *Annual Report of the Department of Marine Fisheries*, Natural History Report, Appendix 18. Can. Lobster Comm., 1910, pp. 368–394.

11. J. D. Pringle, D. G. Robinson, G. P. Ennis, and P. Dubé, An overview of the management of the lobster fishery in Atlantic Canada. *Can. J. Fish. Aquat. Sci., Manuscr. Rep.* **1704**:103 (1983).

12. A. Campbell and D. S. Pezzack, Relative egg production and abundance of berried lobsters, *Homarus americanus*, in the Bay of Fundy and off southwestern Nova Scotia. *Can. J. Fish. Aquat. Sci.* **43**:2190–2196 (1986).

13. R. N. Venning, *The Marine and Fisheries Committee and Lobster Industry*, Ann. Rep. Dept. Mar. Fish., Spec. Appended Report. II. Canada Annual Reports, Marine and Fisheries, Ottawa, 1910, Sess. Pap. No. 22.

14. W. A. Found, The Lobster Fishery of Canada. In *Sea-Fisheries of Eastern Canada*. Mortimer Co., Ottawa, Ont., 1912.

15. W. A. Found, *Memorandum for Royal Commission to Investigate Fishery Conditions and Requirements of the Martime Provinces Including the Magdalen Islands*. Queens Printer, Ottawa, Canada 1927, p. 51.

16. D. G. Robinson, History of the lobster fishery on the eastern shore of Nova Scotia. *Can. J. Fish. Aquat. Sci., Tech. Rep.* **954**:8–23 (1980).

17. W. Templeman, Embryonic developmental rates and egg-laying of Canadian lobsters. *J. Fish. Res. Board Can.* **5**:71–83 (1940).

18. H. C. Perkins, Development rates at various temperatures of embryos of the northern lobster (*Homarus americanus* Milne-Edwards). *Fish. Bull.* **70**:95–99 (1972).

19. A. Campbell, Migratory movements of ovigerous lobsters, *Homarus americanus*, tagged off Grand Manan, eastern Canada. *Can. J. Fish. Aquat. Sci.* **43**:2197–2205 (1986).

20. W. Templeman, Embryonic developmental rates and egg-laying of Canadian lobsters. *J. Fish. Res. Board Can.* **5**:71–83 (1940).

21. G. P. Ennis, Swimming ability of larval American lobsters, *Homarus Americanus*, in flowing water. *Can. J. Fish. Aquat. Sci.* **43**:2177–2183 (1986).

22. J. S. Cobb, Shelter-related behavior of the lobster, *Homarus americanus. Ecology* **52**:108–115 (1971).

23. J. S. Cobb, Review of the habitat behavior of the clawed lobsters (*Homarus* and *Nephrops*). *C.S.I.R.O., Div. Fish. Oceanogr., Circ.* **7**:143–157 (1977).

24. R. A. Pottle and R. W. Elner, Substrate preference behavior of juvenile American lobsters, *Homarus americanus*, in gravel and silt-clay sediments. *Can. J. Fish. Aquat. Sci.* **39**:928–932 (1982).

25. L. L. Stewart, The seasonal movements, population dynamics and ecology of the lobster, *Homarus americanus* (Milne Edwards), off Ram Island, Connecticut. Ph.D. Thesis, University of Connecticut, Storrs, 1972, p. 112.

26. R. H. Cooper and J. R. Uzmann, Ecology of juvenile and adult *Homarus*. In J. S. Cobb and B. F. Phillips, Eds., *Biology and Management of Lobsters*, Vol. II. Academic Press, New York, 1980, pp. 97–142.

27. D. G. Wilder, The growth rate of the American lobster (*Homarus americanus*). *J. Fish. Res. Board Can.* **10**:371–412 (1953).

28. A. Campbell, Growth of tagged American lobsters, *Homarus americanus*, in the Bay of Fundy. *Can. J. Fish. Res. Board Can.* **40**:1667–1675 (1983).

29. A. Campbell and D. G. Robinson, Reproductive potential of three American lobster (*Homarus americanus*) stocks in the Canadian maritimes. *Can. J. Fish. Aquat. Sci.* **40**:1958–1967 (1983).

30. J. Atema, A review of sexual selection and chemical communication in the lobster, *Homarus americanus*. *Can. J. Fish. Aquat. Sci.* **43**:2283–2390 (1986).

31. W. Templeman, Further contributions to mating in the American lobster. *J. Biol. Board Can.* **2**:223–226 (1936).

32. P. J. Dunham and D. Skinner-Jacobs, Intermolt mating in the lobster (*Homarus americanus*). *Mar. Behav. Physiol.* **5**:209–214 (1978).

33. S. L. Waddy and D. E. Aiken, Multiple fertilization and consecutive spawning in large American lobsters, *Homarus americanus*. *Can. J. Fish. Aquat. Sci.* **43**:2291–2294 (1986).

34. F. Herrick, Natural history of the American lobster. *Fish. Bull.* **29**:149–408 (1911).

35. A. Campbell, D. E. Graham, H. J. MacNichol, and A. M. Williamson, Movements of tagged lobsters released on the continental shelf from Georges Bank to Baccaro Bank, 1971–73. *Can. J. Fish. Aquat. Sci., Tech. Rep.*, **1288**:16 (1984).

36. A. Campbell and A. B. Stasko, Movements of tagged American lobsters, *Homarus americanus*, off southwestern Nova Scotia. *Can. J. Fish. Aquat. Sci.* **42**:229–238 (1985).

37. A. Campbell and A. B. Stasko, Movements of lobsters (*Homarus americanus*) tagged in the Bay of Fundy, Canada. *Mar. Biol.* **92**:393–404 (1986).

38. D. S. Pezzack and D. R. Duggan, Evidence of migration and homing of lobsters (*Homarus americanus*) on the Scotian Shelf. *Can. J. Fish. Aquat. Sci.* **43**:2206–2211 (1986).

39. A. G. Huntsman, Natural lobster breeding. *Bull. Biol. Board Can.* **5**:1–11 (1923).

40. J. F. Caddy, The influence of variations in the seasonal temperature regime on survival of larval stages of the American lobster (*Homarus americanus*) in the southern Gulf of St. Lawrence. *Rapp. P-V. Reun., Cons. Int. Explor. Mer* **1975**:204–216 (1979).

41. A. Campbell, Aspects of lobster biology and fishery in the upper reaches of the Bay of Fundy. *Can. J. Fish. Aquat. Sci., Tech. Rep.* **1256**:469–488 (1984).

42. R. I. Perry and P. C. F. Hurley, Circulation and potential ichthyoplankton dispersal in the Gulf of Maine, Browns and Georges Bank areas. *Can. Atl. Fish. Sci. Advis. Comm. (CFSAC), Res. Doc.* **86/37** (1986).

43. D. F. Bumpus and L. M. Lauzier, Surface circulation on the continental shelf off eastern North America between Newfoundland and Florida. *Ser. Atlas Mar. Environ., Am. Geogr. Soc., Folio* **7**:92. (1965).

43a. B. Butman, J. W. Loder, and R. C. Beardsley, The seasonal mean circulation on Georges Bank: Observation and theory 125–138. In R. B. Backus and D. W. Bourne Eds., *Georges Bank*. MIT Press, Cambridge, MA, 1987 593 pp.

43b. D. A. Greenberg, Modelling the mean barotropic circulation in the Bay of Fundy and Gulf of Maine. *J. Phys. Oceanogr.* **13**:886–904 (1983).

43c. P. C. Smith, The mean and seasonal circulation off southwest Nova Scotia. *J. Phys. Oceanogr.* **13**:1034–1054 (1983).

43d. D. A. Brooks, Vernal circulation in the Gulf of Maine. *JGR, J. Geophys. Res.* **90**:4687–4705 (1985).

44. A. B. Stasko and A. Campbell, An overview of lobster life history and fishery in southwestern Nova Scotia. *Can. J. Fish. Aquat. Sci., Tech. Rep.* **954**:208–224 (1980).

45. A. B. Stasko, Inshore-offshore lobster stock interaction: An hypothesis. *Can. Atl. Fish. Sci. Advis. Comm. (CFSAC), Res. Doc.* **78/37**:10 (1978).

46. M. J. Dadswell, A review of the decline in lobster (*Homarus americanus*) landings in Chedabucto Bay between 1956 and 1977 with an hypothesis for a possible effect by the Canso Causeway on the recruitment mechansim of eastern Nova Scotia lobster stocks. *Tech. Rep.—Fish. Mar. Serv. (Can.)* **834**:113–144 (1979).

47. G. C. Harding, K. F. Drinkwater, and W. P. Vass, Factors influencing the size of American lobster (*Homarus americanus*) stocks along the Atlantic coast of Nova Scotia, Gulf of St. Lawrence, and Gulf of Maine: A new synthesis. *Can. J. Fish. Aquat. Sci.* **40**:168–184 (1983).

48. A. B. Stasko and D. J. Gordon, Distribution and relative abundance of lobster larvae off southwestern Nova Scotia, 1977–1978. *Can. J. Fish. Aquat. Sci., Tech. Rep.* **1175**:23 (1983).

49. L. M. Dickie, Fluctuations in abundance of the giant scallop, *Placopecten magellanicus* (Gemlin), in the Digby area of the Bay of Fundy. *J. Fish. Res. Board Can.* **12**:797–865 (1955).

50. V. C. Anthony, Review of lobster mortality estimates in the United States. *Can. J. Fish. Aquat. Sci., Tech. Rep.* **932**:17–25 (1980).

51. A Campbell, A review of mortality estimates of lobster populations in the Canadian maritimes. *Can. J. Fish. Aquat. Sci., Tech. Rep.* **932**:27–35 (1980).

52. G. P. Ennis, Annual variations in standing stock in a Newfoundland population of lobsters. *North Am. J. Fish. Manage.* **3**:26–33 (1983).

53. J. M. Flowers and S. B. Saila, An analysis of temperature effects on the inshore lobster fishery. *J. Fish. Res. Board Can.* **29**:1221–1225 (1972).

54. R. L. Dow, The clawed lobster fisheries. In J. S. Cobb and B. F. Phillips, Eds., *The Biology and Management of Lobsters*, Vol. II. Academic Press, New York, 1980, pp. 265–316.

55. F. L. Orach-Meza and S. B. Saila, Application of a polynomial distributed lag model to the Maine lobster fishery. *Trans. Am. Fish. Soc.* **107**:402–411 (1978).

56. M. J. Fogarty, Temperatuare–yield relationships for the Maine American lobster (*Homarus americanus*) fishery: A time series analysis approach. *NAFO SCR Doc.* **84/VI/57**:23 (1984).

56a. D. W. McLeese and D. G. Wilder, The activity and catchability of the lobster (*Homarus americanus*) in relation to temperature. *J. Fish. Res. Board Can.* **15**:1345–1354 (1958).

57. J. F. Caddy, Modelling stock-recruitment processes in Crustacea: Some practical and theoretical perspectives. *Can. J. Fish. Aquat. Sci.* **43**:2330–2344 (1986).

58. W. H. Sutcliffe, Jr., Correlations between seasonal river discharge and local landings of American lobster (*Homarus americanus*) and Atlantic halibut (*Hippoglossus hippoglossus*) in the Gulf of St. Lawrence. *J. Fish. Res. Board Can.* **30**:856–859 (1973).

59. D. W. McLeese, Effects of temperature, salinity and oxygen on the survival of American lobster. *J. Fish. Res. Board Can.* **13**:247–272 (1956).

60. B. F. Phillips and A. N. Sastry, Larval ecology. In J. S. Cobb and B. F. Phillips, Eds., *The Biology and Management of Lobsters*, Vol. II. Academic Press, New York, 1980, pp. 11–57.

61. D. G. Robinson, Consideration of the lobster (*Homarus americanus*) recruitment overfishing hypothesis, with special reference to the Canso Causeway. *Tech. Rep.— Fish. Mar. Serv. (Can.)* **834**:77–99 (1979).

62. W. G. Wharton and K. H. Mann, Relationship between destructive grazing by the sea urchin, *Stronglocentrotus droebrachiensis*, and the abundance of American lobster, *Homarus americanus*, on the Atlantic coast of Nova Scotia. *Can. J. Fish. Aquat. Sci.* **38**:1339–1349 (1981).

63. R. J. Miller, Seaweeds, sea urchins, and lobsters: A reappraisal. *Can. J. Fish. Aquat. Sci.* **42**:2061–2072 (1985).

64. W. E. Ricker, Stock recruitment. *J. Fish Res. Board Can.* **11**:559–623 (1954).

65. T. J. H. Beverton and S. J. Holt, On the dynamics of exploited fish populations. *Fish. Invest.—Minist. Agric., Fish. Food (G.B.) (Ser. 2)* **19**:533 (1957).

66. M. J. Fogarty and J. S. Idoine, Recruitment dynamics in an American lobster (*Homarus americanus*) population. *Can. J. Fish. Aquat. Sci.* **43**:2368–2376 (1986).

67. D. J. Scarratt, Abundance and distribution of lobster larvae (*Homarus americanus*) in Northumberland Strait. *J. Fish. Res. Board Can.* **21**:661–680 (1964).

68. D. J. Scarratt, Abundance, survival, and vertical and diurnal distribution of lobster larvae in Northumberland Strait, 1962–63, and their relationships with commercial stocks. *J. Fish. Res. Board Can.* **30**:1819–1829 (1973).

69. A. Campbell and J. Brattey, Egg loss from the American lobster, *Homarus americanus*, in relation to nemertean, *Pseudocarcinonemertes homari*, infestation. *Can. J. Fish. Aquat. Sci.* **43**:772–780 (1986).

70. J. E. Stewart, The lobster disease, gaffkemia, in relation to fisheries management. *Int. Counc. Explor. Sea, Shellfish Comm.* **F:16**:1–8 (1978).

71. L. W. Botsford and D. E. Wickham, Behavior and age-specific, denstiy-dependent models and the northern California dungeness crab (*Cancer magister*) fishery. *J. Fish. Res. Board Can.* **35**:833–843 (1978).

72. V. C. Anthony and J. F. Caddy, Eds., Proceeding of the Canada-U.S. Workshop on status of assessment science for N.W. Atlantic lobster (*Homarus americanus*) stocks. *Can. J. Fish. Aquat. Sci., Tech. Rep.* **932**:186 (1980).

73. A. Campbell and D. R. Duggan, Review of the Grand Manan lobster fishery with an analysis of recent catch and effort trends. *Can. J. Fish Aquat. Sci., Tech. Rep.* **997**:20 (1980).

74. A. Campbell, Growth of tagged lobsters (*Homarus americanus*) off Port Maitland, Nova Scotia, 1944–80. *Can. J. Fish. Aquat. Sci., Tech. Rep.* **1232**:10 (1983).

75. J. E. Paloheimo, Estimation of catchabilities and population sizes of lobsters. *J. Fish. Res. Board Can.* **20**:59–88 (1963).

76. A. Campbell, Movements of tagged lobsters released off Port Maitland, Nova Scotia, 1944–80. *Can. J. Fish. Aquat. Sci.*, *Tech. Rep.* **1136:**41 (1982).

77. T. W. Rowell, Sampling of commercial catches of invertebrates and marine plants in the Scotia-Fundy region. *Can. J. Fish. Aquat. Sci.*, *Spec. Publ.* **66:**52–60 (1983).

78. J. D. Pringle and R. E. Duggan, Latent lobster fishing effort along Nova Scotia's Atlantic coast. *Can. Atl. Fish. Sci. Advis. Comm. (CAFSAC), Res. Doc.* **84/56:**23 (1984).

79. M. B. Schaefer, A study of the dynamics of the fishery for yellowfin tuna in the eastern tropical Pacific Ocean. *Bull. Int. Am. Trop. Tuna Comm.* **2:**245–285 (1957).

80. G. D. Marchesseault, S. G. Saila, and W. J. Palm, Delayed recruitment models and their application to the American lobster (*Homarus americanus*) fishery. *J. Fish. Res. Board Can.* **33:**1779–1787 (1976).

81. A. L. Jensen, Assessment of the Maine lobster fishery with surplus production models. *North Am. J. Fish. Manage.* **6:**63–68 (1986).

82. S. B. Saila and G. D. Marchesseault, Population dynamics of clawed lobsters. In J. S. Cobb and B. F. Phillips, Eds., *The Biology and Management of Lobsters*, Vol. II. Academic Press, New York, 1980. pp. 219–241.

83. J. S. Cobb and D. Wang, Fisheries biology of lobster and crayfishes. In D. E. Bliss, Ed., *The Biology of Crustacea*, Vol. X. Academic Press, New York, 1985, pp. 167–247.

84. G. P. Ennis, Canadian efforts to assess yield per recruit in lobsters. *Can. J. Fish. Aquat. Sci.*, *Tech. Rep.* **932:**45–49 (1980).

85. M. J. Fogarty, Assessment of yield per recruit for the American lobster (*Homarus americanus*). *Can. J. Fish. Aquat. Sci.*, *Tech. Rep.* **932:**37–44 (1980).

86. J. F. Caddy, Approaches to a simplified yield-per-recruit model for Crustacea, with particular reference to the American lobster, *Homarus americanus*. *Fish. Mar. Serv. Manuscr. Rep. (Can.)* **1445:**14 (1977).

87. J. F. Caddy, Notes on a more generalized yield-per-recruit analysis for crustaceans, using size-specific inputs. *Fish. Mar. Serv. Manuscr. Rep. (Can.)* **1525:**40 (1979).

88. A. Campbell, Application of a yield and egg-per-recruit model to the lobster fishery in the Bay of Fundy and southwestern Nova Scotia. *North Am. J. Fish. Manage.* **5:**91–104 (1985).

89. A. Campbell, Implications of size and sex regulations for the lobster fishery of the Bay of Fundy and southwestern Nova Scotia, *Can. J. Fish. Aquat. Sci.*, *Spec. Publ.* **92:**126–132 (1986).

90. R. C. A. Bannister and J. T. Addison, The effect of assumptions about the stock-recruitment relationship on a lobster (*Homarus gammarus* L.) stock assessment. *Can. J. Fish. Aquat. Sci.* **43:**2353–2359 (1986).

91. L. W. Botsford and C. R. Hobbs, Static optimization of yield per recruit with reproduction and fishing costs. *Fish. Res.* **4:**181–189 (1986).

92. J. D. Pringle, The human factor in fishery resource management. *Can. J. Fish. Aquat. Sci.* **42:**389–392 (1985).

93. N. Caputi and R. S. Brown, Relationship between indices of juvenile abundance and recruitment in the western rock lobster (*Panulirus cygnus*) fishery. *Can. J. Fish. Aquat. Sci.* **43:**2131–2139 (1986).

7 THE WEST AUSTRALIAN ROCK LOBSTER FISHERY: RESEARCH FOR MANAGEMENT

B. F. Phillips

CSIRO Marine Laboratories, North Beach, Western Australia, Australia

R. S. Brown

Western Australian Marine Research Laboratories, North Beach, Western Australia, Australia

1. INTRODUCTION

The fishery for the western rock lobster *Panulirus cygnus* George, is one of the major rock lobster fisheries in the world and the most valuable single-species fishery in Australia, accounting for approximately 35% of the country's gross income from fisheries products (Morgan 1). Australia is the largest exporter of rock (spiny) lobster in the world. The range of *P. cygnus* extends from North West Cape (21°44′S) to just south of Cape Leeuwin (34°24′S) and includes the Abrolhos Islands (Fig. 1). The majority of the 750 fishing boats are concentrated between Mandurah in the south, Kalbarri in the north, and the Abrolhos Islands (Fig. 1). The boats vary in length from 5 to 28 m (the majority being 8–15 m) and operate 76,000 pots (traps) during a $7\frac{1}{2}$ month coastal season (November 15–June 30) and a $3\frac{1}{2}$ month Abrolhos Island season (March 15–June 30). Pots of traditional design are used—stick or cane "beehive," which are used mainly in the southern area, and wooden "batten" pots, which were formerly used in the northern area but are now used throughout the fishery (Morgan and Barker 2). Pots, the dimensions of which have been fixed (Bowen and Hancock, Chap. 16), are set individually and pulled each day depending on the catch rate and weather.

The fishery has developed from being underexploited in the 1930s when approximately 250,000 kg of lobster was caught for the local market, through an infant export industry in the late 1940s producing 1.0 million kg in 1948, to an average of approximately 10 million kg/yr over the last 10 years to 1984–1985 (Table 1), worth in excess of A$130 million to the fishermen in 1985. More than 90% of the catch is processed as frozen tails at processing establishments at eight coastal

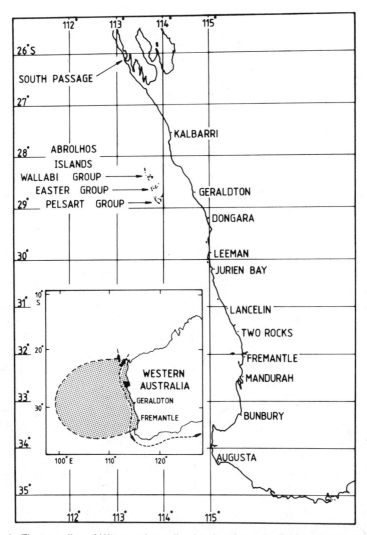

Figure 1. The coastline of Western Australia showing the major fishing centers and the 1° blocks that are used in the fishermen's monthly returns to record catch and fishing effort. The insert shows Western Australia and the southeastern Indian Ocean and indicates the Leeuwin Current (– – –) superimposed on the known extent of the larval distribution of *Panulirus cygnus*, shown as a shaded area (after Phillips 10).

locations and exported to the United States. Sales of live rock lobsters to Japan have recently become important and a small market exists for whole cooked lobsters in France and other European countries. See Sheard (3) for details of the fishery between 1944 and 1961 and Bowen (4) and the annual report series by Morgan and Barker (5) and Brown and Barker (6) for more recent information. For a review of the fisheries regulations see Bowen and Hancock (Chap. 16).

TABLE 1   The Catch, Pot Lifts (Nominal Fishing Effort), Catch per Pot Lift, and Effective Fishing Effort[a]

Season	Catch (millions kg)	Pot Lifts (millions)	Catch per Pot Lift (kg)	Effective Effort (millions of effective pot lifts)
1944–45	0.263	0.240		0.143
1945–46	0.528	0.380		0.228
1946–47	0.875	0.570		0.343
1947–48	1.009	0.680		0.405
1948–49	2.293	1.220		0.731
1949–50	2.909	1.780		1.068
1950–51	3.225	1.860		1.115
1951–52	3.907	2.320		1.391
1952–53	3.465	2.460		1.472
1953–54	4.231	3.000		1.795
1954–55	5.027	3.430		2.050
1955–56	4.842	3.590		2.149
1956–57	4.983	4.110		2.462
1957–58	6.111	4.330		2.591
1958–59	8.060	5.400		3.230
1959–60	8.726	6.630		3.968
1960–61	7.790	6.310		3.777
1961–62	8.481	8.510		5.700
1962–63	9.176	10.010		7.500
1963–64	8.141	7.613		4.648
1964–65	7.485	6.812	1.098	4.798
1965–66	8.120	7.301	1.112	5.036
1966–67	8.631	7.480	1.153	5.147
1967–68	9.853	7.806	1.262	5.173
1968–69	8.078	8.861	0.911	4.292
1969–70	6.918	8.408	0.822	5.771
1970–71	8.013	9.589	0.835	7.888
1971–72	8.171	9.956	0.820	7.536
1972–73	6.809	9.136	0.745	7.187
1973–74	6.780	9.864	0.687	7.127
1974–75	8.260	10.255	0.805	8.035
1975–76	8.720	10.258	0.850	8.063
1976–77	9.312	11.148	0.835	8.368
1977–78	10.052	10.596	0.948	9.265
1978–79	11.394	10.842	1.050	9.366
1979–80	10.709	10.724	0.998	8.350
1980–81	9.937	10.857	0.914	8.583
1981–82	10.551	11.255	0.937	9.721
1982–83	12.423	11.624	1.068	10.298
1983–84	10.585	11.214	0.943	[b]
1984–85	8.973	11.598	0.773	[b]

[a]Using the method of Morgan (33).
[b]Not yet available.

The main thrust of management measures has been to contain the expansion in fishing effort (Bowen 4, 7; Hancock 8; Morgan and Barker 9). Since 1963 the fishery has been subject to limited entry with the objectives of optimal utilization of the resource, reasonable economic return to the fishermen, and orderly exploitation to allow minimum conflicts among professional fisherman and between professional and amateur fishermen.

The successful achievements of this policy have been based on research programs leading to an understanding of the biology, life history, and population dynamics of *P. cygnus* and in defining general production models, yield per recruit, stock–recruitment relationships, and predictive models for recruitment and total catch, based on indexes of prerecruit abundance and puerulus settlement, respectively. The exact scope of the research has varied with time; however, the Commonwealth Scientific and Industrial Research Organisation's Division of Fisheries Research has taken a broad interpretation of its research role, but by agreement has aimed at studying the section of the life cycle from hatching of the larvae until recruitment to the fishery. The Fisheries Department of Western Australia researches the commercial stock and the operations of the fishery.

2. LIFE HISTORY

The life cycle of the rock lobster is complex and includes a 9–11-month oceanic phase (Phillips 10). The nektonic puerulus stage completes the oceanic cycle by swimming across the continental shelf and settling in the shallow limestone reef areas along the coast (Phillips 10). The settled puerulus stage molts into a small juvenile rock lobster; carapace length (CL) = 0.8 cm. The peak settlement of puerulus occurs between September and January, following hatching in the previous December–February period (Phillips et al. 11). Based on the time of hatching of the phyllosoma larvae, Morgan et al. (12) adopted January 1, the midpoint of the hatching period, as an arbitrary "birthday," so that the age of newly settled pueruli is 0+ to 1+ yr, and they are all taken as 1 yr old on January 1.

During their first year as benthic juveniles they are believed to be solitary and they hide in the weed beds. The later development of gregarious behavior at about 1.5 yr of age results in gradual movement to reefs. During daylight the older juveniles are found under the limestone reefs but at night they forage for food in the surrounding sea grass beds (Joll and Phillips 13).

When the juveniles are between 4 and 6 yr old, they emigrate offshore as "whites" from the shallow reef areas to the outer continental shelf into depths of 30–150 m (George 14). The whites phase occurs regularly in November–January each year and is made up of immature, very pale, newly molted animals. Whites are composed mainly of 4- and 5-yr-old juveniles, with the majority being 4 yr old (Morgan et al. 12). The mean size of the whites is around 76 mm CL, the minimum legal size for the fishery (George 14; Chittleborough 15).

The minimum legal size for this fishery is generally less than the size at first breeding of females, though this size varies for different localities along the coast

(Morgan 16). Along the mainland coast juveniles occur in shallow, nearshore waters (< 10 m deep) whereas the subadult and adult populations are located in waters from about 10 to 150 m deep (Morgan et al. 12). Stratification with depth is not as apparent at the Abrolhos Islands (which rise steeply from the outer continental shelf) and, although juveniles are restricted to shallow waters, breeding adults are present in both shallow and deep water (Chittleborough 17).

Reproductive maturity of females generally occurs 12–24 months after the attainment of legal size (i.e., 6–7 yr of age) (Morgan et al. 12). Mating takes place in July–August and the female carries the black spermatophore on the sternal plates of the cephalothorax until October–December when the eggs are extruded, fertilized, and become attached to the pleopods. The eggs are carried on the pleopods for up to 9 weeks, depending on water temperature, before hatching (Chittleborough 17).

3. DATA COLLECTION

3.1. Commercial Fishing Data

3.1.1. Catch and Fishing Effort

There are three major sources of catch and fishing effort data available from the fishery.

(a) Monthly returns. One of the license stipulations is that all fishermen provide details of the weight of their monthly landed catch (kg) by 1° statistical blocks (Fig. 1) and the average number of pots pulled per day. The fishing effort data are available in two different forms for the periods 1944–1945 to 1963–1964 as person-days and for 1964–1965 to the present time as pot lifts (Sheard 3; Morgan and Barker 9; Brown and Barker 6).

(b) Daily returns. These are in the form of research logbooks, which are completed on a voluntary basis by 20–30% of the 750 boats. They give daily records of catch (kg); fishing effort (number of pots pulled) by 10′ latitude transects for depths of 0–18 m, 18–37 m, 37–55 m, 55–73 m, and over 73 m; soak time; numbers of berried females caught (regulations require their return to the sea); and estimates of the number of octopus caught and undersized rock lobsters handled (undersized must be returned to the sea). Information on the distance offshore that the fishing occurred and wind, swell, and current conditions are also recorded. Logbook data are available from 1965–1966.

(c) Processors' monthly returns. It is mandatory for processors to provide information on the weight of live catch supplied by each fisherman and details of sales of processed rock lobsters by weight and grade; that is, number of cartons of each of the four different weights that are used, and the type of tails by seven different grade (weight) categories that are packed in each of the cartons (Brown and Barker 6). Figures are available in various forms from 1967.

3.1.2. Commercial Vessel Monitoring

Technical staff measure a sample of 400 rock lobsters each month of the season from four depth categories (0–18 m, 18–37 m, 37–55 m, and greater than 55 m) at each of four coastal locations (Dongara, Jurien, Lancelin, and Fremantle, Fig. 1). Carapace length, sex, breeding state, and so on, are recorded and the results published annually, for example, Morgan and Barker (5) and Brown and Barker (6).

3.1.3. Other Information

(a) On occasion, measurements of the size and sex ratios of rock lobsters are made at various processing plants along the coast.

(b) Economic studies of the fishery are undertaken intermittently (1974 and 1978) by the Commonwealth Department of Primary Industry (Meany 18).

3.2. Biological Data

3.2.1. Puerulus Settlement

Phillips (19) showed that the last oceanic stage, the puerulus, of *P. cygnus* could be captured using collectors composed of artificial seaweed moored at the surface within the protection of the coastal reefs. Subsequent studies by Phillips and Hall (20) have shown that the catches from these collectors provide a measure of the relative strength of settlement from year to year.

The collectors are checked monthly after each new moon period when most pueruli settled. All settlement takes place at the puerulus stage. The western rock lobsters are removed from the collectors either as pueruli or after they have molted into very small post-puerulus juveniles. Initially, collectors were placed at four sites along the coast, but it was found that the predictions could be made using collectors from only one of these sites, Seven Mile Beach (Phillips 21).

3.2.2. Abundance of Juveniles

The densities of age groups (ages 2–6 yr) have been measured on shallow test reefs at Garden Island (1965–1977) and Seven Mile Beach (1967 to the present time), using the single census trap–mark–recapture method described by Chittleborough (15). These test reefs are adjacent to the collectors used to catch the puerulus stage.

3.2.3. Abundance of Recruits to the Fishery

During late November of each year, large numbers of immature, newly molted, pale colored rock lobsters emigrate into deeper water from the shallow water inshore reefs (which in some cases are inaccessible to fishermen) where they have spent the previous 4 or 5 yr. This offshore movement normally lasts through December and part of January, and in all lasts about 6 weeks. Because they are newly molted, their food requirements are high (Chittleborough 22) and consequently their catchability by baited pots is high (Morgan 23). During this emigration the fishermen catch large quantities of these animals which are locally known as the whites (George 14). Although a small number of animals undergo two or perhaps three white phases in their life cycle, the white phase generally occurs only once during an individual's lifetime (George 14), and this enables the migrating whites to be equated with the recruits to the fishery.

Estimates of the abundance of the potential emigrants from the shallow reefs have been made by Chittleborough (15), although a better measure of their abundance is available from the peak catch rates (measured as the catch per pot lift) of the commercial fishery during November–January, when most of the commercial catch consists of "white" rock lobsters.

3.2.4. Spawning Stock

Since the evaluation by Morgan et al. (12) of a stock and recruitment relationship for *P. cygnus*, considerable research has been undertaken to review the data on which the estimates of spawning stock abundance were made and to obtain more information on the breeding biology of *P. cygnus*.

The review of the data on the abundance of the spawning stock from the daily catch rate of spawning females recorded by fishermen in their research logbooks showed that the early years of data (1966–1972) had been collected using different methods, making them difficult to compare with data from 1973 onward. Therefore the data points at the right side of the figure relating spawning stock puerulus settlement (Fig. 2; Morgan et al. 12, p. 481), cannot be used. Obviously a curve of a different shape would be produced if only the data points from 1973 onward are used.

In addition to the inconsistencies found in the logbook data a recently initiated program to study in detail various aspects of the breeding biology of *P. cygnus* has found that the peak of the breeding season may not necessarily be reflected in the catch rates from logbooks, as it can occur either outside the period the commercial fishery operates (i.e., prior to November 15) or before the fishing fleet has moved from the inshore grounds to the 20–30 fathom depth range (usually early December). The importance of the Abrolhos Islands to the spawning stock is also being reassessed. Its contribution may be much greater than previously believed owing to the large proportions of undersized females that breed there and the fact that a significant proportion of females may be spawning twice (C. F. Chubb, personal communication).

In summary, the relationship produced by Morgan et al. (12) that involved data on the spawning stock should be used with some reservation.

3.2.5. Other Biological Information

A great many research programs have been conducted on many aspects of life history, biology, and population dynamics of the western rock lobster. Extensive reviews of this literature are found in Cobb and Phillips (24), Phillips et al. (25) (synopsis), Hancock (8), and Morgan et al. (12).

4. GROWTH AND MORTALITY RATES

4.1. Growth

Several techniques have been used to estimate growth rates of *P. cygnus*, including progression of size frequency modes (Chittleborough 15), growth of animals held in the laboratory (Chittleborough 17; Phillips et al. 26), and mark–recapture data (Chittleborough 17; Morgan 27).

Morgan (27) found that the von Bertalanffy (28) growth function provided the best description of growth in adult *P. cygnus*; Chittleborough (17) also was able to represent adequately the growth of juvenile *P. cygnus* by the von Bertalanffy curve. However, the adequacy of the von Bertalanffy growth function as a realistic description of the growth process in palinurids has been questioned by Saila et al. (29) and Morgan (30).

A reevaluation of the data used by Morgan (27) and also unpublished data of B. F. Phillips (R. Maller, personal communication) suggest that all von Bertalanffy estimates so far obtained by the mark-recapture method are suspect, mainly because of the inadequacies in sampling distribution of the data and the lack of knowledge of the exact age of the lobsters when marked.

A research program using a microtagging technique to mark post-puerulus stage animals is being initiated to overcome these problems.

4.2. Mortality Rates

4.2.1. Natural Mortality (M)

(a) Exploited phase: Bowen and Chittleborough (31) made the first estimate of natural mortality (*M*) for the exploited phase of *P. cygnus* when they produced their preliminary assessment of the stock of the western rock lobster. They used catch and effort data over the 5-month (March 15–August 15) fishing season at three different island groups of the Abrolhos Islands (Fig. 1) for the period between 1950 and 1963. The values of *M* calculated were 0.222, 0.731, and 0.781 for the inner area (i.e., inside the fringing coral reefs) for the Easter, Pelsart, and Wallabi groups, respectively. However, their estimates were based on changes in uncorrected catch per unit of fishing effort during the season and hence are subject to influence by both recruitment and changes in availability (catchability), which Morgan (23, 27, 32–34) found to be important in estimating values of *M* for *P. cygnus*. Morgan (27) produced estimates of *M* for *P. cygnus* by obtaining values of total mortality (*Z*) using the equation

$$Z = F + M = qf + M$$

where F = fishing mortality, q = catchability coefficient, and f = total fishing effort during period T. When considered on an annual basis this equation can be fitted by a linear regression whose slope is equal to q and whose intercept on the Y axis is equal to M, the annual rate of natural mortality. The value of M was determined by Morgan (27) to be around 0.23. Morgan also felt that the best values of Z could be calculated using the average carapace length of rock lobsters in the grade categories (B to E) of rock lobster that are processed and packed for export.

(b) Juvenile phase: Chittleborough (15) calculated the natural mortality of an unfished population of juvenile *P. cygnus* on shallow nursery reefs at Garden Island, off Fremantle (Fig. 1), by multiple mark–recapture experiments between January and August (the period during which very little migration occurs) over 5 years (1965–1969). Chittleborough's average value for the period was 0.71, which is greater than that found by Morgan for commercially exploited animals. Chittleborough attributed this to the increased number of molts they undergo as juveniles relative to the adults. Chittleborough (15) indicated that most mortality occurs at molting and that natural mortality within these juvenile populations was density dependent.

Chittleborough and Phillips (35) continued the study of juvenile densities and mortalities at Garden Island and expanded them to Jurien Bay and Seven Mile Beach, just north of Dongara (Fig. 1). Estimates of natural mortality between 1965 and 1974 for Garden Island ranged from a high of 1.516 in 1967 and a low of 0.244 in 1970 (Chittleborough and Phillips 35). Morgan et al. (12) provide the natural mortality rates for juveniles on the Seven Mile Beach nursery reefs between 1970 and 1978 as part of their study of the stock and recruitment relationships in *P. cygnus*. These values ranged from a high of 1.934 in 1974 to a low of 0.466 in 1972. Chittleborough (15) and Chittleborough and Phillips (35) reported high levels of density-dependent mortality in the juveniles but the close correlation between the levels of puerulus settlement and subsequent catch suggest that this may be less important than previously believed. Part of the mortality that was observed was probably due to emigration.

4.2.2. Total Mortality (Z)

Bowen and Chittleborough (31) provided the first estimates of total mortality (Z) for *P. cygnus*. They used the decline in catch per unit of effort to obtain values of Z of between 0.72 and 2.40 from three separate inner reef areas of the Abrolhos Islands fishery between 1950 and 1964.

Morgan (27) further refined these estimates by adjusting for changes in catchability during the year (Morgan 32), which allowed the calculation of effective effort (Morgan 27, 33). Because of the problems associated with changes in catchability Morgan (16, 27) preferred to use length composition data to estimate Z. He stated that in general as the total mortality increases, the length composition and the mean size of the population will change. Under steady-state conditions length and mean size are influenced by total mortality rate and growth rate. Using the grade category data available from processed rock lobsters, Morgan (16, 27) calculated values of Z for the fishery to be 0.87–1.01 (1971–1973).

4.2.3. Fishing Mortality (F)

Bowen and Chittleborough (31) calculated values of fishing mortality (F) for the Abrolhos Islands fishery from 1943 to 1964 ranging from 0.011 at the Wallabi group in 1943 (an unexploited population) to 1.817 at the Easter group in 1964 (a fully exploited population). However, the catchability coefficient (q) they used to calculate F was assumed to be constant, whereas Morgan (27, 32, 33) has shown that it is not constant throughout the season. Morgan (27, 32) investigated the use of tag returns from experiments primarily designed to investigate growth, movement, and catchability to obtain values of F, but decided that this method, based on that of Ricker (36), did not meet the assumptions required to produce accurate estimates; for example, tagged animals were more likely to be captured than untagged animals, thus leading to overestimates of Z and F. Morgan (27, 33) estimated F for *P. cygnus* by examining the changes in Z over a wide range of effort between the late 1960s and mid 1970s—the values of 0.64–0.78 obtained currently provide the best estimates available.

5. MODELS

5.1. Stock Assessment

A number of models have been used to describe the *P. cygnus* fishery. Bowen and Chittleborough (31) produced the first model and concluded from it that of the 63.6 million kg of the original standing stock available before intensive exploitation, only 15.9 million kg remained by 1963. They showed that the annual exploitation rate, based largely on the heavily exploited Abrolhos Islands, had risen sharply from 1944–1945 to the early 1960s, when it leveled off at about 60%. They stated that recruitment to the fishery was being significantly reduced due to the mortality of sublegal size animals and proposed a sustainable catch of 7.3 ± 0.9 million kg based on stable recruitment and fishing effort.

Morgan (16, 27, 33) described in detail the Schaefer (37) model, GENPROD model, Fox (38) model PRODFIT, and the delayed recruitment model used to estimate the maximum equilibrium yield and the fishing effort at which the yield is maximized for the *P. cygnus* fishery. The maximum equilibrium yields obtained from the four models above were 8.6, 8.1, 8.5, and 8.0 million kg, respectively, and the effort required to obtain these yields was 5.9, 6.0, 5.6, and 5.8 million units of effective effort (pot lifts), respectively (Morgan 33). Although no statistical tests could be applied to the models to determine which provides the best description of the fishery, Morgan (34) discussed subjectively the advantages and disadvantages of each. He concluded that the four yield models considered provide at best a crude representation of the dynamics of the exploited stock. The models do, however, provide a fit to the observed catch and fishing effort data and so provide a good description of the relationship between catch and fishing effort up to the values of effort so far encountered. Caution should be employed, however, in using the models in a predictive way to estimate equilibrium catches at effort values beyond those already encountered, because there is little evidence of the decline in catches that the models predict with increasing effort (Morgan 34).

Hancock (8) provided a review of the models Morgan (27, 30, 34) used to describe the fishery and added three additional data points (i.e., 1976–1977, 1977–

1978, and 1978–1979) to the data set and with some refinement to Morgan's (1, 16, 34) method fitted a Schaefer curve using the equilibrium approximation of Fox (38), to the new data set. From these Hancock obtained a maximum sustainable yield (MSY) of 9.3 million kg produced with a fishing effort of 7.2 million effective pot lifts, compared to Morgan's estimate (above) of 8.6 million kg with 5.9 million effective pot lifts.

Figure 2 is an update on Hancock's Schaefer curve. The data have been updated to include catch and effective effort figures from 1979–1980 to 1982–1983 (Table 1). With the additional data points included the MSY has increased to 9.8 million kg obtained with an effective effort of 8.2 million pot lifts. It was not considered valid to include data from 1983–1984 and 1984–1985, which relate to a fishing area with boundaries that are not consistent with those used for previous data. Effective effort will therefore have to be recalculated for all previous years before refitting the Schaefer curve to the total data set.

See also Section 7.1.5 for proposed changes in the calculation of effective effort.

5.2. Recruitment Prediction

5.2.1. Recruitment Predictions Using Juveniles

Caputi and Brown (39) have shown that recruitment to the fishery can be predicted 1 and 2 yr in advance by using the abundance of 3- and 4-yr-old animals obtained

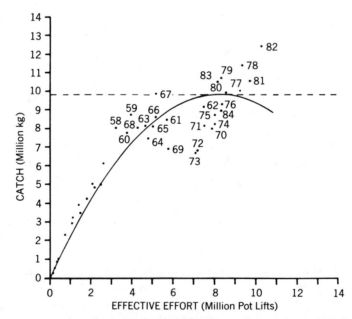

Figure 2. Annual catches related to the effective fishing effort, using the method described by Hancock (8, p. 218). Seasons are represented by 58 = 1958–1959, and so on. Data from 1944–1945 to 1982–1983 have been used (Table 1).

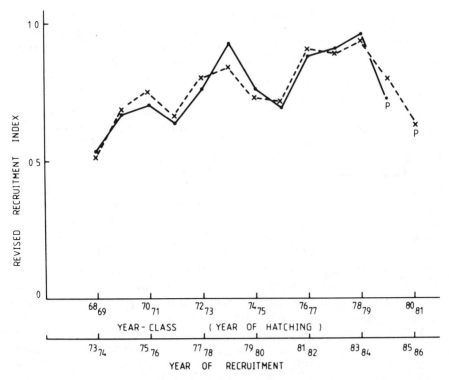

Figure 3. The relationship between the revised recruitment index (——) and its predicted value (– – –) from the multiple regression analysis using both the 3- and 4-yr-old male juvenile indexes (redrawn from Caputi and Brown (39) with permission from Canadian Journal of Fisheries and Aquatic Science).

from length-frequency data from the commercial rock lobster monitoring program that began in 1971–1972 (see Section 3). The 3-yr-old index predicts lowered recruitment for 1985–1986 and 1986–1987, with the 4-yr-old index also predicting a reduction in 1985–1986. Figure 3 shows the close relationship between the predicted recruitment (from a multiple regression using both the 3- and 4-yr-old indexes) and the actual recruitment (multiple $r = 0.97$—Caputi and Brown 39).

5.2.2. Catch Predictions Using Pueruli

Predictions of the total commercial catch of the western rock lobster using data on the presence of puerulus settlement have been successfully made 4 yr in advance. The predictive system, which has operated for 13 yr, has been recognized by industry and government as an important tool in the management of the fishery.

Collectors have recently been placed at seven other sites between South Passage and Augusta (see Fig. 1) to find out whether there is any spatial or temporal variation in puerulus settlement over the whole range of the fishery.

Since 1980, data from Seven Mile Beach have been used to predict both the annual white fishery 4 yr later and to predict the total catch of the fishery in the same year. The relation between the level of puerulus settlement and the level of recruits to the fishery 4 yr later calculated by Morgan et al. (12) was obtained by fitting the Ricker (40) stock–recruitment relationship. The levels of puerulus settlement at Seven Mile Beach and of recruits to the fishery 4 yr later are shown in Figure 4. Data obtained since the paper by Morgan et al. (12) have increased the scatter of the data, but good fits of the models of both Ricker (40) and Beverton and Holt (41) were obtained by linear regression on log-transformed versions of the data (Phillips 21). For both models the variance accounted for on the transformed scale was 86%.

The commercial catch of any year is composed of a number of year classes, but because the white fishery catch is about 40% of the total it strongly influences the total catch. The levels of puerulus settlement at Seven Mile Beach plotted against the total commercial catch of the fishery 4 yr later are shown in Figure 5. However, a clear time trend is apparent in the catch data from 1969 to 1978 (Table 1), with a progressive increase in the catch that leveled off in the final years. A strong relationship between the level of puerulus settlement and the total catch is confirmed by a linear regression of the data, with allowance being made for the trend in the increasing catch ($r = 0.862$, $p < .001$) (Phillips 21).

In the first year in which the prediction system was used (1969–1970) the number of pueruli on the collectors was exceptionally low. It was predicted that low catch levels would be likely in 1972–1973 and even lower levels likely in the 1973–1974

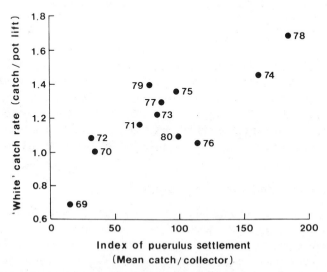

Figure 4. Indexes of annual puerulus settlement at Seven Mile Beach for 1969 to 1980 and catch rates of *Panulirus cygnus* whites, 4 yr later. The year shown is the year of puerulus settlement.

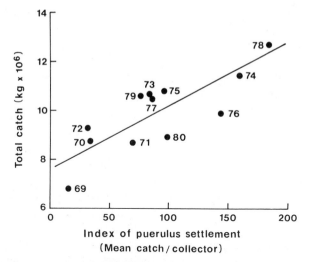

Figure 5. Indexes of annual puerulus settlement at Seven Mile Beach 1969 to 1980 and the total commercial catch of *Panulirus cygnus* 4 yr later. The fitted regression is $y = 7.64 + 0.0261x$. The year shown is the year of puerulus settlement.

season. The prediction was borne out by the catches of these two years, the white season of 1973–1974 being the poorest on record. Since then, the predictions have proved remarkably accurate at about $\pm 13\%$ for the total catch.

In 1979, for example, it was predicted that the total catch in 1983–1984 would be 10.8×10^6 kg. The actual catch was 10.6×10^6 kg. Based on the demonstrated relationships in the data to 1980 and on the level of puerulus settlement in 1982 (Fig. 6, modified and updated from Hancock (8) and using only settlement data

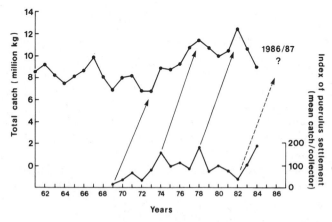

Figure 6. The annual indexes of puerulus settlement at Seven Mile Beach from 1969 to 1984 and the total commercial catch of *Panulirus cygnus* from 1961/1962 to 1983/1984 (modified and updated from Hancock (8)).

from Seven Mile Beach) it is predicted that the total western rock lobster catch in the 1986–1987 season should be in the region of 7.72×10^6 kg. This prediction assumes that both the level of effort and the environmental conditions will be similar to those experienced over the past decade. Because the level of puerulus settlement in 1983–1984 (Fig. 6) was substantially higher than for several past years (and 1984–1985 was the highest ever recorded), the catches can be expected to increase again after the 1986–1987 season.

5.2.3. Use of Predictions

Predictions of the annual catch and of the white fishery, made using levels of puerulus settlement and juvenile abundance estimates, are useful to both the fisheries management and the organizations involved with the fishing industry (fishermen, fishing companies, processors, and banks). Fishermen find the information useful when contemplating new investments, such as buying a new vessel or purchasing additional pot entitlements, as do banks when deciding on loans for these investments. Data on puerulus settlement and juvenile levels are now routinely considered in the estimates of likely future fishing success given by the Industry Advisory Committee to fishermen and the State Government Minister responsible for the fishery.

The predictive system is, however, most valuable for fisheries management. It enables those responsible to anticipate years of reduced catches and to judge whether they are caused by a natural fluctuation in abundance rather than by overfishing of the resource. Although the level of the breeding stock is related initially to the level of settlement at the puerulus stage, it is then affected by the fishing pressure on the adults after they enter the fishery. Strong fishing pressure, combined with a low level of puerulus settlement such as that in 1982–1983, could possibly reduce the size of the breeding stock below the critical level required to maintain recruitment to the fishery.

6. SOCIOECONOMIC CHANGES

The decision in 1962 to limit the number of boats and pots and to stabilize the industry immediately caused a value to be put on a rock lobster license that previously had no value (Morgan and Barker 9). In 1971 the price of a 100-pot license was between A$15,000 and A$20,000 (i.e., $160–$200/pot) and by 1985 the price had risen to A$400,000 to A$450,000. Fishermen now entering the fishery have to obtain and service large loans.

Concurrent with the advent of strict licensing control, the price paid to fishermen for their catch began to rise. In 1962 fishermen received A$0.84/kg, whole weight (Morgan and Barker 9) and by 1973 the price had risen to A$1.20/kg (Morgan and Barker, 2) and more than A$15.00/kg by 1985 (E. H. Barker, personal communication).

These two factors along with steadily rising running costs have been the main pressures on fishermen to increase their fishing effort in order to maximize their share of the catch. The consequent increase in fishing effort has caused major problems for the industry as management and government try to implement ways of offsetting the increases to reduce the exploitation on the stocks. See Bowen and Hancock (Chapter 16) for further details.

7. PRIORITIES AND APPROACHES FOR RESEARCH AND FUTURE MANAGEMENT

7.1. Research

7.1.1. Recruitment

Individual projects within the CSIRO programs have been integrated into a synthesized approach to understanding the recruitment processes of the western rock lobster. Biological oceanographers now believe that three major controlling factors determine the survival of larvae: (1) the degree of predation, (2) the type and amount of food available, and (3) where and how the larvae are dispersed. The interplay of these factors must be understood in detail before larval survival (and ultimate commercial harvest) can be accurately predicted. The oceanic larval phase of the western rock lobster is followed by a juvenile phase. After 4 or 5 yr on the coastal reefs the juveniles recruit to the fishery. In their case, the three controlling factors that determine survival are thought to be, again, predation, food, and availability of suitable shelter.

Current studies include:

1. The measurement of spatial differences in the distribution and density of potential prey organisms, within an area inhabited by dense populations of juvenile rock lobsters, to determine whether the feeding behavior of rock lobsters is related to the availability of food.
2. The measurement of seasonal differences in the availability of potential prey organisms, and the amount of secondary production, to determine whether seasonal fluctuations in food supply could affect the growth or patterns of distribution of the juveniles.
3. Evaluation of the mode and rate of trophodynamic transfer of primary production to higher trophic levels, directly to the post-puerulus and early juvenile stages of the western rock lobster, and indirectly via other invertebrates in the coastal system. Plant particulate material and microbial processes are a special focus.
4. Spatial and temporal settlement of the puerulus stage along the coast of Western Australia over the range of the fishery and behavior of the puerulus stage.
5. Accurate estimates of juvenile densities, aging, and mortality from settlement to 5 yr of age.

6. Evaluation of the factors that affect juvenile mortality, productivity, and distribution, especially predation, cannibalism, food abundance, and shelter.

7. The role of the circulation of the southeastern Indian Ocean in the return of western rock lobster larvae to the coastal reefs of the west coast.

7.1.2. Stock Enhancement

Brown et al. (42) undertook a $3\frac{1}{2}$-yr study (1977–1980) of the fishery-induced mortality of undersized rock lobsters. On average 18 million undersized rock lobsters (<76 mm carapace) are caught, handled, and returned to the sea by fishermen each season despite the use of a 54 × 305 mm escape gap in each trap. Depending on the handling procedure fishermen use, the undersized experience various degrees of exposure, displacement, and damage, which adversely affect their survival and growth (Brown and Caputi 43–45; Brown et al. 42). As exposure time increases, so does the mortality of the undersized (Brown and Caputi 43—Fig. 6, p. 117).

The results of the laboratory and field trials (Brown and Caputi 43), showed that the poor handling methods used by rock lobster fishermen produced a mortality of 14.6% of all undersized handled. This translated to a loss of commercial production of 11.4% worth approximately A$15 million to the industry at 1984–1985 prices.

A 2-yr (1981–1982) education program was undertaken to disseminate the results of the research to the rock lobster industry. This resulted in a significant improvement in handling methods, although 52.2% of fishermen still used a handling method that could cause significant mortality (Brown and Dibden 46).

Field trials were initiated to test various combinations and sizes of escape gap to determine the best configuraion for reducing the numbers of undersized retained in the trap, while still maintaining the catch of legal size. Brown and Caputi (45) reported that reductions of up to 70% could be obtained by increasing the number and width of escape gaps.

7.1.3. Review of Spawning Stock Estimates

A complete reassessment of the stock–recruitment relationships will examine not only the relationship between puerulus settlement and spawning stock but also the environmental factors that may play an important role in the relationship.

7.1.4. Refining Recruitment Predictions

The ability to predict the level of recruitment to the fishery one and two years in advance using the catch rate of prerecruits (3- and 4-yr olds) is proving to be a very important management tool. Therefore additional research will be undertaken to refine the growth and age parameters used in the predictions to enable a more accurate assessment of what year/size classes will recruit to the fishery in a particular season. From preliminary studies it would appear that growth rates at various locations along the west coast are different (Morgan, 27); therefore animals from different locations may recruit to the fishery at different ages and at different times of the year.

In addition to the problems experienced with differences in growth rates, it has been recommended to industry that there should be an increase in the number of the escape gaps (i.e., from one to three of 54 × 305 mm) to reduce the catch rate and hence handling of undersized rock lobsters. This means that the catch rate of prerecruits will be reduced, thus producing different indexes of abundance that will not be directly comparable to previous indexes. Research is currently in progress to quantify these changes to enable adjustment of the old indexes to make them compatible with the present data set.

7.1.5. Fishing Mortality and Effective Fishing Effort

At present it would appear that fishing mortality could be underestimated from a number of causes:

(a) The current measure of fishing effort is the pot lift, which does not take into account the increases in fishing efficiency brought about by changes in technology, the use of large pots in some seasons, and fishing practice. Therefore current figures (Table 1) are an underestimate of the actual effort being expended. Information is currently being collected to examine the impact of these changes on the fishing effort. It is expected that some changes will be made to the values given in Table 1.

(b) Catchability (vulnerability) estimates derived by Morgan (27, 32, 33) and used to calculate effective fishing effort were based on research conducted at the Abrolhos Islands, an area different from the coastal fishery. Morgan (27) stated that it is not known if these values reflect those of the coastal area and therefore additional research is required to clarify this. If new values are obtained for coastal areas, the catchability indexes currently used will be revised and new values of effective effort computed. In addition, the fishing area component used by Morgan (27, 33) in the calculation of effective fishing effort will also be changed to delete the 1° blocks that may be biasing results (i.e., those in which infrequent fishing occurs). These changes will result in a revision of the effective effort values given in Table 1, especially those in the latest years.

(c) Data on the grade categories of processed rock lobster tails (Section 3.1) have previously been used to obtain values of natural (M) and total mortality (Z) (Morgan 27). A program is currently underway to obtain more detailed information from processors on the grades they pack for export and the local market. In addition to this a factory survey is being undertaken at three of the major coastal locations (Fremantle, Jurien, and Geraldton—Fig. 1) to obtain detailed information on the whole weight/tail weight/carapace length relationship. This will allow more accurate partitioning of the total catch into the various grade categories and produce more reliable estimates of M and Z.

7.2. Management

7.2.1. Improving Recruitment

Recommendations based on the results of the research undertaken by Brown and Caputi (43–45), Brown et al. (42), and Brown and Dibden (46) on the fishery-induced mortality of undersized rock lobsters have been implemented by management. The first recommendation was to legislate to make it an offense for fishermen to take a trap on board while still sorting rock lobsters from the preceding trap. Under this regulation undersized can be held on board a boat only for a maximum of 5 min before being returned to the sea. The second recommendation was to increase the number of escape gaps from one of 54 × 305 mm to three gaps of 54 × 305 mm.

With both these recommendations incorporated in the legislation for the 1986–1987 season, fewer undersized will be taken by traps, and most of those that are taken on board will be exposed for a shorter period (less than 5 min) before being returned to the water. This will significantly improve the survival of prerecruits and will increase the total yield from the fishery by an estimated 11.4% (Brown and Caputi 43).

7.2.2. Reducing Fishing Effort

Effective fishing effort has been rising inexorably (Table 1) in the fishery despite the seemingly tight limited-entry framework with which it is surrounded (see Bowen and Hancock, Chap. 16 for details). It has been realized for many years that the fishery would be biologically more stable and economically more profitable if fishing effort could be reduced by as much as 30% (Bowen 4; Morgan 1, 16, 30; Hancock 8).

Based on the index of abundance of prerecruits and the index of settlement of puerulus the recruitment and total catch of the 1986–1987 rock lobster season are predicted to be the lowest recorded since 1973–1974. It is not currently felt that the spawning stock is in danger of being overfished. However, with the present very high levels of effort operating in the fishery, the low abundance of recruits that are expected in 1986–1987 could be exploited so heavily that the subsequent breeding stock (1–2 yr later—Morgan et al. 12) may be reduced to such an extent that future puerulus settlement and hence recruitment to the fishery could be adversely affected. The abundance of the spawning stock would have to fall below those so far experienced in the fishery, because it is currently considered (B. F. Phillips and A. F. Pearce, personal communication) that environmental factors govern the strength of puerulus settlement at current levels of breeding stock.

To ensure that the recruitment in 1986–1987 and the subsequent spawning stock are not adversely affected, the Rock Lobster Industry Advisory Committee, which is made up of representatives from the fishing and processing sectors and government, has recommended that pot numbers be reduced by 10% from the beginning of the 1986–1987 season. This should produce an immediate reduction in effective effort (though fishermen will fish harder to offset the loss—Bowen and Hancock see Chap. 16) and hence reduce the exploitation of the vulnerable new recruits, the spawning stock, and indeed the entire rock lobster population.

8. CONCLUSIONS

The western rock lobster fishery is arguably the best-managed rock lobster fishery in the world. Cooperation between government, fishermen, and researchers is a feature of the manangement program and there is constant communication between all levels of the industry.

The management program is based on a strong background of research information. The arrangement by which two different organizations, that is, the CSIRO and the Fisheries Department, collaborate to supply this information is unusual but has been successful.

The fishery is closely regulated. The effect of legislative changes often makes data over different time periods difficult to standardize. However, it is significant that factors such as the minimum legal size and the design of the unit of fishing effort (the pot) have remained essentially unchanged over a long period. Because the majority of the catch are recruits taken before maturity, pressure on the stock is considerable. It is only by mutual cooperation of all sections of the industry that it has been possible to maintain the fishery at the current high levels of production.

REFERENCES

1. G. R. Morgan, Increases in fishing effort in a limited entry fishery—the western rock lobster fishery 1963–76. *J. Cons. Int. Explor. Mer.* **39(1):**82–87 (1980).

2. G. R. Morgan and E. H. Barker, The western rock lobster fishery 1972–73. *West. Aust., Dep. Fish., Wildl. Rep.* **15:**1–22 (1974).

3. K. Sheard, *The Western Australian Crayfishery 1944–61.* Patterson Brokensha Pty. Ltd., Perth, Western Australia, 1962, 107 pp.

4. B. K. Bowen, Spiny lobster fisheries management. In J. S. Cobb and B. F. Phillips, Eds., *The Biology and Management of Lobsters*, Vol. I. Academic Press, New York, 1980.

5. G. R. Morgan and E. H. Barker, The western rock lobster fishery 1975–76. *West. Aust., Dep. Fish., Wildl. Rep.* **33:**1–20 (1979).

6. R. S. Brown and E. H. Barker, The western rock lobster fishery 1982–83. *West. Aust., Dep. Fish., Wildl. Rep.* **70:**1–123 (1985).

7. B. K. Bowen, Management of the western rock lobster (*Panulirus longipes cygnus*) George. *Proc. Indo-Pac. Fish. Counc.* **14(II):**139–153 (1971).

8. D. A. Hancock, Research for management of the rock lobster fishery of Western Australia. *Proc. Annu. Gulf Caribb. Fish. Inst.* **33:**207–229 (1981).

9. G. R. Morgan and E. H. Barker, The western rock lobster fishery 1961–71. *West. Aust., Dep. Fish., Wildl. Rep.* **55:**1–41 (1982).

10. B. F. Phillips, The circulation of the southeastern Indian Ocean and the planktonic life of the western rock lobster. *Oceanogr. Mar. Biol.* **19:**11–39 (1981).

11. B. F. Phillips, B. A. Brown, D. W. Rimmer, and D. D. Reid, Distribution and dispersal of the phyllosom larvae of the western rock lobster, *Panulirus cygnus*, in the southeastern Indian Ocean. *Aust. J. Mar. Freshwater Res.* **30:**773–783 (1979).

12. G. R. Morgan, B. F. Phillips, and L. M. Joll, Stock and recruitment relationships in *Panulirus cygnus*, the commercial rock (spiny) lobster of Western Australia. *Fish. Bull.* **80(3):**475–486 (1982).

13. L. M. Joll and B. F. Phillips, Natural diet and growth of juvenile western rock lobsters *Panulirus cygnus* George. *J. Exp. Mar. Biol. Ecol.* **75:**145–169 (1984).

14. R. W. George, The status of the white crayfish in Western Australia. *Aust. J. Mar. Freshwater Res.* **9:**537–545 (1958).

15. G. R. Chittleborough. Studies on recruitment in the western rock lobster *Panulirus longipes cygnus* George: Density and natural mortality of juveniles. *Aust. J. Mar. Freshwater Res.* **21:**131–148 (1970).

16. G. R. Morgan, Population dynamics of spiny lobsters. In J. S. Cobb and B. F. Phillips, Eds., *The Biology and Management of Lobster*, Vol. I. Academic Press, New York, 1980.

17. G. R. Chittleborough, Breeding of *Panulirus longipes cygnus* George under natural and controlled conditions. *Aust. J. Mar. Freshwater Res.* **27:**499–516 (1976).

18. T. F. Meany, *The Western Australian Rock Lobster Fishery. A Report of an Economic Survey*, Fish. Rep. 33. Economic Analysis Section, Fisheries Division, Department of Primary Industry, Canberra, Australia, 1981, 84 pp.

19. B. F. Phillips, A semi-quantitative collector of the puerulus larvae of the western rock lobster *Panulirus longipes cygnus* George (Decapoda, Palinuridea). *Crustaceana* **22(4):**147–154 (1972).

20. B. F. Phillips and N. G. Hall, Catches of puerulus larvae on collectors as a measure of natural settlement of the western rock lobster *Panulirus cygnus* George. *CSIRO, Rep. Div. Fish. Oceanogr.* **98:**1–18 (1978).

21. B. F. Phillips, Prediction of commercial catches of the western rock lobster *Panulirus cygnus. Can. J. Fish. Aquat. Sci.* **43:**2126–2130 (1986).

22. R. C. Chittleborough, Environmental factors affecting growth and survival of juvenile western rock lobsters *Panulirus longipes* (Milne-Edwards). *Aust. J. Mar. Freshwater Res.* **26:**177–196 (1975).

23. G. R. Morgan, Aspects of the population dynamics of the western rock lobster *Panulirus cygnus* George. II. Seasonal changes in catchability coefficient. *Aust. J. Mar. Freshwater Res.* **25:**249–259 (1974).

24. J. S. Cobb and B. F. Phillips, Eds., *The Biology and Management of Lobsters*, Vols. I and II. Academic Press, New York, 1980.

25. B. F. Phillips, G. R. Morgan, and C. M. Austin, Synopsis of biological data on the western rock lobster *Panulirus cygnus* George, 1962. *FAO Fish. Synop.* **128:**1–64 (1980).

26. B. F. Phillips, N. A. Campbell, and W. A. Rea, Laboratory growth of early juveniles of the westen rock lobster, *Panulirus longipes cygnus. Mar. Biol. (Berlin)* **39:**31–39 (1977).

27. G. R. Morgan, Aspects of the population dynamics of the western rock lobster and

their role in management. Ph.D. Thesis, University of Western Australia, 1977. University of Western Australia Nedlands Western Australia.

28. L. von Bertalanffy, A quantitative theory of organic growth. *Hum. Biol.* **10:**191–213 (1938).

29. S. B. Saila, J. H. Annala, J. L. McKoy, and J. D. Booth, Application of yield models to the New Zealand fishery for rock lobster *Jasus edwardsii* (Hutton). *N. Z. J. Mar. Freshwater Res.* **13(1):**1–11 (1979).

30. G. R. Morgan, Population dynamics and management of the western rock lobster fishery. *Mar. Policy* **4:**52–60 (1980).

31. B. K. Bowen and G. R. Chittleborough, Preliminary assessments of stocks of the Western Australian crayfish, *Panulirus cygnus* George. *Aust. J. Mar. Freshwater Res.* **17:**93–121 (1966).

32. G. R. Morgan, Aspects of the population dynamics of the western rock lobster, *Panulirus cygnus* George. I. Estimation of population density. *Aust. J. Mar. Freshwater Res.* **25:**235–248 (1974).

33. G. R. Morgan, Trap response and the measurement of effort in the fishery for the western rock lobster. *Rapp. P.-V. Reun., Cons. Int. Explor. Mer.* **175:**197–203 (1979).

34. G. R. Morgan, Assessment of the stocks of the western rock lobster *Panulirus cygnus* using surplus yield models. *Aust. J. Mar. Freshwater Res.* **30:**355–363 (1979).

35. G. R. Chittleborough and B. F. Phillips, Fluctuations in year class strength and recruitment in the western rock lobster. *Aust. J. Mar. Freshwater Res.* **26:**317–328 (1975).

36. W. E. Ricker, Hand book of computations for biological statistics of fish populations. *Bull. Fish. Res. Board Can.* **119:**1–300 (1958).

37. M. B. Schaefer, A study of the dynamics of the fishery for yellowfin tuna in the eastern tropical Pacific Ocean. *Bull. Int. Am. Trop. Tuna Comm.* **2(6):**245–285 (1957).

38. W. W. Fox, Fitting the generalized stock production model by least squares and equilibrium approximation. *Fish. Bull.* **73(1):**23–36 (1975).

39. N. Caputi and R. S. Brown, Prediction of recruitment in the western rock lobster (*Panulirus cygnus*) fishery based on indices of juvenile abundance. *Can. J. Fish. Aquat. Sci.* **43(11):**2131–39 (1986).

40. E. W. Ricker, Computation and interpretation of biological statistics of fish populations. *Bull. Fish. Res. Board Can.* **191:**1–382 (1975).

41. R. J. H. Beverton and S. J. Holt, On the dynamics of exploited fish populations. *Fish. Invest.—Minist. Agric. Fish. Food (G.B.) (Ser. 2)* **19:**1–533 (1957).

42. R. S. Brown, N. Caputi, J. Prince, and J. Jenke, Fishery induced mortality of undersize western rock lobster. *West. Aust. Fish. Dep., Rep.* (in press).

43. R. S. Brown and N. Caputi, Factors affecting the recapture of undersize western rock lobster *Panulirus cygnus* George returned by fishermen to the sea. *Fish. Res.* **2:**103–128 (1983).

44. R. S. Brown and N. Caputi, Factors affecting the growth of undersize western rock lobster *Panulirus cygnus* George returned by fishermen to the sea. *Fish. Bull.* **83(4)** (1985).

45. R. S. Brown and N. Caputi, Improvement of recruitment of the western rock lobster (*Panulirus cygnus* George) by enhancing survival and growth of undersize captured and returned by fishermen to the sea. *Can. J. Fish. Aquat. Sci.* **43(11):**2236–42 (1986).

46. R. S. Brown and C. J. Dibden, A re-examination of the methods used for handling undersize western rock lobsters. *West. Aust. Fish. Dep., Rep.* **78:**1–38 (1987).

8 NORWAY LOBSTERS IN THE IRISH SEA: MODELING ONE COMPONENT OF A MULTISPECIES RESOURCE

Keith M. Brander and David B. Bennett
Ministry of Agriculture, Fisheries and Food
Lowestoft, Suffolk, England

1. INTRODUCTION AND FISHERY DESCRIPTION

The Irish Sea is a relatively small area of some 45,000 km^2, virtually surrounded by the British Isles (Fig. 1). Most of the area is less than 50 m deep, with the maximum depth of more than 200 m occurring in the North Channel. In terms of annual fish yield it is one of the least productive areas around the British Isles, and this may be linked to low recruitment (Brander and Dickson 1).

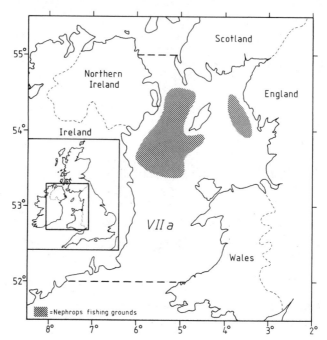

Figure 1. Irish Sea (ICES Division VIIa) Norway lobster (*Nephrops norvegicus*) fishing grounds.

Otter trawling for demersal species, such as cod, whiting, sole, plaice, and Norway lobster, is the main Irish Sea fishery, exploited by vessels from the United Kingdom, Ireland, France, and Belgium. The first two countries land the majority of the total fish and shellfish production, which has averaged just over 80,000 tonnes during the last 10 yr (Fig. 2). Demersal fish make up the major component of the landings but the most valuable single species landed is a shellfish, the Norway lobster (Fig. 3).

Total international landings have increased over the past 30 yr, reaching a peak in the mid-1970s, owing to large landings of pelagic species, particularly herring. Demersal fish landings reached a peak in 1981–1982 of just over 50,000 tonnes (Fig. 2) as a result of increased fishing effort and good cod recruitment (International Council for the Exploration of the Sea 2). Norway lobster landings topped 10,000 tonnes for the first time in 1979, and then slumped dramatically in 1980 because of marketing problems. Landings have recently recovered to around the 10,000-tonne level. There are two main fishing grounds for Norway lobster in the Irish Sea, the small fishery off northwest England prosecuted mainly by English boats, and the much larger ground in the western Irish Sea between the Isle of

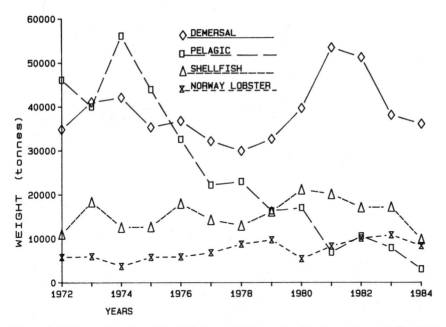

Figure 2. Various categories of the Irish Sea total international landings (tonnes). (Shellfish exclude Norway lobster.)

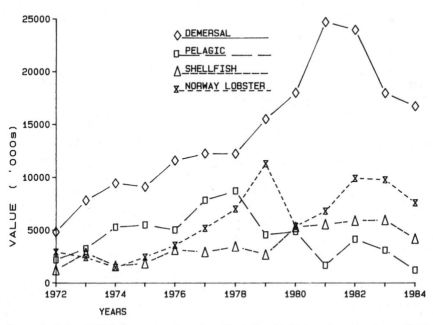

Figure 3. Value (£'000) of various categories of Irish Sea total international landings. (Shellfish exclude Norway lobster.)

Man and Ireland (Fig. 1). In this area vessels from Northern Ireland and the Irish Republic land 83% of the total Irish Sea Norway lobster landings.

The distribution of Norway lobsters is confined to areas where suitable muddy substrate occurs, and the relatively large extent of such substrate in the Irish Sea, compared with other sea areas around the British Isles, explains the high proportion of Norway lobster in the total landings from the Irish Sea. Recent studies have also shown that the stratified water mass overlying the main Norway lobster ground in the western Irish Sea is an area of enhanced biological activity (Fogg et al. 3). A large proportion of the total Irish Sea fish landings, particularly of cod and whiting, come from this area.

The Irish Sea is wholly within the European Economic Community's (EEC) exclusive economic zone (EEZ), and management is by EEC regulations enforced by member countries within their own waters. The main methods of regulating the fisheries are total allowable catches (TAC) and technical measures covering mesh and minimum landing sizes. Scientific advice on predicted catch levels and technical regulations is provided to the EEC by the International Council for the Exploration of the Sea (ICES), an independent scientific organization. Five fish stocks in ICES Division VIIa, cod, whiting, sole, plaice, and herring, are currently managed by annual TAC; Norway lobster has recently become a TAC species, but the TAC covers a much larger area, ICES subarea VII. The demersal fish stocks are further regulated by a 70-mm mesh (75-mm double twine) and various minimum landing sizes. A smaller mesh of 60 mm* is permitted for Norway lobster trawling and there is a minimum landing size (MLS) of 25-mm carapace length (85-mm total length, 46-mm tail length). This small-mesh fishery also takes a considerable by-catch of small whitefish, which are discarded.

The implicit management objective for the Irish Sea is to optimize yields while maintaining adequate spawning stocks, but because of the mixed nature of the main trawl fisheries and the biological interactions between species, it is very difficult to determine how the objective can be achieved. TAC regulations for the main species, based on single-species assessment, are clearly not sufficient to prevent excessive fishing effort and serious overfishing in some cases. For example, Brander (4) has shown that the disappearance of the common skate (*Raia batis*) from the Irish Sea is an inevitable consequence of the fisheries directed at other species.

The difficulties of incorporating technical and biological interactions into usable multispecies models are well documented, but a start has been made in investigating these interactions for Norway lobsters in the Irish Sea and the results are already being used in formulating management policy. Our approach is piecemeal: identifying obvious major interactions and incorporating them in models that are kept as simple as possible. Two principal models are described in this chapter. The first is a quarterly yield-per-recruit model, which incorporates seasonal data on growth, availability, and exploitation for Norway lobster. The second is a

---

*Mesh size changed to 70 mm as from 1 July 1986

multispecies yield model, which incorporates technical and biological interactions between Norway lobster and cod, the predominant predator.

The first model is used, along with length cohort analysis; LCA (Jones 5), mainly to examine the consequences of changes in mesh size and minimum landing size on yield per recruit of Norway lobster. It is not a multispecies model, but is used to complement the results of mesh assessments of fish species (principally whiting), which are caught as a by-catch in the small-mesh Norway lobster fishery. Thus in considering the case for changing the mesh sizes, we can estimate the short-term and long-term effects on both Norway lobsters and whitefish.

The second model is described in greater detail elsewhere (Brander and Bennett 6), but the principal features affecting management strategy for Norway lobster are discussed here. Although it is a relatively simple multispecies extension of standard yield per recruit modeling, it can be seen that even when it is limited to two species (Norway lobster and cod) the resultant complexity is not easy to interpret.

2. BIOLOGICAL AND FISHERIES INFORMATION REQUIRED FOR ASSESSMENTS

The Norway lobster (*Nephrops norvegicus* L.) is a benthic burrowing decapod crustacean with a relatively straightforward life history. Spawning occurs in autumn when the fertilized eggs are attached to the female's pleopods and incubated externally for 8–9 months. Peak larval production in the Irish Sea is mid-May, and settlement takes place after three pelagic larval stages lasting about 40 days (Nichols et al. 7). Larval surveys in 1982 and 1985 have been used to estimate stock biomass of Norway lobsters in the Irish Sea (Nichols et al. 7).

Norway lobsters live in burrows in muddy substrates, and are available for capture by trawls only when they are on the surface of the seabed (Chapman 8). Their emergent behavior has been shown to be related to light levels, and in the Irish Sea catch rates are higher at dawn and dusk. Ovigerous females tend to stay in their burrows during the incubation period and are fully available only in the summer. Differential mortality on males and females can be expected, and this may result in unbalanced sex ratios (ICES 9).

Female Irish Sea Norway lobsters first attain sexual maturity at about 20-mm CL (carapace length), and males do so at 26-mm CL (Farmer 10). Fifty percent maturity of females is at 23-mm CL, with 100% mature at about 30-mm CL (R. P. Briggs, personal communication). About 90% of mature females are berried annually (Thomas 11; Farmer 10).

Growth rate is a critical input to any assessment. The inability to age crustaceans directly means that growth data are scarce. The discontinuous growth of Norway lobsters comprises two components, molt increment and molt frequency.

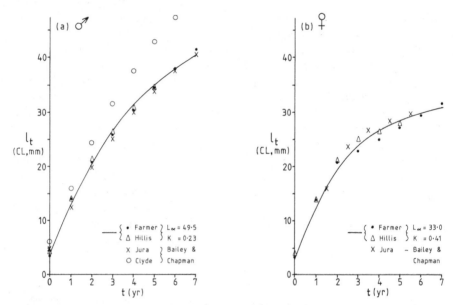

Figure 4. Irish Sea Norway lobster von Bertalanffy growth curves calculated from Farmer (12) and Hillis (13), with Bailey and Chapman's (14) Scottish data for comparison; (a) males, (b) females.

Farmer (12) and Hillis (13), using polymodal analysis of catch length distributions and aquarium studies, have estimated growth of Irish Sea Norway lobsters (Fig. 4). Recent studies in Scotland by Bailey and Chapman (14) have shown that growth rates differ from area to area and may be related to density and/or food availability. Irish Sea Norway lobster growth seems to be similar to the Scottish high density slower growth rate stock at Jura (Fig. 4). There are sexual differences in the growth rates and biometric relationships (ICES 15; Bennett 16) that necessitate assessments of males and females separately whenever possible. For convenience in assessment models the von Bertalanffy growth model has often been used, although it does not give the best fit for discontinuous crustacean growth (Fig. 4). Molting occurs mainly in the summer, but juveniles, which have a higher molt frequency than adults, molt in spring and autumn (Farmer 12). Using Farmer's (12) molt increment data, plus knowledge of molting period and reproductive cycles, an estimate of mean growth in terms of molt increments and timing of molts is given in Figure 5.

   Stomach content analyses of many species of fish in the Irish Sea have identified cod (*Gadus morhua*) as the main predator on Norway lobsters (Armstrong 17, 18; Brander 19; Fannon and Hillis 20; Boyd 21; Patterson 22; Symonds and Elson 23). Brander and Bennett (6) estimated an instantaneous coefficient of natural

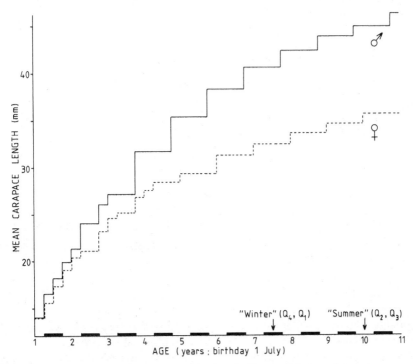

Figure 5. Average growth data for male and female Norway lobsters showing mean molt increments and timing of molts.

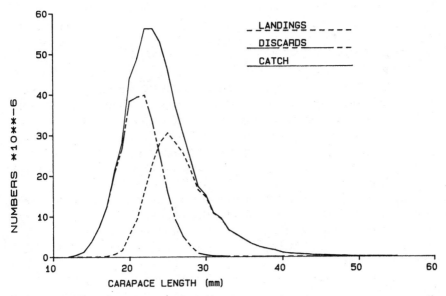

Figure 6. Length compositions, 1980–1984, of the Northern Irish Norway lobster catch, discards, and landings (R. P. Briggs, personal communication).

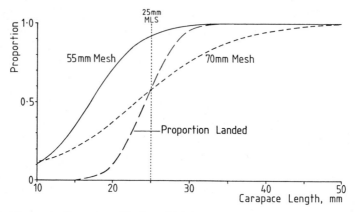

Figure 7. Mesh selection ogives, Northern Irish landing ogive, and Minimum Landing Size (MLS) for Irish Sea lobsters.

mortality of 0.3/yr, which constituted a predation mortality of 0.2/yr due to cod (at average cod biomass levels—see the Appendix) and a basal natural mortality of 0.1/yr. This is consistent with Morizur's (24) summary of Norway lobster natural mortality data from "quasi-unexploited stocks," which gave a range of values of 0.14–0.39/yr (0.28 ± 0.07, $\bar{x}$ ± sd).

Exploitation of Norway lobsters in the Irish Sea begins at about 12-mm CL; mean selection length of the 55-mm average mesh size in use is about 17 mm (Bennett 25). Length compositions of the catch, discards, and landings are available since 1980 from the Northern Ireland fishery (Fig. 6; R. P. Briggs, personal communication); some limited data are also available from Ireland and England (ICES 2). Considerable discarding of small Norway lobsters takes place to satisfy market requirements, rather than because of enforcement of minimum landing size. The Northern Irish landing ogive has a mean selection length of about 25-mm CL, with some retention of undersized lobsters (<25-mm CL), but also with discarding of legal-sized ones (Fig. 7). With up to half the catch being discarded, assessments of Norway lobsters must take account of discard mortality. Experiments to attempt to determine discard mortality suggest a range of 50–75% (ICES 2).

In the western Irish Sea Norway lobsters are mainly caught in a directed fishery, though there is a small by-catch from the larger-mesh cod fishery. Switching of intended target species is quite common. Fisheries on both sides of the Irish Sea are seasonal, with about three-quarters of the catch being taken in the second and third quarters of the year.

Some amplification of particular aspects of this biological and fisheries information will be found in the sections dealing with single-species and multispecies assessments (see Sections 3–5).

3. SINGLE-SPECIES NORWAY LOBSTER ASSESSMENTS

Although Norway lobsters have recently come under TAC regulation, the TAC levels are not based on the kind of short-term catch forecasting techniques used for other species. In fact, the main reason for setting a TAC was to provide a tonnage figure, which could be allocated to EEC member states when Spain and Portugal joined the Community. Short-term catch forecasting for Norway lobster is hindered by our inability to estimate age composition of the catch and by lack of catch-per-unit-effort data until very recently. There are also no estimates of recruiting year classes as yet.

The available data on growth rate and length composition of the catch can be used in length cohort analysis (LCA) (Jones 5) in order to estimate average population numbers and fishing mortality. This model can also be used to calculate yield per recruit and to estimate the effects of a mesh change.

Bennett (26) recently developed a quarterly yield-per-recruit model (QYPR) which incorporates the major biological and technical features applicable to Norway lobsters into a simple modification of the standard Beverton and Holt (27) yield-per-recruit (Y/R) model. Biological and technical factors may be incorporated on a seasonal basis, allowing the inclusion of information on the seasonal pattern of fishing, discontinuous growth during the molting season, and availability to predation and exploitation in relation to seasonal behavior patterns. The model also includes an exploitation pattern with a logistic mesh selection ogive, and fishermen selection using knife-edged minimum landing size or a logistic landing ogive, both with a discard mortality rate. Changes in the level of fishing mortality, mesh size and/or minimum landing size, and discarding practice can be included in predictions of yield per recruit.

The LCA and QYPR models have been applied to Norway lobsters in the Irish Sea to evaluate what effect an increase in mesh size would have on Norway lobster yields.

The inputs for a mesh assessment of Irish Sea Norway lobsters are detailed in Table 1, following a more general discussion in Section 2. Some of these inputs are common to both the models used. A range of values has been used for natural mortality and for discard mortality to examine the sensitivity of the models to these two less well estimated parameters. Both models are quite sensitive to natural mortality, but are less sensitive to discard mortality (Figs. 8 and 9). Likely values of natural mortality of 0.3/yr and discard mortality of 0.5 have been used to assess the impact of a mesh increase from the current 55-mm mesh in use to a proposed 70 mm. The current (1980–1984) mean fishing mortality of $F = 0.34$/yr was obtained from the LCA (ICES 2). At this level of fishing mortality the QYPR model shows little effect upon Y/R of an increase in mesh size (Fig. 10). A 70-mm mesh gives the highest Y/R, but the increase over the current 55-mm mesh is only 4%. At higher levels of fishing mortality there are larger gains in Y/R for mesh increases. The LCA shows a similar small gain in Y/R with a 70-mm mesh

TABLE 1 Main Inputs for Assessments of Irish Sea Norway Lobsters Using Quarterly Yield-per-Recruit (QYPR), Length Cohort Analysis (LCA), and Multispecies Models (MSM)[a]

	QYPR	LCA	MSM
Growth: von Bertalanffy growth curves			
Male: $K = 0.23$, $L_\infty = 49.5$	—	✔	—
Female: $K = 0.41$, $L_\infty = 33.0$	—	✔	—
Sexes combined	—	—	✔
Mean growth at age (Fig. 5)	✔	—	—
Length–weight relationship			
Male: $W = 0.000285*CL**2.936$	✔	✔	—
Female: $W = 0.000287*CL**2.923$	✔	✔	—
Sexes combined	—	—	✔
Natural mortality: range used with $M = 0.3$/yr,	✔	✔	—
Proportion modulated quarterly:			
Males + immature females:			
0.2, 0.3, 0.3, and 0.2	✔	—	—
Mature females: 0.10, 0.23, 0.37, and 0.10	✔	—	—
Basal rate = 0.1/yr + calculated predation mortality	—	—	✔
Predation: functional feeding relationship (Brander and Bennett 6)	—	—	✔
Fishing mortality: estimated by the model	—	✔	—
Range used with quarterly proportions of 0.13, 0.31, 0.43, and 0.13	✔	—	—
Terminal $F = 0.5$	—	✔	—
Range 0–1 used	—	—	✔
Availability			
Males + immature females fully available	✔	—	—
Mature females: 0.10, 0.55, 1.0, 0.10	✔	—	—
Mesh selection			
Mesh size (mm) 55 60 70 80			
Selection factor 0.31 0.32 0.33 0.34	✔	✔	✔
Selection range (50%) 7.1 9.1 13.2 17.3	✔	✔	✔
Length composition: 1980–1984 catch, discards, and landings, Northern Ireland (Fig. 6)	—	✔	—
Landing selection: estimated directly from length distributions	—	✔	—
Male: $L_{50} = 24.6$-mm CL SR = 4.1	✔	—	—
Female: $L_{50} = 24.8$-mm CL SR = 4.2	✔	—	—
Sexes combined	—	—	✔
Discard mortality			
Range 0–100%	✔	✔	—
100%	—	—	✔
Stock–recruit relationship: virtually constant	—	—	✔
Maturity: female 50% at 25-mm CL	✔	—	✔

[a]For cod inputs to MSM see Brander and Bennett (6).

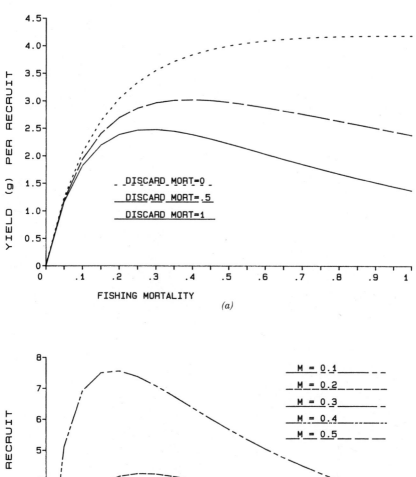

Figure 8. Yield-per-recruit curves for Irish Sea Norway lobsters showing the effects using the QYPR model of a range of (*a*) discard mortality and (*b*) natural mortality at the current mesh size of 55 mm.

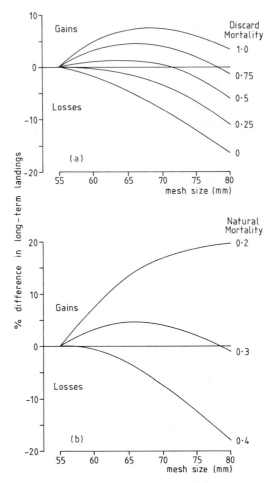

Figure 9. Effects of mesh changes on long-term Irish Sea Norway lobster landings with various levels of (*a*) discard mortality and (*b*) natural mortality using a length cohort analysis (LCA).

(Fig. 11). Short-term losses can be estimated from the LCA, and these indicate that with an increase to 70-mm mesh the short-term loss would be 18% in the first year and 11% in the second, reducing to 3% by the fifth year (Fig. 11).

Thus although there are still some doubts over the inputs to either model for assessing the effects of a mesh increase on Irish Sea Norway lobsters, the best available estimates predict small long-term gains, though some short-term losses could be expected. Any increase in mesh size would result in stock biomass increases (Fig. 11). These results are discussed and put into the management context in Section 6.

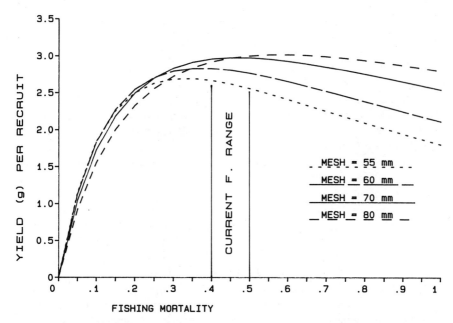

Figure 10. Effects of mesh changes on *Y/R* of Irish Sea Norway lobsters estimated using the QYPR model with a discard mortality of 0.75 and natural mortality of 0.3.

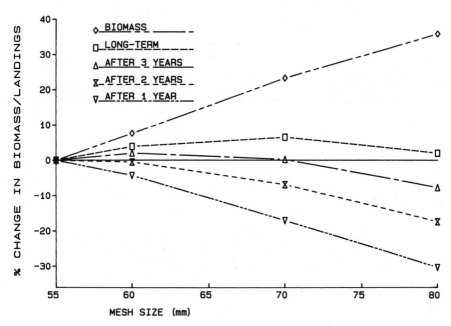

Figure 11. Short-term and long-term effects of mesh changes on Irish Sea Norway lobster landings and biomass using the LCA model with a discard mortality of 0.75 and natural mortality of 0.3.

4. ASSESSING THE EFFECT OF THE NORWAY LOBSTER FISHERY ON WHITEFISH

Considerable quantities of whitefish are caught as a by-catch while fishing for Norway lobsters using 60-mm mesh size and, because of the small mesh used, significant quantities of undersized fish are caught and discarded. The by-catch regulations allow up to 60% of protected whitefish to be landed. Whiting (*Merlangius merlangus*) is the main by-catch species, but cod, dabs (*Limanda limanda*), plaice (*Pleuronectes platessa*), and hake (*Merluccius merluccius*) are also caught and discarded in substantial quantities (Briggs 28). It is estimated that 4225 t of whiting were discarded in 1984 from the Irish Sea Norway lobster fishery (ICES 2); this represents 36% of the landed weight of whiting. Nearly all the discards are undersized (Fig. 12, MLS-27 cm), and few of them survive (Briggs 28). Any reduction in discarding would increase the yield of whiting. ICES (2) calculated that the *Y/R* of whiting would increase by 55% at the present level of fishing mortality, and the spawning biomass by 57%, if the Norway lobster mesh was increased to 70 mm, because virtually all whiting discarding would be eliminated.

Reductions in the by-catch of small fish can also be brought about by modifying

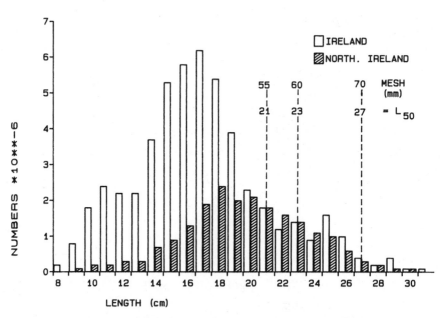

Figure 12. Length compositions of 1984 whiting discards in the Irish Sea Norway lobster fishery, with mesh size and mean selection length (L_{50}) indicated. (Selection factor = 3.8, MLS = 27 cm.)

the trawl. Recent direct observations of fish behavior and fishing gear, using divers and towed underwater vehicles with still and television cameras, show how such modifications can be tested (Main and Sangster 29). Several workers have demonstrated that fish and Norway lobsters can be separated by various means such as a horizontal panel of netting (Main and Sangster 29; Ashcroft 30, 31; Charuau 32), twin cod ends (Symonds and Simpson 33), and diagonal baffles (Hillis 34, 35). Bennett (36) discusses the degree of separation of whiting and Norway lobsters that can be achieved. This would then allow the use of different mesh sizes in the upper and lower parts of the trawl, permitting a choice of mesh that optimized Norway lobster catches while minimizing catches of undersized finfish, particularly whiting.

Although only whiting has been modeled in detail when examining the effect of the Norway lobster fishery on fish, it is clear that other species such as cod would also benefit to a lesser extent. There are, however, two more aspects that need to be looked at when considering the management options for the mixed fishery as a whole. The first is the relative value of the different components, because Norway lobster is far more valuable per unit weight than most fish. The second is the effect of biological interactions between fish and Norway lobsters and in particular the predation of fish. Both of these aspects are included in the model of cod and Norway lobster that is described in the next section.

5. MODELING TECHNICAL AND BIOLOGICAL INTERACTIONS BETWEEN NORWAY LOBSTER AND COD

The use of multispecies rather than single-species models to provide guidance on fisheries management is essential when the dynamic behavior of the group of species being managed is significantly affected by technical and biological interactions. Multispecies models are used for two purposes: (1) to determine whether the inclusion of technical and biological interactions is likely to cause significant changes in dynamic behavior; and (2) to provide a more credible forecast of stock dynamics than the single-species models. The first of these is fairly easy and has always been recognized as a potential shortcoming of single-species models (e.g., Beverton and Holt 27). The second is very difficult for many reasons to do with model specification, parameter estimation, and complex behavior of model systems.

The dynamics of Norway lobster and its fishery in the Irish Sea might be expected *a priori* to be affected by changes in its major predator, the cod. In the interests of parsimony, and because cod account for as much as 88% of the total predation on Norway lobster (Symonds and Elson 23), a model has been constructed that includes only those two species. Details of the construction, inputs

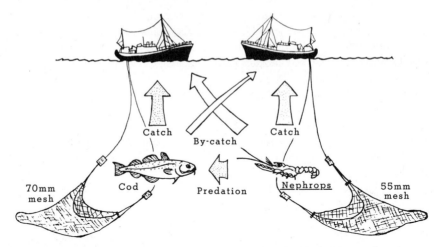

Figure 13. Schematic representation of the biological and technical interactions between Norway lobster and cod.

(Table 1), sensitivity and output of this model are given in Brander and Bennett (6), and a representation of the main features is shown in Figure 13. It is a steady-state yield model developed by Shepherd (37) and incorporates growth, mortality, maturity, stock–recruit relationship, selectivity, by-catch levels, discarding, predation, and unit value.

Because the model is derived directly from single-species *Y/R* models and uses many of the same inputs (Table 1) it shows the consequences of incorporating technical and biological interactions very clearly. Figure 14 shows the yield of Norway lobster as a function of fishing mortality in the cod and Norway lobster fisheries. As fishing mortality on cod increases, the biomass of cod is reduced and so the predation mortality on Norway lobster also declines. This is equivalent to the effect of a reduction in natural mortality on the curves in Figure 8*b*. The relationship between cod biomass and predation mortality is discussed in the Appendix. The multispecies model predicts high yields of Norway lobster if the fishing mortality on cod is kept high. The management advice for Irish Sea cod now takes this interaction into account and the TAC for cod has been set high, partly in order to keep the cod biomass down.

The results of this model are thus already influencing the management strategy, but there is little doubt that they have been accepted for reasons of political expediency as much as because they are scientifically credible. The results are consistent with our knowledge of feeding and average population levels, but the model only looks at the steady state, and if the predation effect is as strong as we have estimated then there should be observable short-term effects as well. We know that cod population biomass has fluctuated by a factor of 2 in recent years

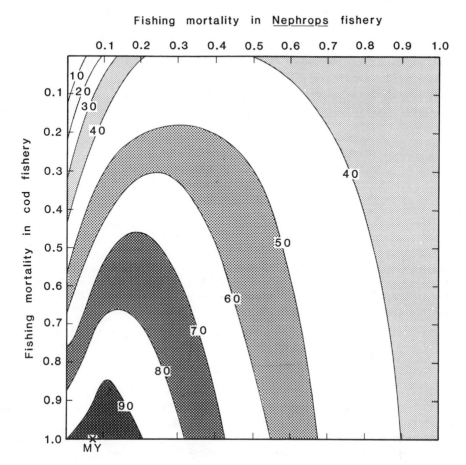

Figure 14. Contour plot of Norway lobster yield as a function of fishing mortality in each of the cod and Norway lobster fisheries. The contours are percentiles of maximum yield (MY) of 4k tonnes.

and this should result in linked fluctuations in the Norway lobster population. If it could be shown that these short-term fluctuations in population levels are linked in a way that is also consistent with the relationships incorporated in the model, then this would greatly enhance its credibility.

Unfortunately the time series of data on Norway lobster population biomass is short and not very reliable, and efforts are being made to improve this. A preliminary look at English and Welsh catch-per-unit-effort data, which are a measure of population biomass, suggests that the Norway lobster biomass responds inversely to changes in cod biomass, with a lag of 1 yr (Fig. 15). This result is encouraging, but not definitive.

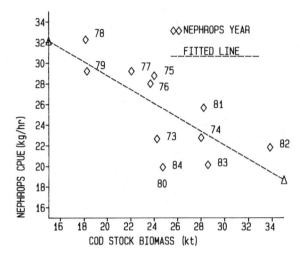

Figure 15. Relationship between Norway lobster catch per unit effort (kg/hr) and cod stock biomass. The 1980 observation was omitted from the fitted regression line, $y = 42.23 - 0.672$ x, $r = -0.736$ ($p < .001$) because abnormal market factors prevailed in 1980. The cod stock biomass is lagged by 1 yr.

6. DISCUSSION

The Y/R and mesh assessments for Norway lobster (Section 3) show the consequences of varying fishing mortality and mesh size, but the results are sensitive to the values of natural mortality used. Feeding studies and the multispecies model (Section 5) show that predation mortality by cod is a major component of natural mortality on Norway lobster and changes in cod biomass therefore have a great effect on the Norway lobster stock. The management strategy for Norway lobster cannot be considered in isolation from the management strategy for cod and, by estimating the joint value of the two species in the two fisheries, the multispecies model provides guidance on their joint management. Broadly the results suggest that the joint value would increase if fishing mortality was higher in the cod-directed fishery and lower in the Norway-lobster-directed fishery. In fact, the highest value is obtained by fishing cod to extinction and catching Norway lobsters in the large-mesh fishery, but this result lies outside the domain of feasibility or validity of the model (Brander and Bennett 6).

The cod stock is probably not affected very much by the present small-mesh Norway lobster fishery, but the whiting stock certainly is affected (Section 4). Therefore the management strategy for Norway lobster has to take account of the consequences on the whiting stock. In particular, the benefits in terms of yield per recruit of Norway lobster from an increase in mesh size are small and uncertain, but the benefits to whiting are considerable. One problem with trying to balance the potential benefits for several interacting species and fisheries is that different groups of fishermen or even different countries may benefit from different strategies. It is fortunate that most fishermen participating in the Irish Sea mixed fisheries are able to switch between species and would therefore be able to benefit from any joint optimum value.

7. CONCLUSIONS

1. In spite of shortcomings in the time series of data on Norway lobster in the Irish Sea and the absence of age information it is possible to provide guidance on management strategy using single-species Y/R and mesh assessment models.

2. However, the dynamics of the Norway lobster population are greatly affected by fluctuations in its major predator, the cod, and a multispecies model that includes technical and biological interactions between them has therefore been developed. This model provides a consistent interpretation of the average state of the cod and Norway lobster populations, and recent results indicate that it may also be consistent with short-term fluctuations in the two populations. Further tests in this direction would greatly enhance its credibility.

3. The results from the multispecies model and from assessments of by-catches in the small-mesh Norway lobster fishery show that management strategy for Norway lobster cannot be considered in isolation. These results are already having an influence on management strategy for the mixed fisheries of the Irish Sea.

REFERENCES

1. K. M. Brander and R. R. Dickson, An investigation of the low level of fish production in the Irish Sea. *Rapp. P.-V. Reun., Cons. Int. Explor. Mer* **183**:234 (1984).

2. International Council for the Exploration of the Sea (ICES), Report of the Irish Sea and Bristol Channel Working Group. *ICES, C. M.* 1985/Assess:**10**:175 (1985) (mimeo).

3. G. E. Fogg, B. Egan, G. D. Floodgate, S. Hoy, D. A. Jones, J. Y. Kassab, K. Lochte, E. I. S. Rees, S. Scrope-Howe, C. M. Turley, and C. J. Whitaker, Biological studies in the vicinity of a shallow-sea tidal mixing front. *Philos. Trans. R. Soc. London, Ser. B* **310**:407 (1985).

4. K. M. Brander, Disappearance of common skate *Raia batis* from Irish Sea. *Nature (London)* **290**:48 (1981).

5. R. Jones, Assessing the long-term effects of changes in fishing effort and mesh size from length composition data. *ICES, C. M.* 1974/F:**33**:7 (1974) (mimeo).

6. K. M. Brander and D. B. Bennett, Interactions between Norway lobster (*Nephrops norvegicus* (L.)) and cod (*Gadus morhua* L.) and their fisheries in the Irish Sea. *Can. Spec. Publ. Fish. Aquat. Sci.*, **92**:269 (1986).

7. J. H. Nichols, D. B. Bennett, D. J. Symonds, and R. Grainger, Estimation of the stock

size of adult *Nephrops* from larvae surveys in the western Irish Sea in 1982. *ICES, C. M.* 1983/K:**6**:7 (1983) (mimeo).

8. C. J. Chapman, Ecology of juvenile and adult *Nephrops*. In J. S. Cobb and B. F. Phillips, Eds., *The Biology and Management of Lobsters*, Vol. II. Academic Press, New York, 1980, p. 143.

9. International Council for the Exploration of the Seas (ICES), Report of the *Nephrops* Working Group. *ICES, C. M.* 1984/K:**4**:82 (1984) (mimeo).

10. A. S. D. Farmer, Reproduction in *Nephrops norvegicus* (Decapoda: Nephropidae). *J. Zool.* **174**:161 (1974).

11. H. J. Thomas, The spawning and fecundity of the Norway lobster (*Nephrops norvegicus* L.) around the Scottish coast. *J. Cons. Perm. Int. Explor. Mer* **29**:221 (1964).

12. A. S. D. Farmer, Age and growth in *Nephrops norvegicus* (Decapoda: Nephropidae). *Mar. Biol. (Berlin)* **23**:315 (1973).

13. J. P. Hillis, Growth studies on the prawn, *Nephrops norvegicus. Rapp. P.-V. Reun., Cons. Int. Explor. Mer* **175**:170 (1979).

14. N. Bailey and C. J. Chapman, A comparison of density, length composition and growth of two *Nephrops* populations off the west coast of Scotland. *ICES C. M.* 1983/K:**42**:10 (1983) (mimeo).

15. International Council for the Exploration of the Sea (ICES), Report of the *Nephrops* Working Group. *ICES, C. M.* 1982/K:**3**:p. 42 (1982) (mimeo).

16. D. B. Bennett, Irish Sea *Nephrops* biometrics, with particular reference to tails. *ICES, C. M.* 1983/K:**10**:6 (1983) (mimeo).

17. M. J. Armstrong, The feeding ecology of a demersal fish community over a muddy substrate off the west coast of the Isle of Man. Ph.D. Thesis, University of Liverpool, Isle of Man, 1979, p. 182.

18. M. J. Armstrong, The predator–prey relationships of Irish Sea poor cod (*Trisopterus minutus* L.), pouting (*T. luscus*), and cod (*Gadus morhua* L.). *J. Cons. Int. Explor. Mer* **40**:135 (1982).

19. K. M. Brander, On the application of models incorporating predation in the Irish Sea. *ICES, C. M.* 1981/G:**29**:11 (1981) (mimeo).

20. E. Fannon and J. P. Hillis, Studies on cod prey with special attention to *Nephrops norvegicus. ICES, C. M.* 1981/G:**51**:18 (1981) (mimeo).

21. R. Boyd, The feeding of cod in the northwest Irish Sea. *ICES, C. M.* 1983/G:**6**:5 (1983) (mimeo).

22. K. R. Patterson, Some observations on the ecology of the fishes of a muddy sand ground in the Irish Sea. Ph.D. Thesis, University of Liverpool, Isle of Man, 1983, p. 180.

23. D. J. Symonds and J. M. Elson, The food of selected fish species on *Nephrops* grounds in the western Irish Sea. *ICES, C. M.* 1983/K:**8**:14 (1983) (mimeo).

24. Y. Morizur, Estimation de la mortalité pour quelques stocks de Langoustine, *Nephrops norvegicus. ICES, C. M.* 1982/K:**10**:19 (1982) (mimeo).

25. D. B. Bennett, A review of Irish Sea *Nephrops* mesh selection. *ICES, C. M.* 1984/K:**5**:6 (1984) (mimeo).

26. D. B. Bennett, Use of a quarterly yield-per-recruit model for a *Nephrops* assessment. *ICES, C. M.* 1985/K:**37**:10 (1985) (mimeo).

27. R. J. H. Beverton and S. J. Holt, On the dynamics of exploited fish populations. *Fish. Invest. (London) (Ser. 2)* **19:**533 (1957).

28. R. P. Briggs, The discarded by-catch of the Northern Ireland *Nephrops* fishery. *ICES, C. M.* 1983/K:**23:**9 (1983) (mimeo).

29. J. Main and G. I. Sangster, A study of a multi-level bottom trawl for species separation using direct observation techniques, *Dept. Agric. Fish. Scott. Fish. Res., Rep.* **26:**17 (1982).

30. B. Ashcroft, *Trials of Prawn/fish Separator Trawl and Survey of New Prawn Grounds in South Central North Sea ICEA Area IVb*, Tech. Rep. 230. Sea Fish Industry Authority, Hull, U.K., 1983, p. 20.

31. B. Ashcroft, *Trials of Fish/prawn Separator Trawls on Fishing Grounds in the Irish Sea ICES Area VIIa*, Tech. Rep. 253. Sea Fish Industry Authority, Hull, U.K., 1984, p. 23.

32. A. Charuau, Expérimentation d'un chalut séparant la Langoustine (*Nephrops norvegicus*) du poisson. *ICES, C. M.* 1985/B:**38:**6 (1985) (mimeo).

33. D. J. Symonds and A. C. Simpson, Preliminary report on a specially designed *Nephrops* trawl for releasing undersized roundfish. *ICES, C. M.* 1971/B:**6:**4 (1971) (mimeo).

34. J. P. Hillis, Experiments with a double cod-end *Nephrops* trawl. *ICES, C. M.* 1983/B:**29:**9 (1983) (mimeo).

35. J. P. Hillis, Further experiments with a double cod-end *Nephrops* trawl. *ICES, C. M.* 1984/K:**36:**9 (1984) (mimeo).

36. D. B. Bennett, The problems of small mesh *Nephrops* trawls in the Irish Sea—are separator trawls the answer? *ICES, C. M.* 1985/K:**38:**9 (1985) (mimeo).

37. J. G. Shepherd, A promising method for the assessment of multispecies fisheries. *ICES, C. M.* 1984/G:**4:**11 (1984) (mimeo).

APPENDIX: THE RELATIONSHIP BETWEEN COD BIOMASS AND PREDATION MORTALITY ON NORWAY LOBSTER

In the multispecies model an instantaneous coefficient of predation mortality on each age of Norway lobster is calculated from the product of cod biomass at age, a prey size preference function, and a coefficient of predation mortality per gram of predator. The average cod biomass at age is estimated by virtual population analysis (ICES 2). Details of the prey size preference function and the method of estimating predation mortality per gram are given by Brander and Bennett (6).

The actual levels of predation mortality on each age of Norway lobster are shown as a function of cod biomass in Figure 16. Because the prey size and age composition are different for each age group of cod, the exact shape of this figure depends on the cod age composition, which depends in turn on the relative mortality in the two fisheries. The prey size preference function is set so that there is no predation on 0- and 1-group Norway lobsters, because cod have not been observed to feed on them. In addition to the predation mortality shown, a "basal natural mortality" component of 0.1 is added to give the total natural mortality. The natural

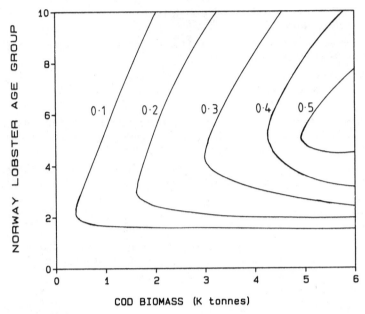

Figure 16. Contours of predation mortality by cod on Norway lobsters related to cod biomass and age group of Norway lobsters.

mortality (0.3) used in the QYPR and LCA models may be compared with the total natural mortality at age generated in the MSM at current average cod biomass, which is:

Age	2	3	4	5	6	7	8	9	10
M	0.27	0.28	0.26	0.24	0.22	0.20	0.18	0.17	0.16

Because these are obtained by trial-and-error adjustment of the predation parameter rather than by a formal estimation procedure, they cannot be regarded as "better" estimates of natural mortality.

Nevertheless, they illustrate that if predation mortality is calculated in this way then natural mortality on Norway lobster is not constant with age and depends on the level of cod biomass. This in turn means that answers to questions such as the optimal mesh size for Norway lobster depend on cod biomass.

9 MANAGEMENT OF A CYCLIC RESOURCE: THE DUNGENESS CRAB FISHERIES OF THE PACIFIC COAST OF NORTH AMERICA

Richard D. Methot*
Southwest Fisheries Center
La Jolla, California

Present address: Northwest & Alaska Fisheries Center, Seattle, Washington

1. INTRODUCTION

Fisheries for the Dungeness crab, *Cancer magister*, range from San Francisco, California to Kodiak Island, Alaska, and date from the mid-1800s. Average annual landings during the period 1978–1983 were about 18,000 t (metric tons) with a landed value of more than $50,000,000. Historically, these fisheries have not been stable. Landings at San Francisco collapsed in the early 1900s, recovered, then collapsed again in 1960. Fisheries in northern California through Washington have been highly cyclic, with a period of 10 yr (Fig. 1). Alaskan fisheries continue to develop; a stock in the lower Cook Inlet region of Alaska was first harvested in 1973 (1). Each of these fisheries is regulated primarily by sex, size, and season: only males may be retained (except in British Columbia), a minimum size limit is applied to these males, and the molting season for males is closed to commercial fishing. These management measures preserve the reproductive potential of the population, but do not address its cyclic nature. The questions facing crab biologists and managers are: is more precise management possible and what would be the gains?

2. DEVELOPMENT OF THE FISHERIES

The fishery in the San Francisco area began in the 1860s with fishermen using hoop nets—small, baited, mesh-covered rings—to capture crabs (2). Landings increased during the 1880s and attained a level of 2000 t by 1900. This fishery was perceived to be depleting the resource because catch rates were declining and the fishing area was expanding. Regulation by size, sex, and season was enacted during 1897–1905, but landings continued down to a low of 100 t in 1915. Landings in the San Francisco area recovered in the late 1920s and gradually increased to a peak of 4000 t in the 1956–1957 fishing season. The fishery collapsed after the 1960–1961 season and has averaged only 242 t per season since 1974.

Fisheries in northern California (3), Oregon (4), and Washington (5) have similar histories. Although landings in Oregon date from 1889, the major expansion of these three fisheries occurred during the 1930s because of development of crab traps (pots) and removal of marketing restrictions (2). Today, the cylindrical, mesh-covered, baited traps typically are 1 m in diameter and have two entrance tunnels and two escape ports. In California, typical crab vessels are converted trollers (salmon and albacore) or trawlers (bottom fish and shrimp) and are 11–12 m in length. Presently, hoop nets are only used substantially by recreational fishermen.

Landings in northern California, Oregon, and Washington have had coherent cycles with a period of about 10 yr. The cycle's greatest amplitude is in northern California, where the harvest was 145 t during the low season of 1973–1974 and 11,400 t in the record 1976–1977 season (3). These fisheries are heavily exploited; about 70% of legal-sized male crabs have been harvested annually in northern California (6, 7), and in recent years, 80–90% of this harvest occurred during the first month of the season (3).

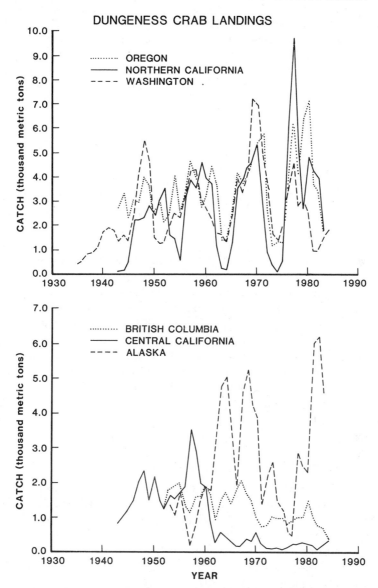

Figure 1. Time series of annual Dungeness crab harvest in each of six geographic regions. In California, Oregon, and Washington the values are for seasonal harvest (December through late summer) and are recorded in the second year of the season.

Crab fisheries in British Columbia and Alaska are more localized because of the complex coastline north of Washington. The British Columbia catch is divided among eight fisheries that have fluctuated rather independently of each other (8). None of these exhibit the 10-yr cycle found to the south. In southeastern Alaska a fishery began before 1915 and regulations were established by 1921 (9), but most Alaskan Dungeness crab fisheries did not expand until the 1960s. Historically, effort in Alaskan Dungeness crab fisheries has increased during periods of poor catch in northern California–Washington. A large increase occurred during the early 1980s as Dungeness crab catch in Washington and king crab catch in Alaska declined (9). If crab abundance has been cyclic in Alaska, the pattern would have been masked by fluctuations in the relatively low level of effort.

3. LIFE HISTORY

3.1. Reproduction

Dungeness crabs typically are found in shallow water. The fishery occurs almost entirely at depths less than 100 m and often within the 30-m isobath. Tagging studies indicate no barrier to crab movement along the northern California to Washington coast, but most do not move far; 19–46% were recovered within 2 km of the release site (10, 11). Both sexes move inshore prior to the spring mating season.

Mating in Dungeness crabs occurs between hard-shelled males and recently molted, soft-shell females. Fertilization occurs as eggs are extruded in about October. Females store sperm and are capable of fertilizing eggs without molting and mating the previous spring (12). Females carry an egg mass containing up to 2 million eggs under their abdomen, but the largest females may have reduced fecundity (12). During brooding, eggs are vulnerable to predation by the nemertean worm, *Carcinonemertes errans* (13). Brooding duration is temperature dependent (14), and hatching occurs in December–January in California and later to the north.

3.2. Larvae and Juveniles

Larvae are planktonic and develop through five zoeal and one megalopal stage before settling (Fig. 2). Zoeal stages drift offshore, then megalopae are found close to shore (15, 16). Zoeae are distributed over the upper tens of meters of the water column, but megalopae are highly surface associated (16) and may be found clinging to floating objects, including the hydroid, *Velella* (17). Salmon prey on the planktonic megalopae (18).

Settlement and metamorphosis to the first crab instar occurs during late spring in central California and through late summer in British Columbia. Stomach contents of predatory fish indicate that most settlement occurs within about 10 km of shore (18). Juvenile crabs are very abundant in bays and estuaries and may be a major contributor to ocean fisheries (19). Although juvenile abundance has been estimated in San Francisco Bay (20), Humboldt Bay (21), and Grays Harbor (22), companion studies on the open coast (20, 23) are insufficient to estimate the relative contribution of bay-reared juveniles to the adult population on the open coast. Bay-reared juveniles benefit from faster growth and separation from most adult Dungeness crabs, which are known to prey on juvenile crabs. Juvenile crabs move out

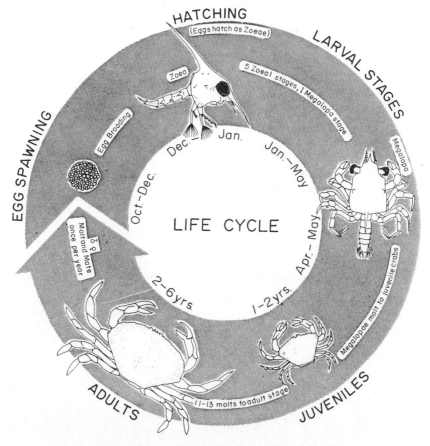

Figure 2. Summary of crab life history in California (from Wild and Tasto (36)).

of San Francisco Bay, California after about 1 yr (24) and out of Grays Harbor, Washington after 2 yr (22).

3.3. Growth

Overall growth of crustaceans depends on growth per molt and molt frequency. Studies of Dungeness crab growth indicate a difference in juvenile growth between bay-reared and ocean-reared individuals, latitudinal variation in growth, and declining molt frequency of larger adults. These patterns introduce uncertainty in growth estimates, which affects calculation of potential yield per recruit and the time lag for density-dependent population feedback. Size and growth is measured as carapace width just anterior to, and excluding, the tenth anterolateral spines.

Growth per molt has been studied by tagging and laboratory studies (9, 12, 24–26). Newly settled crabs are 7.5 mm and their first molt is approximately 3 mm

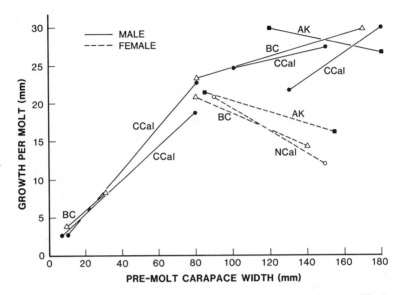

Figure 3. Growth per molt in Dungeness crabs. Growth of small crabs does not differ between the sexes. Inflection in growth rate occurs at size of sexual maturity (about 105 mm). Data sources: central California (24, 25); northern California (12); British Columbia (26); and Alaska, (9).

(Fig. 3). Growth per molt is similar for immature males and females, and increases to 22 mm at the eleventh molt. Both sexes mature at about 105 mm (25, 27), which they attain 2.5 yr posthatch in the San Francisco area and 3.5 yr in Washington. However, males that molt to 105 mm in the spring may not reproduce that year, because mating occurs between hard-shell males and smaller, recently molted females. Growth per molt is approximately 28 mm for adult males and 15 mm for adult females. There is no apparent latitudinal pattern in growth per molt.

Overall growth of juvenile crabs has been described by following progression of size modes (20–30). A summary of these studies (Fig. 4) indicates faster growth in bays compared to nearby open coast, and a north–south gradient in growth. Reduced size of young juveniles on the coast of Washington and in British Columbia is partly due to later settlement rather than slow growth. The difference in growth between bay and ocean juveniles affects assignment of age classes for older crabs in the ocean because some of these ocean crabs were juveniles in bays. The importance of this growth differential depends on the relative abundance of bay-reared juveniles in the ocean, which is not sufficiently known.

The latitudinal trend in juvenile growth (Fig. 4) suggests that most central California males are legal size (159 mm) at age 3 posthatch, in northern California at age 4, and at least age 4 in Washington and British Columbia. Published estimates of age at recruitment have lesser latitudinal pattern: Poole (25) estimated 4 yr in central California and Butler (26) estimated 3–4 yr in British Columbia. In northern California, Warner (29) suggested that 28–60% of 3-yr olds were

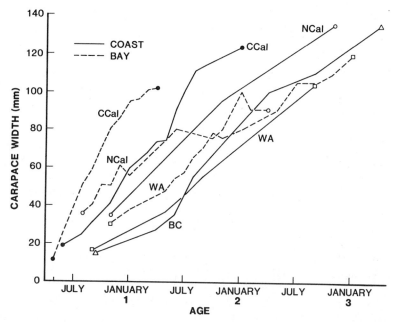

Figure 4. Growth of juvenile Dungeness crabs determined from seasonal progressions of size modes. Growth is faster to the south and in bays. Data sources: central California (20, 24, 25); northern California (21, 28, 29); Washington (22, 23, 27); and British Columbia (26, 30).

recruited, and Botsford (28) concluded a smaller contribution at age 3 and approximately equal contributions at ages 4 and 5.

The variability in the above estimates is largely due to lack of knowledge of annual molt probability for adult crabs. Tag–recapture data and incidence of carapace fouling in northern California (12) and in southeastern Alaska (9) indicated that annual molt probability for 110–139-mm females was 75–80%, then declined to effectively zero for crabs larger than 155 mm. Molt probability of females at 140–149 mm was only slightly less in Alaska than in California. Samples from the size-specific fishery are biased in favor of individuals that molt into the legal-size range (31), so true molt probability for 135–145 mm females may be only half the observed probability (12).

Tag–recapture data in southeastern Alaska (9) indicated that annual size-specific molt probability of males declined from 100% at 130–139 mm to 10% at 160–169 mm. Ten percent of crabs greater than 160 mm did not molt in two consecutive years. Similar data are not available from other areas, but carapace fouling and mating marks on males' legs indicated that 85–90% molted annually in central California (25) and 75–100% molted annually in northern California (32). Unfortunately, these data have not been presented on a size-specific basis. Few legal-sized male crabs escape the intense California fishery, so carapace condition data primarily represents those crabs that molt from about 140 mm into the >159-mm size range available to the fishery. Therefore, molt frequency of California male crabs larger than 170 mm is unknown.

3.4. Mortality

Jow (33) estimated male natural mortality from tag returns. During the 1962–1963 fishing season, the fishery harvested 84%, 15% died of natural causes, and only 1% survived. It seems likely that natural mortality is somewhat greater than 15% because of uncertainty in the report rate of recaptured tags and imprecision caused by the high fishing mortality. Botsford and Wickham (34) and McKelvey et al. (35) used an instantaneous natural mortality rate of about 0.2/yr in their models of crab population dynamics. Recapture of tagged females (12) indicated 50% survival for 125–140 mm females, and only 8–14% survival for females larger than 155 mm. Females are not retained by the fishery and presumably have low fishery-caused mortality.

Exploitation by the fishery has been great. Gotshall (6) examined decline in catch per trap within fishing seasons and concluded that 63–87% of legal-sized males were harvested annually during the 1966–1967 to 1971–1972 seasons. McKelvey et al. (35) related these exploitation rates to number of traps fished and concluded that the increasing trend in northern California catch during 1943–1978 was consistent with increasing effort (number of traps). Methot and Botsford (7) estimated preseason abundance from the decline in catch per effort relative to cumulative catch within each fishing season in northern and central California. In central California, 92% of the available, legal-sized males were harvested annually during 1951–1956. In northern California, annual exploitation rate has varied with the 10-yr cycle in catch (Fig. 5). Notably, catch per effort was at a high, saturated level throughout the second high catch year of each cycle (1957, 1966, 1977). Preseason abundance could not be estimated in these years, but it must have been high enough to allow substantial survival into the following fishing season. Previous studies could not have detected these anomalies because Gotshall (6) started his study the year after the 1965–1966 abundance peak, and McKelvey et al. (35) assumed constant catchability (no trap saturation).

4. CURRENT REGULATIONS

Dungeness crab fisheries are regulated primarily by sex, size, and season. These conservation measures date from 1897–1905 in central California and were intended to protect the reproductive potential of the stock and to reduce wastage. In all areas except British Columbia, only male Dungeness crabs may be landed by commercial fishermen. Some states allow sport fishermen to take females. The size limit is 159-mm (6.25-in.) carapace width in California, Oregon, and Washington, and 165-mm (6.5-in.) in Alaska. In British Columbia, the limit is 159 mm and includes the spines.

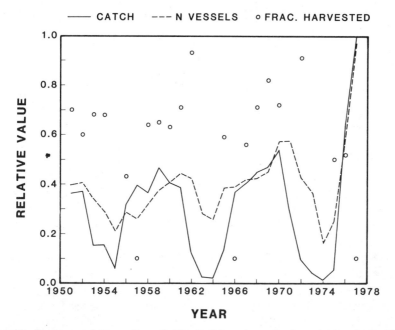

Figure 5. Response of the northern California fishery to cycles in crab abundance (38). *N* vessels is the maximum number of vessels landing crab in any month. Catch and *N* vessels are scaled to their maximum value to illustrate cycles only. Fraction harvested is the ratio of catch to initial abundance in years with significant declines in catch per effort. The years 1957, 1966, and 1977 are set at low values because catch per effort was at a high saturated level.

Escape ports are required in all crab traps to permit free passage of sublegal-sized male crabs and females, which usually are smaller than the male size limit. The minimum diameter of these ports is 108 mm (4.25 in.) in California, Oregon, and Washington; 100 mm in British Columbia; and 110 mm (4.375 in.) in Alaska. Recent studies in central California indicated that 110-mm (4.375-in.) escape ports were most effective at releasing crabs smaller than 159 mm (36). Other regulations regarding traps are limits on number of traps per fisherman in some areas, maximum size of the trap entrance in Alaska to discourage illegal fishing of king crabs, destruct devices to prevent continued fishing by lost traps, and identification on trap buoys.

Conservation of male reproductive stock seems to be adequately achieved with the current size limit. In California, males mature about 2.5 yr posthatch, so they typically have two breeding seasons before recruiting at age 4 yr posthatch. Most males larger than the size limit are harvested each season, so few males larger than the size limit are in the breeding stock. Therefore, this heavily exploited fishery regulated by size limit has potential for genetic selection for slow growing traits.

An accurate assessment of yield per recruit cannot be made because of uncer-

TABLE 1. A Hypothetical Growth Model for Dungeness Crabs in Northern California

Age (yr)	2	3	4	5	6
Molts/yr	2	1	0.5	0.1	0.1
Mean size (mm)	93	143	170	170 (50%)	170 (25%)
				198 (50%)	198 (45%)
					226 (5%)
Mean weight (g)	149	489	788	994	1123
Instantaneous weight growth	1.19	0.48	0.23	0.12	

tainties in adult growth (molt frequency) and natural mortality rate. It is instructive, however, to consider the magnitude of relevant quantities. A hypothetical model of growth (Table 1) indicates that if instantaneous natural mortality is 0.2–0.5/yr, then the current size limit of 159 mm seems appropriate because it causes recruitment at approximately the age at which instantaneous weight growth is equal to natural mortality.

Crab fishing seasons were designed primarily to reduce fishing on recently molted male crabs. These crabs have poor meat yield if landed and probably poor survival if released after capture. In California, Oregon, and Washington, molting occurs during late summer and autumn, with a south–north gradient. The central California fishery opens the second Tuesday of November. Because of the free movement of fishermen between northern California, Oregon, and Washington, management has been simplified by having a uniform December 1 opening date. However, Washington prohibits landing of soft-shell Dungeness crabs, so they monitor shell condition and may delay opening until December 15. The season closing dates range from June 30 in central Caliifornia to September 30 in Washington, but each state may prolong fishing until significant molting begins. In British Columbia and Alaska, crab fishing seasons vary locally and are less tightly linked to seasonal molting. Typical seasons are May to February.

Area closures typically were implemented to reserve crabs in bays and estuaries for sport fishermen and to reduce conflicts between crab fishermen and other commercial fishermen. In California, bottom trawlers are excluded from 3 miles of shore; this reduces crab mortality and trawl damage to and from crab traps. In Oregon, there is no nearshore restriction, but Demory (4) noted that trawlers fish offshore during crab season. Wild and Tasto (36) suggested exclusion of trawlers from the entire Gulf of the Farallones (San Francisco area) during the summer crab molting season to reduce mortality on vulnerable soft-shell crabs.

5. POPULATION DYNAMICS

Studies of population dynamics have focused on two issues; cycles in the northern California fishery, and collapse of the central California fishery. Biotic and abiotic factors have been investigated, and there is no consensus on the cause of either phenomenon (32, 37). Some hypotheses have been discounted. Botsford et al. (38) argued that cycles were not due to the lagged response of a major predator, humans. Botsford et al. (39) suggested that predation on megalopae by salmon also was not responsible. However, Thomas (40) suggested that salmon predation may be inhibiting recovery of the central California stock. Botsford (37) concluded that the egg predator worm, *Carcinonemertes errans* (13), was not the sole cause of the cycles. The worm, however, is still common in central California and may be inhibiting recovery of that stock, although this possibility was discounted by Wild and Tasto (36). Also, pollution in San Francisco Bay probably did not cause the decline of the central California crab stock (36).

5.1. Cyclic Abundance

The cyclic nature of crab stocks along the northern California to Washington coasts is well established (41). The catch cycles are due to changes in abundance and not simply changes in fishing effort (38). In each region, autocorrelation of catch reaches a minimum of about -0.5 at a lag of 4–5 yr (significant at 5% level with 30-yr time series), then rises to a maximum at the cycle period of 9–10 yr (41). This 9–10 yr period is also apparent in the frequency of years with saturated catch per effort in northern California: 1957, 1966, and 1977. The cycles are coherent along the coast with maximum cross-correlation at a lag of 0 years. This is not surprising considering the lack of geographic boundaries along this section of the coast.

Changes in fishing effort (38) tend to lag behind the crab population by about 1 yr. Within each high catch period, the number of vessels and the fraction of available crabs that are landed have increased (Fig. 5). In the first low-catch year, many vessels enter the fishery, but most leave before April (Fig. 6). During low-catch periods, the price for landed crabs doubles. Better prediction of the cycles may allow the fishery to track the population better and achieve a greater profit. More importantly, better understanding of the cause of the cycles may allow manipulation of the fishery to dampen the cycles in northern California–Washington.

5.2. Abiotic Factors

The abiotic factors best correlated with fluctuations in crab catch are ocean water temperature (42) and wind-driven surface water transport (41). Wild et al. (42) demonstrated reduced hatching success at high temperature and a coincidence between the collapse of central California recruitment and the extreme warm water of the 1957–1958 El Niño. The central California stock is near the species' southern limit, so it is not surprising that the high abundance in this area depends on particular environmental conditions. Examination of environmental conditions during the earlier collapse (about 1910) may provide a clue to the causal mechanism.

Johnson et al. (41), building on earlier analyses of the relation between upwelling and recruitment by Botsford and Wickham (43), found that southward

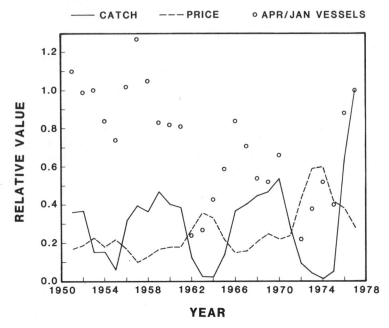

Figure 6. Response of price and within-season effort shifting to cycles in northern California crab catch. Apr/Jan vessels is the ratio of the number of vessels fishing in April and January. Price (dollars) is ex-vessel value adjusted by the consumer price index. Catch is expressed in relative units.

wind strength in the spring was significantly positively correlated with crab catch 4 yr later in northern California, Oregon, and Washington. The likely scenario is that early larval stages diffuse offshore as they are carried north by the Davidson Current. The megalopae, which tend to be found near the surface, are then transported south and onshore by southeastward winds later in the spring. Wild et al. (42) noted that peaks in northern California water temperature coincided with troughs in catch 4 yr later. The wind and temperature hypotheses are consistent: reduced southward winds should be correlated with negative anomalies in southward water transport and with positive anomalies in water temperature. Although crab catch is well correlated with environmental variation, tremendous preseason abundances that occur in the second high catch year of each cycle (7) are not well explained by environmental data (37, 41).

5.3. Biotic Factors

The role of biotic factors in cyclic crab populations has been studied with age-specific, density-dependent models (34, 35). Simply, a population can have an endogenous cycle if recruitment declines sufficiently rapidly as abundance of older age groups increases. The period of the cycles will be approximately twice the difference in mean age between the life stage being affected and the life stage causing the effect.

The effect of a fishery is to decrease the mean age and abundance of the population. Decreased mean age reduces the period of the cycles. If most age groups are available to the fishery, the decreased abundance should dampen the cycles (34, 35), but if only the older age groups are fished, then the population can be destabilized (34). Thus it is possible that unfished Dungeness crab populations are stable, and that the fishery on age 4+ males is causing the cycles. A major goal of the models is to explore this and other possibilities.

Botsford and Wickham (34) and McKelvey et al. (35) differ regarding the most likely density-dependent mechanism for the cyclic phenomenon, largely because both models necessarily included several unmeasured parameters. Botsford and Wickham (34) proposed cannibalism at the time of settlement as a likely factor, but cycles in their model had a period of only about 6 yr. Subsequent studies indicated that variation in growth (28) and a lagged response of the fishery to a change in abundance (38) can increase the period of the model's cycles to more closely match observed 10-yr periods. The model of McKelvey et al. (35) had 10-yr cycles if the effect occurred before settlement and was caused only by females; they discounted the cannibalism mechanism and male involvement because these factors produced unrealistically short cycles. Subsequent field studies (12) have provided much information on female life history parameters for these models.

The structure of the McKelvey et al. (35) model implies either that effective egg production is reduced at high female abundance, or that larval mortality is related to the *initial* number of eggs. In terms of timing, high cannibalism at the time of settlement seems indistinguishable from this latter possibility. Females probably dominate the cannibalistic stock, because the abundance of large males is greatly reduced by the fishery that occurs while larvae are planktonic. Newly settled crabs must move through a field of predatory fish (18) and adult crabs to reach bays and nearshore habitat typically inhabited by juveniles. Offshore settlement, caused by adverse environmental conditions, would prolong this period of onshore movement and increase mortality. Recent investigations by Botsford (37, 44) incorporate density-dependent recruitment and environmental factors.

The management response to a cyclic population depends on the cause of the cycles. If the cycles are strictly environmentally driven, then the management response could be to attempt to forecast the fluctuations. Better forecasts could prevent excess effort from entering the fishery in the first low-catch year, and encourage increased effort when a large year class is recruited. If the cycles are endogenous, then a change in the age/sex structure of the harvest could stabilize the fishery. A third possibility is that the cycles are due to a weaker environmental factor perturbing a population that is barely stable because of a density-dependent mechanism (44). This effect would be greatest if the environmental events tended to have a periodicity similar to the period of the population's natural cycle. In this case, a size-selective fishery could increase the stability of the population and its resilience to environmental perturbation.

5.4. Female Harvest

Harvest of female crabs has been suggested to potentially increase yield (36), stabilize the cycles (35), and reduce the stock's sensitivity to environmental perturbation (44). In practice, the marketing of smaller males and females will pose a formidable problem, and the large female harvest necessary to affect the stock's dynamics would be difficult to achieve. For example, McKelvey et al. (35) found increased model stability for a female harvest beginning at age 2, whereas the current male harvest begins at about age 4. The female size limit would have to be much lower than the current male size limit of 159 mm, because females rarely grow this large. U.S. fishermen and processors may not accept a female harvest because of lower meat yields, and may be reluctant to allow landing of females carrying eggs. A female harvest seems most acceptable after egg hatching in January (which is the effective end of the major male fishing season) and before molting and mating in the spring.

6. CURRENT MANAGEMENT PROBLEMS

Regulation by size, sex, and season protects the reproductive potential and yield per recruit of Dungeness crab populations north of San Francisco, California, but the level of effort in the fisheries is too high for economic efficiency. Even if the cycles cannot be stabilized, the status of the fishery could be improved. Overcapitalization in trawl fisheries for bottom fish and shrimps along the U.S. west coast has caused an increase in crab fishing effort during the early part of the season and 80–90% of northern California's annual landings are taken during December (3). The increased level of crab fishing effort has caused an expansion of fishing grounds and increased conflict with trawlers and shipping lanes.

As crab abundance and catch rate decline within the season, larger vessels move to more profitable fisheries. Price increases as the supply of fresh crab dwindles (northern California price increased from $0.90 to $1.70/lb within the 1982–1983 season). This increase in price allows small vessels to continue to fish at low catch rates. Warner (3) suggested that medium-sized vessels have difficulty competing with large vessels during the early part of the season, and they cannot afford to operate later when catch rate declines. Demory (4) suggested that some control of crab fishing effort would result in reduced competition among fishermen and a more even distribution of effort, landings, and price within each season.

In British Columbia and Alaska, Dungeness crab fishing is minor relative to salmon, herring, and king crabs, and overcapitalization is not yet perceived as a problem. A different problem in British Columbia is that the timing of fishing seasons does not closely match molting cycles. Jamieson (8) suggested that imposition of a lower price for recently molted crabs would be a simple means to discourage their harvest. In Alaska, effort historically has varied inversely with catch along the west coast. But in recent years the level of effort has increased sufficiently to raise concern among managers. Dungeness crab is an increasing share of total crab catch in Alaska (1), and Dungeness crab effort now seems sufficiently high to reduce stock abundance significantly (9). Other problems noted in Alaska are heavy predation by otters in isolated areas and mortality of soft-shell crabs by drift gill nets fishing in shallow water (43).

Any attempt to manage effort in the west coast Dungeness crab fishery must account for the fleet's diverse nature and its tendency to participate in other fisheries (45, 46). In 1981, 28% of the crabs in Washington–California were landed by vessels earning more than $10,000 and obtaining most of this income from a species other than crabs (Table 2) (47). Fishermen involved in the 1981 crab fishery also account for significant fractions of landings in other west coast fisheries: salmon, 13%; albacore, 19%; groundfish, 12%; shrimps, 37%; and herring, 8%. The movement of vessels in and out of the Dungeness crab fishery is affected by levels of recent harvests and prices (Figs. 5 and 6).

Within-season movement of vessels between crab, salmon, and albacore fisheries was modeled by Fletcher et al. (46). The model indicated that trip limits on crab fishermen were not equitable because of the large range in crab vessel capacity. Modeled net revenues for the crab fleet increased with a limited-entry program, but only when fleet size was reduced substantially so that the length of the effective season approached the 8-month duration of the current open season. Interactions between crab and salmon regulations were small because the primary crab, salmon, and albacore seasons are separated in time by 2–4 months. These interactions became important only when effort limitation increased the effective duration of the crab season.

Expansion of the above model to include trawl fisheries seems important if effort control in the crab fishery is to be seriously considered. Groundfish and herring fisheries are the only substantial alternatives to crab fishing during winter months, and the shrimp fishery opens earlier in spring than the salmon fishery. Currently, large groundfish trawlers only fish crabs while crab abundance is high at the start of the season. Effort control in the crab fishery may reduce conflict and increase the crab fleet's net revenues, but may severely restrict opportunities for these large multipurpose vessels.

There are several possibilities for a more even seasonal distribution of the fishery. Trip limits for each vessel have been used in the groundfish fishery, but

TABLE 2. Distribution of 1981 Crab Landings among Vessels Landing Some Crabs in 1981 (47)

	<$10,000 Income (All Species)	≥$10,000 Income	
		Principally Crabs	Principally Other Species
Vessels (no.)	511	451	476
Mean vessel length (m)	9.1	11.7	12.8
% of crab landed	6.3	65.5	28.2

large vessels are adversely affected. Limited licensing to reduce fleet size may achieve the desired result, but equitable selection of vessels seems difficult. Transferable individual fishermen's quotas (48) seem a feasible alternative. After an initial distribution of shares among potential crab fishermen, shares could be bought and sold. At the beginning of each season, management authorities would allocate an amount of crabs to each share. Each fisherman could then decide when in the season to harvest crabs allocated to shares owned.

7. CONCLUSIONS

Management of the cyclic Dungeness crab resource relies on regulation by size, sex, and season to protect reproductive output of the stock and to achieve a reasonable yield per recruit. Examination of the biological basis for these regulations suggests that the male size limit is at an appropriate level, and that the special protection of females is not necessary. Molt frequency of adult males and natural mortality are the least well-known parameters. Models of this cyclic resource have explored the roles of environmental variation and density-dependent factors. Although no single cause of the cycles is unambiguously supported, some models suggest that a change in the size–sex structure of the harvest could dampen the cycles.

In response to these large cycles in abundance, the fishery experiences great inter- and intra-annual changes in price and effort. The west coast fishing fleet contains a large pool of potential crab fishermen because of the high proportion of multipurpose vessels. Many vessels enter the crab fishery in December, then shift to other fisheries as legal-sized crab abundance declines and other opportunities arise.

Another management consideration is control of effort to achieve a more even seasonal distribution of catch. Harvesting 80–90% of the season's total during December creates much competition among fishermen and results in large stores of frozen crab and wide seasonal changes in price. Whatever is done to manage crab effort will affect distribution of effort in other fisheries. In the context of a multispecies fishery with staggered seasons, the current situation of pulse fishing Dungeness crab in December may be a rational policy for the overall fleet.

REFERENCES

1. M. F. Merritt, The Lower Cook Inlet Dungeness crab fishery from 1964–1983. *Alaska Sea Grant Rep.* **85-3:**85–96 (1985).
2. W. A. Dahlstrom and P. W. Wild, A history of Dungeness crab fisheries in California. *Calif. Dep. Fish Game, Fish. Bull.* **172:**7–24 (1983).
3. R. W. Warner, Overview of the California Dungeness crab, *Cancer magister*, fisheries. *Alaska Sea Grant Rep.* **85-3:**11–26 (1985).

4. D. Demory, An overview of Oregon Dungeness crab fishery with management concepts for the future. *Alaska Sea Grant Rep.* **85-3**:27–32 (1985).

5. S. Barry, Overview of the Washington coastal Dungeness crab fishery. *Alaska Sea Grant Rep.* **85-3**:33–36 (1985).

6. D. W. Gotshall, Catch-per-unit-of-effort studies of northern California Dungeness crabs (*Cancer magister*). *Calif. Fish Game* **64**:189–199 (1978).

7. R. D. Methot, Jr. and L. W. Botsford, Estimated preseason abundance in the California Dungeness crab (*Cancer magister*) fisheries. *Can. J. Fish. Aquat. Sci.* **39**:1077–1083 (1982).

8. G. S. Jamieson, The Dungeness crab, *Cancer magister*, fisheries of British Columbia. *Alaska Sea Grant Rep.* **85-3**:37–60 (1985).

9. T. M. Koeneman, A brief review of the commercial fisheries for *Cancer magister* in southeast Alaska and Yakutat waters, with emphasis on recent seasons. *Alaska Sea Grant Rep.* **85-3**:61–76 (1985).

10. D. W. Gotshall, Northern California Dungeness crab (*Cancer magister*), movements as shown by tagging. *Calif. Fish Game* **64**:234–254 (1978).

11. N. Diamond and D. G. Hankin, Movements of adult female Dungeness crabs (*Cancer magister*) in northern California based on tag recoveries. *Can. J. Fish. Aquat. Sci.* **42**:919–926 (1985).

12. D. G. Hankin, N. Diamond, M. Mohr, and J. Ianelli, Molt increments, annual molting probabilities, fecundity and survival rates of adult female Dungeness crabs in northern California. *Alaska Sea Grant Rep.* **85-3**:189–206 (1985).

13. D. E. Wickham, Predation by the nemertean *Carcinonemertes errans* on eggs of the Dungeness crab (*Cancer magister*). *Mar. Biol. (Berlin)* **55**:45–53 (1979).

14. P. W. Wild, The influence of seawater temperature on spawning, egg development, and hatching success of the Dungeness crab, *Cancer magister. Calif. Dep. Fish Game, Fish. Bull.* **172**:197–214 (1983).

15. R. G. Lough, Larval dynamics of the Dungeness crab, *Cancer magister*, off the central Oregon coast, 1970–71. *Fish. Bull.* **74**:353–376 (1976).

16. P. N. Reilly, Dynamics of Dungeness crab, *Cancer magister*, larvae off central and northern California. *Calif. Dep. Fish Game, Fish. Bull.* **172**:57–84 (1983).

17. D. E. Wickham, The relationship between megalopae of the Dungeness crab, *Cancer magister*, and the hydroid, *Velella velella*, and its influence on abundance estimates of *C. magister* megalopae. *Calif. Fish Game* **65**:184–186 (1979).

18. P. N. Reilly, Predation on Dungeness crabs, *Cancer magister*, in central California. *Calif. Dep. Fish Game, Fish. Bull.* **172**:155–164 (1983).

19. D. A. Armstrong and D. R. Gunderson, The role of estuaries in Dungeness crab early life history: A case study in Grays Harbor, Washington. *Alaska Sea Grant Rep.* **85-3**:145–170 (1985).

20. R. N. Tasto, Juvenile Dungeness crab, *Cancer magister* studies in the San Francisco Bay area. *Calif. Dep. Fish Game , Fish. Bull.* **172**:135–154 (1983).

21. D. W. Gotshall, Relative abundance studies of Dungeness crabs, *Cancer magister*, in northern California. *Calif. Fish Game* **64**:24–37 (1978).

22. B. G. Stevens and D. A. Armstrong, Distribution, abundance, and growth of juvenile Dungeness crabs, *Cancer magister*, in Grays Harbor estuary, Washington. *Fish. Bull.* **82**:469–483 (1984).

23. K. R. Carrasco, D. A. Armstrong, D. R. Gunderson, and C. Rogers, Abundance and growth of *Cancer magister* young-of-the-year in the nearshore environment. *Alaska Sea Grant Rep.* **85-3:**171–184 (1985).

24. P. C. Collier, Movement and growth of post-larval Dungeness crabs, *Cancer magister*, in the San Francisco area. *Calif. Dep. Fish Game, Fish. Bull.* **172:**125–134 (1983).

25. R. L. Poole, Preliminary results of the age and growth study of the market crab (*Cancer magister*) in California: The age and growth of (*Cancer magister*) in Bodega Bay. In *Symposium on Crustacean, Proceedings*, Part 2. Mar. Biol. Assoc., India, 1967, pp. 543–567.

26. T. H. Butler, Growth and age determination of the Pacific edible crab (*Cancer magister*) Dana. *J. Fish. Res. Board Can.* **18:**873–891 (1961).

27. F. C. Cleaver, Preliminary results of the coastal crab (*Cancer magister*) investigation. *Wash. State Dep. Fish.*, Biol. Rep. **94A:**47–82 (1949).

28. L. W. Botsford, Effect of individual growth rates on expected behavior of the northern California Dungeness crab (*Cancer magister*) fishery. *Can. J. Fish. Aquat. Sci.* **41:**99–107 (1984).

29. R. W. Warner, Age and growth of male Dungeness crabs, *Cancer magister*, in northern California. *Alaska Sea Grant Rep.* **85-3:**185–188 (1985).

30. D. C. G. MacKay and F. W. Weymouth, The growth of the Pacific edible crab, *Cancer magister* Dana. *J. Biol. Board Can.* **1:**191–212 (1935).

31. N. Diamond and D. G. Hankin, Biases in crab tag recovery data. *Alaska Sea Grant Rep.* **85-3:**341–356 (1985).

32. D. G. Hankin, Proposed explanations for fluctuations in abundance of Dungeness crabs: A review and critique. *Alaska Sea Grant Rep.* **85-3:**305–326 (1985).

33. T. Jow, California cooperative crab tagging study. *Annu. Rep. Pac. Mar. Fish. Comm.* **16/17:**51–52 (1965).

34. L. W. Botsford and D. E. Wickham, Behavior of age-specific, density-dependent models and the northern California Dungeness crab (*Cancer magister*) fishery. *Can. J. Fish. Aquat. Sci.* **35:**833–843 (1978).

35. R. McKelvey, D. Hankin, K. Yanosko, and C. Syngg, Stable cycles in multistage recruitment models: An application to the northern California Dungeness crab (*Cancer magister*) fishery. *Can. J. Fish. Aquat. Sci.* **37:**2323–2345 (1980).

36. P. W. Wild and R. N. Tasto, Eds., Life history, environment, and mariculture studies of the Dungeness crab, *Cancer magister*, with emphasis on the central California fishery resource. *Calif. Dep. Fish Game, Fish. Bull.* No. 172 (1983).

37. L. W. Botsford, Population dynamics of the Dungeness crab (*Cancer magister*). *Can. J. Fish. Aquat. Sci., Spec. Publ.* **92:**140–153 (1986).

38. L. W. Botsford, R. D. Methot, Jr., and W. E. Johnston, Effort dynamics of the northern California Dungeness crab (*Cancer magister*) fishery. *Can. J. Fish. Aquat. Sci.* **40:**337–346 (1983).

39. L. W. Botsford, R. D. Methot, Jr., and J. E. Wilen, Cyclic covariation in the California king salmon, *Oncorhynchus tshawytscha*, silver salmon, *O. kisutch*, and Dungeness crab, *Cancer magister*, fisheries. *Fish. Bull.* **80:**791–801 (1982).

40. D. H. Thomas, A possible link between coho (silver) salmon enhancement and a decline in central California Dungeness crab abundance. *Fish. Bull.* **83:**682–691 (1985).

41. D. F. Johnson, L. W. Botsford, R. D. Methot, Jr., and T. C. Wainwright, Wind stress and cycles in Dungeness crab (*Cancer magister*) catch off California, Oregon and Washington. *Can. J. Fish. Aquat. Sci.* **43**:838–845 (1986).

42. P. W. Wild, P. M. W. Law, and D. R. McLain, Variations in ocean climate and the Dungeness crab fishery in California. *Calif. Dep. Fish Game, Fish. Bull.* **172**:175–188 (1983).

43. L. W. Botsford and D. E. Wickham, Correlation of upwelling index and Dungeness crab catch. *Fish. Bull.* **73**:901–907 (1975).

44. L. W. Botsford, Effects of environmental forcing on age-structured populations: Northern California Dungeness crab (*Cancer magister*) as an example. *Can. J. Fish. Aquat. Sci.* **43**:2345–2352 (1986).

45. R. D. Methot, Management of Dungeness crab fisheries. *Can. J. Fish. Aquat. Sci., Spec. Publ.* **92**:326–334 (1986).

46. J. J. Fletcher, W. E. Johnston, and R. E. Howitt, Modeling a fishery characterized by uncertain resources and a multi-purpose, heterogeneous fishing fleet—the Eureka Dungeness crab fishery. *Alaska Sea Grant Rep.* **85-3**:287–304 (1985).

47. A. Kimker, Overview of the Prince William Sound management area Dungeness crab fishery. *Alaska Sea Grant Rep.* **85-3**:77–84 (1985).

48. D. G. Moloney and P. H. Pearse, Quantitative rights as an instrument for regulating commercial fisheries. *J. Fish. Res. Board Can.* **36**:859–866 (1979).

10 THE FLORIDA STONE CRAB FISHERY: A REUSABLE RESOURCE?

Nelson M. Ehrhardt and Victor Ricardo Restrepo
Rosenstiel School of Marine and Atmospheric Science
University of Miami
Miami, Florida

1. INTRODUCTION

The stone crab, *Menippe mercenaria* (Say 1891), is a benthic species of the family Xanthidae with juveniles inhabiting bays and estuaries (Manning 1) and adults moving offshore to depths of 54 m (Bullis and Thompson 2). The species occurs along the southeast coast of the United States from North Carolina to Florida, and through the Gulf of Mexico (Williams 3) into the Caribbean Sea (Karandeyva and Silva 4). Within this geographical range, *M. mercenaria* supports important fisheries in the United States and Cuba.

Stone crabs are noted for possessing two powerful and disproportionately large claws relative to their body size. In spite of their asymmetric size, these claws are morphologically very similar (Rathbun 5; Williams 3; Cheung 6), and constitute approximately one-half of the body weight in adult individuals (Sullivan 7). This characteristic makes stone crabs a valuable fishery resource mainly for their claws.

The U.S. stone crab fishery is directed to the exploitation of claws only, and it is largely restricted to areas off the west coast of Florida, where stock densities are believed to be the greatest owing to more favorable habitat conditions (Bert et al. 8). During the period 1983–1985, some 300 commercial fishing vessels operated more than 400,000 traps, landing an average of 1.76 million lb of claws per year with an estimated value of over $6.9 million (Phares 9). These statistics place the stone crab fishery among the most important fisheries in the state of Florida.

Stone crab stocks are jointly managed by the Florida Marine Fisheries Commission within state territorial waters (0–9 nm) and by the Gulf of Mexico Fishery Management Council in the Fishery Conservation Zone (9–200 nm). Management regulations for the greater part correspond to those of the state of Florida and are contained in the laws and regulations of the state and in a Fishery Management Plan developed by the Gulf Council. One of these regulations declares that claws are the only legally harvestable portion and that declawed crab bodies must be returned to the water. The primary reason for this measure is that stone crabs, like other brachiurans, possess the ability to regenerate appendages after losing them. It is thought therefore, that those crabs surviving fishing operations might contribute to future catches after regenerating new claws.

Survival of declawed stone crabs and regeneration of claws have been demonstrated by Savage and Sullivan (10), Davis et al. (11), and Sullivan (7). As in other crustaceans, the growth of *Menippe mercenaria* is characterized by the frequency of molting and by the increase in size per molt. Both these factors are related to age, because smaller, thus younger crabs experience shorter molt cycles and greater increases in size relative to older crabs (Mootz and Epifanio 12; Savage and Sullivan 10). Because claw regeneration is related to the molt cycle, it is expected that smaller crabs, if exploited, should be able to regenerate more biomass during their lifetime than older individuals. This condition allows for an interaction between the level of fishing intensity and future levels of regenerated claw biomass, which is of interest to both stone crab managers and commercial fishermen.

The objective of this study was to investigate the effects of various harvest management strategies on the long-term sustainable yield per recruit of a self-renovating and partially self-regenerating stone crab claw population.

2. THE FISHERY

Stone crabs have been recognized as a delicacy since the middle of the nineteenth century. Fishery development, however, was severely affected by poorly developed marketing systems. In fact, the supply of stone crabs exceeded the demand until 1962. Landings in south Florida reached an equivalent of 4680 lb claws in 1895 and 22,000 lb in 1919 (Schroeder 13). Landings from the Gulf of Mexico were less than 50,000 lb claws/yr until the 1950s, but catches increased significantly from 250,000 lb in 1962 to more than 500,000 lb in 1968 (Fig. 1). This

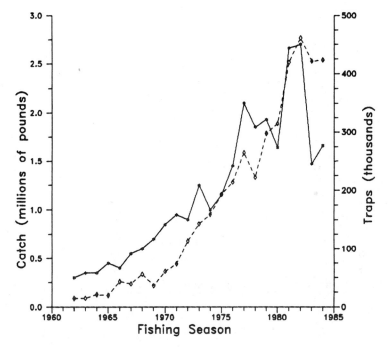

Figure 1. Trends in catch (——) and fishing effort (– – –) for the stone crab fishery from the 1962–1963 fishing season to the 1984–1985 fishing season.

expanding trend continued through the 1982–1983 fishing season, when landings reached 3 million lb of claws. Subsequent to that time, landings have declined to 1.85 million lb and 1.6 million lb in the 1983–1984 and 1984–1985 seasons, respectively (Gulf of Mexico Fishery Management Council 14).

Stone crab traps are the only fishing gear operated in the fishery. Typically, these are wooden traps with an entrance at the top that is not to exceed a size of 10.2 × 16.5 cm. The design and dimensions of other parts of the trap may be modified, however. Most traps measure about 40 × 40 × 28 cm or 53 × 36 × 28 cm in length, width, and height, respectively. Traps are baited with 1–3 lb of trash fish, remnants, or rawhide. Baited traps are usually set in parallel lines following bottom depth contours with traps spaced 30–90 m apart. Soak time varies from a few days to 3 weeks, and fishing trips last 1 day.

Fishing effort, measured in number of traps operated during a 7-month fishing season (October 15–May 15), has increased from 14,600 traps in the 1962–1963 season to 113,300 traps in 1972–1973 and peaked at 461,000 traps during the 1982–1983 fishing season (Fig. 1).

For the purpose of protecting egg-bearing females during peak summer spawning months, a closed fishing season was implemented from May 16 to October 14. Also a minimum claw size (propodus length) (Fig. 2) of 70 mm (2.75 in.) was established because of marketing convenience, and to allow individuals to reach sexual maturity and spawn before being recruited to the fishery. Several other

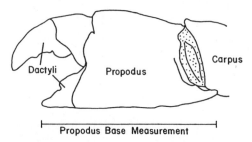

Figure 2. Diagram of a stone crab claw showing the propodus base measurement used to define minimum harvestable claw size.

measures related to harvesting practices have also been promulgated. These are all oriented toward enhancing crab survival prior to declawing. There are no fishing effort limitations or catch quotas imposed upon the fishery.

3. DESCRIPTION OF THE MODEL

The concept of yield per recruit (Beverton and Holt 15; Ricker 16) has been widely applied to examine the effectiveness of minimum size regulations on equilibrium yield responses of exploitable fish populations. A fishery with a successful minimum size regulation will increase total yield per recruit as a result of a higher natural rate of biomass growth in comparison to an unregulated fishery. Although the theory of yield per recruit was developed for scale fish fisheries, it has also been applied to invertebrate stocks. In this last instance, some of the basic biological assumptions underlying this theory may be invalidated (Hancock 17), because the available knowledge makes it clear that the discontinuous growth pattern characterizing crustaceans cannot be represented by the simpler formulations used with fish species. In this sense, a more realistic description of invertebrate population dynamics may be obtained with the Ricker model because it can accommodate allometric growth representations as well as time-specific mortality and growth rates.

The contrasting difference of the exploited dynamics of stone crab stocks requires specific model configurations such that yield per recruit may be assessed under different management alternatives of minimum claw size and varying catch and fishing mortality rates. With this purpose in mind, a discrete-time, age-structured, dynamic pool model was developed that incorporates harvestable biomass of appendages (claws) and delayed regenerated claw biomass as a dynamic function of fishing effort levels. Specific model components are described below.

3.1. Growth

The increase in body size at molting and the frequency of molting are two aspects of the study of discontinuous growth processes and age determination in crustaceans. The increase in body size at molting of stone crabs was estimated by relating postmolt to premolt size in a Hiatt growth diagram similar to that suggested by Mauchline (18, 19), and frequency of molting was obtainable by relating the time required to molt (intermolt period) and the premolt size. For this purpose information on molting of stone crabs kept in the laboratory was obtained from the Florida Department of Natural Resources and from published accounts (Savage 20; Cheung 21; Savage and McMahan 22; Savage and Sullivan 10).

Hyperbolic relationships were fitted to data relating postmolt (PO) to premolt (PR) carapace widths (CW) of male and female stone crabs (Fig. 3). Analyses of covariance demonstrated significant differences in growth between males and females greater than 40 mm CW, but these differences were not significant among younger individuals. Because females mature for the first time at an approximate size of 40 mm CW (Bert et al. 23), the observed sexual dimorphism may be the result of differences in the amount of energy allocated for growth and reproduction between the two sexes. Resulting equations are as follows:

$$PO = \frac{PR}{0.0015PR + 0.7234} \quad \text{(mature males)}$$

$$PO = \frac{PR}{0.0019PR + 0.7463} \quad \text{(mature females)}$$

$$PO = \frac{PR}{0.0005PR + 0.7904} \quad \text{(immature males and females)}$$

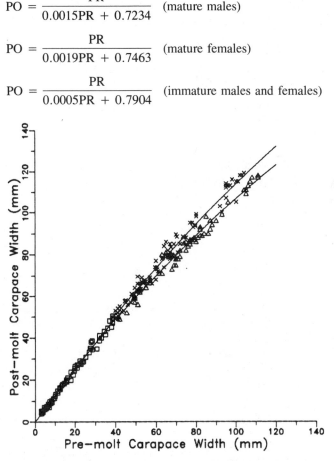

Figure 3. Premolt carapace width and postmolt carapace width relationships for juvenile (□), male (×), and female (△) stone crabs.

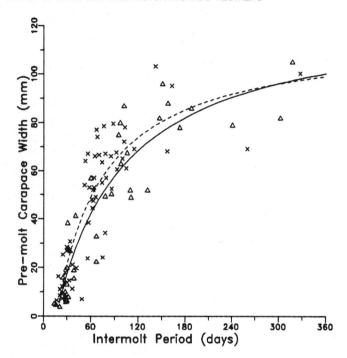

Figure 4. Relationships between pre-exuvial carapace width and intermolt period for male (×) and female (△) stone crabs. (---) Fitted function for males; (——) fitted function for females.

Similar relationships were found for data relating premolt carapace width to intermolt period (IP) in days, for each sex (Fig. 4) and expressed as

$$PR = \frac{IP}{(0.0062\ IP + 0.2551)} - 47.0047$$

$$PR = \frac{IP}{(0.0065\ IP + 0.4546)} - 28.7529$$

Individual body growth was modeled by constructing a size at age history using the above relationships and starting from an arbitrarily selected size of 2.5-mm CW and an age of 0.09 yr.

Claw (propodus) size at age was modeled using average carapace width at age obtained as above, and claw size–body size (CW) relationships reported by Savage and Sullivan (10). Claw size at age relationships for minor (pincer) and major (crusher) claws are presented by sexes in Figure 5. These indicate that males attain the legal size of first capture (70-mm propodus length) at an age of about 1.8 yr, whereas females reach this size 1 yr later.

According to several declawing experiments (Savage and Sullivan 10; Davis et al. 11), adult stone crabs are able to regenerate approximately 70% of the original

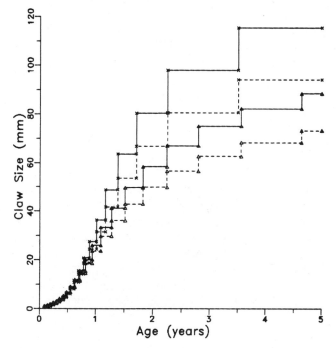

Figure 5. Claw size at age curves for male (×) and female (△) stone crabs. (——) crushers; (---) pincers.

claw size in the first molt and reach 100% of that size in the subsequent molt. This growth rate was used in conjunction with premolt sizes and molt rates to create a delayed regenerated claw recruitment schedule for different sizes at first recruitment.

Data on length and weight of claws collected during commercial fishing operations were used to estimate a common weight–length relationship for either claw and given by

$$W \text{ (g)} = 0.00083[L \text{ (mm)}]^{2.62}$$

3.2. Mortality

Stone crab traps are not size selective. As such, they can capture and retain individuals over a wide range of sizes that may extend from well below the carapace width generating the legal minimum claw size of 70 mm, to the largest individuals in the population. These traps, however, are significantly more efficient in capturing males (Bert et al. 23).

Owing to the peculiar stock utilization practices of the fishery, instantaneous fishing mortality rates (F_d) are a function of crab handling. Consequently, fishing mortality rates are equated with mortality rates associated with declawing. On the other hand, instantaneous catch rates (F_c) are defined as the product of the catchability coefficient (q) and total effective fishing effort (f), and F_c equates to the traditional definition of fishing mortality rate only when there are no survivors from the declawing operation. Thus instantaneous fishing mortality rate was expressed as

$$F_d = qf(1 - S_d)$$

where S_d represents the probability that a crab survives declawing.

A catchability coefficient of 0.00011 per trap pulled was estimated by Ehrhardt et al. (24), whereas survival rates due to declawing have been estimated to range from 0 to 10% in tagging experiments with declawed crabs (Ehrhardt et al. 24), from 53 to 78% during declawing experiments carried out in the laboratory (Davis et al., 11), and from 25 to 97% in the commercial fishery (Bert et al. 8). Savage et al. (25) put the number of regenerated claws landed at 9.95% of the total landings based on an examination of the stridulatory patterns in the claws, which are altered during regeneration.

Instantaneous natural mortality rates of 1.60 and 1.61/yr were reported by Ehrhardt et al. (24) estimated from fishing success methodologies and from an experimental group of stone crabs kept in the laboratory. Both estimates are very close in spite of the differences in the methodologies employed. These high estimates may be expected because stone crabs are characterized by both cannibalistic and aggressive behaviors and because they reach terminal molt at about 5 yr (Restrepo 26).

3.3. Yield per Recruit

In the stone crab yield-per-recruit framework, the rate of yield at time h is assumed to be a direct function of the instantaneous catch rate F_c and the biomass of harvestable claws present. This biomass is originated from individuals that have not been previously captured and declawed in the fishery, and hence are defined as "natural" clawed stock, as well as from those that have regenerated new claws following commercial declawing and are defined as "regenerated" cohorts. Claw biomass from "regenerated" cohorts is recruited following schedules defined by molting rates at age of the individuals and time of capture. The amount of regenerated claw biomass is a function of the abundance of "regenerated" cohorts and as such is a delayed function of both F_c and S_d levels.

Because of the seasonality of the fishery and the size-specific growth rates, yield per recruit was computed by a stepwise numerical integration procedure similar to that suggested by Paulik and Bayliff (27). For this purpose the fishable life-span was divided into monthly time segments (h) and the total life-span after "natural" recruitment was defined as $T = t_T - t_r$, where t_T = terminal age and t_r = age at first recruitment. Ages at first, second, and third "regenerated" recruitments were designated as t_i, t_{ij}, and t_{ijk}, after first (i), second (j), and third (k) declawing in the fishery, respectively.

In equilibrium, the age structure of the stock remains unchanged over time. Therefore, yield in biomass from the entire stock at time h was assumed to be equivalent to the yield from a single "natural" cohort (Y_N) and from several "regenerated" cohorts (Y_R) during their lifetimes. A simplified diagram showing yield contributions from all cohorts is presented in Figure 6.

Then the yield for the entire stock was calculated by summing the contributions from each cohort ($Y_N + Y_R$) as

$$Y = \sum_{h=1}^{T} \left\{ Y_h + \sum_{m=v}^{T} \left[Y_{hm} + \sum_{n=p}^{T} \left(Y_{hmn} + \sum_{s=q}^{T} Y_{hmns} \right) \right] \right\}$$

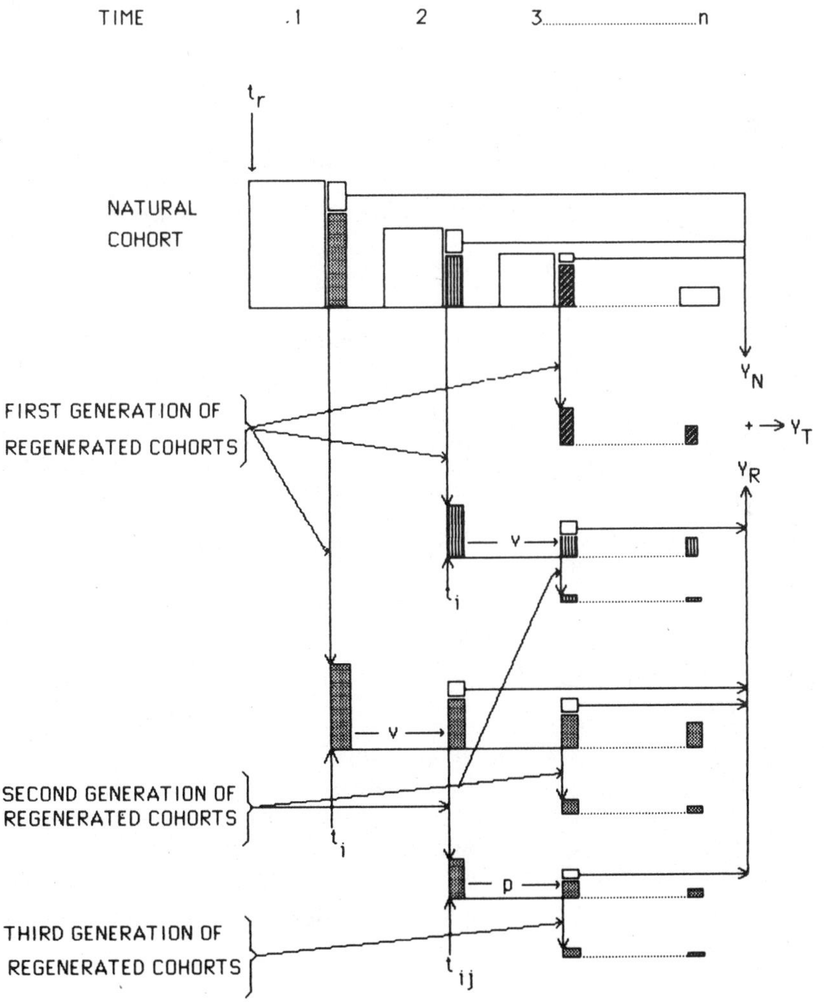

Figure 6. Diagram of yield from natural and regenerated stone crab cohorts.

Here

$$Y_N = \sum_{h=1}^{T} Y_h = \sum_{h=1}^{T} C_h w_h$$

where w_h = average weight of claws from empirical growth curves during
month h, after initial recruitment to the fishery,

$C_h = N_{(N)h} u_h$ = number of crabs from natural cohort N caught in month
h,

$u_h = F_{c,h}[1 - \exp(-Z_h)]/Z_h$ = exploitation rate in month h,

$Z_h = (F_{c,h} + M_h)$ = total instantaneous mortality rate during month h
(equal to M during months closed to fishing),

$N_{(N)h} = R_N \exp(-\sum_{l=1}^{h-1} Z_l)$ = number of survivors to begin month h,
and

R_N = number of "natural" recruits.

Also

$$Y_R = \sum_{h=1}^{T} \sum_{m=v}^{T} \left\{ Y_{hm} + \sum_{n=p}^{T} \left[Y_{hmn} + \sum_{s=q}^{T} \left(Y_{hmns} \right) \right] \right\}$$

where $v = t_i - t_r + 1$ = month at regenerated recruitment after "natural" recruitment of cohort i,

$p = t_{ij} - t_r + 1$ = month at regenerated recruitment after "natural" recruitment of cohort j,

$q = t_{ijk} - t_r + 1$ = month at regenerated recruitment after "natural" recruitment of cohort k.

Yield from "regenerated" cohort i, captured during month $h(i \doteq h)$ and again
during month m, was defined as

$$\sum_{h=1}^{T} \sum_{m=v}^{T} Y_{hm} = \sum_{h=1}^{T} \sum_{m=v}^{T} C_{hm} w_{hm}$$

where w_{hm} = average weight of regenerated claws from empirical growth
curves of cohort (i) during month m after second recruitment,

$C_{hm} = N_{(h)m} u_m$ = number of crabs from cohort (i) caught in month m
after second recruitment,

$N_{(h)m} = R_h \exp(-\sum_{l=1}^{m-1} Z_l)$ = number of survivors in "regenerated"
cohort (i) to begin month m after second recruitment,

$R_h = (C_h S_d) \exp(-\sum_{l=1}^{L1} M_l)$, and

$L1 = t_i - t_r - h$.

Yield from "regenerated" cohort (i, j), captured during month h $(i = h)$ and again during months $m(j = m)$ and n, was defined as

$$\sum_{h=1}^{T} \sum_{m=v}^{T} \sum_{n=p}^{T} Y_{hmn} = \sum_{h=1}^{T} \sum_{m=v}^{T} \sum_{n=p}^{T} C_{hmn} w_{hmn}$$

where w_{hmn} = average weight of regenerated claws from empirical growth curves for cohort (i, j) during month n after third recruitment,

C_{hmn} = $N_{(hm)n} u_n$ = number of crabs from cohort (i, j) caught in month n after third recruitment,

$N_{(hm)n}$ = $R_{hm} \exp(-\Sigma_{l=1}^{n-1} Z_l)$ = number of survivors in "regenerated" cohort (i, j) to begin interval n after third recruitment,

R_{hm} = $(C_{hm} S_d) \exp(-\Sigma_{l=1}^{L2} M_l)$, and

$L2 = t_{ij} - t_r - h$.

Yield from "regenerated" cohort (i, j, k), captured during month h $(i = h)$ and again during months m $(j = m)$, $n(k = n)$, and s, was defined as

$$\sum_{h=1}^{T} \sum_{m=v}^{T} \sum_{n=p}^{T} \sum_{s=q}^{T} Y_{hmns} = \sum_{h=1}^{T} \sum_{m=v}^{T} \sum_{n=p}^{T} \sum_{s=q}^{T} C_{hmns} w_{hmns}$$

where w_{hmns} = average weight of regenerated claws from empirical growth curves for cohort (i, j, k) during month s after fourth recruitment,

C_{hmns} = $N_{(hmn)s} u_s$ = number of crabs from cohort (i, j, k) caught in month s after fourth recruitment,

$N_{(hmn)s}$ = $R_{hmn} \exp(-\Sigma_{l=1}^{s}Z_l)$ = number of survivors in "regenerated" cohort k to begin interval s after fourth recruitment,

R_{hmn} = $(C_{hmn} S_d) \exp(-\Sigma_{l=1}^{L3} M_l)$, and

$L3 = t_{ijk} - t_r - h$.

Many more "regenerated" cohorts may be formed as a function of t_r and claw regeneration schedules. In this study only three regenerations were considered for the minimum claw size limits used in the analyses, since the formation of a greater number of cohorts in the exploited stock is unlikely.

4. THE MODEL APPLIED

The yield-per-recruit model incorporating delayed catch rates and declawing survival rates was used to evaluate the resilience of stone crab stocks to generate claw biomass at different levels of exploitation and minimum claw size limits. The analyses were performed for males and females separately to include the effects of sexual dimorphism on yield per recruit estimates. Each analysis assumed a constant natural mortality rate (M_h) of 0.13 per month and constant recruitment. In addition, a constant monthly catch rate $(F_{c,h})$ was applied during any fishing season (October 15–May 15) throughout the exploitable life-span of a cohort. However, catch rates were varied from 0.07 to 0.50/month in successive runs, to create harvesting situations under two arbitrarily specified claw size limits of 40- and 70-mm propodus length. These lengths were selected because they represent the size corresponding to first maturity and the actual legal size enforced, respectively. Declawing survival fractions of 0, 10, and 50% were adopted in the analyses, because this range is believed to include the most likely rates of survival encountered in the fishery. Relevant growth data were obtained from empirical growth curves developed for this species in previous sections.

Assessment of minimum claw size limits and catch rates focused on the percent gain in yield per recruit under different claw regeneration and declawing survival schemes. Hence the analyses included dynamic yield per recruit estimates for harvesting strategies relative to yield per recruit at 0% declawing survival $(F_c = F_d)$.

Figures 7 and 8 show the estimated yield per recruit for females and males expressed as a function of the decision variables F_c, l_r, and S_d. In all instances

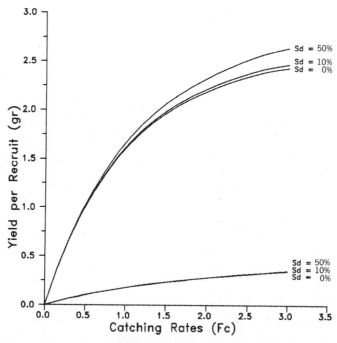

Figure 7. Yield per recruit as a function of catching rates F_c and declawing survival S_d for female stone crabs. Upper curves are for l_r = 40 mm and lower curves are for l_r = 70 mm.

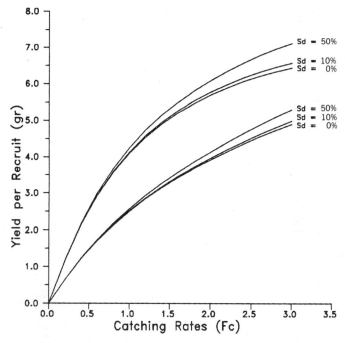

Figure 8. Yield per recruit as a function of catching rates F_c and declawing survival S_d for male stone crabs. Upper curves are for $l_r = 40$ mm and lower curves are for $l_r = 70$ mm.

yield per recruit is low as a consequence of the high natural mortality rate estimated for this species. The results also show a significant difference in yield per recruit between sexes because of the effect of sexual dimorphism on the age of first recruitment, t_r, for the two values of l_r selected.

Among females, yield per recruit at $l_r = 70$ mm and for any F_c level is very much reduced as a result of an excess of natural mortality over growth. For the same reason, declawing survival, S_d, does not play an important role in furthering claw biomass regeneration (Figs. 7 and 9). For values of F_c between 0.30 and 3.00, a reduction in minimum claw size limit to 40 mm resulted in modest gains of yield to the fishery on the order of 1.0–2.5 g per recruit. For that same range of F_c, yield from regenerated claws become important if at least 50% of the crabs survive declawing.

Yield per recruit among males recruited at $l_r = 70$ mm, although low, is significantly greater than that found for females within the same F_c range. This difference is explained by the larger growth rates and the relatively more reduced effect of natural mortality over growth among males than females of the same size.

As with females, substantial gains in yield per recruit can be achieved by lowering the minimum claw size of males to 40 mm (Fig. 8). Also, net gains in yield per recruit as a consequence of regenerated claw biomass become significant only at a 50% declawing survival (Fig. 10). These gains are slightly more impor-

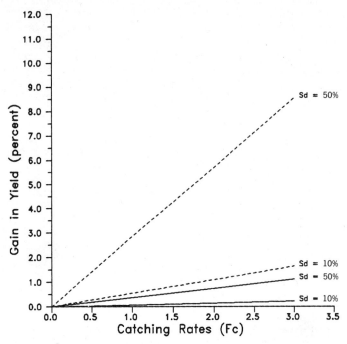

Figure 9. Net gain in yield due to claw biomass regeneration under varying S_d and I_r levels for female stone crabs. (‒‒‒) I_r = 40 mm; (——) I_r = 70 mm.

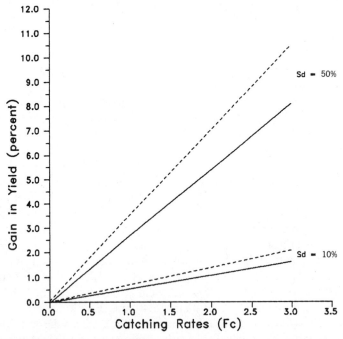

Figure 10. Net gain in yield due to claw biomass regeneration under varying S_d and I_r levels for male stone crabs. (‒‒‒) I_r = 40 mm; (——) I_r = 70 mm.

tant when minimum claw size is 40 mm, thus indicating an increased opportunity for younger individuals to contribute to claw biomass regeneration over an extended time period of their exploitable life-span.

In conclusion, the imposed minimum propodus length of 70 mm seems to be too large, thus decreasing yield to the fishery. This condition is more notorious among females, which contribute only slightly toward the overall yield of the stock, a fact that is also observed in the fishery. Also, the actual minimum claw size considerably restricts the possibility of incorporating regenerated claw biomass, as molting periods become more extended with age. However, the success of decreasing the size limit depends heavily on the cooperation from fishermen, for it was demonstrated that net gains in yield become significant at higher levels of declawing survival. Thus if 50% declawing survival can be achieved through good fishing practices at either 40- or 70-mm claw length, then a 10 or 5% increase in yield from regenerated claw biomass could be expected. At current exploitation levels this is translated into 65,000–117,000 lb of claws per season added to the actual landings. As these results show, under properly observed magement schemes the stone crab stocks off Florida can be categorized as a unique reusable resource.

REFERENCES

1. R. B. Manning, Some growth changes in the stone crab, *Menippe mercenaria* (Say). *Q. J. Fla. Acad. Sci.* **23**(4):273–277 (1961).

2. H. R. Bullis and J. R. Thompson, Collections by the exploratory fishing vessels *Oregon, Silver Bay, Combat*, and *Pelican* made during 1956 to 1960 in the southwestern North Atlantic. *U.S. Fish Wildl. Serv., Spec. Sci. Rep.-Fish.* 510, 130 p (1965).

3. A. B. Williams, The decapod crustaceans of the Carolinas. *Fish. Bull.* 65(1), 298 p (1965).

4. O. G. Karandeyeva and A. Silva, Intensity of respiration and osmoregulation of the commercial crab, *Menippe mercenaria* (Say) from Cuban coastal waters. *Investigations of the Central American Seas* (translated from Russian). Published for the Smithsonian Institution and National Science Foundation by the Indian National Scientific Document Center, New Delhi, India, 1973, pp. 292–310.

5. M. J. Rathbun, The Crancoid crabs of America of the families Euryalidae, Portunidae, Atcelecyclidae, Cancridae, and Xanthidae. *Bull.—U.S. Natl. Mus.* 152, 609 p (1930).

6. T. S. Cheung, A biostatistical study of the functional consistency in the reversed claws of adult male stone crabs, *Menippe mercenaria* (Say). *Crustaceana* **31**(2):137–144 (1976).

7. J. R. Sullivan, The stone crab, *Menippe mercenaria*, in the southwest Florida USA fishery. *Fla. Mar. Res. Publ.* 36, 37 p (1979).

8. T. M. Bert, R. E. Warnes and L. D. Kessler, The biology and Florida fishery of the stone crab, *Menippe mercenaria* (Say), with emphasis on Southwest Florida. *Univ. Fla. Sea Grant Program, Tech. Pap.* 9, 82 p (1978).

9. P. L. Phares, Analysis of stone crab assessment data for 1979 to 1985 and evaluation of the logbook data collection system. *Natl. Mar. Fish. Serv., Southeast Fish. Cent., Contrib.* ML1-85-27, 25 p (1985).

10. T. Savage and J. R. Sullivan, Growth and claw regeneration of the stone crab, *Menippe mercenaria. Fla. Mar. Res. Publ.* 32, 23 p (1978).

11. G. E. Davis, D. S. Baughman, J. D. Chapman, D. MacArthur, and A. C. Pierce, *Mortality Associated with Declawing Stone Crabs*, Menippe mercenaria, South Florida Research Center Report T-522. Everglades National Park, Homestead, FL, 1979. 23 p.

12. C. A. Mootz and C. E. Epifanio, An energy budget for *Menippe mercenaria* larvae fed *Artemia* nauplii. *Biol. Bull. (Woods Hole, Mass.)* **146**:44–55 (1974).

13. W. C. Schroeder, *Fisheries of Key West and the Clam Industry of Southern Florida.* Report of the U.S. Commissioner of Fisheries for 1923. U.S. Comm. Fish., 1924, Appendix 12.

14. Gulf of Mexico Fishery Management Council, Environmental assessment and supplemental regulatory impact review and initial regulatory flexibility analysis and amendment number 3 to the Fishery Management Plan for the Stone crab fishery in the Gulf of Mexico, Tampa, FL, 1986.

15. R. J. H. Beverton and S. J. Holt, *On the dynamics of exploited fish populations. Fish. Invest.—Minist. Agric. Fish. Food (G.B.) (Ser. 2)* 19, 533 p (1957).

16. W. E. Ricker, Computation and interpretation of biological statistics of fish populations. *Bull., Fish. Res. Board Can.* 191, 382 p (1975).

17. D. A. Hancock, Population dynamics and management of shellfish stocks. *Rapp. P.-V. Reun., Cons. Int. Explor. Mer* 175(8): 8–19 (1979).

18. J. Mauchline, The Hiatt growth diagram for crustacea. *Mar. Biol. (Berlin)* **35**:79–84 (1976).

19. J. Mauchline, Growth of shrimps, crabs and lobsters—an assessment. *J. Cons., Cons. Int. Explor. Mer* 37(2) 162–169 (1977).

20. T. Savage, Effect of maintenance parameters on growth of the stone crab, *Menippe mercenaria* (Say). *Spec. Sci. Rep.—28, Fla., Dep. Nat. Resour., Mar. Res. Lab.* 28, 19 p (1971).

21. T. S. Cheung, The environmental and hormonal control of growth and reproduction in the adult female stone crab, *Menippe mercenaria* (Say). *Biol. Bull. (Woods Hole, Mass.)* **136**:327–346 (1969).

22. T. Savage and M. R. McMahan, Growth of early juvenile stone crabs, *Menippe mercenaria* (Say, 1819). *Spec. Sci. Rep.—Fla. Board Conserv.* 21 17 p (1968).

23. T. M. Bert, J. Tilmant, J. Dodrill, and G. E. Davis, *Aspects of the Population Dynamics and Biology of the Stone Crab* (Menippe mercenaria) *in Everglades and Biscayne National Parks as Determined by Trapping*, Report SFRC-86/04. National Park Service, South Florida Research Center, Homestead, FL, 1986. 77 p.

24. N. M. Ehrhardt, D. J. Die, and V. R. Restrepo, Assessment of a stone crab *Menippe mercenaria* stock in the Everglades National Park, Florida. *Proc. 1st Stone Crab Symp., 1986* (1986).

25. T. Savage, J. R. Sullivan, and C. E. Kalman, An analysis of the stone crab, *Menippe mercenaria*, landings on Florida's west coast, with a brief synopsis of the fishery. *Fla. Mar. Res. Publ.* 13, 37 p (1975).

26. V. R. Restrepo, On the growth character of the stone crab *Menippe mercenaria. Proc. 1st Stone Crab Symp., 1986* (1986).

27. G. J. Paulik and W. H. Bayliff, A generalized computer program for the Ricker model of equilibrium yield per recruitment. *J. Fish. Res. Board Can.* **24**:249–259 (1967).

11 CRAB FISHERIES AND THEIR MANAGEMENT IN THE BRITISH ISLES

ERIC EDWARDS
Shellfish Association of Great Britain
London, England

1. INTRODUCTION

Several species of crabs are found around the British coasts but the most important one fished for human consumption is the European edible crab, *Cancer pagurus*. This crab is particularly abundant in the coastal waters of northwest Europe, being found in large numbers where the seabed is rocky.

Although edible crabs are exploited in most European countries, the most important crab fisheries are found off Norway, around the coasts of Scotland, and along the northeast and south coasts of England (1). A large fishery for *Cancer* also occurs around France, mainly off Brittany, and around the Channel Isles. Smaller crab fisheries are found off Spain, Portugal, and Ireland.

In Britain, one of the main crab-fishing countries in Europe, about 9000 tonnes of edible crab, with a first-sale value of about £4 million, is landed every year. This figure represented around 10% of the total value of British shellfish landings during each of the years from 1980 to 1985.

Crabs are caught in baited traps, the design of which varies around the coast. The catch, which is landed alive, is utilized in various ways: a large proportion is processed and the meat extracted, or crabs are exported alive to various markets on the Continent. In some areas the crab catch is cooked and sold locally, especially during the holiday season. There is also a growing trade abroad in frozen, whole cooked crabs.

Given the economic importance to Britain of crab fisheries, it is not surprising that most of the biological studies on *Cancer pagurus* have been carried out by British scientists. This included the work of Professor A. Meek (2, 3), who studied the northeast of England crab stock from 1895 to 1925, and Dr. H. C. Williamson, who studied the Scottish fishery during the same period (4, 5).

More recent studies included work by my colleagues and myself when employed by the Ministry of Agriculture, Fisheries and Food at their Fisheries Laboratory, Burnham-on-Crouch. This research program, undertaken between 1960 and 1976, was part of a stock management survey covering the English crab fisheries, which included tagging experiments to evaluate growth (6–8) and migrations and observations on the biology of *Cancer* (1, 9).

This chapter describes the present British crab fishery and the measures currently used to manage the stocks. It also includes a description of the crab's life cycle—emphasizing various factors in the biology and behavior of this crustacean that affect catch levels or are important to the fishery and its proper management.

2. THE PRESENT DAY FISHERY

2.1. Landings and Value

During the past 5 yr the average annual U.K. crab catch has been around 9500 tonnes, with a first-sale value of £4 million (Table 1). Ports in England land the highest crab catches; on average 70% of the U.K. crab catch being landed in

TABLE 1   Crab Landings and Their Value in Scotland, England, and Wales, 1980–1984

Year	England and Wales[a] (tonnes)	Scotland (tonnes)	Northern Ireland (tonnes)	Total United Kingdom (tonnes)	Value (£1000)
1984	9070	4952	4	14,026	8422
1983	7310	4004	4	11,318	6373
1982	5404	3161	2	8,567	4246
1981	7104	2637	4	9,745	4258
1980	7147	2530	6	9,683	3968

[a] The figures for England and Wales included about 500–1000 tonnes of spider crabs (*Maia*) during this period.

Source: Ministry of Agriculture, Fisheries & Food.

England since 1978. Landings at Welsh ports are minimal and the remainder of the U.K. catch, 25–30% over that period, has been landed in Scotland.

The main English crab fishing grounds are off south Devon, around Cornwall, and along the south coast from Portland Bill to Dungeness in the English Channel (Fig. 1). Other important fishing areas are found along the east and northeast coasts of England including grounds off Norfolk, Yorkshire, Durham, and Northumberland (1).

In Scotland there are important crab fisheries on the east and west coasts and around the islands of Orkney and Shetland (10). Recently fishing has been expanded around the Outer Hebrides, where largely unexploited stocks offer scope for new developments.

2.2. Seasonality

In Britain some boats fish for crabs throughout the year, but the main season is from April to November. Regional differences in catch levels occur; on the east coast of England peak catches occur in May and June, but in the southwest of England catches are low in the spring but rise from July to November (Fig. 2). In Scotland, the most important months for crabs are from May to October but some landings are made throughout the year.

Crab landings in all areas are affected by various factors, such as weather or crab behavior—including molting and migrations—and by the seasonal emphasis on the capture of other fish or shellfish species. For example, in most areas lobsters are caught with the same type of traps and, because lobsters command a much higher price than crabs, fishermen may then concentrate on lobsters at times when they are most available.

The weather in particular controls the level of crab landings. Because crabs, like other crustaceans, are cold-blooded animals, they take their body temperature from their surroundings. Therefore, in cold water they are less active and their feeding is reduced; catches during the winter and early spring are poor. Landings

Figure 1. Principal crab fishing ports in Britain and Ireland.

can also be affected by stormy weather when boats cannot leave harbor to haul
their traps.

2.3. Fishing Gear

Crabs are caught in baited traps, sometimes called crab pots, set in coastal waters.
In an industry that is distributed along most of the British coastline, tradition and

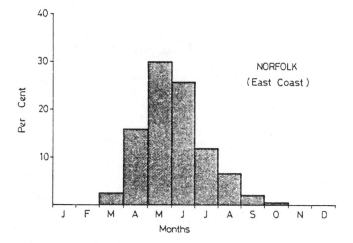

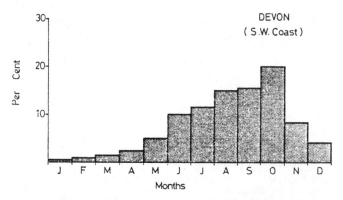

Figure 2. Monthly distribution of the crab catches in Norfolk and Devon (based on landings in the early 1980s).

individual preferences have influenced the pattern of gear design and fishermen use a variety of different traps (1). However, the two main types of trap used are the creel (Fig. 3), fished mainly on the east coast of England and all around Scotland, and the inkwell pot, which is mainly used on the south and southwest coasts of England and in parts of Wales.

Crabs have been caught in traps for generations. As early as 1874 it was reported by Holdsworth (11) that "crab and lobster fishing employ a large number of men at particular seasons. They are caught in a creel or cage trap, also called a "pot.' "

Creels are mainly constructed of wood and netting and usually have a base diameter of 24 in. × 18 in. (61 cm × 46 cm) and a height of 12 in. (30 cm). Three or four half hoops of hazel, cane, or sometimes plastic tubing support the netting, which is nowadays mostly synthetic material. The pot has two openings

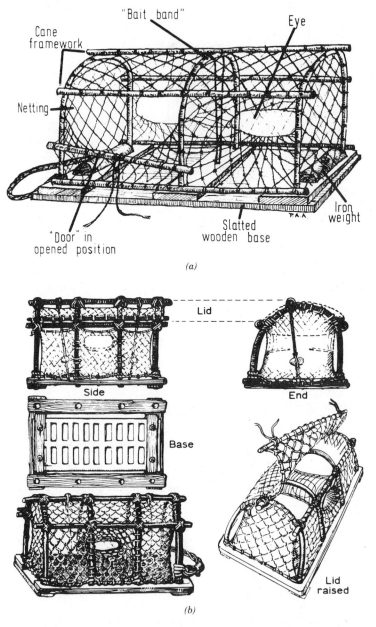

Figure 3. (a) A Yorkshire crab and lobster pot. (b) A Norfolk crab creel.

or *eyes*—each constructed in the form of a short tunnel and entering the pot from opposite sides, each eye having a diameter of about 5 in. (12.7 cm). Pieces of iron, concrete, or chain lashed to the base are used to weight the pot to prevent movement on the seabed. The bait, which is usually pieces of fish, is held between toggles in the *bait band*, which is a double length of stretched twine fitted from the top of the pot to the wooden base. These crab pots are usually made by the fishermen themselves, although manufactured frames are now available.

Variation in creel design is common; the Norfolk creel has four hoops rather than the three used in Scotland and the northeast of England (1). Sometimes larger creels are used in Scotland for crabbing, but the basic creel design is similar in all parts of the United Kingdom (10).

Circular pots—known as inkwell pots—have been used in southwest England and in some parts of Wales for centuries. In days gone by these were made from woven willow branches cut from specially planted willow stands, but today, although the shape of the pot has changed little, durable modern materials are used to make this type of crab pot.

Today fishermen can buy welded tubular polythene frame pots which they then cover with netting. In these pots the entrance is a wide opening at the top which is fitted with a plastic or glass fiber neck. The fish used as bait is held firmly to the neck by strong rubber bands or a skewer. This change to more modern designs has led to a subsidiary industry springing up making pots, pot frames, pot entrances, and also plastic swivels, which prevent the gear getting tangled in stormy weather. All this makes the crab fishery in southwest England more important as a source of employment than is generally realized.

2.4. Catching and Handling Methods

Crab fishing occurs at varying depths and distances from the shore. In Scotland and along the east coast of England crabbing usually involves daily excursions to the grounds, which may be found just outside the harbor entrance or up to 10 miles from port and within 2 and 6 miles of the coast.

In other parts of the country, particularly the English Channel area, the best crabbing grounds are often 20–30 miles offshore, and although most fishermen try to land their catch daily, some vessels are equipped with "vivier" storage tanks in which the catch can be kept alive for several days.

The fishing methods have remained unchanged for generations; most fishermen prefer fresh to stale bait for crabs and the pots are most effective if baited daily. Most crab fishermen work their traps in strings or "fleets" of between 20 and 80 pots set on the sea bottom at intervals of about 10 fathoms (18 m) along a main rope that has a buoyed anchor or sinker attached to each end. A string of pots should not have more creels than can comfortably be accommodated and baited, and so on, on the deck of the boat. For example, a 30-ft (9-m) boat may work 200 traps, whereas a 50-ft (15-m) boat can handle up to 500.

The traps are usually set on selected crab grounds, left to fish overnight, and then hauled the next day if weather allows. After hauling, the catch is removed, the traps rebaited, and the gear reset. At one time the traps were hauled by hand; today, however, mechanical capstans worked off the main engine or by hydraulic power are widely used to haul the pots up from the seabed to the surface. Mechanization—by the use of pot haulers—has helped to increase the number of grounds that can be exploited, because areas with strong tides and deep water can now be worked much more easily.

Crab boats are usually skipper-owned and are sometimes crewed by members of the same family. Large boats work on a share basis, with perhaps several of the crew owning part of the gear. In addition, several of the larger U.K. shellfish merchants now own and operate their own vessels.

Boats of a variety of classes and lengths, between 12 and 60 ft (4–18 m), are used to catch crabs. Half-decked boats from 30 to 35 ft (10 m), crewed by two men, are popular. There is, however, no restriction at present in Britain on the size of boat that can land crabs, and sometimes large trawlers will sell boxes of whole crabs or crab claws taken while fishing for demersal species such as cod, plaice, or haddock.

Experienced fishermen handle live crabs as little as possible; after careful removal from the traps they are checked to ensure they meet the legal landing size and then packed right side up and close together in boxes to prevent movement and fighting. Mortality is further reduced if the crabs are kept out of the hot sun and drying winds.

2.5. Commercial Utilization

After being removed from the traps the day's catch is packed alive into boxes or baskets on deck. A few crab boats have seawater storage tanks but in general most catches are transported back to port dry. Once taken ashore—usually on the same day as captured—the crabs are sold either by auction or by a prearranged agreement with a merchant or processor (12).

The methods of marketing vary. In some areas crabs are sent in large quantities to processing factories, where they are cooked by boiling in vats and then the meat is extracted and frozen. Picking out the meat is still mainly done by hand but mechanical extraction of the body and leg meat is nowadays more common in the U.K. trade. The most valuable meat is the white muscle extracted from the claws; the brown meat—which is mainly the liver and reproductive tissues—is used to make paste and crab spreads (12).

In contrast, some fishermen boil their own catch and have stalls on the seafront of seaside towns or villages where they sell crabs and other shellfish to holiday-makers. Merchants also buy crabs, boil or "dress" them, and distribute them fresh or frozen to inland towns and cities in England.

In recent years, expanded outlets for live crab in France and Spain have helped develop a major export business to the Continent. Good demand has increased prices, and many crab fishermen along the south coast of England sell most of their catch to shellfish exporters, who send them away alive by "vivier" lorries, each holding 3–5 tonnes of crabs and other shellfish in aerated, chilled seawater.

The growing demand for convenience foods has also affected our crab trade; some U.K. producers now sell ready-prepared crab dishes which sell as high-value products. Some of these products contain the cheaper white meat recovered by mechanized means from the small legs and body of the crab. But by incorporating it with other foods, for example, crab stuffed with plaice, new seafoods have been developed that attract the modern housewife who is uncertain how to handle a live crab.

3. THE LIFE CYCLE OF THE CRAB

The edible crab *Cancer pagurus* is one of the largest invertebrates found around Europe, and around Britain crabs and lobsters often inhabit the same areas where they compete for food and shelter.

Despite its reputation as a scavenger, the edible crab feeds mainly on living food, including small fish and marine worms. Shellfish such as mussels and barnacles are easily crushed in its powerful claws. It is extremely voracious and has a keen sense of smell that helps to find its food on the sea bottom.

3.1. Growth

Crabs, like all crustaceans, are covered with a hard rigid exoskeleton made of calcified chitin, and in order to grow, crabs have to cast these outer shells by molting or ecdysis. During the first few years of life, when growth is rapid, a crab molts several times a year, but by the time it becomes sexually mature molting may occur only once a year or even less frequently (1, 7).

In Britain the main molting period is from July to October each year. At this time the crabs move close inshore and seek shelter among the rocks. Molting commences when the hard outer shell splits along a precise line dividing the upper and lower halves, and the soft crab inside slowly backs out through the gap.

The crab, now in the soft-shelled condition, absorbs water and swells, increasing in size across the back by as much as 20–30% in one molt (6, 7). On average, a 3.5-in. (88-mm) male will reach 4.5 in. (115 mm) in one molt and 5.75 in. (146 mm) in the next. After a molt has been completed the new shell slowly hardens and the crab's body does not increase in size again until the next molt. Research has shown that crabs that molt do not return to the final stage of calcification for at least two, or sometimes three, months after ecdysis has taken place. Hardness is judged by pressing the shell to determine its flexibility. Soft-shelled crabs easily break whereas a good-quality crab with a hard shell cannot be broken. While the new shell is hardening the crab is called "soft-shelled" and by law it must be returned to the sea (6).

3.2. Breeding

The sex of a crab can easily be determined; the female or "hen" crab has a broad abdomen, whereas the male or "cock" crab has a narrow abdomen which fits tightly to the body, as shown in Figure 4. The claws of the male are also larger than those of a female of the same size.

Mating occurs in inshore waters during the summer, immediately after the female crab has molted and while it is in the soft-shelled condition (13). Prior to the molt, and for a period of up to a fortnight after, the female is attended by a hard-shelled male, which is attracted to the female by a scent or pheromone she exudes at this breeding time. Immediately after the female has cast her shell, mating takes place and the male sperm are introduced into the female's two sperm sacs. One supply of sperm may fertilize two or more batches of eggs in subsequent

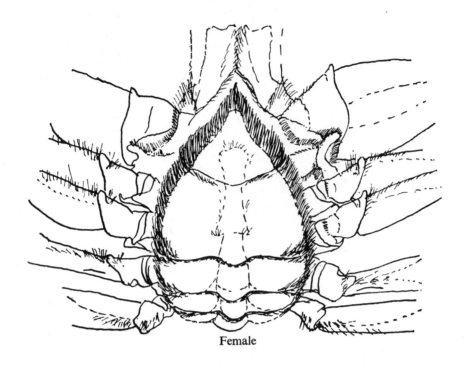

Female

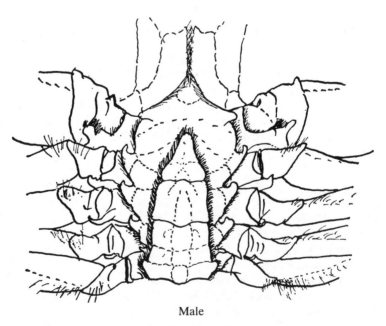

Male

Figure 4. The external abdominal features of male and female crabs.

years, and the majority of females that mate in July or August will spawn, that is, carry eggs, in November or December of the same year, but in some cases spawning is delayed until the next winter. Crabs usually select a soft seabed for spawning, often in deep water (14), and the eggs remain attached to the "swimmerets" on the abdomen of the parent for about 7 months (1). A crab with eggs is sometimes called a berried crab, and the number of eggs carried can vary from 0.5 million on a 5-in. (127-mm) crab to 3 million on a 7-in. (178-mm) one.

3.3. Hatching and Larvae

In the spring and summer following spawning, berried females move inshore, where the eggs hatch. Hatching times vary for the different stocks of crabs around our coasts but the main periods are all between May and September (1, 4).

The young larval crabs that emerge from the eggs have a shrimp like appearance and live among the free-floating plankton in the surface water layers for about 1 month (4, 5). This period is a dispersal phase, because the microscopic larvae can be transported considerable distances by water movements from where they first hatched. During this period they grow through several stages and by the time the larval crab has reached a size of 2.5 mm it settles to the seabed as the first postlarval stage. As described in the next section, the transport of crab larvae to an area where they will survive is aided by directed migrations by adult females.

3.4. Migrations

Although generations of fishermen have known that crabs move about on the coastal fishing grounds, the full extent of these migrations was never properly defined until marine scientists undertook extensive tagging studies in the 1960s and 1970s (1, 15, 16).

In these experiments, suture tags which are not lost when the crab casts its shell, were attached to large numbers of crabs of both sexes. Each tag was numbered and a record was kept of the size, sex, and shell condition of each crab marked in this way. Fishermen who caught tagged crabs were rewarded if they returned the crabs and also provided information on the date and position of recapture (17).

In 1965, in a major east coast tagging experiment, 3400 tagged crabs were released in the Norfolk, Yorkshire, and Northumberland crab fisheries. Just over a third were later recaptured by fishermen and the results added considerably to the available knowledge on crab movements along the east coasts of England and Scotland (1, 18).

For example, in the Yorkshire experiments a total of 135 female crabs were recaptured, and of these 40% were found to have moved distances of 20 miles or more. Nearly all these migrations were in a northerly direction, as shown in Figure 5, the crabs being caught in North Yorkshire, Durham, Northumberland, or Scotland. The longest straight-line distance moved was by a female released near Flamborough Head, Yorkshire, which was recaptured off Berwick (Northumber-

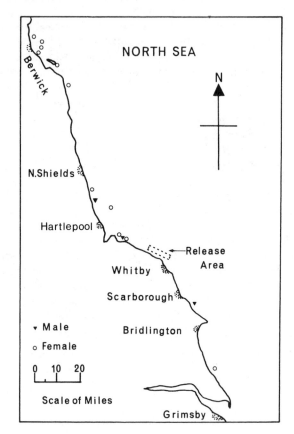

Figure 5. Migrations of recaptured suture-tagged crabs, released off north Yorkshire in May 1965, that had moved 20 miles or more.

land) 16 months later, having moved 163 miles. For male crabs the results were different; only 5% of those recaptured had moved distances of 20 miles or more. Tagging studies in southwest England (19) again showed that some crabs, particularly females, make extensive movements, mainly west or southwest down the English Channel. The maximum distance covered was 155 miles but the results confirmed extensive migratory movements with crabs moving from one part of the Channel to another.

These migration patterns undertaken by *Cancer pagurus* are associated with the crab's breeding behavior. The long-distance movements along the east coast of Britain and in the English Channel, identified by tagging experiments, are believed to be part of an offshore movement in the autumn, when the eggs are laid, followed by a return to shallower inshore waters in spring, when the eggs are hatched. Tagging experiments in Scottish waters have also confirmed that female crabs undertake considerable migrations, often in excess of 100 miles. Male crabs rarely move far from the point of release.

3.5. Meat Yield and Size

The cooked meat, which makes up the edible portions of the crab, consists of the white muscle from the claws, legs, and body cavities, and the brown meat that comes from the reproductive and digestive tissues of the body (12). Tests have shown that the total meat yield from the edible crab varies around the United Kingdom, and even between grounds in the same locality. Yields are affected by the winter fasting and by the molting and breeding cycle.

More white meat is extracted from male crabs, which have larger claws than females, whereas more brown meat is available in females, especially during the breeding season in late summer and autumn when the reproductive tissues are ripe. Male crabs also yield more meat in total than females of similar size.

The size of crabs caught around Britain is not uniform. In general, edible crabs grow to a larger size in southwest England than anywhere else in the country (9). In contrast, crabs along the east coast, especially off Norfolk and Yorkshire, have a smaller average size (15, 21, 22).

Growth rates have an effect on the size of crabs caught. Scientific studies using tagged crabs to evaluate growth have shown that the size a crab increases at a molt does not vary between the east and southwest coasts of England. However, in Devon and Cornwall crabs molt more frequently than in the Norfolk and Yorkshire fisheries (7). Furthermore, the studies showed that in the southwest, males molt more frequently than females (hence the large cock crabs in the area); the opposite is the case for east and northeast crab stocks. A valid reason for this difference has not yet been established.

3.6. "Black Spot" Shell Disease

Sometimes the quality of crabs is affected by black lesions on the shell, which is usually associated with pitting and a general discoloration of the exoskeleton. This naturally occurring condition in crabs—called "black spot" in the trade,—is caused by bacteria that attack the shell chitin. This shell necrosis or black spot is normally more evident in older animals and its incidence is greater in lightly fished crab populations (23).

In some areas crabs may be so badly affected that the shell is pitted and even underlying soft tissues are discolored. Then the commercial value is so reduced in these diseased crabs that fishermen will either kill or reject them. The microorganisms that attack the shell are widely distributed in the marine environment but are not harmful in any way to humans.

4. REGULATING THE CRAB FISHERY

4.1. History of Legislation

Management of the crab fishery in Britain is achieved mainly by control of the minimum landing size. In addition, the landing of berried (egg-carrying) and soft-shelled (recently molted) crabs is prohibited. There are no national restrictions on length of the fishing season, the number of traps worked, or the size of vessel.

Regulation of the English crab fishery first began in 1876 when the Crab and Lobster Fisheries (Norfolk) Act first introduced a minimum landing size for crabs of 4.25-in. (108-mm) width. This early act also restricted the sale of female crabs or lobsters carrying spawn (ovigerous). These conservation measures, which applied only to the Norfolk coast, were recommended by the eminent marine biologist Frank Buckland, who was concerned about the large-scale killing of small crabs and the destruction of crabs and lobsters carrying spawn (19).

Buckland continued his studies on the crab fisheries of England and a comprehensive report, prepared with his colleague Walpole, was published in 1877 (24). Their recommendations led to the Fisheries (Oysters, Crabs and Lobsters) Act 1877, which introduced statutory regulations for the whole of England, Wales and Scotland. These new conservation measures—the first of their kind for the British shellfish industry—included a minimum landing size for *Cancer* of 4.25 in. (108 mm) measured across the broadest part of the carapace, the protection of crabs carrying spawn, and the protection of soft-shelled crabs, that is, those that had recently molted and have a low meat yield, which improves, because their shell hardens, if they are returned to the sea unharmed.

In 1894, the Sea Fisheries Regulation Act authorized local Sea Fisheries Committees to impose further bylaws for the protection of the crab fishery. These committees introduced local bylaws regarding the taking of crabs, and therefore there was some variation between the eleven sea fisheries areas in England and Wales. For example, along the east coast of England the tendency has been to prohibit the use of soft-shelled or juvenile crabs for bait in the long-line fishery for cod, and in parts of the south and southwest of England, where crabs are larger than on the east coast, the legal landing size was raised on a local basis to 5 in. (127 mm).

No further attempts to regulate the crab fisheries were made during the 74 yr between 1877 and 1951, when it was agreed to raise the crab landing size from 4.25 in. (108 mm) to 4.5 in. (115 mm) in an attempt to give better protection to the stocks. This was introduced with the Sea Fishing Industry (Crabs and Lobsters) Order 1951, which increased the national minimum size; the protection of berried and soft-shelled crabs was retained as specified under the old 1877 Act.

To date, the setting of the legal landing size (i.e., first at 4.25 and later at 4.5 in. shell width), has been mainly based on the commercial requirement for better-quality crabs on the market.

In the 1960s scientific studies I carried out for the Ministry of Agriculture, Fisheries & Food—on east coast crab stocks (15, 21, 25, 26) concluded that the 4.5-in. size limit was appropriate for the Norfolk and Yorkshire crab fisheries. However, in the late 1960s, fears that the important southwest crab grounds were being overexploited led to an intensive study by Bennett and Brown (27). The program, which took 5 yr to complete, included observations at sea aboard commercial vessels and at processing factories to record the size and sex composition of the crab catch on different grounds, and tagging experiments—using the suture tag, which is not lost at molting—to determine growth and migrations and general observations on the biology of the crab [7]. Considerable effort was spent on collection of catch statistics, including a logbook scheme to evaluate catch-per-unit effort in the fishery.

Bennett (28) carried out a yield assessment aimed at determining the ideal minimum crab landing size for the southwest fishery using information on catch and effort, population structure, growth, and mortalities. Molt increments, annual molt frequency, and hence annual growth, were determined by suture tagging experiments. Using likely values for mortality parameters in a simple yield model, the probable effects on yields per recruit were determined for various minimum sizes. It was determined that an increase in the minimum size of both males and females would increase the yield, and a recommendation was made that the minimum landing size for females should be raised to 5.5-in. (140 mm) carapace width and for males to 6.5-in. (160 mm) carapace width.

Bennett and Brown (27) proposed fishery management actions that would implement two fundamental changes to the U.K. fishery regulations. These were:

(a) It would be necessary to consider the management of crab fisheries on a regional rather than national basis.

(b) Different minimum sizes would be required for male and female crabs if the maximum sustainable yield for each were to be achieved.

Despite support from trade associations and fishermen themselves, the recommendation to raise the legal landing size—based on a regional size limit proposed in 1976—was not implemented until 1986, when it was introduced on April 4 in the form of the Undersized Crabs Order 1986. This legislation introduced regional minimum sizes for edible crabs (*Cancer*) landed in the United Kingdom from British fishing boats.

The new order reflected the scientists' concern over overexploitation of some stocks by raising the legal landing size from a uniform 115 mm to 160 mm for males and 140 mm for females in certain areas (i.e., Devon and Cornwall); while other parts of the country still retain 115-mm limits, or have a carapace length of 140 mm for both males and females (south coast of England). Legislation has also been introduced to ban the landings of crab claws.

4.2. Management and Development of Crab Stocks

I have been involved in programs to consider both the conservation and further exploitation of crab stocks in England, Scotland, and Ireland (26). What is surprising is that although records (1) are available to show that English and certain Scottish crab stocks have been exploited since the 1800s, most stocks appear to have stood up reasonably well to years of fishing, although there are few hard data on stock sizes and dynamics.

Tagging has shown that there is considerable interchange of adults between different populations, and we know that larval dispersal by tides and currents is wide. Also, female crabs in the egg-bearing stage are rarely taken in the traps. These behavioral factors all have helped to maintain stock levels.

Despite the fact that mobile crab stocks seem to be able to cope with varying levels of fishing mortality over a period of years, I still believe that stock management is essential. In fact, management of a crab or other shellfish stock may be required for a variety of reasons, although in practice the most pressing justifications are poor or falling catch rates and overcapitalized fishing fleets. Nowadays, with many stocks being developed or fully developed, the role of the fisheries manager assumes greater importance. Sophisticated theories of fisheries management can be used to considerable advantage when there is adequate biological information plus accurate data on landings and fishing effort. In practice, however, all too often only limited data are available, or it would take too long, or be too costly to collect *all* the data required to fit a dynamic model.

Even so, it is dangerous to make fishery management recommendations without some knowledge of the exploited stock. We need to know something about the biological parameters of a shellfish stock. This should include the following.

The size of the population and its distribution, are important data, bearing in mind that these factors may vary throughout the year and from year to year.

Mortality rates, which include losses from both natural and fishing mortality, must be estimated—a parameter so hard to quantify with shellfish. Typical values for mortality rates are around $Z = 0.5$–0.6, $F = 0.4$–0.5, and M in the vicinity of 0.1. In all cases, a slightly lower mortality rate was noted for females (28).

Growth rates must be known. With crustaceans such as crabs and lobsters, simple tagging techniques are not applicable owing to shell casting, but in recent years basic growth data have been collected by using specialized tags, which survive the molt and provide data on molt increment and frequency.

Recruitment is a parameter that is now receiving a great deal of attention from fishery scientists. Two main themes are as follows:

(a) What are the natural factors that lead to fluctuations in recruitment from year to year? Is it possible to forecast the abundance of recruits that will enter the fishable stocks?

(b) Is it possible to increase the level of recruitment by protecting the breeding stock of ovigerous females or by creating better conditions for the survival of the young shellfish?

Four additional biological features that are of extreme importance to management are the *longevity* or the life-span of a species, its *catchability* or vulnerability to fishing gear, and the nature and magnitude of any *migration* in and out of a fishery. The *fecundity* of a species, as related to its potential for producing young, will also be of special importance.

All the above factors must be studied by the fisheries biologist considering management recommendations, although in practice many decisions on conservation will have to be made using a more simple approach, because several of the above parameters are not easy to obtain for shellfish stocks.

4.3. Fishery Regulations

Earlier in this chapter the management regulations used to protect British crab stocks for more than 100 years were described. Most of these were based on some elementary biological knowledge plus economic considerations.

Let us, however, consider on a broad basis the type of controls that potentially could be used to manage exploited stocks of shellfish such as crabs, and achieve a sustained yield.

1. Control on the size of shellfish landed by a minimum legal landing size or by gear selectivity (escape gaps in traps) that allows the juveniles to escape and remain free to grow to a size providing a higher yield.

2. Control and adjustment of fishing mortality either by regulating the level of fishing effort or the efficiency of the gear or by limiting the quantity of the catch.

3. Improvement of recruitment by ensuring that crabs are not caught before they reach maturity. Protection of ovigerous females or restriction of fishing by means of closed seasons or areas to protect breeding stocks. With some high-priced species it may be possible in the future to increase production by culture techniques or stock enhancement.

4. Habitat improvement, especially pollution control and the provision of sanctuaries or extra shelters such as artificial reefs to increase production.

4.4. Management and Industry

The aim of fisheries management is to improve the yield from a fishery and to prevent overfishing. However, most fishermen accept that catches vary year by year and are affected by factors beyond their control (e.g., weather, size of normal recruitment).

The concept of maximum sustainable yield still receives support in some quarters, although in reality economic forces have greater impact than biological ones. It is not surprising that most fisheries administrators, the industry, and scientists now accept that decisions on management policy must also take into account the fishermen and their economic situation. Furthermore, policy makers must accept that it is often difficult for fishermen to comply with new management regulations unless this is the case.

As explained earlier in this chapter, most rules or regulations for controlling fisheries arise out of certain theoretical considerations that relate stock abundance to the level of fishing mortality. It is implicit in the various regulations that when the fisherman is told to cease fishing or put the shellfish back, his future harvest will be better, or at least no worse, than if he had kept and landed them. Although this is fine in theory, much money is spent by our governments under the term enforcement—to force the fisherman to conserve—for his future benefit. These costs also must be included in the economic equation for the fishery.

This chapter has tried to explain how fairly simple regulatory measures have helped to conserve the U.K. crab fishery over a period of many decades. I have emphasized the need to combine biological knowledge with the economic situation, especially market demand.

It is essential to have cohesive programs that involve fishery scientists, economists, and representatives from the industry—because there is a joint responsibility to see that management measures will be practical and effective— and that are supported by the majority of fishermen, processors, and others in the shellfish industry.

REFERENCES

1. E. Edwards, *The Edible Crab and its Fishery in British Waters.* Fishing News Books, Farnham, Surrey, England, 1979, 142 pp.

2. A. Meek, *The Crab and Lobster Fisheries of Northumberland.* Report of the Northumberland Sea Fisheries Committee, 1904.

3. A. Meek, *The Migrations of Crabs*, Report No. 2, Dove Marine Laboratory, Cullercoats, Northumberland, 1914, pp. 13–20.

4. H. C. Williamson, Contributions to the life-history of the edible crab (*Cancer pagurus*, Linn). *Rep. Fish. Board Scotl.* **18(3)**:77–143 (1900).

5. H. C. Williamson, Contributions to the life-histories of the edible crab (*Cancer pagurus*) and of other Decapod Crustacea. *Rep. Fish. Board Scotl.* **22 (3)**:100–140 (1904).

6. E. Edwards, Observations on the growth of the edible crab (*Cancer pagurus*). *Rapp. P.-V. Reun., Cons. Int. Explor. Mer* **156**:62–70 (1965).

7. D. B. Bennett, Growth of the edible crab (*Cancer pagurus* L.) off south-west England. *J. Mar. Biol. Assoc. U.K.* **54(4)**:803–823 (1974).

8. D. A. Hancock, and E. Edwards, Estimation of annual growth in the edible crab (*Cancer pagurus* L.). *J. Cons., Cons. Int. Explor. Mer*, **31**:246–264 (1967).

9. C. G. Brown and D. B. Bennett, Population and catch structure of the edible crab (*Cancer pagurus*) in the English Channel. *J. Cons., Cons. Int. Explor. Mer* **39(1)**:88–100 (1980).

10. H. J. Thomas, Lobster and crab fisheries in Scotland. *Mar. Res.*, **8**:1–107 (1958).

11. E. W. H. Holdsworth, *Deep-sea Fishing and Fishing boats.* Edward Stanford, London, 1–74.

12. E. Edwards and J. Early, Catching, handling and processing the edible crab. *Torry Advis. Note* **26**:1–17 (1967).

13. E. Edwards, Mating behaviour in the European edible crab (*Cancer pagurus* L.). *Crustaceana* **10** (Pt. I):23–30 (1966).

14. A. E. Howard, The distribution and behaviour of ovigerous edible crabs (*Cancer*

pagurus) and consequent sampling bias. *J. Cons.*, *Cons. Int. Explor. Mer* **40**:259–261 (1982).

15. E. Edwards, The Yorkshire crab stocks. *Lab. Leafl.—Fish. Lab. Burnham-on-Crouch-(New Ser.)* **17**:1–34 (1967).

16. D. B. Bennett and C. G. Brown, Crab (*Cancer pagurus*) migrations in the English Channel. *J. Mar. Biol. Assoc. U.K.* **63**:371–398 (1983).

17. E. Edwards, The use of the suture-tag for the determination of growth increments and migrations of the edible crab (*Cancer pagurus*). *ICES, C.M.* Doc. 42 (1964) (mimeo).

18. E. Edwards, Further observations on the growth of the edible crab (*Cancer pagurus*). *ICES, C.M.* Doc. 17 (1966) (mimeo).

19. F. Buckland, *Report on the Fisheries of Norfolk Especially Crabs, Lobsters, Herring and the Broads.* H.M.Stationery Office, London, 1–75.

20. J. Mason, The Scottish crab tagging experiments, 1960-61. *Rapp. P.-V. Reun., Cons. Int. Explor. Mer.* 156 (12) (1965).

21. E. Edwards, The Norfolk crab fishery. *Lab. Leafl.—Fish. Lab. Burnham-on-Crouch (New Ser.)* **12**:1–23 (1966).

22. C. G. Brown, Norfolk crab investigations. *Lab. Leafl.—Fish. Lab. Lowestoft* (New Series) **30**:1–12 (1975).

23. P. A. Ayres and E. Edwards, Notes on the distribution of "Black Spot" shell disease in Crustacean fisheries. *Chem. Ecol.* **1**:125–130 (1982).

24. F. Buckland and S. Walpole, *Reports on the Crab and Lobster fisheries of England and Wales, of Scotland and of Ireland.* H.M.Stationery Office, London, 1877.

25. D. A. Hancock, Yield assessment in the Norfolk fishery for crabs (*Cancer pagurus*). *Rapp. P.-V Reun., Cons. Int. Explor. Mer* **156**(13):81–93 (1965).

26. E. Edwards, A contribution to the Bionomics of the Edible Crab (*Cancer pagurus*) in English and Irish Waters. Ph.D. Thesis, National University of Ireland, 1971.

27. D. B. Bennett and C. G. Brown, The crab fishery of south-west England. *Lab. Leafl.—Fish. Lab. Lowestoft* **33**:1–11 (1976).

28. D. B. Bennett, 1979. Population assessment of the edible crab (*Cancer pagurus* L.) fishery off south-west England. *Rapp. P.-V. Reun., Cons. Int. Explor. Mer* **175**:229–235 (1979).

12 NORTHWEST ATLANTIC SNOW CRAB FISHERIES: LESSONS IN RESEARCH AND MANAGEMENT

Richard F. J. Bailey
Department of Fisheries and Oceans
Institut Maurice-Lamontagne
Ste-Flavie, Québec, Canada

and
Robert W. Elner
Department of Fisheries and Oceans
Halifax, Nova Scotia, Canada

1. INTRODUCTION

The snow crab, *Chionoecetes opilio*, is one of the most widely exploited decapod crustaceans, in terms of scale and distribution of landings, in the world. Various nations harvest snow crabs in the deep, cold waters of the northwest Atlantic, the northern Pacific, and in the Sea of Japan. What is presently the major fishery for snow crabs developed off Atlantic Canada over the past 20 years, and is the subject of this chapter (Fig. 1).

Initially, snow crabs were an exasperating nuisance to Canadian groundfisherman who were tired of cleaning ''spiders'' off their nets. Eventually, however, the commercial potential of this by-catch was recognized, and a directed trap fishery was initiated in 1966 after exploratory vessels discovered extensive grounds in the Gulf of St. Lawrence. Subsequently, the fishery increased steadily in importance as further commercial concentrations were located in the Gulf, around Cape Breton Island, and off the coast of Newfoundland (1). Currently, the snow crab is the fourth most valuable species landed in eastern Canada, after cod, lobsters, and scallops. All the commercial grounds are on mud or sand–mud substrates at depths of 60–200 m where bottom temperatures remain between −1 and 4°C year round. Because female snow crabs do not attain an acceptable size, commercial exploitation is focused exclusively on larger males.

The Canadian snow crab fishery is conducted by approximately 400 vessels using baited traps as their exclusive gear type (Fig. 2). Vessels less than 14 m long exploit nearshore grounds around the Gulf of St. Lawrence and Cape Breton

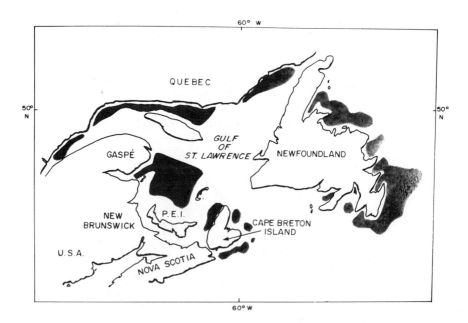

Figure 1. Snow crab fishing areas (shaded) in eastern Canada.

Figure 2. Hauling a commercial snow crab trap at sea.

Island, where 1-day trips are feasible. These vessels may also pursue other commercial species, such as lobster or cod, depending on the season and potential profits. However, the majority of snow crab landings are made by vessels 14–21 m long that often fish snow crabs exclusively and are capable of making up to 3-day trips onto offshore grounds. Offshore vessels in the Gulf of St. Lawrence and most inshore vessels around Cape Breton Island use large rectangular traps. The basic pattern, borrowed from the Alaska king crab fishery, consists of a steel frame covered by polypropylene netting with two entrances on opposite sides. The initial dimensions of $1.5 \times 1.5 \times 0.6$ m were increased to $1.8 \times 1.8 \times 0.9$ m in the early 1980s. Owing to handling and depth considerations, fishermen off Newfoundland and the north shore of the Gulf of St. Lawrence favor Japanese-type conical top-entry traps, with a 1.2 m base diameter, 0.65 m height. Conical traps may be fished in strings of up to 60, whereas the rectangular traps are set individually. For regulation purposes, one rectangular trap is judged equivalent to two conical traps, reflecting the approximate fishing power ratio (2). Trap soak times commonly vary between 1 and 3 days.

Biological assessments are performed annually for most Canadian Atlantic snow crab management areas. Data from commercial catch sampling (3–6), research cruises, official sales slip records of landings, mark–recapture studies, and fishermen's logbooks may all be considered in the assessment procedures. Currently, however, trends in catch-per-unit-effort (CPUE) indexes, gleaned from the mandatory logbooks, and landing records, are usually the primary consideration (7–10). Although research and management procedures have generally followed established fisheries methodology, the population dynamics of the northwest Atlantic snow crab resource remains enigmatic. The apparently inconsistent responses of the resource to management and the convoluted evolution of biological understanding of the snow crab therefore form the "lessons" of this chapter.

2. MANAGEMENT

Early in the development of the snow crab fishery, the Canadian government, in collaboration with industry introduced some basic management measures, which were essentially dictated by economic considerations. Undamaged live snow crabs, large enough to be processed easily, were required. To achieve this, fishing with potentially destructive gear such as trawls was forbidden and only traps were allowed. Recently molted soft-shelled crabs were to be avoided or returned to the sea when caught. In addition to having a low survival rate, soft-shelled crabs give a reduced meat yield of poor quality (11). Regulations prevented females from being landed owing to their small size and relatively short legs. Initially, only males larger than 102 mm across the carapace (carapace width, CW) were permitted to be landed, for this was considered the minimum size profitable to handle. However, with experience, industry became able to process smaller crabs and the minimum size limit was decreased to 95-mm CW.

The initial set of regulations had the additional advantage of being biologically justifiable. Female brood stock was entirely protected. Males were considered to be mature and able to participate in reproduction well before their recruitment to the fishery (12). Finally, soft-shelled crabs were allowed to harden before being retained, thus maximizing their meat yield.

In the mid-1970s, effort restrictions were introduced in order to control overcapitalization in the snow crab fishery. Access to the resource was controlled by a licensing policy and the number of traps fished per boat was limited. The yields and the markets being favorable, snow crabs became a lucrative harvest for the licensed participants; soon less fortunate fishermen exerted pressure to enter the fishery.

Management concern then shifted toward allocation problems. Inshore zones were created and limited numbers of local fishermen were granted exclusive access. To maximize the number of participants, crab fishing in these zones was deemed as supplementary to other fisheries. Thus fishermen were not to depend solely on crab for their annual income. Short seasons were implemented in some areas and fewer traps per boat than on the major fishing grounds were permitted. In a few zones, a total allowable catch (TAC) was set and each boat was allocated an individual quota.

On the major fishing grounds, regulations were expanded to cope with new problems as they arose. With increasing exploitation, soft-shelled crabs became difficult to avoid. A closed season during the molting period was finally found to be the only effective tool to reduce the landing of these commercially undesirable crabs. The at-sea sorting of crabs below the minimum legal size was greatly improved by the introduction of a minimum mesh-size measure that allowed escape of undersized crabs from the traps. Finally, precautionary TACs were set when effort controls were not considered sufficient to curtail sharp increases in fishing intensity.

The actual management decisions pertaining to snow crabs are made through the Federal Minister of Fisheries and Oceans or his representatives, the regional Directors General. The Atlantic coast of Canada is divided into four administrative regions. Each regional Director General, with his staff, is responsible for the development and the implementation of a management plan for the snow crab fisheries under his jurisdiction.

Before any important management decisions are made, the industry is consulted through Regional Snow Crab Advisory Committees. These committees meet at least once a year to debate current and proposed regulations, allocation of the resource, and any difficulties encountered by the fishery. Members represent fishermen, processors, and provincial and federal governments. The results of their deliberations are communicated for consideration by the fisheries managers.

A formal scientific body, the Canadian Atlantic Fisheries Scientific Advisory Committee (CAFSAC), is requested to generate annual advice for consideration by the advisory committees and fisheries managers. CAFSAC members are mostly government biologists familiar with resource assessment. The CAFSAC Steering Committee issues scientific advice on snow crabs, based on the work of a specialized Invertebrates Subcommittee. Working papers, prepared by snow crab biologists, are presented and discussed during meetings of this subcommittee.

Ideally, managers seek precise estimates from CAFSAC on resource size, rate of exploitation, and expected annual production for each management unit. Furthermore, they would like to know when molting will occur and what maximum annual catch could be sustained without endangering the perennial nature of the fishery. Of course, such information is not available for each management unit. Consequently, controversies within advisory committees are often created and are difficult to resolve without solid scientific arguments. Thus the development of management plans is usually an arduous exercise and CAFSAC is under pressure to provide whatever guidance it can. The advice is generated, but usually at the expense of several unverified assumptions, on the basis that a professional, usually conservative "guesstimate" is often more expedient than no estimate at all. However, confidence in the CAFSAC process may be compromised when these assumptions are subsequently proved wrong, or when the initial scientific foundation, sometimes assumed definitive by CAFSAC's clients, is modified by new information.

3. ASSESSMENT METHODOLOGIES AND EVOLVING KNOWLEDGE

Biological advice, from both stock assessments and more general scientific studies, has always been regarded as critical for snow crab management. Preliminary work

(12–17) established the initial biological foundations for management. The subsequent strategy has been to adopt simple but usable assessment techniques and, while acknowledging their deficiencies, apply them routinely until new approaches can be developed to elucidate aspects of crab biology and population dynamics. Models developed for the management of finfish stocks have been discarded, mostly because the input parameters are either unknown or inappropriate for snow crabs (e.g., age and growth) and the required long-term quality data series are unavailable. The following section presents the techniques applied with variable success and describes the evolving state of biological knowledge.

3.1. The Leslie Method

Most quantitative assessments of snow crabs are made with the Leslie method (18). The technique relies on a declining trend in CPUE with effort over the fishing season. Because most snow crab grounds are under relatively high exploitation, this drop in CPUE over a season is generally significant. Mean CPUE values, obtained from fishermen's logbooks, are plotted against the cumulative catch for the whole fishing fleet, for regular time intervals through the fishing season. A regression line is then fitted to the data by the equation

$$\text{CPUE}_t = qB_0 - q \sum C_t \tag{1}$$

where CPUE_t is the mean catch rate over the time period t,
$\quad\quad\;\; \sum C_t$ is the cumulative catch up to the middle of the same period,
$\quad\quad\;\; q$ is catchability, the fraction of the commercial biomass captured by a single trap haul, estimated by the slope of the regression line, and
$\quad\quad\;\; qB_0$ is the intercept on the y axis.

The initial commercial biomass (B_0), estimated by dividing the intercept (qB_0) by the catchability (q), corresponds to the intercept of the regression line on the cumulative catch axis (Fig. 3). Thus in theory, CPUE would drop to zero if all the initial biomass was caught. Inherent assumptions of the model are that no significant input of biomass or losses due to causes other than fishing occur during the season and that catchability remains constant over time.

The rate of exploitation is the ratio of the season's landings over the initial commercial biomass. The biomass (B_t) present at any given period can be found from

$$B_t = \frac{\text{CPUE}_t}{q} \tag{2}$$

Molting periods, noticeable by increases in the frequency of soft-shelled crabs and by concurrent rises in mean CPUE reflecting an increase in biomass, are excluded from the regression per se. However, equation 2 sometimes permits an estimate

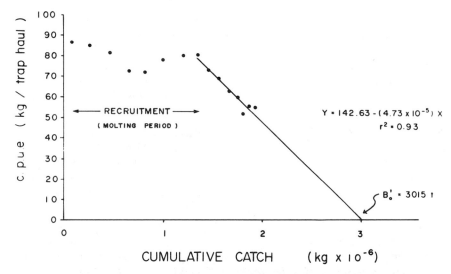

Figure 3. Estimation by Leslie's method of the snow crab harvestable biomass in north-western Cape Breton Island for 1978. A molting period during the initial part of the fishing season was eliminated from the regression. The estimated initial biomass was
$$CPUE_1/q = 87/4.73 \times 10^{-5} = 1839 \text{ t}.$$
Therefore, recruitment contributed 1176 t in a total harvestable biomass of 3015 t.

of biomass increases from growth and recruitment through molting during the fishing season (19). In such cases, the exploitable biomass during a fishing season will become larger than the initial biomass.

In practice, effort is not standardized for soak time in the assessments because logbook data are not usually amenable to such resolution. Nonetheless, when the potential biases of soak time on Leslie analysis were investigated (7, 20), the estimates generated using only defined soak times were close to those from pooled data from the whole range of soak times. Comparisons with Petersen estimates from mark–recapture data sometimes have corroborated the Leslie estimates but sometimes have also stressed the weaknesses of the methods (19, 21, 22). Moreover, it has been pointed out by Morgan (23) that both these methods can be similarly biased.

Overall, although Canadian snow crab biologists acknowledge the deficiencies of the Leslie technique, they continue to utilize it, especially because no practical effective alternative has been developed. Detailed accounts of the inherent assumptions and practicality of the Leslie technique are found in references 24–26.

Ideally, commercial biomass and recruitment histories can be built up over time for snow crab in a given management unit by using Leslie analyses. Indeed, such a chain of estimates has facilitated the stabilization of a fishery off the northwest coast of Cape Breton Island. The level of harvesting was set by TACs approximating the estimated mean annual biomass increases resulting from growth and recruitment in previous years (27). This management strategy has been successful because the annual replenishment of the resource on the fishing grounds has been relatively stable, representing approximately 50% of the exploitable biomass. However, for most other management units a history of reliable Leslie estimates has not been achieved, or the recruitment dynamics have proved too unstable to permit a realistic mean annual TAC to be instigated.

Scientific advice by CAFSAC on the status of the crab resource is generally made in reference to a target rate of exploitation of 50–60%. The initial justification for this target level was its effectiveness in giving stable yields off northwest Cape Breton. However, further rationale is that higher rates of exploitation result in a greater dependence on the vagaries of recruitment, an increased incidence of soft-shelled, recently recruited crabs, and lower average catch rates. For most crab fisheries, management adjustments are made on the basis of previous years' estimated rate of exploitation in comparison with the target level. The approach is a posteriori and does not account for potential regional and annual variability in production levels. In fact, the 50–60% target rate of exploitation suitable for northwest Cape Breton was too high for the eastern Cape Breton crab grounds, where productivity was so low that the fishery collapsed (8, 27).

3.2. Tagging and Alternative Methods

Biomass and rate of exploitation estimates obtained from mark–recapture studies have been used to supplement, or occasionally, replace Leslie estimates (19, 21, 22, 28). Three types of tags have been used in Canadian snow crab studies: (1) A spaghetti-like vinyl tube tied around the cephalothorax, printed with instructions for return and individually numbered (29), is the most visible tag but it cannot be retained through a molt. (2) The T-bar tag, also made of vinyl and printed with similar instructions and numbering, although less visible to fishermen, has some potential for retention through a molt (30). (3) A small magnetic tag implanted internally (28) carries limited information and is recovered by magnetic detection equipment. A recent study (G. V. Hurley, unpublished report) has shown that although retention through a molt is better than for the T-bar tag, the magnetic tag can be lost or cause limb deformities. Despite their inherent problems, the tags have generated some information on growth and movement patterns (29–31). In general, the high cost in labor and materials have relegated tagging studies to a secondary position in snow crab research and assessment. Attempts are still being made to develop more effective tagging techniques for long-term growth and movement studies.

Biologists have tried to develop alternative methodologies for estimating the abundance of snow crab. Trawls, traps, submarines, and video and still cameras have all been tested. For instance, the effective fishing area of a trap has been determined by a mark–recapture technique (32–34) and by comparison with photographic estimates of crab density (35). Density estimates have been used to tentatively assess resource size on some fishing grounds (17, 35–38). Although various methodologies are still under development, underwater cameras may ultimately prove the most reliable technique to measure snow crab density (39).

3.3. Size Frequencies

Historical data on size frequencies and mean carapace widths from port and at-sea catch sampling are often used in the assessments (40). Shifts in the mean carapace width and relative abundance of large individuals reflect the combined action of recruitment, growth, and exploitation. The interpretation is generally based on a qualitative examination of the data. A gradual disappearance of the largest size classes suggests that exploitation levels are outpacing growth into these size classes. A sudden increase in the relative abundance of the small size classes, on the other hand, is a good indication of a recruitment pulse entering the fishery.

A more objective approach to studying changes in size-frequency distributions, the factorial analysis of correspondence, led to the development of a working hypothesis on recruitment processes (41). This method clearly identified the existence of two annual pulses of recruitment into the fishery. Based on the proposition that not all prerecruits are on the fishing grounds prior to their molt to commercial size in the summer, the hypothesis postulates that those already on the grounds would enter the fishery in early autumn following the molt period, and the remainder would enter the next spring after a period of migration. Tagging studies have been initiated and may help to test the hypothesis.

Growth at molt can be estimated from modes in the size-frequency distributions. In his study on juveniles, Robichaud (42) describes these modes as different molt classes and, following Hiatt's method (43), he obtained the following growth equation:

$$L_{n+1} = 1.351L_n + 0.671$$

where L_n and L_{n+1} are premolt and postmolt carapace widths in mm, respectively. Such modes are also identifiable in some populations of commercial-size crabs (44). Differences in size between successive molt classes were used to estimate size increments ranging from 32% at a premolt carapace width of 71 mm, to 20% at a premolt width of 118 mm. In comparison, Miller and Watson (16) observed an average size increment of 18.4% in crabs kept in captivity. Unfortunately, modes are not always visible in the size frequencies of commercial-size crabs. The variability in the individual growth rates may indeed be expected to blur the presence of such modes at larger sizes.

3.4. The Soft-Shell Problem

Although shell condition is routinely monitored as an indicator of growth and recruitment, molting also has product quality implications. As indicated previously, recently molted crabs, termed white crabs by the industry, are unacceptable for processing. To avoid landings of white crabs, managers require a practical technique for indexing shell hardness, a knowledge of the time required by crabs to regain an acceptable commercial condition after molting, and finally a method to forecast the molting period.

Devising a quantitative yet practical index for shell hardness continues to prove difficult. Presently, effective definitions for shell hardness in general and white crabs in particular remain unresolved. Snow crabs are currently categorized as soft-, intermediate-, or hard-shelled on the basis of coloration and resistance to thumb pressure applied across the propodus of the chela (45). However, such a subjective index is prone to considerable individual bias; effective quantification is particularly arbitrary when crabs with intermediate-stage shells are present. Although other indexes—setal development (46–48), hemolymph protein concentration (49), relative water content (38), and shell breaking point (50, 51)—may be more accurate, they all require sophisticated equipment and trained personnel and are impractical for use on a large scale.

Observations of snow crabs in captivity (52) and tagged individuals in their natural habitat (53) indicate that a newly molted crab takes up to 2 months to recover to an acceptable commercial condition, yielding at least 20% of fresh meat. Consequently, a fishing season closure must last at least 2 months. On most grounds, the actual molting period is confined to 1 or 2 months in the summer. However, this period may show considerable intra- and interground variability in starting date and duration. In the southwest Gulf of St. Lawrence, actual fishing has been postulated to have caused increased recruitment, and hence a white crab problem (54). In contrast, low bottom temperatures may be related to the absence of molting off Newfoundland in certain years (D. Taylor, personal communication). Factors such as photoperiod and intraspecific communication by pheromones are other potential factors controlling the onset of molting (48). Certainly the biological basis for molting in snow crabs requires further probing before effective molt forecasting can be achieved and the white crab problem resolved.

3.5. Population Modeling

Simple approaches to modeling the population dynamics of snow crabs have been attempted and failed. The first approach, under the assumption that the resource productivity is mostly dictated by fishing mortality, was to fit production models for the southwest Gulf management area (55). The Graham–Schaeffer logistic model and the Gulland–Fox exponential model were fitted to the data from this major fishery for the years 1968–1977. Because the fishery was obviously not stabilized, equilibrium conditions were approximated by Gulland's method (56) with a 2-yr averaging of effort. The results indicated that maximum sustainable yield (MSY) had been attained and that further increases in effort would not increase landings but, rather, would negatively affect CPUE and hence the profitability of the fishery.

Subsequent performance by the fishery in the southwest Gulf proved that the indications from the production models were false. From 1977 to 1980, total landings increased by 56% from 9450 t to 14,757 t. Landings reached a maximum of 28,543 t in 1982 before they were finally stabilized by a TAC of 26,000 t in 1984. Further testing with production models was abandoned when the quality of the effort and catch data came under serious questioning in recent years. It has also been suggested (40) that the discovery of all the fishing grounds in the southwest

Gulf was not completed before the mid-1970s. Hence relative abundance indexes from 1968 to 1977 would be biased owing to an expanding resource base with virgin grounds being incorporated into the fishery.

Various hypotheses to interpret the dynamics of the resource in the southwest Gulf have been proposed (27). The increases in relative abundance since 1977 can reflect a reaction triggered after the entire resource was being exploited. Waiwood and Elner (54) proposed that commercial harvesting acted as the "trigger" by generating a regular, critical drop in biomass to prevent intraspecific competition from restricting recruitment.

An alternative hypothesis to explain the increases in abundance of the resource in the southwest Gulf is based on the interaction with cod (*Gadus morhua*), a predator of juvenile snow crabs (57). A relationship was found between a year's crab catch and the average biomass of cod 3–6 yr previously. Other averaging periods were tested but the fits were poorer. Recent evidence suggests that crabs recruit to the fishery at approximately 6 yr old, and are subject to most fish predation during their first 3 yr (42). Thus increased catches of crab from 1976 to 1981 could be accounted for by improved recruitment resulting from a decrease in cod abundance in the mid-1970s. A downward trend in crab abundance predicted to start in 1982 was not perceived until 1987 when landings fell seriously. The applicability of the model is still debated and further investigation is required to elucidate this apparent lag in response.

3.6. Maturity, Mating, and Terminal Molt

Initial studies on snow crabs in eastern Canada suggested that the reproductive potential of the resource was unaffected by exploitation (14, 58). All females examined above 60-mm CW, and some as small as 47-mm CW, were mature (12) (Fig. 4). The female's molt to maturity, into a size range of 47–95-mm CW, was postulated to be a terminal molt, as seems to be the case for all majid crabs. Females were observed to mate in a soft-shelled condition soon after their terminal molt in late winter, then extrude the fertilized eggs onto their pleopods. Watson (15) also demonstrated that females were able to extrude subsequent batches of eggs, fertilized by spermatophores stored in their spermathecae. Because females could not be induced to mate in a hard-shelled condition, Watson implied that mating could occur only soon after the female molt to maturity. Recently, however, females in a hard-shelled condition have been observed re-mating, in both the field (59) and laboratory (R. W. Elner, personal observations).

Male snow crabs can mature at a small size just as females do. Males as small as 51-mm CW have been observed with ripe gonads (12). Based on morphometric measurements and on the inspection of the vasa deferentia in 133 hard-shelled individuals, Watson (12) concluded that males attain maturity at sizes between 50- and 80-mm CW (50% maturity at 57-mm CW). Because males can reach a size of 150-mm CW, it appeared logical to conclude that they, unlike females, continue to molt after maturity. As will be seen below, recent evidence now indicates that this conclusion is probably incorrect.

Observations of snow crab mating behavior in the laboratory demonstrated that

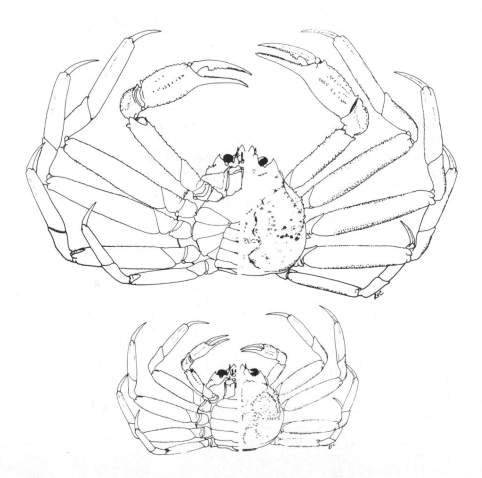

Figure 4. Dorsal and ventral illustrations of adult male (top) and female (bottom) snow crab, *Chionoecetes opilio* (drawing by L.V. Colpitts).

males can be polygamous and can mate with larger females at a relatively small size (12, 15). In comparison (Fig. 5), males observed in copulatory embraces in nature were always larger than their female partners (59). Hence small mature males may be restricted in their mating activities by competition from larger males. However, when the abundance of large males is reduced by fishing, the smaller males may increase their breeding activity, as was suggested by Butler (60) for *Cancer magister*.

All these observations combine to support the contention that snow crab reproductive output should be effectively immune to exploitation. Indeed, even on highly exploited fishing grounds, 90–100% of mature females are usually brooding eggs (58, 61, 62). Although a reduction in egg production is now evident in the collapsed fishery off eastern Cape Breton Island, the cause does not appear to be fisheries-

Figure 5. Mating pair of snow crabs in their natural surroundings in Bonne Bay, Newfoundland (photograph by G. P. Ennis and G. King).

related (8, 27, 63, 64). Barren females and females with reduced egg clutches became more apparent only after the fishery experienced a series of recruitment failures; hence it is unlikely that egg production was a factor in the initial collapse. Rather, the changes in fecundity probably reflect the natural process of large numbers of individual females becoming reproductively senile with age, a phenomenon that is normally masked on grounds that experience annual influxes of freshly mature females. Jewett (65) observed a similar reduction in egg-bearing females on nonexploited snow crab ground off Alaska but suggested the phenomenon was environment related.

In the past, incidental observations led some biologists to speculate about the existence of a terminal molt in male snow crabs. Some males that have been tagged were recaptured after up to 6 yr at large without showing growth (D. A. Robichaud, unpublished data). On the basis of survey data on shell condition of snow crabs from a Newfoundland bay, Miller and O'Keefe (45) suggested that a relatively large number of males less than 95-mm CW minimum legal size had reached a terminal molt; further, they argued that crabs of legal size probably also attain a terminal molt but are caught before their shells age. In a mark–recapture experiment using tags lost at ecdysis, more tagged crabs with old than new hard shells were recovered after a molting period (19). Hence Bailey's initial assumption, that crabs with old hard shells had been in intermolt for a longer time and thus were more likely to molt and lose their tag than newer-shelled crabs, had to be reversed.

Similarly, Miller and Watson (16) observed that only crabs with relatively new shells molted in the laboratory and crabs with old encrusted shells did not.

Males of other majid crab species are known to reach a terminal molt at maturity (66–69). The terminal molt has been related to a degeneration of the Y-organs (70–72), a pair of endocrine glands responsible for the secretion of a molting hormone. Ongoing research indicates that in both male and female snow crabs, the Y-organs have a degenerated aspect in mature individuals and that the molting hormone can be detected in juvenile crabs but not in adults. Furthermore, reexamination of the allometric relationship between crab CW and chela dimensions have shown that some male crabs are still morphometrically immature at a size up to 120-mm CW (73). Data presented by Powles (13) and Coulombe et al. (74) also suggest the existence of large immature males. Such data have provided further evidence that male snow crabs probably reach a terminal molt at maturity and are in accord with other Majidae.

At present, although the evolution of our understanding of snow crab biology modifies the biological basis to management, there is no clear indication of how management strategy should be adapted. Whereas previously all males were assumed to have the opportunity to mate before entering the fishery, it is now apparent that some juvenile males are being exploited and cannot contribute to reproduction. Perhaps large mature males that escape the fishery are capable of mating with most of the available females. If not, the small mature males probably contribute to a greater extent than on virgin grounds. Such a scenario could have profound long-term implications for the fishery if the size reached at terminal molt is determined genetically. A heavily exploited population might eventually breed only crabs smaller than legal size. Similarly, we have no appreciation of how the new knowledge on female multiple mating behavior relates to the reproductive potential of an exploited population. One can argue that egg production has continued to be high on most grounds, but will reproductive output be maintained under the current fishing regime?

A further consequence of a terminal molt in males relates to yield-per-recruit considerations. In a resource where many of the recruited individuals have ceased growing, maximum yield per recruit will be achieved at a very high fishing mortality. Nevertheless, this may not be an economically viable situation because catch rates would be lower, fishing expenses higher, and the fishery highly vulnerable to recruitment fluctuations.

3.7. Recruitment and Stock Delineation

A corollary to the assumption of a reproductive potential immune to exploitation is that the number of larvae produced each year is the same as under virgin conditions. Studies on larval ecology are important to appreciate the naturally induced component of variability in the population. Larvae are hatched from mid-April to June (14) and develop through three planktonic larval stages before settling to the bottom in September (75, 76). Their survival within the Baie-des-Chaleurs has been estimated at approximately 0.19%, which represents 28–79 small juveniles per breeding female each year (76). These juveniles are found on the same muddy substrates as the adults and are subjected to predation by cod and skate for at least 3 yr (42). Later-stage juveniles are rare on the muddy habitat and are found mainly on heterogeneous substrates in shallower waters (74). There are indications of a migration to the deeper muddy habitat after reaching puberty; this is when most large males would recruit into the fishery. Subsequent movements seem to be random and largely restricted to distances less than 10 km (29, 31). Nevertheless, a spring breeding migration has been observed in Bonne Bay, Newfoundland (59, 77).

The delineation of stocks, defined as intraspecific groups of individuals exhibiting unique phenotypic, genotypic, or biological attributes, remains of universal concern to fisheries managers (78). Morphometric, meristic, and fecundity analyses have indicated that snow crabs from four Atlantic grounds are phenotypically distinct and probably represent separate phenotypic stocks (63). However, the electrophoretic analysis showed that snow crabs from grounds each side of Cape Breton Island could not be distinguished and may represent a single "genetic" stock. Larval drift with prevailing currents from the Gulf of St. Lawrence to the Atlantic side of Cape Breton provides the most probable mechanism for the genetic mixing. Surprisingly, there is a widely different degree of resilience to exploitation and response to the same management strategy between the two Cape Breton grounds; hence a phenotypically and/or genotypically defined stock is not necessarily a useful management tool. Davidson et al. (63) suggested that stocks may be subdivided into more meaningful management units that reflect intrastock characteristics such as growth and recruitment patterns.

4. CONCLUSIONS

The development of scientific knowledge on snow crabs in eastern Canada was largely dictated by management needs. A series of preliminary studies probed the reproductive biology of the species and concluded that the management measures seemed biologically sound. Subsequent research effort was aimed at measuring stock size, rate of exploitation, and recruitment fluctuations. Fishery managers needed such information to elaborate an optimal fishing strategy. When required to provide advice for snow crab management, scientists were frequently confronted with the absence of satisfactory research results. Consequently, they often relied on simplified tools and assumptions in preparing their advice. Although the practical working context of fisheries frequently justifies such an approach, the Atlantic snow crab experience now suggests that scientists should be more aware of potential risks. What should remain working hypotheses sometimes evolve into accepted truth. Preconceptions and paradigms must be confronted, as well as any other assumption used in elaborating scientific advice. Unfortunately, the fishery scientist is often overwhelmed by the necessity to follow and refine established management approaches. Strong determination is required to consider and test different ones.

For the future, efforts should be made to develop increased insights into snow crab biology and the fishery. When possible, hypotheses should be scientifically tested. Nevertheless, we recognize that fisheries problems are often multifactorial by nature and difficult to solve in a clear scientific fashion. Given that uncertainties will always remain, it is the duty of scientists to inform managers properly about their nature.

REFERENCES

1. R. W. Elner, Overview of the snow crab *Chionoecetes opilio* fishery in Atlantic Canada. In *Proceedings of the International Symposium on the genus Chionoecetes.* Alaska Sea Grant Program, University of Alaska, Fairbanks, 1982, pp. 3–19.
2. R. Dufour, Rendements comparatifs et sélectivité de trois types de casiers à crabes des neiges. *Can. Atl. Fish. Sci. Advis. Comm. (CAFSAC), Res. Doc.* 84/1 (1984).
3. R. F. J. Bailey, A review of the sampling of snow crab (*Chionoecetes opilio*) catches in the south-western Gulf of St. Lawrence. *Can. Spec. Publ. Fish. Aquat. Sci.* **66:**77–81 (1983).
4. D. G. Parsons, E. G. Dawe, G. P. Ennis, K. S. Naidu, and D. M. Taylor, Sampling of commercial catches for invertebrates in Newfoundland. *Can. Spec. Publ. Fish. Aquat. Sci.* **66:**39–51 (1983).
5. T. W. Rowell, Sampling of commercial catches of invertebrates and marine plants in the Scotia–Fundy region. *Can. Spec. Publ. Fish. Aquat. Sci.* **66:**52–60 (1983).
6. P. Lamoureux, P. Dubé, P. E. Lafleur, and J. Fréchette, Problématique de l'échantillonnage du crabe des neiges (*Chionoecetes opilio*) et du homard (*Homarus americanus*) et analyse du système d'échantillonnage au Québec. *Can. Spec. Publ. Fish. Aquat. Sci.* **66:**279–290 (1983).
7. R. F. J. Bailey and R. Cormier, Évaluation du stock de crabe des neiges exploité par le Nouveau-Brunswick dans le sud-ouest du Golfe Saint-Laurent. *Can. Atl. Fish. Sci. Advis. Comm. (CAFSAC), Res. Doc.* 83/54 (1983).
8. R. W. Elner and D. A. Robichaud, Assessment of the fishery for snow crab off the Atlantic coast of Cape Breton Island in 1985. *Can. Atl. Fish. Sci. Advis. Comm. (CAFSAC), Res. Doc.* 86/10 (1986).
9. D. M. Taylor and P. G. O'Keefe, Analysis of the snow crab, *Chionoecetes opilio*, fishery in Newfoundland for 1984. *Can. Atl. Fish. Sci. Advis. Comm. (CAFSAC), Res. Doc.* 85/93 (1985).
10. R. Dufour, Évaluation des stocks de crabe des neiges (*Chionoecetes opilio*) de l'estuaire et du nord du Golfe St-Laurent. *Can. Atl. Fish. Sci. Advis. Comm. (CAFSAC), Res. Doc.* 85/13 (1985).
11. P. J. Ke, B. Smith-Lall, and A. B. Dewar, Quality improvement investigations for Atlantic queen crab (*Chionoecetes opilio*). *Can. Tech. Rep. Fish. Aquat. Sci.* **1002:**1–74 (1981).
12. J. Watson, Maturity, mating and egg laying in the spider crab, *Chionoecetes opilio*. *J. Fish. Res. Board Can.* **27:**1607–1616 (1970).
13. H. Powles, Distribution and biology of the spider crab *Chionoecetes opilio* in the

Magdalen Shallows, Gulf of St. Lawrence, *Fish. Res. Board Can., Manuscr. Rep.* No. **997**:1–106 (1968).

14. J. Watson, Biological investigations on the spider crab *Chionoecetes opilio*, Proceedings of the Meeting on Atlantic Crab Fishery Development. *Can. Fish. Rep.* **13**:24–47 (1969).

15. J. Watson, Mating behavior in the spider crab, *Chionoecetes opilio. J. Fish. Res. Board Can.* **29**:447–449 (1972).

16. R. J. Miller and J. Watson, Growth per molt and limb regeneration in the spider crab, *Chionoecetes opilio. J. Fish. Res. Board Can.* **33**:1644–1649 (1976).

17. R. J. Miller, Resource underutilization in a spider crab industry. *Fisheries* **2**:9–12, 30 (1977).

18. P. H. Leslie and D. H. S. Davis, An attempt to determine the absolute number of rats on a given area. *J. Anim. Ecol.* **8**:94–113 (1939).

19. R. F. J. Bailey, Analysis of the snow crab population in northwestern Cape Breton, 1978. *Can. Atl. Fish. Sci. Advis. Comm. (CAFSAC), Res. Doc.* 78/41 (1978).

20. R. F. J. Bailey and R. Cormier, Review of snow crab resources in western Cape Breton (area 1 and 7) for 1982. *Can. Atl. Fish. Sci. Advis. Comm. (CAFSAC), Res. Doc.* 83/55 (1983).

21. R. W. Elner and D. A. Robichaud, Assessment of the Cape Breton inshore fishery for snow crab, 1980. *Can. Atl. Fish. Sci. Advis. Comm. (CAFSAC), Res. Doc.* 81/40 (1981).

22. D. M. Taylor and P. G. O'Keefe, Assessment of snow crab (*Chionoecetes opilio*) stocks in Newfoundland in 1980. *Can. Atl. Fish. Sci. Advis. Comm. (CAFSAC), Res. Doc.* 83/3 (1983).

23. G. R. Morgan, Aspects of the population dynamics of the western rock lobster *Panulirus cygnus* George. I. Estimation of population density. *Aust. J. Mar. Freshwater Res.* **25**:235–248 (1974).

24. W. E. Ricker, *Computation and interpretation of biological statistics of fish populations. Bull., Fish. Res. Board Can.* **191**:149–161 (1975).

25. G. A. F. Seber, *The Estimation of Animal Abundance*, 2nd ed. Macmillan, New York, 1982, pp. 297–308.

26. R. F. J. Bailey, Overview of the Leslie fishing success method as an assessment tool for snow crab stocks. *Can. Atl. Fish. Sci. Advis. Comm. (CAFSAC), Res. Doc.* 83/85 (1983).

27. R. W. Elner and R. F. J. Bailey, Differential susceptibility of Atlantic snow crab, *Chionoecetes opilio*, stocks to management. *Can. Spec. Publ. Fish. Aquat. Sci.* **92**:335–346 (1986).

28. R. F. J. Bailey and R. Dufour, Field use of an injected ferromagnetic tag on the snow crab (*Chionoecetes opilio* O. Fabr.). *J. Cons. Int. Explor. Mer* **43**:237–244 (1987).

29. J. Watson, Tag recaptures and movements of adult male snow crabs *Chionoecetes opilio* (O. Fabricius) in the Gaspé region of the Gulf of St. Lawrence, *Tech. Rep.—Fish. Res. Board Can.* **204**:16 (1970).

30. D. M. Taylor, A recent development in tagging studies on snow crab, *Chionoecetes opilio* in Newfoundland: Retention of tags through ecdysis. In *Proceedings of the International Symposium on the Genus Chionoecetes*. Alaska Sea Grant Program, University of Alaska, Fairbanks, 1982, pp. 405–417.

31. J. Watson and P. G. Wells, Recaptures and movements of tagged snow crabs (*Chionoecetes opilio*) in 1970 from the Gulf of St. Lawrence, *Tech. Rep.—Fish. Res. Board Can.* **349:**12 (1972).

32. J.-C. Brêthes, R. Bouchard, and G. Desrosiers, Determination of the area prospected by a baited trap from a tagging and recapture experiment with snow crabs (*Chionoecetes opilio*). *J. Northwest Atl. Fish. Sci.* **6:**37–42 (1985).

33. R. J. Miller, Comments on paper by Brêthes et al. (1985). *J. Northwest Atl. Fish. Sci.* **6:**173–174 (1985).

34. J.-C. Brêthes, Responses to comments of R. J. Miller. *J. Northwest Atl. Fish. Sci.* **6:**174–175 (1985).

35. R. J. Miller, Density of the commercial spider crab, *Chionoecetes opilio*, and calibration of effective area fished per trap using bottom photography. *J. Fish. Res. Board Can.* **32:**761–768 (1975).

36. D. M. Taylor, W. R. Squires, and P. G. O'Keefe, An alternate methodology for estimating snow crab (*Chionoecetes opilio*) populations in commercially-fished areas. *Can. Atl. Fish. Sci. Advis. Comm. (CAFSAC), Res. Doc.* 83/1 (1983).

37. F. Coulombe, Estimation de la biomasse exploitable du crabe des neiges du sud-ouest de golfe du Saint-Laurent pour l'année 1984: Utilisation de l'aire efficace de pêche d'un casier. *Can. Atl. Fish. Sci. Advis. Comm. (CAFSAC), Res. Doc.* 84/88 (1984).

38. P.-E. Lafleur, R. F. J. Bailey, J.-C. Brêthes, and P. Lamoureux, Le crabe des neiges (*Chionoecetes opilio* O. Fabricius) de la Côte-Nord de l'estuaire et du Golfe du Saint-Laurent. *Trav. Pêch. Québec* No. 50 (1984).

39. G. Y. Conan and D. R. Maynard, Estimates of snow crab (*Chionoecetes opilio*) abundance by underwater television. *J. Appl. Ichthyol.* **3:**158–165 (1987).

40. P. Lamoureux and P.-E. Lafleur, The effects of exploitation on snow crab populations of the southwestern Gulf of St. Lawrence, between 1975 and 1981. In *Proceedings of the International Symposium on the Genus Chionoecetes.* Alaska Sea Grant Program, University of Alaska, Fairbanks, 1982, pp. 443–481.

41. R. Bouchard, J.-C. F. Brêthes, G. Desrosiers, and R. F. J. Bailey, Changes in size distribution of the snow crabs (*Chionoecetes opilio*) in the southwestern Gulf of St. Lawrence. *J. Northw. Atl. Fish. Sci.* **7:**67–75 (1986).

42. D. A. Robichaud, Écologie du crabe des neiges (*Chionoecetes opilio*) juvénile au large des côtes nord-ouest du Cap Breton et ses interactions avec la morue (*Gadus morhua*) et la raie (*Raja radiata*), M. Sc. Thesis, University of Moncton, New Brunswick, Canada, 1985.

43. R. W. Hiatt, The biology of the shore crab, *Pachygrapsus crassipes* Randall. *Pac. Soc.* **2:**135–213 (1948).

44. R. Greendale and R. F. J. Bailey, Résultats d'inventaires du crabe des neiges (*Chionoecetes opilio*) dans l'estuaire et le Golfe Saint-Laurent. *Rapp. Tech. Can. Sci. Halieut. Aquat.* **1099F:**1–40 (1982).

45. R. J. Miller and P. G. O'Keefe, Seasonal and depth distribution, size and molt cycle of the spider crabs, *Chionoecetes opilio, Hyas araneus,* and *Hyas coarctatus* in a Newfoundland bay. *Can. Tech. Rep. Fish. Aquat. Sci.* **1003:**1–18 (1981).

46. P. Drach and C. Tchernigovtzeff, Sur la méthode de détermination des stades d'intermue et son application générale aux crustacés. *Vie Milieu* **18:**595–610 (1967).

47. M. Moriyasu and P. Mallet, Reading molt stages of the snow crab (*Chionoecetes opilio*) by observation of maxilla. *J. Crust. Biol.* **6:**709–718 (1986).

48. M.-J. O'Halloran, Moult cycle changes and the control of moult in male snow crab, *Chionoecetes opilio*, M. Sc. Thesis, Dalhousie University, Nova Scotia, Canada, 1985.

49. J. Barlow and G. J. Ridgway, Changes in serum protein during the molt and reproductive cycles of the american lobster (*Homarus americanus*). *J. Fish. Res. Board Can.* **26:**2101–2109 (1969).

50. K. Ito, Ecological studies on the edible crab, *Chionoecetes opilio* O. Fabricius in the Japan sea. I. When do female crabs first spawn and how do they advance into the following reproductive stage? *Bull. Jpn. Sea Reg. Fish. Res. Lab.* **17:**67–84 (1967) (Fish. Res. Board Can. Transl. Ser. No. 1103).

51. K. Ito, Ecological studies on the edible crab, *Chionoecetes opilio* (O. Fabricius) in the Japan sea. III. Age and growth as estimated on the basis of the seasonal changes in the carapace width frequencies and the carapace hardness. *Bull. Jpn. Sea Reg. Fish. Res. Lab.* **22:**81–116 (1970) (Fish. Res. Board Can. Transl. Ser. No. 1512).

52. J. Watson, Ecdysis of the snow crab, *Chionoecetes opilio. Can. J. Zool.* **49:**1025–1027 (1971).

53. D. M. Taylor, P. G. O'Keefe, and G. Marshall, Recovery of tagged soft-shelled snow crabs (*Chionoecetes opilio*) from Bonavista Ray, Newfoundland. *Proc. Nat. Shellfish. Assoc.*, Abstr. (in press).

54. K. W. Waiwood and R. W. Elner, Cod predation of snow crab (*Chionoecetes opilio*) in the Gulf of St. Lawrence. In *Proceedings of the International Symposium on the Genus* Chionoecetes. Alaska Sea Grant Program, University of Alaska, Fairbanks, 1982, pp. 499–520.

55. R. F. J. Bailey, Status of snow crab (*Chionoecetes opilio*) stocks in the Gulf of St. Lawrence. *Can. Atl. Fish. Sci. Advis. Comm. (CAFSAC), Res. Doc.* 78/27 (1978).

56. J. A. Gulland, *Fish Stock Assessment: A Manual of Basic Methods*, Vol. 1. Wiley (Interscience), New York, 1983, p. 70.

57. R. F. J. Bailey, Relationship between catches of snow crab, *C. opilio* (O. Fabricius) and abundance of cod (*Gadus morhua* L.) in the southwestern Gulf of St. Lawrence. In *Proceedings of the International Symposium on the Genus* Chionoecetes. Alaska Sea Grant Program, University of Alaska, Fairbanks, 1982, pp. 485–497.

58. R. W. Elner and D. A. Robichaud, Observations on the efficacy of the minimum legal size for Atlantic snow crab, *Chionoecetes opilio. Can. Atl. Fish. Sci. Advis. Comm. (CAFSAC), Res. Doc.* 83/63 (1983).

59. D. M. Taylor, R. G. Hooper, and G. P. Ennis, Biological aspects of the spring breeding migration of snow crabs, *Chionoecetes opilio*, in Bonne Bay, Newfoundland (Canada). *Fish. Bull.* **83:**707–711 (1985).

60. T. H. Butler, Maturity and breeding of the Pacific edible crab, *Cancer magister* Dana. *J. Fish. Res. Board Can.* **17:**641–646 (1960).

61. R. W. Elner and C. A. Gass, Observations on the reproductive condition of female snow crabs from N.W. Cape Breton Island, November 1983. *Can. Atl. Fish. Sci. Advis. Comm. (CAFSAC), Res. Doc.* 84/14 (1984).

62. F. Coulombe, Étude comparative de la fécondité de deux populations de crabe des neiges (*Chionoecetes opilio*) de la Côte-Nord du Saint-Laurent. *Can. Atl. Fish. Sci. Advis. Comm. (CAFSAC), Res. Doc.* 84/55 (1984).

63. K. Davidson, J. C. Roff, and R. W. Elner, Morphological, electrophoretic, and fecundity characteristics of Atlantic snow crab, *Chionoecetes opilio*, and implications for fisheries management. *Can. J. Fish. Aquat. Sci.* **42**:474–482 (1985).

64. R. W. Elner and D. A. Robichaud, Assessment of the 1984 fishery for snow crab off the Atlantic coast of Cape Breton Island. *Can. Atl. Fish. Sci. Advis. Comm. (CAFSAC) Res. Doc.* 85/5 (1985).

65. S. C. Jewett, Variations in some reproductive aspects of female snow crabs, *Chionoecetes opilio*. *J. Shellfish Res.* **1**:95–99 (1981).

66. G. Teissier, Croissance des variants sexuels chez *Maia squinado*. *Trav. Stn. Biol. Roscoff* **13**:93–130 (1935).

67. R. G. Hartnoll, The biology of spider crabs: A comparison of British and Jamaican species. *Crustaceana* **9**:1–16 (1965).

68. G. Vernet, C. Bressac, and J.-P. Trilles, Quelques données récentes sur l'organe-Y (glande de mue) des crustacés décapodes. *Arch. Zool. Exp. Gen.* **119**:201–225 (1978).

69. J.-C. Chaix, Le cycle biologique et quelques aspects de la reproduction du crabe oxyrhynque *Acanthonyx lunulatus* (Risso, 1816) (Crustacea Decapoda Oxyrhyncha). *Tethys* **9**:17–22 (1979).

70. D. B. Carlisle, On the hormonal inhibition of moulting in decapod crustacea. II. The terminal anecdysis in crabs. *J. Mar. Biol. Assoc. U.K.* **36**:291–307 (1957).

71. J.-C. Chaix, J.-P. Trilles, and G. Vernet, Dégénérescence de l'organe-Y chez les mâles pubères d'*Acanthonyx lunulatus* (Risso) (Crustacea, Decapoda, Oxyrhyncha). *C.R. Hebd. Seances Acad. Sci.* **283**:523–525 (1976).

72. G. Vernet-Cornubert, Influence de l'ablation des pédoncules oculaires sur la mue, la ponte et les caractères sexuels externes de *Pisa tetraodon* Pennant. *Bull. Inst. Oceanogr.* No. 1186 (1960).

73. G. Y. Conan and M. Comeau, Functional maturity and terminal molt of male snow crab, *Chionoecetes opilio*. *Can. J. Fish. Aquat. Sci.* **43**:1710–1719 (1986).

74. F. Coulombe, J.-C. F. Brêthes, R. Bouchard, and G. Desrosiers, Ségrégation édaphique et bathymétrique chez le crabe des neiges, *Chionoecetes opilio* (O. Fabr.), dans le sud-ouest du golfe du Saint-Laurent. *Can. J. Fish. Aquat. Sci.* **42**:169–180 (1985).

75. K. Davidson, Stock delineation and larval taxonomy of the snow crab *Chionoecetes opilio* O. Fabricius (Decapoda, Brachyura, Majidae) in Atlantic Canada. M.Sc. Thesis, University of Guelph, Ontario, Canada, 1983.

76. M. Lanteigne, Distribution spatio-temporelle des larves de crabe appartenant aux genres *Chionoecetes* et *Hyas*, dans la baie des Chaleurs, Canada, M.Sc. Thesis, University of Moncton, New Brunswick, Canada, 1985.

77. R. G. Hooper, A spring breeding migration of the snow crab (*Chionoecetes opilio*) into shallow water in Newfoundland. *Crustaceana* **50**:257–264 (1986).

78. P. E. Ihssen, H. E. Booke, J. M. Casselman, J. M. McGlade, N. R. Payne, and F. M. Utter, Stock identification: Materials and methods. *Can. J. Fish. Aquat. Sci.* **38**:1838–1855 (1981).

REVIEWS AND SPECIAL TOPICS

13 THE MANAGEMENT OF COASTAL PENAEID SHRIMP FISHERIES

S. Garcia
Food and Agriculture Organization
Rome, Italy

1. INTRODUCTION

Shrimp are one of the world's most valuable fishery resources. The present total landings are estimated to be around 1.8×10^6 tons, and an additional demand for 200,000 tons is foreseen by 1990 (Sribhibhadh 1). Tropical penaeid shrimp landings amount to about 700,000 tons (Gulland and Rothschild 2) and their high value and strong demand on the markets of the richer countries (United States, Japan, Europe) were powerful incentives to the development of shrimp fisheries in the 1960s and 1970s. From 1977 to 1981, for instance, the landings increased by 22% and the value by about 60% in terms of U.S. dollars (International Trade Center 3). The high prices for shrimp on export markets have stimulated rapid development, leading in many cases to excessive effort, even with regulatory measures, which have usually been unable to prevent overinvestment, excessive production costs, low or even negative economic returns to the country, and perhaps an overall reduction in total catch value. Most potentially productive areas are now being exploited, and no major increase in landings of capture fisheries can be foreseen. As shrimp prices continue to rise and oil prices are presently decreasing (1986), there is a risk of additional fishing effort being injected into fisheries where conspicuous conflicts have already appeared between small-scale and offshore industrial shrimp fisheries, between shrimp and finfish fisheries, and between shrimp aquaculture and fisheries for the postlarval resources and for the markets. Most fisheries are now in a situation of economic overfishing, shrimp fisheries are the major source of conflict and problems in the tropical zone, and recruitment problems are presently receiving increased attention (Penn 4; Penn and Caputi 5; Garcia 6, 7). Considerable progress has been made in understanding the essential biological aspects of shrimp resources in the past decade and, although the data may not always be as complete as one would wish, an extraordinary amount of information has been accumulated since the pioneering work of Boerema (FAO/UN 8), Gunter (9), and Gulland (10) on shrimp fisheries management. A first review was prepared in 1981 by Garcia and Le Reste (11) and since then three major workshops have been held (Gulland and Rothschild 2; Rothlisberg et al. 12; FAO 13) and management plans for shrimp have been prepared in many countries.

This chapter identifies the major management issues, objectives, and approaches in shrimp fisheries. It does not deal with gathering of information for decision makers, management organizations, or practical implementation because of lack of space and because these aspects either do not differ markedly from other fisheries and are treated extensively elsewhere or are poorly documented (e.g., practical implementation in small-scale fisheries).

2. MANAGEMENT ISSUES

Although shrimp fisheries have developed in drastically different socioeconomic contexts (from Australia and the United States to Malaysia or Senegal) a limited

number of issues for management are common to all of them (see also Poffenberger
14).

2.1. Biological and Economic Overfishing

It is usually assumed that in trawl offshore shrimp fisheries, catches increase more
slowly than fishing effort and that some maximum catch (maximum sustainable
yield, MSY) and value are reached for some intermediate level of effort. It is also
accepted that the maximum economic yield (MEY) is obtained at some level of
effort lower than the level corresponding to MSY. Most shrimp fisheries around
the world are common property resources, even when the coastal countries have
claimed exclusive economic zones, for access is still open to nationals. The usual
competition among participants for a greater share of the common resource has
often led to an uncontrolled increase of effort to the point where the economic rent
is dissipated and often the economic situation of the fishery is very critical, some
fishermen no longer being able to cover their capital costs and even sometimes
their operational costs. Governments intervene by providing subsidies, soft loans,
tax reductions, and so on, which usually aggravate the problem, leading possibly
to overall economic losses to the country. The Gulf of Mexico shrimp fishery of
the United States, for instance, valued at more than $400 million annually, was
considered to be in a state of "economic overfishing" by Neal (15). The numerous
analyses made in the 1970s (Greenfield 16; Griffin and Beatie 17; Blomo et al. 18)
confirmed this, gave a good analysis of the situation, and stressed that the crisis
was exacerbated by rising fuel costs, and sustained by the fact that the marginal
yield in shrimp fisheries was still higher than in potential alternative fisheries
because of high prices. A management plan was initiated in 1976 and has been
implemented since 1981. Many measures are being enforced (sanctuaries, seasonal
closures, etc.) but the limitation and reduction of effort proposed by Rounsefell
(19) has not been adopted as a strategy and the present situation is still largely one
of overcapitalization (Leary 20). It is worth mentioning also that although in
Australia limited entry was enforced in many cases since the inception of the
fisheries a decade or more ago, the present situation is not yet totally satisfactory
and excess of effort has not been entirely avoided (Bowen and Hancock 21). A
similar situation of economic overfishing is encountered in most fisheries around
the world, as shown in the various case studies presented at the Key West meeting
in 1981 (Gulland and Rothschild 2); the wealthy situation of the shrimp fishery in
Saudi Arabia due to sole ownership (R. Willmann, personal communication) is a
noteworthy exception.

2.2. Optimization of Yield Per Recruit (Growth Overfishing)

Shrimp are very fast growing animals. Seasonal and age-specific fishing patterns
have marked consequences on annual yield in weight and in value. One of the key
issues in present-day shrimp fishery management is to determine the most appro-

priate age at first capture and the fishing pattern to reach a specific economic objective or a given shrimp market. This implies the use of bioeconomic yield-per-recruit modeling with preseason surveys, mesh-size regulations, closed areas, and seasonal or temporary closed seasons (see Section 4.1). The problem here is a trade-off between immediate loss of small shrimp catch and future gains in weight and value from the survivors. In many fisheries intensive exploitation starts much too early, leading to growth overfishing, and the solution is not always simple because there are conflicts in the use of the resource.

2.3. Conflicts

Problems of resource allocation in shrimp fisheries are similar to those in other fisheries and have international, intranational, and interstate territory or community aspects. These have been considered by FAO (13); I deal here with only two specific issues, that is, the conflict between artisanal and industrial fisheries and the conflict with aquaculture. The potential conflict between shrimp fisheries and finfish fisheries is dealt with in Section 2.6.

Conflicts between industrial and small-scale fisheries cannot be overstated and have been particularly acute in Southeast Asia and India (Unar and Naamin 22; Naamin and Martosubroto 23; Silas et al. 24). Shrimp are often exploited by small-scale commercial or sport fisheries inshore as well as trawl industrial fisheries offshore. The overall input and benefits for the combined fisheries varies with the intensity of fishing inshore and offshore, and decisions are needed on optimal fishing patterns and allocation between the various fishing sectors involved sufficiently acceptable to be enforceable. Direct conflicts also exist for space, and small-scale fishermen complain about destruction of their gear by trawlers as well as conflicts for the market when trawlers land large quantities of coastal shrimp at low price. Some answers can be found in setting appropriate closed seasons and areas (see Section 4.1). In some cases, however, more drastic measures have been taken, such as total banning of trawling in Indonesia (Naamin and Martosubroto 23) or total elimination of artisanal fisheries in Cuba (Perez et al. 25).

Conflicts with shrimp culture are increasing in some countries. Extensive aquaculture is growing very rapidly; the present world production, not known precisely, varies between 35,000 and 80,000 tons according to various sources (Pedini, cited by Sribhibhadh 1; Lawrence 26) and great potential for expansion is said to exist. The production could reach 200,000 tons in the mid-1990s (Lawrence 26), and over-optimistic figures of 400,000 tons can be found in the literature. These cultured shrimps compete with wild shrimp on the market and also for growing space and postlarval seed. The largest production of cultured shrimp is obtained by large-scale extensive aquaculture often using littoral nursery areas (mangroves, marshlands) for growing space and wild postlarval seed. According to Lawrence (26) 95% of the present commercial shrimp production in the western hemisphere depends on collection of postlarvae from natural sources, and this situation has created concern regarding the potential effect on capture fisheries. The shortage of wild seed in Ecuador recently and the overall decrease of the capture fisheries have increased this concern. The problem is not easy to address and how to model it as juvenile natural mortality is still unclear.

2.4. Variability and Uncertainty

Shrimp fisheries exploit essentially one year class. The annual yield is therefore largely a function of the importance of the annual level of recruitment and the latter is widely influenced by environmental conditions. The consequence is that annual catches vary from year to year either randomly or, more probably, following long-term autocorrelated oscillations. This fact has many consequences on stock assessment, modeling, effort control, and management strategy. Garcia (27) has stressed the existence of year-to-year variations in shrimp production due to environmental conditions, leading to difficulties in establishing an appropriate production model to assess the present state of the stock and to estimate MSY when only short time series are available. Year-to-year variability also renders difficult the use of annual catch quotas for effort regulation. When exceptional year classes enter the fishery, or exceptional prices are obtained, higher than average profitability is generated and it has been shown in the Gulf of Mexico (Rounsefell 19; Poffenberger 14) that these years were usually followed by pulsed increases in boat numbers. Variable resources are therefore less prone to economic self-regulation and lead more easily to heavy overfishing and overcapitalization (Csirke and Sharp 28; Garcia 29). In order to reduce the uncertainty of the production and management sector, it is necessary to elaborate predictive models to foresee the coming year's production a few months ahead and allocate fishing time and effort accordingly. These models are based either on environmental factors such as rainfall or temperature in a given critical seasonal period or on preseason indexes of recruitment. Most of them have to prove their effectiveness, but Leary (20) indicates that the Laboratory of Galveston predicted effectively the 1982 and 1983 annual catches in Texas with a precision of 1% on the basis of bait shrimp fisheries catch rates used as a prerecruitment index. It is often argued that, because of the short life-span of shrimp, forecasts will always give too short a lead time to be really useful. This belief was, however, rejected by the industry in Australia, which stated that 1–6 weeks of lead time is largely sufficient to enable cost-effective deployment of the fleets (FAO 13). The question of forecasting models will not be elaborated further here; a review is available in Garcia and Le Reste (11). It should also be briefly noted that when the ability to predict cannot be developed at a reasonable cost the solution consists in evaluating and including uncertainty in the models (Sissenwine 30) as well as developing flexible and efficient reactive management systems.

When recruitment is highly variable and annual production only loosely linked to stock size and effort level, it seems necessary to optimize the fishery on a pre-recruit basis (Gulland 10), regulating mesh size, but above all implementing closed areas and closed seasons. This concept is largely followed in the U.S. management plan for the Gulf of Mexico (Poffenberger 14) and in some Australian fisheries (Bowen and Hancock 21). It appears, however, that without a definite limitation on the level of fishing effort, this sort of "fine-tuning management" is bound to meet with difficulties (see Section 4.1.3).

2.5. Recruitment Overfishing

This issue has been neglected for a long time in shrimp fisheries and the accepted paradigm was that because of the high fecundity of shrimp and the importance of inshore nurseries in determining cohort survival, shrimp stocks were unlikely to be exploited intensively enough to cause recruitment problems and that economic factors probably would limit effort to below the level critical for shrimp stock reproduction. Most of the evidence given in the past to demonstrate the existence of stock–recruitment relationships can be interpreted as artifacts owing to the short life-span of shrimp and the autocorrelation in environmental variations (Garcia 6). Penn and Caputi (5) presented some evidence that in a small and well isolated stock off an arid zone, recruitment might be affected by fishing. It is of course obvious that at some high level of effort problems of recruitment are to be encountered, although at levels of exploitation of up to 70–80% no effort–recruitment relationship was encountered in northern Australia stocks (Staples et al. 31). Garcia (7) in examining the reproduction mechanisms of the shrimp populations showed that selective fishing out of the main cohorts by the perfectly aimed shrimp trawl fishery could lead to severe disturbance of the delicate mechanism developed by shrimp through their evolution to cope with a highly seasonal environment. The stock–recruitment question remains open and is certainly worth more attention than it has received in the past.

Two sets of management measures address the recruitment overfishing issue by trying to improve, on the one hand, larval survival and estuarine carrying capacity (see Section 4.4 on stock enhancement and habitat conservation in nurseries), and, on the other hand, the spawning stock size (see Sections 4.1.3 on seasonal fishing closures or 4.2 on overall effort regulations).

2.6. Multispecies Management

Many different species of shrimp with different distributions exist in the waters of any tropical country. Shrimp fisheries tend to begin on white coastal shrimps of the genus *Penaeus* caught during the day. They develop progressively with additional night fishing on brown and tiger shrimps as the effort increases. As overall profitability decreases further they tend also to develop later on more coastal small shrimp of the genus *Xyphopenaeus*, *Trachypenaeus*, *Lithopenaeus*, *Metapenaeus*, and so on, of smaller size and value.

Most of the shrimp fisheries tend therefore to evolve into exploiting a mixture of shrimp species. Management is complicated if these species have different population parameters and value. Compromises must be found regarding optimum mesh size, closed seasons, closed areas, and so forth. Clark and Kirkwood (32), for instance, elaborated a model for two species and two types of boats and used it to define optimal space–time allocation of effort and fleet composition.

Management is further complicated when the accompanying finfish species are taken into consideration. Shrimp are only one element (a major one in value but a minor one in weight) of the fish assemblage available for exploitation on tropical shelves and one of the important characteristics of shrimp fishing is the importance of by-catch and discards, amounting to about 2,700,000 tons and 1,400,000 tons, respectively (Gulland and Rothschild 2). The discards are usually dead when returned to the sea. Because the issue has two facets, the by-catch can be better used or substantially reduced. The question of a better use of discards was debated at a special Technical Consultation in Georgetown, Guyana (FAO/IDRC 33).

Although there are technical problems to keep large amounts of by-catch on freezer shrimpers, the better utilization of discards is essentially an economic problem and discards are more important in some areas, for example, the Gulf of Mexico or the Arafura Sea (Irian Jaya, Indonesia), where markets for "trash" species are more limited than in others, such as India or Senegal, where such markets have developed. The landings of large quantities of trash fish can compete on the market with the fish landed by small-scale fisheries. If the quantities presently returned to the bottom are utilized, the biological problem is to know whether the resulting decrease in food on the bottom will have any effect on shrimp production. The preliminary study by Sheridan et al. (34) indicates that this effect is negligible.

The problem of by-catch is also one of conflict with the finfish fisheries. Shrimpers use smaller mesh than other trawlers (about 40–50 mm stretched) and accidentally capture juveniles of fish species targeted by other trawl, small-scale, and sport fisheries in coastal areas. These gear interactions are not receiving sufficient attention at the moment.

Shrimp fisheries can include nonshrimp species as secondary targets. Lhomme and Garcia (35) showed in Senegal that the proportion of trips aimed solely at shrimp decreased from about 100% in 1969 to only 25% in 1978 as high-value species such as soles, kingklips, and croakers were progressively added to the target list. Haysom (36) indicates an increase in pressure by shrimpers in Australia on sea snakes (for leather), pipefishes (for aphrodisiacs), sand crab, and whiting, leading in the two latter cases to conflict with professional crab fishermen and sport fishermen, respectively. Bowerman (37) indicates that by-catches of whiting helped fishermen to face unfavorable economic exploitation of shrimp, and Walker (38) refers to mixed shrimp/scallop fishery. In Malaysia fairly high economic revenues are achieved by sales of pelagic species taken in the high-opening bottom trawl used for shrimp fisheries (F. T. Christy, Jr., personal communication). Because more species are kept at each trip, it becomes necessary to consider these fisheries as single multitarget fisheries (Gulland and Garcia 39), and to manage them as such. In some instances (e.g., Thailand) shrimp are even considered as a valuable by-catch. A major difficulty arises because the added value to the catch produced by the commercialization of the by-catch or secondary target species may allow fishing effort to remain economical and to develop well beyond the optimum economic level for the highest value species, possibly leading to biological disruption (Penn and Caputi 5).

The solution to reduction of by-catch or by-catch mortality, still at an experimental stage, may be in better trawl selectivity and sorting devices; many attempts have been made, especially with nonpenaeid shrimp with trawls equipped with selection panels (FAO 40) and by-catch excluder devices (Naamin and Sujastani 41). Special on-the-deck "fish-friendly" sorting devices allowing the return of discards to the sea alive have also been developed (Boddeke and Verbaan 42; Boddeke 43). The problem is partly technical because of gear complexity, but also economic because of additional gear costs and potential loss of shrimp escaping with the by-catch. It is also biological because of the predation possibly added on shrimp from potential predators returned alive to the bottom. This last problem might be secondary because few fish caught in shrimp trawls are big enough to eat the accompanying shrimp anyway. Their survival should therefore not affect significantly shrimp predation or natural mortality. Pauly (44) has argued that in the Gulf of Thailand prerecruit survival of shrimp has increased with decreased fish abundance, but his results are still to be confirmed because most prerecruit mortality occurs in estuaries where predators cannot be affected by industrial offshore trawling and also because of bias in the computational procedures used (Garcia 6; Bailey 45). The whole issue remains open, therefore.

3. MANAGEMENT OBJECTIVES

The theory of fisheries management usually states that successful management is based on a clear definition of the objectives and their ranking (e.g., Gulland 10). It is, however, also generally recognized that these objectives are rarely clearly defined in reality and are at best expressed as a list of broad and often conflicting goals. The reason for this discrepancy might be that management is not a flow of information, decisions, and controls coming from the administration, advised by scientists, and aimed at manipulating the catching sector, a unidirectional process in which clear and ranked a priori objectives would certainly be the key to success. In practice, the fishery sector is a complex system comprising many subsystems with their own time scales, objectives, and pressure groups (fishermen, boat owners, processors, retailers, consumers, politicians, scientists, administrators, etc.). Ideally, management in this context must offer mechanisms allowing the fishery to evolve toward an acceptable compromise on basic objectives with minimum constraint and maximum consensus. This requires from the managing authority consultations and negotiations with all parties concerned and, unfortunately, a key-condition to successful negotiations for each of these parties is to avoid a priori statement of the true objectives (Brewer 46).

Therefore the objectives cannot always be as explicitly and precisely stated as the advisors would wish, and they will have to accept the difficult task of providing advice under a range of options that will most probably "be significantly tempered in the real world of decision by the inclusion of increased numbers of players or participants and by the presence of institutional mechanisms whose primary function is to temper and enrich the decision context" (Brewer 47).

Among the broad range of objectives retained for shrimp fisheries management the following have been noted. The long-term conservation of the resources is usually given top priority, at least rhetorically (Beddington and Rettig 48; FAO 49; Bowen and Hancock 21). The maximization of physical yield as retained, for example, in the U.S. Gulf of Mexico shrimp fishery (Leary 20) is a very traditional one. However, this simple concept of MSY has been repeatedly criticized since the 1960s (Christy and Scott 50; Gulland 51; Larkin 52; Sissenwine 53) and a review of many of the biological, technical, and socioeconomic arguments against MSY as an objective are given in Garcia and Le Reste (11, pp. 169–172). The maximization of other benefits such as economic rent, revenue incomes, foreign currency earnings, and employment are also often retained, although the whole concept of maximization has in fact been criticized because it is clear now that very rarely do managers search for a maximum of any single output that a fishery can generate (Gulland 54). Because of the complexity of the fishery system and the diversity of the objectives of its various components, some viable compromise will be looked for, and the objective mix and ranking considered ''acceptable'' is likely to change with time as the fishery evolves. Decrease in production costs, improvement of socioeconomic conditions of small-scale fishermen, or protection of sports on recreational fishermen are also mentioned. The better distribution of benefits is a sensitive issue because any attempt to change the established distri-bution pattern is likely to generate resistance and conflicts. However, it has to be tackled particularly when a disadvantaged social group must be protected. Other objectives include better use of by-catch, improvement of overall biological production by protecting juveniles, and last but not least, the reduction of conflicts. This objective has probably been, and still is, one of the top priorities of fisheries management, and it has sometimes been said that fisheries administrations too often try to solve conflicts reactively, instead of managing the fisheries. One could, of course, argue that the high level of uncertainty resulting from the complexity of the fishery system, the lack of appropriate data or models, the physical and economical environmental variability, and even the institutional uncertainty (Hannesson 55) make rational forecasting very difficult and often force manage-ment authorities to consider the fishery system as a black box and to monitor its state and stress through the violence and number of complaints, aiming at reducing them as much as possible. John Pope coined this as the ''maximum sustainable whinge'' strategy. Considering the potential cost of conflicts, this strategy still has a significant role to play in the future as conflicts increase in shrimp fisheries (see Section 2.3).

4. MANAGEMENT STRATEGIES AND TECHNIQUES

These are usually classified into two not entirely distinct categories: the regulation of catch–age composition and the regulation of fishing effort. In practice, a mix of tools from these two categories is needed for successful management. In addition, shrimp are particularly amenable to nursery habitat management and conservation, because year-class strength is largely determined by survival of larvae and postlarvae in littoral fringe areas.

4.1. Regulation of Catch–Age Composition

The underlying family of analytical models are based on the yield-per-recruit concept. Methods considered under this heading are aimed at reducing mortality on small sizes in the hope of improving production to the extent that potential gains in weight and value through growth of survivors will compensate for the immediate losses due to their delayed capture. This can usually be obtained by regulating mesh sizes or minimum size limits on landings. Because shrimp are fast-growing animals with seasonal recruitment, similar effects can also be obtained by regulating the fishing season and establishing seasonal or permanent closed areas; the main issue is to determine the optimum time–space allocation of fishing effort.

Because these measures exert no influence on fleet size they cannot prevent excessive investments and fishing costs. In fact these measures are more largely used because they are easier to implement than effort regulations. They are more acceptable to fishermen because they have no obvious redistributive effects; that is, they do not extensively change the traditional distribution of wealth, though in fact they may actually change it (see below). However, because they do not address the main cause of the overfishing problem (an excess of effort) they cannot be expected to solve it.

4.1.1. Regulation of Mesh Size

The selection process in shrimp is not very effective and the selection range covers a large part of the life-span. Lhomme (56) showed that the selection curves of mesh sizes from 40 to 70 mm (stretched) overlap widely. An appreciable increase in length at first capture would therefore usually involve unacceptable immediate losses. Al-Hossaini et al. (57) concluded that the 50% retention length is not well related to mesh size, perhaps because of the amount of discards and trash usually taken in shrimp fisheries. However, as noted by Garcia and Le Reste (11) the regulation of mesh size can in theory be useful because the long-term gains can be obtained in the same year, without short–term losses, and the unit value of shrimp increases rapidly with size, so that gains in value are therefore potentially higher than gains in weight. In addition, a slightly wider mesh size could also help to reduce by-catch (while affecting shrimp catches very little) and thereby potentially improving the potential yield of coastal finfishes (see Section 2.6).

Mesh-size regulation is often complicated by the fact that shrimp fisheries tend to exploit a mix of shrimp species with different population parameters and market value. Adjusting the mesh size to the most profitable stock (usually the larger species) leads to underexploitation of the smaller ones. When the main associated species is fish (as in the fishery for *Parapenaeus longirostris*, sea breams, and hake in Morocco), optimizing for shrimp, the high-value species, leads to overfishing of finfish.

The selectivity of push nets, stake nets, traps, bamboo weirs, and other types of estuarine devices can, in theory, be modified to allow juveniles to escape. In bamboo weirs, for example, the spacing between the bamboo lattices can be regulated (Le Reste and Marcille 58). However, owing to the artisanal production of such lattices the possibility of enforcement seems rather poor. In addition, because the target is often migrating juveniles, the likelihood of any "long-term" effect to compensate for immediate losses does not exist and it is doubtful that a consensus can be obtained.

Consistent program of control of mesh size are usually necessary in the ports and at sea because fishermen can circumvent the mesh size regulation by using a different mesh size; lining the cod end with a finer mesh size, inside or outside; superimposing two layers of authorized mesh size reducing by about half the actual escapement openings; attaching heavy weights to the cod end; trawling fast or changing the mesh hanging ratio; or lining the regular cod end externally with a larger mesh but making this tighter than the cod end, impeding the cod end from expanding normally and keeping its mesh size closed.

Mesh sizes are regulated in most shrimp fisheries. However, the age at first capture in shrimp is determined not only by mesh size but also by the distribution of effort in time (in relation to the recruitment period) and space (in relation to nurseries or shallow depths). Garcia et al. (59) have shown in French Guiana that the size at first capture had decreased with time as effort increased, despite the fact that the mesh size had apparently not changed. Experience shows that mesh-size regulations are difficult to enforce on depleted resources and that the use of small mesh sizes by fishermen is often the consequence of overfishing (excess of effort), not its primary cause (Garcia 60), leading to the need to address the problems of excess effort and excessively small mesh size simultaneously and not considering mesh-size regulation as a viable second best alternative to effort regulation. In the case of sequential trawl fisheries operating successively inshore and offshore, the implementation of a common mesh size (or single age at first capture) may lead to a transfer of fishable biomass from the inshore sector of the fishery to the offshore sector and this may be a source of conflict. This point is also discussed in Section 4.1.3 on closed seasons.

4.1.2. Minimum Landing Size

This regulation is intended to render the fishing unattractive in areas where small shrimp are abundant, to make the regulation of mesh size more effective, and to reduce the temptation to evade the mesh-size regulation. It is usually agreed that this method is useless when used alone, and it is difficult to enforce when large adult shrimp are mixed with juveniles. This is the case, for example, in the Gulf of Mexico, where adult white shrimp are mixed with juveniles of brown shrimp. Small shrimp are fished and discarded dead. Rounsefell (19) reported that up to

80% of the catch was discarded. In cases where a minimum landing size has been applied it was often under pressure from processors or dealers with marketing problems and not for serious biological reasons. The regulation has, in fact, been abolished in Texas in the new management plan and replaced by a total seasonal closure (Leary 20). Ruello (61) had already recommended the abolition of such a measure in Australia.

4.1.3. Seasonal Closures

The greater part of the very shallow trawlable areas of bays, lagoons, and littoral fringes are occupied by shrimps migrating toward deeper waters for spawning. Here shrimp size varies seasonally in relation to the seasonal migration cycle, and in order to improve the yield per recruit, they must be protected from fishing up to a certain size. The "ideal" optimal size at first capture can be determined by yield-per-recruit analysis by weight and value. Other outputs can be considered as well, if required (e.g., employment and fuel consumption). The trade-off is between catching small low-value shrimp at low operating cost inshore, and catching bigger and higher-value shrimp at higher operating costs offshore. For instance, in artisanal fisheries shrimp can be caught by push nets in lagoon fringes with very little gear and equipment, or later on in the deeper channels during migration, using stake nets, canoes, and outboard engines, or later still in coastal waters using "baby trawlers" or drift nets, and finally, in offshore waters using sophisticated multi-rig freezer trawlers. Because shrimp migrate continuously to deeper waters, there is, for each of these types of exploitation and depth strata, an optimum size at first capture below which there is "local" growth overfishing, and above which there are important losses by natural mortality and migration.

If we consider the whole life cycle, there is also an optimum size at first capture and fishing pattern to achieve the overall highest value per recruit from a stock, but this may involve totally banning some fishing methods, totally protecting some areas, and regulating seasonal fishing in estuaries and at sea. The optimum fishing regime evidently will depend on the objectives retained for the fishery. The highest possible total market value or foreign exchange earnings might be obtained by promoting offshore trawling and licensed foreign fishing for export. The higher level of employment and lower fuel consumption might be obtained, on the contrary, by promoting small-scale fishing. In most cases compromises will have to be found between conflicting objectives of this sort. Another objective frequently retained for closed season management is the improvement of the spawning potential of the stock by closing the fishery either during the recruitment period or during the spawning period.

The problem can generally be considered at two levels: (1) determination of the "best" closed season, on the average, and (2) fine tuning of the opening date and duration of the fishing season from year to year in order to optimize results according to small changes in the recruitment parameters.

The determination of the average "best" season can easily be made using a yield-per-recruit model. The Thompson and Bell model (in Ricker 62) is particularly useful because it offers a time-discrete representation of the life cycle and is easy to explain to fisheries administrators. It has been used, for example, by Garcia and van Zalinge (63) in Kuwait and Willmann and Garcia (64) for the Guyanas–Brazil fisheries. The impact of a seasonal closure depends on its dates and duration, as well as on the overall level of effort and the seasonal pattern of catchability (Garcia 65; Sluczanowski 66). When an inshore and offshore fishery operate sequentially, the overall results of a closure must be considered and the total catch or value, as well as other economic benefits, depend also on the respective effort levels in both fisheries (Garcia 65; Grant and Griffin 67; Nichols 68; Somers 69; Willmann and Garcia 64).

Clearly in sequential fisheries for shrimp, the closed seasons in the inshore bays and offshore coastal areas must be coordinated for optimum results, and their effects may involve not only overall improvement in the fisheries output, but possibly also a change in resource allocation. Garcia (65), Blomo et al. (18), and Nichols (68) have shown that the total output in terms of tonnage or value is not greatly affected by changes in effort in inshore and offshore fishing. The main effect is on allocation and is generated through migration of biomass (and therefore value) from the inshore to the offshore fishery given that the stock may no longer be accessible to inshore fishing if it reopens, as for example in the Kuwait bay fishery for *Penaeus semisulcatus* (FAO 70). This can be a source of social unrest in some countries. A similar problem exists when migration during the closed season transfers some benefits of a closed season to another country. This may be the case between Texas and Mexico (Leary 20) and between Senegal and Guinea-Bissau (Lhomme and Garcia 35).

Once the "average" appropriate period for a closed fishing season is defined, it is possible to fix it definitively. However, shrimp are highly sensitive to year-to-year changes in the coastal environment which vary the onset of spawning, timing of larval recruitment inshore, subsequent growth and survival, dates of the migration of the main cohorts from lagoons, and so on. In such situations fishermen have reservations about arbitrary dates, and it may be necessary to set seasons annually, based on preseason surveys. This has been applied in the Gulf of Mexico (Ingle 71; Ford and St. Amant 72) and in Australia (Ruello 61; Bowen and Hancock 21). The optimal opening date is forecast by extrapolating observed growth rates of the main cohorts to the date at which 50% or 75% of the cohort is at optimum size. Applying this technique can be complicated when two or more species with different life cycles occur together. The problem is discussed by Ford and St. Amant (72) and a solution is proposed by Eldridge and Goldstein (73).

High seas closures to improve yield per recruit have met with variable success. In Texas waters the standing stock has been theoretically increased by 30–36% and the benefits reached apparently 6–9% of the annual predicted catch (Nichols 68). A longer closed season would produce higher benefits but probably at a higher cost. According to Rackowe (cited by Sribhibhadh 1), however, the Texas closure had other negative effects such as a decline in product quality because vessels and plants had difficulties in handling very large catches over a short period and the percentages of small shrimp landed rose, obviously contrary to one of the original objectives of the measure.

In Australia (Gulf of Carpentaria, St. Vincent Gulf) management by flexible closed seasons has been very successful and, confirming the earlier statement by Ruello, the benefits are said to pay largely for the costs incurred (Somers 69; Bowen and Hancock 21). Sluczanowski (66) rightly mentions, however, that the likelihood of successful fine tuning of management depends largely on the precison of the parameters used in forecasting and that in Spencer Gulf, Australia, 90% of the optimum results can already be obtained with the average parameter. It can probably be added that the cost–benefit ratio of fine tuning depends on the year-to-year variability of the recruitment parameters. This author also stresses that the losses incurred by suboptimal management increase rapidly with effort. At high levels of effort fine tuning might be essential, but the precision required will be obtained only at high additional research cost. During the discussions organized by the 2nd Australian National Prawn Seminar (Rothlisberg et al. 12) the industry directly involved in undertaking the preseason surveys in the Gulf of Carpentaria declared that the direct cost of the surveys needed for fixing the flexible dates was about A$200,000/yr, and the benefits reached A$2,000,000. It is worth noting that in Australia the closed-season management system operates on the basis of flexible dates within the framework of a limited entry, contrary to what happens in the United States. Blomo et al. (18) underlies that "fine-tuning of the closed season could involve the analysis of optimization of the use of various boat sizes at various depth ranges (probably also during various time periods)." Sluczanowski adds that optimization of management of a complex of stocklets with slightly different recruitment parameters may lead to the need for different closure dates in different subareas.

A very important conclusion of bioeconomic simulations obtained by Blomo et al. (18) and confirmed by Sluczanowski (66) is that the likely upper limit that fishing effort can reach during the open season must be known in order to calculate the optimum fishing dates. Only in the case of a "sole owner" situation as in some Australian limited-entry fisheries or in Saudi Arabia this condition can be strictly fulfilled. This confirms the general statement made earlier (Section 4.1.1) that regulation of size at first capture is likely to be inefficient if total effort level is not properly controlled, and is perfectly in line with optimization theory. In fact, Clark (74) stated "achievements of satisfactory levels of economic efficiency is probably impossible unless some form of exclusive 'property rights' or appropriate substitute can be established with respect to the fishery resource."

In the Persian Gulf, a regional 5-month closed season was implemented from 1980 to 1982 after a major decrease in total catches believed to be linked with excessive fishing effort. Subsequently Morgan and Garcia (75) showed that the long-term decrease in recruitment had no relationship with the increase in effort and suggested that environmental causes were most likely responsible. The closed season was later on reduced to 3 months in Kuwait and flexible dates were recommended. Fishing for a secondary species (*Metapenaeus affinis*) has recently been allowed during the closed season for the main target, *Penaeus semisulcatus*, in specially defined areas (Abdul Ghaffar and Mathews 76). In Saudi Arabia the closed season has been lengthened to 6 months (February 1 to July 31), that is, beyond the legal requirements, by the sole owner fishing company operating in the country, with substantial profits (R. Willmann, personal communication.).

In situations of heavy overfishing it is very often proposed by fishermen to close fishing during the spawning season to help conserve enough spawning potential for reproduction. If fishing is concentrated on juveniles at recruitment it can be easily shown that unless spawning corresponds to a particularly marked schooling behavior (which would drastically increase fishing mortality), it would not be very fruitful to close fishing on spawners once the main cohorts have been decimated by excessive fishing. The fecundity per recruit at a given level of exploitation is in this case increased more significantly by protecting juveniles during the recruitment period (Garcia 65; Garcia and van Zalinge 63).

Moreover, the assured positive effect of protecting spawners or increasing spawning potential per recruit relies on the assumption that there is a stock–recruitment relationship (SRR) and that increasing spawning stock size will increase recruitment and subsequent overall biomass; this has still to be convincingly demonstrated (see Section 2.5). It is, however, obvious that below some level of stock size, problems on stock reproduction are to be expected, and that once a stock or its main cohorts have been driven to excessively low levels of abundance by fishing or by a combination of intensive fishing and adverse environmental conditions, the question of whether or not there is indeed an SRR is not the most relevant. The depleted spawning stock should in that case be enhanced if only to give the stock a higher probability of recovering when environmental conditions improve (Csirke and Sharp 28).

4.1.4. Other Closed Periods

Bowen and Hancock (21) refer to moon closures. In Western Australia fishing is periodically closed for 10 nights around the full moon in order to prevent harvesting of a significant proportion of soft newly molted prawns, to restrict fishing during periods of low catchability (sic), and to reduce effective effort.

4.2. Regulation of Total Fishing Effort

The concept is to reduce fishing mortality (and economic inputs) to improve stock size, yields, and benefits. The annual fishing mortality by trawling can be defined by

$$F = \sum_{i=1}^{n} q_i f_i = \sum_{i=1}^{n} q_i p_i t_i$$

or by

$$F = Y/B$$

where q = catchability coefficient; p is the individual fishing power of a vessel; t the fishing time; $i = 1, \ldots, n$, the number of vessels; Y the annual yield; and B the stock biomass. Fishing mortality can therefore be reduced by reducing either the power of each vessel, p (gear and vessel limitations), the fishing time t (by catch quotas, institutional reduction of fishing time, moon closures, weekend closures, etc.), or the number n of boats allowed in the fishery (limited entry).

4.2.1. Limitation of Vessel Fishing Efficiency

This can be done by controlling gear or vessel characteristics. The limitation of gear characteristics is a particularly relevant option for shrimp fisheries where technological progress has been very significant. The shift from single rig (one trawl per boat) to double rig (two trawls), triple rig, and twin rig (four trawls/ boats) has produced a significant increase in efficiency (see Garcia and Le Reste 11, p. 34 for a short review). Progress in sorting on board has also contributed to the increase in efficiency. As a consequence total fishing pressure has increased faster than nominal effort. Many shrimp fisheries around the world still present potential for increased efficiency if no gear limitation is implemented, although fishing effort might already be excessive. Gear size is regulated in some countries as well as the number of trawls allowed on each trawler (Ruello 61; Bowen and Hancock 21). Such an imposed decrease in efficiency adds, however, to the cost of fishing and is hardly acceptable from an economic point of view.

Management techniques in shrimp fisheries include such vessel restrictions as limits on overall boat length and/or engine power, especially when effort is excessive, whether or not effort is limited and boat replacement policy is implemented (see Walker 38), for example, in southern Australia). Vessels above a given size, tonnage, or horsepower can then be prohibited, and the shift to a preferred boat size can be accelerated by economic incentives such as special soft loans or subsidies for a particular size. The boat replacement policy is facilitated if the limited factor is quantified in units, which are transferable. A new boat can then be entered into the fishery only if the equivalent amount of units have been bought out (see Bowen and Hancock 21) for an example). Subsidies to promote a given type of vessel can have perverse economic (see Section 4.5) and technological effects.

4.2.2. Limitation of Fleet Activity

In critical situations where excessive fishing power exists, fishing time may be reduced by setting catch quotas with or without formally limiting the overall fleet size.

Institutional fleet immobilization by which fishing is restricted some days per week (e.g., weekend closures) or per month (moon closures, see Section 4.1.4) have also been used. Clark (77) indicates that catch quotas alone cannot limit fleet growth and only "replace overfishing by overcapacity." When short-lived animals such as shrimp, squid, and anchovy are involved, regulation by annual total catch quota is even worse. In practice the individual race for catching as much as possible before the overall quota is taken leads to increased fishing power and concentration of fishing earlier and earlier in the season. The same annual catch in weight is taken in a progressively shorter season, and is composed of more and more younger shrimps. Fishing mortality continues to increase and nothing prevents investments from becoming excessive. This policy may result in very high peak catches in a short period of the year, creating problems and high costs of storage. The overall effort could be best distributed by quotas for shorter periods (quarterly or by month). Such regulations would be difficult to apply because they necessitate a good control on landings; this is not always possible, particularly in developing countries. Leary (20) indicates that in the Gulf of Mexico inshore fishery daily catch limits are implemented. Another solution apparently not yet used on tropical shrimp, the transferable individual fisherman quotas (Christy 78; Moloney and Pearse 79), could in theory avoid excessive costs of fishing.

4.2.3. Limitation of Access to the Fishery

None of the measures discussed earlier can really avoid biological or economic overfishing when used alone. At best, improved state of stocks and profitability is temporary and the competition between fishermen for appropriation of the newly created rent rapidly leads to its dissipation (see introduction to Section 2). It is important to note that the extension of jurisdiction has not changed the issue for developing countries in the absence of regulated access of citizens to the EEZ resources.

Direct regulation of the level of exploitation by controlling the fleet size and the horsepower is designed to minimize such difficulties to prevent biological overfishing, but also to reduce costs, if effective. Walker (38) indicates, however, that in Australia "even in limited entry regimes there is a tendency to fish prawns to their biological maximum and beyond." A review of the problem of effort regulations in general can be found in Stokes (80), FAO (49), and Beddington and Rettig (48). The limited-entry system has advantages and disadvantages and its chances of success depends on many complicated factors.

In relation to shrimp stocks, Gulland (10) noted that limited entry should be implemented early in the development of fisheries because it is much easier to stop development of a fleet than to reduce it after the crisis has started, although it is probably difficult to convince fishermen of the need for limited entry when earnings are still very high. However, in some Australian shrimp fisheries limited entry has been in force from the beginning of the fishery in 1960 and according to Hundloe (81) and Bowen and Hancock (21) excessive fishing effort and overcapitalization are presently a problem in most fisheries whether or not limited entry was implemented at an early stage. According to Meany (82), however, limited entry has limited/restricted the overcapitalization process. The reason for failure to contain total effort in limited-entry shrimp fisheries lies merely in a rapid increase in effort just prior to closing the fishery when this measure was made public, and subsequently a progressive increase in fishing power by fleet upgrading to the authorized engine or size limit.

Limiting entry consists in limiting the number of fishing permits. It must involve estimation and monitoring of boat performances in order to detect subtle changes in fishing efficiency—''seepage effects''—stemming from technical innovation (see Section 4.2.1) or improvement in the space–time distribution of fishing operations. The number of licenses may have to be adjusted periodically to compensate for the increase in efficiency, for instance, by implementing a buyback scheme, possibly funded from the revenues generated by selling fishing rights. It is usually considered that the efficiency of such a mechanism is improved when the fishing rights are transferable through the market (Clark 77). Logbooks that can be required in exchange for licenses could be an excellent source of data for monitoring the fleet activity and efficiency.

One of the most controversial features of limited entry is that it may result in large rents. If they accrue to the license holders, they add considerable value to the license. This surplus value reached a $150,000 in southern Australia (Slucza-nowski 66) and can be cashed by the original first license holder when selling his right (windfall gain). Subsequently the surplus value becomes part of the capital costs to the new entrants. Whether or not this surplus value should accrue to the license holder or be extracted by the governments is a matter of philosophy.

Limited entry may result in social tension if the returns to the group of ''privileged'' license holders are much greater than in comparable employment opportunities and results in great pressure from those prohibited entry to the fishery. Introduction of appropriate license fees can allow the state to recuperate part of the surplus value created by management, discourages further applications for entry, and provides financing for other governmental initiatives such as the promotion of development of a priori unfavorable areas This last possibility has been used in Australia (Hancock 83).

It has been argued that limited entry creates a discrimination against outside fishermen, may allow inefficient fishing methods to continue (Bowen 84), and discourages technical advances by creating a quasi-monopoly (Gulland 10). According to Hancock (83) this has not been the case in Australia, where the techniques have advanced rapidly.

Limited entry to one fishery tends to lead to transfers of the excess efforts into neighboring fisheries or stocks and it usually becomes necessary to limit entry in all neighboring fisheries in order to control such transfers.

In extensive artisanal fisheries, limited-entry management is considered less appropriate, particularly in developing countries where it creates social and economic problems. The identification of fishermen is a significant deterrent to direct attempts of direct control of effort, although licensing of artisanal fishermen has received attention in some areas (FAO 13). In any case, close coordination between authorities in charge of management and development is particularly needed in this sector.

4.3. Closed Areas

Areas can be permanently or seasonally closed to fishing. Permanently closed areas or sanctuaries are widely used to protect both very small shrimp from capture and shrimp habitat from degradation by trawling gears. They aim, therefore, at optimization of the yield per recruit in value (as with seasonal closures) and conservation of critical habitats. They have been maintained, for instance, in various states of Australia (Bowen and Hancock 21) and in the Gulf of Mexico (Leary 20). They can be limited to small areas of marshland or extended into the open sea, as in Florida where the permanent sanctuary extends from the Everglades National Park to 18 m deep offshore. The prohibition of fishing can be reinforced by a regulation on minimum commercial size. In extensive deltas in developing countries these measures might be difficult to enforce because of the difficulty of access and also because small shrimps are often commercialized in a dry powdered form and sold as condiment at a high price.

Some areas are only seasonally closed to exploitation in relation to seasonal fishing closures for which the rationale for this measure is similar. The area closed could be only part of the area of distribution of the species (e.g., where the concentration of juveniles is highest) or the whole area of distribution of the species. In Texas, before 1981, only the waters under state jurisdiction (9 miles) were closed to fishing. The measure had to be replaced after this date by a closure of the entire Fishery Conservation Zone of Texas because of difficulties in enforcing the partial area closure. In Kuwait (Mathews 85) the closure concerns only the major area of distribution of the most important species whereas the rest of the fishing zone remains open to exploitation for secondary species.

Areas are sometimes closed to some type of fishing in order to allocate the resource to a particular socioeconomic stratum of fishermen. This is the case for coastal zones prohibited to trawling and reserved for passive small-scale gear. In countries with excessive trawling capacity, encroachment on the coastal closed area is usual and leads to permanent conflict with small-scale fishermen (see Section 2.3).

4.4. Estuarine Habitat Management and Stock Enhancement

The concept underlying these management techniques is to maintain and improve the reproductive potential of the stock by increasing larval survival through aquaculture and seeding of postlarvae and young juveniles, or to maintain or increase the carrying capacity of the nursery by habitat enhancement or conservation.

The first of these approaches (larval or postlarval seeding) has been followed in some countries for various species. It has been applied widely in Japan for shrimp (Hiroko 86; Doi et al. 87; Hasegawa et al. 88) and also in Kuwait (Mathews 85). It is usually a by-product of aquaculture and popular with managers and fishermen although its biological and economic efficiency has still to be convincingly demonstrated.

Nursery habitat protection and enhancement is more likely to lead to successful results. The very early stages of the life cycle occur in the intertidal zone of estuaries, in fringing creeks, mud banks, tidal swamps, mangroves, eelgrass meadows, and so on. It has been shown that the potential of a shrimp resource is proportional to the amount of habitat available in the nursery (Turner 89; Barrett and Gillespie 90, 91). The importance of the mangrove area available was shown by Martosubroto and Naamin (92). Doi et al. (87) have shown that the shrimp production in the Seto Inland Sea decreased progressively as a direct function of the amount of estuarine land reclaimed. Habitat conservation is therefore a very important component of shrimp stock management and identified nurseries must be protected from pollution, deforestation, land reclamation, damming, and housing development. One usual problem is that the modification of the estuaries by various user groups is not under the control of the authority managing the fishery. Intensive dredging of lagoon inlets changes the salinity regime in the nurseries and may have important consequences on shrimp stocks, composition, or abundance (Ewald 93). Kurata (94, 95) has proposed actively to manage the nursery areas by constructing artificial tidelands to improve postlarval survival.

4.5. Monetary Measures

These measures, used either for management or for development, are powerful tools for influencing fishermen's behavior, and general discussions on their advantages and drawbacks can be found in Clark (77) or Beddington and Rettig (48). Monetary measures have decisive effects on effort levels even though they are often implemented without explicit reference to this effect and their direct use to regulate effort can be difficult (Crutchfield 96). The literature on shrimp fisheries regarding these measures is very limited indeed.

Taxation of fishery inputs seem an obvious way of reducing fishing effort by reducing profits and incentives for expansion, particularly in artisanal fishing. However, the effects depend on the supply response of the fishery and specifically on the price elasticity of the demand for the species; that is, it depends on how far the fishermen can pass on the increased costs to the consumer (Lawson and Robinson 97). Baisre (98) indicates that price manipulations were used in concert with spatial and seasonal regulations to reallocate effort among alternative shrimp stocks after the 1979 hurricane in Cuba.

Subsidies have opposite effects. They can be used, for example, to divert excess effort to underexploited species or influence fleet size composition. In the latter case they can have powerful distorting effects and, according to Hundloe (81), they have increased costs, added to the burden of overcapacity, and led to a boat size structure that might not be ideal in Australian shrimp fisheries. This is particularly important when it is considered that the duration of a profitable fishing season depends on boat size (Penn and Hall, cited by Bowen and Hancock 21). Fishery development measures often have more effect on effort than those of management, and a coordination of action is needed between these two domains of fisheries. Development subsidies should be temporary and suppressed as soon as the aim is reached. Financial resources extracted through license fees could in turn be used for development and excess effort can be channeled.

High license fees can also be considered as a way of accruing part of the rent of the fishery to the country. They would reduce the surplus value attached to the fishing right and therefore the surplus capital cost of new entrants to the fishery.

REFERENCES

1. A. Sribhibhadh, International shrimp marketing situation. In P. C. Rothlisberg, B. J. Hill, and D. J. Staples, Eds., *Proceedings of the Second Australian National Prawn Seminar*. Simpson, Halligan, Brisbane, Australia, 1985, pp. 297–304.

2. J. A. Gulland and B. J. Rothschild, Eds., *Penaeid Shrimps—Their Biology and Management*. Fishing News Books, Farnham, Surrey, England, 1984, 308 pp.

3. International Trade Center, *Shrimps: A Survey of the World Market*. UNCTAD/GATT, Geneva, 1983, 273 pp.

4. J. W. Penn, The behaviour and catchability of some commercially exploited penaeids and their relationship to stock and recruitment. In J. A. Gulland and B. J. Rothschild, Eds., *Penaeid Shrimps—Their Biology and Management*. Fishing News Books, Farnham, Surrey, England, 1984, pp. 173–186.

5. J. W. Penn and N. Caputi, Stock-recruitment relationships for the tiger prawn, *Penaeus esculentus*, fishery in Exmouth Gulf, Western Australia and their implications for management. In P. C. Rothlisberg, B. J. Hill, and D. J. Staples, Eds., *Proceedings of the Second Australian National Prawn Seminar*. Simpson, Halligan, Brisbane, Australia, 1985, pp. 165–173.

6. S. Garcia, The stock recruitment relationship in shrimps—Reality or artifacts and misinterpretation? *Oceanogr. Trop.* **(1):**25–48 (1983).

7. S. Garcia, Reproduction, stock assessment models and population parameters in exploited penaeid shrimp populations. In P. C. Rothlisberg, B. J. Hill, and D. J. Staples, Eds., *Proceedings of the Second Australian National Prawn Seminar*. Simpson, Halligan, Brisbane, Australia, 1985, pp. 139–158.

8. FAO/UN, Report to the Government of the Republic of Panama on the Panamanian shrimp resources. Based on the work of L. K. Boerema, FAO/ETAP marine fisheries biologist. *Rep. FAO/ETAP* **1423:**1–28 (1961) (issued also in Spanish).

9. G. Gunter, Principles of shrimp fishery management. *Proc. Annu. Gulf Caribb. Fish. Inst.* **18:**99–106 (1966).

10. J. A. Gulland, Some introductory guidelines to management of shrimp fisheries. **IOFC/DEV/72/24:**1–12 (1972).

11. S. Garcia and L. Le Reste, Life cycles, dynamics, exploitation and management of coastal penaeid shrimp stocks. *FAO Fish. Tech. Pap.* **203:**1–215 (1981).

12. P. C. Rothlisberg, B. J. Hill, and D. J. Staples, Eds., *Proceedings of the Second Australian National Prawn Seminar*. Simpson, Halligan, Brisbane, Australia, 1985, 361 pp.

13. FAO, Report of the FAO/Australian Workshop on the management of penaeid shrimp/prawns in the Asia Pacific region. *FAO Fish. Rep.* **323:**1–19 (1985).

14. J. R. Poffenberger, An economic perspective of problems in the management of penaeid shrimp fisheries. In J. A. Gulland and B. J. Rothschild, Eds., *Penaeid Shrimps: Their Biology and Management*. Fishing News Books, Farnham, Surrey, England, 1984, pp. 299–306.

15. R. A. Neal, The Gulf of Mexico research and fishery on penaeid prawns. In P. C. Young, Ed., *Proceedings of the First Australian National Prawn Seminar*. Gov. Publ. Serv., Canberra, 1975, pp. 1–8.

16. J. E. Greenfield, The economics of shrimp production and marketing. *Proc. Annu. Gulf Caribb. Fish. Inst.* **27**:129–137 (1975).

17. W. L. Griffin and B. R. Beatie, Economic impact of Mexico's 200 mile offshore fishing zones on the United States Gulf of Mexico shrimp fishery. *Land Econ.* **54(1)**:27–37 (1978).

18. V. Blomo, K. Stokes, W. Griffin, W. Grant, and J. Nichols, Bioeconomic modelling of the Gulf shrimp fishery: An application to Galveston Bay and adjacent offshore areas. *South. J. Agric. Econ.* July Issue:119–125 (1978).

19. G. A. Rounsefell, Management of bankruptcy in the Gulf shrimp industry. *Proc. Annu. Gulf Caribb. Fish. Inst.* **27**:111–124 (1975).

20. T. R. Leary, Review of the Gulf of Mexico management plan for shrimps. In P. C. Rothlisberg, B. J. Hill, and D. J. Staples, Eds., *Proceedings of the Second Australian National Prawn Seminar*. Simpson, Halligan, Brisbane, Australia, 1985, pp. 267–274.

21. B. K. Bowen and D. A. Hancock, Review of penaeid prawn fishery management regimes in Australia. In P. C. Rothlisberg, B. J. Hill, and D. J. Staples, Eds., *Proceedings of the Second Australian National Prawn Seminar*. Simpson, Halligan, Brisbane, Australia, 1985, pp. 247–265.

22. M. Unar and N. Naamin, A review of the Indonesian shrimp fisheries and their management. In J. A. Gulland and B. J. Rothschild, Eds., *Penaeid Shrimps—Their Biology and Management*. Fishing News Books, Farnham, Surrey, England, 1984, pp. 104–110.

23. N. Naamin and P. Martosubroto, Effect of gear changes in the Cilacap shrimp fishery. Paper presented at the Fourth Session of IPFC/SCORRAD, Jakarta, Indonesia, 23–29 August 1984. *FAO Fish. Rep.* **318**:25–32 (1984).

24. E. G. Silas, M. J. George, and T. Jacobs, A review of the shrimp fisheries of India: a scientific basis for the management of the resources. In J. A. Gulland and B. J. Rothschild, Eds., *Penaeid Shrimps—Their Biology and Management*. Fishing News Books, Farnham, Surrey, England, 1984, pp. 83–103.

25. A. Perez, R. Puga, and J. Rodriguez, The stock assessment and management of Cuban shrimp stocks. In C. P. Mathews, Ed., Proc. shrimps and finfisheries management workshop, 9–11 October 1983. *Kuwait Inst. Sci. Res. (KISR)* **1366**:48–119 (1984).

26. A. L. Lawrence, Marine shrimp culture in the western hemisphere. In P. C. Rothlisberg, B. J. Hill, and D. J. Staples, Eds., *Proceedings of the Second Australian National Prawn Seminar*. Simpson, Halligan, Brisbane, Australia, 1985, pp. 327–336.

27. S. Garcia, A note on environmental aspects of penaeid shrimp biology and dynamics. In J. A. Gulland and B. J. Rothschild, Eds., *Penaeid Shrimps—Their Biology and Management*. Fishing News Books, Farnham, Surrey, England, 1984, pp. 268–271.

28. J. Csirke and G. D. Sharp, Eds., Report of the expert consultation to examine changes in abundance and species composition of neritic resources. *FAO Fish. Rep.* **291(1)**:1–102 (1984).

29. S. Garcia, The problem of unstable resources management. CECAF/ECAF Ser. 84/28:1–34 (1984).

30. M. P. Sissenwine, The uncertain environment of fishery scientists and managers. *Mar. Res. Econ.*, **1(1)**:1–30 (1984).

31. D. J. Staples, W. Dall, and D. J. Vance, Catch prediction of the banana prawn, *Penaeus merquiensis*, in the south-eastern Gulf of Carpentaria. In J. A. Gulland and B. J. Rothschild, Eds., *Penaeid Shrimps—Their Biology and Management*. Fishing News Books, Farnham, Surrey, England, 1984, pp. 259–267.

32. W. C. Clark and G. P. Kirkwood, Bioeconomic model of the Carpentaria prawn fishery. *J. Fish. Res. Board Can.* 36(11):1304–1312 (1979).

33. FAO/IDRC, *Fish By-Catch—Bonus from the Sea*. Report of a technical consultation on shrimp by-catch utilization held in Georgetown, Guyana, 27–30 October 1981. FAO/IDRC, 1982, 163 pp.

34. P. F. Sheridan, J. A. Browder, and J. E. Powers, Eds., Ecological interactions between penaeid shrimps and bottomfish assemblages. In J. A. Gulland and B. J. Rothschild, Eds., *Penaeid Shrimps—Their Biology and Management*. Fishing News Books, Farnham, Surrey, England, 1984, pp. 235–253.

35. F. Lhomme and S. Garcia, Biologie et exploitation de la crevette penaeide au Sénégal. In J. A. Gulland and B. J. Rothschild, Eds., *Penaeid Shrimps—Their Biology and Management*. Fishing News Books, Farnham, Surrey, England, 1984, pp. 111–141.

36. N. M. Haysom, Review of penaeid prawn fisheries of Australia. In P. C. Rothlisberg, B. J. Hill, and D. J. Staples, Eds., *Proceedings of the Second Australian National Prawn Seminar*. Simpson, Halligan, Brisbane, Australia, 1985, pp. 195–203.

37. M. Bowerman, ''By-catch'' pays the bill for Clarence prawners. *Aust. Fish*. July:14–16 (1984).

38. R. H. Walker, Australian prawn fisheries. In J. A. Gulland and B. J. Rothschild, Eds., *Penaeid Shrimps—Their Biology and Management*. Fishing News Books, Farnham, Surrey, England, 1984, pp. 36–48.

39. J. A. Gulland and S. Garcia, Observed patterns in multispecies fisheries. *Life Sci. Res. Rep*. 32:155–190 (1984).

40. Food and Agriculture Organization (FAO), Expert Consultation on selective shrimp trawls. Ijmuiden, the Netherlands, 12–14 June 1973. *FAO Fish. Rep.* **139**:1–71 (1973).

41. N. Naamin and T. Sujastani, *The By-Catch Excluder Device*. Paper presented at the FAO/Australian workshop on the management of penaeid shrimp/prawns in the Asia Pacific region. Kooralbyn Valley, Queensland, Australia, 1984, 20 pp.

42. R. Boddeke and A. Verbaan, Het mechanisch transport voor de spoelsort eet machine. *Visserij* **29(6)**:386–396 (1976).

43. R. Boddeke, *Report of Consultancy on the By-Catch in Senegalese Shrimp Fisheries*. Paper to be presented at the FAO Expert Consultation on selective shrimp trawl development, Mazathon (Mexico), November 1986.

44. D. Pauly, An attempt to estimate the stock-recruitment relationship of Gulf of Thailand penaeid shrimps. *Trans. Am. Fish. Soc.* **111(1)**:13–20 (1982).

45. P. B. Bailey, Pre-recruit mortality in the Gulf of Thailand shrimp stocks. *Trans. Am. Fish. Soc.* **133**:403–406 (1984).

46. G. Brewer, The wider dimensions of management uncertainty in world fisheries. In R. M. May, Ed., *Exploitation of Marine Communities*. Springer-Verlag, Berlin and New York, 1984, pp. 275–285.

47. G. Brewer, Managing fisheries, some design considerations. *FAO Fish. Circ.* **718:**85–104 (1979).

48. J. R. Beddington and B. R. Rettig, Approaches to the regulation of fishing effort. *FAO Fish. Tech. Pap.* **243:**1–39 (1983).

49. Food and Agriculture Organization (FAO), Report of the expert consultation on the regulation of fishing effort (fishing mortality). *FAO Fish. Tech. Pap.* **289:**1–34 (1983).

50. F. T. Christy, Jr. and A. Scott, *The Common Wealth in Ocean Fisheries. Problems of Growth and Economic Allocation.* Johns Hopkins Press for Resources for the Future Inc., Baltimore, MD, 1965, 281 pp.

51. J. A. Gulland, Manual of methods for fish stock assessment. Part I. Fish population analysis. *FAO Man. Fish. Sci.* **4:**1–154 (1969).

52. P. A. Larkin, An epitaph for the concept of maximum sustainable yield. *Trans. Am. Fish. Soc.* **106(1):**1–11 (1977).

53. M. P. Sissenwine, Is MSY an adequate foundation for optimum yield? *Fisheries* **3(6):**22–24, 37–42 (1978).

54. J. A. Gulland, Introductory guidelines to shrimp management: Further thoughts. In J. A. Gulland and B. J. Rothschild, Eds., *Penaeid Shrimps—Their Biology and Management.* Fishing News Books, Farnham, Surrey, England, 1984, pp. 290–299.

55. R. Hannesson, Fisheries management and uncertainty. *Mar. Resour. Econ.* **1(1):**89–97 (1984).

56. F. Lhomme, Biologie et dynamique de *Penaeus duorarum notialis* au Sénégal. Partie 1. Séléctivité. *Doc. Sci. Cent. Rech. Oceanogr. Dakar-Thiaroye* **63:**1–30 (1978).

57. M. Al-Hossaini, A. R. Abdul Ghaffar, and M. El-Musa, Gear selection in Kuwait's shrimp fishery and its effect on *P. semisculcatus* recruitment. A preliminary study. Proceedings, shrimp and finfisheries management workshop, 9–11 October 1983. *Kuwait Inst. Sci. Res. (KISR)* **1366:**398–422 (1984).

58. L. Le Reste and J. Marcille, Réflexion sur les possibilités d'aménagement de la pêche crevettière à Madagascar. *Bull. Madagascar* **320:**1–15 (1973).

59. S. Garcia, E. Lebrun, and M. Lemoine, Le recrutement de la crevette *Penaeus subtilis* en Guyanne française. *Rapp. Tech. ISTPM* **9:**1–43 (1984).

60. S. Garcia, Seasonal trawling bans can be very successful in heavily overfished areas: The Cyprus effect. *Fishbyte* **4(1):**7–12 (1986).

61. N. V. Ruello, Biological research and management of prawn fisheries in New South Wales. In P. C. Young, Ed., *Proceedings of the First Australian National Prawn Seminar.* Aust. Gov. Publ. Serv. Canberra, 1975, pp. 222–233.

62. W. E. Ricker, Computation and interpretation of biological statistics of fish populations. *Bull. Fish. Res. Board Can.* **191:**1–382 (1975).

63. S. Garcia and N. P. van Zalinge, Shrimp fishing in Kuwait: Methodology for a joint analysis of the artisanal and industrial fisheries. In *Assessment of the Shrimp Stocks off the West Coast of the Gulf Between Iran and the Arabian Peninsula,* FAO/UNDP Proj. FI/DP/RAB/80/015. FAO, Rome, 1982, pp. 119–142.

64. R. Willmann and S. Garcia, A bioeconomic model of sequential artisanal and industrial shrimp fisheries. *FAO Fish. Tech. Pap.* **270:**1–49 (1986).

65. S. Garcia, Biologie et dynamique des populations de crevettes roses, *Penaeus duorarum*

notialis (Perez-Farfante, 1967) en Côte d'Ivoire. *Trav. Doc. ORSTOM* **79:**1–271 (1977).

66. P. R. Sluczanowski, Modelling and optimal control: A case study based on the Spencer Gulf prawn fishery for *Penaeus latisullatus* Kishinouye. *J. Cons., Cons. Int. Explor. Mer* **41:**211–225 (1984).

67. W. Grant and W. Griffin, Bioeconomic model of the Gulf of Mexico shrimp resources. *Trans. Am. Fish. Soc.* **108:**1–13 (1979).

68. S. Nichols, Impact of the combined closures of the Texas territorial sea and FCZ on brown shrimp yields. 1984, 18 pp. Unpublished Manuscript.

69. I. F. Somers, Maximizing value per recruit in the fishery for banana prawns, *Penaeus merguiensis* in the southeastern Gulf of Carpentaria. In P. C. Rothlisberg, B. J. Hill, and D. J. Staples, Eds., *Proceedings of the Second Australian National Prawn Seminar.* Simpson, Halligan, Brisbane, Australia, 1985, pp. 185–191.

70. Food and Agriculture Organization (FAO), *Assessment of the Shrimp Stocks of the West Coast of the Gulf Between Iran and the Arabian Peninsula,* FAO/UNDP Proj. FI/DP/RAB/80/015. FAO, Rome, 1982, 163 pp.

71. R. M. Ingle, Intermittent shrimp sampling in Apalachicala Bay with biological notes on regulatory applications. *Proc. Annu. Gulf Caribb. Fish. Inst.* 5 (1956).

72. T. B. Ford and L. S. St. Amant, Management guidelines for predicting brown shrimp, *Penaeus aztecus,* production in Louisiana. *Proc. Annu. Gulf Caribb. Fish. Inst.* **23:**149–164 (1971).

73. P. J. Eldrige and S. A. Goldstein, The shrimp fishery of the South Atlantic United States: A regional management plan. *Ocean Manage.* **3:**87–119 (1977).

74. C. W. Clark, Optimization theory and fishery management. *FAO Fish. Circ.* **718:**43–48 (1979).

75. G. R. Morgan and S. Garcia, The relationship between stock and recruitment in the shrimp stocks of Kuwait and Saudi Arabia. *Oceanogr. Trop.* **17(2):**133–137 (1982).

76. A. R. Abdul Ghaffar and C. P. Mathews, *Metapenaeus affinis* fishery in the open sea during the closed season. Proceedings, shrimp and finfisheries management workshop, 9–11 October 1983. *Kuwait Inst. Sci. Res. (KISR)* **1366:**155–162 (1985).

77. C. W. Clark, *Bioeconomic Modelling and Fisheries Management.* Wiley, New York, 1985.

78. F. T. Christy, Jr., *Fisherman Quotas: A Tentative Suggestion for Domestic Management,* Occas. Pap. Ser., No. 19. Law of the Sea Institute, University of Rhode Island, Kingston, 1973, 6 pp.

79. D. G. Moloney and P. H. Pearse, Quantitative rights as an instrument for regulating commercial fisheries. *J. Fish. Res. Board Can.* **36(7):**859–866 (1979).

80. R. L. Stokes, Limitation of fishing effort: an economic analysis of options. *Mar. Policy* October:289–301 (1979).

81. T. S. Hundloe, The financial and economic health of the northern prawn fishery and the effect of shipbuilding bounties. In P. C. Rothlisberg, B. J. Hill, and D. J. Staples, Eds., *Proceedings of the Second Australian National Prawn Seminar.* Simpson, Halligan, Brisbane, Australia, 1985, pp. 289–295.

82. T. F. Meany, *Limited Entry in the Western Australian Rock Lobster and Prawn Fisheries: An Economic Evaluation.* Paper presented at the Symposium on policies for

economic rationalization of commercial fisheries, Powell River, British Columbia, August 1978.

83. D. A. Hancock, The basis for management of the West Australian prawn fisheries. In P. C. Young, Ed., *Proceedings of the First Australian National Prawn Seminar.* Govt. Publ. Serv., Canberra, 1975, pp. 252–269.

84. B. K. Bowen, The economic and sociological consequences of license limitation. In P. C. Young, Ed., *Proceedings of the First Australian National Prawn Seminar.* Aust. Govt. Publ. Serv., Canberra, 1975, pp. 270–276.

85. C. P. Mathews (Ed.), Final report. Proceedings. Shrimp and finfisheries management workshop, 9–11 October 1983. Kuwait, *Kuwait Inst. Sci. Res.*, MB-48/KISR 1366, Vol. 1:239 p., Vol. 2:240–600 (1984).

86. H. Hiroko, The studies on some ecological and physiological characteristics of artificial seedlings of prawns, *Penaeus japonicus* Bate. *Bull. Nansei Reg. Fish. Res. Lab.* **6:**59–84 (1973).

87. T. Doi, K. Okada, and K. Isibashi, Environmental assessment on survival of Kuruma prawn, *Penaeus japonicus*, in Tideland. I. Environmental conditions in Saizyo tideland and selection of essential characteristics. *Bull. Tokai Reg. Fish. Res. Lab.* **76:**37–52 (1973).

88. A. Hasegawa, Economic effectiveness of liberation of Kuruma prawn seedlings. *Bull. Tokai Reg. Fish. Res. Lab.* **83:**7–23 (1975).

89. E. Turner, Intertidal vegetation and commercial yields of penaeid shrimp. *Trans. Am. Fish. Soc.* **106(5):**411–416 (1977).

90. B. B. Barrett and M. C. Gillespie, Primary factors which influence commercial shrimp production in coastal Louisiana. *Tech. Bull. La. Wildl. Fish. Comm.* **9:**1–29 (1973).

91. B. B. Barrett and M. C. Gillespie, Environmental conditions relative to shrimp production in coastal Louisiana. *Tech. Bull., La. Wildl. Fish. Comm.* **15:**1–22 (1975).

92. P. Martosubroto and N. Naamin, Relationships between tidal forests (mangroves) and commercial shrimp production in Indonesia. *Mar. Res. Indonesia* **18:**81–86 (1977).

93. J. J. Ewald, The Venezuelan shrimp industry. *FAO Fish. Rep.* **57(3):**765–774 (1969).

94. H. Kurata, Certain principles pertaining to the penaeid shrimps seedling and seeding for the farming of the sea. *Bull. Nansei Reg. Fish. Res. Lab.* **5:**33–74 (1972).

95. H. Kurata, Artificial tideland and its effects on prawn breeding. *Yamaha Fish. J.* **2:6** (1977).

96. J. A. Crutchfield, Economic and social implications of the main policy alternatives for controlling fishing effort. *J. Fish. Res. Board Can.* **36(7):**742–752 (1979).

97. R. Lawson and M. Robinson, Artisanal fisheries in West Africa. Problems of management and implementation. *Mar. Policy* October:279–290 (1983).

98. J. A. Baisre, Resultados en la administracion scientifica de las pesquerias de camarones (*Penaeus* spp.) y langostas. *FAO Fish. Rep.* **278**(Suppl.):120–142 (1983).

14 PERFORMANCE AND SELECTIVITY OF TRAP FISHERIES FOR CRUSTACEANS

Jay S. Krouse

Department of Marine Resources
West Boothbay Harbor, Maine

1. INTRODUCTION

Throughout the world traps of various materials and designs are employed extensively by commercial and recreational fishermen to catch clawed and spiny lobsters, crabs, crayfish, and to a much lesser extent, shrimp (see review by von Brandt 1). These crustacean trap fisheries are not only some of the world's most valuable, but as one might expect, some of the more heavily exploited. For instance, in 1984 the crab and lobster catch of North America alone weighed 197,447 metric tons and was valued at U.S.$519,145,000, or 16% of the total value of marine catches (Fisheries of the United States, 1984, and Canadian Fisheries Statistical Highlights,

1984). Moreover, Canadian and U.S. scientists estimate 70–90% of all legal-sized American lobsters, *Homarus americanus*, entering the fishery are harvested each year (Cobb 2; Pringle et al. 3), and similarly high fishing mortalities for spiny lobster populations have been reported by Bowen (4).

In all likelihood, the high unit value of most crustaceans is the main incentive for technological advances such as larger, faster, and more powerful boats, hydraulic haulers, and electronic equipment. These enable fishermen not only to fish more traps but to do so more efficiently. The effective fishing effort of most trap fisheries will thus continue to escalate, further taxing many already overexploited shellfish fisheries. Scientific management, based on a wide array of biological, sociological, and economic data, is therefore essential to ameliorate this situation and to ensure the future well-being of the stock. Accordingly, many researchers have studied various aspects of trap performance and selectivity for purposes of evaluating fishing power and catchability; understanding this aspect is necessary for realistic stock assessments with catch and effort data, and also provides knowledge of how a trap "works" relative to the target species. As a result, the biologist can recommend trap modifications to improve and, in certain cases, control gear efficiency and overall catch characteristics.

In the following pages those factors affecting the fishing characteristics of baited traps such as the mechanical size-selective properties related to trap design, lath, and mesh spacings; entrance location, shape, and size; and behavioral and physiological considerations are examined. Furthermore, this chapter discusses how many decapod crustacean trap fisheries have realized conservation and cost-saving benefits from various regulations based on an understanding of the trap capture process.

2. TRAP FISHERY DEVELOPMENT

Over the years, technological advances in vessels, electronics, and gear have all contributed in varying degrees and ways to the evolution of shellfish trap fisheries. To illustrate the types of changes that have occurred and are indeed typical of many crustacean fisheries, the trap fishery for American lobster along the Maine coast is briefly reviewed.

By the mid- to late-nineteenth century, when lobsters had become less abundant and more difficult to catch, fishermen converted from hand gathering, spears, and hoop nets to the more efficient lath pot (Fig. 1). This forerunner of today's lobster trap was aptly described by Herrick (5), the first scientist to study extensively the biology and fishery of the American lobster, as follows:

> The principle of the modern lobster trap is that of the old-fashioned rat trap adapted for taking an aquatic animal with as keen a scent as the rodent, but with far duller wits. The device is undoubtedly of great antiquity, but as modified and applied for the lobster it is apparently not over 200 years old. It was introduced to this country (United States) from Europe, where . . . it was first applied in this way by the Dutch in 1713, and was adapted from the eelpot then in use.

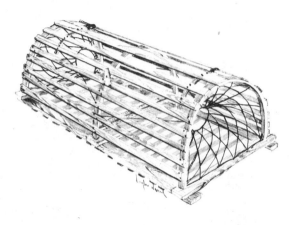

Figure 1. Typical single-chambered lath trap used in the American lobster, *Homarus americanus*, fishery from the mid-nineteenth century until well into the 1900s (from *Scribner's Monthly*, June 1981, courtesy of Maine Maritime Museum, Bath, Maine).

Figure 2. Modern-day single- and double-parlor lath traps used in the American lobster, *Homarus americanus*, fishery (photographs by David Parkhurst). (*a*) Wire traps; (*b*) wooden traps.

Although there have been several improvements in trap construction materials (synthetic twine, net, vinyl-coated wire, etc.) and design, certainly the most significant modification has been the incorporation of an inner chamber (parlor), which greatly enhances trap efficiency (Fig. 2). Once lobsters enter the parlor, escape is very difficult. With this increase in retention capability, fishermen could set their traps for longer periods without a reduction in catch and consequently were enticed to fish more traps. In fact, today many Maine lobstermen are using traps equipped with double parlors, usually arranged in series, or at opposite ends of the trap.

Until the early 1900s and then again through the depression years between World Wars I and II (Dow 6), fishermen used rowboats and sailing dories to deploy their traps (then averaging 50–100 traps per fisherman), in close proximity to shore. Subsequently, as the result of 20th-century technology such as the development of the gasoline- and eventually diesel-powered engine, the power winch, and finally the hydraulic trap hauler, and sophisticated electronic equipment (loran, radar, depth recorder, and marine radio), lobstermen had not only the means to fish more traps, but could distribute this gear over a considerably larger area. Whereas in 1900 there were 3105 licensed Maine lobstermen fishing 327,000 traps close to shore, in 1985 7879 fishermen (a 154% increase) used about 2 million traps (a 512% increase), or an average of 365 traps per boat (J. S. Krouse, unpublished data). Although the spatial distribution of fishing effort has not been defined, it is not uncommon for today's inshore lobstermen to fish 10–20 miles offshore.

Similar to the Maine lobster fishery, technological advances have influenced significant change and development in other major crustacean fisheries including the Western Australian spiny lobster, *Panulirus cygnus* (Bowen 4) and the California Dungeness crab, *Cancer magister* (Dahlstrom and Wild 7), and in many cases these advances, in conjunction with economic factors, have caused excessive fishing on a finite resource. In describing innovations affecting the Canadian lobster industry, aside from the gasoline engine and pot hauler, Pringle et al. (3) wrote,

> Other significant developments were the introduction of the parlour traps early in the century and the echo sounder in the post war era . . . electronic gear such as radio, loran, and radar, as well as larger, faster, more reliable, and more comfortable boats. Thus they are a more efficient predator of lobsters than their predecessors. The drastic decline in landings (in certain districts) . . . may be due to recruitment overharvesting.

The excessive number of traps set along with the increased hauling frequency are attributable at least in part to technological improvements, which have in many instances led not only to growth and recruitment overfishing, but also to problems such as higher cull incidence, predation, within-trap cannibalism, and reduced growth associated with the entrapment, handling, and subsequent release of high numbers of sublegal-sized animals (Hunt et al. 8; Brown and Caputi 9, 10; Davis 11; Joll 12; Krouse 13; Krouse and Thomas 14). Compounding this situation are the durable synthetic trap materials such as nylon and vinyl-clad wire mesh used for trap entrances and coverings and wood preservatives, which significantly prolong the effective fishing life of traps. If a trap is severed from its surface buoy, an event more likely today with the high trap densities and boat traffic, it may continue to fish for an extended period, lobsters being attracted to entrapped fish, other invertebrates, conspecifics, or the shelter offered by the trap, thereby removing animals from the fishable stock without contributing to yield. Studies on fishing by derelict traps are reported on by Sheldon and Dow (15) and Pecci et al. (16), and reviewed by Smolowitz (17) for *H. americanus*, by High and Worlund (18) for the king crab, *Paralithodes camtschatica*, by Casey and Wesche (19) for the blue crab, *Callinectes sapidus*, and by High (20) for the Dungeness crab, *Cancer magister*. To reduce the deleterious effects of ''ghost'' fishing, beginning in 1990, traps used for *H. americanus* in Maine will be required to have escape (ghost) openings (3.75 × 3.75 in.) covered by a natural fiber or a panel fastened with degradable links; in Oregon some type of escape mechanism is also required in *C. magister* traps. Conversely, Miller (21) observed that the snow crab, *Chionoecetes opilio*, was attracted only by the trap's original bait; subsequently the modest potential loss of crab did not warrant any gear alterations.

Considering the intense fishing pressure on most crustacean fisheries, in association with destructive factors that may cause unnecessary damage and waste to the resource, fishery managers are advised by Brown and Caputi (10), in their study of the effects of fishing practices on sublegal-sized spiny lobsters caught and released by Western Australian fishermen, of

> the need for constant monitoring of the fishery and the updating of management procedures and regulations to ensure the stability and viability of the rock lobster stock. . . . An important component of management of a fully exploited stock is to reduce waste, e.g. by predation and poor handling techniques.

3. FACTORS AFFECTING CAPTURE BY TRAPS

Whether a crustacean is attracted to, enters, and is retained by a trap depends on a series of environmental, physiological, and behavioral factors along with mechanical characteristics of the trap. These factors are illustrated in a flow diagram depicting the sequential events of the trap capture process (Fig. 3) developed by Bennett and Brown (22) in the assessment of the effect of the immersion time on catch. Further modifications to this capture scenario were presented by Caddy (23) and Brêthes et al. (24). An understanding of the multifaceted capture process is beneficial to the fisherman in allowing for more efficient gear design and effective fishing strategy, and to the biologist for accurate interpretation and assessment of catch and effort data and the regulation of mechanical trap characteristics to minimize resource waste.

3.1. Bait

Owing to well-developed chemoreceptory senses, crustaceans are adept at distant detection and eventual location of the source of food odors (Phillips et al. 25).

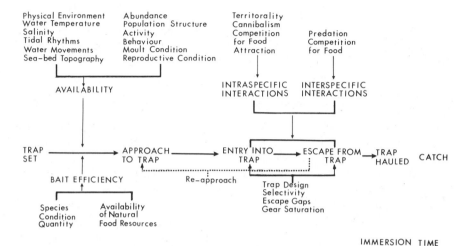

Figure 3. A flow diagram showing the major factors affecting the capture process of crustaceans with baited traps (from Bennett and Brown 22).

This olfactory response, which is affected by environmental (temperature, salinity, etc.) and physiological (molt and reproductive condition) factors, has long been exploited by fishermen through the use of various species of fish carcasses, invertebrates, farm animal by-products such as hides and hocks, and synthetic bait to lure their quarry to traps (Bowen 4; Brown and Barker 26; Dahlstrom and Wild 7; Dow 6; Phillips et al. 25; Prudden 27; Savage et al. 28; Van Engel 29).

The kind of bait selected by fishermen is primarily related to availability, cost, perceived attractiveness, and tradition. Certainly this choice is significant: bait trials conducted by Dow and Trott (30) on American lobsters and Thomas (31) on European lobsters, *Homarus gammarus*, showed that redfish, *Sebastes marinus*, and plaice, *Platessoides platessa*, are more efficient bait than other species tested. Similarly, traps baited with fresh rather than salted fish catch more crabs (Cargo 32; Edwards and Meaney 33; Kinnear 34; Otwell et al. 35; Thomas 31).

Aside from the influence bait type has on capture success, fishing practices also affect bait performance: for instance, the amount of bait used in a trap, its location, frequency of change, and baiting method (bait bag/line, or perforated container) (Bjordal 36; Simpson 37). Miller (38) found traps with 3 kg of bait caught more snow crabs, *Chionoecetes opilio*, than traps with 1 kg of bait. Miller (39) also reported that trap saturation, that is, reduction of catch rate with increasing catch, could be minimized but not eliminated by utilizing larger quantities of bait, fishing larger traps, and reducing escapement. Depending on bait size and consistency, hagfish, *Myxine glutinosa*, and scavenging amphipods are capable of devouring bait within 24 hr (J. S. Krouse, personal observation) and thus there may be advantages to using perforated containers. However, traps with unprotected bait made larger catches, probably owing to a less restricted bait odor trail (Bjordal 36). Likewise, Miller (40) noted higher levels of trap saturation for *C. productus* in traps with exposed bait.

As Hancock (41) points out in his study of attraction and avoidance responses of marine invertebrates to various baits, an understanding of this behavior should be useful in developing a selective artificial bait, formulated to attract target species and repel unwanted animals. Knowledge of the intraspecific avoidance displayed by many crustaceans has also been put to practice in certain fisheries. For example, Western Australian lobstermen reportedly keep traps free of any spiny lobster remains (Hancock 41), but some Maine lobstermen string *Cancer* crabs on the bait line to minimize incidental crab catches (J. S. Krouse, unpublished data); this practice is supported by findings of Chapman and Smith (42), who showed reduced catches of the European crab, *Cancer pagurus*, in traps baited with a conspecific crab. Although in a derelict pot study, king crabs were not attracted to pots containing dead crabs, it was not clear whether this was an avoidance response (High and Worlund 18).

Besides the conventional baiting methods characteristic of most trap fisheries, live decapods serve as attractants in a few fisheries. Bonded south Florida fishermen are permitted to bait traps with living sublegal-sized spiny lobsters, *Panulirus argus*, to decoy their larger gregarious counterparts into the trap. Even though this practice eliminates the purchase of bait, it is not without cost, for a study by Hunt et al. (8) showed that excessive handling, on-deck exposure, and confinement of sublegals all cause mortality and reduce growth, resulting in a significant economic loss to the fishery. In the South Carolina blue crab fishery, Bishop et al. (43) report that because of the attraction of pubertal-molt female crabs to mature males, fishermen can successfully catch premolt females in peeler pots baited with large males at certain times and locations.

An understanding of the area influenced by a baited trap is necessary in order to estimate population size with catch data. This has been studied by Brêthes et al. (24) using tag–recapture experiments, Eggers et al. (44) with catch data, Miller (45) with crab densities from bottom photography, and Caddy (23) with calculations on olfactory response thresholds and diffusion rates. Brêthes et al. (24) caution that because of the pronounced effect of biological factors that are rarely measurable (e.g., molt and reproduction behavior, migration) and hydrology (namely currents) on catch success, it is difficult at best to delimit accurately the area effectively fished by a trap. Consequently, stock assessments derived from baited trap catch data alone must be carefully scrutinized.

For fishermen deploying strings of multiple traps the spacing between traps is critical to the catch rates. Traps too close together will compete with each other. Sinoda and Kobayasi (46) and Williams and Hill (47) found traps spaced at 33- and 50-m intervals caught fewer crabs (*Chionoecetes japonica* and *Scylla serrata*) than traps set further apart.

3.2. Trap Characteristics

After a decapod crustacean locates a trap, entry and finally retention are affected by behavioral interactions as well as design features such as trap shape and size; mesh and lath spacing; entrance size, shape, and location relative to the bait; and anti-escape devices and escape ports. Trapping experiments by Miller (39, 40, 48, 49) with the crabs *C. irroratus*, *C. productus*, and *Hyas araneus* demonstrated the effect that entrance location has on trap efficiency. Highest catches occurred in traps with the entrances aligned along the bait odor plume, thereby increasing the crab's success in quickly finding the opening. For those traps with hard-to-find entrances, as the number of crabs searching for the opening increased so did agonistic encounters, causing many animals to flee (Miller 48, 49). SCUBA observations of trap-related behavior of *H. americanus* by Auster (50) revealed that some of the larger lobsters were able to "bridge" the gap between the entrance port and bait, thereby consuming bait from the safety of the entrance funnel before departing the trap. In order to prevent this "funnel feeding" activity the distance separating entrance ring and bait should be increased. Suitability of entrance location is also related to behavioral considerations; for instance, Stasko (51) found that *H. americanus* would not enter traps with an entrance mounted on top, although the same traps captured *C. irroratus*.

Trap entrances equipped with various types of antiescape mechanisms can effectively minimize escape and significantly improve trap performance. Triggers on side-entry traps have been shown to inhibit egress of *C. productus* (Miller 49), *C. magister* (High 20), and *C. pagurus* (Mason 52). For top-entry traps, Miller (40) found that entrances with plastic collars deterred escape of *C. irroratus*, and Stasko (51) reported that top-mounted metal openings caught more *C. irroratus* than side-entry traps.

In contrast to the escape-inhibiting features of many crab trap entrances, nonescape ports are not generally found in lobster traps, but rather fishermen in pursuit of the clawed lobsters *H. americanus* and *H. gammarus* use traps with usually one and sometimes two inner chambers or "parlors" which are known to enhance retention and trap efficiency appreciably (Fig. 2). Lovewell et al. (53) noted that after a 1-day set traps with parlors caught twice as many *H. gammarus* as single-chambered traps. In fact, to restrict gear efficiency Western Australian fishermen are not permitted to have inner chambers in pots of the rock lobster, *Panulirus cygnus*, fishery.

The success of ingress and egress of homarid lobsters from traps has been shown in trapping studies of Thomas (54) and Pecci et al. (16) to be affected by head design, features including mesh size, slope, opening size and shape, and height above bottom. Spurr (55) concluded that a properly designed parlor head with a long steep slope and small opening near the top of the trap significantly improved efficiency.

Miller (39) reported a relationship between trap size and gear saturation that has two contributing components: escape and reduced entry. With larger traps Miller observed higher saturation levels of *Cancer productus* resulting in better catches. He attributed this to the high packing density of *C. productus*, which produced a larger catch with an increase in trap size. Similarly, Munro (56) noted there were higher yields of fish with larger Antillean fish traps, because escape was determined to be inversely proportional to trap volume.

The ready availability, improved durability, longevity, and competitive cost of synthetic trap covering material such as plastic, steel, and twine meshes have been responsible for extensive gear modifications in many fisheries. For instance, many Maine lobstermen, who have traditionally fished wooden traps, over the past decade have switched to wire traps. Of nearly 2 million traps used in the Maine inshore lobster fishery in 1985, about 48% were of wire (J. S. Krouse, unpublished data). Even though the relative fishing power of wire traps is unknown, on the basis of the gear's rise in popularity with many discerning fishermen (some of the finest gear technologists) and from an inconclusive study by Acheson (57) in which wire traps caught more lobsters than wood traps, the overall efficiency of wire traps appears to be at least comparable to that of wooden gear.

Mechanical size selection of traps is chiefly a function of the greatest dimension of the trap's covering (mesh size, lath spacing, etc.), presence of escape ports, and entrance opening size. Early trap selectivity studies of *H. americanus* by the Canadian scientists Templeman (58) and Wilder (59–62) revealed the effect of entrance ring diameter on catch size, and the inverse relation between lath spacing and numbers of sublegal lobsters caught. Following this, many investigators (Brown 63, 64; Dybern 65; Fogarty and Borden 66; Krouse 67; Krouse and Thomas 14; Nulk 68; Pecci et al. 16; Smolowitz 69; Weber and Briggs 70) examined the effects of escape vents on gear selectivity in clawed lobster fisheries. The results of these studies conclusively document the effectiveness of vents in allowing the escape of most sublegal lobsters and in enhancing the catch rate of legal lobsters; effects that Fogarty and Borden (66) attributed to gear saturation. Other benefits of escape vents include reductions in the number of animals injured by handling, intraspecific aggression during trap confinement, and in time spent by fishermen in sorting catch; escape vents would also deter the illicit trade in sublegal sized lobsters and the number of animals imprisoned and dying in ghost traps.

Despite the use of escape gaps in the Western Australian rock lobster fishery, Brown and Caputi (9, 10) reported that many undersized lobsters (estimated 16–20 million annually) are still caught. Subsequently these may be injured or, depending on the extent of displacement from home range and deck exposure time, become more prone to predation upon return to the bottom. The authors recommended an increase in vent size along with an educational program to alert fishermen to the importance of vents in quickly returning undersize lobsters to the sea.

Although escape gaps have been shown by Bowen (71) and Ritchie (72) to minimize catches of small spiny lobsters (*Jasus edwardsii* and *Panulirus cygnus*), Crous (73) and D. E. Pollock (personal communication) contended that escape ports are ineffective in certain Cape rock lobster, *Jasus lalandii*, fisheries off South Africa. Unlike the trap-related behavior of the majority of crustaceans, most Cape rock lobster escape occurs while the trap is being hauled (D. E. Pollock, personal communication). Crous (73) suggested that legal-sized lobsters block the escape openings at this time, and Pollock cited the same mechanism for shorts in high recruitment areas. Consequently, South African boats employ grid sorters on deck to reduce handling time. More recently, D. E. Pollock (unpublished data) determined that the sorting time for undersized animals could be further reduced by increasing the mesh size of the trap covering to 100 mm (stretched mesh).

For those fishermen pursuing both lobsters and crabs, guided by experiments of Stasko (51) with traps having entrances of varied sizes, shapes, and locations, a circular escape port was demonstrated by Krouse (67) to be effective in retaining harvestable sized *Cancer borealis* and *C. irroratus* while permitting the egress of sublegal lobsters. Circular escape rings have also been shown by Eldridge et al. (74), High (20), and Jow (75) to reduce trap catches of undersized Dungeness (*C. magister*), Alaska king (*P. camtschatica*), and blue crabs (*C. sapidus*). Moreover, Weber and Briggs (70) found that catches of market-size black sea bass (*Centropristis striata*), an important by-catch of the New York lobster fishery, were similar in number and size for unvented traps and traps with 58-mm circular ports. Miller (21) reported that the catch of small snow crabs, *C. opilio*, could be minimized by regulating trap mesh size.

In Alaska, where traps are used interchangeably during the noncurrent seasons for king and tanner crabs (*Chionoecetes spp.*), tunnel entrance size is regulated to achieve species selectivity (Miller 76). During the season for tanner crab (the smaller of the two species), tunnel "eyes" are not allowed to exceed 5 in. in height, whereas during the king crab season, openings must be at least 5 in. Williams et al. (77) observed that "tanner boards" placed across the upper half of tunnel opening in side entry traps not only excluded king crabs but also reduced the incidental catch of halibut, *Hippoglossus stenolepis*.

Koike and Ogura (78) found that the size composition of pink shrimp, *Pandalus borealis*, catches varied with trap mesh size, whereas the size selectivity of the Beni-zuwai crab, *Chionoecetes japonica*, was related to both entrance and mesh size. The influence of mesh size is depicted in the selectivity curves derived by Sinoda and Kobayasi (46).

3.3. Availability

The catchability of crustaceans with traps is known to be affected by various environmental, physiological, and behavioral factors as discussed by Auster (50), Caddy (23), Morgan (79), and Skud (80), who emphasized the importance of understanding such processes when interpreting catch and effort data. For example, cyclic fluctuations of water temperatures have been shown by Dow (81, 82) to be positively correlated with the abundance and availability of the clawed lobster (*H. americanus*, *H. gammarus*, and *Nephrops*), and McLeese and Wilder (83) reported the catchability of *H. americanus* to be a linear function of temperature. In his analysis of Canadian, European, and U.S. lobster catch trends Dow (6) concluded that seasonal changes in the English *H. gammarus* fishery (Hepper 84) are similar to those in the New England *H. americanus* fishery, where the catch peaks in August and September occur in association with warmer temperatures, increased molting (greater recruitment), and metabolic rates, resulting in greater foraging activity and enhanced vulnerability. Similarly, temperature and salinity along with physiological factors were shown by Morgan (85) to influence the vulnerability of rock lobster *Panulirus cygnus*, causing seasonal changes in catchability. By using mark and recapture methods on the portunid crab, *Scylla serrata*, Williams and Hill (47) determined that 66% of the variation in monthly catches was attributable to temperature and the occurrence of newly molted crabs. Passano (86) and Chittleborough (87) reported a reduction or even a cessation in decapod feeding during the molt or premolt period, whereas shortly after ecdysis, Ennis (88) found *H. americanus* to display high feeding levels to ensure and hasten physiological recovery from the molt.

Differential vulnerability to traps is also known to be related to size and sex in association with behavioral differences. The unsuccessful attempts of researchers (Elner 89; Krouse 90) to collect small juvenile *H. americanus* in small mesh research traps are best explained by the unavailability of lobsters less than 40-mm carapace length owing to their cryptic behavior (Cooper and Uzmann 91). Temporal variations in size and sex composition of trap catches of *Cancer borealis* and *C. irroratus* have been mainly attributed to behavioral responses to molting and reproduction (Krouse 92). Moreover, Bert and Stevely (93) reported similar variations in trap catches of the Florida stone crab, *Menippe mercenaria*.

3.4. Intraspecific and Interspecific Interactions

Whether a crustacean enters a trap and then subsequently escapes at least in part depends on intraspecific and interspecific behavioral interactions such as competition for food and space, predation, cannibalism, agonistic encounters, and attraction and avoidance responses. In trap experiments with *Cancer irroratus, C. productus,* and *Hyas araneus,* Miller (40, 48, 49) demonstrated that the crab's success in trap entry may be limited by agonistic encounters that are likely to occur when trap entrances are hard to find and crabs become concentrated outside the trap, and also by intimidation of crabs about to enter the trap by crabs entrapped. Auster (50) observed *H. americanus* to burrow beneath traps and suggested high levels of this activity might deter other individuals from approaching the trap.

Pronounced differences in sex ratios of trap-caught *Cancer pagurus* were attributed by Bennett (94) not only to temporal changes in the catchability of male and female crabs, but also to the possibility that the more abundant sex might exclude its co-gender by competing more successfully for food and space both in and outside the trap. In a series of experiments, traps stocked with different combinations and densities of *C. irroratus, C. borealis,* and *H. americanus,* Richards et al. (95) discovered that the probability of both trap entry and escape is affected by intraspecific and interspecific interactions, including predator avoidance, and competition for preferred positions; both leading to trap saturation. Catch rates of *H. americanus* and of both crab species were lower in traps containing lobsters, whereas for traps stocked with crabs there was no effect on the lobster catch, but the catch of *C. borealis* was greater. Finally, the catchability of certain decapods can be affected by either attraction or avoidance reactions to living or dead conspecifics (reviewed by Hancock 41; Chapman and Smith 42) (see Section 3.1).

3.5. Immersion Time

The duration of time between setting a trap and hauling it is recognized to have a marked effect on catch. Accordingly, numerous authors (Bennett 94; Bennett and

Brown 22; Caddy 23; Auster 50, 96; Skud 80) have emphasized that before catch-per-unit-effort (CPUE) data are used, it is necessary to understand the influence soak time has on the catch rate of the species of interest, because this relation is species and fishery dependent. For example, Thomas (97) and Skud (80), investigating the inshore and offshore *H. americanus* fisheries, and Austin (98) those for *Panulirus argus*, described a nonlinear catch rate that moved toward an asymptote. In contrast, Mason (52) with *Cancer pagurus* and G. R. Morgan (personal communication) with *Panulirus longipes cygnus* noted that catches were unaffected by set-over time. Of course, the effect of immersion time may change with improvements in gear efficiency as illustrated by the English fishery for *C. pagurus* and *H. gammarus*, where parlor pots replaced the traditional single chambered creels during the 1960s. Lovewell et al. (53) and Kinnear (34) reported that catches with parlor traps increased with soak time, whereas with creel traps, maximum lobster catches resulted from daily hauling (Bennett 94; Bennett and Lovewell 99). Bennett and Kinnear attributed this chiefly to escape, because the creel pot has no antiescape mechanism. Other factors causing a reduction in the CPUE with soak time may be the loss of bait, along with its attractiveness, and trap saturation, resulting in behavioral interactions (see Section 3.3).

For a detailed review of the effects of soak time on the catches of various crustaceans, along with the different mathematical treatments to standardize and validate CPUE data for the accurate assessment of population parameters, the reader is referred to studies of Auster (50, 96), Austin (98), Bennett and Brown (22), Caddy (23), Morgan (85), and Skud (80).

4. CONCLUDING COMMENTS

A complete understanding of the many factors and influences on the capture process for decapod crustaceans and their interactions with baited traps is a complex task and, as such, requires an imaginative, integrated, and often labor-intensive and costly research program. Fortunately, the importance of this information for management purposes is widely recognized for evaluating the effectiveness of gear restrictions, assessing population parameters with catch and effort data, and advising fishermen of improved gear design and fishing strategies, and has received considerable research attention. Nevertheless, with today's technological advances and their impact on increasing gear design efficiency and fishing methods, as applied to heavily exploited shellfish fisheries, it is of paramount importance to maintain research programs that monitor fisheries developments, as well as any associated changes in gear selectivity and performance.

REFERENCES

1. A. von Brandt, *Fish Catching Methods of the World*. Fishing News Books, Farnham, Surrey, England, 1984.
2. J. S. Cobb, The American lobster. The biology of *Homarus americanus*. R.I., Univ., *Mar. Tech. Rep.* **48**:1–32 (1976).

3. J. D. Pringle, D. G. Robinson, G. P. Ennis, and P. Dube, An overview of the management of the lobster fishery in Atlantic Canada. *Can. J. Fish. Aquat. Sci., Manuscr. Rep.* **1701:**1–103 (1983).

4. B. K. Bowen, Spiny lobster fisheries management. In J. S. Cobb and B. F. Phillips, Eds., *The Biology and Management of Lobsters*, **Vol. II.** Academic Press, New York, 1980, pp. 243–264.

5. F. H. Herrick, Natural history of the American lobster. *Fish. Bull.* **29:**149–408 (1911).

6. R. L. Dow, The clawed lobster fisheries. In J. S. Cobb and B. F. Phillips, Eds., *The Biology and Management of Lobsters*, **Vol. II.** Academic Press, New York, 1980, pp. 265–316.

7. W. A. Dahlstrom and P. W. Wild, A history of Dungeness crab fisheries in California. *Calif. Dep. Fish Game, Fish Bull.* **172:**7–23 (1983).

8. J. H. Hunt, W. G. Lyons, and F. S. Kennedy, Jr., Effects of exposure and confinement on spiny lobster, *Panulirus argus*, used as attractants in the Florida trap fishery. *Fish. Bull.* **84:**69–76 (1986).

9. R. S. Brown and N. Caputi, Factors affecting the recapture of undersize western rock lobster *Panulirus cygnus* George returned by fishermen to the sea. *Fish. Res.* **2:**103–128 (1983).

10. R. S. Brown and N. Caputi, Factors affecting the growth of undersize western rock lobster, *Panulirus cygnus* George, returned by fishermen to the sea. *Fish. Bull.* **83:**567–574 (1985).

11. G. E. Davis, Effects of injuries on spiny lobster, *Panulirus argus*, and implications for fishery management. *Fish. Bull.* **78:**979–984 (1981).

12. L. M. Joll, The predation of pot-caught western rock lobster (*Panulirus longipes cygnus*) by octopus. *West. Aust., Dep. Wildl., Rep.* **29:**1–58 (1977).

13. J. S. Krouse, Incidence of cull lobsters, *Homarus americanus*, in commercial and research catches off the Maine coast. *Fish. Bull.* **74:**719–724 (1976).

14. J. S. Krouse and J. C. Thomas, Effects of trap selectivity and some population parameters on size composition of the American lobster, *Homarus americanus*, catch along the Maine coast. *Fish. Bull.* **73:**862–871 (1975).

15. W. W. Sheldon and R. L. Dow, Trap contributions to losses in the American lobster fishery. *Fish. Bull.* **73:**449–451 (1975).

16. K. J. Pecci, R. A. Cooper, D. C. Newell, R. A. Clifford, and R. J. Smolowitz, Ghost fishing of vented and unvented lobster, *Homarus americanus*, traps. *Mar. Fish. Rev.* **40:**9–43 (1978).

17. R. J. Smolowitz, Trap design and ghost fishing: Discussion. *Mar. Fish. Rev.* **40:**59–67 (1978).

18. W. L. High and D. D. Worlund, Escape of king crab, *Paralithodes camtschatica*, from derelict pots. *NOAA Tech. Rep.*, NMFS **SSRF-734:**1–11 (1979).

19. J. F. Casey and A. E. Wesche, Effect of derelict crab pots on future commercial harvest investigated by tidewater administration. *Md. Dep. Nat. Resour., Tidewater Fish. News* **14:**1–2 (1981).

20. W. L. High, Escape of Dungeness crabs from pots. *Mar. Fish. Rev.* **1184:**19–23 (1976).

21. R. J. Miller, Resource underutilization in a spider crab industry. *Fisheries* **2:**9–13 (1977).

22. D. B. Bennett and C. G. Brown, The problems of pot immersion time in recording and analysing catch-effort data from a trap fishery. *Rapp. P.-V. Reun., Cons. Int. Explor. Mer* **175:**186–189 (1979).

23. J. F. Caddy, Some considerations underlying definitions of catchability and fishing effort in shellfish fisheries, and their relevance for stock assessment purposes. *ICES, C.M.* **1977/K:18:**1–22 (1977).

24. J. C. Brêthes, R. Bouchard, and G. Desrosiers, Determination of the area prospected by a baited trap from a tagging and recapture experiment with snow crabs (*Chionoecetes opilio*). *J. Northwest Atl. Fish. Sci.* **6:**37–42 (1985).

25. B. F. Phillips, J. S. Cobb, and R. W. George, General biology. *In* J. S. Cobb and B. F. Phillips, eds., *The Biology and Management of Lobsters*, **Vol. I.** Academic Press, New York, 1980, pp. 1–82.

26. R. S. Brown and E. H. Barker, The western rock lobster fishery 1982–1983. *West. Aust., Dep. Fish. Wildl., Rep.* **70:**1–23 (1985).

27. T. M. Prudden, *About Lobsters.* Cumberland Press, Freeport, ME, 1973.

28. T. Savage, J. R. Sullivan, and C. E. Kalman, An analysis of stone crab (*Menippe mercenaria*) landings on Florida's West Coast, with a brief synopsis of the fishery. *Fla. Dep. Nat. Resour.* **13:**1–37 (1975).

29. W. A. Van Engel, The blue crab and its fishery in Chesapeake Bay. Part 2. Types of gear for hard crab fishing. *Commer. Fish. Rev.* **24:**1–10 (1962).

30. R. L. Dow and T. T. Trott, *A Study of Major Factors of Maine Lobster Production Fluctuations.* Dep. Mar. Resour., Augusta, ME, 1956.

31. H. J. Thomas, The efficiency of fishing methods employed in the capture process of lobsters and crabs. *J. Cons., Cons. Int. Explor. Mer* **18:**333–350 (1953).

32. D. G. Cargo, Maryland commercial fishing gears. III. The crab gears. *Chesapeake Biol. Lab., Educ. Ser.* **36:**1–18 (1954).

33. E. Edwards and R. A. Meaney, Crab fishing trials at Schull, Co. Cork. *Resour. Dev. Fish. Dev. Div., Ir. Sea Fish.* Note 3:1–12 (1968).

34. J. A. M. Kinnear, The effect of trap immersion time on the catch of lobsters in two Scottish fisheries. *ICES, C.M.* 1983/K:**43:**1–8 (1983).

35. W. S. Otwell, J. Bellairs, and D. Sweat, Initial development of a deep-sea crab fishery in the Gulf of Mexico. *Fla. Sea Grant Coll. Program, Rep.* **61:**1–29 (1984).

36. J. A. Bjordal, Factors affecting pot-fishing of Norway lobster (*Nephrops norvegicus*, L.) and pink shrimp (*Pandalus borealis*, Kroyer), studied in fishing and behavior experiments. Thesis, Dep. Agric. Fish. for Scotland, 1979 (Translation No. 2124, 32 pp.), University of Bergen.

37. A. C. Simpson, Effort measurements in the trap fisheries for crustacea. *Rapp. P.-V. Reun., Cons. Int. Explor. Mer* **168:**50–53 (1975).

38. R. J. Miller, How many traps should a crab fisherman fish? *North Am. J. Fish. Manage.* **3:**1–8 (1983).

39. R. J. Miller, Saturation of crab traps: Reduced entry and escapement. *J. Cons., Cons. Int. Explor. Mer* **38:**338–345 (1979).

40. R. J. Miller, Design criteria for crab traps. *J. Cons., Cons. Int. Explor. Mer* **39:**140–147 (1979).

41. D. A. Hancock, Attraction and avoidance in marine invertebrates—Their possible role in developing an artificial bait. *J. Cons., Cons. Int. Explor. Mer* **35**:328–331 (1974).

42. C. J. Chapman and G. L. Smith, Creel catches of crab, *Cancer pagurus* L. using different baits. *J. Cons., Cons. Int. Explor. Mer* **38**:226–229 (1978).

43. J. M. Bishop, E. J. Olmi, III, and G. M. Yianopoulos, Efficacy of peeler pots for capture of premolt blue crabs. *Trans. Am. Fish. Soc.* **113**:642–654 (1984).

44. D. M. Eggers, N. A. Rickard, D. G. Chapman, and R. R. Whitney, A methodology for estimating area fished for baited hooks and traps along a ground line. *Can. J. Fish. Aquat. Sci.* **39**:448–453 (1982).

45. R. J. Miller, Density of the commercial spider crab, *Chionoecetes opilio*, and calibration of effective area fished per trap using bottom photography. *J. Fish. Res. Board Can.* **32**:761–768 (1975).

46. M. Sinoda and T. Kobayasi, Studies on the fishery of Zuwai crab in the Japan Sea. VI. Efficiency of the Toyama Kago (a kind of crab trap) in capturing the Beni-zuwai crab. *Bull. Jpn. Soc. Sci. Fish.* **35**:948–956 (1969).

47. M. J. Williams and B. J. Hill, Factors influencing pot catches and population estimates of the portunid crab *Scylla serrata. Mar. Biol. (Berlin)* **71**:187–192 (1982).

48. R. J. Miller, Crab (*Cancer irroratus* and *Hyas araneus*) ease of entry to baited traps. *Tech. Rep.—Fish. Mar. Serv.* **771**:1–8 (1978).

49. R. J. Miller, Entry of *Cancer productus* to baited traps. *J. Cons., Cons. Int. Explor. Mer* **38**:220–225 (1978).

50. P. J. Auster, Factors affecting catch of American lobster (*Homarus americanus*), in baited traps. *Univ. Conn. Sea Grant Program, Tech. Bull Ser.* **CT-SG-85-1**:1–41 (1985).

51. A. B. Stasko, Modified lobster traps for catching crabs and keeping lobsters out. *J. Fish. Res. Board Can.* **32**:2515–2520 (1975).

52. J. Mason, The efficiency of the Gourdon crab creel. *Rapp. P.-V. Reun., Cons. Int. Explor. Mer* **156**:95–97 (1965).

53. S. R. J. Lovewell, A. E. Howard, and D. B. Bennett, The efficiency of parlour pots for catching crabs and lobsters. *ICES, C.M.* **1979/K**:13:1–5 (1979).

54. H. J. Thomas, Creel selectivity in the capture of lobsters and crabs. *J. Cons., Cons. Int. Explor. Mer* **24**:342–348 (1959).

55. E. Spurr, Lobster research project: Final report of 3-105-R, July 1969–June 1971. Fish Div., New Hampshire Fish Game Dep., Concord (unpublished manuscript), 1972, 22 pp.

56. J. L. Munro, The mode of operation of Antillean fish traps and the relationships between ingress, escapement, catch and soak. *J. Cons., Cons. Int. Explor. Mer* **35**:337–350 (1974).

57. J. M. Acheson, Factors influencing productivity of metal and wooden lobster traps. *Maine Sea Grant, Tech. Rep.* **63**:1–37 (1980).

58. W. Templeman, Investigations into the life history of the lobster (*Homarus americanus*) on the west coast of Newfoundland, 1938. *Newfoundland Dep. Nat. Resour., Res. Bull. (Fish.)* **7**:1–52 (1939).

59. D. G. Wilder, The effect of lath spacing and size of fishing ring on the catch of lobster traps. *Fish. Res. Board Can., Prog. Rep. Atl. Coast Stn.* **34**:22–24 (1943).

60. D. G. Wilder, Wider lath spaces protect short lobsters. *Fish. Res. Board Can., Atl. Biol. Stn. Circ.* **G-4:**1 (1945).

61. D. G. Wilder, The protection of short lobsters in market lobster areas. *Fish. Res. Board Can., Atl. Biol. Stn. Circ.* **G-11:**1 (1948).

62. D. G. Wilder, The lobster fishery of the southern Gulf of St. Lawrence. *Fish. Res. Board Can., Gen. Ser. Circ.* **24:**1–16 (1954).

63. C. G. Brown, Trials with escape gaps in lobster and crab traps. *ICES C.M.* **1978/K:**7:1–6 (1978).

64. C. G. Brown, The effect of escape gaps on trap selectivity in the United Kingdom crab (*Cancer pagurus* L.) and lobster (*Homarus gammarus* (L.)) fisheries. *J. Cons., Cons. Int. Explor. Mer* **40:**127–134 (1982).

65. B. I. Dybern, Catch of lobster in traps with escape vents of 40 × 300 and 45 × 300 mm. *Medd. Havsfiskelab. Lysekil* **267:**1–13 (1980).

66. M. J. Fogarty and D. V. D. Borden, Effects of trap venting on gear selectivity in the inshore Rhode Island American lobster, *Homarus americanus*, fishery. *Fish. Bull.* **77:**925–933 (1980).

67. J. S. Krouse, Effectiveness of escape vent shape in traps for catching legal-sized lobsters, *Homarus americanus*, and harvestable-sized crabs, *Cancer borealis* and *Cancer irroratus. Fish. Bull.* **76:**425–432 (1978).

68. V. E. Nulk, The effects of different escape vents on the selectivity of lobster traps. *Mar. Fish. Rev.* **40:**50–58 (1978).

69. R. J. Smolowitz, Trap design and ghost fishing: An overview. *Mar. Fish. Rev.* **40:**1–8 (1978).

70. A. M. Weber and P. T. Briggs, Retention of black sea bass in vented and unvented lobster traps. *N.Y. Fish Game J.* **30:**67–77 (1983).

71. B. K. Bowen, Preliminary report on the effectiveness of escape gaps in crayfish pots. *West. Aust., Dep. Fish Fauna, Rep.* **2:**1–9 (1963).

72. L. D. Ritchie, Crayfish pot escapement gap survey, November 1965–January 1966. *Fish. Tech. Rep. N.Z. Mar. Dep.* **14:**1–23 (1966).

73. H. B. Crous, A comparison of the efficiency of escape gaps and deck grid sorters for the selection of legal-sized rock lobsters *Jasus lalandii. Fish. Bull., S. Afr.* **8:**5–12 (1976).

74. P. J. Eldridge, V. G. Burrell, Jr., and G. Steele, Development of a self-culling blue crab pot. *Mar. Fish. Rev.* **41:**21–27 (1979).

75. T. Jow, Crab trap escape-opening studies. *Pac. Mar. Fish. Comm.* **5:**49–71 (1961).

76. R. J. Miller, North America crab fisheries: Regulations and their rationales. *Fish. Bull.* **74:**623–633 (1976).

77. G. H. Williams, D. A. McCaughran, S. H. Hoag, and T. M. Koeneman, II. A comparison of Pacific halibut and tanner crab catches in (1) side-entry and top-entry crab pots and (2) side-entry crab pots with and without tanner boards. *Int. Pac. Halibut Comm., Tech. Rep.* **19:**15–35 (1982).

78. A. Koike and M. Ogura, Selectivity of meshes and entrances of shrimp traps and crab traps. *J. Tokyo Univ. Fish.* **64:**1–11 (1977) (in Jpn., Engl. abstr.) (Translation by Translation Bureau, Multilingual Services Division, Department of the Secretary of State of Canada).

79. G. R. Morgan, Trap response and the measurement of effort in the fishery for the western rock lobster. *Rapp. P.-V. Reun., Cons. Int. Explor. Mer* **175**:197–203 (1979).

80. B. E. Skud, Soak time and the catch per pot in an offshore fishery for lobsters (*Homarus americanus*). *Rapp. P.-V. Reun., Cons. Int. Explor. Mer* **175**:190–196 (1979).

81. R. L. Dow, Relationship of sea surface temperature to American and European lobster landings. *J. Cons., Cons. Int. Explor. Mer* **37**:186–190 (1977).

82. R. L. Dow, Effects of sea surface temperature cycles on landings of American, European, and Norway lobsters. *J. Cons., Cons. Int. Explor. Mer* **38**:271–272 (1978).

83. D. W. McLeese and D. G. Wilder, The activity and catchability of the lobster (*Homarus americanus*) in relation to temperature. *J. Fish. Res. Board Can.* **15**:1345–1354 (1958).

84. B. T. Hepper, An apparent relationship between catch per unit of effort and temperature in the English lobster fishery. *Shellfish Benthos Comm., ICES C.M.* **K**:8 (1971).

85. G. R. Morgan, Aspects of population dynamics of western rock lobster, *Panulirus longipes cygnus* George. 2. Seasonal changes in catchability coefficient. *Aust. J. Mar. Freshwater Res.* **25**:249–259 (1974).

86. L. M. Passano, Molting and its control. In T. H. Waterman, Ed., *The Physiology of Crustacea*, **Vol. I.** Academic Press, New York, 1960, pp. 473–536.

87. R. G. Chittleborough, Environmental factors affecting growth and survival of juvenile western rock lobsters. *Aust. J. Mar. Freshwater Res.* **26**:177–196 (1975).

88. G. P. Ennis, Food, feeding, and condition of lobsters, *Homarus americanus*, throughout the seasonal cycle in Bonavista Bay, Newfoundland. *J. Fish. Res. Board Can.* **30**:1905–1909 (1973).

89. R. W. Elner, Lobster gear selectivity—A Canadian overview. *Can. J. Fish. Aquat. Sci., Tech. Rep.* **932**:77–83 (1980).

90. J. S. Krouse, Maturity, sex ratio, and size composition of the natural population of American lobster, *Homarus americanus*, along the Maine coast. *Fish. Bull.* **71**:165–173 (1973).

91. R. A. Cooper and J. R. Uzmann, Ecology of juvenile and adult clawed lobsters, *Homarus americanus, Homarus gammarus*, and *Nephrops norvegicus. Circ.—C.S.I.R.O., Div. Fish. Oceanogr. (Aust.)* **7**:187–208 (1977).

92. J. S. Krouse, Distribution and catch composition of Jonah crab, *Cancer borealis*, and rock crab, *Cancer irroratus*, near Boothbay Harbor, Maine. *Fish. Bull.* **77**:685–693 (1980).

93. T. H. Bert and J. M. Stevely, Aspects of the population dynamics of stone crabs (*Menippe mercenaria*) as determined by remote vs. in situ observation: A comparative study. *Spec. Rep. NMFS—Fla. Dep. Nat. Resour.* (1982), 13 pp.

94. D. B. Bennett, The effects of pot immersion time on catches of crabs, *Cancer pagurus* L. and lobsters, *Homarus gammarus* (L.). *J. Cons., Cons. Int. Explor. Mer* **35**:332–336 (1974).

95. R. A. Richards, J. S. Cobb, and M. H. Fogarty, Effects of behavioral interactions on the catchability of American lobster, *Homarus americanus*, and two species of *Cancer* crab. *Fish. Bull.* **81**:51–60 (1983).

96. P. J. Auster, The utility of computing CPUE at each level of effort in the lobster trap fishery. *NAFO Sci. Counc. Stud.* **10**:53–56 (1986).

97. J. C. Thomas, An analysis of the commercial lobster (*Homarus americanus*) fishery

along the coast of Maine, August 1966 through December 1970. *NOAA Tech. Rep.,* NMFS **SSRF-667:**1–57 (1973).

98. C. B. Austin, Incorporating soak time into measurement of fishing effort in trap fisheries. *Fish. Bull.* **75:**213–218 (1977).

99. D. B. Bennett and S. R. J. Lovewell, The effects of pot immersion time on catches of lobsters *Homarus gammarus* (L) in the Welsh coast fishery. *Fish. Res. Tech. Rep.* (U.K., *Dir. Fish. Res.*) **36:**1–4 (1977).

15 THE POPULATION BIOLOGY OF DECAPODS

J. Stanley Cobb
Department of Zoology
University of Rhode Island
Kingston, Rhode Island

and

John F. Caddy
Food and Agriculture Organization
Rome, Italy

1. INTRODUCTION

In this chapter we review recent developments in our understanding of the population biology of commercially important marine decapods. Although an apparently homogeneous group, the decapods encompass a wide variety of adaptations built around a common crustacean biology, and as noted in the Preface, particular emphasis is placed in this chapter on those aspects of their population dynamics that reflect life-history characteristics of decapods, their behavior, and fisheries. For more detailed reviews of conventional models of fish stock assessment, the reader is referred to general texts on population dynamics and stock assessment, such as those of Ricker (1) and Gulland (2), and to Garcia and Le Reste (3) and Morgan (4) for applications of generalized models to penaeid and lobster fisheries, respectively. Commercially important decapods have a high value in developed countries, and are thus heavily fished. They often are exported (also by developing countries), generating foreign currency earnings within the country of origin. Because of their generally high unit value, considerable research effort has been directed toward understanding the ecology and population dynamics of many decapods, and significant progress made toward elucidating some of the special features of this important group of organisms.

This chapter is organized around several themes: First we briefly consider the commonalities and contrasts in general biology, then proceed to describe the individual parameters that influence population dynamics. A consideration of population parameters is followed by a brief review of single-species fishery models as they apply to decapods, ending with a brief overview and speculation on possible ecological approaches to multispecies modeling including these crustaceans.

2. DECAPOD BIOLOGY

The commercially important decapods consist of three major groups: shrimps (Penaeidae, Pandalidae, and to a lesser extent, Crangonidae), crabs (Brachyura and Anomura), and lobsters (Nephropidae and Palinuridae). Most species are benthic as adults, and all have pelagic larvae. Generally, penaeids are limited to tropical and warm temperate waters, whereas pandalids are boreal in distribution. Clawed lobsters are generally cold temperate animals, although some species are found at great depths in tropical waters, and spiny lobsters range from tropical to temperate waters. Crabs are distributed world wide. All groups inhabit depths from tidal to abyssal (and occasionally above the tide marks), but commercial fisheries are generally limited to species found at depths of less than 200 m.

Life-spans range from ≤ 1 to 1.5 yr (many penaeids) to more than 50 yr (*Homarus* spp.). Growth is continuous, although size increase is accomplished almost entirely by molting. Courtship and mating have not been described for most species, but it is likely that many have some sort of courtship that includes both tactile and chemical communication. Spermatophores are stored by the female for periods of days (penaeids, some crabs) to a year or more (lobsters), whereas the Anomura fertilize their eggs immediately upon transfer of the spermatophores. In all but the penaeids, which lay demersal eggs, the eggs are carried on the ventral surface of the abdomen on the pleopods. Egg development time varies widely. In lobsters, for example, some species of *Panulirus* carry the eggs for as little as 4 weeks, whereas in *Homarus* eggs are attached to the female for 10–12 months. The duration and form of the larval stages are similarly varied. The sole commonality seems to be a pelagic postlarval stage that is similar to the adult form. In some, the postlarval stage is an adaptation for return to nursery sites. Many shrimps and crabs have a juvenile stage spent in shallow estuarine areas.

Ecological adaptations vary greatly, but there are some common themes. All the commercially important species are in relatively high abundance and may be among the key species in their benthic communities. Many decapods are nocturnal, emerging from the sediment or from burrows or lairs to feed only at night. Most are carnivores or detritivores. Migrations are common and often are correlated with specific periods in the life cycle. For instance, many penaeid species move to deeper water as they enter the adult phase, or a contranatant spawning migration occurs as in the spiny lobster *Panulirus ornatus*.

3. INDIVIDUAL PARAMETERS IMPORTANT TO POPULATION DYNAMICS

3.1. Growth

Growth describes the relationship between size (length or weight) and age (units of weeks to years). Age has to be determined indirectly from size after the construction of some sort of age–size key based on data from modal analysis or tagging experiments. This is clearly not an ideal situation, but hard parts used for aging fish, for example, are lost each time a decapod molts. Because size increase occurs almost entirely at the time of molt,* the measured growth rate is made up of two components, size increase at molt and the time interval between molts. These two variables can be quite labile and often respond differently to variations in the environment, so ideally should be analyzed separately for long-lived decapods (Hartnoll 5). For short lived species, molting is frequent enough that continuous growth functions are often used. Factors such as crowding, food deprivation, injury, autonomy, or other stress can influence either or both components.

---

*As for other organisms, body mass increases in a more or less continuous fashion throughout the intermolt. For smaller species with flexible exoskeletal plates, some increase in length can occur between molts due to "stretching" of the intersegmental integument (Boddeke 6).

Usually the intermolt interval is the first to be affected. After puberty, females generally grow less rapidly than males (see references in Hartnoll 5), presumably because gonadal growth and parental care (egg carrying) divert more resources in females than in males, and because eggs are attached to the exoskeleton, so that parental care prolongs the intermolt period.

A Hiatt growth diagram is often used to represent the relationship between size increment and size at molt, usually plotted as premolt size against postmolt size (Hiatt 7; Kurata 8; Hartnoll 5). Mauchline (9) suggested plotting percent increment against size, and this allows a different perspective on the data. Figure 1 shows that there are wide variations among decapods in their growth strategies. In general the percent size increment decreases with size, but this is not universal (Hartnoll 5). Intermolt period generally increases with increasing size, so the older the animal is, the longer the time between molts. Some crabs show a terminal molt at or around maturity, which may be atypical of previous increments in size (e.g., Conan and Comeau 10); further growth does not then occur. Many factors can affect growth rate, usually by influencing the intermolt interval. These include injury, temperature, food supply, photoperiod, and social interactions (Hartnoll 5; Cobb and Wang 11). Additionally, after maturity, molt frequency of some shrimp may be independent of size, depending rather on extrinsic events such as lunar periodicity (Boddeke et al. 12; Racek 13). One effect of maturation on growth can be a change in the slope of the growth curve at maturity (Caddy 14), particularly for females of species that hold eggs under the abdomen for protracted periods.

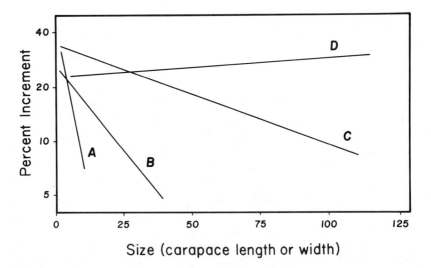

Size (carapace length or width)

Figure 1. The relationship between percent increment at molt and size (carapace length or carapace width) in representatives of four genera of decapods: *A, Rithropanopeus; B, Pachygrapsus; C, Homarus D, Callinectes*. In general, percent increment decreases with increasing size, but *Callinectes* provides an exception (Redrawn from Hartnoll 5).

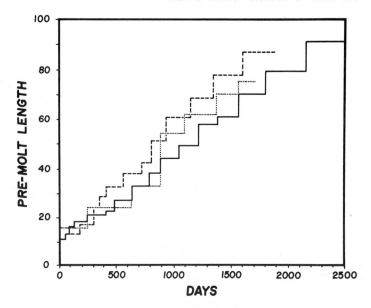

Figure 2. Decapod growth strategies illustrated by size increase of three individual *Homarus americanus* over time. A faster growing individual can be distinguished. Of the two growing at approximately the same rate, one molts frequently with small increments, the other less often with large increments (M. J. Fogarty, personal communication).

A matter of considerable importance to anyone attempting to separate cohorts is the degree of individual variability in growth. The coefficient of variation in size increment increases with age (Hartnoll 5). Variable growth strategies between individuals have been demonstrated in lobsters (M. J. Fogarty, U.S. National Marine Fishery Service, personal communication), in which an individual shows consistently large or consistently small growth increments (Fig. 2); this may be linked to variations in size reached at maturity.

The discontinuous nature of size increase and lack of any way to determine age accurately makes estimation of growth rates difficult. The data used generally are of three types: length–frequency, often from the commercial catch, size increase during time at large from marking to recapture, or a combination of separate estimates of molt increment and molt interval. Length frequencies can be used most effectively in short-lived species in which cohorts are identifiable, such as shrimp. In longer-lived species such as lobsters, variability in growth among individuals can result in a disappearance of clear peaks marking cohorts. In environments with marked seasonal variation in temperature, this could even result in plurimodal age classes if for some members of a cohort, the "temperature sum" needed to complete molting is not attained by the end of the warm-water season (Caddy 15). Additional variability may be introduced by breeding seasons, multiple spawnings, temporal or spatial variation in food availability, and so on. In particular, migration of mature shrimps may occur at or around maturity, and be scheduled in relation to the molt cycle (Boddeke et al. 12). Various methods have been suggested for the analysis of length–frequency data, especially if size modes can be considered equivalent to age classes. Several techniques to split samples with many peaks into component age classes (e.g., Cassie 16; Tanaka 17; Bhattacharaya 18), lead to the analysis of the progression of modes of age classes over several years. New computer-based adaptations of these approaches promise to improve approaches to size–frequency analysis greatly (e.g., Sparre 19).

Mark and recapture methods are conceptually simple, but they are also time consuming and expensive. Various tags are available for crustaceans. Some tags are retained through ecdysis if implanted at the rear of the carapace where splitting occurs on molting. Various assumptions must be met (Gulland 2); for example, that marking does not change the growth rate, which may be a considerable problem in many species (Stewart and Squires 20; Garcia and Le Reste 3).

An estimation of molt frequencies and increments can be combined to determine growth rate as outlined by Mauchline (21) and applied to lobsters (Aiken 22; Cooper and Uzmann, 23), shrimps (Boddeke et al. 12), and crabs (Hankin et al. 24). Data are often obtained from tagging, but the proportion of a population molting may also be estimated by morphological changes (e.g., setagenesis, Aiken 25; Vigh and Fingerman 26) over the molt cycle. Because increment is relatively invariant except under severe stress, average increment from laboratory studies may be used.

Three models are generally used to describe growth: the von Bertalanffy growth function, the Gompertz function, and the logistic. All describe asymptotic curves. When comparisons are made among the models (e.g., Parrack 27) the von Bertalanffy model generally fits the data best (as determined by examination of the residual sums of squares), especially for short-lived species, although Saila et al. (28) found their empirical model to be superior for spiny lobsters. The von Bertalanffy model historically has been the choice of fishery biologists because the parameters of this model are easily used in the Beverton and Holt yield formation, but this justification is less relevant now that more flexible and realistic formulations can be easily handled by small computers.

In fact, the von Bertalanffy growth function does not always do a good job of describing growth over the entire life-span. This is particularly true for long-lived species, or where growth rate changes markedly or effectively stops after maturity, as in some crabs that exhibit a terminal molt (e.g., *Chionoecetes opilio*, Conan and Comeau 10). Empirical growth models that combine the separate elements of increment and frequency to derive the growth curve have significant advantages for slow-growing species and for species with an effective cessation of molting at some stage in the life history. Examples of this approach are found in Campbell (29), Bennett (30), Saila et al. (28), and Fogarty (31). Among other advantages, they allow the construction of a length- or weight-based model, rather than an age-based model, as is necessary for the traditional growth formulations. Additionally, impacts of maturity, spawning, season, terminal molt, or other factors can be readily incorporated.

3.2. Maturity

Age and size at sexual maturity are important variables for fisheries biologists because they not only mark the start of reproductive activity, but also suggest an appropriate minimum legal size for retention, a regulation frequently used in decapod fisheries.

A variety of measures have been used to determine maturity in females. Among them are ovarian development (size and/or color of ova), presence of sperm in the sperm receptacle, the presence of eggs under the abdomen, the relative width of the abdomen, and the presence of "cement" glands and/or nonplumose setae on the pleopods. Some of these cannot be used in penaeids because their eggs are shed at spawning. Most of the pandalid shrimps are protandrous hermaphrodites and thus present special problems. Less attention is generally paid to male sexual maturity. Measures include the presence of mature sperm, the development of testes or vas deferens, relative size or weight of chelipeds, first walking legs, or mating behavior. Techniques have been developed to use ontogenic changes in relative growth of body parts as an indication of the onset of sexual maturity. Somerton (32) showed how to identify the inflection point in growth rate in crabs that is associated with maturity. A similar procedure was suggested by George and Morgan (33) for changes in relative growth of the first legs of the male spiny lobster.

Size or age at sexual maturity may not be constant for a species over its whole geographical range. An extreme example of this is *Homarus americanus*, in which Bay of Fundy females mature at a size of 100 mm carapace length (Groom 34), nearly twice that of females in western Long Island Sound (Briggs and Mushacke 35).

A problem arises if one wishes to distinguish between functional and physiological maturity. To our knowledge this has been addressed only in lobsters and in the majid crab, *Chionoecetes opilio* (Conan and Comeau 10). Male *H. americanus* mature physiologically (as indicated by vas deferens size and the presence of mature sperm) long before they are behaviorally "interested in mating" (Aiken and Waddy 36). At a size when 100% of female *H. americanus* appear mature according to criteria of cement gland presence (Aiken and Waddy 37), tagging studies have shown that only 40–60% of the physiologically mature females actually extruded eggs (Ennis 38). These differences could have a considerable effect on calculations of population fecundity and on female growth rate, as well as on management decisions based on size at maturity.

The special problems in the dynamics of populations of hermaphroditic pandalid shrimps are a function of a highly variable life cycle. For example, not all individuals of *Pandalus montagui* in Maine start life as males, though for these, transition to the female gender occurs after 1 or 2 yr (Chapter 3). Maturity seems to be size dependent rather than age dependent, but total fecundity clearly is a function of how many years are spent as a female (Stevenson and Pierce 39). *Pandalus montagui* in the North Sea grow more rapidly and show a higher proportion of females than does the Gulf of Maine population (Allen 40). At the extreme north of the range (North Laborador), transition to females in very cold water may be delayed beyond the average life-span.

The problem presented by variable, hermaphroditic life cycles could be a real concern to decapod fishery biologists (see Fox 41). It is easy to predict population crashes upon overfishing because the largest animals are females, and the highly variable nature of the male–female transition may lead to inaccurate predictions from generalized yield models. The pandalid example is only one example, but an extreme lesson in the importance of a full understanding of the biology of the species being managed.

3.3. Fecundity

The measurement of the reproductive potential of a population is based on an estimate of individual fecundity, which is usually taken as the mean number of eggs produced individually per female. In all decapods but the penaeids, measuring fecundity is relatively easy because maturing eggs are borne externally. The range in fecundity is staggering, from the low hundreds of eggs per brood in some lobsters and pandalids, to 3×10^6 or more in some crabs. Simply counting externally carried eggs may be misleading unless the time of estimation is standardized, because egg loss during incubation is known to occur in some lobsters and crabs (Cobb and Wang 11; Wickham 42) and may happen in other decapods. Egg loss can be due to faulty attachment of nonfertile eggs, disease, or parasitism. Because the loss may be large (up to 36% in *H. americanus*, Perkins 43) and entire broods may be lost after attachment (Ennis 38), we suggest estimates of egg numbers be made only shortly before hatching is expected.

The relationship between body size and number of eggs carried has been studied in many species. In some the relationship seems linear (e.g., *Cancer irroratus*, Reilly 44) or linear-segmented, whereas in others an exponential fit to the data has been used, and in at least one species (*Lithodes covesi*), an asymptotic relationship has been described (Somerton 45). Of five studies on *Homarus* fecundity, two used a linear relationship and three an exponential one (Aiken and Waddy 36), but this may be a function of the fitting procedure chosen. Jamieson (46) pointed out that in Dungeness crabs and red king crabs the fecundity–size relationship varies both between populations and over time.

Determination of individual fecundity is necessary to an estimation of the reproductive potential (population fecundity) of a size class or population. This is a function of the frequency of spawning, the number of eggs produced, and the proportion of the population represented by the size class under consideration (Aiken and Waddy 36; Kanciruk and Herrnkind 47). Following Caddy (15, 48), Campbell and Robinson (49) developed an egg-per-recruit model for *H. americanus* that expressed reproductive potential per molt group (m) as

$$E_m = N_m P_m Ex_m F_m$$

where E = total number of eggs produced in molt group m
N = number of females in m

P = proportion of sexually mature females in m
Ex = proportion of females extruding eggs in m
F = fecundity of females in m at time of egg hatching

The total egg production, E_T, from the first (partially mature) group n, to the oldest group max, in the population is then simply

$$E_T = \sum_{m=n}^{max} E_m$$

Relative fecundity is usually highest for females of intermediate size classes, which, although producing fewer eggs per female, are larger in number and spawn more frequently than larger females (Campbell and Robinson 49; Berry 50). Gregory et al. (51) found the age group of new recruits to the spiny lobster fishery to have the highest relative fecundity because fishing removes most of the larger females. Many warm-water decapods spawn more than once annually, and this must be taken into account when determining population fecundity. Clearly population fecundity is based not only on the fecundity of each size class, but also on the relative abundance at size and the mating frequency.

Similar calculations can be made of the mean fecundity per recruit resulting under a given fishing regime (Caddy 48; Fogarty 31), and can be useful in testing the implications of a given change in size limit or mortality on population fecundity. Note, however, that this does not automatically change survival to recruitment in a proportional way (see Section 7).

3.4. Sex Ratio

Gender is determined genetically in decapods in much the same way as it is in mammals. Hermaphroditism (generally protandric) is found in most pandalids and in some anomurans. Environmental factors usually do not affect sex determination, but parasitic castration of males before puberty will inhibit expression of secondary sexual characteristics. In the pandalids, environmental quality may affect growth rate and the age (and perhaps size) at which the gender change occurs. The effect of this on the sex ratio in *Pandalus jordani* has been explored by Charnov et al. (52). Generalizations about sex ratios at the species level should be made with caution because local variation from a ratio of $1:1$ may be caused by differential mortality, migration or habitat selection, and restricted nutrition (Wenner 53). Sex ratios are highly variable in lobsters (Ennis 54; Briggs and Mushacke 35; Lyons et al. 55) crabs, and shrimps (Garcia and Le Reste 3; Stevenson and Pierce 39).

Size-specific changes in sex ratio have been well documented (Wenner 53) as due to a variety of factors including hermaphroditism, differential growth, and or longevity/mortality. In most decapods growth rates change at puberty. Female growth rate may then decrease markedly, causing the growth curve to become more inflected, because molting may then occur only in alternate years with egg bearing. Thomas (56), Saila and Flowers (57), and Skud (58) have modeled this problem. They all suggest that changes in size-specific sex ratios can be used to estimate mortality where other measures are not available.

4. STOCK IDENTITY

When a population or stock is modeled, the assumption is normally made that the individuals within the stock are uniform. As Gulland (2) points out, this is a necessary simplifying assumption but it is important that the assumption be verified. In practice, a stock is often defined operationally as a production or management unit about which conclusions can be made while ignoring differences within the group and exchanges with other groups (Larkin 59; Gulland 2), or biologically, as a genetically discrete population. This unit stock assumption can be made more freely, of course, if spawner–recruit relationships are not being explored. Working with a biologically isolated or genetically discrete population is the ideal (Ihssen 60), but a mixture of stocks with limited interchange is probably more usual, though the problem of stock identity has been ignored for most decapod stocks. Ihssen et al. (61) cite seven methods of stock separation, with criteria ranging from biochemical to population parameters. We consider some of those here.

Decapods generally have very low levels of genetic variation as measured by electrophoretic techniques, the only exceptions seeming to be species occurring over broad areas and in widely different environments (Hedgecock et al. 62; Nelson and Hedgecock 63; Mulley and Latter 64; Bulnheim and Scholl 65). Eight populations of *Homarus americanus* were surveyed by Tracey et al. (66), who found differentiation between several populations at one gene locus. They interpreted this, along with earlier migration studies, to mean that American lobsters are divided into local populations with little cross breeding. Smith et al. (67) found little variation among populations of *Jasus edwardsii* or between *J. edwardsii* and *J. novaehollandiae*. Nine species of Australian penaeids showed little intraspecific polymorphism (Mulley and Latter 68). On the other hand, Hedgecock et al. (62) found considerable divergence between geographically separate populations of the freshwater prawn *Machrobrachium rosenbergii*.

Conclusions drawn from biochemical data should be confirmed by other types of biological data. The American lobster is a case in point. Despite the conclusion of Tracey et al. (66) that the inshore and offshore populations are discrete, evidence from tagging studies (Cooper and Uzmann 69; Fogarty et al. 70) shows that a large proportion of the population migrates back and forth between the two areas annually. Additionally, the larvae are planktonic, spending 3–5 weeks in surface waters, raising the possibility that larvae released in one area may become recruits in another. Allendorf and Phelps (71) show that significant allelic divergence may be found even when there is considerable exchange between subpopulations, so differences in alleles among populations need not be interpreted as evidence for a high degree of isolation.

Evidence other than allelic divergence is frequently used for stock separation in decapods. Movements of individuals between areas can be monitored by use of artificial tags or natural markers such as stable carbon isotope ratios (Fry 72) or parasites (Uzmann 73). A long pelagic larval phase suggests that the opportunity for dispersal is present, and has been used as part of the argument for interpreting the Dungeness crab fishery, for instance, as a single stock (Soule and Tasto 74). A considerably more complex situation seems to exist with the spiny lobster *Panulirus argus* in the Caribbean. The adults and long-lived larvae (more than 6 months in the plankton) are distributed throughout the region, and preliminary biochemical evidence has indicated distinct subpopulations (Menzies 75). However, evidence from studies of larval distributions and current patterns gave no evidence for local recruitment in any of the populations. Morphometric criteria also are used for stock separation. A multivariate analysis of measurements on 16 body parts of the American lobster suggested differences between inshore and offshore stocks in *H. americanus* (Saila and Flowers 76). Farmer (77) combined regional differences in body proportions of the Norway lobster (*Nephrops norvegicus*) with differences in relative infestation of a parasite to suggest isolation of subpopulations. Fournier et al. (78) introduced a statistical technique based on a maximum-likelihood procedure to determine the proportion of several stocks in a mixed-stock fishery. Several very different types of evidence can be used, including morphometric, meristic, and electrophoretic traits.

Rather than attempting to define subpopulations biologically, Campbell and Mohn (79) took a pragmatic approach, defining a stock as "a group of individuals that sustains itself over time and that responds in a similar way to environmental changes within a discrete geographic area." Such a definition does not demand reproductive isolation or genetic discontinuity, but approaches more directly the functional concept of a stock that a fisheries manager must deal with. They used several multivariate techniques to analyze a 90-year period in landings data from the Canadian Maritimes lobster fishery and found at least three main stock areas. For penaeid shrimp, current practice is generally to assume that although the smallest population unit might correspond to a given lagoon or estuarine nursery area, most offshore fisheries on adults are likely to be sustained by recruitment from several such areas, and in turn, contribute larvae to them.

It would seem reasonable for fisheries managers to continue the pragmatic definition of stock advocated by Campbell and Mohn (79), and it may not be vitally important, unless stocks are severely overexploited, to know the extent of gene flow from adjacent populations. Of primary importance are characteristics such as growth rate, size at maturity, and indexes of population size and recruitment. If these can be reliably discriminated between geographical areas, they should allow the corresponding stocks to be managed separately, whether or not there is evidence for genetic discreteness.

5. MORTALITY

One of the critical problems in the study of the population dynamics of decapods is the estimation of mortality rates. As with growth, the problem for crustaceans is compounded by the lack of age-based data. The data used in the estimation of instantaneous mortality rates often come from the commercial fishery; thus the total mortality rate (Z) is a sum of the rates M (natural mortality) and F (mortality due to fishing), with migration being a frequent source of bias. The usual case is that the total mortality rate Z and either F or M are estimated separately, and the third mortality component found by difference. (In unfished populations, of course, M may be estimated directly, although it is then likely to be somewhat higher than for the exploited stock: see Section 5.2).

5.1. Total Mortality Rate, Z

Two general classes of methods are used for estimating this parameter. First, those that make steady-state assumptions, and second, those that follow a cohort throughout its life-span.

In the first category are methods such as catch curve analysis and estimates based on mean (e.g., Beverton and Holt 80), or median size or age statistics (Hoenig et al. 81). These approaches assume equilibrium conditions, and that mortality, growth, and recruitment are constant over the interval; assumptions that may be approximated to if data are pooled over a number of years when mortality rates have been fairly constant. These estimators are liable to bias, but may provide useful indexes of overall mortality rate.

Second are methods such as modal analysis and sequential cohort analysis, which follow a given cohort or class in time and determine the rate of decline in numbers. If cohorts can be identified and the size of each cohort followed, then mortality can be estimated by the ratio of the natural logarithms of numbers before and after a time interval if a constant catchability coefficient (q) can be assumed. If so, the catch rates before and after the time interval can be substituted for population numbers in the above equation (Gulland 2). However, because q may change with size due to differential vulnerability to gear through mesh size selectivity or behavioral changes, this can be a hazardous assumption to make (e.g., for trap fisheries). The use of length–frequency catch curves to estimate Z may also suffer from problems due to size-related changes in emigration or catchability, which can bias apparent values of Z.

Catch curve analysis can be used to develop mortality estimates by considering the age structure of a population, although since most decapods have some migratory component or segregation between mature and immature individuals, care should be taken in ensuring that all components of the population are correctly weighted in pooling the size frequencies taken throughout the range of the stock. Each cohort will have been exposed to one year's more mortality than the next younger group, so that the ratio of their abundances is a measure of annual survival. In decapods, where absolute age is rarely known, length–frequency data can be used (e.g. Hancock 82; Annala 83) if the annual increase in size at each initial size is known.

Conan (84) and Caddy (15, 85) suggest the use of the molt period as the appropriate biological time unit in decapods with multiannual life-spans. Rather than being expressed in days or years, this is expressed in biologically relevant units of "physiological time"; that is, one unit is the time to complete one intermolt period (irrespective of the variations in real duration with season, age, sex, etc). In this approach, the log frequency of catches at size by locality (and ideally by sex) are pooled over several (steady-state) years, then segmented using size increment at molt data to obtain

$$Z_m = (\log_e N_m - \log N_{m+1}) / Dt_m$$

where Dt_m is the mean intermolt period for the mth molt (years) and Z_m is the annual mortality rate.

Calculating total mortality rate on an annual basis requires more information. Thus tagging studies can give the intermolt period Dt for a range of arbitrary premolt sizes, and observations on animals pre- and postecdysis can provide the increment. Annual values of abundance at size can then be stepped off in intermolt increments, up through the size range of the pooled sample to estimate Z's at size.

5.2. Natural Mortality Rate, M

The natural mortality rate is a parameter that is difficult to estimate directly, despite its importance. It is usually expressed annually, and ranges systematically with longevity from around 0.1 in long-lived lobsters and crabs to $M = 1.5$ or even $2.0+$ in short-lived penaeid shrimps. Garcia and Le Reste (3) pointed out that for heavily exploited, short-lived species caution must be used in correlating longevity with natural mortality within a group, especially if the species is migratory.

Estimating M directly without data on fishing effort and total mortality for a series of years is problematic. Literature on the same species in other areas is not uncommonly used for preliminary yield calculations, or the calculation of yield per recruit for a reasonable range of guesses of M may be resorted to (and the most conservative prediction adopted!). If annual effort and mortality data are available, the classical approaches of either Silliman or Paloheimo (for details, see Ricker 1), may be used to solve the equation

$$Z_t = q\bar{f}_t + M$$

for M, where Z_t and $\bar{f}_t$ are total mortality rate and mean fishing effort over two or more periods. If in addition to the mean annual fishing effort, estimates of (Z_t/K) are available from the Jones and Van Zalinge (86) method for example, for two or more stable periods, these could be used to solve for M/K (Caddy 87): this ratio may be used directly as input to the Beverton and Holt (88) yield tables.

The above methods suffer, however, from the assumption of constant M, which Munro (89) argued was inappropriate, given that predation is the major cause of natural mortality in *Panulirus argus*. He pointed out that for lightly fished areas Z approximates to M.

Various authors (e.g., Wickins 90; Caddy 48; Fogarty 31) have suggested that because the period of ecdysis is a time of high risk for many crustaceans, the probability of natural death may be highest at this time. If this assertion is true (and we must admit that there have been few field studies aimed so far at investigating it), then the probability of natural death should be directly related to molt frequency. If this hypothesis is true, it would imply that M is high for juveniles and drops off to a more constant, lower rate for older animals; a situation suggested on other grounds by, for example, Cushing (91) for marine fish. Further reasons for supposing that such a relationship applies to crevice-dwelling organisms are discussed in a later section, but research on the relationship between molt frequency and M should clearly be a high priority for further work in crustaceans.

Most decapods show cryptic behavior during at least some stage in their life histories, and except for those species that burrow in soft substrates, they seek shelter in crevices in outcrops and reefs, or among epifauna and flora, for at least part of the diurnal or tidal cycle. Although crevices or burrows may become the focus for a variety of activities such as molting, mating, and feeding, or simply for the conservation of energy, a reduction in the natural mortality rate probably was a selective force in the evolution of crevice-dwelling behavior. The number of available crevices has been considered as one factor limiting population density of lobsters (e.g., Howard 92) and has led to the suggestion that density-dependent mortality is important (Addison 93). Competition for crevices also involves interspecific competition, however, and in other ways M is not a simple result of density dependence. In qualification to models such as those described above, which make M a function of predator abundance, it seems likely that the "size spectrum" and number of available crevices will prove factors that influence the size frequency of the survivors where this special habitat is limiting. Individuals unable to find suitable crevices will have a higher probability of being preyed upon (Richards and Cobb 94).

One of the features of the natural environment that has long been recognized but only recently quantified is that natural surfaces tend to be "infinitely dissected." For most natural surfaces, the total number of crevices present usually drops off with increasing size. Surfaces showing this characteristic have been referred to as "fractal" by Mandelbrot (95), and this theory has been used to explain the size spectrum of terrestrial arthropods in natural vegetation (Morse et al. 96). Such a feature of natural surfaces has obvious relevance for investigations on decapod species that rely on crevices of a suitable size in outcrops, coral reefs, or other epifauna and flora for protection from predators. Because for most of these organisms the rate of instantaneous growth also slows with size, this should reduce the frequency with which larger shelters have to be located by older animals.

A simple modeling approach assuming no density-dependent growth and migration, and assuming that predation is restricted to cryptic organisms displaced from their niches, seems to demonstrate this, and leads to what can be referred to as the "musical chairs" hypothesis of recruitment (Caddy 14). Here growing individuals compete for a continually diminishing number of crevices in a fractal surface, and M may then be expected to drop off rapidly with size and age. For heavily fished commercial sizes, it seems improbable that limitation of available shelter is the

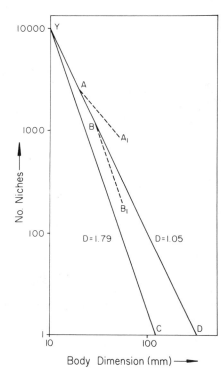

Figure 3. Illustration of how population bottle-necks for a crevice-dwelling decapod might result from an environment showing fractal characteristics (two examples here shown for arbitrary values of D, the fractal dimension. D can, however, change over different size ranges: (the two changes in this theoretical example hypothesized to occur due to (A) a natural change in topography and (B) habitat enhancement. (Modified from Caddy 158.)

sole factor determining population size; nonetheless, the fractal concept helps to explain how "bottlenecks" in recruitment, or other limits to standing stock sizes, might be mediated, or even alleviated by habitat changes. Direct measurement of the fractal dimension D of a natural surface is far from simple, however, and discontinuities may occur at different scales (Mark 97) and could provide the physical basis for bottlenecks in recruitment referred to by various authors (e.g., Conan 98, and Fig. 3).

5.3. Fishing Mortality Rate, F

Fishing mortality is estimated conventionally from a knowledge of Z and M (given $F = Z - M$), by cohort analysis using ages (e.g., Abramson and Tomlinson 99), or sizes (e.g., Jones 100; Schumacher and Tiews 101), by means of mark–recapture studies (see Morgan 4), by swept-area methods (102, 103), or by use of the relationship

$$F = C/\overline{B}$$

where C = catch and B = mean biomass (derived from surveys). Because these approaches are also widely applicable to exploited resources other than crustaceans, the reader is referred to standard texts in population dynamics (1, 2).

6. FECUNDITY AND SURVIVAL RATE

Under unexploited conditions, the female fecundity and the individual probability
of survival for a species bear a reciprocal relationship to one another. Given stable
conditions, one might expect a relationship between numbers of spawners (S) and
recruits (R), such that on average, one female replaces itself by producing sufficient eggs so that one female recruit survives a generation later. This can be
expressed by the general relationship

$$(R/S) \exp - \sum_{i=1}^{n} Z_i Dt_i = 1$$

where mortality (Z_i) and duration (Dt_i) of stages $i = 1, 2 \ldots, n$ of multistage life
histories characteristics of decapods, are known for each stage. This is rarely the
case, of course, but those values that are known, or for which magnitudes can be
reasonably guessed, can be expressed in a "life table." Given the above stability
conditions, this can be used to estimate an order of magnitude for remaining
unknowns (e.g., the mortality rate in larval and postlarval stages), in order that
the above equality is satisfied. This type of segmented life history can also be
modelled by matrix algebra (see Vaughan and Saila 104). In the form of a simple
life table for females (usually assuming 50% eggs become females), the general
characteristics shown in the table are evident.

Time	Interval	(No. Survivors by Stage)
t_0		1 female spawns producing $E/2$ eggs that become female
	Dt_1	
t_1		$N_1 = 0.5E \exp(-Z_1 Dt_1)$ eggs survive larval life to benthic-stage females
	Dt_2	
t_2		$N_2 = N_1 \exp(-Z_2 Dt_2)$ female postlarvae reach size at recruitment
	Dt_3	
t_3		$N_3 = N_2 \exp(-Z_3 Dt_3) = 1$ female reaches maturity and spawns

Obviously, this is likely to be a simplification of the situation, and does not
take into account a range of phenomena, including multiple spawning, migratory
stages, discard mortality, and so forth, or a variety of fishery-related time intervals.
Some of these variables can be incorporated in a life table (see, e.g., Caddy 14,
which by oversight assumed that all eggs result in females!). Life tables can be

used as a method of sensitivity analysis (e.g., in evaluating the impact of discarding berried females, or changing fishing intensity) or in deciding on priorities in life-history investigations. These may lead to a more formal model of an exploited decapod population, such as that described by Fox (41).

7. STOCK AND RECRUITMENT RELATIONSHIPS

Many authors have identified the need to elucidate the relationship between stock and recruitment (variously referred to as the stock–recruitment or spawner–recruit relationship, SRR) as one of the critical problems in fishery management, for the simple reason that if predictive, such a relationship would be of great use to fisheries managers. However, it was not many years ago that Hancock (105) commented that there was very little information about the SRR for any crustacean stock. It was felt at that time that identification of the nature of the SRR might lead to an improved understanding of the underlying ecological dynamics. We seem to be in the paradoxical situation today that for a number of crustaceans, the major strides made in understanding population biology give us our best clues about the likely SRR that applies.

As we saw earlier in discussing life tables, the relationship between numbers of spawners and numbers of their progeny entering the fishable population is far from easily determined in marine populations. On top of measurement error, the effects of "environmental noise" on survival rates usually obscures the simple deterministic picture fishery managers would like to see. The nature of this environmental noise and variability in the SRR can be attributed to many factors, some abiotic, others biotic (Ricker 1; Sissenwine 106; Fogarty and Idione 107); a great deal of biology is subsumed within the apparently simple relationship between stock and recruit. In fact, in the few cases where the SRR could be plotted for crustaceans (and for obvious reasons, these are largely annual species such as penaeid shrimps; for a discussion of this, see Garcia 108), the points typically form a "cloud" from which it is often difficult to select an optimal model on purely statistical grounds.

Traditional two-parameter stock–recruitment models take two forms; overcompensatory (Ricker 109) and asymptotic (Beverton and Holt 110). More flexible models with three or more parameters have been developed more recently (e.g., Paulik 111; Shepherd 112; Tang 113), and have been applied to theoretical modeling of lobster populations (Bannister and Addison 114), but most of the discussion can still be centered around these simple two-parameter versions. These models use data on, or indexes of, stock size (ideally, the size of the spawning stock) and of recruit or immediate prerecruit abundance. When a stock–recruitment relationship is postulated, it is often presumed that the shape of the curve presents some evidence about the existence and nature of density-dependent population-regulating mechanisms. Both models incorporate the notion of density-dependent mortality. Ricker (109) developed the following relationship:

$$R = CS \exp{-(DS)}$$

where R = recruitment, S = abundance of spawning stock, and C and D are constants (notation as in Gulland 2). This gives a dome-shaped curve, reaching a maximum at some intermediate stock size. In contrast, the asymptotic curve described by Beverton and Holt (110) is

$$R = S / (A + BS)$$

where A and B are constants.

The dynamics underlying these two curves was clear when they first were proposed. Ricker was describing a situation (a salmon population), in which density-dependent negative interractions occur between adults and their offspring, involving a limited spawning area with destruction or reduced survival of excess eggs or offspring, or cannibalism by adults on eggs or fry. Beverton and Holt's curve implies a rather less extreme reduction in survival to recruitment at high densities, suggesting, for example, a limited carrying capacity with regard to food or habitat (implying competition at some early life history stage?). In practice, the two relationships are often fitted to a particular data set, and the one that describes the data best is reported. However, some knowledge of the ecological or behavioral mechanism likely to be underlying the relationship could profitably be used in deciding which theoretical curve to apply to the data; it is not clear to us that this has been done in any of the stock–recruitment relationships published for decapods. Cannibalism is unlikely because of female guarding of eggs, the pelagic nature of most decapod larvae, and habitat segregation of juveniles and adults in many species. There is one possible mechanism that could lead to a domed spawner–recruit curve, namely, where egg parasitism is extreme at high densities of spawners, especially if spawning is in aggregations. However, most considerations seem to suggest an asymptotic relationship, unless the data or specific a priori reasons can be cited for preferring the domed model.

Only a few publications have documented the existence of a stock–recruitment relationship in decapods, and in a major review, Garcia (108) urged caution in accepting the existence of one simple, deterministic relationship for any shrimp population. Nevertheless, for some shrimp fisheries, evidence (e.g., Staples 115; Penn and Caputi 116) is somewhat more convincing. Fogarty (117) and Fogarty and Idione (107) reanalyzed a 15-yr series of data on larval abundance and adult stock size for *Homarus americanus* in the Northumberland Strait, and found an asymptotic relationship between the final larval stage and a weighted average of stock size 5–7 yr later. No density-dependent relationships were found between the four pelagic stages, and these authors suggested that some form of density-dependent population regulating mechanism must be operating between the time of the final larval stage and recruitment to the fishable stock. The asymptotic nature of the relationship may explain the observed relative stability of the lobster stock, as well as its resilience to continued high exploitation rates. A steep ascending slope of the left side of the curve suggested that relatively high levels of fishing mortality might be sustained without a demonstrable decrease in recruitment. An alternative hypothesis suggested at the 1978 St. Andrews lobster workshop

(Anthony and Caddy 118), is that refugia of less heavily exploited segments of the population may be able to supply recruitment for the entire population. As Fogarty and Idione (107) point out, if an apparent population stability is due to density-dependent mechanisms it has important implications for fishery management.

It is clear, however, that even if a SRR is developed from historical data series, it cannot usually throw light on the mechanisms underlying it. In an analysis of the above data set, Caddy (119) demonstrated that survival rate of larvae produced throughout the season is not uniform: poor survival for this stock occurred with low and high summer surface temperatures and for larvae produced late in the season, and was usually optimal for early-produced larvae and in those summers with intermediate temperatures. Here, intermediate surface temperatures are associated with reduced stratification and improved vertical mixing of the highly stratified water masses in the Northumberland Strait. This may imply that vertical migration is important for completion of the larval life history, and shows that the causative mechanisms underlying simple SRRs can be very diverse. Stock–recruitment models that include more than two parameters have been proposed to account for other variables (Paulik 111, Tang 113) and to allow the curve to take nearly any form. The flexibility this imparts is useful, but given the high inherent variability, relationships with multiple parameters pose serious problems in fitting SRRs, particularly when applied without much knowledge of the biology that drives them.

There has been some progress in incorporating environmental parameters into SRRs. Thus Penn and Caputi (116 and Chapter 5) fitted SRRs to a population of the banana prawn using multiple regression techniques and incorporating cyclone effects as historic rainfall data. Cyclones (rainfall) were a dominant factor in two of the 14 years of the study; in other years recruitment appeared to be a direct function of the spawning stock levels of the preceding year.

Paulik (111) proposed the synthesis in a graphical model of interstage relationships and this has obvious potential (see Fig. 4) for decapod life-history stages (egg, pelagic larvae, benthic juveniles, adults). A similar approach was used by Staples (115), and a matrix algebra approach by Vaughan and Saila (104). Fogarty and Idione (107) demonstrated linear relationships between successive larval stages, and an asymptotic relationship between the final larval stage and subsequent recruitment to the stock. They assumed linear relationships between parental stock and production of first-stage larvae, and between recruitment and spawning stock size (Fig. 4a). Each of the four quadrants represents one of the interstage relationships, and the trajectory of the life cycle can be followed (dotted line) to project changes in population levels from any chosen initial stock size. The form of the SRR chosen is critical to predicting the amount and direction of changes, as illustrated in Figure 4b, where a hypothetical Ricker-type curve rather than an asymptotic curve was chosen for comparison. The situation soon becomes quite complex if any other relationships are nonlinear.

As a simplifying assumption that may be useful for conceptualizing the impact of SRRs on management of the stock, SRRs could be divided into two segments: the ascending left-hand limb, and the flat asymptotic section (Beverton and Holt),

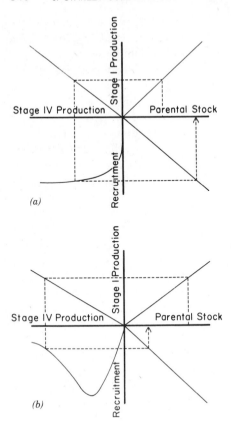

(a)

(b)

Figure 4. Graphical approach to a multistage recruitment model. (a) The lobster *Homarus americanus* (adapted from Fogarty and Idione 107); (b) hypothetical replacement of the asymptotic relationship by an overcompensatory curve.

or the descending right limb (Ricker). Once they are partitioned in this way S–R curves becomes more tractable. For the manager concerned with recruitment overfishing in a heavily exploited fishery, the concern is to estimate roughly where the ascending limb intersects the flat or descending section of the curve. This is most easily visualized if the curve is replaced by two intersecting straight lines, and the point of intersection may be called $S(\text{lim})$; see Figure 5. Above the stock size $S(\text{lim})$, recruitment is more or less constant for the Beverton–Holt type of curve. Garcia (108) and Caddy (14) proposed using a ramp function approximating this (Fig. 5). Although such a simplistic approach has some utility in conceptualizing management implications of SRRs, $S(\text{lim})$ is unlikely to be estimated with great precision from field data. As noted earlier, the research challenge is to determine the nature and relative importance of density-dependent and independent factors on the relationship between stock and recruitment, and if at all possible, to elucidate mechanisms.

Given the problems outlined above, it might be questioned if formulation of SRRs is a necessary exercise, or one that reflects reality. At a fundamental level, the answer to the second question must be yes, in that little recruitment is possible if only a few spawners are present, and alternatively, individual survival rate must

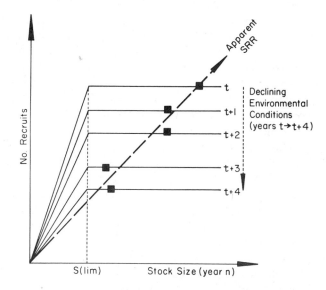

Figure 5. Beverton and Holt SRR shown as a family of ramp functions varying with environmental conditions, for an annual penaeid shrimp stock (adapted from Garcia 108). The straight-line segments are postulated to occur at critical stock size S(lim). The apparent SRR shown by the dashed line could be an artifact of deteriorating conditions over the period t to t + 4 yr.

be low at high spawning stock sizes, or the population would run away to infinity. Such relationships must therefore be taken into account in modeling the population, even if they are difficult to measure in the field. Whether it is a fruitful exercise to spend a great deal of research effort in developing the data base for such a relationship is questionable, especially because for most exploited stocks, the range of stock sizes that would allow proper fitting of such a relationship is currently restricted to the left of any inflection point. However, the necessary data series of annual spawners and recruits have other valuable uses, as explained in the next section.

8. FORECASTING STOCK ABUNDANCE AND CATCHES FROM RECRUITMENT

As we have seen, predicting the number of recruits from the stock size of spawners presents major problems for most crustaceans. In large part this is because environmental factors play a major role in determining spawning success and larval survival, and usually overwhelm any SRR present. This may occur to such an extent that as noted by Marchessault et al. (120), with the relatively high intensities of fishing and low stock sizes characteristic of exploited crustaceans, recruitment becomes almost entirely a function of fluctuations in environmental variables.

More success has been achieved in practice with what may be considered the inverse of the above relationship; namely, predicting stock size and catch from various indexes of recruitment of the same cohort earlier in its life history. Generally speaking, this type of prediction is less liable to catastrophic effects than predicting the number of offspring surviving from the number of spawners a generation earlier. It can also be extremely useful to the industry in planning future investment strategy, as well as to fisheries managers. Chapter 5 refers to some attempts at short-term forecasting for penaeid shrimps, where estimated juvenile abundance on inshore or estuarine nursery areas and/or the amount of freshwater runoff into these areas can become useful indexes of likely catch offshore later in the season. The importance of recruitment indexes is also recognized (e.g., for crabs; Jamieson 46 and Chapter 9), but the most extensive experience to date has been in the the Western Australian rock lobster fishery (Chapter 7), where a number of forecasting techniques have been developed that successfully indicate the order of magnitude of future catches for an animal that enters the fishery 5 yr after settlement. An asymptotic relationship seems to best fit the relationship between the index of puerulus settlement and the abundance of new recruits entering the fishery 4 yr later (Phillips 121), but this forecast can be backed up later, still 1–2 yr before these animals enter the commercial fishery, by one of several recruitment indexes (RIs) based on juvenile abundance in commercial catches (Caputi and Brown 122). These may be fitted by a power curve:

$$\text{catch} = a \, (\text{RI})^b$$

where a and b are constants. In this case (see also Pollock 123), environmental factors seem to have a limited effect once settlement has occurred, with little evidence of density-dependent bottlenecks between settlement and recruitment to the fishery.

9. YIELD MODELS FOR AN EXPLOITED POPULATION

When a model of the dynamics of a population is developed for a commercially important species, the goal of the modeling exercise is to predict accurately changes in population abundance as a result of various fishing strategies. There are two basic approaches: the first either implicitly or explicitly incorporates the dynamics of growth, recruitment, and mortality—the conceptual approach; the second uses statistical analysis of a time series of population estimates and one or more environmental factors—the empirical approach. Both conceptual and empirical models have been used to study the dynamics of exploited decapod populations.

9.1. Empirical Models

Fluctuations in abundance are a characteristic of all populations. An extreme example may be that of the Dungeness crab, which in northern California has undergone major, apparently cyclic changes in abundance (see Chapter 9). Attempts to explain the fluctuations have used correlations with upwelling indexes, sunspot activity, water temperature, predators, and the density-dependent effects of one life stage on another (reviewed in Hankin 124; Jamieson 46).

A number of environmental parameters have been used as correlates in empirical models of decapod populations including water temperature, river discharge, rainfall, salinity, wind speed and direction, upwelling, and primary production (see reviews by Garcia and Le Reste 3; Cobb and Wang 11; Jamieson 46). When there are interannual variations in yield not related to fishing effort, attempts can be made to correlate them with one or more environmental fluctuations in a predictive model. The model, in that it examines only the statistical relationship between abundance and other factors, essentially treats the situation as a ''black box.'' Thus it is predictive only in the sense that we assume that relationships extracted from past data will hold in the future. Although biological principles may be inferred from the model, they are not necessarily explicit in it. For instance, a correlation between catches of *Cancer magister* and sunspot activity may exist (Love and Westphal 125), but what biological mechanism (if any) is operating, is unclear. The simplest approach is to use only catch, or catch and effort data for time series analyses, as was done for *Jasus edwardsii* from New Zealand by Saila et al. (28). A more sophisticated approach is represented in an analysis of effort dynamics in the Dungeness crab fishery by Botsford et al. (126). They used the observation that fishing mortality varies in a regular pattern with regard to catch (Methot and Botsford 127), suggesting that the dynamics of fisherman and market behavior are important to understanding the dynamics of the population, and found a lagged response of harvest rate to subsequent changes in abundance.

Frequently the longest time series of environmental data available is that for water temperature. A number of investigators have used this in the American lobster fishery (Dow 128; Flowers and Saila 129; Saila and Marchessault 130; Fogarty 31). Fogarty (31) used both autoregressive–integrated moving average (ARIMA) approaches and transfer function models (Box and Jenkins, 131) to analyze 56 yr of data from the Gulf of Maine. Forecast accuracy for the final year (held in reserve) of the data was quite good. Lags predicted by the model were 0 yr (immediate effects, presumably due to changes in activity, therefore catchability) and 6 yr (the approximate time from larva to legal size). The analysis thus suggests that biological factors are important influences on catch levels.

Penaeid shrimp are short-lived; thus it should be easier to make correlations with environmental factors as well as predictions because only environmental factors in the preceding 12–18 months must be considered (Staples et al. 132). *Penaeus mergiuensis*, an Australian shrimp from the Gulf of Carpentaria, shows considerable interannual variation in abundance. Staples et al. (132) incorporated several environmental variables with commercial catch statistics and emigration rates in a forward stepwise multivariate regression analysis. The rainfall variable accounted for more than 80% of the catch variation, and was the single predictor variable used in the equation

$$C = 2.626R - 950.5$$

where C = catch (tonnes) and R = spring plus summer rainfall (mm).

Many of the empirical analyses point to critical periods in the life cycle. Often this appears to be early in life (May 133). Such factors as egg predation in Dungeness crabs (Wickham 42), larval mortality in blue crabs (Sulkin and van Henklen 134), and post-settling mortality (Fogarty 31), are possibilities. Nonetheless, all life stages are potentially critical (Jamieson 46), as was suggested when the red king crab fishery collapsed, apparently due to high mortality among the larger prerecruits.

9.2. Conceptual Models of Exploited Populations

Two types of conceptual models are used to describe the dynamics of fished populations, surplus production and analytic models. Traditional surplus yield models take the black box approach, using only catch and effort data in a single compensatory population growth model. The analytic approach explicitly considers processes such as individual growth, mortality, and recruitment. A much more detailed understanding of basic biology and population dynamics is necessary for the second approach.

More recently, several data-economical approaches have been described that offer promise for modeling crustaceans that incorporate aspects of both of the above, though these have not yet been applied to decapods.

9.2.1. Surplus Production Models

Because of difficulties in aging, crustacean stock evaluations, particularly in the early stages of investigation of a fishery, have relied heavily on monitoring catch and effort data in assessing the state of the stocks. In their original simplistic form, the category of models referred to as surplus production models reduce to "control curves" or plots of annual catch on annual fishing effort, with the intention of following the way that resource harvest changes with fishing pressure. All other population parameters and processes are undefined and considered to be occurring within a black box, into which the only input is fishing effort and from which the only output is catch.

Surplus production models are based on the concept that at equilibrium, growth plus recruitment must equal mortality. Production over and above that needed for replacement can be considered surplus, and thus harvested. Harvesting reduces intraspecific competition as well as shifting the age composition to younger, faster growing individuals. This creates a "surplus production" that, it is presumed, may be harvested without detrimental effects. The black box nature of the model becomes clear when you consider that the constant g in the general form of the production equation

$$dB/dt = g(B) - qfB$$

incorporates the aggregated effects of recruitment, growth, and natural mortality on the mean biomass B, which at equilibrium ($dB/dt = 0$) are considered to be equal to the catch, defined as $C = qfB$, where B = biomass, g = compensatory growth function, q = catchability coefficient, and f = fishing effort. The original formulation of Graham (135) expressed equilibrium yield in terms of biomass; that is,

$$Y_E = rB(B_\infty - B)/B_\infty$$

where r and B_∞ (the virgin biomass) are the two parameters of the logistic formulation. Other formulations express equilibrium yield or catch rate in terms of standard fishing effort. We will not go further into the theory of production modeling further here, but refer the reader to Gulland (2) for more details of equilibrium approaches.

Catch–effort plots have been used as rough and ready "back of the envelope" visualizations of the way that a resource responds to fishing, and still have some utility for this purpose if it is remembered that yield generally exceeds equilibrium values as effort increases, and vice versa as effort declines, and that these "departures from equilibrium" are more extreme for dynamic fisheries (e.g., Mohn 136) and for long-lived species. Gulland (2) and Hancock (137) set out some of the underlying assumptions and considerations to be borne in mind in fitting and interpreting production models. Clearly, although these are objective procedures, a knowledge of the fishery and resource biology is a precondition to their correct use and interpretation.

Three forms of production function have been fairly widely used: logistic (Schaefer 138), exponential (Fox 139), and several generalized versions (e.g., Pella and Tomlinson 140; Shepherd 141) that allow for different shapes of the production function (e.g., Fig. 6). The only data needed for production modeling are a time series of catch or biomass estimates and related fishing effort. The stock

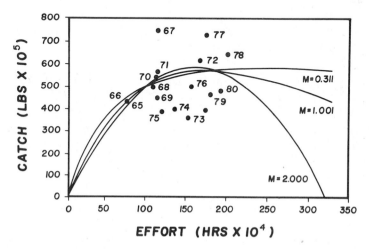

Figure 6. Surplus production model for the brown penaeid shrimp in the Gulf of Mexico. The three curves are produced by different values of the shape parameter M. (From Brunen-meister 142.)

biomass $B = C/qf$, is considered to be measured by catch per unit effort (variously expressed as U or CPUE). In a conventional approach, after adjustment for departure from equilibrium conditions, CPUE is plotted against effort and one of the above models is fitted to the data (Fig. 6) (142). Despite an apparent simplicity in mathematical form and assumptions, the literature on production modeling has proliferated greatly as workers on population dynamics and fishery economics have attempted to improve, or correct for departures from, the assumptions underlying the simple models. In the main, the problems with simple production modeling together with approaches adopted to solving them, can be summarized as follows:

(a) *Time delays.* The effect of fishing on the maternal generation in one year, and hence on the contribution to yield from their offspring t_r years later, is ignored in the simple model. This problem was addressed for lobster stocks, however, by Marchessault et al. (120) and more generally by Fogarty and Murawski (143).

(b) *Departures from equilibrium* due to rapid fisheries development and, later, stock collapses, can only be approximately accounted for by ''adjustments'' to the models or to the input data series. The method of Gulland (144) is still used, and with Fox (145) incorporates the idea that stock abundance (as measured by catch per unit effort U) should be related to the average effort exerted over the mean number of years T, that recruited individuals spend in the fished population; thus from (Gulland 2):

$$U_t = C/\overline{f_t} \qquad \text{where } \overline{f_t} = (1/T) \sum_{i=0}^{T-1} f_{t-i}$$

Some workers have questioned the efficacy of simple adjustments for departure from equilibrium, especially because analytical information on mortality rates is necessary to calculate the precise correction for departure from equilibrium. This has been the main reason why models incorporating some knowledge of growth and mortality parameters (e.g., Deriso 146; Schnute 147; Kimura et al. 148) have been developed.

(c) *Catchability constant?* One major deficiency of simple models is that they assume that fishing effort is proportional to the fishing mortality rate F (i.e., the catchability coefficient q is constant) over a wide range of densities, whereas q is often density dependent. The gear type used and the unit of effort selected can be critical here, as can changes in fishing technology and boat type if not taken into account.

In theory, systematic variations in the catchability q may be allowed for

1. By ''calibrating'' fishing effort against F using tagging experiments, cohort analysis, or direct estimation of biomass. (In this case, however, enough data are then usually available to permit more flexible analytical procedures!)
2. Or by using a production function that traces flexible shapes (e.g., Pella and

Tomlinson 140; Shepherd 141), so that systematic changes in q are in effect "internalized" into the model's black box along with the other analytical parameters.

The latter is a dubious procedure however, because changes in q due to fishermen learning more efficient methods in the development phase of the fishery will not be automatically cancelled out if effort then declines. One example of the uncertainties involved is the widespread use of the Fox (139) production function for fitting catch and effort for penaeid shrimp stocks. Here, production typically reaches a plateau after a certain level of effort, and declines relatively slowly with subsequent increases of effort. The fact that overall production declines slowly with effort beyond maximum sustainable yield (MSY) is one reason for the overcapitalization typical of industrial shrimp fisheries (see, e.g., Chapter 13). Whether this is due to a decline in q with increasing effort (i.e., a progressive difficulty in finding concentrations of shrimp near port with increasing effort?) or to other factors such as an underlying Beverton–Holt SRR and/or a similarly flat-topped yield-per-recruit curve is rarely clear, and production models alone give little insight into processes going on in the population.

The assumption of constant catchability may be particularly troublesome for fisheries in which gear may be saturated at high densities, as is usually the case in many trap fisheries, especially since there must be a wide variation in annual effort over the time series for the model to be fitted successfully. It is also clear that catchability can vary with environmental variables such as temperature (McLeese and Wilder 149) and other factors such as density (Morgan 150; Penn 151).

Two brief examples can be given to illustrate the pitfalls of the constant q assumption for crustaceans; one for a trawl (shrimp) fishery, the other for a trap fishery for lobsters. With respect to penaeid shrimps, engine horsepower may increase in the course of fleet development; or for multirigged boats, the number of trawls towed and/or their width. More subtle "learning" effects due to improving skipper skills, or better navigational or depth-finding equipment, all require continual effort calibration. For trap fisheries, factors other than effort also affect catchability, and the problem of calibrating trap fishing effort must be reviewed periodically (see Chapter 14) because traps rely on active behavioral responses for their success. These in turn depend on ambient temperature, sex, size, and the point in the molting and spawning cycle, as well as the saturation characteristics of the gear (see, e.g., Morgan 152; Boutillier 153).

It was the difficulty of calibrating effort units in many fisheries, and current improvements in our ability to estimate mortality rates from size data, that prompted the development of production models in which total mortality rate is substituted for fishing effort in the production function (Csirke and Caddy 154). Adjustments for departures from equilibrium are also possible with this approach (Caddy 87), but should be less pronounced for multiage species if mortality estimates are based on all age classes in the population. Production modeling could be extended to use other indexes as indirect measures of the impact of fishing in a similar "control curve" mode (Caddy 87). Thus for crustacean fisheries, the percentage of newly molted individuals in the catch should qualify here, because it normally increases under exploitation as the mean age decreases, given that young individuals generally molt more frequently than the older animals that dominate the unexploited stock. This index also has an advantage for species with a terminal molt (e.g., some crabs) over mean size as an index of mortality rate, for mean size may not decline much until the stock is fully exploited. Other potential measures of the degree of exploitation are the sex ratio (when males and females are not equally catchable), and the percent immature animals in the catch. Using biological indexes of population stress will also likely be less costly than effort estimation and calibration. With experience, crustacean fisheries might come to be regulated in part to maintain some biological characteristic of catches within a specified range that has proved from past experience to be associated with optimal yields and stock sizes, as opposed to relying exclusively on some difficult-to-measure and calibrate effort unit.

(d) *Environmental effects* (biotic and oceanographic) on yield are not explicitly considered in most production models, though the annual recruitment variations they cause are a principal reason why high variability around the fitted line makes it difficult to determine the appropriate model to be fitted. Similar considerations apply here to those described for spawner–recruit relationships.

9.2.2. Analytical Models

These models incorporate information on growth, age (or size) structure, natural mortality rates, and harvesting strategy, into population dynamics of the species. They, of course, require more data to be fully refined, and hence generally are used only in developed fisheries where there is a good biological data base. The results are often expressed initially as yield per recruit (Y/R), especially if the nature of the parent stock–recruitment relationship is not known. From experience, however, excessive reliance on Y/R analysis can be hazardous if the actual number of recruits is declining, and developing some measure of annual recruitment is strongly recommended.

A number of approaches to Y/R calculation have been suggested, such as those of Thompson and Bell (155) and Beverton and Holt (110), and these are reviewed in standard texts (1, 2). In this section we consider only some special problems in their application to decapods.

Analytical models offer an age-structured approach to population dynamics, and were first developed in fisheries for finfish assessment, where individual age can be determined. Because this is not directly possible in decapods, population size structure is usually substituted for age structure. This is not an entirely satisfactory exchange for reasons discussed in Sections 3 and 4. The transposition of continuous growth functions to organisms showing stepwise increases in body dimensions also poses some problems when yield modeling is begun. Thus the Beverton and Holt formulation for yield per recruit using parameters from the von Bertalanffy growth function (VBGF) to describe growth has been widely applied in crustacean Y/R analysis (Garcia and Le Reste 3; Cobb and Wang 11), but cannot easily incorporate other biological information such as that on molt stage and fecundity.

Various more detailed approaches offer promise where more extensive information is available: for example, a model incorporating size-specific growth rates and a stochastic representation of lobster growth was developed for *Homarus americanus* (Fogarty 31), and a similar approach was suggested toward modeling complex choices in the life history of crustaceans (Caddy 14), which proposes use of the biologically relevant intermolt time units mentioned earlier. In the Fogarty model, molt-dependent natural mortality rate was included in the formulation based on the assumption that mortality was higher for recently molted individuals. The model was applied separately to males and females because of different growth rates (females grow more slowly), molting frequency (females do not molt in the year they bear eggs), and fishing strategy (ovigerous females may be protected). A flow chart of the model for females is shown in Figure 7, and is the same for males, without the path for egg bearing. The model was used to explore the effects of changes in minimum legal size and of fishery mortality rates. Fogarty (31) found that a major reduction in F from its current estimated 1.5–2.0 level was necessary to affect yield in any meaningful way, but changes in minimum legal size could give significant increases in yield; the patterns for males and females were very similar. One cautionary result of this simulation is clear: at high levels of fishery

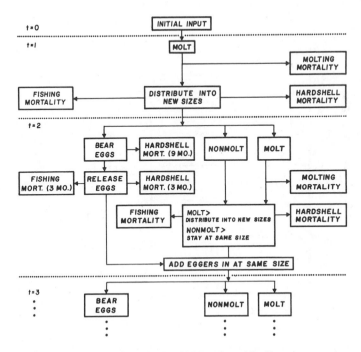

Figure 7. Flow chart for an analytical yield model developed for *Homarus americanus*. (From Fogarty 31.)

mortality (over $F = 1.0$) it may be impossible to use size composition data from the fishery to determine changes in fishery mortality, a conclusion also supported on a different basis by Addison (93). This possibility poses a considerable problem in assessing many decapod fisheries.

For many commercial crustaceans, the unit weights of different sized individuals on landing do not have the same market value. This is most particularly true for shrimps, whose value per unit weight increases markedly with size, and for lobsters, where it may decline for very large individuals. Partly as a consequence, in addition to gear selection effects, F is not independent of age or size. Rothschild and Brunenmeister (156) developed a Beverton and Holt model for three species of shrimps in the Gulf of Mexico which showed that the eumetric fishing curves were very sensitive to changes in natural mortality (Fig. 8) and suggested an appropriate size at first capture. They went on to discuss the problem of discards of small shrimp. The proportion discarded varies with value: when abundance is low (so that value is high) smaller shrimps are kept; when the situation is reversed, small shrimps are discarded. In a situation such as this the standard models may supply misleading management advice and eumetric curves such as Figure 8 may suggest a too-small size at first capture. The use of models in which growth is represented by value rather than tissue accumulation may allow better analysis for management.

In addition to yield or value it is often very relevant to examine fecundity per

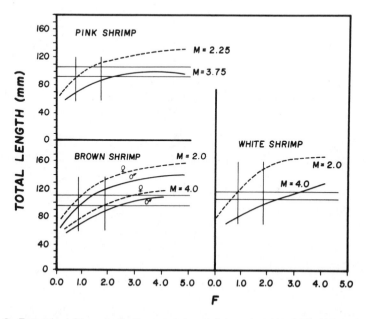

Figure 8. Eumetric yield curves for three species of shrimps from the Gulf of Mexico. Approximate lengths at recruitment are shown by horizontal lines. Fishing mortality lies between 1.0 and 2.0 as shown by the vertical lines. (From Rothschild and Brunenmeister 156).

recruit, particularly in situations where recruitment overfishing is suspected. Calculating egg production per molt group has been suggested (Caddy 85): other formulations use size-structured models (Campbell 29; Ennis 157; Fogarty 31) to estimate fecundity per recruit.

10. AN ECOSYSTEM APPROACH

The models considered so far treat the fished populations as if they existed in isolation; only interaction with fishermen (as predators) being considered. Additionally, the models all assume equilibrium or steady-state conditions. Neither assumption conforms to reality. Fluctuations in physical and biological environments, as well as in fishing mortality, are the rule rather than the exception. In general, the fisheries manager must make decisions about levels of fishery effort given only single-species models that submerge variability in their structure. Recent developments in single-species production and analytic models have probably now become more sophisticated than can be justified by the amount and quality of data usually available, or by their ability to explain the phenomena faced by fisheries managers. Despite this mild caveat, single-species models will continue to be applied, and the circumstances when they are most applicable will become more evident as long time series of data accumulate on a number of key indexes. In the rest of this section, however, we begin to explore some simple concepts from multispecies fisheries and a food web approach to the problem of fisheries management. No pretense is made that these approaches offer easy solutions, but they appear to us to offer great promise for the future.

Although it is dangerous to consider a multispecies fishery as simply the sum of a series of fisheries for the individual species, there is some interest initially in considering the overall curve of production for all species as the composite of the production curves for the individual species in the catch. If we ignore interactions between species and just look at the past catch record and size or age composition, a value for $F(MSY)$, the fishing mortality required to reach a single-species MSY, can in theory be derived for each species. Overfishing, and possibly severe overfishing, is likely to occur on a species whose $F(MSY)$ is less than that for the whole system. This occurs even if the fishery is managed so that there is no evidence for overfishing from the multispecies yield. This is the case for many multispecies penaeid shrimp fisheries, (e.g., Caddy 158, and Fig. 9), but also includes finfish taken incidentally by shrimp fisheries. A discussion of some of the yield implications of shrimp fishing on fish by catch is given in Caddy (159). Now if in addition we make the reasonable assumption that there are predatory or competitive interrelationships among species, it is clear that overfishing one species, possibly to extinction, could have severe repercussions on the others via relationships unaccounted for in the model. More seriously, the way that fisheries for other species are prosecuted that are linked trophically to the target species (often a decapod) can affect the value for $F(MSY)$ for the latter. Caddy and Sharp (160) explore this type of consideration in more detail. When dealing with multi-

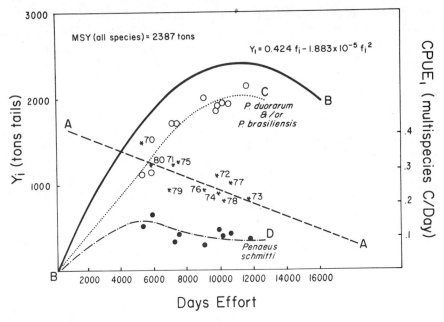

Figure 9. Simple descriptive approach to dissection of a multispecies production curve; in this case for the Nicaraguan Caribbean shrimp fishery (from Caddy 158). [B-B is a regression fit: B-C and B-D fitted by eye]

species fisheries it makes sense to be concerned with the nature of the ecological interactions among the species even if these are not always explicitly considered in a formal modeling context.

The use of structural diagrams for the analysis of marine food webs (e.g., Paine 161) has a long and illustrious history (see Hardy 162) although a rigorous approach (e.g., Steele 163) is only relatively recent. Cobb et al. (164) suggested a structural diagram for the three-species system of large decapods found in the sandy–rocky subtidal of Narragansett Bay, Rhode Island. Two species of *Cancer* apparently compete with each other and with *Homarus* for shelter space. Generalized bottom-feeding predators such as the tautog, *Tautoga onitis* feed on all three species (Fig. 10*A*). We can speculate that addition of fishermen to the system (Fig. 10*B*) causes a decrease in the size of the lobster component, releasing competitive pressure on the crabs, and allowing their population size to increase. Further manipulations can be carried out, for instance, adding size categories to the model (Fig. 10*C*) or adding fishing pressure on another species (Fig. 10*D*). The predictions here are qualitative, but given the observed interrelationships, they indicate the likely directions of change. Loop analysis (Levins 165) has been suggested for studies on multispecies fisheries (Saila and Parrish 166) and species introductions (Li and Moyle 167), but the results are not always satisfactory (Cobb et al. 164). In part, this is because the relative strengths of each link can be critical, and because simple qualitative models such as Figure 10 may not be complex enough to allow the model to be balanced. These structural models should perhaps be regarded as

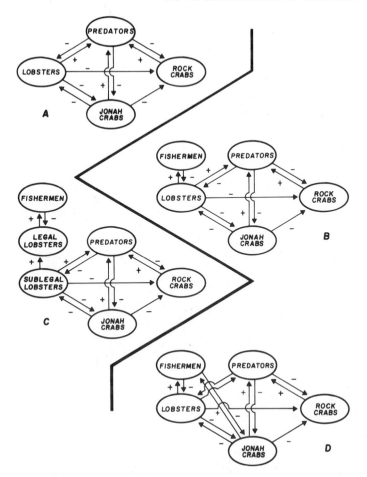

Figure 10. Structural model of the species group, lobsters, Jonah crabs, and rock crabs in Narragansett Bay, RI. Arrows with a + indicate a positive effect on population size in the direction indicated, arrows with a −, a negative. Thus a pair + − indicates a predator–prey link, and − − a competitive interraction. An unidentified common predator is added for realism. (A) The situation as understood from field and laboratory studies. (B) Fishing mortality added to A. (C) As for B, but size constraints on harvesting imposed. (D) As for B, but fishing on Jonah crabs also.

"thought experiments," not as predictive; they are most useful in describing the present state of knowledge and for planning further experiments. In some systems more quantitative food web data are available, as is the case for the Gulf of Mexico shrimp fishery. Flint and Rabelais (168, and Fig. 11) and Sheridan et al. (169) have developed material transfer and energy flow models using long data bases on interactions between elements of their models. The goal of Flint and Rabelais (168) was to determine the trophodynamics of the ecosystem and to examine the role of penaeids in it. They found that benthic infaunal production was insufficient

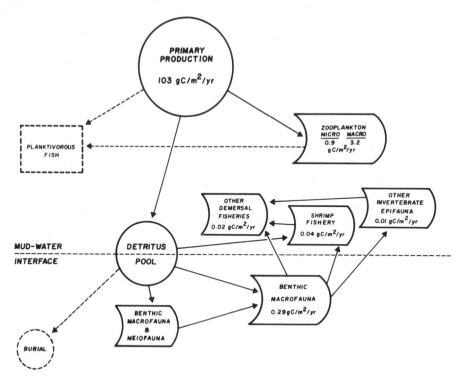

Figure 11. Energy flow diagram for the shrimp fishery of the south Texas continental shelf. (From Flint and Rabelais 168.)

to support the demersal component (shrimps, fish, and other epifauna) without a large detrital component to balance the carbon budget, and suggested that inputs of detritus and discards from the shrimp fishery are what allows the productivity of the demersal fauna to be so high. This system contrasts markedly with the North Sea (Steele 163) and Scotian Shelf (Mills and Fournier 170), where important demersal fisheries are also located. A detailed review (Sheridan et al. 169) of the problem of discards from the shrimp fishery in the trophodynamic context suggested that a reduction by one-half in discards would reduce shrimp stocks by 8–25%, depending on the method of discard reduction.

Other conceptual tools are also becoming available. A biomass budget developed for French Frigate Shoals in the Hawaiian Islands (Polovina 171) included growth, mortality due to fishing and predation, and nonpredator mortality, for species components interacting trophically. Although essentially an equilibrium or static representation of a system, this approach is useful in getting a feel for quantitative interactions, and a computer program, ECOPATH, was developed (Polovina and Ow 172). Nixon (173) pointed out that freshwater ecologists have been successful in linking fisheries yields with indexes of primary production, and attempted with some success to make the same correlation for coastal lagoon ecosystems, and compared them with other shallow marine systems. Because juvenile shrimps can be an important seasonal component of coastal lagoons, this type of analysis is of interest to fisheries biologists working on penaeids. Finally, in some very special situations the interactions may be simple enough to lend themselves to in-depth analysis. This seems to be the case on the west coast of South Africa where the spiny lobster *Jasus lalandii* is the most abundant benthic predator and feeds almost exclusively on the ribbed mussel *Aulacomya ater*, the dominant species in the benthos. The size-frequency distribution of *A. ater* is bimodal; large mussels cannot be opened by any but the largest lobsters, small mussels find refuge from predation among the large ones. Thus it is possible for the population of lobsters to be food limited even where mussel biomass is large (Pollack 174; Pollack et al. 175; Griffiths and Siderer 176). Given that fishing has cropped the largest lobsters from the population, what would be the effect of changing minimum legal size in the fishery? This question was addressed elegantly by Siderer et al. (177) using a mathematical model incorporating the population structure, growth and mortality rates of lobsters and mussels, as well as the results of behavioral experiments on prey choice in lobsters. A decrease in minimum legal size from 90 to 70 mm CL was predicted to result in increases of mussel biomass and lobster yield, but decreases in lobster density and population fecundity.

11. SOME PERSPECTIVES

Our knowledge of the special nature of decapod population dynamics has increased enormously in the past 15 years. A better understanding of growth processes and larval ecology has been particularly helpful. Considerably more sophisticated models are now being used for analysis of fishery and environmental data. New perspectives on the relationship between stock and recruitment will spur studies on the ecology of early life stages. The major challenge was and continues to be the development of conceptual approaches appropriate to decapods in a field dominated by models developed for finfish stocks.

The description of growth provides an example of the change from finfish to decapod paradigms. The discontinuous nature of growth in decapods is not represented by continuous models such as the Gompertz or von Bertalanffy functions. They continue to be applied to growth data because they do a reasonably good job and, more importantly, because the parameters of the von Bertalanffy function are used in the traditional Beverton and Holt yield-per-recruit model. However, the development of empirical growth descriptors that use molt increment and molt interval allows a better interpretation of the data and the development of simulation models with a much greater degree of biological realism.

Most decapod fisheries are relatively recent; few have reliable data for a long enough time series to make predictive models based on environmental parameters. Nevertheless in some fisheries unsophisticated time series analysis, often using only one environmental parameter, yields surprisingly good predictions. Often catch and effort statistics are all that are available, and although the increasing sophistication of production modeling is remarkable, it does not always appear that energy expended on making traditional models better—for instance, by adding delay terms to production models—always compensates for poor data and unaddressed sources of bias. New approaches that incorporate biological reality into simulation models are needed.

Age cannot be determined for individual crustaceans, which may remain at the same size for relatively long periods of time. Demographic models based on population structure by size or molt group are now beginning to be used, especially since many ecological processes are size- rather than age-dependent. Mortality is a good example; predation and fishing are both size selective, thus the use of a size-specific mortality function in demographic models is important when considering effects of fishing or predation on population size distribution (Kirkpatrick 178). Similarly, fecundity, growth, feeding behavior, and habitat selection all are size dependent, as is intraspecific competition (Hyatt 179). Although one would predict that size is also important in determining the outcome of interspecific competition, this is not always the case (Richards and Cobb 94). Feeding behavior and habitat selection may be of more importance than is generally appreciated in understanding the distribution and abundance of populations. Cannibalism by large *Cancer magister* on young-of-the-year crabs has been suggested (Botsford and Wickham 180) but questioned because size-specific habitat distribution makes encounter unlikely (Hankin 124). Choice of substrate type by the European lobster is based on the size and availability of suitable crevices that can act as shelter as well as current speed (Howard and Bennett 181). Structural complexity of seagrass beds affords refuge from predation to some but not all species preyed upon by *Penaeus duoarum* (Leber 182). Finally, the fractal nature of substrates (Caddy 14) may explain how changes in natural mortality rate with size result from modulations by available habitat of the effects of predation.

Molting is a phenomenon that dominates the life processes of decapods. If mortality is higher at the time of ecdysis (as has been suggested by Ennis and Akenhead 183, and others), then a term relating mortality to molt frequency should be built into demographic models. Many decapods can mate only when the female has recently molted, severely constraining the timing of reproduction, and perhaps creating as yet unrecognized problems with mate supply in species where, because of their larger size, only males are fished. A knowledge of the stage in the molt cycle may also provide useful demographic information for the fishery biologist. Although age determination of specimens is not possible, the stage (e.g., A through D of Drach 184) in the molt cycle can be determined by examining setal hairs (e.g., Aiken 25). This can be particularly useful when estimates of the proportion of the population molting are needed, or in characterizing patterns of larval distribution (Hatfield 185).

Commercially important decapods are often the most abundant, and frequently the largest benthic invertebrate predator in their environment. Through predation and competition they probably play important roles in the organization and dynamics of the benthic communities of which they are a part. Decreases in population abundance of decapods due to fishing are sure to affect those communities, but how is not at all sure. Some attempts have been made at multispecies and ecosystem modeling, but these must be developed in greater detail. Specific hypotheses can be formulated and tested. An example of this approach is the sudden appearance of barren grounds devoid of normal kelp cover in the subtidal of eastern Canada. Breen and Mann (186) postulated that overfishing lobsters released predation pressure on sea urchins, which in turn overgrazed kelp. Testing this hypothesis and corollaries showed the relationships to be far more complex than originally imagined (Mann 187), and that lobster predation probably is not involved in the regulation of population size in sea urchins (Miller 188; Vadas et al. 189). In fact, mass mortalities of urchins apparently due to a pathogen have allowed the regrowth of kelp beds (Miller 190) without a dramatic change in the size of the lobster population. Nevertheless, the initial simple and testable hypothesis had considerable implications for the lobster fishery and an impact on research strategy.

The testing of hypotheses, as Larkin (191) so eloquently stated, should be an important part of fisheries ecology. However, direct tests require population manipulation in the field, and this is difficult in marine systems. Crayfish populations in small freshwater lakes can be subjected to experimental fishing (Momot and Gowing 192; Momot 193), and one of the most important lessons learned by Momot was that even with an apparently long time series of data (8 yr on a crayfish with a 2-yr life-span), population responses to exploitation were unclear. He concluded that real progress in understanding how exploitation affects populations will involve "the accumulation of costly, unfashionable, well documented, long-term experimental mangement studies." One of the problems we face in understanding the dynamics of marine resources is our growing realization (e.g., Caddy 194) that we are not dealing with stable systems in equilibrium: both fishery and ecological interactions, as well as the physical environment and development dynamics of the fishing industry, all mean that systems must be considered responsive to a variety of inputs, even if their effects show up only after appropriate lag periods. Our ability to model the dynamics of exploited populations has clearly outstripped our ability to test the reality of these models. Fisheries ecologists wishing to do this in marine systems face even greater challenges than their colleagues working in lakes, but we seem to be learning that modeling simple but biologically realistic subsystems that can be joined in a flexible fashion may be one way to proceed. Nevertheless, it appears to be time to start making testable hypotheses that are amenable to field experimentation and comparative study. This may require the closure of large study areas for extensive periods of time, for it is our opinion that the next major step forward in the understanding of decapod population dynamics will come from experimental manipulation of populations in the field.

ACKNOWLEDGMENTS

The senior author thanks the members of the Crustacean Population Dynamics Seminar, University of Rhode Island, for their help and enthusiasm. He also thanks Michael Clancy and Tracey MacKenzie for technical support and the U.R.I. Sea Grant Program for financial support.

REFERENCES

1. W. E. Ricker, Computation and interpretation of biological statistics of fish populations. *Bull., Fish. Res. Board Can.* **191**:1–382 (1975).

2. J. A. Gulland, *Fish Stock Assessment: A Manual of Basic Methods*, Vol. 1. Wiley, New York, 1983, 222 pp.

3. S. M. Garcia, and L. Le Reste, Life cycles, dynamics, exploitation and management of coastal penaid shrimp stocks. *FAO Fish. Tech. Pap.* **203**:1–215 (1981).

4. G. R. Morgan, Population dynamics of spiny lobsters. In J. S. Cobb and B. F. Phillips, Eds.), *The Biology and Management of Lobsters*, Vol. II. Academic Press, New York, 1980, pp. 189–217.

5. R. G. Hartnoll, Growth. *In* L. G. Abele, Ed., *The Biology of Crustacea*, Vol. II. Academic Press, New York, 1982, Chap. 3.

6. R. Boddeke, The seasonal migration of the brown shrimp *Crangon crangon. Neth. J. Sea Res.* **10(1)**:103–130 (1976).

7. R. W. Hiatt, The biology of the lined shore crab, *Pachygrapsus crassipes* Randall. *Pac. Sci.* **2**:135–213 (1948).

8. H. Kurata, Studies on the age and growth of Crustacea. Bull. Hokkaido reg. *Fish. Res. Lab.* **25**:1–115 (1962).

9. J. Mauchline, The Hiatt growth diagram for crustacea. *Mar. Biol. (Berlin)* **35**:79–84 (1976).

10. G. Y. Conan, and M. Comeau, Functional maturity and terminal molt of show crab, *Chionoecetes opilio. Can. J. Fish. Aquat. Sci.* **43**:1710–1719 (1986).

11. J. S. Cobb and D. Wang, Fisheries biology of lobsters and crayfishes. In A. J. Provenzano, Ed., *The Biology of Crustacea*, Vol. X. Academic Press, New York, 1985, Chap. 3, pp. 167–247.

12. R. Boddeke, R. Dijkema, and M. E. Siemelink, The patterned migration of shrimp populations: A complete (comparative) study of *Crangon crangon* and *Penaeus brasiliensis. FAO Fish. Rep.* **200**:31–49 (1976).

13. A. D. Racek, Prawn investigations in eastern Australia. *Res. Bull. State Fish. N.S.W.* **6**:1–57 (1959).

14. J. F. Caddy, Modelling stock–recruitment processes in crustacea: Some practical and theoretical perspectives. *Can. J. Fish. Aquat. Sci.* **43**:2330–2344 (1986).

15. J. F. Caddy, Approaches to a simplified yield-per-recruit model for crustacea, with particular reference to the american lobster, *Homarus americanus. Fish. Mar. Serv. Manuscr. Rep. (Can.)* **1445**:1–14 (1977).

16. R. M. Cassie, Some uses of probability paper for the graphical analysis of polymodal frequency distributions. *Aust. J. Mar. Freshwater Res.* **5**:513–522 (1954).

17. S. Tanaka, A method of analysing polymodal frequency distributions and its application to the length distribution of the porgy, *Taius tumifrons* (T. and S.). *Fish. Res. Board. Can.* **19**:1143–1159 (1962).

18. C. G. Bhattacharaya, A simple method of resolution of a distribution into Gaussian components. *Biometrics* **23(1):**115–135 (1967).

19. P. Sparre, Computer programs for fish stock assessment. Length-based Fish Stock assessment for Apple II computers. *FAO Fish. Tech. Pap.* No. 101 (Suppl. 2.) 218 p. (1987).

20. J. E. Stewart, and H. J. Squires, Adverse condition as inhibitors of ecdysis in the lobster *Homarus americanus. J. Fish Res. Board Can.* **25:**1763–1774 (1968).

21. J. Mauchline, Growth of shrimps, crabs and lobsters—An assessment. *J. Cons. Cons. Int. Explor. Mer* **37(2):**162–169 (1977).

22. D. E. Aiken, Molting and growth. In J. S. Cobb and B. F. Phillips, Eds., *The Biology and Management of Lobsters*, Vol. I. Academic Press, New York, 1980, pp. 91–163.

23. R. A. Cooper, and J. Uzmann, Ecology of juvenile and adult *Homarus*. In J. S. Cobb and B. F. Phillips, Eds., *The Biology and Management of Lobsters*, Vol. II. Academic Press, New York, 1980, pp. 97–142.

24. D. G. Hankin, N. Diamond, M. Mohr, and J. Ianelli, Molt increments, annual molting probabilities, fecundity, and survival rates of adult female Dungeness crab in Northern California. *Alaska Sea Grant Rep.* **85-3:**189–206 (1985).

25. D. E. Aiken, Proedysis, setal development, and molt prediction in the American lobster (*Homarus americanus*). *J. Fish. Res. Board Can.* **30:**1337–1344 (1973).

26. D. A. Vigh, and M. Fingerman, Molt staging in the fiddler crab *Uca pugilator. J. Crustacean Biol.* **5:**386–396 (1985).

27. M. L. Parrack, Aspects of brown shrimp, *Penaeus aztecus*, growth in the northern Gulf of Mexico. *Fish. Bull.* **76(4):** 827–837 (1978).

28. S. B. Saila, J. H. Annala, J. L. McKoy, and J. D. Booth, Application of yield models to the New Zealand rock lobster fishery. *N.Z.J. Mar. Freshwater Res.* **13:**1–11 (1979).

29. A. Campbell, Growth of tagged American lobsters, *Homarus americanus*, in the Bay of Fundy. *Can. J. Fish. Aquat. Sci.* **40:**1667–1675 (1983).

30. D. B. Bennett, Growth of the edible crab (*Cancer pagurus* L.) off south-west England. *J. Mar. Biol. Assoc. U.K.* **54:**803–823 (1974).

31. M. J. Fogarty, Population dynamics of the American lobster (*Homarus americanus*). Ph.D. Thesis, University of Rhode Island, Kingston, 1986, 277 pp.

32. D. A. Somerton, A computer technique for estimating the size of sexual maturity in crabs. *Can. J. Fish. Aquat. Sci.* **37:**1488–1494 (1980).

33. R. W. George and G. R. Morgan, Linear growth stages in the rock lobster (*Panulirus versicolor*) as a method for determining size at first physical maturity. *Rapp. P.-V. Reun., Cons. Int. Explor. Mer* **175:**182–185 (1979).

34. W. Groom, *Lobster Project, 1977, Gran Manan Island*, Inf. Notice 1. New Brunswick, Dep. Fish., Bay of Fundy Fishermen, 1977, pp. 1–10.

35. P. T. Briggs, and F. M. Mushacke, The American lobster in western Long Island Sound. *N. Y. Fish Game* **26:**59–86 (1979).

36. D. E. Aiken and S. L. Waddy, Reproductive biology. In J. S. Cobb and B. F. Phillips, Eds., *The Biology and Management of Lobsters*, Vol. I. Academic Press, New York, 1980, Chap. 4, pp. 215–276.

37. D. E. Aiken and S. L. Waddy, Cement gland development, ovary maturation, and reproductive cycles in the American lobster, *Homarus americanus. J. Crustacean Biol.* **2:**315–327 (1982).

38. G. P. Ennis, Comparison of physiological and functional size–maturity relationships in two Newfoundland populations of lobsters *Homarus americanus. Fish. Bull.* **82:** 244–249 (1984).

39. D. K. Stevenson, and F. Pierce, Life history characteristics of *Pandalus montaqui* and *Dichelopandalus leptocerus* in Penobscot Bay, Maine. *Fish. Bull.* **83:**219–233 (1985).

40. J. A. Allen, Observations on the biology of *Pandalus montaqui* (Crustacea: Decapoda). *J. Mar. Biol Assoc. U.K.* **43:**665–682 (1963).

41. W. W. Fox, Jr. A general life history exploited population simulator with pandalid shrimp as an example. *Fish. Bull.* **71(4):**1019–1028 (1973).

42. D. E. Wickham, Carcinonemertes errans and the fouling and mortality of eggs of the Dungeness crab, *Cancer magister. J. Fish. Res. Board Can.* **36:**1319–1324 (1979).

43. H. C. Perkins, Egg loss during incubation from offshore northern lobsters (Decapoda: Homaridae). *Fish. Bull.* **69:**451–453 (1971).

44. P. N. Reilly, The biology and ecology of juvenile and adult rock crabs, *Cancer irroratus* Say, in southern New England waters. M.S. Thesis, University of Rhode Island, Kingston, 1975, 146 pp.

45. D. A. Somerton, Contribution to the life history of the deep sea king crab, *Lithodes covesi* in the Gulf of Alaska. *Fish. Bull* **79:**259–269 (1981).

46. G. S. Jamieson, Implications of fluctuations in recruitment in selected crab populations. *Can. J. Fish. Aquat. Sci.* **43:**2085–2098 (1986).

47. P. Kanciruk and W. Herrnkind, Autumnal reproduction in *Panulirus arqus* at Bimini, Bahamas. *Bull. Mar. Sci.* **26:**417–432 (1976).

48. J. F. Caddy, Notes on a more generalized yield-per-recruit analysis for crustaceans, using size-specific inputs. *Fish. Mar. Serv. Manuscr. Rep. (Can.)* **1525:**1–40 (1979).

49. A. Campbell and D. G. Robinson, Reproductive potential of three American lobster (*Homarus americanus*) stocks in the Canadian Maritimes. *Can. J. Fish. Aquat. Sci.* **40:**1958–1967 (1983).

50. P. F. Berry, The biology of the spiny lobster, *Panulirus homarus* (Linnaeus), off the east coast of southern Africa. *Oceanogr. Res. Inst. Invest. Rep. (Durban)* **28:**1–75 (1971).

51. D. E. Gregory, Jr., R. L. Labisky, and C. L. Coombs, Reproductive dynamics of the spiny lobster *Panulirus arqus* in South Florida. *Trans. Am. Fish. Soc.* **111:**575–584 (1982).

52. E. L. Charnov, D. W. Gotshall, and J. G. Robinson, Sex ratio: Adaptive response to population fluctuations in Pandalid shrimp. *Science* **200:**204–206 (1978).

53. A. M. Wenner, Sex ratio as a function of size in marine crustacea. *Am. Nat.* **106:**321–350 (1972).

54. G. P. Ennis, Size–maturity relationships and related observations in Newfoundland populations of the lobster *Homarus americanus. Can. J. Fish. Aquat. Sci.* **37:**945–956 (1980).

55. W. G. Lyons, D. G. Barber, S. M. Foster, F. S. Kennedy, and G. R. McLano, The spiny lobster, *Panulirus arqus*, in the middle and upper Florida keys: Population structure, seasonal dynamics and reproduction. *Publ.—Fla., Dep. Nat. Resour., Mar. Res. Lab.* **38:**1–38 (1981).

56. H. J. Thomas, Observations on the sex ratio and mortality rates in the lobster (*Homarus vulgaris* Edw.). *J. Cons., Cons. Int. Explor. Mer* **20:**295–305 (1955).

57. S. B. Saila and J. M. Flowers, Geographic morphometric variation in the American lobster. *Syst. Zool.* **18:**330–338 (1969).

58. B. E. Skud, The effect of fishing on size composition and sex ratio of offshore lobster stocks. *Fiskeri dir. Skr., Ser. Havunders.* **15:**295–309 (1969).

59. P. A. Larkin, *The Stock Concept and Management of Pacific Salmon.* Univ. of British Columbia Press, Vancouver, B.C., 1972, 213 pp.

60. P. E. Ihssen, Physiological and behavioral genetics and the stock concept for fisheries management. In *Fish Genetics—Fundamentals and Implications to Fish Management.* Great Lakes Fish. Comm., Ann Arbor, MI, 1977, pp. 27–30.

61. P. E. Ihssen, H. E. Brooke, J. M. Casselman, J. M. McGlade, N. R. Payne, and F. M. Utter, Stock identification. Materials and methods. *Can. J. Fish. Aquat. Sci.* **38:**1838–1855 (1981).

62. V. Hedgecock, D. J. Stelmach, K. Nelson, M. E. Lindenfelser, and S. R. Malecha, Genetic divergence and biogeography of natural population of *Macrobrachium rosenbergii. Proc. Annu. Meet.--World Maric. Soc.* **10:**873–879 (1979).

63. K. Nelson, and D. Hedgecock, Enzyme polymorphism and adaptive strategy in the Decapod crustacea. *Am. Nat.* **116:**238–280 (1980).

64. J. C. Mulley, and B. D. H. Latter, Geographic differentiation of eastern Australian prawn populations. *Aust. J. Mar. Freshwater Res.* vol. **32 (6):**889–906 (1981).

65. H. P. Bulnheim, and A. Scholl, Genetic variation between geographic populations of the amphipods *Gammarus zaddachi* and *Gammarus salinus. Mar. Biol. (Berlin)* **64:**105–115 (1981).

66. M. L. Tracey, K. Nelson, D. Hedgecock, R. A. Shleser, and M. L. Pressick, Biochemical genetics of lobsters: Genetic variation and the structure of American lobster (Homarus americanus) populations. *J. Fish. Res. Board Can.* **32:**2091–2101 (1975).

67. P. J. Smith, J. L. McKoy, and P. J. Machin, Genetic variation in the rock lobsters *Jasus edwardsii* and *Jasus novaehollandiae. N. Z. J. Mar. Freshwater Res.* **14(1):**55–63 (1980).

68. J. C. Mulley, and B. D. H. Latter, Genetic variation and evolutionary relationships within a group of thirteen species of penaeid prawns. *Evolution (Lawrence, Kans.)* **34:**904–916 (1980).

69. R. A. Cooper and J. R. Uzmann, Migrations and growth of deep-sea lobsters, *Homarus americanus. Science* **171:**288–290 (1971).

70. M. J. Fogarty, D. V. D. Borden, and H. J. Russell, Movements of tagged American lobster, *Homarus americanus*, off Rhode Island. *Fish. Bull.* **78:**771–780 (1980).

71. F. W. Allendorf and S. R. Phelps, Use of allelic frequencies to describe population structure. *Can. J. Fish Aquat. Sci.* **36:**1587–1514 (1981).

72. B. Fry, Natural stable carbon isotope tag traces Texas shrimp migrations. *Fish. Bull.* **79:**1–5 (1981).

73. J. R. Uzmann, Use of parasites in identifying lobster stocks. *J. Parasitol.* 56(Suppl. II):349 (1970).

74. M. Soule and R. N. Tasto, Stock identification. Studies on the Dungeness crab *Cancer magister. Calif. Dep. Fish Game., Fish. Bull.* **172:**39–42 (1983).

75. R. A. Menzies, Biochemical population genetics and the spring lobster recruitment problem: An update. *Proc. Annu. Gulf. Caribb. Fish. Inst.* **33:**230–243 (1981).

76. S. B. Saila and J. M. Flowers, A simulation study of sex ratios and regulation effects with the American lobster, *Homarus americanus. Proc. Annu. Gulf Caribb. Fish. Inst.* **18:**66–78 (1965).

77. A. S. D. Farmer, Relative growth in *Nephrops norveqicus* (L) (Decapoda: Nephropidae). *J. Nat. Hist.* **8:**605–620 (1974).

78. D. A. Fournier. T. D. Beacham, B. E. Riddell, and C. A. Basack, Estimating stock composition in mixed stock fisheries using morphometric, meristic and electrophoretic characteristics. *Can. J. Fish. Aquat. Sci.* **41:**400–408 (1984).

79. A. Campbell and R. H. Mohn, Definition of American lobster stocks for the Canadian Maritimes by analysis of fishery-landing trends. *Trans. Am. Fish. Soc.* **112:**744–759 (1983).

80. R. J. H. Beverton and S. J. Holt, A review of methods for estimating mortality rates in exploited fish populations with special reference to sources of bias in catch sampling. *Rapp. P.-V. Run., Cons. Int. Explor. Mer* **140:**67–83 (1956).

81. J. M. Hoenig, W. D. Lawing, and N. A. Hoenig, Using mean age, mean length and median length data to estimate the total mortality rate. *ICES, C.M.* 1983/**D:23:**1–11 (1983).

82. D. A. Hancock, Yield assessment in the Norfolk fishery for crabs (*Cancer pagurus*). *Rapp. P.-V. Reun., Cons. Int. Explor. Mer* **156:**81–94 (1965).

83. J. H. Annala, Mortality estimates for the New Zealand rock Lobster, *Jasus edwardsi. Fish. Bull.* **77:**471–480 (1979).

84. G. Y. Conan, A growth model of *Nephrops norvegicus* from Biscay Bay in function of periodicity of molt. *ICES,* C.M. 1975/**K:**10, 21 p (mines) (1975).

85. J. F. Caddy, Size frequency analysis for crustacea: Moult increment and frequency models for stock assessment. Kuwait Shrimp Workshop, KISR, Manama, Kuwait, November 1985.

86. R. Jones and N. S. Van Zalinge, Estimates of mortality rate and population size for shrimp in Kuwait waters. *Kuwait Bull. Mar. Sci.* **2:**273–288 (1981).

87. J. F. Caddy, Stock assessment in data-limited situations—The experience in tropical fisheries and its possible relevance to evaluation of invertebrate resources. *Can. J. Fish. Aquat. Sci., Spec. Publ.* **92:**379–392 (1986).

88. R. J. H. Beverton and S. J. Holt, Tables of yield functions for fishery assessment. *FAO Fish. Tech. Pap.* 38 (Rev. 1): 1–49 (1966).

89. J. L. Munro, The biology, ecology, and bionomics of Caribbean reef fishes. VI. Crustaceans (Spiny lobsters and crabs). *Res. Rep. Zool. Dep. Univ. West Indies* 3 (pt. V) (1):1–57 (1974).

90. J. F. Wickins, Prawn biology and culture. *Oceanogr. Mar. Biol.* **14:**435–507 (1976).

91. D. H. Cushing, The natural mortality of the plaice. *J. Cons., Cons. Int. Explor. Mer* **36(2):**150–157 (1975).

92. A. E. Howard, Substrate controls on the size composition of lobster (*Homarus gammarus* L.) populations. *J. Cons. Cons. Int. Explor. Mer* **39:**130–133 (1980).

93. J. T. Addison, Density-dependent mortality and the relationship between size composition and fishing effort in lobster populations. *Can. J. Fish. Aquat. Sci.* **43(11):**2360–2367 (1986).

94. R. A. Richards and J. S. Cobb, Relative body size and competition for shelter between lobsters (*Homarus americanus*) and Jonah crabs (*Cancer borealis*). *Can. J. Fish. Aquat. Sci.* **43:**2250–2255 (1986).

95. B. Mandelbrot, *Fractals: Form, Chance and Dimension.* Freeman, San Francisco, CA, 1977.

96. D. R. Morse, J. A. Lawson, M. M. Dodson, and M. H. Williamson, Fractal dimension of vegetation and the distribution of arthropod body lengths. *Nature (London)* **314:**731–733 (1985).

97. D. M. Mark, Fractal dimension of a coral reef at ecological scales: A discussion. *Mar. Ecol.: Prog. Ser.* **14:**293–294 (1984).

98. G. Y. Conan, Summary of session 5: Recruitment enhancement (International workshop on lobster recruitment). *Can. J. Fish. Aquat. Sci.* **43:**2384–2390 (1986).

99. N. J. Abramson and P. K. Tomlinson, An application of yield models to a California ocean shrimp population. *Fish. Bull.* **70(3):**1021–1040 (1972).

100. R. Jones, An analysis of a *Nephrops* stock using length composition data. *Rapp. P.-V. Reun., Cons. Int. Explor. Mer.* **175:**259–269 (1979).

101. A. Schumacher and K. Tiews, On the population dynamics of the brown shrimp (*Crangon crangon* L.) off the German coast. *Rapp. P.-V. Reun., Cons. Int. Explor. Mer.* **175:**280–286 (1979).

102. D. L. Alverson and W. T. Pereyra, Demersal fish explorations in the Northeastern Pacific Ocean—An evaluation of exploratory fishing methods and analytical approaches to stock size and yeild forecasts. *J. Fish. Res. Board Can.* **26:**1985–2001 (1969).

103. FAO/UNDP, Assessment of the shrimp stock of the west coast of the Gulf between Iran and the Arabian Peninsula. *UNDP/FAO Programme for Fishery Development in the Gulf,* FI:RAB/80/015. FAO/UNDP, Rome, 1982, 163 pp.

104. D. S. Vaughan and S. B. Saila, A method for determining mortality rates using the Leslie Matrix. *Trans. Am. Fish. Soc.* **105(3):**380–383 (1976).

105. D. A. Hancock, The relationship between stock and recruitment in exploited invertebrates. *Rapp. P.-V. Reun., Cons. Int. Explor. Mar* **164:**113–131 (1973).

106. M. P. Sissenwine, Why do fish populations vary? In R. May, Ed., *Exploitation of Marine Communities.* Springer-Verlag, Berlin, 1984, pp. 59–109.

107. M. J. Fogarty and J. S. Idione, Recruitment dynamics in an American lobster (*Homarus americanus*) population. *Can. J. Fish Aquat. Sci.* (1986) Vol. **43(11):**2368–2376.

108. S. Garcia, The stock–recruitment relationship in shrimps: Reality or artefacts and misinterpretations? *Oceanogr. Trop.* **18:**25–48 (1983).

109. W. E. Ricker, Stock and recruitment. *J. Fish. Res. Board Can.* **11:**559–623 (1954).

110. R. J. H. Beverton and S. J. Holt, On the dynamics of exploited fish populations. *Fish. Invest. (London) (Ser. 2)* **19:**1–533 (1957).

111. G. J. Paulik, Studies of the possible form of the stock–recruitment curve. *Rapp. P.-V. Reun., Cons. Int. Explor. Mer* **164:**302–315 (1973).

112. J. G. Shepherd, A versatile new stock–recruitment relationship for fisheries, and the construction of sustainable yield curves. *J. Cons., Cons. Int. Explor. Mer* **40:**67–75 (1982).

113. A. Tang, Modification of the Ricker stock recruitment model to account for environmentally induced variation in recruitment with particular reference to the blue crab fishery in Chesapeake Bay. *Fish. Res.* **3:**13–21 (1985).

114. R. C. A. Bannister and J. T. Addison, Effect of assumptions about the stock–recruit-

ment relationship on a lobster (*Homarus gammarus*) stock assessment. *Can. J. Fish. Aquat. Sci.* **43(11)**:2353–2359 (1986).

115. D. J. Staples, Modelling the recruitment processes of the banana prawn, *Penaeus merquinsis*, in the southeastern Gulf of Carpentaria, Australia. In P. C. Rothlisberg, B. J. Hill, and D. J. Staples, Eds., *Proceedings of the Second Australian National Prawn Seminar*. Simpson, Halligan, Brisbane, Australia, 1985, pp. 175–184.

116. J. W. Penn and N. Caputi, Spawning stock–recruitment relationships and environmental influences on the tiger prawn (*Penaeus esculentus*) fishery in Exmouth Gulf, Western Australia. *Aust. J. Mar. Freshwater Res.* **37**:491–505 (1986).

117. M. J. Fogarty, Distribution and relative abundance of American lobster, *Homarus americanus*, larvae: A review. *NOAA Tech. Rep. NMFS SSRF*-**775**:3–8 (1983).

118. V. C. Anthony and J. F. Caddy, Proceedings of the Canada–US workshop on status of assessment science for N.W. Atlantic lobster (*Homarus americanus*) stocks. St. Andrews, N.B., Oct. 24–26, 1978). *Can. Fish. Aquat. Sci., Tech. Rep.* **932**:1–186 (1980).

119. J. F. Caddy, The influence of variations in the seasonal temperature regime on survival of larval stages of the american lobster (*Homarus americanus*) in the Southern Gulf of St. Lawrence. *Rapp. P.-V. Reun., Cons. Int. Explor. Mer* **175**:204–216 (1979).

120. G. D. Marchessault, S. B. Saila, and W. J. Palm, Delayed recruitment models and their application to the American Lobster (*Homarus americanus*) fishery. *J. Fish. Res. Board. Can.* **33**:1779–1789 (1976).

121. B. F. Phillips, Prediction of commercial catches of the western rock lobster, *Panulirus cygnus*. *Can. J. Fish. Aquat. Sci.* **43**:2126–2130 (1986).

122. N. Caputi and R. S. Brown, Relationship between indices of juvenile abundance and recruitment in the Western Rock Lobster (*Panulirus argus*) fishery. *Can. J. Fish. Aquat. Sci.* **43**:2131–2139 (1986).

123. D. E. Pollock, Review of the fishery for and biology of the Cape lobster, *Jasus lalandii* with notes on larval recruitment. *Can. J. Fish. Aquat. Sci.* **43**:2107–2117 (1986).

124. D. G. Hankin, Proposed explanations for fluctuations in abundance of Dungeness crabs: A review and critique. *Proc. Symp. Dungeness Crab Biol. Manage. 1985* (1985), pp. 305–326.

125. M. S. Love and W. V. Westphal, A correlation between annual catches of Dungeness crab, *Cancer magister*, along the west coast of North America and mean annual sunspot number. *Fish. Bull.* **79**:794–795 (1981).

126. L. W. Botsford, R. D. Methot, Jr., and W. E. Johnston, Effort dynamics of the Northern California Dungeness Crab (*Cancer magister*) Fishery. *Can. J. Fish. Aquat. Sci.* **40**:337–346 (1983).

127. R. D. Methot and L. W. Botsford, Estimated pre-season abundance in the California Dungeness crab (*Cancer magister*) fishery. *Can. J. Fish. Aquat. Sci.* **39**:1077–1083 (1982).

128. R. L. Dow, Cyclic and geographic trends in sea water temperature and abundance of American lobster. *Science* **164**:1064–1067 (1969).

129. J. M. Flowers and S. B. Saila, An analysis of temperature effects on the inshore lobster fishery. *J. Fish. Res. Board Can.* **29**:1221–1225 (1972).

130. S. B. Saila and G. Marchessault, Population dynamics of clawed lobsters. In J. S.

Cobb and B. F. Phillips, Eds., *The Biology and Management of Lobsters*, Vol. II. Academic Press, New York, 1980, Chap. 6, pp. 219–241.

131. G. E. P. Box and G. M. Jenkins, *Time Series Analyses: Forecasting and Control*, rev. ed. Holden-Day, San Francisco, CA, 1976, 575 pp.

132. D. J. Staples, W. Dall, and D. J. Vance, Catch prediction of the banana prawn *Penaeus merguiensis*, in the south-eastern Gulf of Carpentaria. In J. A. Gulland and B. J. Rothschild, Eds., *Penaeid Shrimps—Their Biology and Management*. Fishing News Books, Farnham, Surrey, England, 1984, pp. 259–267.

133. R. C. May, Larval mortality in marine fishes and the critical period concept. In J. H. S. Blaxter, Ed., *The Early Life History of Fish*. Springer-Verlag, New York, 1974, pp. 3–19.

134. S. D. Sulkin and W. van Henklen, Larval recruitment in the crab *Callinectes sapidus* Rathbun: An amendment to the concept of larval retention in estuaries. In V. Kennedy, Ed., *Estuarine Comparisons*. Academic Press, New York, 1982, pp. 459–475.

135. M. Graham, Modern theory of exploiting a fishery and its application to North Sea Trawling. *J. Cons.; Cons. Int. Explor. Mer* **10**:264–274 (1935).

136. R. K. Mohn, Bias and error propogation in logistic production models. *Can. J. Fish. Aquat. Sci.* **37**:1276–1283 (1980).

137. D. A. Hancock, Population dynamics and management of shellfish stocks. *Rapp. P.-V. Reun., Cons. Int. Explor. Mer* **175**:8–19 (1979).

138. M. B. Schaefer, Some aspects of the dynamics of populations important to the management of the commercial marine fisheries. *Bull. Int. Am. Trop. Tuna Comm.* **1**:27–56 (1954).

139. W. W. Fox, Jr., An exponential surplus yield model for optimising exploited fish populations. *Trans. Am. Fish. Soc.* **99(1)**:80–88 (1970).

140. J. J. Pella and P. K. Tomlinson, A generalized stock production model. *Bull. Int. Am., Trop.-Tuna Comm.* **13**:419–496 (1969).

141. J. G. Shepherd, A family of general production curves for exploited populations. *Math. Biosci.* **59**:77–93 (1982).

142. S. L. Brunenmeister, Standardization of fishing effort and production models for brown, white and pink shrimp stocks fished in U.S. Waters of the Gulf of Mexico. In J. A. Gulland and B. J. Rothschild, Eds., *Penaeid Shrimps—Their Biology and Managment*. Fishing News Books, Surrey, England, 1984, pp. 187–210.

143. M. J. Fogarty and S. W. Murawski, Population dynamics and assessment of exploited invertebrate stocks. *Can. J. Fish. Aquat. Sci., Spec. Publ.* **92**:228–244 (1986).

144. J. A. Gulland, Fishing and the stocks of fish off Iceland. *Fish. Invest.—Minist. Agric., Fish. Food. (G. B.) (Ser. 2)* **23(4)**:1–52 (1961).

145. W. W. Fox, Fitting the generalised stock production model by least-squares and equilibrium approximation. *Fish Bull.* **73(1)**:23–36 (1975).

146. R. B. Deriso, Harvesting strategies and parameter estimation for an age structured model. *Can. J. Fish. Aquat. Sci.* **37**:268–282 (1980).

147. J. Schnute, A general theory for analysis of catch and effort data. *Can. J. Fish. Aquat. Sci.* **42**:414–429 (1985).

148. D. K. Kimura, et al. Generalized stock reduction analysis. *Can. J. Fish. Aquat. Sci.* **41**:1325–1333 (1984).

149. D. W. McLeese and D. G. Wilder, The activity and catchability of the lobster

(*Homarus americanus*) in relation to temperature. *J. Fish. Res. Board Can.* **15:**1345–1354 (1958).

150. G. R. Morgan, Assessment of the stocks of the Western Rock Lobster *Panulirus cyanus* using surplus production models. *Aust. J. Mar. Freshwater Res.* **30:**355–363 (1979).

151. J. W. Penn, The behavior and catchability of some commercially exploited penaeids and their relationship to stock and recruitment. In J. A. Gulland and B. J. Rothschild, Eds., *Penaeid Shrimps—Their Biology and Management* Fishing News Books, Farnham, Surrey, England, 1984, pp. 173–186.

152. G. R. Morgan, Trap response and the measurement of effort in the fisheries for the Western Rock Lobster. *Rapp. P.-V. Reun., Cons. Int. Explor. Mer* **175:**197–203 (1979).

153. J. A. Boutillier, Fishing effort standardisation in the British Colombia prawn (*Pandalus platyceros*) trap fishery. *Can. J. Fish. Aquat. Sci.* **92:**176–181 (1986).

154. J. Csirke and J. F. Caddy, Production modelling using mortality estimates. *Can. J. Fish. Aquat. Sci.* **40:**43–51 (1983).

155. W. F. Thompson and F. H. Bell, Biological statistics of the Pacific halibut fishery. 2. Effect of changes in intensity upon total yield and yield per unit of gear. *Rep. Int. Pac. Halibut Comm.* **8:**1–49 (1934).

156. B. J. Rothschild and S. L. Brunenmeister, The dynamics and management of shrimp in the Northern Gulf of Mexico. In J. A. Gulland and B. J. Rothschild, Eds., *Penaeid Shrimps—Their Biology and Management.* Fishing News Books, Farnham, Surrey, England, 1984, pp. 145–172.

157. G. P. Ennis, An assessment of the impact of size limit and exploitation rate changes in a Newfoundland lobster population. *North Am. J. Fish. Manage.* **5:**86–90 (1985).

158. J. F. Caddy, 1985. Displaying multispecies information in relation to fishing intensity: A sampling and production modelling approach. *FAO Fish. Rep.* 327(Suppl.): 286–290 (1985) (Doc. FIP/R327 Suppl.).

159. J. F. Caddy, Management of shrimp fisheries. In *Fish By-Catch; Bonus from the Sea*, Report of a technical consultation on shrimp by-catch utilization held in Georgetown, Guyana, 27–30 Oct 1981. IDRC, Ottawa, Ontario, Canada, 1981, pp. 120–123.

160. J. F. Caddy and G. D. Sharp, An ecological framework for marine fishery investigations. *FAO Fish. Tech. Pap.* **000:**1–151 (1987).

161. R. T. Paine, Food webs: Linkage, interaction strength and community infrastructure. *J. Anim. Ecol.* **49:**667–685 (1980).

162. A. C. Hardy, The herring in relation to its animate environment. I. The food and feeding habits of the herring with special reference to the east coast of England. *Fish. Invest. (London) (Ser. 2)* **7(3)** (1924).

163. J. H. Steele, *The Structure of Marine Ecosystems.* Harvard Univ. Press, Cambridge, MA, 1974, 128 pp.

164. J. S. Cobb, D. Wang, R. A. Richards, and M. J. Fogarty, Competition among crabs and lobsters and its possible effects in Narragansett Bay, R.I. *Can. J. Fish. Aquat. Sci., Spec. Publ.* **92:**282–290 (1986).

165. R. Levins, Evolution in communities near equilibrium. In M. L. Cody and M. Diamond, Eds., *Ecology and Evolution of Communities* 16–50. Harvard Univ. Press, Cambridge, MA, 1975.

166. S. B. Saila and J. D. Parrish, Exploitation effects upon interspecific relationships in marine ecosystems. *Fish. Bull.* **70(2):**383–393 (1972).

167. H. Li and P. B. Moyle, Ecological analysis of species introductions into aquatic ecosystems. *Trans. Am. Fish. Soc.* **110:**772–782 (1981).

168. R. W. Flint and N. N. Rabelais, Gulf of Mexico shrimp production: A food web hypothesis. *Fish. Bull.* **79:**737–748 (1981).

169. P. F. Sheridan, J. A. Browder, and J. E. Powers, Interaction with other species, ecological interactions between penaeid shrimp and bottom fish assemblages. In J. A. Gulland and B. J. Rothschild, Eds., *Penaeid Shrimps—Their Biology and Management.* Fishing News Books Farnham, Surrey, England, 1984, pp. 235–252.

170. E. L. Mills and R. O. Fournier, Fish production and marine ecosystems of the Scotian Shelf, eastern Canada. *Mar. Biol. (Berlin)* **54:**101–108 (1979).

171. J. J. Polovina, Model of a coral reef ecosystem. Part 1. ECOPATH and its application to French Frigate Shoals. *Coral Reefs* **3:**1–11 (1984).

172. J. J. Polovina and M. D. Ow, An approach to estimating an ecosystem box model. *Fish Bull.* **83(3):**457–460 (1985).

173. S. W. Nixon, Nutrient dynamics primary production and fisheries yields of lagoons. *Oceanol. Acta* **5:**356–371 (1982).

174. D. E. Pollack, Predator–prey relationships between the rock lobster *Jasus lalandii* and the mussel *Aulacomya ater* at Robben Island on the Cape west coast of Africa. *Mar. Biol. (Berlin)* **52:**347–356 (1979).

175. D. E. Pollack, G. L. Griffiths, and L. J. Seiderer, Predation of rock lobsters on mussels. *S. Afr. J. Sci.* **75:**562 (1979).

176. C. L. Griffiths and L. J. Siderer, Rock lobsters and mussels—limitations and preferences in a predator–prey interaction. *J. Exp. Mar. Biol. Ecol.* **41:**95–109 (1980).

177. L. J. Siderer, B. D. Hahn, and L. Lawrence, Rock-lobsters, mussels and man? A mathematical model. *Ecol. Modell.* **17:**225–241 (1982).

178. M. Kirkpatrick, Demographic models based on size, not age, for organisms with indeterminate growth. *Ecology* **65:**1874–1884 (1984).

179. G. W. Hyatt, Qualitative and quantitative dimensions of crustacean aggression. In S. Reback and D. W. Dunham, Eds., *Studies in Adaptation.* Wiley, New York, 1983, Chap. 5, pp. 113–139.

180. L. W. Botsford and D. E. Wickham, Behavior of age-specific, density dependent models and the northern California Dungeness crab (*Cancer magister*) fishery. *J. Fish. Res. Board Can.* **35:**833–843 (1978).

181. A. E. Howard and D. B. Bennett, The substrate preference and burrowing behavior of juvenile lobsters (*Homarus gammarus*) (L.). *J. Nat. Hist.* **13:**433–438 (1979).

182. K. M. Leber, The influence of predatory decapods, refuge, and microhabitat selection on seagrass communities. *Ecology* **66:**1951–1964 (1985).

183. G. P. Ennis and S. A. Akenhead, A model and computer program used to assess yield per recruit in Newfoundland lobster stocks. *Can. Atl. Fish. Sci. Advis. Comm. (CAFSAC), Res. Doc.* **78/30:**1–13 (1978).

184. P. Drach, Mue et cycle d'intermue chez les crustaces decapodes. *Ann. Inst. Oceanogr. (Paris)* **19:**103–391 (1939).

185. S. E. Hatfield, Intermolt staging and distribution of Dungeness crab, *Cancer magister,* megalopae. *Calif. Dep. Fish Game, Fish. Bull.* **172:**85–96 (1983).

186. P. A. Breen and K. H. Mann, Changing lobster abundance and the destruction of kelp beds by sea urchins. *Mar. Biol. (Berlin)* **34:**137–142 (1976).

187. K. H. Mann, Invertebrate behaviour and the structure of marine benthic communities. In R. M. Sible, and R. H. Smith Eds., *Behavioural Ecology*, Blackwell, London, 1985, pp. 227–246.

188. R. J. Miller, Seaweeds, sea urchins, and lobsters: A reappraisal. *Can. J. Fish. Aquat. Sci.* **42:**2061–2072 (1985).

189. R. L. Vadas, R. W. Elner, P. E. Garwood, and I. G. Babb, Experimental evaluation of aggregation behavior in the sea urchin *Strongylocentrotus droebachiensis* a reinterpretation. *Mar. Biol. (Berlin)* **90:**433–488 (1986).

190. R. J. Miller, Sea urchin pathogen: A possible tool for biological control. *Mar. Ecol.: Prog. Ser.* **21:**169–174 (1985).

191. P. A. Larkin, Fisheries management—An essay for ecologists. *Annu. Rev. Ecol. Syst.* **9:**57–73 (1978).

192. W. T. Momot and H. Gowing, Results of an experimental fishery on the crayfish, *Orconectes virilis*. *J. Fish. Res. Board Can.* **34:**2056–2066 (1977).

193. W. T. Momot, Production and exploitation of the crayfish, *Orconectes virilis*, in northern climates, *Can. J. Fish. Aquat. Spec. Sci., Publ.* **92:**154–167 (1986).

194. J. F. Caddy, An alternative to equilibrium theory for management of fisheries. *FAO World Conf. Fish. Manage. Dev.* Prep. 1, Sess. 2/Panel 2.2:1–44 (1983).

16 EFFORT LIMITATION IN THE AUSTRALIAN ROCK LOBSTER FISHERIES

B. K. Bowen
Fisheries Department
Perth, Western Australia, Australia

and

D. A. Hancock
Marine Research Laboratories, Waterman
Perth, Western Australia, Australia

1. INTRODUCTION

Many countries have now moved toward tight management controls of fisheries to limit either fishing effort or catch, or in some cases both. This has been the consequence of both the implementation of international access arrangements and the recognition that many of the major fished stocks are fully exploited, or already overexploited. In Australia the major stocks of rock lobsters are fully exploited and are subject to license limitation. All major Australian prawn stocks now appear to be fully or excessively exploited and managers face the problems of reducing the fishing effort (Bowen and Hancock 1). At different stages of development each has become subject to some form of limited-entry arrangements. Other invertebrate fisheries throughout Australia are similarly managed by some form of license limitation, including scampi, scallops, abalone, pearl oysters, and edible oysters, which also involves security of land tenure. In addition, pearl oyster growers in the northwest of Australia are restricted to quotas of shell from the wild stocks, and catches of squid by foreign operators in eastern Australian waters are subject to a total allowable catch (TAC).

However, when arrangements for limited-entry fisheries for rock lobsters and prawns were commenced in Western Australia in 1963 there was little experience to go on, and there were lessons to be learned and some mistakes to be made and corrected.

In particular, in the retrospect of more than 20 years of management of Western Australian (W.A.) rock lobster and prawn fisheries under a system of license limitation, it has become clear that even where the basic philosophies of managers and fishing industry representatives are in fundamental agreement, in practice there emerges a battle of wits between the legislators and the practical fishermen. History has shown that through the highly competitive nature and ingenuity of the fisherman, the spirit of management by conservation of stocks may be continuously undermined and, as one set of controls progressively becomes less effective, it has to be supplemented with additional measures in the same or a different direction. It has been seen how legistators must be specially sensitive to the potential for evasion and circumvention of controls—controls that are the price paid for limited entry and that are essential to prevent progressively, but not always successfully, the expansion of fishing effort to an undesirable level that may eventually well exceed the original constraints set for the fishery. Inevitably such controls are limiting to personal efficiency and therefore may be specially frustrating to individual fishermen, especially at times of low abundance. This had led to a change in the industry's approach to the decision-making process, and industry is now seeking more direct representation on industry–government management advisory committees to ensure that sector interests are taken into account. The majority of industry members usually do not welcome a reduction in fishing capacity.

However, in any lucrative fishery the alternative of management by regulation without controls on fishing effort or catch cannot be expected to prevent overexploitation in the long term, and the stipulation of an annual maximum catch or quota presents its own problems. These and the advantages and problems of limiting fishing effort are explored here with reference to the limited-entry fisheries for rock lobster in Australia, with major emphasis on the Western Australian experience which is most familiar to us.

2. THE WESTERN ROCK LOBSTER FISHERY

Bowen (2) listed the following conditions appropriate for license limitation.

(a) The product from the fisheries resource is a luxury item such that the price paid has a potential to rise at a higher rate than the rate of inflation.

(b) The return per unit of capital investment is higher than the return in other fisheries in the same area, or higher than that for other investment opportunities, even taking into consideration the risk factor involved.

(c) The fleet is operating over a unit stock of fish.

(d) The resource is highly vulnerable to the fishing gear being used.

(e) It is desirable to maintain high economic returns to fishermen in order to encourage exploitation of a new area.

He commented that the first four criteria were certainly operating in the Western Australian rock lobster fishery in 1963.

The history of development of the fishery for western rock lobster (*Panulirus cygnus*) in Western Australia is well represented in Figure 1. Although rock lobsters were caught for sale before the turn of the century (Saville-Kent 3), it was not until the 1940s that catches became significant. Sheard (4) reported that where 270,000 kg were taken by 42 men in 1944–1945, the catch had reached 8.7 million kg by 1959–1960. By 1977–1978 the catch exceeded 10 million kg, with a record

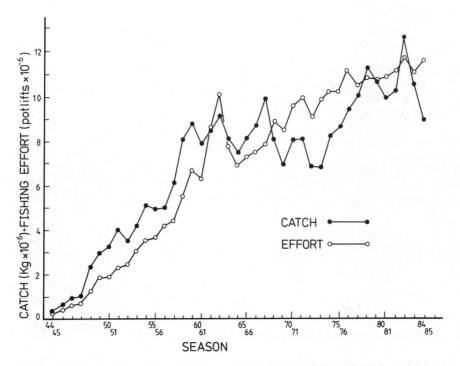

Figure 1. Annual catch and nominal (uncorrected) fishing effort in the W.A. rock lobster fishery 1944–1945 to 1984–1985.

12.4 million being taken in 1982–1983. These catches were largely paralleled by increases in fishing effort (Fig. 1). The fishery is now pursued to the limits of the quite discrete distribution of the western rock lobster, that is, from shallow water to the edge of the continental shelf between 21° and 34°S approximately.

The raw data presented in Figure 1 must be examined against the background of management controls introduced as the fishery developed. These have been listed in Table 1.

The situation that led to the introduction of limited entry in 1963 is described in Figure 2 (Bowen 5). During the early 1960s the W.A. government was faced with a common property resource that was being fished increasingly by professional and fractionally by amateur rock lobster fishermen (Hancock 6). The rapid expansion of a fishery with unrestricted access had led to levels of fishing effort (Figs. 1 and 2) that appeared to have exceeded that required for maximum sustainable yield (MSY) (Fig. 3) (Gulland 7) and therefore were well in excess of maximum economic yield (MEY) (Fig. 4) (Panayotou 8). Figure 2 shows clearly how in the three seasons up to 1962–1963 the nominal fishing effort (pots in the water) had doubled with no commensurate increase in catch.

At that time there were no limited-entry fisheries in Australia, nor was this management measure generally in use outside of Australia (Bowen 9). However,

TABLE 1   History of Important Regulatory Measures

1887	Minimum legal weight equivalent to 76-mm carapace length
	System of various seasonal closures commenced
1899	Taking off berried females prohibited
1962	Closed seasons: Coastal—Aug. 15 to Nov. 14 Incl.
	Abrolhos—Aug. 15 to March 14 Incl.
1963	Commencement of limited-entry fishery
	Number of boats limited
	Number of pots fixed at 3/ft of boat length
1965	No increase in pots per vessel
	No increase in vessel length
1966	Escape gaps (2 in.) introduced
1972	Escape gap increased to $2\frac{1}{8}''$ (54 mm)
1973	Multiple entrances to pots prohibited
	Spearguns etc. prohibited
1978	Season to end June 30, not Aug. 14
1979	Flexibility in size of replacement boats permitted
	Amendments to Fisheries Act
	Fish purchases to be recorded
	Fishery and logbook returns confidential
	Updates penalties for offences
1981	Pots to be 1 m or less in length or diameter
	Pots not to have side entrances
1983	Tighter regulations for amateur fishermen
1984	Further controls on dimensions of pots

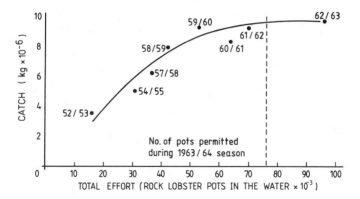

Figure 2. Line, fitted by eye, relating annual catches and total (uncorrected) fishing effort for 8 yr from the western rock lobster fishery (after Bowen 5).

fishermen were becoming concerned that the trend toward reduced catches per boat and per pot, which was implicit in the data in Figure 2, might eventually cause their operations to become unprofitable. In consequence, there were overtures by many fishermen to the government, through the then Fishermen's Advisory Committee, to introduce measures to prohibit the introduction of additional boats and to limit the number of pots per boat. The basic requirement for a regulation to be effective, especially when there is a heavy enforcement requirement, is the philosophical support of the fishing industry. Similar requests were being made to limit the number of prawn boat licenses in the W.A. Shark Bay fishery; in this case as a preventive measure against the development of excessive fishing effort and the accompanying risk to capital investment, in a newly developing fishery.

In the rock lobster fishery, approximately 4 months notice was given that those vessels engaged in the rock lobster industry as of March 1, 1963 would constitute the rock lobster fleet, and to the present date no further licenses have been granted; though the number of boats, for various reasons, has fallen from 815 to 764 in 1985. It should be noted that the 1963 procedure did not involve some of the problems that are a feature of present-day fisheries being considered for limited entry, where there is a need for a reduction in the number of boats permitted to fish because of perceived excessive fishing effort. In 1963, to reduce the effective fishing effort, the measure taken was to control the number of catching units, that is, the rock lobster pots (Bowen 9). Detailed submissions were received from the fishing industry, with the result that the number of pots was fixed at three per foot of boat length, with a maximum of 200. This resulted in a reduction in the total number of pots from approximately 97,000 in the 1962–1963 season to 76,000 during 1963–1964; a number that has remained virtually unchanged to this day. The effect of this can be seen in Figure 1, which shows the reduction in both catch and fishing effort after 1962–1963; and later, in Figure 5, which shows the same reduction in fishing effort for a relatively small fall in catch.

From this successful beginning, the western rock lobster fishery has developed

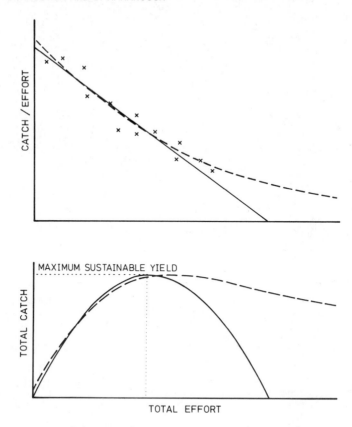

Figure 3. Possible relationships between effort and catch per unit of effort and corresponding relationships between effort and catch (after Gulland 7).

into a highly productive and well maintained fishery, but this has not been achieved without considerable administrative and inspection costs. Despite arrangements being agreed upon by government and leaders of the fishing industry, these have resulted from the personal attitudes and competitive spirit of individual fishermen that have repeatedly capitalized on any loophole for increasing the amount of fishing effort within the limited-entry framework.

Moreover, it must not be assumed that the granting of a privileged and exclusive license will imply strict adherence to the rules. In fact, the reverse seems to have been the case in the early years of limited entry with flagrant overpotting and retention of legally undersized animals by some operators, which needed to be controlled by an elaborate and expensive enforcement system. Such infringements are perhaps surprising in view of the possibility of losing a license that attracts a resale value in the hundreds of thousands of dollars. Similarly, flouting the regulations by licensed prawn boats, for example illegally fishing in the nursery areas, is more likely to reflect the attitudes of some hired crew rather than the owners, processors, and leaders of the fishing industry. Although in no way condoning

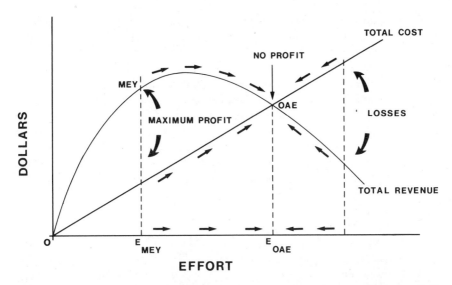

Figure 4. Relationship between total revenue and total cost showing the long-term equilibrium (OAE) that would result from an open access fishery (adapted from Panayotou 8).

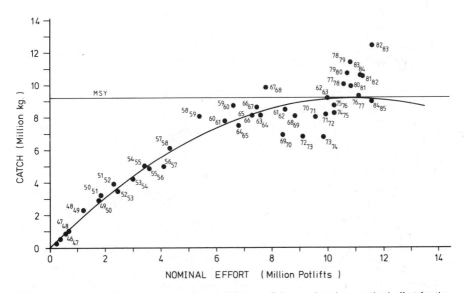

Figure 5. Annual catches in the western rock lobster fishery related to nominal effort for the period from 1944–1945 to 1984–1985. The Schaefer curve was calculated from a linear regression of observed catch per pot lift (1946–1947 to 1984–1985) against the weighted average of nominal effort in the corresponding season and two preceding seasons, with weighting factors of 3, 2, and 1, respectively.

such departures from the established rules, we must recognize that it is within the catching sector that the competitive spirit is likely to be manifest and difficult to keep within bounds when it becomes necessary to contain fishing effort.

Figure 5 shows how despite the firm intent of measures introduced in 1962–1963 to stabilize and control fishing effort and therefore rate of exploitation, the level of fishing effort has risen steadily, with current levels now some 70% more than those in 1964. Although the annual catch increased to a record 12.4 million kg in 1982–1983 as a consequence of a series of above-average years of recruitment, believed environmentally controlled, recent catches have fallen drastically to resemble levels characteristic of the early 1960s. The high catch levels led to a feeling of optimism in the fishery, which tended to overshadow the fact that they were at the cost of greater fishing intensity, higher operating costs, and increased exploitation rates.

The yield curve (Schaefer 10) fitted to the data in Figure 5, though it has no predictive value, provides a useful description of events in the fishery. It is a representation that may not be well received by fishermen because its inevitable descending right limb may be interpreted as deliberately pessimistic for management purposes. Nevertheless it is a useful research and management tool, which allows the relationship between catch and fishing effort to be viewed in its proper perspective. Clearly the relativity between data points on the graph will determine its shape, which will have a direct bearing on the interpretation of health of the stocks. Any increase in catch rates resulting from the discovery of as yet unfished grounds, and in effective fishing effort due to greater crew efficiency, improved fish finding (color sounders, sonar), faster, larger boats, and so on must be recognized, but can seldom be quantified in making adjustments to the effort scale on the yield curve.

In addition, validation and where possible, adjustment of values of catch and of fishing effort in the western rock lobster fishery proved to be necessary following claims of inaccuracies resulting from illegal and unrecorded landings of undersized lobsters, changes in the type and quantity of bait, and use of pots in excess of the number authorized (overpotting) earlier in the history of the fishery. However, from the results of a questionnaire requesting personal and confidential information on these matters, it was concluded that the magnitude of such efforts would not substantially affect the conclusions.

The level of fishing effort, which in itself requires careful surveillance, must be considered in conjunction with the rate of exploitation. This is a concept that is less easily understood by the fishing industry, and Figures 6 and 7 are given here in order to underline the principles involved. It can be seen with the current annual exploitation rate (E) of around 60%, and after adjustment for losses from natural causes (M or natural mortality), there remains only a small proportion of the stock available for exploitation after its first, recruited year in the fishery. Because exploitation rate (E) is equal mathematically to

$$[F/(F + M)] \, (1 - e^{-Zt})$$

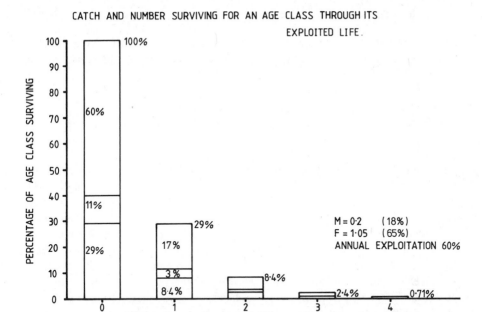

Figure 6. Relationship between percentage losses due to natural causes (*M*) and fishing (*F*) with an exploitation rate of 60%.

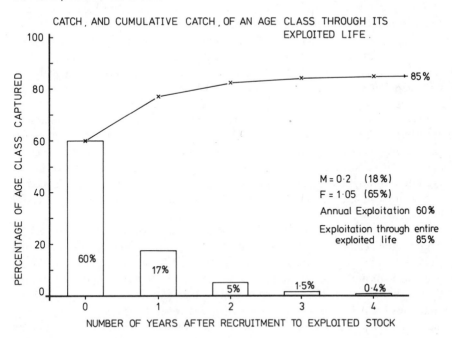

Figure 7. Relationship between annual exploitation rate (60%) and cumulative fishing of a year class through life (85%) using data from Figure 6.

(Ricker 11) [where F, M, and Z are instantaneous coefficients of fishing, natural, and total ($Z = F + M$) mortality, respectively, during time t], the implication is that of all those that die (i.e., 100% during the lifetime of a year class and when $E = F/Z$), more than 80% are taken by fishing, a very high figure that does not leave much room for management maneuvers or natural disaster.

The growth of catching capacity of the fleet might have been prevented by a stricter vessel replacement policy or a total quota on catching, but these have their own built-in disadvantages, to be discussed below.

Bowen (9) commented that because in 1963 the Western Australian government was feeling its way in the administrative realities of a limited-entry policy, it was not surprising that additional rules had to be introduced to complement the initial decision. Because there was no boat replacement policy, a number of fishermen with small vessels made plans to build larger vessels and use the additional pots allowed under the three pots per foot of boat length rule. In consequence, from May 1965, no additional pots were permitted even though a new vessel could be longer. This policy, however, while allowing bigger boats to be built, did not permit any greater catching power to compensate for the increased capital investment, with the result that some fishermen used more pots than their entitled number in order to meet vessel repayments. Bowen (9) concluded that this was one of the causes of considerable overpotting in the industry. In December 1965, an amended boat replacement policy limited the size of a new boat to that of the boat replaced, with the exception that for reasons of boat safety at sea, a boat of less than 7.6 m could be replaced with one up to 7.6 m.

It was not until 1979 that some flexibility was allowed in the size of replacement boats, according to a formula that established a minimum and maximum overall length based on the pot entitlement of the boat to be replaced. Although this permitted some increases in fishing effort where larger boats were more efficient in moving pots from place to place, the arrangement did not allow for any increase in the number of pots in the industry. However, for the first time fishermen now had some flexibility in their business decisions, because they could now, but within strict limits, change the size of their boats and their pot entitlement by sale or purchase of pot entitlements. This would prove important to the effective implementation of any later consideration given to comprehensive measures for effort reduction, such as percentage reductions in individual pot entitlements or a vessel buyback scheme.

However, prior to 1979 other avenues for increasing fishing effort were beginning to give cause for concern. The contribution of improved efficiency and technology (color echo sounders, sonar, etc.) had been noted but could not be quantified nor impeded. In addition, there was a marked trend toward greater effectiveness of pots as catching units by increasing their size and introducing additional entrances. In 1973, following research that confirmed their enhanced efficiency, the number of entrances was regulated to one, followed in 1981 by a prohibition on side entrances. Also in 1981 the diameter or length of a pot was limited to 1 m on the understanding that oversized pots caught significantly better, particularly at times of high abundance, for example, early in the season; and that this measurement would not exclude any pots of a size traditionally used in the fishery. This measure, while preventing a massive swing toward excessively large pots, did generate, perhaps surprisingly considering supposed loyalties to tradition, an element of increase in average pot size, as replacements were made up to the permitted 1 m. Moreover, fishermen were quick to capitalize on a prescribed single length measurement that did not restrict the area of oblong-based (batten) pots or the height of either beehive or batten pots, and in 1984 it became necesary to regulate the volume of each pot.

The experience with shortening the season by 6 weeks from 1978, against a background of escalating effort, showed that it is not possible to predict precisely the consequence of effort controls. It had been anticipated from the 1976–1977 season that the reduction in pot lifts would be 9%, but in the first year, owing to increased activity during the shortened season, a reduction of only 4.95% was achieved; a figure that declined until the original level had been resumed after 5 years. Although fishermen claimed that the measure had achieved nothing, the considered view was that it had in fact reduced the expected rate of increase in fishing effort.

The western rock lobster fishery is entering a crucial phase of its existence. Management measures so far have been based on a good data set, with the ability to predict, within limits, future stock levels 4 yr ahead (Hancock 6) from quantitative data on first settlement of 1-yr-old puerulus larvae on artificial collectors (Phillips and Hall 12; see also Chapter 7 by Phillips and Brown). These predictions have more recently been supported by forecasts of recruitment (at 5 yr of age) based on indexes of 3- and 4-yr-old juveniles in the monitored catch of representative commercial vessels (Caputi and Brown 13). However, unless and until the scientists can show beyond reasonable doubt that current high levels of fishing effort are not only prejudicial to the profitability of the fishery, but that they are reducing the spawning stock to levels that may result in lowered recruitment, neither administrators nor fishermen can be expected to be receptive to drastic controls that would be perceived to be a threat to their livelihood. Continued "nervousness" that because fishing effort has increased over many years without commensurate increase in catch is no match against pressures to maximize on investments and to service massive loans. Indeed, European Economic Community fisheries managers found that the concept of precautionary TACs for fisheries where it was deemed necessary to avoid the danger of excessive effort, was quite unacceptable to fishermen who moreover continued to subvert measures to control catches in fisheries even when the danger had been well established. Unquestionable biological evidence of overfishing can transform the administrator's role from one of giving priority to industry's requirements, to one of a mandatory responsibility to initiate the much stronger, and unpalatable, procedures required to restore and preserve continuity of stocks. Such was the case in Western Australia's Exmouth Gulf prawn fishery in which, following the novel and clear demonstration of recruit overfishing in a prawn fishery (Penn and Caputi 14), it proved necessary to limit both effort and total catch, accompanied by an acceptable and practical vessel buyback scheme (Bowen and Hancock 1; Penn, Hall and Caputi, Chapter

5 of this book). In the W.A. rock lobster fishery, although such a buyback scheme has always been the favored mechanism for reducing fishing effort, the costs involved have proved daunting in recent years to both the government and the fishing industry. This is due to the fact that the resale value of licenses is based on a pot transfer price currently around A$4000, so that to reduce the fleet of 76,000 pots by only 10% would cost in excess of A$30 million.

It is a common experience in fisheries management that a disaster must be perceived before there is a sympathetic response from industry, and not always then. Clearly a continuing, effective, and clear dialogue between managers, scientists, and fishermen will play a vital role in the acceptance and implementation of management philosophies and procedures. Such a mutual understanding is imperative when facing critical events in the history of a fishery; one such being the forecast of poor catches for the 1986–1987 season in the W.A. rock lobster fishery. The impact of this situation has however, been lessened by the conclusion that recruit levels to date have been governed by environmental rather than biological controls; that is, no apparent recruitment overfishing (see also Chapter 7 by Phillips and Brown). Moreover, any advance warning of symptoms of growth overfishing; that is, a reduction of average size of individuals in the catch, was for many years not apparent from catches graded for size by processors, though more recent observations (R. S. Brown and N. G. Hall, personal communication) suggest that such reassurance is no longer valid.

The case for reducing fishing effort in the western rock lobster fishery, although recognizing growing apprehension about the likely biological effects of the high levels of fishing effort, must therefore rest on the high rates of exploitation generated by insidiously increasing levels of both recorded and unquantifiable fishing effort, which have generated catches in excess of those perceived to constitute the MSY. It is a well-known fact that MSY, which is so often a target for fisheries management, whenever reached, inevitably causes problems. First, there is a potential for overshooting MSY and therefore greatly exceeding the MEY (Fig. 4), and second, there are inherent difficulties in pulling back from excessive fishing effort. In this case however, these were not sufficiently convincing arguments to lessen the industry's determination to maximize on catching opportunities.

The immediate concern, however, is about the consequences to future spawning stocks when poor conditions for settlement coincide with currently high levels of fishing effort. A proposed remedial reduction in the number of pots in the fishery has so far met with a response similar to that to shortening the season—marked resistance to its implementation, and promises to circumvent its effectiveness if and when it is introduced. It is fortunate that the western rock lobster is such a robust species and has survived the continuing onslaught of fishermen in a state of full exploitation for more than 20 yr.

Meanwhile, as well as documenting fishery-related events, research has been aimed at identifying and quantifying other elements that could contribute to error in yield models and, where appropriate, recommending measures for avoiding unnecessary losses from the stock or catch. For example, an amateur component of the catch was identified but was concluded to represent no more than 2% of the commercial catch (Norton 15). Octopuses are significant predators of rock lobster trapped in pots—to the extent of 1% of the value of the commercial catch in 1975, but to a much less significant extent under natural conditions (Joll 16). Fishermen, through careless handling, wasted 15% of the annual catch of undersized in 1978–1979 (Brown and Caput 17); a loss that has been reduced to some extent by intensive publicity, but that could be further reduced by increasing the size (from 54 to 55 mm) and number (from one to three) of escape gaps (Brown and Caputi 18).

3. DATA AND RESEARCH BASE—ESSENTIAL INGREDIENTS

Hancock (6) presented a full account of the sources of data and conduct of research programs, without which a proper understanding of this important fishery would not have been possible, and which provide the basis for management measures and documenting their consequences. Data sources include mandatory monthly returns of catches, together with the magnitude and area distribution of fishing effort. The latter, together with the record of seasonal distribution of catches, allows conversion from the nominal description of fishing effort or number of pots lifted (Fig. 5), to a measure of effective fishing effort that is essential for any year-to-year comparisons required for yield modeling (Phillips and Brown, Chapter 7). Fishermen's returns can be notoriously inaccurate, a fact that may be capitalized upon when research data are challenged, and it is therefore necessary to have means of validating industry-based information. This is provided not only by regular and careful scrutiny, but also by figures provided in Rock Lobster Processors' Monthly Returns, as well as from Research Log Sheets. The latter are completed by about one-quarter of rock lobster boats. The view has been held that compulsory log sheets would be subject to the same sort of inaccuracies leveled at monthly returns, but it is a source of disappointment that although the information in research log sheets is provided voluntarily, there is often a reluctance to respond or continue at times of actual or anticipated unwelcome management action. Such a scheme therefore requires regular supervision, dialogue, and feedback of information. Since 1971, factory measurements of commercially sized individuals have been superseded by regular monitoring of the activities of commercial vessels at sea to provide data on the undersized component of the catch and on spawning individuals; neither of which may be legally landed.

The wide-ranging and continuous research programs on the western rock lobster for more than 30 yr have been described by Hancock (6) and others, as well as in the preceding paragraphs. Perhaps the most significant contribution to meaningful management, short of establishing a scientific basis for a stock and recruitment relationship, has been the provision of at least a qualitative means of predicting future stock levels. At the outset this was of considerable advantage in defusing industry panic resulting from poor catches in the early 1970s, which could be shown to have resulted from poor larval settlement 4 yr previously, but which was followed by improved settlement in subsequent years. More recently, indexes of larval (puerulus) settlement (Phillips 19) have provided a 4-yr lead time in predicting catches, which has been supported latterly by predictions based on the representation of 3- and 4-yr-old (lead time 1 yr) juveniles in commercial catches monitored at sea by research staff (Caputi and Brown 13). This capability has been further enhanced by the credible explanation provided by oceanographers of recruit levels geared to the degree to which warm currents penetrate southward into rock lobster areas each year. In the meantime, work on the distribution and magnitude of annual settlement of puerulus larvae has been expanded, together with an intensive program to quantify the annual production of larvae from available spawning stock. From these, a better understanding should be obtained of the important stock and recruitment relationship.

4. THE SOUTHERN ROCK LOBSTER FISHERY

There are certain similarities, but some differences from the above account, in the management of the southern rock lobster (*Jasus novaehollandiae*), which is fished in South Australia, Tasmania, and Victoria, and to a smaller extent in New South Wales and Western Australia.

Although the western rock lobster fishery is managed by a simple bilateral arrangement between the state of Western Australia and the commonwealth government of Australia, the southern rock lobster fishery is divided into five zones: two (western and eastern) are managed by Victoria, two (northern and southern) by South Australia, and one by Tasmania, all in collaboration with the commonwealth government. The appropriateness of the management regimes is under constant review.

License limitation for the southern rock lobster was introduced into Tasmania (1967), South Australia (1967–1968), and Victoria (1976). The experience in South Australia to some extent paralleled that for the western rock lobster; the fishery commenced before the turn of the century, but did not become significant until after World War II. Spurred on by good prices and a buoyant export market, effort escalated rapidly and in the 18 yr to 1966–1967, multiplied by a factor of 29 (Copes 20), during which catch per pot lift declined from 10.4 to 0.9 kg. In 1967–1968 licensing restrictions prevented further entry to the fishery, and pots were allocated in relation to boat length. In 1968 there were 421 licensed boats using 22,740 pots (Copes 20). By 1980 there were 367 licensed rock lobster boats in South Australia (324 in Tasmania and 199 in Victoria). The reduction in South Australia was brought about by amalgamation of licenses resulting through the transfer of pot authorizations within the industry. Earlier attempts to contain and reduce fishing effort through extended seasonal closures did not prove successful. Experience showed that unless the extensions are drastic, effort reduction is not achieved, because a compensatory increase occurs during the open months by pulling pots more than once a day, or fishing on days that would not normally be fished. This is best illustrated in the South Australian southern zone rock lobster fishery, where before 1978, when five out of the winter months May to October were open to fishing, there were 600,000 pot lifts. In 1979, when only three of these months were open to fishing, the number of pot lifts was 622,000, and in 1980 with 2 months open to fishing there were 658,000 pot lifts. Similar compensatory fishing effort increases were described for Western Australia earlier in this chapter.

These effort reduction proposals were considered to be economically advantageous to the industry—the revenue would be spread among the fleet without commensurate increase in operating costs—and Hunt (21) explored ways of stimulating license amalgamation, together with the options of reducing the number of pots and a buyback scheme. The buyback scheme was rejected by the fishing industry, and in 1984 there was a 15% across-the-board pot number reduction in the southern zone fishery of South Australia and at the same time a new ceiling of 80 pots per boat. This and other effort reduction proposals were designed to promote the economic well-being of fishermen. Although the fishery is based on high density stocks, is fully exploited, and has remained relatively stable since the introduction of license limitation in 1968 (Lewis 22), Hunt (21) referred to the deteriorating economic state of the fishery, and concluded that the transfer value of pots increased by more than 60% as a result of the 15% reduction in pot numbers, a ceiling of 80 pots per boat, and changes in the international exchange rates that significantly increased the price received per kilogram of rock lobster and therefore the free market value of a pot. It is significant that the proposal for effort reduction in the South Australian rock lobster fishery came from the Professional Fisherman's Association. South Australia is now assessing the impact of effort reduction measures to date, and the industry and the Department of Fisheries are considering further options to greater improve the viability of the fishery.

Victoria went through a similar experience following a buildup of effort and deterioration of the economic state of its rock lobster fisheries during the 1970s (R. H. Winstanley, personal communication). Finally in 1982, after many years of consultation between government and industry, a package was introduced that removed the pot number/vessel length restrictions, enabling pots to be freely transferable between license holders in each management zone, in the the western zone, subject to pot forfeit on transfer; that is, loss of 1-in-5 on part transfer and 1-in-20 when a complete vessel pot authorization is transferred. Also in the western zone, there was a once-only reduction of pots by 5–15% based on original pot entitlements, which were related to vessel length. The immediate impact of this on the 106 licensed boats was to reduce the total pot fleet from 6508 to 5745. In 1984 the 1-in-5 forfeit was extended to the eastern zone, together with minimum pot entitlements for both zones and a maximum pot entitlement for the earlier zone. The effectiveness of pot reduction measures in Victoria has yet to be assessed.

Tasmania to date has not introduced across-the-board pot reductions, but has moved toward allowing pot entitlements to be split for redistribution but subject to 15% reduction in a boat's entitlement before sale, in order to offset the expected increase in overall efficiency.

The readiness with which the fishing industry based on the southern rock lobster accepted effort reduction appears to stem from the lower return on capital than from the western rock lobster fishery, and the fact that fishermen were experiencing a steady decline in net incomes to a level of real concern. Moreover, the level of dependence of rock lobster fishermen on the resource in Western Australia is higher compared with the southern rock lobster fishery, where there is more diversification into alternative or complementary fisheries.

A serious feature stemming from the aim of directly or indirectly reducing the number of vessels fishing for rock lobster, is the additional fishing pressure that will be placed on alternative fisheries that are mostly already fully exploited. In Western Australia, where an endorsement to catch "wet fish", that is, scalefish and shark, by line is attached to a rock lobster license, such an endorsement may not be separated from the rock lobster license. If a vessel leaves the rock lobster industry, all fishing licenses are removed.

5. CONCLUSIONS

A basic management requirement of a fully exploited fish resource is a mechanism for preventing the growth in fishing capacity of the fleet or in some cases reducing that capacity. An alternative to capacity control is a change from input controls to output controls involving some form of quotas.

Output controls using quotas are theoretically the most direct approach when there is a clear need to control the exploitation rate. However, there are three major problems involved: first, establishing an equitable method of dividing the allowable catch among the participants; second, setting the annual catch to take account of variable recruitment; and third, auditing the catch of each participant. Auditing the catch is a relatively simple matter when a small number of large animals is taken and the outlets are few, but it would become a major exercise in a rock lobster fishery involving 800 vessels operating along a coastline of several hundred miles, and having numerous market outlets. It is therefore not surprising that input controls have remained the favored management method for rock lobsters.

Fishing capacity will continue to increase gradually through a combination of advances in fish-finding technology, vessels, and fishing gear. History has also shown that any measures to curb the rate of exploitation, whether or not they gain philosophical acceptance by the fishing industry, will engender a reaction aimed at maintaining catches. Even where compensatory increases in fishing effort are initiated by only a minority of fishermen, this will inevitably undermine the impact of those measures.

This tendency for compensatory fishing will be compounded in a fishery where the transfer value of whole or part licenses has become inflated, and fishermen see additional fishing as the only way to service financial arrangements. Furthermore, it has been demonstrated that even where a downturn in catches can be predicted with reasonable certainty, it is hard to win acceptance for unpalatable emergency measures.

It is imperative, therefore, that a mechanism be established at the start of effort limitation which sets out a method of fishing capacity reduction to counterbalance the increases that inevitably occur. This requires a clear definition of the unit capacity to be used in the fishery.

In prawn fishing off northern Australia the unit of capacity combines a measure of the "under deck volume" and the engine horsepower. However, in the rock lobster fisheries around Australia the unit of fishing capacity is the rock lobster pot.

Two mechanisms for capacity reduction can be adopted. First, in rock lobster fisheries, the number of pots can be reduced across the board to reduce the capacity of the total fleet. Second, when a replacement vessel is introduced, a further pot reduction can be required to offset to some extent the increase generated by the greater efficiency of the new vessel. In the northern prawn fishery, which has established the unit of fishing capacity as measured by the dynamics of the vessel rather than by gear controls, the reduction of fishing capacity can take place only when a vessel is replaced. In both the rock lobster and prawn fisheries, in order to maintain the existing capacity, the operator will therefore need to purchase additional units as available within the unit ceiling currently in force for the fishery.

A third method of reduction, but one for which there is little experience to date, is the requirement that on vessel replacement, the operator be required to purchase and remove from the fishery another vessel together with its fishing units so that the number of fishing units is reduced by one.

Whichever method of effort reduction is decided upon, it is necessary that it be integrated into the development of a management program, so that industry can take it into account when making business decisions. This was not done in the early days of license limitations (1963) in the western rock lobster fishery, and later has led industry to be critical of management planning. Any fishery embarking on a limited-entry system of management ought therefore to have these principles well established from the outset. Direct buyback schemes, although simple in concept, may be excessively costly or slow to implement, and have usually been deferred in Australia in favor of the more indirect means of removing units of fishing capacity piecemeal as described.

The arrangements described for reducing fishery capacity clearly can apply only in fisheries in which management procedures allow for transfer, by sale, of fishing authorizations, and in all of Australia's rock lobster and prawn fisheries this can be based, as a commercial transaction, on the current market values of pots or vessel units. In fisheries where this is not the rule, or in others, for example scallop fisheries, that do not lend themselves to unitization, measures will have to be considered that in some way remove complete vessels, or individual fishermen (e.g., in Australian abalone fisheries) from the industry; or which redistribute a reduced number of licenses periodically through some system of tender, lottery, and so on.

The fact that most Australian limited-entry fisheries involve transfer values attaching to pots or vessels usually leads to two major disadvantages. First there is always a tendency for the average pot value to be inflated well beyond its annual catching capacity, which exacerbates the need to maximize catching potential; and second, it is very difficult for new entrants to raise the necessary capital to buy into the fishery.

Fisheries management schemes that involve effort control will become increasingly important as fisheries are progressively subjected to higher rates of exploitation. Their successful introduction and ongoing review require close liaison between research personnel, managers, and representatives of the fishing industry. However, in the process, the research staff should ensure that their role remains one of providing good research data and interpretation of those data. The representatives of industry are in a particularly difficult position because on the one hand they will understand the research data and the need for controls on fishing capacity, and on the other hand they will be aware that neither themselves, nor those they represent, desire to reduce their catching ability. The managers have to take advice from both the research staff and the fishing industry representatives, and endeavor to obtain an equitable solution to the problem of growth in fishing capacity without overlooking their management responsibility. When action is required the manager must act to ensure the objectives of fisheries management, which in broad terms can be listed as maintenance of the stock, economic performance, and social equity.

ACKNOWLEDGMENTS

The authors wish to acknowledge the valuable advice of Messrs. R. Brown, N. Caputi, and N. Hall of Western Australia and Mr. R. Lewis of South Australia. Mr. I. Lethbridge drew the figures.

REFERENCES

1. B. K. Bowen and D. A. Hancock, Review of penaeid prawn fishery management regimes in Australia. In P. C. Rothlisberg, B. J. Hill, and D. J. Staples, Ed., *Proceedings of the Second Australian National Prawn Seminar*. Simpson, Halligan, Brisbane, Australia, 1985, pp. 247–265.
2. B. K. Bowen, The economic and sociological consequences of license limitation. In P. C. Young, Ed., *Proceedings of the First Australian National Prawn Seminar*. Aust. Govt. Publ. Serv., Canberra, 1975, pp. 270–274.
3. W. Saville-Kent, *The Naturalist in Australia*. Chapman & Hall, London, 1897, 302 pp.
4. K. Sheard, *The Western Australian Cray Fishery 1944–1961*. Paterson Brokensha, Perth, Australia, 1962, 107 pp.
5. B. K. Bowen, Management of the western rock lobster (*Panulirus longipes cygnus*) George. *Fish. Ind. News Serv., West. Aust., Dep. Fish. Wildl., Perth* **4:**20–34 (1971).
6. D. A. Hancock, Research for management of the rock lobster fishery of Western Australia. *Proc. Annu. Gulf Caribb. Fish. Inst.* **33:**207–229 (1981).
7. J. A. Gulland, The concept of the maximum sustainable yield and fishery management. *FAO Fish. Tech. Pap.* **70:**1–13 (1968).

8. T. Panayotou, Management concepts for small-scale fisheries: Economic and social aspects. *FAO Fish. Tech. Pap.* **228:**1–53 (1982).

9. B. K. Bowen, Spiny lobster fishery management. In J. S. Cobb and B. F. Phillips, Ed., *The Biology and Management of Lobsters*, Vol. 2. Academic Press; New York, 1980, pp. 243–264.

10. M. B. Schaefer, A study of the dynamics of the fishery for yellowfin tuna in the eastern tropical Pacific Ocean. *Bull. Int. Am. Trop. Tuna Comm.* **2:**245–285 (1957).

11. W. E. Ricker, Handbook of computations for biological statistics of fish populations. *Bull., Fish. Res. Board Can.* **119:**1–300 (1958).

12. B. F. Phillips and N. G. Hall, Catches of puerulus larvae on collectors as a measure of natural settlement of the western rock lobster *Panulirus cygnus* George. *CSIRO Div. Fish. Oceanogr. Rep.* **98:**1–18 (1978).

13. N. Caputi and R. S. Brown, Relationship between indices of juvenile abundance and recruitment to the western rock lobster (*Panulirus cygnus*) fishery. *Can. J. Fish. Aquat. Sci.* (in press).

14. J. W. Penn and N. Caputi, Stock–recruitment relationships for the tiger prawn, *Penaeus esculentus*, fishery in Exmouth Gulf, Western Australia, and their implications for management. In P. C. Rothlisberg, B. J. Hill, and D. J. Staples, Eds., *Proceedings of the Second Australian National Prawn Seminar*. Simpson, Halligan, Brisbane, Australia, 1985, pp. 165–173.

15. P. N. Norton, The amateur fishery for the western rock lobster. *West. Aust., Dep. Fish. Wildl., Rep.* **46:**1–108 (1981).

16. L. M. Joll, The predation of pot-caught western rock lobster (*Panulirus longipes cygnus*) by octopus. *West. Aust., Dep. Fish. Wildl., Rep.* **29:**1–58 (1977).

17. R. S. Brown and N. Caputi, Factors affecting the recapture of undersize western rock lobster *Panulirus cygnus* George returned by fishermen to the sea. *Fish. Res.* **2:**103–128 (1983).

18. R. S. Brown and N. Caputi, Conservation of recruitment of the western rock lobster (*Panulirus cygnus*) George by improving survival and growth of undersize captured and returned by fishermen to the sea. *Can. J. Fish. Aquat. Sci.* (in press).

19. B. F. Phillips, Prediction of commercial catches of the western rock lobster *Panulirus cygnus* George. *Can. J. Fish. Aquat. Sci.* (in press).

20. N. Copes, *Resource Management for the Rock Lobster Fisheries of South Australia* (mimeo). A report commissioned by the Steering Committee for the Review of Fisheries of the South Australian Government, 1978, 295 pp.

21. C. Hunt, *Management Issues in the Catching Sector of the "N" Zone Rock Lobster Industry*, Vol. 9. SAFIC (the magazine of the Department of Fisheries and the Australian Fishing Industry Council, South Australian Branch), 1985, pp. 5–8.

22. R. Lewis, *Southern Rock Lobster (Jasus novaehollandiae)—A Review of the Zone S Fishery*, Vol. 5. SAFIC (the magazine of the Department of Fisheries and the Australian Fishing Industry Council, South Australian Branch), 1981, pp. 31–43.

PART II
MOLLUSCAN AND OTHER INVERTEBRATE HARVESTS

OVERVIEW OF ASSESSMENT AND MANAGEMENT OF MARINE MOLLUSKS AND INVERTEBRATES OTHER THAN CRUSTACEANS

Part I of this book draws attention to the range and scale of resources of marine crustaceans that support human activities. The diversity of other marine invertebrates exploited by humans is even more striking; the chapters that follow describe fisheries for representatives of three classes of the phylum Mollusca and for members of at least two other phyla, the Coelenterata (Chapter 28) and Echinodermata (Chapter 29). Part II must also for the moment stand as representative of the general principles that might apply to management of those other groups such as the Annelida and Chordata that are also exploited by humans, usually on a small scale, but that for reasons of space or the absence of a good information base are not touched on here.

USES OF MARINE INVERTEBRATES OTHER THAN FOR FOOD

Many considerations discussed in Part I for crustacean fisheries also apply here, but are not repeated at great length in this overview. An important new theme that begins to be developed in Part II however, is that, although as for most crustaceans, these invertebrates command high unit prices, not all of them are exploited for food. Other reasons for their exploitation may be referred to here, namely, the use of invertebrates for personal adornment (see Chapters 19 and 28) and to supply the interests of collectors of beautiful objects such as marine shells (see also Chapters 18 and 24). Three other applications, which are not discussed at length but are mentioned for completeness, are the use of some marine invertebrates as a source of pharmaceuticals and for educational purposes and research.

Two additional groups of invertebrates, particularly among the Gastropoda and

Echinodermata, are also important as predators, vectors for diseases, or parasites on other organisms of value to humans: these are touched on briefly in several chapters, but a full treatment is beyond the scope of this volume.

THE CONSEQUENCE OF A SEDENTARY LIFE

With the exception of the Cephalopods, where Chapters 26 and 27 provide only a minor indication of the importance and potential of these invertebrates, we are generally dealing in these chapters with resources that are sedentary or that have limited mobility, and to a significant extent this characteristic dominates any discussion of their population biology and management. The importance of spatial factors is one key consequence of this in modeling exploited populations, where the "dynamic pool" assumption (implying perfect mixing within the exploited stock), is not an appropriate one to make in most cases. In much more practical ways, the sedentary nature of most of these organisms can lead to exclusive territorial rights being assigned for their harvesting and intensive culture.

SPECIES AND SOCIAL INTERACTIONS

The gross ecological simplification implied by the term single-species management was referred to in several of the crustacean chapters, and inevitably recurs in a number of chapters in Part II. Again, societal choice should precede an assessment of any single one-species option: questions such as those raised in Chapter 17, for instance, whether sea otter populations should be controlled in order to permit human harvest of a luxury seafood and to maintain the unique life style of abalone divers, cannot be answered by biologists. Biologists do, however, have the responsibility of outlining the choices involved and their consequences, and to the extent possible, quantifying them so that conflicting courses of action can be evaluated by society.

The scale and variety of resources being exploited are usually, rightly or wrongly, taken into account in deciding on the need for management action. Because few fisheries for gastropods (other than abalone) or for representatives of other invertebrate taxa are as large as those for bivalves and crustaceans, their management has generally lagged behind, often with serious consequences.

In the case of the Caribbean queen conch described in Chapter 18, the importance of this large gastropod to island states has led to the current state of resource depletion in most areas. Belated attempts to understand the population dynamics of this gastropod may offer some hope of stock replenishment and better management of these locally important resources. Although conch fisheries can probably be managed largely on a single-species basis, research on resolving problems of defining stock units along an island chain will be assisted greatly by collaborative efforts and the exchange of management information between different islands and fisheries administrations. The shells of the conch, and of gastropods such as trochus (which has a parallel importance for some Pacific islands), as well as of bivalves such as the pearl oyster, where the "shell stock" is of importance, all have economic relevance, for example, in the mother-of-pearl trade (Chapter 19). For pearl oysters this is of course, in addition to the value of the pearl itself, whose production is now largely induced by artificial means in cultivated pearl oysters. Chapter 19 touches on the importance of managing the harvesting of rarities in the shell trade and some of the problems faced in doing so, and discusses similar problems now being encountered in the souvenir shell trade, where again value and rarity go hand in hand. This dichotomy poses difficult problems, particularly for developing countries that need to profit from resources that may themselves be ecologically important, for example, as gastropod predators important for controlling epidemics of undesirable species such as the crown-of-thorns starfish.

RESOURCE ALLOCATION PROBLEMS AND THE IMPACT OF TECHNOLOGY

Conflicting societal choices often play a decisive role in managing some fisheries, even to the extent of preventing courses of action that are in the interests of all parties, particularly if the resource lies across areas of multiple or conflicting jurisdiction. Chapter 20 describes this situation, where local jurisdictions conflict in the management of an estuarine fishery for oysters, and some of the consequences that can result from it.

Technical innovation driven by market demand can lead to uncontrolled growth in many fisheries. This is the situation often encountered in harvesting sedentary resources that may require only a small technological innovation in order to become vulnerable to intensive exploitation. Stock depletion from the perspective of the resource and overcapitalization and subsequent economic loss from that of the entrants to the fishery are the usual consequences of uncontrolled introduction of new technology. Chapters 21 and 22 illustrate this well, with two independent accounts of the effects of introducing a highly efficient method of harvesting bivalves, the hydraulic dredge, into the offshore clam fisheries of the eastern seaboard of the United States and of the Italian Adriatic, respectively. Both case studies show the extreme vulnerability of sedentary resources to overexploitation, and the way that societal choices are reduced or made more complex when the development of an appropriate management framework is outpaced by technological innovation.

The theme of societal choice in the face of development innovations leads us back to the question touched on in the preface, namely, that many (especially sedentary) invertebrates, and particularly mollusks, are the focus of marine farming or mariculture, as it is commonly referred to now. For a number of these species, the transition from wild harvesting to cultivation is fueled both by uncontrolled harvest of the wild stock and by unfulfilled market demand and rising prices, leading to new technologies of cultivation. Potentially at least, it is possible to farm marine organisms successfully without having an explicit knowledge of growth and mortality rates and how they determine production. However, to do so with any degree of flexibility in the face of changes in environmental conditions, predator abundance, and with differential rates of recruitment and population density, seems to call for a more widespread familiarity with resource assessment procedures than generally seems to have been the case in the past. To some extent also, this is one of the challenges that is currently posed to invertebrate stock assessment workers, namely, to review existing theory and extend it to deal with other important areas such as those involved in modeling species dynamics and bioeconomics under extensive culture conditions. Chapter 23 deals with a rather unique situation, the Mutsu Bay scallop industry, where hanging and bottom culture coexist with a long-standing monitoring of recruitments. This example shows that recruitment success can, in fact, be directly modified by experimentally manipulating spawning stock size; a concrete result that in most fisheries is still largely a subject of speculation on the so-called spawner–recruit relationship. Other insights, such as those into the impact of predator control on the natural mortality rate of their invertebrate prey, or the importance of prey density in determining predator population size, seem feasible as a result of a controlled study of culture systems in the sea, but do not yet seem to have attracted the attention they deserve from mathematical ecologists. Rehabilitation of depleted stocks is another area where aquaculture and wild harvest meet, and Chapters 18 and 24 touch on this aspect.

How to choose research priorities and select methods of assessment and management of invertebrates that properly reflect the main features of the biology of the resource are the main themes of this book. These questions are tackled in Chapter 25, in which I review current developments and future possibilities for research on the dynamics of molluscan populations, in the particular case of scallop fisheries. Chapter 30 extends this theme to include a review of existing work on sedentary mollusks, which seems to reveal that a number of promising directions are possible for future research on these organisms that could throw considerable light on a whole range of factors still largely outside the range of experimental manipulation for most fishery populations. These include an investigation of density-dependent factors in growth and mortality, stock–recruitment relationships and intraspecific competition, and predator–prey interactions. The possibilities for future theoretical and practical advances in this area are highly promising.

CASE STUDIES

A / *GASTROPODS*

17 THE CALIFORNIA ABALONE FISHERY: PRODUCTION, ECOLOGICAL INTERACTIONS, AND PROSPECTS FOR THE FUTURE

Mia J. Tegner
Scripps Institution of Oceanography
La Jolla, California

1. INTRODUCTION

Abalones, prosobranch gastropods of the family Haliotidae, are both highly prized shellfish and an important component of the ongoing debate over management of California kelp forest resources. The kelp forests of California, generally organized around the giant kelp, *Macrocystis pyrifera*, form a highly productive ecosystem that supports commercial harvest of the kelp itself, plus abalones, sea urchins, spiny lobsters, rock crabs, and various finfishes (1). This ecosystem is undergoing transition with the recovery of the sea otter, *Enhydra lutris*, a species hunted nearly to extinction in the eighteenth and nineteenth centuries. Voracious predators with a high metabolic rate, sea otters have caused a major realignment of benthic invertebrate biomass and distribution within the kelp forests of central California (see Fig. 1). Within their established range, otter foraging clearly precludes commercial fisheries for abalones and sea urchins (2–6).

Macrocystis forests, which cover most rocky subtidal habitats, especially in southern California, are characterized by large standing stocks of kelp with high rates of growth, production, and turnover (1). The resulting supply of algal drift, which is both abundant and relatively predictable, probably contributed to the success of the Haliotidae in California. There are eight abalone species (Table 1) five of which are sufficiently large and abundant to support commercial fisheries (9, 10). These fisheries, however, are likely to have resulted from release in predation pressure when sea otters were eliminated from most of their range, which once stretched from central Baja California in an arc up the west coast of North America to the northern Japanese archipelago. Following protection in 1911, the small seed colony that survived along the rugged Big Sur coast began to recover (5) and by 1983, about 1300 sea otters occupied 320 km of central California (1). As their range expanded, the mammals came into direct conflict with the abalone fishery, a conflict that continues today.

2. HISTORY OF THE FISHERY

Archaeological evidence indicates that California Indians exploited a variety of marine resources, and abalone shells and otter bones are both abundant in middens throughout the state (11). Humans on Santa Rosa Island were hunting marine mammals, fishing, and collecting a large number of red abalones as early as 6750 years ago. There is considerable controversy over the nature and extent of abalone fishing and whether Indian predation on otters was sufficient to limit populations of this competitor for food. As the Indian population grew, human predation probably caused a decline in the abundance of both sea otters and shellfish in densely populated coastal areas late in the prehistoric period (11).

Modern fisheries date to the arrival of Chinese immigrants in the 1850s (9). Using only rowboats and gaffs, this thriving industry harvested 1700 tonnes of primarily green and black abalones in 1879. Alarmed coastal counties passed laws prohibiting the take of abalones from shallow waters in 1900. The fishery then shifted to the Japanese, who had diving capabilities. Their virtual monopoly of hard-hat diving techniques enabled them to dominate the commercial fishery until World War II, when many were placed in detention centers. Caucasian divers took over after the war (9).

Abalone landings in this century (Fig. 2) reflect three major trends: (1) a south-

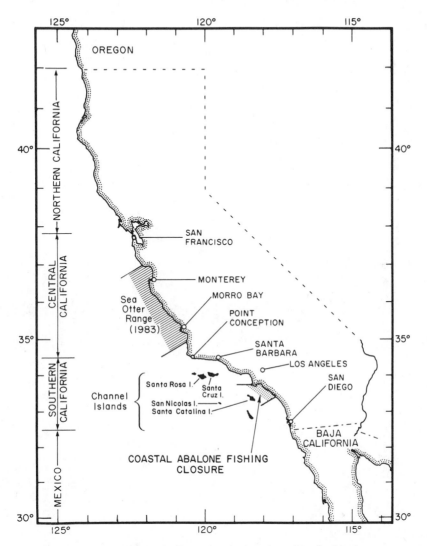

Figure 1. Map of California illustrating the locations described in the text.

ward shift of the major fishing grounds, (2) episodic changes in species composition, and (3) a sharp decline in total landings beginning in 1969 (12). The fishery began with red abalones in Monterey, then gradually shifted to Morro Bay, Los Angeles, and Santa Barbara. The mainland coast south of Point Conception and the Channel Islands were reopened to commercial fishing in 1943; this made available the smaller and less desirable pink, green, and white abalones. Reduction in the red abalone legal size limit in 1959 and the green size limit in 1969, and legislative action in 1970 and 1971 that permitted the drying, canning, and export of black abalones led to peaks in the landings of these species. Conversely, an

TABLE 1 California Abalones[a]

Species	Common Name	Range	Depth Range	Maximum Size (mm)	Minimum Legal Size (mm) (sport/commercial)
Haliotis rufescens	Red	Sunset Bay, OR to Bahia Tortugas, Baja California	Low intertidal (north of Point Conception) to more than 180 m; major concentrations shallower than 11 m in northern CA.: 8–25 m in southern CA	>280	178/197
H. corrugata	Pink	Point Conception, CA to Bahia Tortugas, Baja California	Low intertidal to 60 m: major concentrations 6–24 m	254	152/159
H. fulgens	Green	Point Conception, CA to Bahia Magdalena, Baja California	Low intertidal to 18 m: major concentrations 1.5–8 m	254	152/178
H. cracherodii	Black	Coos Bay, OR to Cabo San Lucas, Baja California	Mid-intertidal to 6 m: predominantly intertidal	203	127/146
H. sorenseni	White	Point Conception, CA to Bahia Tortugas, Baja California	Deep-water species found from 5 to 50 m, major concentrations 25–30 m	254	152/159
H. wallalensis	Flat	British Columbia to La Jolla, CA; rare south of Carmel, CA	Subtidally to 25 m	178	102/102
H. kamtschatkana	Pinto	Sitka, Alaska to Point Conception, CA	Shallow subtidal to 15 m, submerges with decreasing latitude	152	102/102
H. assimilis[b]	Threaded	Point Conception, CA to Bahia Tortugas, Baja California	Found from 3 to 40 m: major concentration 25–30 m	153	102/102

[a]Compiled from Cox (7) and various California Department of Fish and Game sources.
[b]Some authors prefer *H. kamtschatkana assimilis* (7, 8).

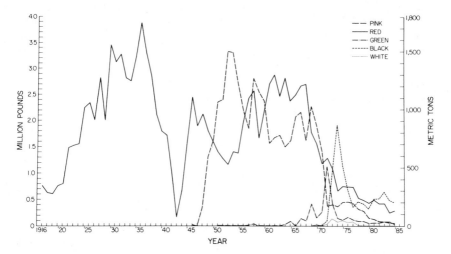

Figure 2. California commercial abalone landings. The California Department of Fish and Game began keepings records in 1916. A specific category for white abalones was not included on landing receipts after 1977. Compiled from various CDFG sources.

increase in the pink abalone minimum size in 1970 led to a sharp reduction in landings. With the depletion of mainland stocks and the introduction of faster boats, the fishery shifted increasingly to the Channel Islands near Santa Barbara. From 1951 to 1968, total landings exceeded 1800 tonnes/yr. With the complete exploitation of accumulated stocks, loss of some very productive areas owing to range expansion of the sea otter, environmental degradation of some mainland sites, and closure of some island sites, total landings have since declined to less than 400 tonnes/yr. This decline was also associated with a large increase in fishing pressure as evidenced by the introduction of more practical dive gear and faster boats, a greater than 300% increase in the commercial fleet between 1951 and 1974, and a similar increase in sport diving (12, 13).

3. LIFE HISTORY AND FISHERY MANAGEMENT

The natural history of abalones seems to facilitate overfishing, a problem in every producing country in the world (10). Abalones are slow growing and long-lived, and recruitment may be unpredictable (14, 15). As a result, population size–frequency distributions are often strongly skewed with an accumulation of old individuals. These animals live in shallow water near stands of their algal food, in relatively predictable and accessible locations, and movement is minimal. Their high unit value warrants extensive searching and the abundance of legal-sized animals is often quite low (13).

An abalone attaches to a rock with its muscular foot; fishermen insert a flat bar under the foot to pry the animal off the substrate. Poor bar placement, movement by the abalone, or irregular substrates often lead to incidental harvest damage, referred to as "bar cuts." These shellfish bleed profusely and laboratory tests showed nearly 60% mortality of red abalones with a 12-mm cut (13); in the presence of predators, the mortality rate may be considerably higher. As stocks decrease, fishermen pick more short animals to sort out the few legal-sized individuals. The effects of bar cutting are apparent in the population size–class structure of pink and green abalones, which depict a reduction in animals within 6 mm of minimum legal size (13), and in the size–frequency distribution of red abalone shells (Fig. 3).

4. ABALONE ECOLOGY

California abalones feed primarily on algal drift, fragments of macrophytes moved by currents and surge; foraging on attached plants is rare (9). They feed preferentially on *Macrocystis* and other laminarians; minor variations in preference appear to reflect the habitat where each is found (16, 17). A common feature of abalone growth is its variation with habitat and food quality. Breen (18) found that maximum size, but not abundance, was determined by local food availability, both the quantity and species composition. Transplant experiments showed that stunted

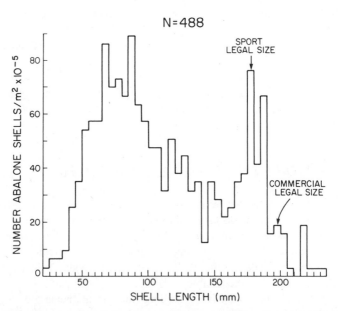

Figure 3. Red abalone shells collected from Johnson's Lee, Santa Rosa Island, in July, 1980. Peaks in the size–frequency distributions of shells just smaller than sport and commercial minimum legal sizes are evidence of bar cut mortality.

individuals retain the potential for good growth when moved to more favorable habitat (15). Manipulations of protected black abalone stocks on Santa Cruz Island demonstrated food limitation in dense populations (19).

Red and purple sea urchins, *Strongylocentrotus franciscanus* and *S. purpuratus*, are the other conspicuous invertebrates that specialize in feeding on drift. The apparent sea urchin population explosion in the 1960s, after fishing reduced abalone standing stocks, suggested that urchin population growth may have been caused in part by release from competition with abalones (20). Examining competition between red urchins and red abalones, Tegner and Levin (21, 22) found that abalones exert weak negative effects on the urchins and concluded that predation was more likely to regulate urchin populations. Spiny lobsters (*Panulirus interruptus*) and a wrasse (*Semicossyphus pulcher*) probably limited sea urchin populations in southern California from the demise of the otter in the mid 1800s until their populations were severely reduced by fishing more than a century later (22). When food is limiting, sea urchins are likely to have a negative impact on other grazers. The echinoids attack attached plants when drift becomes scarce; en masse they are capable of destroying entire stands of kelps, and their subsequent substrate scraping can prevent kelp recruitment (16). A red sea urchin fishery developed in the 1970s (23) and has reduced the incidence of destructive grazing episodes.

Sea urchins also benefit abalones. Encrusting coralline algae induce settlement and metamorphosis in red abalone larvae and are a source of nutrition for small juveniles (24). Urchins maintain patches of coralline algae free from overgrowth, and juvenile abalones and urchins find protection from predators under the spine canopy of the adult urchins. In some areas with little other appropriate habitat (generally small rocks with less than 0.1 m^2 bottom surface area), this is the major habitat for young of the year abalones (25).

Sea otters have a dramatic effect on the distribution, abundance, and size structure of abalone populations. Because of their large body size and caloric content, these mollusks are highly preferred prey (26, 27). Sea otters often use rocks to break the abalones' hold on the substrate (4, 26). Within the established otter range, nearly all abalones are confined to crevices inaccessible to otters, and the average abalone size is less than half as large as in populations outside the otter range (4). Where the habitat is appropriate, abalones persist; the population at Hopkins Marine Life Refuge, within the established otter range for many years, has exhibited stable abalone density, size distribution, and species composition for a 10-yr period. The stability of this population under high predation pressure appears to depend on both high recruitment rates and the crevice refuge afforded by the granodiorite habitat (4). However, siltstone substrates with few deep interstices may not support abalones in the presence of otters (2). The change in red abalone density at Point Estero, part of a commercial bed north of Morro Bay that withstood heavy fishing pressure for nearly 40 yr before sea otter reoccupation, is shown in Figure 4. The decline affected all size classes and the bed was abandoned by the commercial fishery in 1971 (28).

A variety of other species also prey on abalones; different predators are important in different areas and on different size classes of abalones (4, 9, 17, 25, 29).

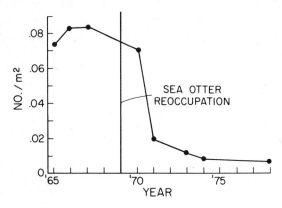

Figure 4. Red abalone density before and after sea otter reoccupation of Point Estero, central California (3).

These include octopuses (*Octopus* spp.), rock (*Cancer* spp.) and sheep (*Loxorhyncus grandis*) crabs, spiny lobsters, various starfishes, bat rays (*Myliobatis californica*), and cabezon (*Scorpaenichthys marmoratus*). Octopuses, for example, appear to have minimal effect on large abalones, for we found no red shells larger than 125 mm with drill holes, but their intense predation on juvenile red abalones may be limiting the recovery of this species on the Palos Verdes Peninsula near Los Angeles (25, 29). It is clear that none of these predators can affect abalone populations to the same degree as sea otters.

Storms can be an important source of abiotic mortality. After particularly violent storms in central and northern California, many abalones frequently wash up on the shore. Although freshwater runoff and siltation associated with the storms can be lethal, Cox (9) interprets the broken condition of most of the shells as evidence for mortality by wave-induced rock rolling.

Macrocystis is itself harvested for alginate production, economically one of the most important kelp forest resources. Harvesting, restricted to the canopy no lower than 1.2 m below the surface, is regulated to ensure minimal impact on the community; considerable research has not demonstrated adverse effects (1). Harvesting may cause some variability in the availability of drift to grazers but, given the general abundance of this resource, the effects are not likely to be important.

More serious impacts on abalone food resources are caused by El Niños and pollution. El Niños, defined in terms of episodic events in the eastern tropical Pacific Ocean, are also associated with pronounced warming of the ocean off California. Depleted of nutrients, the warm waters cause loss of *Macrocystis* canopies, considerable reductions of standing stock, and varying degrees of mortality (30). During the major El Niño of 1957–1959, when many California kelp forests were virtually destroyed, Cox (9) reported that abalone growth practically ceased, body tissues appeared to shrink, and there was a shortage of new animals entering the fishery. Abalone gonads did not increase in size during the regular spawning season and many animals were probably unable to spawn. We

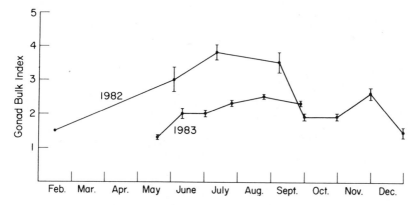

Figure 5. Green abalone gonad development at Golden Cove, on the Palos Verdes Peninsula, 1982–1983. The gonad bulk index is a visual index developed by George Lockwood. The scale varies from 1, a spawned out condition, to 6: spontaneous spawning has been observed at 3.5 or above. The curve for 1982 shows the normal two peaks of green abalone spawning (31). Storms in early 1983 destroyed the kelp forest and the abalones were without food until the kelp recruited in May: the abalones apparently missed the first spawning period. Bars indicate ± 1 standard error.

were following the gonad development of transplanted green abalones (Fig. 5) (31) when winter storms associated with the 1982–1983 El Niño decimated kelp populations on the Palos Verdes Peninsula. The increase in gonad size was delayed until after the kelp recruited in May, and the animals apparently missed the normal late spring/early summer spawning.

Relatively stable prior to 1940, the kelp forests off the metropolitan centers of southern California began to regress near major sewage outfalls in 1945 and the deterioration progressively affected beds at increasing distances from the discharges. The 1957–1959 El Niño compounded the problems, and by the early 1960s, coastal kelp forests were at an all-time low (32). Unpublished California Department of Fish and Game (CDFG) records suggest that abalone landings from the Palos Verdes Peninsula decreased in parallel with the decline in the *Macrocystis* canopy. Although the exhaustion of accumulated stocks doubtlessly contributed to the decrease in landings, reciprocal transplant experiments with abalones from Santa Catalina Island indicated that the Palos Verdes animals were starving (33). Subsequent improvements in waste water management, kelp restoration projects, and sea urchin control measures (including the urchin fishery) have contributed to the recovery of coastal kelp populations in southern California (32).

5. MONITORING STOCKS AND LANDINGS

California requires official receipts (pink tickets) of commercial landings to be issued by licensed seafood dealers (Fig. 6). These receipts include the name of the fisherman, permit and boat registration numbers, the numbers of each species

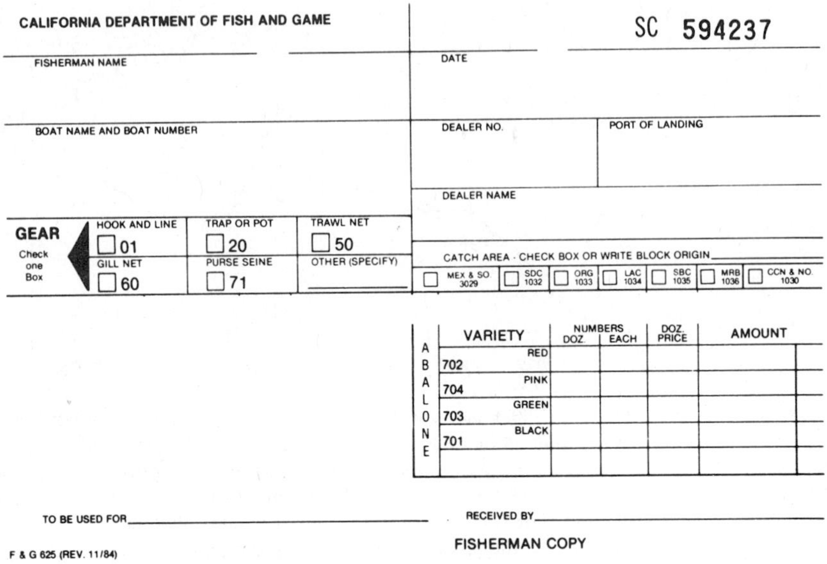

CALIFORNIA DEPARTMENT OF FISH AND GAME   SC 594237

| FISHERMAN NAME | DATE |
| | |

| BOAT NAME AND BOAT NUMBER | DEALER NO. | PORT OF LANDING |
| | DEALER NAME | |

GEAR
Check one Box

HOOK AND LINE	TRAP OR POT	TRAWL NET
☐ 01	☐ 20	☐ 50
GILL NET	PURSE SEINE	OTHER (SPECIFY)
☐ 60	☐ 71	

CATCH AREA - CHECK BOX OR WRITE BLOCK ORIGIN____

| ☐ MEX & SO 3029 | ☐ SDC 1032 | ☐ ORG 1033 | ☐ LAC 1034 | ☐ SBC 1035 | ☐ MRB 1036 | ☐ CCN & NO. 1030 |

	VARIETY	NUMBERS DOZ.	EACH	DOZ. PRICE	AMOUNT
A	RED				
B	702 PINK				
A	704				
L	GREEN				
O	703 BLACK				
N	701				
E					

TO BE USED FOR____   RECEIVED BY____

F & G 625 (REV. 11/84)   **FISHERMAN COPY**

Figure 6. Landings record. This illustrates the portion of the official receipt relevant to abalones. These are issued by licensed seafood dealers and submitted to CDFG.

received, and the block (blocks are $0°10'$ square) or county where the catch was taken. Standard factors are used to convert numbers to weight (34). The landing data (the weight and value of monthly landings by port) are tabulated and published by the CDFG (34), usually with a several year delay. Catch per unit effort could be calculated in terms of pounds per landing (trips vary from 1 to 3 days) per permit but this has not been done.

The sport fishing take is incompletely monitored. Commercial passenger fishing vessels are required to record all fish taken, but much sport fishing is done from the shore or off private boats. The CDFG periodically surveys the recreational catch in different areas. Such surveys in the 1960s and 1970s documented the crash of the sport fishery as the sea otter range expanded (28). Combined flight censuses and simultaneous shore interviews on the north coast, which has been reserved for sport fishing since 1945 (12), led to estimates that the recreational take of red abalones in this area is now larger than the statewide commercial catch of this species (35).

It is not feasible to estimate stock sizes regularly by direct survey. Suitable habitat is too extensive, and abalones are too patchily distributed to make large-scale surveys practical. Furthermore, sampling for most species requires diving, which is very labor intensive. CDFG biologists monitor the block landing data for changes in harvest by area, but this analysis is limited by the scale and accuracy of the reporting. Occasionally direct surveys are made of specific areas, for example, in response to requests for access by commercial fishermen, the CDFG plans to resurvey sites on the north coast in the fall of 1986 (35). Breen (15) evaluated several approaches for back-calculating virgin stocks in British Columbia a few years after that fishery began.

6. GROWTH AND MORTALITY STUDIES

It does not appear that California abalones can be aged by counting rings and, given the variability of growth among habitats, size (measured as maximum shell length) is poorly related to age (9, 10). Nevertheless, because of their sedentary life style and ease of tagging, growth is one of the better-known aspects of abalone biology. Growth is typically determined from size–frequency distributions for smaller animals, although this is problematic for red abalones, which may spawn year round (36), and from tag returns for larger sizes. For most haliotids, growth is adequately described over a broad size range by the von Bertalanffy growth curve (14). However, growth may be linear until the animals reach sexual maturity in some cases (17). CDFG tagging studies of several populations in southern California indicate that sport fishing size (Table 1) is reached after about 8 yr in three of the major species, and commercial size after a minimum of 10 yr for greens and pinks and 15 yr for red abalones (13, 37).

Breen (15) reviewed natural instantaneous mortality estimates, derived for different abalone species in different locations using a variety of techniques, and found most values to be relatively low ($M = 0.05$–0.30). There is evidence that mortality rates may be higher in California. Tutshulte (17), working with sublegal sized animals, estimated annual finite mortality rates of 0.5 for pinks and 0.6 for green abalones at Santa Catalina Island ($M = 0.7$ and 0.9, respectively), and Deacon (38) estimated a natural finite mortality rate of 0.4 per year for all California pinks ($M = 0.5$). Hines and Pearse (4) used several indirect methods to estimate annual turnover (finite mortality) rates ranging from 0.3 to 1.0 for a population of red and flat abalones at Hopkins Marine Life Refuge. Although these latter estimates are somewhat crude, they do suggest that abalone populations within the sea otter range are highly dynamic. Doi et al. (39) estimated natural instantaneous mortality rates of 0.35 and 0.43 per year for two populations of pink abalones in Baja California. Because the range of most California species extends through different zoogeographic zones with different suites of predators, mortality rates may vary considerably between habitats.

7. FISHERY MODELS

Yield models have not yet been applied to California abalones. The CDFG assumed that an appropriate size limit, large enough to allow several spawnings of these highly fecund animals but small enough that the size would be achieved within a reasonable number of years following recruitment, could maintain the stocks (13). Burge et al. (13) felt that if the size limit was so large that only a few faster growing animals could be harvested, fishing would lead to stunted populations with poor growth potential. Thus the goals of size limit changes have been improvement of biomass yields and the economic efficiency of exploitation (40).

If there is no relationship between stock size and subsequent recruitment, abalones could be managed to obtain the best yield from a cohort (15). The weakness of yield-per-recruit analysis is the lack of information regarding the effect on reproduction of a given level of exploitation, for recruitment failure is the most common cause of fishery collapse (41, 42). Furthermore, abalone larvae have a short planktonic phase and Shepherd (reported in Ref. 40) found a positive relationship between the biomass of abalones over 4 yr old (the onset of effective egg production) and subsequent recruitment of 1-yr-olds in an isolated substock in South Australia. Genetic studies of wild populations of Japanese abalones suggest high levels of inbreeding, which are also consistent with limited larval dispersal (43, 44). Thus Sluczanowski (40) and Breen (15) have developed yield models that take into account egg production per recruit. The models indicate that size limits that achieve high yield-per-recruit efficiencies may have led to very poor lifetime egg production per recruit. Egg production can be increased substantially by relatively minor reductions in yield, especially by periodic closures (40). Sluczanowski's (40) model has the advantage for managers that the period between successive fishing visits to each substock is an index of exploitation that is usually more easily measured and controlled than fishing mortality (F). In the absence of egg–recruit data for abalones, Breen (15) suggests a method using an appropriate minimum size as the major regulatory tool so that yield per recruit is maximized and some arbitrary level of egg production is maintained.

8. CURRENT FISHERIES REGULATIONS

The sharp decrease in landings that began in 1969 (Fig. 1) led to an expanded CDFG research program in 1973 and new regulations were implemented by 1977 (13, 45). The regulations, a compromise between CDFG, the industry, sport fishermen, and environmentalists, limited entry to the commercial fishery and further reduced access for all fishermen. It was widely accepted that there were too many commercial fishermen; by 1975 there were 383 commercial permits with an annual turnover rate around 50%. The new law restricted permits to those who held them in 1976, with an additional 5% to be chosen by lottery from among qualified applicants. The law required a permit holder to land at least 4530 kg or make 20 landings per year to keep the permit. The goal was to reduce the number of divers to 200 (a number derived from political considerations) by attrition. By eliminating less experienced divers, the loss associated with bar cuts was expected to drop. To further reduce this problem, sport fishermen were required to keep the first four (the bag limit) legal abalones they picked (45).

Additional regulations governing the commercial fishery include a prohibition against fishing in February and August, size limits (Table 1), gear requirements including surface-supplied air, minimum depth and area restrictions, a permit fee, and restriction of fishing to the period between half an hour before sunrise to half an hour after sunset. A quota for black abalones restricts fishermen to no more than 20 dozen of this species per trip. Sport fishermen are required to have a license, a bar of specified dimensions, and a fixed-caliper measuring gage. Sport fishermen north of a location near Monterey in central California are prohibited from using scuba and have different seasons from those in the south. Area closures are less restrictive than in the commercial fishery and sport size limits are generally smaller (Table 1), a factor that may effectively preclude commercial fishing in some popular locations. All abalones must be landed alive and attached to the shell to enforce size limits.

In spite of the major shift in regulations in the 1970s, the commercial catch has continued to fall (Fig. 1). New legislation in 1985 reduces the number of permits to 100 by further attrition. When the number of permits is 100 or less (there were 130 permits in the 1985–1986 license year), the commercial size limits for red, pink, and green abalones will drop 6 mm and a daily quota of 10 dozen of these species in combination will become effective. Some depth restrictions will also be eliminated. The goal of this legislation is to utilize better the growth potential of these species and to prevent a short-lived bonanza in landings (13). Lower size limits and quotas were proposed by CDFG in 1975 but dropped in the face of strong opposition by the industry at that time (45).

9. SOCIOECONOMIC CONSIDERATIONS

Several socioeconomic aspects of the California abalone fishery are unusual. Cicin-Sain, Moore, and Wyner (12) studied the demographic characteristics of commercial fishermen the year before limited entry began. The divers were young (median age of 32 years), most made a poor economic return from fishing (median annual gross income was $6750), and 64% had additional sources of income. Noneconomic benefits, including the unique life style and the independence, contributed to the decision to stay in the fishery but poor economic return was cited as the major reason for the high attrition rate (12).

With the exception of black abalones, virtually all California production is consumed within the state, but domestic price levels are almost completely set by conditions in international markets (46). Only a portion (22% in 1975) (12) of the California market is satisfied by domestic production; the rest is imported, primarily from Mexico. Ex-vessel prices were relatively stable in real terms during the 1950s and early 1960s but almost doubled between 1967 and 1978, the period when production plummeted. Nevertheless, Deacon (46) found that the increase in California prices coincided almost precisely with the decline in the value of the U.S. dollar in relation to the currencies of other abalone-consuming nations, particularly the Japanese yen.

Cicin-Sain (47) summarized the social polarization that has developed from the shellfish–sea otter conflict:

Sea otter protectionist groups have tended to view those who fished as greedy rapists of the marine environment, interested only in their short term economic gain. Divers, on the other hand, have often viewed sea otter protectionists as members of an elitist class who care more about the rights and fates of animals than about the rights and fates of humans.

Despite evidence to the contrary, otter protectionists blame humans alone for all decline of abalone stocks (12). The high price ($77/kg retail in December, 1985) exacerbates the issue; abalones are clearly a luxury product, not a basic food item.

10. PROSPECTS FOR STOCK ENHANCEMENT

At the time that limited entry was instituted, the CDFG also recommended a mariculture program to enhance abalone production. This became something of a motherhood issue; there was universal support and willingness to pay additional fees for that purpose (12, 45). The Japanese pioneered seeding hatchery-reared juveniles to augment natural recruitment; their efforts were the model for California. Abalone seeding is practiced on a large scale in Japan; in 1979, 44 culture centers produced more than 11 million seed for release into the sea (48). Although biologists report experimental survival rates ranging from less than 1 to 80% (49), the success of the massive national effort is hard to judge. Kafuku and Ikenoue (48) suggest that the number released is too low to affect the total catch, and they document a slow decline in landings from 1970 to 1979.

Six large-scale seeding experiments, testing red and green seed under a variety of conditions, were conducted at island and mainland sites in southern California. In each case the estimated 1- or 2-yr survival rate, based on recovery of live seed, was 2.8% or less (25, 29). Intensive predation pressure on both newly planted and adapted seed, especially by octopuses, appeared to be the major cause of the poor survival (29). Laboratory experiments demonstrated behavioral differences between hatchery-reared and native abalones of the same sizes that significantly increased susceptibility to predation (25). Thus the Japanese approach of seeding roughly 1-yr-old (2–3 cm) abalones does not appear to be feasible for California conditions. Morse (50) suggested the use of miniseed (3–5 mm) as an alternative approach. Using transfer substrates, large numbers could be transplanted at a very low cost (relative to older seed) and the considerably shorter time in the hatchery might reduce or eliminate the behavioral problems. This approach remains to be tested.

In 1977, the legislature also closed a stretch of the severely depleted southern California mainland coast to all abalone fishing to allow natural recovery (Fig. 1). Red abalone populations on the Palos Verdes Peninsula responded; pink and green abalones, historically the more abundant species, did not. Nearby Channel Islands had good breeding stocks, but the short larval life of the green abalone (51) suggested that larval transport between isolated populations was not common. A drift tube study (52) supported this hypothesis and stressed the importance of local brood stock. To provide local sources of larvae, 4500 reproductively mature green abalones were transplanted into sites within the closed area where the drift tube data suggested a high probability of larval retention. When the Palos Verdes Peninsula was recensused 3 yr later, we found a dramatic increase in the abundance of green abalones up to 3 yr old; more than 90% of the sample ($N = 348$) was in this size range, a pattern not found at off-site controls (53). Although brood stock transplants may be an effective approach for abalone enhancement, the approach is expensive, limited by the availability of brood stock, and may not be feasible for pink abalones, the species most susceptible to bar cut damage (13). To extend the approach one step further, we are experimenting with the release of veliger larvae, which are at a developmental stage very close to settlement. Larvae are being released inside kelp forests where the drag from *Macrocystis* plants greatly reduces coastal current velocities (54), thus increasing the potential for larval retention. Because larval production is inexpensive and minimal effort is required for their release, this could be a highly cost-effective approach to abalone enhancement.

11. THE FUTURE

There are many problems facing the California abalone fishery, some of which are amenable to scientific research and some of which require political resolution. Highest priority should be given to yield-per-recruit modeling that takes egg production per recruit into account. Sainsbury (14) and Fournier and Breen (55) have recently developed methods for estimating mortality rates from size frequency data. Mortality rates of red abalones (in different habitats) represent the only major category of information required for the yield-per-recruit models described previously (15, 40) that is not presently available for this species, for example. The results of such modeling exercises could put management of the fishery on a much more scientific basis.

Whether the new size-limit quota legislation can stabilize the decline in landings or lead to a long-term increase in production remains to be seen. Deacon (38) notes that after accumulated stocks of pink abalones were largely depleted (by 1953 or 1954), the population supported a sustained harvest at a lower level until the size limit was raised in 1970. The size increase, which required 2 or more years of additional growth to attain legal size (13), caused a permanent proportional reduction in recruitment to legal size (38). After the number of commercial permits has dropped to 100 or less, the size limit will drop to the pre-1970 level. It may take considerable time for the number of permits to drop as fishermen recognize the potential economic benefits of retaining their licenses; during this period, the decline in landings is likely to continue. Once the size levels are reduced, the daily quota will slow the harvest of the sudden bonanza in stocks. However, the experience of other limited-entry fisheries has been that given sufficient incentive, effort increases dramatically despite limited entry and attempts to control the components of effort (15, 56, 57). Harrison (57) suggests that Australia, which has considerable experience with the effectiveness of limiting entry to abalone fisheries, may have to go to an individual quota system. High prices guarantee continued pressure on the resource.

Other important issues include the question of stock identity, the relationship between stock size and recruitment, and the concomitant problems of severely depleted brood stocks and recruitment failure. The short larval lives (51), the drift tube results, and census data from the Palos Verdes Peninsula (52) strongly suggest that Channel Island and mainland green and probably pink abalone populations are largely separate stocks. This has important implications for fishery management and resource enhancement. If stocks are highly localized, the symptoms of recruitment overfishing could be obscured if fishery statistics are geographically coarse or inaccurate (57). Sluczanowski (40) shows the potential economic benefits of pulse fishing such local stocks, significantly greater egg production with only a marginal decrease in yield. Commercial fishermen favor rotation of open and closed areas, but the CDFG has opposed this in the past because of enforcement problems and the heavy pressure to which newly reopened areas would be subjected (45). Whether on a rotation basis or a complete opening, the industry continues to push for access to the north coast, a move strongly opposed by recreational fishermen.

The prospects for resource enhancement are similarly linked to these issues of stock and recruitment. Area closures will be successful only if there are stocks available to support recovery. It has generally been assumed that reduced recruitment is the primary cause of depleted stocks, but it is not known to what extent mortality at different life-history stages may be limiting populations (6). Three approaches to abalone stock enhancement, brood stock transplants, larval releases, and seeding hatchery-reared juveniles, each make somewhat different assumptions about the limiting life-history stage. The short-term success of the green abalone brood stock transplants on the Palos Verdes Peninsula (53) supports larval supply as the key issue for that species. However, the high mortality rate of both seeded and native red abalone juveniles in the same region (29) indicates that it is not the whole answer. At any rate, seeding does not appear to be the hoped-for panacea (25). Data from the Palos Verdes Peninsula suggest that closures, especially in conjunction with some method to augment natural recruitment, will be successful but not rapid (52).

All this will eventually be moot if the sea otter range continues to expand. No matter what the outcome of the debate over whether it is otters or fishermen that have caused the decline of abalone stocks in central California, it is clear that abalone stocks within the range of the otter will not support a fishery (2–4, 6). Shellfish interests favor zonal management of the otter, and otter protectionist groups favor unlimited range expansion. Resolution of this conflict is hampered by jurisdictional priority. Although the state government is charged with the protection and enhancement of all fish and wildlife, the federal government preempted jurisdiction over otters in 1972 with the passage of the Marine Mammal Protection Act and later the Endangered Species Act. The California population (there are more than 100,000 otters in Alaska) (5), was designated as threatened owing to the possibility of oil spills. Removal of the otter from the threatened category, which could open the way to zonal management, would require translocation of a portion of the California population to a site away from oil spill potential. The federal government proposed San Nicolas Island, a productive abalone bed. From San Nicolas Island, the otters are likely to spread throughout the Channel Islands, the major grounds of the abalone fishery. The debate over sea otter management, which dates back to the early 1960s, continues with no resolution in sight.

ACKNOWLEDGMENTS

I am grateful to D. Parker for supplying considerable CDFG material and to P. Breen, J. F. Caddy, C. Lennert, A. MacCall, D. Parker, P. Parnell, and N. Sloan for their comments on the manuscript. This research was sponsored by the California Department of Fish and Game, the National Marine Fisheries Service, the National Sea Grant College Program under grant number NA80AA-D-00120, and the California State Resources Agency, project numbers R/F-47 and R/F-73. The U.S. Government is authorized to produce and distribute reprints for governmental purposes.

REFERENCES

1. M. S. Foster and D. R. Schiel, The ecology of giant kelp forests in California: A community profile. *U.S., Fish Wildl. Serv., Biol. Rep.* **85:**1–152 (1985).

2. D. J. Miller and J. J. Geibel, Summary of blue rockfish and lingcod life histories: a reef ecology study, and giant kelp, *Macrocystis pyrifera*, experiments in Monterey Bay, California. *Calif. Dept. Fish Game, Fish. Bull.* **158:**1–137 (1973).

3. R. Hardy, F. Wendell, and J. D. DeMartini, A status report on California shellfish fisheries. In B. Cicin-Sain, P. M. Grifman, and J. B. Richards, Eds., *Social Science Perspectives on Managing Conflicts between Marine Mammals and Fisheries.* Marine Policy Program, University of California, Santa Barbara, 1982, p. 328.

4. A. H. Hines and J. S. Pearse, Abalones, shells, and sea otters: Dynamics of prey populations in central California. *Ecology* **63:**1547 (1982).

5. K. W. Kenyon, Sea otter *Enhydra lutris*. In J. A. Chapman and G. A. Feldhamer, Eds., *Wild Mammals of North America*. Johns Hopkins Univ. Press, Baltimore, MD, 1982, p. 704.

6. J. A. Estes and G. R. VanBlaricom, Sea otters and shellfisheries. In J. R. Beddington, R. J. H. Beverton, and D. M. Lavaigne, Eds., *Marine Mammals and Fisheries.* Allen & Unwin, London, 1985, p. 187.

7. B. Owen, J. H. McLean, and R. J. Meyer, Hybridization in the eastern Pacific abalones (Haliotis). *Bull. Los Angeles County Mus. Nat. Hist.: Sci.* **9:**1–37 (1971).

8. D. L. Leighton, The influence of temperature on larval and juvenile growth in three species of southern California abalones. *Fish. Bull.* **72:**1137 (1974).

9. K. W. Cox, California abalones, family Haliotidae. *Calif. Dept. Fish Game, Fish. Bull.* **118:**1–133 (1962).

10. M. G. Mottet, A review of the fishery biology of abalones. *Wash., Dept. Fish., Tech. Rep.* **37:**1–81 (1978).

11. P. L. Walker, California Indians, sea otters, and shellfish: The prehistoric record. In B. Cicin-Sain, P. M. Grifman, and J. B. Richards, Eds., *Social Science Perspectives on Managing Conflicts between Marine Mammals and Fisheries.* Marine Policy Program, University of California, Santa Barbara, 1982, p. 17.

12. B. Cicin-Sain, J. E. Moore, and A. J. Wyner, Management approaches for marine fisheries: The case of the California abalone. *Univ. Calif. Sea Grant Col. Program, Sea Grant Publ.* **54:**1–223 (1977).

13. R. Burge, S. Schultz, and M. Odemar, *Draft Report on Recent Abalone Research in California with Recommendations for Management,* (mimeo). Report to California Fish and Game Commission, California Department of Fish and Game, San Diego, CA, 1975, 62 pp.

14. K. J. Sainsbury, Population dynamics and fishery management of the paua, *Haliotis iris.* I. Population structure, growth, reproduction and mortality. *N. Z. J. Mar. Freshwater Res.* **16:**147(1982).

15. P. A. Breen, Management of the British Columbia fishery for northern abalone (*Haliotis kamtschatkana*). *Can. Spec. Pub. Fish. Aquat. Sci.* **92:**300 (1986).

16. D. L. Leighton, Grazing activities of benthic invertebrates in kelp beds. In W. J. North, Ed., *The Biology of Giant Kelp Beds in California.* Cramer, Lehre, 1971, p. 421.

17. T. C. Tutshulte, *The comparative ecology of three sympatric abalones.* Ph.D. Thesis, University of California, San Diego, 1976, 335 pp.

18. P. A. Breen, Measuring fishing intensity and annual production in the abalone fishery of British Columbia. *Can. Tech. Rep. Fish. Aquat. Sci.* **947:**1–49 (1980).

19. W. J. Douros, *Density, growth, reproduction and recruitment in an intertidal abalone: Effects of intraspecific competition and prehistoric predation.* Master's Thesis, University of California, Santa Barbara, 1985, 112 pp.

20. W. J. North and J. S. Pearse, Sea urchin population explosion in southern California coastal waters. *Science* **167:**209 (1970).

21. M. J. Tegner and L. A. Levin, Do sea urchins and abalones compete in California kelp forest communities? In J. M. Lawrence, Ed., *International Echinoderms Conference, Tampa Bay,* A. A. Balkema, Rotterdam, 1982, p. 265.

22. M. J. Tegner and L. A. Levin, Spiny lobsters and sea urchins: Analysis of a predator-prey interaction. *J. Exp. Mar. Biol. Ecol.* **73:**125 (1983).

23. M. J. Tegner, Multispecies considerations of resource management in southern California kelp beds. *Can. Tech. Rep. Fish. Aquat. Sci.* **954:**125 (1980).

24. A. N. C. Morse and D. E. Morse, Recruitment and metamorphosis of *Haliotis* larvae induced by molecules uniquely available at the surfaces of crustose red algae. *J. Exp. Mar. Biol. Ecol.* **75:**191 (1984).

25. M. J. Tegner and R. A. Butler, Abalone seeding. In K. Hahn, Ed., *Handbook of Culture of Abalones and Other Gastropods.* CRC Press, Boca Raton, FL, (in press).

26. E. E. Ebert, A food habits study of the southern sea otter, *Enhydra lutris nereis. Calif. Fish Game* **54:**33 (1968).

27. R. S. Ostfeld, Foraging strategies and prey switching in the California sea otter. *Oecologia* **53:**170 (1982).

28. P. W. Wild and J. A. Ames, A report on the sea otter, *Enhydra lutris L.,* in California. *Calif. Dept. Fish Game, Mar. Res. Tech. Rep.* **20:**1–93 (1974).

29. M. J. Tegner and R. A. Butler, The survival and mortality of seeded and native red abalones, *Haliotis rufescens*, on the Palos Verdes Peninsula. *Calif. Fish Game* **71:**150 (1985).

30. P. K. Dayton and M. J. Tegner, Catastrophic storms, El Niño, and patch stability in a southern California kelp community. *Science* **224:**283 (1984).

31. M. J. Tegner and P. K. Dayton, El Niño effects on southern California kelp forest communities. *Adv. Ecol. Res.* **17:**243 (1987).

32. K. C. Wilson, P. L. Haaker, and D. A. Hanan, Kelp restoration in southern California. In R. W. Krauss, Ed., *The Marine Plant Biomass of the Pacific Northwest Coast.* Oregon State Univ. Press, Corvallis, 1977, p. 183.

33. P. H. Young, Some effects of sewer effluent on marine life. *Calif. Fish Game* **50:**33 (1964).

34. M. S. Oliphant, California marine fish landings for 1976. *Calif. Dept. Fish Game, Fish Bull.* **170:**1–50 (1979).

35. D. Parker, CDFG, personal communication (1986).

36. J. S. Young and J. D. DeMartini, The reproductive cycle, gonadal histology, and gametogenesis of the red abalone, *Haliotis rufescens* (Swainson). *Calif. Fish Game,* **56:**298 (1970).

37. D. Parker, CDFG, unpublished data (1985).

38. R. T. Deacon, On the economics of fishery dynamics: An age class model of the California abalone fishery. *Univ. Calif., Santa Barbara, Work. Pap. Econ.* **203:**1–41 (1982).

39. T. Doi, S. A. Guzman del Proo, V. Marin, M. Ortiz, J. Camacho, and T. Munoz, Analisis de la poblacion y diagnostico de la pesqueria de abulon amarillo (*Haliotis corrugata*) en el area de Punta Abreojos e Isla Cedros, B.C. *Dir. Gen. Inst. Nac. Pesca, Ser. Cient.* **18:**1–17 (1977).

40. P. R. Sluczanowski, A management oriented model of an abalone fishery whose substocks are subject to pulse fishing. *Can. J. Fish. Aquat. Sci.* **41:**1008 (1984).

41. J. A. Gulland, Can a study of stock and recruitment aid management decisions? *Rapp. P.-V. Reun., Cons. Int. Explor. Mer* **169:**368 (1973).

42. K. J. Sainsbury, Population dynamics and fishery management of the paua, *Haliotis iris*. II. Dynamics and management as examined using a size class population model. *N. Z. J. Mar. Freshwater Res.* **16:**163 (1982).

43. K. Fujino, Genetic studies on Pacific abalone. I. Inbreeding and overdominance as evidenced by biochemical polymorphism in a wild population. *Bull. Jpn. Soc. Sci. Fish.* **44:**357 (1978).

44. Y. Fujio, R. Yamanaka, and P. J. Smith, Genetic variation in marine molluscs, *Bull. Jpn. Soc. Sci. Fish.* **49:**1809 (1983).

45. A. J. Wyner, J. E. Moore, and B. Cicin-Sain, Politics and management of the California abalone fishery. *Mar. Policy* **1:**326 (1977).

46. R. T. Deacon, An Economic Analysis of the California Abalone Fishery and the Experimental Abalone Enhancement Program, *In 1980–1982 Bienn. Rep.*, University of California Sea Grant College, La Jolla, CA, 1982, p. 126.

47. B. Cicin-Sain, Introduction: Exploring conflicts between marine mammals and fisheries. In B. Cicin-Sain, P. M. Grifman, and J. B. Richards, Eds., *Social Science Perspectives*

on Managing Conflicts between Marine Mammals and Fisheries. Marine Policy Program, University of California, Santa Barbara, 1982, p. 1.

48. T. Kafuku and H. Ikenoue, Eds., *Modern Methods of Aquaculture in Japan.* Kodansha Ltd., Tokyo, 1983, Chap. 18.

49. K. Saito, Ocean ranching of abalones and scallops in northern Japan. *Aquaculture* **39:**361 (1984).

50. D. E. Morse, personal communication (1977).

51. D. L. Leighton, M. J. Byhower, J. C. Kelly, G. N. Hooker, and D. E. Morse, Acceleration of development and growth in young green abalone (*Haliotis fulgens*) using warmed effluent seawater. *J. World Maric. Soc.* **12:**170 (1981).

52. Tegner, M. J. and R. A. Butler, A drift tube study of the dispersal potential of green abalone (*Haliotis fulgens*) larvae in the southern California bight: Implications for recovery of depleted populations. *Mar. Ecol. Prog. Ser.* **26:**73 (1985).

53. M. J. Tegner, unpublished data.

54. G. A. Jackson and C. D. Winant, Effect of a kelp forest on coastal currents. *Cont. Shelf Res.* **2:**75 (1983).

55. D. A. Fournier and P. A. Breen, Estimation of abalone mortality rates with growth analysis. *Trans. Am. Fish. Soc.* **112:**403 (1983).

56. G. R. Morgan, Increases in fishing effort in a limited entry fishery—the western rock lobster fishery 1963–1976. *J. Cons. Int. Explor. Mer* **39:**82 (1980).

57. A. J. Harrison, Gastropod fisheries of the Pacific with particular reference to Australian abalone. *Can. Spec. Pub. Fish. Aquat. Sci.* **92:**14 (1986).

18 CONSERVATION AND MANAGEMENT OF QUEEN CONCH (*Strombus gigas*) FISHERIES IN THE CARIBBEAN

Carl J. Berg, Jr.*
Marine Biological Laboratory
Woods Hole, Massachusetts

and

David A. Olsen
Fort Lauderdale, Florida

*Present address. Florida Department of Natural Resources, Bureau of Marine Research, Marathon Florida.

9. Priorities and Approaches for Future Management
    References

1. HISTORY OF THE FISHERY

Few marine snails have received as much attention as the queen conch, *Strombus gigas*, from the Caribbean. Prized for its large, beautiful pink shell (Fig. 1), its pearl, and its delicious meat, this species has been fished from Caribbean waters since prehistoric times. It is only in the past two decades, however, that there developed a sizable commercial fishery for conch. Initially subject only to inter-island trade, the migration of island peoples to the United States brought about the rapid development of a large export market. Now the conch is the subject of numerous symposia and workshops (1–4); the literature on conch has also been reviewed by Randall (5), Berg (6), Brownell and Stevely (7), and Wells et al. (8). Recent annotated bibliographies have been prepared by Darcy (9) and Turnbull (10).

The distribution and abundance of conch have been greatly affected by human fishing pressure. Once common in shallow, even intertidal waters, conch are now caught in progressively deeper or more inaccessible areas. As commercial fishing developed, conch were pulled from deeper water by fishermen using long poles

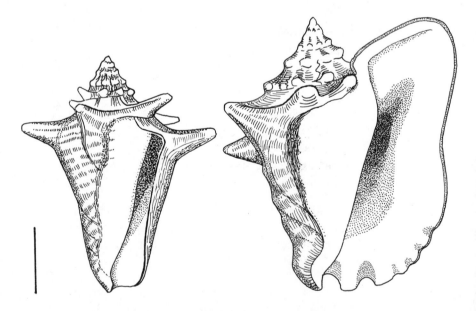

Figure 1. The queen conch, *Strombus gigas*, from the Caribbean. Shell on left is from an immature juvenile without a flared lip; shell on right is from a mature adult. Bar indicates 5 cm.

with hooks or tines on the end (7,11,12). This method has been gradually replaced by the use of diving masks and snorkeling gear, which permits fishing in still deeper waters. Divers using scuba now routinely exploit conch at depths down to 40 m. In addition, depletion of local populations has forced fishermen to replace wooden canoes and skiffs with far-ranging fiber glass boats with powerful engines. In some islands "mother ships," each with numerous small collecting boats, are being used for extended fishing expeditions. The advent of large-scale harvesting has brought with it fishing cooperatives, processing plants, modern freezing technology, and exports to far distant markets.

Brownell and Stevely (7) reviewed the status of conch fisheries throughout the Caribbean, and this was updated by Wells et al. (8) as they placed conch in the IUCN Invertebrate Red Data Book as a "commercially threatened" species. Conch are now protected in Bermuda (13) and Florida (14), with no commercial or private collecting allowed. Because of the increasing importance of this fishery, catch and export data are being reported for new areas, for example, Puerto Rico (15, 16) and Cuba (17). Imports to the United States are closely monitored and have been used as a measure of the status of the fishery (7) for it is believed that those imports represented more than half of the total catch. These data (Fig. 2) show a mean

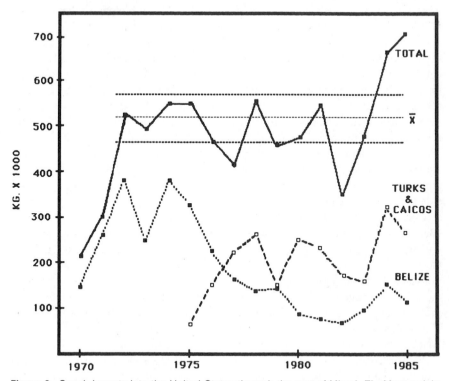

Figure 2. Conch imports into the United States through the port of Miami, FL. Meat weight in thousands of kilograms. Dashed lines indicate ± 10% of the mean value (solid line) for the years 1972–1985.

level of imports of 5.17×10^5 kg for the period 1972–1985, with most values within $\pm 10\%$ of the mean. Imports from the Turks and Caicos Islands and from Belize account for most of U.S. imports. Recent data from Cuba (17), however, show its fishery to be comparable to the total export fishery for the rest of the Caribbean. In 1977 the Cuban fishery landed 23.5×10^5 kg, but this fishing pressure could not be sustained. The fishery crashed and was closed from 1978 to 1982. Landings for 1984 were set at a quota of 7.8×10^5 kg but populations are still decreasing. Because of differences among countries in processing conch meat before weighing, it is difficult to estimate the total number of animals harvested each year. A conservative estimate from 1984 data, based on the realization that not all conch are exported to the United States, would be 10–15 million conch. It is doubtful that the fishery can continue at this level, for fishermen are going farther and deeper to fish, local stocks are becoming depleted, and the few estimates of catch per unit effort (CPUE) available show sharp declines (16, 18–21).

2. GENERAL BIOLOGY AND LIFE HISTORY

Strombus gigas is found in clean waters throughout the Caribbean biological province, where it feeds on algae and detritus. Its general biology is characteristic of most mesogastropods and has been reviewed repeatedly (1, 5–8). Most notable is its determinate shell growth; the shell does not increase in length once a flaring shell lip is produced. Average shell length for adult conch is approximately 20–21 cm, although this varies with area (22). Males are 1–2 cm smaller than females (5, 23–25). The sex ratio in nonbreeding aggregations is 1 : 1 (5, 18, 24, 26). Sexual maturity usually occurs when the conch are roughly 4 yr old, but only after the shell lip is thickened (4 mm); this estimate is based on the presence of mature secondary sex organs (6, 24, 26, 27) and histological analysis of gonads (26, 28). This may be as much as a year after the start of lip formation and entry into the fishery. The time delay between attaining legal length into the Belize fishery (17.8 cm) and becoming sexually mature means that 94% of the population may be fished before they can reproduce (24, 26, 28).

Conch reproduce year round in some areas, as evident by formation of gametes (Belize; 24, 26) or direct observation of copulation (Mexico; J. Ogawa, personal communication). Breeding has been observed to occur from February through October in shallow-water populations (Florida; C. J. Berg, Jr. personal observation) and March through November in deeper water (Virgin Islands; 5, 22). Because there is a bias against observations from deep water and/or winter months, it is probable that at least reduced reproductive activity occurs throughout the year. Hurricanes and winter storms are known to disrupt breeding activity by causing conch to burrow below the sand surface (27, 29).

Female conch are fecund, laying an average of eight egg masses per 6-month breeding season (29), with each mass containing between 3.1×10^5 and 7.5×10^5 larvae (5, 30, and P. Chanley, personal communication). Therefore as many as 6×10^6 larvae may be produced by a female each year. There are no data on age-or size-specific fecundity of conch, nor on natural mortality of larvae, although the latter is assumed to be quite high.

From the egg masses hatch planktonic veliger larvae that are carried by ocean currents. In hatchery conditions the fastest growing take 2 weeks to reach metamorphosis and the nonplanktonic juvenile stage (15, 29), whereas others may take 4–5 weeks (11, 29, 31). Larvae may live much longer (32), but there is no evidence that they would still be competent to metamorphose. Because these planktonic offspring are widely dispersed by currents, it is possible that an island's fishery totally depends upon recruitment of larvae that originated elsewhere. Biochemical/genetic studies of stock differentiation, mediated by such recruitment, are underway (33, 34). Isolated islands, and those "upstream" relative to prevailing currents, may exhibit oceanographic larval retention mechanisms that permit self-sustaining populations. Similarities among downstream conch populations in Belize and the Turks and Caicos Islands suggest considerable larval recruitment from upstream populations throughout the Caribbean (34).

In addition, there may be recruitment into the fishery by juvenile and adult conch that actively migrate up from deep, unfished waters. Hesse (35) observed conch ranging over several kilometers and deep-water populations have been observed at 40–60 m (5, 22, 23, 35) and netted from 160 m (C.J. Berg, Jr., personal observation). The burrowing behavior of adult conch (5, 22, 35) also affects their availability to the fishery.

The combination of high fecundity, extended breeding seasons, and extensive dispersal suggests that recruitment of larval/juvenile conch is prolonged, but because of high larval mortality and changing current patterns, not necessarily constant in time or space (e.g., 23). There is also variability in recruitment of older conch to the fishery, based upon their migratory and burrowing behavior. Taken all together, these factors have a major impact on the size–frequency distributions that are used in calculating population growth and mortality estimates and on the assumptions underlying stock–recruitment models.

3. IMPACTS ON THE RESOURCE

The greatest impact on conch resources of the Caribbean has been from the increased fishing pressure owing to expanding export and local tourist markets. Humans are the greatest predator on conch, for old adult conch appear to have a refuge from predation within their large, heavy shells. Natural predation is heaviest upon smaller conch, with fish, lobsters, crabs, octopus, and even other snails taking their share (5, 36). There is no information concerning competition for natural resources with other species, and removal of conch does not cause a noticeable change in the remaining community structure. Conch are also relatively free of parasites: the only observation being of two conch from Biscayne Bay that each had a pair of nematodes wrapped around their subesophageal ganglia (C.J. Berg, Jr., personal observation), which, because of the ganglia's role in hormone secretion (37), may have affected the conch's reproductive behavior. Finally, habitat destruction by runoff from land development has been observed to kill nearshore populations of strombids in the Pacific (C.J. Berg, Jr., personal observation). This must be considered important for Caribbean populations, but there have been no documented cases. Because conch are no longer common in nearshore waters, habitat destruction, including water pollution, is not as immediate a threat to conch conservation as overfishing.

4. CURRENT METHODS FOR SAMPLING AND SURVEYING POPULATIONS

Because of their ease of collection, most sampling of natural conch populations is simply done with snorkel or scuba gear. Measurement of shell size is by calipers and weighing with a balance. Visual surveys of population density and distribution have been made for stock assessment purposes. These entailed visual censusing while swimming transects (38) or being towed at depth (20). Calculated values based on mean density per habitat type were then extrapolated for the entire island shelf or bank to predict stock size and hence potential yields. Appeldoorn (16) discusses problems inherent in population analyses because of the burrowing behavior of conch and because the habitat of the first year class is unknown and those animals are poorly represented in all samplings.

Measurement of catch and fishing effort is difficult in most areas because of lack of sufficient fisheries personnel and centralized landing sites. Catches are recorded at processing plants in the Turks and Caicos Islands (19, 21), Cuba (17), and Belize (18). Payment is usually based upon pounds landed as the animals are removed from the shell on the fishing grounds. Numbers landed are calculated from approximate values for conch per pound of meat, which can vary appreciably (2–8/lb) depending upon the location fished (39), how well the meat is cleaned, and if juveniles are being taken. On some islands conch are landed in the shell (e.g., St. Lucia), so statistics collectors can also make size and age estimates.

Fishing effort is sometimes included on catch receipts at processing plants, allowing calculation of catch per unit effort (CPUE) (18, 19, 21, 24). Effort is most commonly assessed as days or hours fishing, which does not take into account method, travel time, time spent harvesting other species, and so on. Although man-hours actually spent fishing is a suitable measure, calculations might also include gas consumption or scuba tanks used.

The most difficult measure to obtain, however, is age at capture. Because of the determinate shell growth of conch, growth occurs in two distinct phases. After the flared shell lip is added, all further growth is channeled into shell thickening. Shell length therefore cannot be used as a measure of age for adult animals with thick-lipped shells. As they age, their shells are worn thin and destroyed by boring organisms (5). Rate of lip thickening has been measured using tagged animals (19, 26) and stable isotope ratios (40).

5. GROWTH AND MORTALITY MODELS

5.1. Growth

Growth of conch has been estimated using a variety of techniques and in numerous locations throughout its range (see Table 1). Size–frequency (6, 11, 16, 19, 20, 23), tag–recapture (5, 6, 23, 24, 27, 41), laboratory and field rearing (6, 11, 15, 31, 41–43), and stable isotope data (40) all have been used to calculate age-specific growth rates. Conch exhibit surprising uniformity in growth rates.

Growth rates have been calculated from differences in size–frequency distributions (e.g., 6, 11, 20, 23), which have been analyzed using Harding's probit analysis (44) or Bhattacharya's techniques (45). Both these techniques serve to differentiate size class modes as well as standard deviations for each modal group. In addition, they provide information that can be used to derive abundance within each modal group (and consequently mortality between age classes or cohorts). Use of these techniques requires that the time period between modes is known or can be assumed.

With most animals mean values of modal size groups are used to calculate parameters of the von Bertalanffy growth equation (equation 1) after the method of Walford (46), where the size at time t is plotted against size at time $t + 1$. The natural logarithm of the slope of the regression line is considered to represent the coefficient of catabolism, K, in the equation and the asymptotic maximum size L_∞ is calculated at the point where L_t and L_{t+1} converge. The parameter t_0, a curve-fitting parameter that has been considered to approximate the time at which $L_t = 0$, can be calculated from a regression of the natural logarithm of $(L_\infty - L_t)$ on age:

$$L_t = L_\infty [1 - \exp - K(t - t_0)] \tag{1}$$

The above analysis assumes that modal sizes represent age class cohorts, which may not be valid in areas where spawning occurs evenly throughout the year, as discussed above. Certainly any concentration of larval recruitment in time will create similar peaks. Size class variability can be expected to be greater in areas with prolonged recruitment than in areas where well defined and limited spawning periods occur.

TABLE 1 Growth Parameters from von Bertalanffy Analyses of Conch, *Strombus gigas*, Populations in the Caribbean

Source	Reference	L_∞	K	t_0
St. John, U.S.V.I.	5	26.0	0.516	0
St. Croix, U.S.V.I.	6	20.1	0.589	0
Cuba	23	20.8–38.4	0.287–0.571	−0.05 to +0.13
Turks and Caicos Islands	27	31.8	0.381	−0.08

With conch, shell length–frequency analysis provides accurate results only for the first 3 yr of juvenile shell growth, but not after shell lip formation. Therefore, size–frequency analysis generally provides only two points of shell length or shell thickness for Walford analysis and is of questionable utility for conch data. Analysis of lip thickness has been used to identify age classes in the same manner as shell length of juveniles (20).

Tag–recapture data, however, tend to validate size–frequency estimates of growth rate. Data from tag–recapture studies of natural populations are available from the U.S. Virgin Islands (5), Turks and Caicos Islands (27), Cuba (23), Venezuela (41), and Belize (24). Randall's Virgin Island data (5) and Hesse's (27) unpublished Turks and Caicos Islands data have been analyzed with Fabens' (47) computer program, which calculates the parameters of the von Bertalanffy growth equation in the form

$$L_t = L_\infty \left\{ 1 - b \left[\exp - K(t - t_0) \right] \right\} \tag{2}$$

where b is a parameter related to the size of the animal at birth and is calculated by Faben's program from tag–recapture data and a size at known age. Alcolado (23) used Petersen's direct method with his tag–recapture data to calculate growth rates that were in close agreement with analysis of size class modes from the same area.

There is a surprising degree of uniformity in conch growth data. This can be seen when modal sizes from a variety of sources (Table 2) are plotted over growth curves derived from tag–recapture data (Fig. 3). Modal values lie close to expected values, even though there is no standard "age" when the size data were collected. Some of this effect is shown in Table 2, where Alcolado's (23) modal groups from a number of months are shown. The only major deviations occur when

TABLE 2 Mean Shell Lengths for modal groups of juvenile conch, *Strombus gigas*, populations in the Caribbean[a]

		Modal Group			
Source	Reference	1	2	3	Symbol[a]
St. Thomas, U.S.V.I.	20	9.0	12.6	15.7	1
Turks and Caicos Islands	19	11.5	18.0	20.0	2
Turks and Caicos Islands	27	12.1	17.8	21.2	3
Puerto Rico	6	8.8	12.6	18.0	4
Venezuela	11	7.6	12.8	18.0	5
Cuba	23				
Diego Perez A—October		7.7	13.3	17.0	6
Diego Perez A—December		10.3	14.3	17.3	7
Diego Perez B—May		6.8	12.5	15.5	8
Diego Perez B—August		9.0	13.8	15.8	9
Cayo Anclitas—July		11.5	18.5	21.5	0

[a]Symbols are those used in Figure 3.

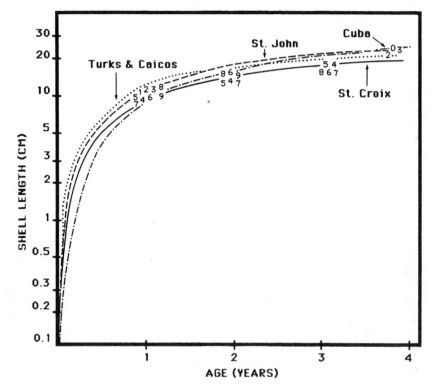

Figure 3. Von Bertalanffy growth curve derived from Randall's tag–recapture data for the Virgin Islands (6). Mean sizes of modal groups from throughout the Caribbean from Table 2.

intensive harvest of young conch has been suspected to account for a reduced year-3 mean size (6, 20) or the under-representation of year-1 conch owing to their cryptic nature. Data for Cuba were selected from Alcolado's Table 2 (23) for areas with little fishing pressure, for large sample size, and for samples where three modal groups were represented.

5.2. Mortality

Mortality occurring in conch populations has been estimated from size–frequency distributions (19, 20, 23), from tag–recapture of natural populations (6, 23, 27, 41, 48) and enclosed natural populations (17, 48), and from laboratory-reared field-released populations (41–43, 49). These studies all provide estimates of annual mortality rates from

$$N_{t+1} = N_t(\exp -Zt) \tag{3}$$

where N is number of individuals at time t and $t + 1$, e is base of natural logarithms, and Z is annual coefficient of total mortality (i.e., the sum of fishing mortality F and natural mortality rate M).

Size-class abundance can be derived either by Bhattacharya's technique (45) or after the method of Olsen (50). In the latter method, the modal size groups and standard deviations are derived by probit analysis. A z-score is calculated by dividing the means for each modal group by the standard deviation for each group. The expected frequency for the upper and lower bonds of the modal interval is obtained from a table of normal probabilities at each calculated z-score. Size-class abundance is calculated by dividing sum of the expected frequencies into the abundance within the modal interval for each size class (50).

Natural mortality M can be derived from the unfished year classes by size–frequency analysis (19, 20) or from protected unfished populations (17). Appeldoorn (16, 42) has compared published values for natural mortality over the complete age–size range of conch and noted the dynamic nature of the relationship and the variability in published estimates. As an example, in Cuba (17) 3200 juvenile conch were placed in a 5000 m^2 enclosure. One year later there were 2365 left alive and the investigators were able to account for 243 deaths. The remaining 592 were considered as having escaped or been stolen. This provides an annual estimate of mortality $M = 0.097$, which is close to the size–frequency derived estimates ($M = 0.115$) of Wood and Olsen (20) for the Virgin islands and Olsen's value of $M = 0.081$ for the Turks and Caicos Islands (19). Ursin (51) observed that natural mortality generally approximated the negative cube root of total weight (grams) for a wide variety of organisms for which direct measurements had been recorded. For conch, the natural mortality of an animal that weighs 616 g would be $M = 0.118$; surprisingly close to the figures reported above. It appears that natural mortality can be approximated at $M = 0.1$ (90.4% annual survival rate).

Appledoorn (16) argued that size–frequency derived estimates of natural mortality based on the abundance of the first two size classes would provide under-estimates of natural mortality owing to the cryptic nature of the young animals. The close agreement mentioned above between size–frequency derived estimates (19, 20), enclosed controlled populations (17), and a widely based theoretic value are considered to validate the size–frequency technique. Higher values for M frequently fail to account for unknown fishing pressure.

Fishing mortality rate F has been estimated to range from $F = 0.586$ in a lightly fished population in the Turks and Caicos Islands (27) to as high as $F = 2.90$ in a heavily fished population in Puerto Rico (6). Fishing mortality is site-specific and should be recalculated for each area under consideration.

6. YIELD MODELS

6.1. Estimation of Yield

Regression equations for converting linear growth measurements into total weight (shell and animal) are available, especially for juvenile and young adult conch (5,

6, 20, 23), and all show a strong correlation. Wood and Olsen (20) calculate total weight (grams) from shell length (centimeters), with $n = 514$, $r = 0.98$, $p \leq .01$, from

$$W = 0.0164L^{3.713} \tag{4}$$

They calculate total weight in grams from shell lip thickness T in millimeters of adult conch, with $n = 135$, $r = 0.52$, $p \leq .01$, from

$$W = 885T^{0.256} \tag{5}$$

They calculate marketable weight MW (grams) from total weight TW (grams) from equation 6, with $n = 102$, $r = 0.95$, $p \leq .01$. Using these regressions, maximum meat weight W_∞ was found to equal 231.6 g (20), which is used in yield-per-recruit (Y/R) analysis.

$$MW = 6.465 + 0.077(TW) \tag{6}$$

Berg (6) also provides equations for calculating total weight and marketable weight from shell length, and marketable weight from total weight, but these were derived only for shells without a flared lip and predict a much higher yield than those of Wood and Olsen (20). Analysis of four adult conch populations from St. Lucia showed only weak correlations ($r = 0.06–0.44$) between shell thickness and marketable weight (Berg, personal observation). Figure 5 of Alcolado (23) illustrates how variable the relationship of total weight to shell length can be among populations, where total weight of a 20-cm-long shell is calculated to range from approximately 700 to 1200 g.

Values of marketable weight for the above studies (6, 20) measured the weight of the animal taken from its shell and with only the viscera removed, as it is sold in native markets. Processing plants remove all organs and skin from the columellar muscle, reducing the effective yield of processed meat to 40% of total animal body weight (viscera attached) as it is purchased directly from the fishermen (19). Olsen (19) estimates a 12% reduction from market weight to processed weight. Care must be taken in comparing landings or exports values to be sure that the meat has been processed to similar levels.

6.2. Yield per Recruit

Because the extensive conch fisheries are found on shallow banks where they are easily accessible to fishing effort, they are extremely susceptible to overfishing. An examination of landing records from most fisheries (e.g., 11, 17, 19, 21, 24) shows that these fisheries seem to cycle between very high landings and virtual elimination of the resource (Fig. 4). It appears obvious that extensive harvest of pre-reproductive individuals can lead to the collapse of any fishery. Y/R analysis (Ricker 52) of conch populations can provide guidelines for management of conch

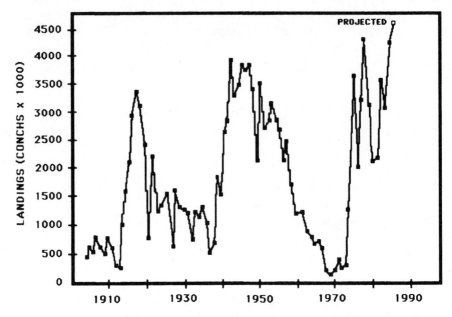

Figure 4. Conch landings (primarily export figures) in the Turks and Caicos Islands from 1904 to present. Values are given in thousands of conchs. The 1985–1986 landing is projected from April to August landings when normally 67% of the total annual figure is landed. See text for details and sources of data.

fisheries, in addition to being used for calculation of maximum sustainable yield (MSY).

Yield per recruit (Y) can be calculated from the Beverton–Holt formulation (52) from

$$Y = FN_0 \, e^{-Mr} W_\infty \left(\frac{1}{2} - \frac{3e^{-2Kr}}{Z + K} + \frac{3e^{-2Kr}}{Z + 2K} - \frac{e^{-3Kr}}{Z + 3K} \right) \qquad (7)$$

where F = rate of fishing
N_0 = number of recruits
M = natural mortality rate
T_r = average age at recruitment to the fishery
$r = T_r - t$,
W_∞ = asymptotic weight at L_∞
Z = total mortality rate ($M + F$)
K = the coefficient of growth from equation 1.

Yield per recruit analysis can suggest the age (or size) at which the fishery can realize the maximum return (in weight) from its resource. Wood and Olsen (20) used this technique to plot yield isopleths and calculated that Y/R is maximized at 80 g when conch are harvested between 3.0 and 5.1 yr of age. Olsen (19) calculated that maximum Y/R of Turks and Caicos Islands conch (100 g) occurs after 2.9 yr of age, even though the average age of recruitment to the fishery (T_r) is 2.78 yr (Fig. 5). Both studies were based on the growth parameters derived by Berg (6) from Randall's data (5). Y/R estimates are greatly affected by both growth and mortality rates. Natural mortality rates of Olsen (19, 20) and Munoz et al. (17) appear reasonable for large conch that have been recruited into the fishery. When equation 7 is calculated with the coefficient of growth (K) varying between 0.4 and 0.6 (the probable range for conch growth on the banks), the maximum Y/R increases from 96 g/recruit at $K = 0.4$ to 127 g/recruit at $K = 0.6$. In addition, at any given level of fishing effort, higher yield is available. This is presented as an indication of the importance of accurate growth parameter estimates.

In cases where extreme fishing pressure lowers the average age at recruitment, the Y/R function responds by providing higher yields at lower ages. In fact, however, with an animal whose populations consist of a relatively small number of year classes, this strategy endangers the resources' reproductive and regenerative capabilities.

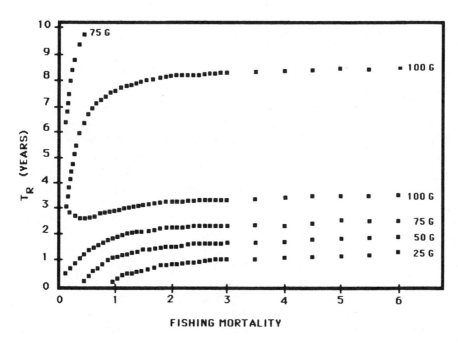

Figure 5. Yield per recruit analysis for *Strombus gigas* using growth and mortality figures given in text. Isopleth values are shown in grams per recruit.

6.3. Maximum Sustainable Yield

The Turks and Caicos conch fishery offers the most complete data source for analysis of the yield potential and population dynamics of an active fishery for *Strombus gigas*. Records exist from 1904 indicating the export of conch from the islands. Effort statistics, given in conch per man-day, have been derived from 1976 through 1985 (19, 21) and can be used to calculate maximum sustainable yield by the surplus production method outlined by Ricker (52). Some CPUE data are also available for the Belize fishery (18, 24).

If the Turks and Caicos fishery can be assumed to be in equilibrium, then the surplus production of the resource available for harvest can be described by a parabolic function of the equilibrium yield (Y_e) and fishing effort (f_e) (52). This function is

$$Y_e = a - b(f_e) \tag{8}$$

where f_e is the fishery effort at equilibrium effort, and a and b are the intercept and slope of a regression of Y_e/f_e (CPUE) on f_e. The maximum sustainable yield (Y_s) can be calculated from

$$Y_s = a^2/4b \tag{9}$$

and the optimum level of fishing effort (f_s) can be calculated from

$$f_s = a/2b \tag{10}$$

Maximum sustainable yield (Y_s) was estimated from equation 9 using the CPUE (number of conch per man-day) from Nardi (21). CPUE from 1982 to the present was derived from a sample of 50 landings per month, taken from the landing tickets (19). The regression of CPUE on total effort (Fig. 6) was significant at the $p =$ .01 level ($r = 0.713$, df $= 11$). The slope of the regression line was $b = -0.0115$ and the intercept $a = 426$. The sustainable yield derived from equation 9 was 3.92 million conch (1.57 million lb) and the optimum level of fishing effort (f_s) was 18,425 man-days (Fig. 7). A review of the landings indicates that the sustainable yield was seldom exceeded in the Turks and Caicos conch fishery until the current decade. The 1985–1986 catch can be projected from the normal annual cycle of landings and preliminary data from April until August. The projected figure is 1.9 million lb (4.75 million conch), substantially in excess of the sustainable yield. The increase in effort appears to be related to a declining lobster resource, which is causing Caicos Island fishermen to spend more effort fishing conch.

Yield per recruit analysis was used by Wood and Olsen (20) to estimate the potential of the two U.S. Virgin Islands shelves. They utilized Berg's (6) growth parameters because size class means approximate that curve, which was also derived from Virgin Island conch. Natural mortality was calculated from the year-1 and -2 size class abundances (see Section 5.2), which were the only unfished

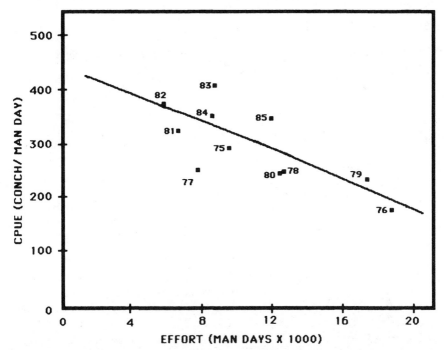

Figure 6. Catch per unit effort (conch/man-day) when plotted against total effort (man-days × 1000) is used to calculate the parameters of the surplus production model and to compute maximum sustainable yield for the Turks and Caicos Islands.

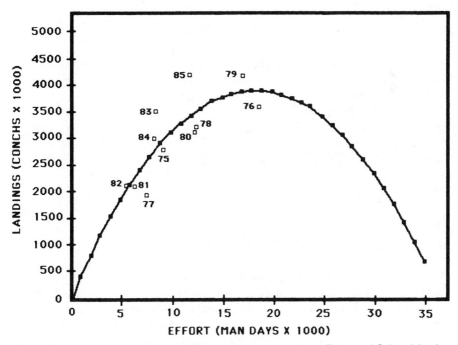

Figure 7. Maximum sustainable yield (thousands of conchs) in the Turks and Caicos Islands. Actual landings are shown with year.

TABLE 3   **Maximum Sustainable Yield Estimates for Conch, *Strombus gigas*, in the Caribbean**

Area	References	Method[a]	MSY (kg/km^2)	Recalculated
Little Bahama Bank	38	VSB	152–263	19
Great Bahama Bank	38	VSB	54–93	6.8
St. Thomas, U.S.V.I.	20	Y/R	101	
St. Croix, U.S.V.I.	20	Y/R	79	
Caicos Bank	19	SPM	116	

[a]VSB, virgin stock biomass (Gulland 53); Y/R, yield per recruit (Ricker 52); SPM, surplus production model (Schaeffer 54).

age classes. The number of recruits present was determined by a visual census technique using scuba and a towed submersible. The standing stock was adjusted to the number of recruits by taking the average estimated age of the censused populations and back-calculating to the age at recruitment (T_r) when maximum Y/R occurred. The results are shown in Table 3.

Smith and van Nierop (38) censused populations of commercially important species on the Little and Great Bahama Banks. Their estimates of sustainable yield (Table 3) were obtained by using census data together with Gulland's (53) virgin stock biomass estimate

$$MSY = 0.5MB_0 \tag{11}$$

where MSY is a function of natural mortality rate (M) and biomass of standing stock (B_0). Gulland (55) has recently suggested that the constant 0.5 in equation 11 may be too high. Smith and van Nierop (38) used Alcolado's (23) and Wood and Olsen's (20) estimates of total mortality, in place of natural mortality, to calculate their sustainable yield estimates. Using Wood and Olsen's value of ($M = 0.12$), the new value for MSY of the Little Bahama Bank is 19 kg/km^2 and 6.8 kg/km^2 for that of the Great Bahama Bank. These values are much closer to the actual yield (11 and 3 kg/km^2, respectively) and underscore the need for accurate estimates of mortality.

7. CURRENT FISHERIES REGULATIONS

Because of heavy overfishing, conch have achieved a "protected" status in Bermuda (13) and Florida (14) and the fishery has been closed for prolonged periods in Bonaire (56) and Cuba (17). In other countries the fishery is regulated through a variety of means (1, 7, 8) including limited entry and licensing (Venezuela, Belize, Bonaire), by prohibiting the use of scuba (French West Indies, Belize, Turks and Caicos Islands), by setting aside closed areas (Bonaire), by setting fishery catch quotas (Belize, Cuba), and by protecting breeding or immature conch through shell or meat size limits (Bahamas, Belize, Turks and Caicos Islands).

The fishery is closed during the summer breeding season in Belize (July–September), but open only during that period in Venezuela (May–November, 41). Pre-reproductive conch are protected by regulations against taking non-flared lipped animals (Bahamas, Belize, Turks and Caicos Islands) and by minimum size limits on the shell (7 in. in Belize and Turks and Caicos Islands, 20 cm in Bonaire). Belize also requires that processed meats weigh more than 3 oz and the Turks and Caicos Islands require that meats measure greater than 4 in. from the base of the operculum to the end of the white meat.

The Organization of Eastern Caribbean States has proposed harmonized regulations for its membership that prohibit the use of scuba, require a shell with a *thickened* flared lip, 18 cm shell length, and 8 oz marketable (not processed) meat weight. The requirement of a *thickened* shell lip is most important, because it allows the animal to become sexually mature before entry into the fishery. By prohibiting the use of scuba, it is hoped that a deep-water reproductive population will be protected.

The major problem with all fisheries regulations in the Caribbean region is that there are no effective means for their enforcement. This is especially true on small islands where everyone is related and close family ties would discourage the arrest or censure of a relative. Where processing plants are operating, monitoring and enforcement are leveled at the owners or the fishing cooperatives. Most important for management, however, is an informed and sympathetic populace. Many islanders are ignorant of the basic biology of conch, thinking that they simply spring up from mines in the ground, and that juvenile conch belong to a different species. Conservation and management education programs are a must for all levels of island society.

8. SOCIOECONOMIC CONSIDERATIONS

Depletion of conch populations near human population centers has necessitated increased capitalization in the form of faster boats, larger engines, and scuba. As a consequence, operating expenses have also increased, which decreases profitability and also decreases national benefits because much of these increased costs are associated with imported goods that must be imported with hard currency.

Economic analysis of the Turks and Caicos Islands fishery (19) indicated that conch and lobster fishermen have gross incomes of approximately U.S. $7700, with approximately $2500 in expenses. Analysis of processing plants on those islands indicated that the simple break-even point (without consideration of capital investment, interest, maintenance, etc.) occurred at a sale price of U.S. $2.50/lb (average sale price was $2.80/lb).

The expansion of the conch fishery into deeper waters through the use of scuba has led to serious health problems from decompression sickness. Fishermen in Puerto Rico and St. Croix (U.S. Virgin Islands) have severely overtaxed available decompression facilities. We estimate that as many as 25% of the conch fishermen using scuba in the Grenadines have suffered from decompression sickness at least once; in some instances it has been fatal.

9. PRIORITIES AND APPROACHES FOR FUTURE MANAGEMENT

Conch are being heavily fished throughout the Caribbean as new and expanding markets increase demand. Plans for effective management must be developed and enacted soon. The summary evaluations and recommendations that came from recent gatherings of conch specialists and fisheries officers (1–3) still hold true.

Because of extensive larval drift, the maintenance of conch stocks in the Caribbean is a multinational problem. Regional harmonized regulations have been proposed to ensure the maintenance of breeding populations. The adoption of these measures should be a priority among Caribbean nations.

Recent evidence suggests that because of deep-water populations and long-term larval drift, depleted populations can recover naturally if totally protected. For isolated and up-current populations such recovery might be slow, as observed for Barbados and Bermuda (34), but for downstream areas it would be faster, as observed for Cuba (17). Perhaps these factors have maintained recruitment to the Belize and Turks and Caicos conch fisheries, in spite of heavy local exploitation.

The enhancement or rehabilitation of conch stocks through aquaculture was proposed over a decade ago (6) and the appropriate technology has been developing ever since (6, 11, 15, 17, 29, 31, 41, 56). Now two large conch hatcheries are producing thousands of juvenile conch annually (57, 58). There are still many problems to be overcome, including excessive juvenile mortality upon reseeding (42, 43, 49, 50), high production costs (49, 57, 59), and the possibility of artificial genetic manipulation of stocks (60). More research and development are needed along these lines before costly attempts at stock rehabilitation are attempted.

ACKNOWLEDGMENTS

We thank members of the fisheries departments throughout the Caribbean for their assistance in gathering samples and providing data, and Katherine Orr for providing unpublished data and reviewing the manuscript. This work was supported, in part, by grant #3638 from the World Wildlife Fund—U.S.

REFERENCES

1. C. J. Berg, Jr., Ed., *Proceedings of the Queen Conch Fisheries and Mariculture Meeting*. Wallace Groves Aquacult. Found., Discovery House, Freeport, Bahamas, 1981, 46 pp.

2. J. B. Higman, Ed., Conch mariculture. *Proc. Annu. Gulf Caribb. Fish. Inst.* **35:**43–135 (1983).

3. F. Williams, Ed., Conch Fisheries Management Workshop. *Proc. Annu. Gulf Caribb. Fish. Inst.* 38 (in press).

4. R. Mann, Ed., *J. Shellfish Res.* **4:**1–74 (1986).

5. J. E. Randall, Contributions to the biology of the queen conch, *Strombus gigas*. *Bull. Mar. Sci. Gulf Caribb.* **14:**246–295 (1964).

6. C. J. Berg, Jr., Growth of the queen conch *Strombus gigas*, with a discussion of the practicality of its mariculture. *Mar. Biol. (Berlin)* **34:**191–199 (1976).

7. W. N. Brownell and J. M. Stevely, The biology, fisheries, and management of the queen conch, *Strombus gigas*. *Mar. Fish. Rev.* **43:**1–12 (1981).

8. S. M. Wells, R. M. Pyle, and N. M. Collins, Queen or Pink Conch. In *The IUCN Invertebrate Red Data Book*. IUCN, Gland, Switzerland, 1983, pp. 79–90.

9. G. H. Darcy, Annotated bibliography of the conch genus *Strombus* (Gastropoda, Strombidae) in the Western Atlantic Ocean. *NOAA Tech. Rep. NMFS SSRF* NMFS SSRF-**748:**1–16 (1981).

10. D. Turnbull, *Strombus gigas* (the Queen Conch) and *Turbinella pyrum* (the Indian Chank): A selected bibliography. *Int. Dev. Res. Cent., Manuscr. Rep.* **42e:**1–15 (1981).

11. W. N. Brownell, C. J. Berg, Jr., and K. C. Haines, Fisheries and aquaculture of the conch, *Strombus gigas* in the Caribbean. *FAO Fish. Rep.* **200:**59–69 (1977).

12. C. O. Hesse and K. Hesse, Conch industry in the Turks and Caicos Islands. *Underwater Nat.* **10:**4–9 (1977).

13. J. Burnett-Herkes, Bermuda. In C. J. Berg, Jr., Ed., *Proceedings of the Queen Conch Fisheries and Mariculture Meeting*. Wallace Groves Aquacult. Found., Discovery House, Freeport, Bahamas, pp. 35–37.

14. J. H. Hunt, Status of queen conch (*Strombus gigas*) management in the Florida Keys, U.S.A. *Proc. Annu. Gulf Caribb. Fish. Inst.* 38 (in press).

15. D. L. Ballantine and R. S. Appeldoorn, Queen conch culture and future prospects in Puerto Rico. *Proc. Annu. Gulf. Caribb. Fish. Inst.* **35:**57–63 (1983).

16. R. S. Appeldoorn, Practical considerations in the assessment of queen conch fisheries and population dynamics. *Proc. Annu. Gulf Caribb. Fish. Inst.* **38** (in press).

17. L. Munoz, P. Alcolado, I. Fraga, and P. Llorente, Status of populations and fisheries of *Strombus gigas* in Cuba, with some results of juvenile rearing in pens. *Proc. Annu. Gulf Caribb. Fish. Inst.* 38 (in press).

18. J. Gibson, 1981. A summary of conch fisheries in Belize. In C. J. Berg, Jr., Ed., *Proceedings of the Queen Conch Fisheries and Mariculture Meeting*. Wallace Groves Aquacult. Found., Discovery House, Freeport, Bahamas, 1981, Appendix 3, 5 pp.

19. D. A. Olsen, 1985. *Fishery Resource Assessment of the Turks and Caicos Islands*, Final report on FAO Project TCI/83/002. FAO, Rome, 1985, 94 pp.

20. R. S. Wood and D. A. Olsen, Application of biological knowledge to the management of the Virgin Islands conch fishery. *Proc. Annu. Gulf. Caribb. Fish. Inst.* **35:**112–121 (1983).

21. G. C. Nardi, An analysis of the queen conch fishery of the Turks and Caicos Islands, with a review of a new, multi-purpose dock receipt. M.S. Thesis, State University of New York, Stony Brook, 1982, 47 pp.

22. M. L. Coulston, R. W. Berey, A. C. Dempsey, and P. Odum, Assessment of the queen conch (*Strombus gigas*) population and predation studies of hatchery reared juveniles in Salt River Canyon, St. Croix, V. I. *Proc. Annu. Gulf Caribb. Fish. Inst.* **38** (in press).

23. P. M. Alcolado, Crecimiento, variaciones morfologicas de la concha y algunas datos biologicos del cobo *Strombus gigas* L. (Mollusca, Mesogastropoda). *Acad. Cienc. Cuba, Ser. Oceanol.* **34**:1–36 (1976).

24. H. L. Blakesley, 1977. *A Contribution to the Fisheries and Biology of the Queen Conch in Belize.* Report of the Fisheries Unit Lab., 1977, 31 pp.; *Am. Fish. Soc., Annu. Meet.* **107**:12 (1977) (abstr.).

25. P. L. Percharde, Notes on distribution and underwater observations on the molluscan genus *Strombus* as found in the waters of Trinidad and Tobago. *Caribb. J. Sci.* **8**:47–55 (1968).

26. B. D. Egan, 1985. Aspects of the reproductive biology of *Strombus gigas*. M.S. Thesis, University of British Columbia, 1968, x + 147 pp.

27. K. O. Hesse, An ecological study of the queen conch, *Strombus gigas*. M.S. Thesis, University of Connecticut, Storrs, 1976, 107 pp.

28. J. Gibson, S. Strasdine, and K. Gonzales, The status of the conch industry of Belize. *Proc. Annu. Gulf Caribb. Fish. Inst.* **35**:99–107 (1983).

29. M. Davis and C. Hesse, Third world level conch mariculture in the Turks and Caicos Islands. *Proc. Annu. Gulf. Caribb. Fish. Inst.* **35**:73–82 (1983).

30. R. Robertson, Observations on the spawn and veligers of conchs (*Strombus*) in the Bahamas. *Proc. Malacol. Soc. London* **33**:164–171 (1959).

31. W. N. Brownell, Reproduction, laboratory culture, and growth of *Strombus gigas, S. costatus* and *S. pugilus* in Los Roques, Venezuela. *Bull. Mar. Sci.* **27**:668–680 (1977).

32. C. N. D'Asaro, Organogenesis, development, and metamorphosis in the queen conch, *Strombus gigas*, with notes on breeding habits. *Bull. Mar. Sci.* **15**:359–416 (1965).

33. C. J. Berg, Jr., K. S. Orr, and J. B. Mitton, Genetic variation in the queen conch, *Strombus gigas*, across its geographic range. Preliminary results. *Biol. Bull. (Woods Hole, Mass.)* **165**:505 (1983).

34. C. J. Berg, Jr., J. B. Mitton, and K. S. Orr, Genetic analysis of the queen conch *Strombus gigas*. I. Preliminary implications for fisheries management. *Proc. Annu. Gulf Caribb. Fish. Inst.* **37**:112–118 (1986).

35. K. O. Hesse, Movements and migration of the queen conch, *Strombus gigas*, in the Turks and Caicos Islands. *Caribb. J. Sci.* **16**:105–107 (1979).

36. D. E. Jory and E. S. Iversen, Queen conch predators: Not a roadblock to mariculture. *Proc. Annu. Gulf Caribb. Fish. Inst.* **35**:108–111 (1983).

37. J. L. Ram, M. L. Ram, and J. P. Davis, Hormonal control of reproduction in *Busycon:* Laying of egg-containing capsules caused by nervous system extracts and further characterization of the substance causing egg capsule laying. *Biol. Bull. (Woods Hole, Mass.)* **162**:360–370 (1982).

38. G. B. Smith and M. van Nierop, Distribution, abundance and potential yield of shallow-water fishery resources of the Little and Great Bahama Banks. *UNDP/FAO Fish. Dev. Proj.* BHA/82/002:1–78 (1984).

39. C. Higgs, 1981. Bahamas. In C. J. Berg, Jr., Ed., *Proceedings of the Queen Conch*

Fisheries and Mariculture Meeting. Wallace Groves Aquacult. Found., Discovery House, Freeport, Bahamas, 1981, pp. 32–34.

40. G. Wefer and J. S. Killingly, Growth histories of strombid snails from Bermuda recorded in their O-18 and C-13 profiles. *Mar. Biol. (Berlin)* **60:**129–135 (1980).

41. R. A. Laughlin and E. Weil, M. Queen conch mariculture and restoration in the Archipielago de Los Roques: Preliminary results. *Proc. Annu. Gulf Caribb. Fish. Inst.* **35:**64–72 (1983).

42. R. S. Appeldoorn and D.L. Ballantine, Field release of cultured queen conch in Puerto Rico: Implications for stock restoration. *Proc. Annu. Gulf Caribb. Fish. Inst.* **35:**89–98 (1983).

43. R. S. Appeldoorn, Growth, mortality and dispersion of juvenile, laboratory-reared conchs, *Strombus gigas* and *S. costatus*, released at an offshore site. *Bull. Mar. Sci.* **37:**785–793 (1985).

44. J. D. Harding, The use of probability paper for the graphical analysis of polymodal frequency distributions. *J. Mar. Biol. Assoc. U.K.* **28:**141–153 (1949).

45. D. Pauly and J. F. Caddy, A modification of Bhattacharya's method for the analysis of mixtures of normal distributions. *FAO Fish. Circ.* **781:**1–16 (1985).

46. L. A. Walford, A new graphic method of describing the growth of animals. *Biol. Bull. (Woods Hole, Mass.)* **90:**141–147 (1946).

47. A. J. Fabens, Properties and fitting of the von Bertalanffy growth curve. *Growth* **29:**265–289 (1965).

48. E. S. Iversen, Feasibility of increasing Bahamian conch production by mariculture. *Proc. Annu. Gulf Caribb. Fish. Inst.* **35:**83–88 (1983).

49. R. S. Appeldoorn, The effect of size on mortality of small juvenile conchs (*Strombus gigas* Linne and *S. costatus* Gmelin). *J. Shellfish Res.* **4:**37–43 (1984).

50. D. A. Olsen and I. G. Koblick, Population dynamics, ecology and behavior of spiny lobsters, *Panulirus argus* of St. John, U.S.V.I. (II) Growth and mortality. *Bull. Natl. Hist. Mus. Los Angeles County* **20:**17–21 (1975).

51. E. Ursin, A mathematical model of some aspects of fish growth, respiration, and mortality. *J. Fish. Res. Board Can.* **24:**2355–2453 (1967).

52. W. E. Ricker, Computation and interpretation of biological statistics of fish populations. *Bull., Fish. Res. Board Can.* **191:**1–382 (1975).

53. J. A. Gulland, *The Fish Resources of the Ocean.* Fishing News (Books), West Byfleet, Surrey, England, 1971, 255 pp.

54. M. B. Schaeffer, Some aspects of the dynamics of populations important to the management of the commercial marine fisheries. *Bull. Int.-Am. Trop. Tuna Comm. Bull.* **2:**247–268 (1954).

55. J. A. Gulland, *Fish Stock Assessment: A Manual of Basic Methods.* John Wiley, New York, 1983, 223 pp.

56. R. R. Hensen, Queen conch management and culture in the Netherlands Antilles. *Proc. Annu. Gulf Caribb. Fish. Inst.* **35:**53–56 (1983).

57. M. Davis, C. Hesse, and G. Hodgkins, Commercial hatchery produced conch seed for the research and grow-out market. *Proc. Annu. Gulf Caribb. Fish. Inst.* 38 (in press).

58. R. Cruz, Avances en la experimentacion de produccion massiva de caracol en Quintana Roo, Mexico. *Proc. Annu. Gulf Caribb. Fish. Inst.* **37:**12–20 (1986).

59. S. E. Siddall, Biological and economic outlook for hatchery production of juvenile queen conch. *Proc. Annu. Gulf Caribb. Fish. Inst.* **35**:46–52 (1983).

60. C. J. Berg, Jr., M. H. Goodwin, W. G. Lyons, J. McVey, and D. Miller, Panel summary and evaluations of the conch workshop. *Proc. Annu. Gulf Caribb. Fish. Inst.* 38 (in press).

19 IMPACTS OF THE PRECIOUS SHELL HARVEST AND TRADE: CONSERVATION OF RARE OR FRAGILE RESOURCES

S. M. Wells
IUCN Conservation Monitoring Centre, Cambridge, England

1. Introduction
2. The Commercial Shell Trade
3. The Ornamental Shell Trade
4. Impacts of Shell Collecting
5. Life Cycles and Growth Rates
6. Management Options
7. Protected Areas
8. Mariculture
   References

1. INTRODUCTION

The shells of mollusks have been highly valued by humans since prehistoric times. They have been used for a variety of purposes including currency, magical and religious symbols, and purely ornamental objects. Ground shells produce a fine lime used in pottery glazes, poultry food additives, and in south Asia for chewing with betel nuts. Where shells are numerous enough to be dredged, they are used for road foundations and building materials. This chapter is concerned only with those that are of economic value because of their esthetic appeal, most of which

443

are tropical species. Despite their importance, particularly in developing countries, they have received very little scientific attention compared with species valued as food.

Detailed records of catches are rarely kept for miscellaneous marine products because the fisheries are often artisanal. FAO statistics (FAO 1) considerably underestimate production. International trade statistics give some indication of exploitation levels but ornamental shells are often included with other commodities (Wells 2). Dealers in the rarer specimen shells operate by mail order or through personal contact (Fig. 1). Total world production is not known, therefore, but some idea of the trade can be obtained from the figures quoted below for individual countries and products. There are two broad categories: "commercial shell" and "ornamental (or precious) shell."

2. THE COMMERCIAL SHELL TRADE

"Commercial shell" refers to mollusks harvested for their nacre or mother-of-pearl. Trochus (*Trochus niloticus*), the gold-lip (*Pinctada maxima*) and black-lip (*P. margaritifera*) pearl oysters, and the green snail (*Turbo marmoratus*), all from the Indo-Pacific, are the main species, although others (*Pinctada* spp., *Haliotis*

Figure 1. Table lamps made of spider shells (*Lambis lambis*), helmet shells (*Cypraecassis rufa*), *Bursa spp.*, and *Charonia spp.* for sale in a Mombasa tourist shop (December 1978).

spp., and freshwater mussels) are sometimes used. Commercial fisheries are found in India (Andamans), Indonesia, Philippines, Australia, Papua New Guinea, New Caledonia, Vanuatu, Fiji, French Polynesia, Solomon Islands, and several other Pacific countries (FAO 1; Wells 2; Carleton 3). Most commercial shell is destined for Southeast Asia, which is the center of the button and shellcraft industries. For example, Japan exports about 75 tonnes of pearl buttons and in 1982, 114 tonnes of shellcraft a year; Taiwan exported 300 tonnes of worked mother-of-pearl in 1981; and the Philippines exported 114 tonnes of worked mother-of-pearl in 1981 (Carleton 3).

Trochus is produced in the greatest quantities (Heslinga and Hillman 4; Heslinga 5). Production declined briefly with the advent of plastics but at the beginning of the 1980s there was an annual demand of about 6000 tonnes (Bouchet and Bour 6). The main suppliers are New Caledonia (exported 1915 tonnes in 1978), Indonesia (exported 1220 tonnes in 1981), Solomon Islands (production of 399 tonnes in 1980), Fiji (exported 316 tonnes in 1980), the Philippines (exported 203 tonnes in 1981), and Vanuatu (exported 68 tonnes in 1981) (FAO 1; Wells 2; Bour and Hoffschir 7). In the South Pacific trochus is the only source of cash income on some of the remote islands (Bouchet and Bour 6; Glucksman and Lindholm 8; Salvat 9). Dockside prices in Palau, for example, have increased by about 500% over the last decade (Heslinga and Hillmann 4) and world trade is valued at U.S. $4 million (Heslinga et al. 10).

The gold-lip and black-lip pearl oysters are used extensively for shellcraft and are valued for their pearls. Exporters include Indonesia (exported 770 tonnes in 1981), the Philippines (exports 200–500 tonnes annually), Australia (produced 73 tonnes in 1979), Japan (produced 48 tonnes in 1980), and Fiji (produced 6 tonnes in 1980). *P. margaritifera* is cultured in French Polynesia, the fishery for its shell having died out earlier this century. There is little information on the green snail, which is highly valued in Korea and China for furniture inlay. The main suppliers are Malaysia (production of 400 tonnes in 1980), Andamans (105 tonnes landed in 1976) (Appukuttan 11), Indonesia (77 tonnes exported in 1981), Papua New Guinea (annual export of 60 imperial tons in early 1960s), the Philippines (32 tonnes exported in 1978), Vanuatu (annual export of 10–20 tonnes), and the Solomon Islands (exported 15 tonnes in 1978).

3. THE ORNAMENTAL SHELL TRADE

The ornamental shell trade includes shells exploited whole for their decorative value, rare and precious shells sought by shell collectors or conchologists, and those used in shellcraft. Shell collecting, a favorite hobby in the nineteenth century, has undergone a recent revival, largely because of growth in the tourist and leisure industries. The trade is centered in the Philippines (Wells 2, 12). Exports rose rapidly from less than 1000 tonnes/yr in the 1960s to more than 3400 tonnes in 1979. In 1981, 2300 tonnes was exported, valued at U.S. $3.5 million. Other important suppliers include Mexico, India, and Haiti. Countries with their own supplies, such as Kenya, now often import ornamental shells if they are cheaper or of better quality or if local stocks have declined. The main consumer is the United States, followed by Japan and Europe. The United States has an estimated 1000 shell dealers; in Florida alone there are about 5000–10,000 retail outlets such as gift shops, hotels, and department stores. As do Philippine exports, U.S. imports reflect world trade, having risen from just over 1000 tonnes in 1960 to nearly 5000 tonnes in 1979, before dropping to 2293 tonnes (valued at just over U.S. $3 million) in 1983.

Some 5000 species are probably involved, 1000 regularly appearing in the Philippine trade alone (Abbott 13; Anonuevo et al. 14). The most popular are the big, colorful gastropods from the Caribbean, Indian Ocean, and Southeast Asia, including cowries (Cypraeidae), helmet shells (Cassidae), cones (Conidae), volutes (Volutidae), and spider shells and conchs (Strombidae). Few countries record trade in individual species but, for example, India exports 3–42 tonnes of cowries annually, and in the 1960s the Maldives was exporting between 20 and 60 tonnes annually.

Shellcraft ranges from high-quality jewelry and carved ornaments to cheap souvenirs for the gift trade. Italy has historically been the center of the cameo industry, which uses helmet shells (Cassidae), cowries (Cypraeidae), and the queen conch, *Strombus gigas*. In 1977, the Bahamas exported 11,000 conchs to Italy but world production figures are not available. Most shells are probably a by-product of the conch food fishery. Other forms of shellcraft have become increasingly popular during the last decade. About 75% comes from the Philippines, where the industry is promoted by the government as a source of employment. Small cowries, dove shells, other small species, mother-of-pearl, broken and subquality shell, and items such as urchin spines are used to make lampshades, decorated boxes, figurines, and jewelry destined for souvenir shops and tourist resorts. In the 1960s, annual imports of shellcraft into the United States were worth U.S. $1 million and by 1981 had risen to $9 million. The windowpane oyster or capiz (kapis) shell, *Placuna placenta*, is of particular value, its white and translucent shells being used for large decorative objects such as lampshades, ornamental boxes, trays, and bowls. The only major commercial fishery is in the Philippines where nearly 3.5 million capiz shell articles were exported in 1979, valued at more than U.S. $2 million (Wells 2, 12; Magsuci et al. 15).

4. IMPACTS OF SHELL COLLECTING

Loss of habitat, particularly for reef-dwelling species, is likely to be a greater threat than overexploitation for most species involved in the shell trade. Nevertheless, depletion does occur on a local basis. In Kenya (Evans et al. 16), Guam (Hedlund 17). Florida (Abbott 13), and the Seychelles (Salm 18), certain species are now extremely scarce on accessible parts of the coast. Scuba and outboard engines permit the exploitation of deep-water populations and more widespread harvesting. Large, sedentary, and easily accessible species with comparatively restricted ranges, such as trochus and giant clams, are vulnerable. The giant triton, *Charonia tritonis*, listed in the IUCN Invertebrate Red Data Book, is widespread, but also potentially threatened by exploitation because of its low densities (Wells et al. 19). Many volutes, such as *Voluta aulica* from the southern Philippines and *Cymbiola rossiniana* endemic to southwest New Caledonia (Bouchet 20), cowries, such as *Cypraea cribellum* and *C. esontropia* from Mauritius, *C. thersites* from South Australia (Coleman 21), and the Hawaiian endemic *C. mauiensis* (Beals 22), and the imperial harp *Harpa costata* from Mauritius have very narrow ranges and are potentially threatened by collectors. Some are rare because of their deep-water habitat and are highly valued on the precious shell market. New collecting techniques permit the collection of comparatively large numbers (Wells 12); the glory-of-the-sea cone, *Conus gloriamaris*, which fetched U.S. $2000 in 1964, now sells for only $200–300. Juveniles of these species are increasingly being taken. Certain species may prove to be vulnerable; for example, the famous golden cowry *C. aurantium* is reported to have become rare on the east coast of the Philippines where it was common in the 1970s (Anonuevo et al. 14).

Commercial shell has been overcollected since early this century in French Polynesia (Salvat 9), Papua New Guinea (Glucksman and Lindholm 8), the Andaman and Nicobar Islands (Rao 23), New Caledonia (Bouchet and Bour 6), Indonesia (Usher 24), Palau (Heslinga and Hillmann 4), Yap, Truk (McGowan 25), Vanuatu (Devambez 26), and the Philippines. For example, in French Polynesia, overcollection of *P. margaritifera* was reported in 1863; by the 1960s eight of the 25 atolls originally producing mother-of-pearl had declining fisheries, and on seven the beds were exhausted (Salvat 9).

Shell collecting also has indirect effects on the habitat. If rocks and coral heads are overturned and left in the sun, the marine organisms they shelter will die; corals may be broken off in search of burrowing species or trampled on; and sandy areas may be disturbed by trawling (Wells 12). Removal of predatory species can affect the ecological balance of the ecosystem. It has been suggested that overcollection of the giant triton has contributed to population explosions of the crown-of-thorns starfish *Acanthaster planci* on coral reefs (Endean and Cameron 27), although this has not yet been proved conclusively.

5. LIFE CYCLES AND GROWTH RATES

The shell trade is labor intensive and generates high economic returns through the added value industries of quality shellcraft and the high prices of precious shells. It should therefore be encouraged if exploitation is directed toward sustainable utilization. For the majority of species, however, the basic data on life cycles, breeding seasons, annual population fluctuations, growth rate, habitat, stock density, and harvesting levels necessary for successful management are lacking. There is clearly a need for much further research in this area perhaps, as suggested by Johannes (28), making use of knowledge accumulated by the collectors themselves.

Most ornamental shells are gastropods, with a wide range of very different life styles; their comparative rareness, cryptic life style, and frequent nocturnal habits make their study very difficult. The mesogastropods include the herbivorous Strombidae, which feed on algae and organic debris in shallow sandy areas; the Cypraeidae, which are omnivores, nocturnal, and found in habitats ranging from shallow reefs to depths of 200 m; and the Cassidae, which feed on echinoderms and other mollusks in sea grass beds. Their free-swimming larvae are largely planktotrophic, often with long pelagic stages and great capacities for dispersal, which may explain the apparent resistance of many species, such as shallow-water cowries (Halsted and Halsted 29) and the popular giant helmet *Cassis cornuta* (Wells 12), to collecting. Neogastropods in trade include the Muricidae, Harpidae, Volutidae, Mitridae, and Conidae, which are active predators, have a slow growth rate, and often have eggs that develop directly into adults.

Trochus, an archaeogastropod, is the only commercially valuable species with a well-documented life history. The larvae are lecithotrophic (yolk bearing), develop rapidly, and metamorphose within about 3 days (Heslinga 30), which may account for the poor dispersal potential and restricted range of this species; it is indigenous only to the Indo-Malaysian area, Melanesia, and Micronesia, although because of its economic importance it has been introduced to various islands further east. Optimum habitat includes a gently sloping bottom, a wide reef-flat exposed at spring low tides, and a subtidal hard substrate at 1–3 m (Bour and Hoffschir 7; Heslinga et al. 10; McGowan 25; Sims 31). Young individuals settle in outer reef-flat intertidal areas and migrate into deeper water as they mature, but are generally not found beyond about 8 m (Heslinga et al. 10). Sand is usually avoided because it inhibits locomotion and the adhesion of the foot, and so are live corals, which feed on larvae. Neither live coral nor loose sand promote the growth of the low filamentous algae that form the main part of the diet.

There are now considerable data on life tables and age classes (Heslinga 5; Bouchet and Bour 6; Bour et al. 32; Smith 33). Growth is variable but it is generally fast in the first 2–3 yr up to a size of 8 cm and then slows down, taking 10 yr for a basal diameter of 12 cm to be reached. Methods of monitoring trochus stock abundance are described by Heslinga et al. (10), Sims (31), and Bour and Hoffschir (7). Bour et al. (34) are now developing a program to calculate stock density using the high-resolution SPOT satellite because trochus habitat appears to be readily recognizable from satellite images.

Apart from pearl oysters, giant clams, and capiz shells, few bivalves play a major role in the shell trade. Although many occur in large shallow beds and are easily accessible, they tend to have rapid growth and recovery rates. For example, capiz shells are found in large "colonies" of more than a thousand individuals in shallow estuary waters, and local populations are regularly depleted (Magsuci et al. 15), but stocks overall in the Philippines show no signs of declining. The biology of many bivalves is well known because of their importance to the edible mollusk fisheries.

6. MANAGEMENT OPTIONS

Attempts to control exploitation of commercial shell have concentrated mainly on trochus, but until recently management strategies were applied with little knowledge of population dynamics and with minimal monitoring of the effects. Minimum size limits at capture (measured across the base of the shell) vary from 7.6 cm in Palau to 9 cm in Vanuatu and 4 in. (10.16 cm) in Guam. In New Caledonia size limits of 9–12 cm have been introduced so that individuals have two reproductive years following maturation at 7 cm, before they are collected. Shells larger than 12 cm are less valuable and are left as a permanent breeding stock, although they may still be taken for food (Bour and Hoffschir 7).

Closed seasons are also variable. In Guam, trochus collection is permitted only in May, June, and July and is prohibited inside the fringing reef; there is a quota of 50 lb of trochus per day for home consumption and a commercial quota is set each season for licensed collectors by the Director of Agriculture (Hedlund 17). In Palau, states or villages may voluntarily stop collecting for one or more years and collecting tends to be limited to a month (usually June) or less, a year (Heslinga et al. 10). In New Caledonia, closed seasons were used in the 1950s but met with little success; rotation of fishing zones was also rejected because this would disrupt local customs and cause problems for sedentary fishermen (Bour and Hoffschir 7).

Attempts are now being made to estimate the sustainable yield of trochus populations. In New Caledonia, in the 1800-km^2 area of lagoon that is considered trochus habitat, annual take is about 200 tonnes, close to the optimum for the current standing stock, estimated at 16,300,000 individuals providing 1700 tonnes of shell. It has been calculated that the lagoon could yield more under appropriate management (Bour and Hoffschir 7). If the harvest is reduced to less than 200 tonnes, by reducing the number of fishermen or by limiting their take to only 2 tonnes/month (assuming current size limits are maintained), the standing stock would increase enough in 5 yr to permit a quota of 400 tonnes. At this stage added-value industries such as button-cutting factories (there are currently two in New Caledonia) could be promoted to generate further income. On Aitutaki, in the Cook Islands, a sustainable yield of almost 0.9 tonnes of shell per kilometer of reef length has been calculated. Sims (31) recommends that a quota should be set at around 30% of the estimated standing stock. With any of these systems, the stocks would have to be monitored continuously and the harvest adjusted as necessary.

Heslinga et al. (10) discuss the system of 16 trochus sanctuaries which were set up in Palau in the 1960s as a result of work by McGowan (25). Because trochus larvae spend only a few days in the plankton, there is a high probability that those that recruit successfully do so within a few days' drift of the point of origin, so that larvae from sanctuaries could help to populate nearby areas. At present the sanctuaries are too numerous and widely scattered to be satisfactorily patrolled, but it is recommended that this strategy would be effective if new areas are designated in suitable habitat accessible for periodic surveillance.

In French Polynesia, regulations were introduced to control the pearl oyster fishery in the last century; these included rotation of lagoons open for diving, declaration of diving seasons, size limits, and quotas, but they were largely unsuccessful because of the difficulty of enforcement (Salvat 9). Currently a major project is directed toward management of the fishery (Coeroli et al. 35). Green snail fishing is largely uncontrolled, but Vanuatu has a minimum size limit of 15-cm basal diameter, and the introduced stocks in Tahiti are protected.

No attempts have been made to devise management plans for ornamental shells, but a number of countries have tried to curb the trade. Enforcement is often difficult. Collecting areas tend to be remote, making the control of closed areas and seasons time consuming and expensive, although some fisheries are naturally restricted to certain seasons by the weather. Australia, Bermuda, and Mauritius have protective legislation for several ornamental species. In Kenya licensed collectors may take only 5 kg of ornamental shells a day and visitors may buy and export 5 kg providing they obtain a license. In Papua New Guinea shells may be marketed only by nationally owned companies (Wells 12).

Collection of *C. tritonis* is prohibited in Fiji, the Seychelles, and Australia, and in Vanuatu there is a minimum size limit of 20 cm. Giant clams are listed in Appendix II of CITES, the Convention on International Trade in Endangered Species of Wild Fauna and Flora, which means that a valid permit is required from the country of export. This provides a means of monitoring trade but the success of CITES in regulating it depends on the extent to which party states can implement the necessary legislation. Strict criteria must be fulfilled for the listing of species, and as discussed earlier the necessary data are usually lacking for species in the ornamental shell trade. There are no controls on the Philippines capiz fishery (Magsuci et al. 15) and, although only shells with an 8–13 cm diameter are of value, the fishery is unselective and wasteful. Despite laboratory studies indicating the contrary, fishermen believe that capiz die after spawning and lose their translucency. Attempts to persuade them to forestall harvesting until after spawning have been unsuccessful.

Kenchington and Hudson (36), Salm and Clark (37), and Wells and Alcala (38) provide general guidelines for the management of coral reefs including recommendations for regulating shell collecting based largely on common sense rather than scientific data. Several malacological and conchological societies already issue instructions to shell collectors, and these are equally applicable to commercial collectors who should be licensed. For the time being such recommendations seem to be the best solution. There are several general principles that should be encouraged. Eggs, juveniles, breeding groups, and specimens with imperfect shells should not be taken. The habitat should be disturbed as little as possible and ecologically sound collecting methods used. Coral should not be touched and stones and loose coral should be returned to their original positions. Tourists and holiday makers, who are often unaware that shells come from living animals, should be discouraged from taking them from the wild. Posters or leaflets displayed in hotels and resorts can play a valuable educational role.

Ornamental shells must be collected live because beach shells usually lack color and shine, but the time-consuming and labor-intensive process of cleaning is often carried out in a haphazard fashion, leading to breakage. However, broken shells can be used for shellcraft rather than thrown away, thus reducing wastage. The use of shells that are by-products of the food fisheries should be promoted, and conversely there should be increased utilization of edible meat from ornamental shells. Shellcraft and other value-added industries that are both economically and environmentally advantageous should be encouraged.

In Papua New Guinea, where the shell trade is government controlled, an attempt was made to develop it along the lines of sustainable utilization. Collectors were issued a booklet advising them on methods that avoid wastage, habitat damage, and overexploitation (Wells 12). This program was not entirely successful owing to lack of follow-up, but a similar scheme is being developed in Fiji (Parkinson 39) and could be incorporated into marine resource management programs in many countries.

7. PROTECTED AREAS

The development of marine protected areas is moving increasingly in the direction of multiple use, with areas zoned for particular uses. Such areas can play an important role in protecting populations of mollusks from which recolonization of adjacent depleted areas can take place. Commercial collection can be permitted in certain zones, whereas other areas are fully protected for their recreational and scientific value (Salm and Clark 37). At present there are few cases where the ornamental shell harvest has been specifically taken into consideration, and none where the benefits can be clearly shown. For example, the reef around the island of Sumilon in the Philippines are fully protected on one side but open to fishing using traditional methods on the other side. Although it was considered that the sanctuary area provided protection for giant clams (Alcala 40), there are no problems with enforcement of the legislation. In the Seychelles, four shell reserves, in which collection of live shells is banned, have been established under the Conservation of Marine Shells Act 1981, but are poorly enforced.

8. MARICULTURE

Mariculture could solve some of the problems of overexploitation, particularly for species that occur in areas where other forms of control are difficult to enforce. Mariculture of *Strombus gigas*, giant clams, and trochus (Heslinga and Hillman, 4) is proving to be economically feasible. The latter is a particularly appropriate species for mariculture. At the Micronesian Mariculture Demonstration Center in Palau, techniques have been developed for rearing larvae and juveniles on a large scale, with the subsidiary aim of using hatchery-reared individuals for restocking depleted reef areas. Adults are easy to collect, and survive well in flowing seawater tanks supporting a rich growth of algae on which they feed outside in full sunlight. Gravid brood stock is available year round; spawning occurs on a monthly cycle, and the larvae have a short planktonic period and require no feeding. Juvenile growth is also rapid, individuals reaching 6 cm in a year with good survival rates. Maturity is reached at 5 cm and a harvestable size of 7.6 cm in 2–3 yr, as in the wild (Heslinga 5). In French Polynesia, hatchery-bred individuals of *P. margaritifera* have been found to be unsuitable for the grafting techniques used in pearl culture, and emphasis is now on spat collection and rearing. Green snail has not yet been cultured but this may well be possible.

In the Philippines, small capiz, below the commercially useful size but collected from the wild, have for many years been broadcast over appropriate shallow habitat and left to reach marketable size. The SEAFDEC Aquaculture Department is now carrying out a capiz resource enhancement project to produce seed shells in hatcheries for reseeding depleted areas. Stocking rate is about 80,000–120,000 pieces per hectare and the seedlings are broadcast at low tide. The culture period is 3–4 months and survival has been estimated at 80–90% (Magsuci et al. 15). Areas suitable for capiz culture are good for mussel and oyster farming, and it appears that the three species can be reared simultaneously.

Pterynotus phyllopterus, a valuable muricid apparently endemic to Guadeloupe and Martinique, is one of the few ornamental species to have been studied with a view to mariculture (Pointier and Lamy 41). Breeding in captivity is considered successful enough for restocking depleted areas and supplying specimens directly to the market. Juveniles reach maturity in 8–16 months and adult size is reached 300–400 days after hatching, rates that are considered high compared with other similar species. However, mariculture of precious shells has yet to be proved economically feasible and must be able to undercut collection of the wild product significantly, to avoid further reductions in wild stocks through stimulation of demand. Captive-reared individuals may eventually be used to reseed depleted natural stocks, although the introduction of cultured stocks to wild habitats outside the species' natural range must be considered very carefully.

REFERENCES

1. FAO, 1980 Yearbook of Fishery Statistics: Catches and Landings, Vol. 50. FAO, Rome, 1981.

2. S. M. Wells, *International Trade in Ornamental Shells*. IUCN Conservation Monitoring Centre, Cambridge, U.K., 1981, 22 pp.

3. C. Carleton, *Miscellaneous Marine Products in the South Pacific: A Survey of the Markets for Specific Groups of Miscellaneous Marine Products*. South Pacific Forum Fisheries Agency, Honiara, Solomon Islands, 1984.

4. G. A. Heslinga and A. Hillmann, Hatchery culture of the commercial top shell *Trochus niloticus* in Palau, Caroline Islands. *Aquaculture* 22:35–43 (1981).

5. G. A. Heslinga, Growth and maturity of *Trochus niloticus* in the laboratory. *Proc. 4th Int. Coral Reef Symp.* 1:39–45 (1981).

6. P. Bouchet and W. Bour, The *Trochus* fishery in New Caledonia. *South Pac. Comm., Fish. Newsl.* 20:9–12 (1980).

7. W. Bour and C. Hoffschir, *Evaluation et gestion de la ressource en Trocas de Nouvelle Calédonie*. Rapport final de Convention ORSTOM/Territoire de Nouvelle Calédonie et dependances, ORSTOM, Noumea, 1985.

8. J. Glucksman and R. Lindholm, A study of the commercial shell industry in Papua New Guinea since World War Two, with particular reference to village production of trochus (*Trochus niloticus*) and green snail (*Turbo marmoratus*). *Sci. New Guinea* **9:**1–10 (1982).

9. B. Salvat, The living marine resources of the South Pacific past, present and future. *Population–Environment Relations in Tropical Islands: The case of Eastern Fiji* MAB Tech. Note 13. UNESCO, 1980, pp. 131–148.

10. G. A. Heslinga, O. Orak, and M. Ngiramengior, Coral reef sanctuaries for Trochus shells. *Mar. Fish. Rev.* **46(4):**73–80 (1984).

11. K. K. Appukuttan, Trochus and Turbo fishery in Andamans. *Seafood Export J.* **11(1):** 41–44 (1979).

12. S. M. Wells, *Marine Conservation in the Philippines and Papua New Guinea with Special Emphasis on the Ornamental Coral and Shell Trade*. Unpublished report to Winston Churchill Memorial Trust, London, 1982.

13. R. T. Abbott, *The Shell Trade in Florida: Status, Trade and Legislation*, Spec. Rep. 3. TRAFFIC (USA), Washington, DC, 1980.

14. M. V. Anonuevo, J. Cabrera, and M. V. Hizon, *A Market Study and Catalogue of Commercially Viable Seashells*, unpublished report. University of the Philippines, Manila, 1982.

15. H. Magsuci, A. Conlu, and S. Moyano-Aypa, The Window-pane Oyster (*Kapis*) fishery of Western Visayas. *Fish. Res. J. Philipp.* **5(2):**74–80 (1980).

16. S. M. Evans, G. Knowles, C. Pye-Smith, and R. Scott, Conserving shells in Kenya. *Oryx* **13(5):**480–485 (1977).

17. S. E. Hedlund, The extent of coral, shell and algal harvesting in Guam waters. *Tech. Rep.—Univ. Guam Mar. Lab.* 37 (1977).

18. R. V. Salm, *Conservation of Marine Resources in Seychelles*. IUCN report to the Government of Seychelles, Morges, Switzerland, 1978.

19. S. M. Wells, R. M. Pyle, and N. M. Collins, *The IUCN Invertebrate Red Data Book*. IUCN Conservation Monitoring Centre, Cambridge, U.K., 1983.

20. P. Bouchet, *Coquillages de collection et protection des recifs*. Report to ORSTOM, Centre de Noumea, New Caledonia, 1979.

21. N. Coleman, Closed season for *Cypraea thersites*. *Aus. Newsl.* [N.S.] **18:**10 (1972).

22. M. Beals, *Cypraea mauiensis* — an endangered species? *Hawaii. Shell News* January:10 (1976).

23. H. S. Rao, On the habitat and habits of *Trochus niloticus* Linn. in the Andaman Seas. *Rec. Indian Mus.* **39:**47–82 (1937).

24. G. Usher, *Coral Reef Invertebrates in Indonesia: Their Exploitation and Conservation Needs*, Rep. Proj. 1688, IUCN/WWF, Bogor, Indonesia, 1984.

25. J. A. McGowan, *The Trochus Fishery of the Trust Territory of the Pacific Islands*. Report to the High Commissioner, U.S. Trust Territory of the Pacific Islands, Saipan, 1958, 46 pp.

26. L. C. Devambez, *Survey of Trochus Reefs in the Central and Southern Groups of the*

New Hebrides. South Pacific Commission, Noumea, New Caledomia, 1959, 7 pp. (unpublished).

27. R. Endean and A. M. Cameron, Ecocatastrophe on the Great Barrier Reef. *Proc. 5th Int. Coral Reef Cong.* **5**:309–314 (1985).

28. R. E. Johannes, *Making Better Use of Existing Knowledge in Managing Pacific Island Reef and Lagoon Ecosystems,* South Pac. Reg. Environ. Programme Top. Rev. 4. South Pacific Commission, Noumea, New Caledonia, 1981.

29. J. P. Halsted and D. C. Halsted, The cowries of Zanzibar. *Bull. East Afr. Nat. Hist. Soc.* **7–8**:67–69 (1980).

30. G. A. Heslinga, Larval development, settlement and metamorphosis of the tropical gastropod *Trochus niloticus. Malacologia* **20(2)**:349–357 (1981).

31. N. Sims, The abundance, distribution and exploitation of *Trochus niloticus* L. in the Cook Islands. *Proc. 5th Int. Coral Reef Cong.* **5**:539–44 (1985).

32. W. Bour, F. Gohin, and P. Bouchet, Croissance et mortalité naturelle des trocas (*Trochus niloticus* L.) de Nouvelle Calédonie. *Haliotis* **12**:71–90 (1982).

33. B. D. Smith, Growth, abundance and distribution of the topshell *Trochus niloticus* on Guam. M.S. Thesis, University of Guam, 1979, 24 pp.

34. W. Bour, L. Loubersac, and P. Rual, Reef thematic maps viewed through simulated data from the future SPOT satellite: Application to the biotope of topshell *(Trochus niloticus)* on the Tetembia Reef (New Caledonia). *Proc. 5th Int. Coral Reef Cong.* **4**:225–230 (1985).

35. M. Coeroli, D. de Gaillande, J. P. Landiet, and D. Coatanea, Recent innovations in cultivation of molluscs in French Polynesia. *Aquaculture* **39**:45–67 (1984).

36. R. A. Kenchington and B. E. T. Hudson, *Coral Reef Management Handbook.* UNESCO Regional Office for Science and Technology for Southeast Asia, Jakarta, Indonesia, 1984.

37. R. V. Salm and J. R. Clark, *Marine and Coastal Protected Areas: A Guide for Planners and Managers.* IUCN, Gland, Switzerland, 1984.

38. S. M. Wells and A. C. Alcala, Collecting of shells and corals. In B. Salvat, (Ed.), *Human Impacts on Coral Reefs: Facts and Recommendations*, Antenne Museum E.P.H.E., French Polynesia, 13–27 (1987).

39. B. J. Parkinson, *The Specimen Shell Resources of Fiji.* South Pacific Commission, Noumea, New Caledonia, 1982.

40. A. C. Alcala, Standing stock and growth of four species of molluscs (family Tridacnidae) in Sumilon Island, Central Visayas, Philippines. *Proc. 4th Int. Coral Reef Symp.* (1981). **2**:757.

41. J. P. Pointier and D. Lamy, Rearing of *Pterynotus phyllopterus*, Mollusca, Muricidae, from Guadeloupe (French West Indies). *Proc. 5th Int. Coral Reef Cong.* **5**:171–176 (1985).

CASE STUDIES

B / BIVALVES

20 THE CHESAPEAKE BAY OYSTER FISHERY: TRADITIONAL MANAGEMENT PRACTICES

Victor S. Kennedy
University of Maryland
Center for Environmental and Estuarine Studies
Cambridge, Maryland

The Chesapeake Bay and its numerous salt-water tributaries contain prolific and valuable oyster beds, probably about equally divided between the two states of Maryland and Virginia. . . . The legislatures of Maryland and Virginia have, at every session for many years, revised and re-revised the laws upon this subject for their respective states; but have always been content to work in the dark, knowing nothing practically, and never seeing the value of obtaining full information upon so important an industry.
  —E. Ingersoll 1881. *The History and Present Condition of the Fishery Industries.*
The Oyster Industry.

1. INTRODUCTION

The oyster fishery in Chesapeake Bay was dominant until recently in the shellfisheries of North America and in the economies of the states of Maryland and Virginia in which the bay is located (Fig. 1). Maryland's catch alone once constituted about 39% of the oysters landed in the United States and more than twice the combined oyster catch of all foreign countries (1). The economic importance of the fishery and its long-term decline have attracted extensive scientific and political attention for more than 100 years. Major publications that provide greater historical detail include references 1–12. Extensive reviews of the Virginia (13) and Maryland (14,15) fisheries have appeared recently because critical reduction of harvests as a result of disease, pollution, reduced recruitment, and overfishing has aroused renewed concern about the future of the resource. These reviews should be consulted for detail on the history of both fisheries, which, necessarily, is presented more briefly here.

In this chapter I compare and contrast the history and management of the oyster resource in different parts of Chesapeake Bay (Fig. 1). There are three major administrative units: the Tidewater Administration of the Maryland Department of Natural Resources (DNR), the Virginia Marine Resources Commission (VMRC), and the Potomac River Fisheries Commission (PRFC). The first two state management agencies have been in existence in one form or another for decades. The third organization, comprising three Maryland and three Virginia members, came into being in 1962 (16) and manages the fishery in the Potomac River. This river divides Maryland from Virginia (Fig. 1) but is Maryland territory to the low-water mark of the Virginia shore. However, the main stem of the Potomac River from the point at which it enters Chesapeake Bay up to the border of the District of Columbia is administered by PRFC (17), whereas the tributaries come under the administration of the respective state resource agencies (DNR, VMRC).

All three resource agencies have had different management strategies over time, and these are examined in this chapter. It will be seen that management by DNR and VMRC has largely been sociopolitical in origin, with modest to no attention paid to the plethora of scientific data and suggestions that have been developed over the past century (15). Management by PRFC has depended more upon scientific insight and has been much less political. However, management by all three agencies has involved the retention of inefficient fishing methods coupled with subsidized shell planting and seed production.

2. THE OYSTER

Crassostrea virginica (Gmelin) is distributed from the Gulf of St. Lawrence along the North American Atlantic and Gulf coasts to the Caribbean. Details of the natural history of the species are extensively reviewed by Galtsoff (18) and Kennedy and Breisch (14). Briefly, this epibenthic filter feeder thrives in estuaries and is capable of surviving turbid conditions by collecting, packaging, and rejecting silt particles

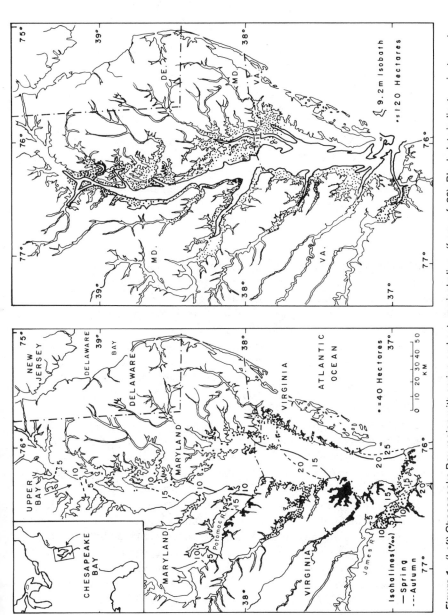

Figure 1. (Left) Chesapeake Bay region with spring and autumn isohalines (from ref. 20). Black dots indicate private oyster grounds. Modified from Alford *Geographical Review*, 1973, with permission from the American Geographical Society (9). (Right) Public oyster grounds (black dots). Modified from J. J. Alford, *Annals of the AAG*, Vol 65, p. 230, Assocation of American Geographers, Washington, D.C., 1975 (10).

in large quantity. In Chesapeake Bay, reproduction occurs mainly from June through September with release and external fertilization of gametes (up to 60 million eggs may be released at one time and up to 250 million in a season by a large female). The planktotrophic larvae develop for 8 or more days depending upon temperature and food. Thereafter, they settle on hard substrate upon which they crawl. If the substrate (usually oyster shell) is attractive, they cement their left valve to it; if not, they swim away to search another area of the bottom. Once attached, the oyster is called a spat. The result of a mass settlement of larvae over time is called a "set." Growth is rapid in the Chesapeake Bay. Seed oyster size (about 2.5–4 cm) may be reached by the end of the first summer by early settling spat. Market size (7.5 cm) is reached in 3 or 4 yr, depending upon location.

3. THE HABITAT

Chesapeake Bay is one of the world's largest estuaries, lying on the temperate east coast of North America and stretching north–south for about 290 km (Fig. 1). It was formed at the end of the Pleistocene by the drowning of the Susquehanna River as sea level rose and water intruded inland to attain a surface area of about 5200 km². Within this sedimentary environment, extensive oyster beds (about 250,000 ha) form the major source of hard substrate (cultch) on which oyster larvae can settle. Without human interference, there was undoubtedly a dynamic ebb and flow between the building up of calcium carbonate reefs by living oysters and the smothering effects of storm- and flood-distributed sediment, with a tendency for oysters to form dense agglomerations in optimal environments. When European explorers first visited the bay, they reported that oysters were present in these optimal regions in immense abundance (19), to the extent that their reefs in shallow waters reached the surface and were navigational hazards. The initial dredging of these oyster "rocks" in the early 1800s was beneficial (2) in that the tightly consolidated virgin reefs were broken up, with oysters and shell spread onto new ground to expand the area of the beds. As a result, the less crowded oysters were probably subject to less competition for planktonic food and therefore became larger and attained better condition. The expanded area of exposed shell may have enhanced spat settlement. However, as dredging and harvesting continued, oyster stocks diminished in abundance and shell debris built up in proportion on the exploited beds (3). In addition, in the absence of cull laws before 1890, little hard substrate was returned to the beds. The resultant lowering of reef height tended to shift the balance in the reef expansion–smothering cycle toward a smothering of overexploited beds and their loss as cultch. Deforestation of bay watersheds as human populations expanded undoubtedly resulted in increased sediment input to the bay, compounding the problem. Today, many formerly productive areas of Chesapeake Bay contain large areas of buried shell.

Chesapeake Bay is prime oyster habitat if suitable hard substrate is available for settlement. Sixty-five percent of the bay is shallow (<9 m; 10), allowing for rapid warming in spring. At about 10°C, feeding commences on the bloom of phytoplankton, enabling mature gametes to be produced by May. Throughout the bay there are a few regions where excellent sets occur, resulting in dependable production of seed oysters. Most seed areas are in secluded low-salinity regions where growth is slow. More open areas provide for rapid growth of market oysters.

Salinity distribution varies seasonally, with the vernal freshwater pulse affecting especially the upper bay and the upper reaches of tributaries (Fig. 1; 20). Below about 10‰ oysters do not reproduce successfully (21). Thus their populations wax and wane in these regions. In Maryland, therefore, the upper bay is not suitable for sustained harvest because freshets or floods inhibit reproduction, recruitment, and growth (22). However, the rest of Maryland's portion of the bay is quite suitable for natural and artificial culture. Maryland's mesohaline (5–18‰) salinities deter predators such as polyhaline (18–25‰) oyster drills (the gastropods *Urosalpinx cinerea* and *Eupleura caudata*) and starfish. Also, disease organisms, specifically *Perkinsus marinus* and *Haplosporidium nelsoni*, are inhibited. In contrast, Virginia's saltier waters allow these predators and pathogens to survive with serious effects on oyster survival and the industry (see below).

4. THE FISHERY

4.1. The Present Situation

The oyster industry in Maryland and Virginia is complex in organization. The basic philosophy of management has been different in each state. Virginia supports production from private leased bottom whereas Maryland discourages such production and PRFC bans it. These strategies have resulted in different types of gear predominating in each management region. Furthermore, in Maryland especially, different management practices have been applied in main-stem bay waters as compared to tributary (county) waters.

Figure 2 presents a general outline of harvesting, processing, and distribution of seed and market oysters in Virginia and Maryland. Seed oysters are harvested by hand tongs from public seed beds such as in the James River in Virginia or in some tributaries on Maryland's Eastern Shore of the bay. Seed oysters may be harvested from private beds by dredges and tongs; such harvesting is more common in Virginia than Maryland. Seed oysters and oyster shells are placed on public grounds in both states. Shells are obtained from shuckers or packers in both states and from the dredging of buried shell in the bay, primarily in Maryland. Private lease operators also have access to public seed beds and purchase shell from various sources to stabilize their beds or to act as cultch. Various fishing gears are used to harvest market oysters from public and leased grounds. Divers operate primarily in Maryland. Dredges are permitted on public grounds in Maryland but only on public management areas and private bottom in Virginia. Harvested oysters are processed in a variety of ways for sale to the consumer.

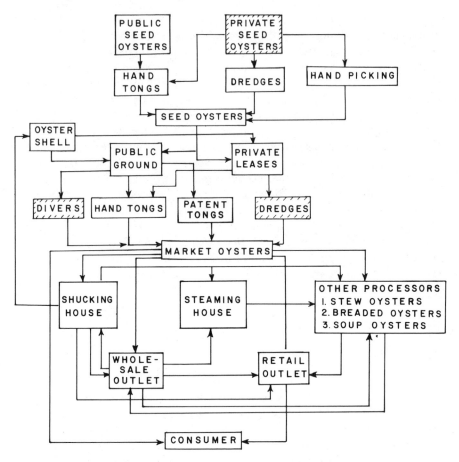

Figure 2. Stages in harvesting, processing, and distribution of seed and market oysters in Virginia and Maryland. Hatched boxes represent the following differences between states (from top of chart and left to right): (1) Private seed oysters are harvested mainly in Virginia, where powered dredges are employed. (2) Divers operate predominantly in Maryland compared with Virginia. (3) Dredges are used on public grounds in Maryland. Figure modified, with permission, from Haven et al. (13).

A variety of boats and fishing gear is involved in the oyster fishery of Maryland and Virginia. Tongers scrape oyster beds with hand tongs (Fig. 3) that consist of two long, flexible wooden shafts (usually 5.5–7 m long) joined in scissor-like fashion toward the ends, to which metal basket rakes are attached. The tonged contents are spilled onto an on-deck *culling board* for sorting. Tongers usually operate from shallow-draft, multipurpose boats (Fig. 3) that can be converted in season for crab fishing (trot-lining and potting). Harvesters are assisted usually by one or more individuals, perhaps another tonger or an individual (sometimes the tonger's wife) who culls the catch by separating market-sized oysters from the mass of undersized oysters, shell, and debris. Culled material must be thrown back

Figure 3. Hand tonging in Maryland's Chesapeake Bay. One person works at culling board while another tongs. Courtesy of Skip Brown and University of Maryland Sea Grant College.

on the oyster bar to serve as cultch and to allow undersized oysters to grow to market size.

Market oysters are usually sold by the bushel on the day of harvest. In the past, *buy boats* (20–30 m long) would sail near a group of tongers to purchase their catch. The decks of these wide-beamed boats (Fig. 4) carried several thousand bushels of oysters, which were transported to a wharf for off-loading at a shucking or packing house. More recently, buy boats have been used less commonly; tongers take their boats to a dock and off-load onto a conveyer belt that carries the catch into a truck owned by a packer or planter.

Patent tongs were developed to fish in waters deeper than hand tongers can reach, although they are sometimes used in shallow waters. They are large and heavy metal tongs (Fig. 4) operated by a winch system. They are raised and lowered by a cable that runs through a block on the vessel's boom and are usually opened and closed by a hydraulic system. Boats may be single or double rigged (one or two sets of tongs, respectively). Harvesting usually involves one (single rig) or two (either rig) people.

Dredges are large, heavy, triangular metal frames with a collection belly and pocket made of metal rings and S-hooks and a toothed lower bar at the mouth that acts as a rake. In the past, dredges weighed hundreds of kilograms and undoubtedly did enormous damage to living oysters and spat as they were dragged over beds. They are now restricted in size and weight in Maryland. Usually two dredges are towed on alternate sides of a boat, with the catch being winched up and dumped on deck for culling. Smaller dredges, called hand scrapes, can be used in certain waters from smaller boats of the size of tonging boats. Engine-powered dredge boats are used in Virginia and on leased bottoms in Maryland. However, sailboats, called skipjacks, are used on public grounds in Maryland (Fig. 5).

The sailing fleet in Maryland is the last all-sail fleet of commercial craft operating in North American waters. The wooden sailboats (about 10–18 m long) are very graceful, sporting clipper bows with carved trail boards and a sharply raked single mast positioned well forward. They have shallow, wide, V-bottomed hulls with a centerboard and a square stern. These sturdy, highly maneuverable vessels have low freeboard to facilitate hoisting dredges on board over a roller. The raked masts make coming about easy and the triangular sail spills wind easily when gusts occur, permitting steady even hauling of the dredge over the bottom. A century ago the bay was fished by more than 1000 sailboats, with many, including schooners, being larger than today's vessels. In the late nineteenth century, smaller, shallow-draft, cheaper skipjacks (the name apparently is an archaic English word meaning "inexpensive yet useful servant") became common. With the decline of the fishery and the banning of dredging in most tributaries, the fleet dwindled and only about 35 vessels are still afloat in Maryland. Most are more than 50 years old and those that are have been placed in the National Register of Historic Places as of 1985.

Most recently, oysters have been collected by divers using scuba or hookah devices delivering air from the boat to the diver below. This is a more efficient system for collecting larger oysters from deeper areas untouched by tongers and

Figure 4. Patent tonger with hydraulic tongs resting on culling board. Note wide-beamed buy boat *Rebecca Forbush* moored in background. Courtesy of Skip Brown and University of Maryland Sea Grant College.

Figure 5. Skipjack under way. Working dredges (on both sides of boat) are out of the water with catches being culled by three oystermen. Note extra dredges, roller on side, and "push boat" or yawl hoisted at the stern on this sailing day. Courtesy of Michael Fincham and University of Maryland Sea Grant College.

from more scattered clumps of oysters than could be fished efficiently by dredge boats or patent tongers.

Table 1 presents a condensed listing of regulations governing the oyster fishery in Maryland, Virginia, and the Potomac River. Although five types of gear are listed, only hand tonging and hand scraping are allowed in the Potomac River. All five methods of fishing are practiced elsewhere in Maryland and Virginia, although diving is uncommon in Virginia.

Generally, entry into the fisheries is virtually unrestricted, being regulated only by one's period of residence in a state and by the requirement of a license fee. Seasons and hours of fishing vary according to gear type and location. For dredging in Virginia, there are few or no restrictions on season, hours, catch limits, or dredge size. Maryland has more restrictions because dredging is allowed on public beds.

Regulations have been aimed at protecting the public stocks by mandating use of the most inefficient gear or by hobbling more efficient gear. The per capita catch using a motorized dredge has been estimated to be more than 10 times the catch of a single hand tonger in the same time period (10). Sail dredges are also more efficient than hand tonging, though probably less so than motor dredges. To counter such efficiency, therefore, the dredging season is limited in length and hours of operation in Maryland, and power can be used only on Mondays and Tuesdays

TABLE 1 Comparative Fishing Regulations Governing the Maryland, Virginia, and Potomac River Oyster Industries, 1985.[a]

	Maryland	Virginia	Potomac River
A. Gear Type			
Hand tonging			
License fee	$50/person	$10/person	$50/person
Season	Sept. 15–March 31	Oct. 1–June 1[b]	Oct. 1–March 31
Hours	Sunrise–sunset	Sunrise–sunset	Sunrise–1500 EST
Daily catch limits			
Per licensee	25 bushels	No limit except in a few areas	No limit
Per boat	75 bushels		No limit
Restrictions	No Sunday tonging	No Sunday tonging	No Saturday or Sunday tonging
Patent tonging			Banned
License fee	$50/person	$35/person	
Season	Sept. 15–March 31	Oct. 1–March 1	
Hours	Sunrise–sunset	Sunrise–1400	
Daily catch limits			
Per licensee	25 bushels	No limit except in a few areas	
Single-rig boat	75 bushels		
Double-rig boat	100 bushels		
Restrictions	No Sunday tonging; certain counties off limits	No Sunday tonging; certain regions off limits; tongs < 100 lb weight, ≤ 4-in. teeth	

(Continued)

TABLE 1 (*Continued*)

	Maryland	Virginia	Potomac River
Dredging			Banned
License fee	$50/person	$50/dredge	
Season	Nov. 1–March 15	No restriction	
Hours	Sunrise–1500	Daylight, except sunrise–1400 in some management areas	
Daily catch limits	150 bushels/boat	None, except in some management areas	
Dredge size		None specified except in one management area	
Rock bottoms	<200 lb weight		
Mud bottoms	≤ 42-in. tooth bar		
	≤ 44-in. tooth bar		
Restrictions	No Sunday dredging; limited to bay waters and a few tributaries; sail only, except Monday and Tuesday when a yawl may push dredge boat	No Sunday dredging; banned on public grounds; legal only in management areas and on leased grounds[c]	
Hand scraping			
License fee	$50/person	$50/scrape	$100/boat and person
Season	Two weeks maximum if inclement weather has disrupted tonging season	No restriction	Nov–Dec, M–W–F; March, M–T–W–Th
Hours	Sunrise–1200	Sunrise–sunset	0800–1200
Daily catch limits		No limit	No limit
Per licensee	10 bushels		
Per boat	30 bushels		

Restrictions	No Sunday scraping; all but two counties off limits	No Sunday scraping; banned on public grounds; legal on leased grounds	No Saturday or Sunday scraping
Diving			
License fee	$50/person		Banned
Season	Sept. 15–March 31	Not regulated per se; considered to be taking oysters by hand, thus coming under hand tonging regulations	
Hours	Sunrise–sunset		
Daily catch limits	30 bushels/boat		
Restrictions	No Sunday fishing; attendant required on board; certain grounds off limits		
B. *Culling Laws*	In all three fisheries shell must be returned to fished bed; maximum market size is 3 in.		
C. *Taxes* (per bushel)			
Unshucked oysters shipped out-of-state	$0.15 inspection tax	$0.20 export tax	—
Oysters fished from public grounds	$0.45 severance tax	$0.50 replenishment tax[d]	$0.50 inspection tax
Tax on purchasers, planters, packers, importers, shippers	—	$0.03 inspection tax	—

[a] Generally, licenses available only to 12-month residents of the appropriate state. EST = Eastern Standard Time; 1 in. = 2.54 cm; 1 lb = 0.45 kg; 1 bushel = about 300–350 oysters.

[b] James River seed area season, Oct. 1–July 1.

[c] Dredges are powered by engines.

[d] James River seed area taxes, $0.05–$0.50/bushel depending on market price of the bushel of oysters.

and then only through use of a *push boat* (Fig. 5). When divers began operating in Maryland in the late 1970s, an uproar ensued because of the efficiency of the method. Nondiving watermen agitated to ban the harvest system entirely, claiming that divers collect the older brood stock which, they believed, maintain recruitment to the oyster fishery. Soon thereafter, divers were limited to collecting oysters of ≥ 10 cm, rather than ≥ 7.5 cm, as was the case with the other harvest methods. That regulation is no longer in effect and diving is now tolerated in Maryland, but with catch and manpower restrictions (Table 1). Also, certain grounds have been made off limits, to be reserved for tongers.

In Maryland it is not uncommon for certain counties to pass laws imposed on state regulations for even more conservative management. For example, although patent tongs are legal, they are banned in certain "county waters" (i.e., in tributaries in the county as opposed to open bay waters where the state has control). As Power (8) notes, such variations have no rational justification but they do prevent uniform management of the resource.

Taxes are applied by all three jurisdictions but they vary in name and amount (Table 1). A bushel of oysters harvested from public grounds and shipped in the shell out of the region would be subject to taxes totaling $0.60, $0.73, and $0.50 for Maryland, Virginia, and the Potomac River, respectively.

4.2. History of the Fishery

It can be said that the regulations summarized above have produced an economically inefficient industry but that depletion of the fishery has not been abated (7), although it may have been slowed. Thus in spite of regulations and legislative attention paid to the oyster industry in the Chesapeake Bay, harvests have declined over the past century (Fig. 6). An understanding of the reasons for the decline in both states requires a brief retracing of the history of management practices of the states (13, 14) and an evaluation of recent disease and pollution problems.

Maryland's earliest oyster-related law was enacted in 1820, prohibiting dredging and out-of-state transport of oysters in ships not wholly owned for the preceding year by a state resident. This law was necessary because earlier in 1811 Virginia had prohibited dredging by nonresidents, thus forcing Connecticut-based dredge schooners to fish in Maryland to obtain oysters. The Connecticut dredges were in Chesapeake waters because the New England oyster grounds to the north had been polluted and overfished. By 1808 it was necessary to harvest Chesapeake Bay oysters to meet market demand in Connecticut (1). With the bay states restricting transport of oysters to residents only, New England tradesmen opened packing houses in Maryland and Virginia in the mid-1830s and took advantage of improved turnpikes (roads) and railway systems to ship canned oysters north, south, and especially, west (23). Harvesting began to involve more Marylanders and Virginians and catches increased greatly (15). As a result of essentially unregulated fishing, the catch peaked in both states in the 1880s.

The ensuing decline in harvests disturbed watermen, processors, and politicians. In Virginia in 1892, Lieutenant Baylor of the U.S. Coast and Geodetic

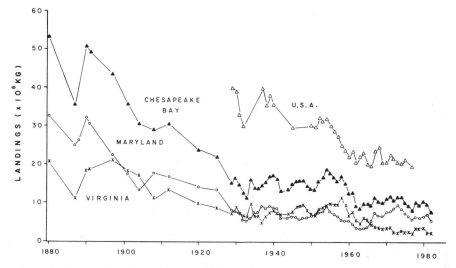

Figure 6. Reported landings of oysters in Chesapeake Bay and the United States. Data from Fishery Statistics of the United States (1950, 1952, 1965 and 1966–1977) and Stagg (24). Data for the United States are more limited because there were only 12 yrs between 1880 and 1952 in which complete data on oyster harvests were available for all states.

Survey was commissioned by the state to survey the oyster grounds (13). He issued a report in 1894 delineating the boundaries of all the natural oyster reefs in Virginia (about 210,000 acres or 84,000 ha) and, with occasional adjustments, the "Baylor Grounds" have remained essentially as he outlined them (even though many beds are now overfished or silted over). In 1906, Maryland's Shellfish Commission sponsored a 6-yr survey of oyster grounds by the Coast and Geodetic Survey's Lieutenant Yates (15). His survey produced 37 m^2 of charts of Maryland's oyster beds; many of these beds no longer exist. Both Baylor's and Yates's surveys have been the basis for defining the locations of public oyster beds in Virginia and Maryland, respectively (Fig. 1). However, within the last decade both states have undertaken new surveys of the public beds because of the loss of overfished beds and the demand of watermen wishing to fish for clams (*Mya arenaria* L. or *Mercenaria mercenaria* (L.)) on abandoned oyster grounds.

In their conclusions concerning the future of the industry, both Baylor and Yates urged the leasing of unproductive or "barren" bottom for private planting purposes. They were not alone in this, for similar recommendations had been made by Winslow (3), Brooks (5), and numerous others during the late nineteenth century when the bay oyster industry was beginning its major decline. However, the results (especially in Maryland) have been limited. Figure 1 (left) reveals that more oyster grounds are leased in Virginia than in Maryland; this is also true in the Potomac River whose south shore is in Virginia. Compare Figure 1 (left and right) to see that the relative lack of private beds in Maryland is not due to a lack of suitable oyster habitat. Even marginal oyster ground can be improved by careful management by leaseholders. For example, private leases have been common on Virginia's lower Eastern Shore of the bay (Fig. 1 left), whereas natural oyster beds are limited (Fig. 1 right) because higher salinity in this region allows predators and disease to penetrate the Bay along that shore. In regions such as this, mortality can be limited to a certain extent by collecting predators mechanically and by transplanting oysters when they have reached seed or even market size but before they succumb to disease.

The difference in the mix of public and private grounds in the two states has had an effect on harvest and gear use (Fig. 7). From 1952 until the mid-1960s (see also Fig. 6), Virginia landed more than 50% of all oysters harvested in both states, with most of that coming from privately leased beds (Fig. 7A), which represent about one-third of the oyster acreage in Virginia (Table 2). In contrast, Maryland landings were predominantly from public beds (Fig. 7A); nevertheless, Maryland's private oyster grounds (~3% of total acreage, Table 2) have yielded an average of 11% of Maryland's total catch (24). It is typical that private grounds outproduce public beds. For example, in 1943, 13% of all U.S. oyster-producing bottoms were privately leased or owned but produced 55% of the total oyster crop (25).

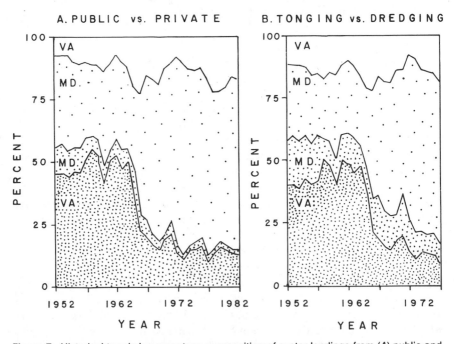

Figure 7. Historical trends in percentage composition of oyster landings from (*A*) public and private grounds harvested by (*B*) tongs or dredges in Maryland and Virginia. Data on private landings and on dredged oysters from Virginia (VA) and Maryland (MD) are presented in the lower two (darker) segments of *A* and *B*, respectively, whereas public harvests and tonged harvests are presented in the upper two (lighter) segments of *A* and *B*, respectively. Data from Stagg (24).

TABLE 2   Comparative Data on Public and Private Oyster Grounds and on Leasing
Policy in Maryland and Virginia, 1985.[a]

	Maryland	Virginia
Approximate extent of all oyster grounds (acres)		
Public	270,000 (97%)	243,000 (69%)
Private	9,000 (3%)	111,000 (31%)
Quantity leased to any one person (acres)		
Bay	5–500	≤ 5000
Tangier Sound	1–100	—
Elsewhere (tributaries)	1–30	≤ 250/assignment; total ≤ 3000
Annual rental costs		
Tributaries	$3.50/acre	$1.50/acre
Bay	$3.50/acre	≥ $0.75/acre
Duration	20 yr initial; 5 yr renewal	10–20 yr, depending on when lease was granted; 10 yr renewal
Restrictions	Natural bars off limits, as is any area within 150 ft (county waters) or 600 ft (bay waters) of a natural oyster or clam bar; illegal in six tidewater counties; no private corporation may lease oyster grounds; Maryland residents only; tonging allowed only on some leased grounds; hand scrapers and power dredges allowed in other limited leased areas; moratorium on new leasing imposed in 1976	Natural bars (Baylor Grounds) off limits; Virginia residents or state-chartered oyster culture corporations only

[a] 1 acre = 0.4 ha. 1 ft = 30 cm.

This is probably due to the better culture and management practices of leaseholders
whose capital is at risk. In spite of evidence of the greater yield possible from
privately leased grounds (usually marginal "barren bottoms" at best), Maryland
has persisted in inhibiting leasing. The history of the consistently strong hostility
of Maryland's watermen and their political representatives to private culture is
outlined in greater detail in Kennedy and Breisch (15).

Because of higher harvests from Virginia's private beds than from public beds, the yield from dredging exceeded that from tonging from 1952 to 1975 with few exceptions (Fig. 7B). Contrarily, tonging has predominated in Maryland because more beds are in the public fishery.

Figure 7A reveals a sharp drop in harvests by Virginia's private leaseholders beginning in the early 1960s. This decline was the result of an invasion or manifestation of a lethal disease in Chesapeake Bay in 1959–1960. This disease, initially dubbed MSX, had virtually destroyed the oyster industry earlier in Delaware Bay (Fig. 1) (26). The disease organism was eventually determined to be a haplosporidan protozoan (27) now designated *Haplosporidium nelsoni*. In Chesapeake Bay where salinities were > 15‰ (predominantly in Virginia, although also in southern Maryland waters) the disease reached epizootic proportions during some years after 1960 (28). Several large processing companies in Virginia that had extensive lease-holdings were crippled or driven out of business (13). Nearly half the leased bottoms were abandoned, resulting in a major decline in total landings (Fig. 6). Lower-salinity waters of Maryland were much less ravaged by this disease, although MSX killed oysters well up into Maryland's portion of the bay during recent drought years. The impact of MSX was exacerbated by the persistent presence of the common polyhaline disease, *Perkinsus* (= *Dermocystidium*) *marinus*, which increases in intensity during dry periods and extended warm autumn seasons.

Since the epizootic, there has been a continued decline or depression of landings and recruitment in Virginia since 1961 (13). This decline has occurred on public and private grounds both in high- and low-salinity regions. In addition to disease, it is thought that pollutants, especially chlorine and chlorine derivatives resulting from sewage treatment and fouling control programs, may be implicated in larval mortality. However, economic factors have also played a role. Haven et al. (13) implicated "rising production costs, stagnant dockside prices, consumer resistance, failure of the industry to adjust to modern production methods, inadequate management by industry and the public sector, and competition from growers and harvesters outside of the State" as contributing causes of the decline in the industry.

Most or all these factors undoubtedly affected Maryland's industry as well. However, Maryland's harvests have risen somewhat since the early 1960s during Virginia's decline (Fig. 6). This has been attributed to Maryland's more extensive public repletion program. This program includes seed transplantation from poor to good growing areas and planting of fresh oyster shell (purchased by the state from shucking houses) and "fossil" shell (dredged from extensive buried reefs) in productive areas (11). From 1960 to 1978, more than 120 million bushels of dredged shell, fresh shell (thought to be more attractive to setting larvae but less available than dredged shell), and seed oysters were planted by the program, at an annual cost of about $1.3 million (D. G. Swartz, University of Maryland Sea Grant College, personal communication). The program has been moderately to heavily subsidized by Maryland taxpayers even though legislation in 1967 and 1968 was enacted in an effort to make it self-sustaining by taxing the industry. Few studies have been performed to determine the efficacy of the repletion program, either from a biological or an economic perspective (11). Recently, however, some evidence has accrued from modeling efforts that indicates that repletion activities did influence production positively in certain regions (29,30; D. G. Swartz, personal communication). Unfortunately, mortality from MSX resurgence in drought years and poor or no settlement in wet years have resulted in declines in Maryland's production since about 1973 (Fig. 6).

Haven et al. (13) examined possible reasons for Virginia's sustained decline in production, as mentioned. As in Maryland, a major biological factor involves lowered setting levels, especially in the seed beds of the James River, once the major source of seed for leaseholders but now producing about one-quarter as many seed oysters as in the 1950s. Reasons for poor settlement are unclear, but it is thought that the MSX-caused reductions in brood stock of adult oysters have led to reductions in the larval populations once produced by these adults (31). If true, this phenomenon may be exacerbated by the extreme sensitivity of oyster larvae to chlorine and its derivatives (13). Another factor affecting Virginia's production may be competition in Virginia markets from oysters imported from southern states. The decline in James River seed beds and losses to MSX led to reluctance by planters to spend money on private culture. More recently, high interest rates and inflation led to a greater perceived risk of private oyster culture in Virginia. Processors have turned to imported oysters to meet consumer demand. In addition, the decade of the 1970s was wet, exacerbating poor reproductive success. This, combined with continued harvesting, has driven oyster population levels even lower in Virginia and Maryland.

5. THE PRESENT AND THE FUTURE

At present, there are few or no dependable statistics collected in either state concerning fishing mortality, natural mortality, recruitment success over the long term, stock size, damage caused by harvesting, and so on. Those statistics that have been collected have been unreliable (11,13). Cabraal (11) found errors in the transfer of data by DNR from records made by buyers in the field (bar name, date, boat and waterman license numbers, bushels caught, price per bushel, buyer license number) to coding sheets, then to computer cards, and finally to magnetic tape. He discovered that data had been erased accidentally on some records, different codes for fishing gear were used in different years, the number of people on board different kinds of fishing boats was arbitrarily set (e.g., five for dredge boats), and deciphering the actual bar location where the oysters were harvested was difficult because of the multiplicity of local names in use. Yet DNR's records are the only source of detailed oyster production information available in Maryland. Aggravating this lack of reliability (which applies also to Virginia) is the penchant for watermen and processors to understate their catch information and any other data that might allow monitoring of their incomes for taxation purposes. Naturally, management of the industry is hindered by lack of such information.

In addition, political pressure in Maryland especially has inhibited attempts to encourage private leasing (Table 2). Thus the state has operated a subsidized repletion program, with management decisions dependent on demands of watermen who put pressure on DNR through their county committees of tongers and dredgers. Similar pressures exist in Virginia. For example, in both states, shell has been planted in places where historical data indicate that limited spat settlement occurs, and at the wrong time for successful settlement. Both states have been pressured to open areas to fishing and to lift harvest limits when it has not been scientifically appropriate to do so. In the Potomac River, the management agency appears to be somewhat less susceptible to such political pressure.

At present, both states are undertaking major examination of the management of their oyster fisheries. Although these examinations are not new—as previously noted, Maryland has sponsored numerous studies over the last 100 yr and has largely neglected the resulting recommendations (15)—there may be more urgency now, because recent oyster recruitment has been so erratic and harvests continue to decline. Deliberations coordinated by DNR in 1985 brought together oystermen, politicians, packers, and scientists. The result was a consensual "White Paper" for managing the fishery. Among the shortcomings of this document were the neglect of either oyster bottom leasing or encouragement of private oyster culture. Instead, the proposed action for 1986 involved banning oystering on Saturday, limiting catches to 50 bushels per tong boat and 100 bushels per dredge boat, and prohibiting the placement of oysters in containers except during unloading at a dock (to prevent the hiding of illegal-sized oysters in baskets filled at sea). These proposed regulations seem insufficient to revitalize Maryland's industry. In addition, in spite of the involvement of oystermen and politicians in production of the White Paper, other oystermen and a local politician opposed the agreed-upon regulations; many did not become law in 1986. A major disease outbreak has since reduced landings to ~0.5 million bushels (1988). Virginia's management plan was completed in 1986. It remains to be seen how substantive it will be.

In their review of Virginia's fishery, Haven et al. (13) made numerous sound recommendations for rehabilitating the industry (most of these recommendations would be useful in Maryland also). They lamented the inattention paid by management, politicians, watermen, and leaseholders to the litany of scientific advice over the past century. Such inattention is discouraging because shellfish, of all marine animals, are among the most readily susceptible to deliberate management. McHugh (32) has stated that

> It appears that reasonable solutions to the problems of the oyster industry will come about only by the route that so many fishery solutions appear to take, when resistance to change is so weakened by disaster and virtual destruction of the industry and the resource that resistance crumbles and the industry finally begs for help.

It is not clear that such an end has been reached yet in Chesapeake Bay.

6. CONCLUSIONS

In spite of the demonstrated resilience of eastern oyster populations in Chesapeake Bay, decades of overfishing and mismanagement, coupled with recent failures of recruitment, have led to historically depressed harvests. The harvest rate exceeds the repletion rate, in spite of gear restrictions. Broodstock and cultch are removed with insufficient attempts to establish adult reserves or to return shucked shell to appropriate regions of the system. The leveling of reefs that originally projected above the surrounding bottom has undoubtedly produced less suitable oyster habitat. This is because reefs are probably (a) less susceptible to effects of suspended sediment, (b) more favorable setting sites, and (c) better growing sites because they project into the food-laden water column.

Recent imposition of measures to minimize sediment runoff from farming and land clearing should help slow smothering of oyster beds and cultch. Further rehabilitation will require brood stock protection, planting seed oysters in good growing regions where the bottom has been stabilized by hard substrate, placing mounds of cultch only in regions with a history of excellent settlement success, and managing by scientific principles rather than in response to political pressure. Such rehabilitation would be facilitated by encouraging private culture, especially in Maryland. Thus the costs of rehabilitation would be shared among numerous leaseholders, not just by state taxpayers. The effort of rehabilitation would also be shared. The public grounds could be restored as larval populations increased with the success of privately operated oyster bottom. Shelling of public grounds could provide suitable substrate for settlement of these pelagic larvae, which are not restricted by property lines. In both states, an understanding of the reasons for continued depressed recruitment is needed, as is encouragement in Virginia of a renewed level of private culture. Finally, dependence on a politicized, heavily subsidized public fishery will have to yield to a moderate mix of public and private production.

ACKNOWLEDGMENTS

I thank J. Andrews (especially), J. DiCosimo, E. Dunnington, D. Haven, M. Leffler, and D. Meritt for criticizing an early draft of this paper, C. Stagg for harvest data, J. Alford for the data of Figure 1, and D. Kennedy for drafting figures. The Maryland Sea Grant Program supported some of the costs of producing this chapter and provided photographs of fishing gear. I am indebted to all for their help. The U.S. Government is authorized to produce and distribute reprints for governmental purposes notwithstanding any copyright notation that may appear herein. Contribution No. 1679HPEL of the Center for Environmental and Estuarine Studies, University of Maryland.

REFERENCES

1. C. H. Stevenson, The oyster industry of Maryland. *Bull. U.S. Fish. Comm.* **(1892):**205–297 (1894).
2. E. Ingersoll, *The History and Present Condition of the Fishery Industries. The Oyster Industry.* U.S. Census Bureau, 10th Census, Department of the Interior, Washington, DC, 1881, 251 pp.

3. F. Winslow, Present condition and future prospects of the oyster industry. *Trans. Am. Fish. Soc.* **13:**148–163 (1884).

4. R. H. Edmonds, The oyster interests of Virginia. In G. B. Goode, Ed., *The Fisheries and Fishery Industries of the United States, 1884–1887*, Vol. 4. U.S. Commission of Fish and Fisheries, Washington, DC, 1884.

5. W. K. Brooks, *The Oyster.* Johns Hopkins Univ. Press, Baltimore, MD, 1891, 225 pp.

6. E. P. Churchill, Jr., *The Oyster and the Oyster Industry of the Atlantic and Gulf Coast. Report of the U.S. Commissioner of Fisheries for 1919*, Washington, DC, 1920, 51 pp.

7. F. T. Christy, The exploitation of a common property natural resource: The Maryland oyster industry. Ph.D. Dissertation, University of Michigan, Ann Arbor, 1964.

8. G. Power, More about oysters than you wanted to know. *Md. Law Rev.* **30:**199–225 (1970).

9. J. J. Alford, The role of management in Chesapeake oyster production. *Geogr. Rev.* **63:**44–54 (1973).

10. J. J. Alford, The Chesapeake oyster fishery. *Ann. Assoc. Am. Geogr.* **65:**229–239 (1975).

11. R. A. Cabraal, *Systems analysis of the Maryland oyster fishery: Production, management and economics*. Ph.D. Dissertation, University of Maryland, College Park, 1978, 318 pp.

12. G. E. Krantz, Oyster propagation in the Maryland portion of Chesapeake Bay. In K. K. Chew, Ed., *Proceedings of the North American Oyster Workshop*, Spec. Publ. No. 1. World Maric. Soc., Baton Rouge, LA, 1983, pp. 159–186.

13. D. S. Haven, W. J. Hargis, Jr., and P. C. Kendall, *The Oyster Industry of Virginia: It Status, Problems and Promise*, Spec. Pap. Mar. Sci. No. 4. Virginia Institute of Marine Science, Gloucester Point, 1978, 1024 pp.

14. V. S. Kennedy, and L. L. Breisch, *Maryland's Oysters: Research and Management*, Sea Grant Program Publ. No. UM-SG-TS-81-04. University of Maryland, College Park, 1981, 286 pp.

15. V. S. Kennedy, and L. L. Breisch, Sixteen decades of political management of the oyster fishery in Maryland's Chesapeake Bay. *J. Environ. Manage.* **16:**153–171 (1983).

16. K. Lasson, A history of Potomac River conflicts. In G. Power, Ed., *Legal Rights in Potomac Waters*, Gen. Publ. 76-2. Interstate Commission on the Potomac River Basin, Bethesda, MD, 1976, pp. 2–35.

17. D. S. Haven, The shellfish fisheries of the Potamac River. In W. T. Mason, Ed., *The Potamac Estuary: Biological Resources, Trends and Options*. Interstate Commission on the Potomac River Basin, Bethesda, MD, 1976, pp. 88–94.

18. P. S. Galtsoff, The American oyster, *Crassostrea virginica* (Gmelin). *Fish. Bull.* **64:**1–480 (1964).

19. J. Wharton, *The Bounty of the Chesapeake. Fishing in Colonial Virginia.* University Press of Virginia, Charlottesville, 1957, 78 pp.

20. A. J. Lippson, *The Chesapeake Bay in Maryland.* Johns Hopkins Univ. Press, Baltimore, MD, 1973, 55 pp.

21. P. A. Butler, Gametogenesis in the oyster under conditions of depressed salinity. *Biol. Bull. (Woods Hole, Mass.)* **96:**263–269 (1949).

22. J. Engle, Commercial aspects of the Upper Chesapeake Bay oyster bars in the light of recent oyster mortalities. *Proc. Natl. Shellfish. Assoc.* 1946:42–46 (1947).

23. A. J. Nichol, *The Oyster-packing Industry of Baltimore. Its History and Current Problems*, Bull. Chesapeake Biol. Lab. University of Maryland, Solomons, 1937, 32 pp.

24. C. Stagg, *An Evaluation of the Information Available for Managing Chesapeake Bay Fisheries: Preliminary Stock Assessments*, Univ. MD., Cent. Environ. Estuarine Stud. Rep. 85-29. Chesapeake Biological Laboratory, Solomons, MD, 1985, 225 pp.

25. P. S. Galtsoff, Increasing the production of oysters and other shellfish in the United States. *U.S. Dep. Int., Fish Wild. Serv., Fish. Leafl.* **22**:1–14 (1943).

26. H. H. Haskin, W. J. Canzonier, and T. L. Myhre, The history of "MSX" on Delaware Bay oyster grounds 1957–1965. *Annu. Rep. Am. Malacol. Union* pp. 20–21 (1965).

27. H. H. Haskin, L. A. Stauber, and T. G. Mackin, *Minchinia nelsoni* n. sp. (Hadosporida: Haplosporidiidae): Causative agent of the Delaware Bay oyster epizootic. *Science* **153**:1414–1416 (1966).

28. J. D. Andrews, and J. L. Wood, Oyster mortality studies in Virginia. VI. History and distribution of *Minchinia nelsoni*, a pathogen of oysters in Virginia. *Chesapeake Sci.* **8**:1–13 (1967).

29. R. E. Ulanowicz, W. C. Caplins, and E. A. Dunnington, The forecasting of oyster harvest in central Chesapeake Bay. *Estuarine Coastal Mar. Sci.* **11**:101–106 (1980).

30. R. A. Cabraal and F. W. Wheaton, Production functions for the Maryland oyster fishery. *Trans. ASAE* **24(1)**:248–251, 254 (1981).

31. J. D. Andrews, Transport of bivalve larvae in James River, Virginia. *J. Shellfish Res.* **3**:29–40 (1983).

32. J. L. McHugh, *Fishery Management*. Springer-Verlag, New York, 1984, pp. 51–63.

21 MECHANIZED SHELLFISH HARVESTING AND ITS MANAGEMENT: THE OFFSHORE CLAM FISHERY OF THE EASTERN UNITED STATES

Steven A. Murawski and Fredric M. Serchuk
National Marine Fisheries Service
Woods Hole Laboratory
Woods Hole, Massachusetts

1. Introduction
2. History and Development of Ocean Clam Fisheries
3. Population Dynamics of the Surf Clam and Ocean Quahog
   3.1. Surf Clam
   3.2. Ocean Quahog
4. Management of Ocean Clam Resources in the EEZ
5. Bioeconomic Overview of Bivalve Fishery Management
   References

1. INTRODUCTION

Oceanic clam fisheries off the eastern United States currently produce landings of 56,000 tonnes of meats with an ex-vessel (first sale) value of U.S. $58 million (Table 1). These figures alone are impressive, but they are even more so when one considers that the entire catch is processed into products worth many times their dockside value (value-added products). Landings and value derived from these

TABLE 1 Total Landings and Ex-vessel Value for Surf Clams and Ocean Quahog Fisheries off the Northeast United States, 1940–1986[a]

Year	Surf Clam		Ocean Quahog		Total	
	Landings	Value	Landings	Value	Landings	Value
1940	599	55	—	—	599	55
1941	NA		NA		NA	
1942	342	47	NA		NA	
1943	406	124	NA		NA	
1944	546	188	293	60	839	248
1945	2,168	630	595	110	2,763	740
1946	NA		649	126	NA	
1947	1,677	424	141	20	1,818	444
1948	1,683	393	108	15	1,791	408
1949	2,417	514	29	5	2,446	519
1950	3,511	766	100	18	3,611	784
1951	5,451	1,188	93	15	5,544	1,203
1952	5,736	1,408	220	43	5,956	1,451
1953	5,749	1,412	125	25	5,874	1,437
1954	5,358	1,458	89	20	5,447	1,478
1955	5,452	1,365	201	47	5,653	1,412
1956	7,245	1,782	175	42	7,420	1,824
1957	8,142	2,240	176	48	8,318	2,288
1958	6,560	1,572	119	36	6,679	1,608
1959	10,537	1,924	43	10	10,580	1,934
1960	11,370	1,713	84	19	11,454	1,732
1961	12,473	1,766	56	12	12,529	1,778

Year						
1962	13,993	2,010	30	7	14,023	2,017
1963	17,499	2,676	47	10	17,546	2,686
1964	17,299	2,619	51	11	17,350	2,630
1965	19,995	3,197	42	11	20,037	3,208
1966	20,459	3,876	41	11	20,500	3,887
1967	18,165	4,352	20	6	18,185	4,358
1968	18,391	4,137	102	29	18,493	4,166
1969	22,483	5,913	290	99	22,773	6,012
1970	30,530	7,730	792	305	31,322	8,035
1971	23,825	6,890	922	345	24,747	7,235
1972	28,740	7,941	635	235	29,375	8,176
1973	37,356	9,867	661	250	38,017	10,117
1974	43,587	12,225	380	146	43,967	12,371
1975	39,436	12,570	588	249	40,024	12,819
1976	22,273	23,344	2,540	1,617	24,813	24,961
1977	23,146	26,442	8,412	5,524	31,558	31,966
1978	17,795	20,901	10,415	6,707	28,210	27,608
1979	15,833	19,273	15,748	10,233	31,581	29,506
1980	17,114	19,107	15,343	10,187	32,457	29,294
1981	20,907	23,466	16,375	10,184	37,282	33,650
1982	22,549	25,963	15,779	10,850	38,328	36,813
1983	25,369	24,914	15,978	10,753	41,347	35,667
1984	31,856	34,334	17,602	11,829	49,458	46,163
1985	32,889	38,877	23,566	15,874	56,455	54,751
1986	35,714	42,613	20,582	15,716	56,296	58,329

[a]Landings in tonnes of meat, value in thousands of U.S. dollars.

mechanized clam fisheries have increased steadily since the late 1970s when stocks of surf clam (*Spisula solidissima*) were at historically low levels and large-scale fisheries for the deeper-water-dwelling ocean quahog (*Arctica islandica*) had not yet been initiated. The current prognosis for the long-term economic viability of these fisheries is good. Both species are relatively long-lived, and exploitation rates are low in comparison to standing stocks. Based on known resources, current production levels of both species can be sustained well into the 1990s. In the case of ocean quahog resources, exploitation rates are so low (an estimated 2% of standing stock per year) that the fishery can be maintained for several decades without significant new recruitment. A rigorous management program adopted initially in 1977 has resulted in sustainable landings remaining at or near historically high levels and considerable economic benefits accruing to a relatively small number of individuals and vertically integrated companies (i.e., companies involved in catching, processing, and marketing these products). The surf clam-ocean quahog fisheries currently are the object of a level of management control unprecedented for U.S. marine fisheries. The fisheries have been described as the most stringently regulated in waters under the jurisdiction of the U.S. federal government.

This chapter examines the historical development of the mechanized clam fisheries and the population dynamics of surf clams and ocean quahogs in relation to exploitation. Biological and economic justifications are reviewed for various restrictive management measures implemented initially to conserve and rebuild declining surf clam stocks and to promote orderly development of the ocean quahog industry. Subsequent management decisions have focused on stabilizing product flow over time from a substantially rebuilt surf clam resource, and equitable allocation to a significantly overcapitalized harvesting sector. We also consider the unique biological characteristics of these and similar bivalve resources with respect to issues of resource productivity, recruitment variability, and density-dependent processes, particularly as they relate to long-term management stratagems.

2. HISTORY AND DEVELOPMENT OF OCEAN CLAM FISHERIES

Yancey and Welch (1) divided the history of the surf clam fishery into three relatively distinct phases: The early period (1870–1942), the developmental period (1943–1949), and the recent period (1950–1965). Subsequent to these we would add: full development and overcapitalization period (1966–1976), and intensive management period (1977–present). Each period is characterized in terms of the status of surf clam utilization, the development of harvesting and processing equipment and methods, and the state of knowledge of the magnitude and distribution of the resource base (Fig. 1).

Although surf clams had been utilized by aboriginal Americans along the Atlantic seaboard for several centuries (Yancey and Welch 1; Parker 2), a formal industry for the species was not initiated until the 1870s. This early fishery was

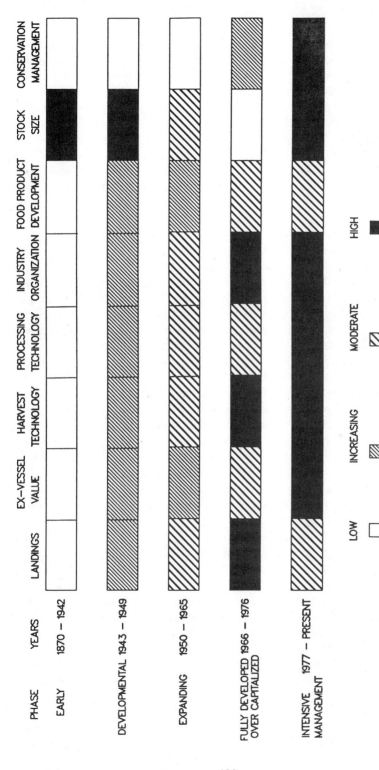

Figure 1. Phases in the development of the surf clam fishery off the northeast United States. Attributes of the fishery, technological development, resource status, and management are categorized by time period.

conducted off southeastern Massachusetts (Fig. 2) and employed rakes and tongs in addition to hand gathering of clams. The earliest use of surf clams was as bait for the hand-line cod fishery, and thus clams were generally not used for human consumption during the early period. Maximum annual production during 1870–1928 was 120 tonnes of salted clam meats (1908). Power dredging techniques were first introduced into the fishery during 1929, in nearshore waters off Long Island, New York (Fig. 2). These early scrape dredges were $\frac{1}{2} - \frac{3}{4}$ m wide with the blade set to dig to about 20 cm (Westman 3). These early dredges relied solely on mechanical factors for substrate penetration, and thus could not be used in hard-packed substrata. Although landings increased during 1929–1942, to an average of 385 tonnes of meat per year (Lyles 4), most of the landings continued to be used as bait.

The period from 1943 to 1949 was important in the development of the modern surf clam fishery because of the advent of the significant utilization of surf clams as food. Early attempts to use the clam meats for human consumption failed, in part owing to the inability to remove the considerable quantities of sand that permeate the meat. Most of the sand was "blasted" into the mantle cavity and viscera during the dredging process (Westman 3). The technological development of mechanical washers, combined with increased wartime protein demands, stimulated a fishery for surf clam human consumption. At the same time, similar circumstances contributed to the initial interest in harvesting of ocean quahog resources, primarily off Rhode Island (Neville 5).

A second major technological development, introduced during 1945, was the use of hydraulic dredges incorporating water jets at the cutting edge of the dredge, instead of the dry or scrape dredges (Yancey and Welch 1). By employing water pressure to loosen the clams from the substrate, harvesting efficiency (CPUE) was increased with a corresponding decrease in the breakage rate of clams, and a reduced incidence of "cut feet" (Westman 3). The latter condition occurs when the clam clamps its valves together without first retracting the large foot muscle, thereby lacerating the foot and potentially reducing yields from the most valuable portion of the clam.

Landings statistics (Table 1) document the substantial development of surf clam and ocean quahog fisheries during the late 1940s. Although interest in the ocean quahog waned after 1948, the surf clam fishery (which had expanded to cover New York and northern New Jersey waters) increased greatly. Improved meat yields per bushel and higher-density (bushels per hour) beds were sought along the New Jersey, Delaware, and Maryland coasts during 1950–1965 (Yancey and Welch 1; Ruggiero 6). During this period landings from New York waters declined as the fishery developed fully in more southerly areas. In 1957, the fleet had expanded to approximately 100 small vessels, primarily owner-operated, but by 1965 the fleet had been reduced by nearly half, with a corresponding increase in vessel size and the advent of vessel ownership by the major clam processing companies (Yancey and Welch 1).

Technological innovation in clam harvesting, shucking, and processing methods was most rapid during 1966–1976. The introduction of large stern-rigged vessels

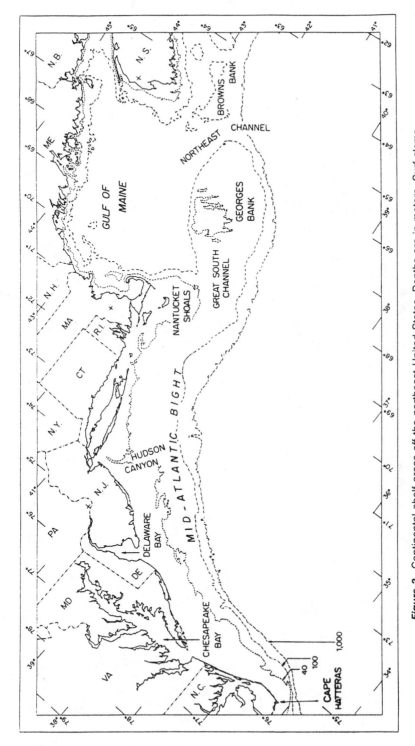

Figure 2. Continental shelf areas off the northeast United States. Depths are in meters. Surf clams and ocean quahog fisheries are conducted primarily from Hudson Canyon to Cape Hatteras, and secondarily in the Nantucket Shoals–Georges Bank area.

Figure 3. A modern stern-rigged surf clam–ocean quahog fishing vessel. Note hydraulic dredge on stern ramp. Vessel length is 29 m; 190 gross registered tons.

Figure 4. Off-loading a "cage" of surf clams from a dredge vessel. The cage has a volume of 32 U.S. bushels (approximately 1.1 kL) and weighs 4000 lb (1.8 tonnes) when full.

(Fig. 3), and less labor-intensive methods for landing and transporting the catch to the processing plants (Fig. 4) were among the most important developments during this phase. Also significant were the introduction of mechanized shucking techniques to replace laborious hand opening and shucking and improved methods for eviscerating and washing the clam meats. The advent of automatic shucking was a critical development, because prior to this innovation, only the largest clams (> 14-cm shell length) were landed, for hand-shucking smaller clams is not efficient. Further improvements were made in the design of dredging and dredge handling systems (Fig. 5). Dredge width also increased markedly in response to less labor-intensive catch handling methods and the need to increase fishery productivity to meet market demands. Biological research was also identified at the time as an important element for the continued viability of the expanding surf clam fishery. Intensive research, initially sponsored by industry, was undertaken to (1) survey for possible new concentrations of harvestable clams, (2) characterize the size composition, growth rate, and recruitment of surf clams, and (3) monitor the production and relative productivity (CPUE) of various clam fishing areas along the Mid-Atlantic coast (Fig. 2). Much of this initial research is summarized by Ropes (7, 8).

The period from 1966 to 1976 was one of substantial flux in the surf clam fishery. The areal distribution of landings shifted from a concentration off the New Jersey coast to Maryland–Delaware waters, and later and most importantly, to a

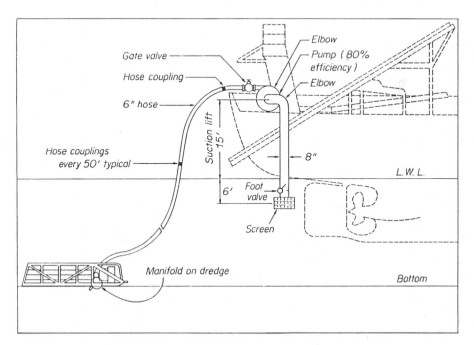

Figure 5. Schematic diagram of a surface-supplied hydraulic dredging system. From Smolowitz and Nulk (25).

large aggregation of clams of a single-year class, off the mouth of Chesapeake Bay, during the early 1970s. Landings stabilized at about 20 thousand tonnes during 1965–1969, in spite of the significant areal shifts by the fishery to virgin resources. The discovery of the large concentration off Chesapeake Bay (1971–1972) resulted in a dramatic increase in landings, CPUE, and fleet size (Table 1, Figs. 6 and 7). By 1974 the fleet had again increased to about 100 vessels, with a large increase in the proportion of vessels in the 100-GRT and greater range (Fig. 7: Ropes, 8). Processing capacity was also increased during the mid-1970s to accommodate increased fleet capacity and landings, particularly from the area off Chesapeake Bay.

Landings of surf clams peaked in 1974 at 44 thousand tonnes (Table 1). The extensive beds of a single year class of clams that contributed to the rapid expansion of the fishery were quickly depleted, and rapid reductions in landings and CPUE ensued (Fig. 6). By 1976 total landings had declined 49% from the peak in 1974. Processing capacity and the combined fishing power of the fleet had increased dramatically however, in response to the stimulus of unconstrained fishing on the local resource off the Chesapeake. Overcapacity in harvesting and processing sectors manifested itself in several ways during the late 1970s. Considerable interest was renewed in ocean quahogs as a market substitute for surf clams. Surf clams had been the preferred species because they occur at shallower depths (Merrill and Ropes 9), produce significantly greater usable meat yields per bushel

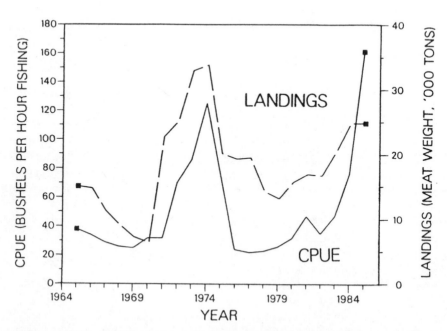

Figure 6. Landings (thousands of tonnes of meats), and CPUE (bushels per hour fishing) for the Mid-Atlantic surf clam fishery conducted in waters more than 3 nautical miles from the coast (the Exclusive Economic Zone, EEZ, under federal government control).

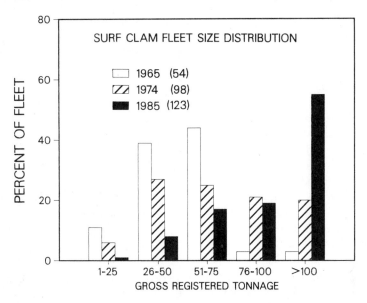

Figure 7. Size distributions of fleet of vessels used to catch surf clams, 1965, 1974, and 1985. Data in parentheses are the numbers of vessels landing surf clams from EEZ waters during the three time periods.

of shell stock, and are more desirable as a finished product. Ocean quahog landings increased slightly during the late 1960s, exceeding wartime production for the first time in 1970 (Table 1). Prior to 1976 virtually all quahog landings were from nearshore Rhode Island waters. It was not until 1976 that a fishery for ocean quahog was developed in the offshore Mid-Atlantic area (off New Jersey and Maryland). Developments in food processing methods rendered the ocean quahog an effective substitute for the increasingly scarce surf clam during the late 1970s (Bakal et al. 10, 11). Ocean quahogs were not, however, a complete substitute in all product forms owing to some significant anatomical differences between the two clams. Traditionally, surf clams had been processed into several product forms including minced clams for chowders and soups, and the large foot muscle was sliced into thin strips for frying. The ocean quahog has a significantly smaller foot muscle than the surf clam, precluding quahogs from being processed for clam strips. The scarcity of the surf clam resource, combined with the lack of a suitable analog for certain products, combined to increase significantly the ex-vessel unit value of surf clams. Average price per kilogram of surf clam meat increased from U.S. $0.28 in 1974 to $1.05 in 1976 (a 275% increase). In spite of declining surf clam landings during the late 1970s, the total ex-vessel value of the fishery more than doubled (Table 1), thereby stimulating the construction of even more large and efficient vessels.

Clearly the surf clam industry was in a crisis mode during the late 1970s. Landings had decreased greatly; harvesting and processing sectors were significantly overcapitalized. New virgin surf clam resources could not be located and prospects for substantial recruitment to areas previously fished were unknown. A widespread and virtually complete natural kill of surf clams off the northern New Jersey coast during the summer of 1976 (Ropes et al. 12) further diminished a tenuous resource base.

Initial attempts at developing a comprehensive management program for surf clams were begun in 1972. At the time, several of the large vertically integrated companies were becoming concerned at the rapid increases in fleet size and landings, particularly with respect to the long-term viability of the resource. Discussions of the necessity for management efforts, and their desirability, were initiated under the auspices of the Atlantic States Marine Fishery Commission (ASMFC). It was not until the passage of federal extended fisheries jurisdiction legislation in 1976 (the Magnuson Fishery Conservation and Management Act), however, that a clear authority and mandate for conservation and management of fisheries resources in offshore waters was established. The development of a comprehensive management plan for surf clam and ocean quahog fisheries in waters under federal jurisdiction [the Exclusive Economic Zone (EEZ), 3–200 nautical miles from the coast] was one of the first tasks of the management councils established under the Magnuson Act. The development of explicit management objectives and strategies for the offshore clam fisheries, and measures used to attain them are discussed in Section 4.

Subsequent to the imposition of restrictive management measures for offshore clam fisheries, the industry carried out substantial development activity exploring fishing areas not traditionally fished. Thus, for example, harvestable surf clam resources have been identified on Georges Bank (Murawski and Serchuk 13), in southern New England (Murawski and Serchuk 14), and in Long Island Sound, New York. Since the adoption of the management program for offshore clams, the surf clam resource has been substantially rebuilt, ocean quahog landings have increased steadily, and new surf clam resources in both offshore and inshore areas have been explored and are now contributing to production.

3. POPULATION DYNAMICS OF THE SURF CLAM AND OCEAN QUAHOG

3.1. Surf Clam

Systematic programs to collect data on the population dynamics of surf clams (and to some extent ocean quahogs) were initiated in the early 1960s (Ropes 8; Merrill and Webster 15). Two major focuses of these activities were: (1) the collection and compilation of fishery statistics (catch, effort by area fished, fleet size and composition) and biological sampling of the catch, and (2) the initiation of comprehensive regionwide surveying to document the distribution and abundance of harvestable sized clams (an industry objective) and prefishery recruits (Serchuk et al. 16). These programs have retained their essential elements for more than two decades, although data collection procedures and sampling design have changed markedly (Murawski and Serchuk 17). Extensive dockside interviews by personnel assigned to the major landing ports have been conducted since 1965. Beginning in 1978, mandatory logbooks have been submitted on a trip basis by all participants in the EEZ fishery. Other specific studies have been undertaken to assess growth rates (Ropes and O'Brien 18; Serchuk and Murawski 19; Murawski and Serchuk 20), length–weight relationships (Murawski and Serchuk 20), and aspects of density-dependent growth and implications for harvest strategies (Murawski and Fogarty 21; Fogarty and Murawski 22).

Ropes (7, 8) and others have extensively documented annual changes in the areal distribution patterns of landings, their size composition, and trends in CPUE for various subareas of the Mid-Atlantic region (Fig. 2). We have combined the estimates of CPUE by subareas, weighting by landings, to derive a single index of vessel performance for the EEZ surf clam fishery, 1965–1985 (Fig. 6). Vessel interviews were not conducted in 1975, and thus data for that year were interpolated. CPUE data from 1978 to 1985 are based on logbook submissions and thus represent a census rather than a subsample of vessel catches (Murawski and Serchuk 17).

Synoptic research vessel surveys of surf clam (and later ocean quahog) resources began in 1965 (Parker and Fahlen 23). Between 1965 and 1986, a series of 17 surveys have been conducted to evaluate the distribution, relative abundance, and size structure of the oceanic clam and quahog populations (Murawski and Serchuk 17; Murawski 24). Prior to 1976, surveys were conducted intermittently, but have since been performed on an annual basis (1976–1984). Earliest surveys (1965–1977) were based on a grid-type design, with stations generally spaced at 10-nautical mile intervals along either LORAN or latitude–longitude lines. Beginning in 1978 the survey was changed to a stratified-random design (Fig. 8), with selection of strata based primarily on depth, and to a limited extent on bottom type. Earlier survey results (pre-1978) have been post-stratified to conform to the stratified-random scheme, and abundance indexes in numbers and weight developed for a number of subareas corresponding to those from which the CPUE data were acquired.

Surveys are performed by allocating a predetermined number of stations (tows) to each stratum. Initially, the number of stations allocated is proportional to stratum area. However, in strata known to contain significant clam resources (as determined from previous surveys and commercial catch data), additional stations are added in order to reduce overall variance of abundance estimates. At each station a 5-min tow is now made with a hydraulic clam dredge 1.5 m (60 in.) wide (Smolowitz and Nulk 25). Surveys prior to 1978 employed smaller dredges and different tow times, and earlier data have been standardized by ratios of current tow time and dredge width to those employed in earlier years (Murawski and Serchuk 20). Survey catches are enumerated and a subsample is measured to record the size distribution of the catch at each station. The total meat weight of clams caught at each station is then computed from length–weight equations (Murawski and Serchuk 20). Mean number and weight per tow (Fig. 9), and number per tow

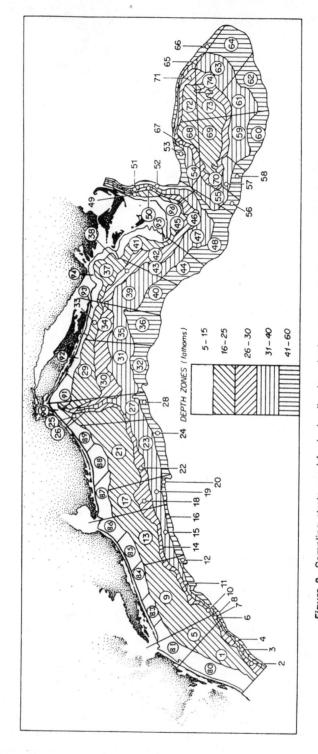

Figure 8. Sampling strata used for hydraulic clam dredge surveys in the Mid-Atlantic and Georges Bank regions off the northeast United States. Stratification is based primarily on depth and to a lesser extent on bottom type. Catch per tow data from stratified random surveys are aggregated for analysis of abundance indexes by subarea (Figs. 9 and 10).

492

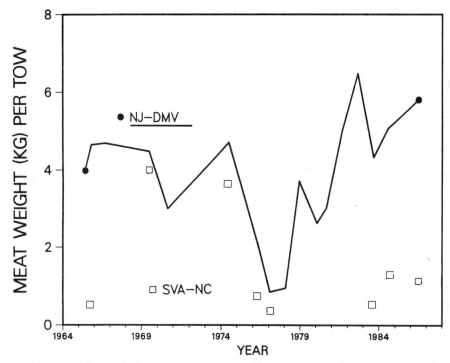

Figure 9. Stratified mean meat weight (kg) per tow research vessel survey indexes for surf clams in two Mid-Atlantic subareas. Indexes are presented for combined New Jersey (NJ) and Delmarva (DMV, off the states of Delaware, Maryland, and Virginia) areas. Indexes for southern Virginia—North Carolina (SVA–NC) are presented separately (open squares) because the region was not surveyed consistently in all years.

at each 1-cm shell length interval (Fig. 10) are computed for each set of survey strata corresponding to assessment subareas.

Both commercial catch rates (CPUE) and research vessel survey data for 1965–1986 document significant changes in the abundance and age structure of surf clam resources in the Mid-Atlantic assessment areas (Figs. 6, 9, 10). During 1965–1970 the resource primarily consisted of large (> 14-cm shell length) clams at moderate to low levels of abundance. Average CPUE during this period was about 30 bushels/hr fishing (Fig. 6), with the fleet primarily comprised of vessels from 26 to 75 GRT (Fig. 7). Resource abundance was relatively stable during 1965–1970, with a low proportion of clam biomass in the southern Virginia–North Carolina region (SVA-NC, Figs. 8 and 9). Total landings declined during 1965–1970 as the fishery searched for clam concentrations in both inshore (< 3 nautical miles from shore) and offshore waters.

The discovery of a large concentration of a single year class of clams off the entrance to Chesapeake Bay (Fig. 2) during the early 1970s had a dramatic effect

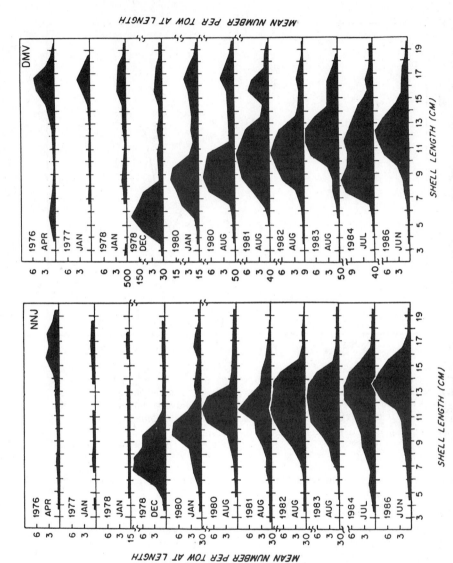

Figure 10. Stratified mean catch per tow (numbers) for surf clams captured in research vessel surveys two Mid-Atlantic subareas, 1976–1984. Data are presented separately for the northern New Jersey (NNJ) and Delmarva (DMV) regions. Survey strata corresponding to each subarea are (Fig. 8): NNJ = 88–90, 25, 21; DMV = 82–86, 9, 10, 13, 14.

494

on surf clam CPUE, landings, and fleet composition. This resource was primarily situated in the SVA–NC region (Fig. 9) and landings were dominated by relatively small clams (the mean shell length in the SVA–NC fishery during 1972–1974 was 13 cm; Ropes 8). CPUE increased rapidly from 28 bushels/hr in 1971 to more than 120 bushels/hr in 1974 (Fig. 6). Research vessel survey indexes for the SVA–NC region peaked in 1974, coincident with the year of maximum landings and highest CPUE (Table 1; Figs. 6 and 9). The fleet size increased and its composition exhibited a shift to larger, more modern vessels during the early and mid-1970s. These shifts were primarily stimulated by the larger vessel hold capacities needed because CPUE had increased greatly, and processing capacity had expanded to accommodate the elevated landings levels. Total fleet size increased 81% from 54 vessels in 1965 to 98 in 1974 (Ropes 8). Vessels in the 76+ GRT range made up about 40% of the fleet in 1974, whereas in earlier years they formed a much smaller proportion of the fleet (Fig. 7). Resource abundance in the SVA–NC area declined rapidly after 1974 as evidenced by reductions in both CPUE and research vessel survey indexes. Landings from the entire Mid-Atlantic surf clam fishery declined from a peak of 44,000 tonnes of meats in 1974 to 22,000 tonnes by 1976. As the resource in the SVA–NC area dwindled, the fishery began to concentrate on the residual stock of relatively large-sized clams in the Delmarva assessment area (Figs. 9 and 10). Average fleet CPUE returned to levels exhibited during 1965–1971; the fleet now, however, significantly expanded in numbers, and was represented by larger, more efficient vessels. The research vessel survey conducted in 1976 indicated a predominance of large, relatively old clams in the major fishing areas off northern New Jersey (NNJ) and Delmarva (DMV, Fig. 10).

During the summer of 1976 a large hypoxic water mass killed a considerable fraction of the extant surf clam resource off NNJ (Ropes et al., 12). The "clam kill" resulted in an historic low in stock abundance in that area, and led to a further concentration of the fleet in the Delmarva assessment region. The entire fishable resource of surf clams in the Mid-Atlantic EEZ thus reached its lowest levels during 1976–1978 after the demise of the SVA–NC fishery, the kill of surf clams off NNJ, and the concentration of fishing on the limited Delmarva resource (Fig. 9). It was in this period that management plans for the EEZ surf clam fishery were first implemented.

Recovery of the Mid-Atlantic surf clam resource in the NNJ and DMV subareas is documented in research vessel and commercial catch statistics (Figs. 6, 9, and 10). Larger than usual 1976 (NNJ) and 1977 (DMV) year classes resulted in the total biomass of the Mid-Atlantic EEZ resource recovering to those levels observed during the mid-1960s (Fig. 9).

It is clear that, apart from the catastrophic mortality occurring in 1976, the instantaneous rate of natural mortality (M) is generally low (<0.1) for surf clams greater than 1 yr of age. Substantial predation mortality (due to crabs and snails) is common for post-settlement juvenile clams (C. McKenzie, personal communication). However, crabs are generally not capable of killing clams greater than ~2 cm. Thus M is thought to be initially very high, but reduced substantially after about 1 yr. The observation of low post-juvenile M is supported by aging studies indicating a maximum life-span of about 30 yr (Ropes and O'Brien 18) and by the persistence of strong year classes in resource surveys when fishing mortality is low (Fig. 10). Number-per-tow indexes for the 1976 and 1977 cohorts did not decline significantly in 9 yr of consecutive surveys (1978–1986, Fig. 10). Considering that fishing occurred during a portion of the 9-yr period on these two year classes, their persistence in surveys strongly indicates a very low total instantaneous mortality rate (Z). Year classes after 1977 have been relatively poor and thus the 1976 and 1977 cohorts will support the bulk of the EEZ fishery through the early 1990s. Given continued presumed low fishing mortality rates, there should be adequate resources to support current landings levels for some years to come based on these known concentrations and the absence of any catastrophic mortality in the area occupied by the stock.

Surf clams are fully capable of spawning at the end of their second year of life (Ropes 26). The production of large year classes (i.e., 1976 and 1977) when spawning stock biomass was low (Fig. 9) and the lack of significant new year classes generated after recovery of the stocks suggests no apparent positive relationship between parental biomass and subsequent recruitment. There may well be an inhibitory effect of the current high stock size on recruitment, but the necessary microscale studies of stock–recruitment dynamics have not been conducted. It has been postulated that the hypoxic event off NNJ during 1976 acted to reduce the abundance of clam predators such as crabs and *Limulus* (Botton and Haskin 27), as well as of clams. The hypoxic conditions subsided during the late summer months, when surf clam larvae settled. Owing to the dearth of predators, small post-settlement clams off NNJ were probably not subjected to normal high predation mortality, and thus a large 1976 year class resulted. However, this mechanism does not apparently explain the appearance of an outstanding 1977 year class off DMV. Although factors influencing year class strength are at this point conjectural, it is clear that outstanding cohorts are indeed rare: we have observed but three strong year classes in two decades of surveying.

The large 1976 cohort off NNJ has exhibited reduced growth in comparison to rates determined from previous aging studies in that area (Serchuk and Murawski 19; Murawski and Fogarty 21). Density-dependent growth rates have obvious implications for management, particularly since biological reference points from yield-per-recruit studies have been used to analyze alternative minimum size options for the fishery (Murawski and Fogarty 21). Minimum size regulations had to be adjusted downward in 1985 because the growth rate of the cohort did not meet expectations. Adjustments to the minimum size were required owing to high levels of at-sea discarding of undersized clams, because only a portion of this cohort had reached minimum legal size (Murawski and Serchuk 17; see Section 4).

CPUE indexes for the Mid-Atlantic FCZ surf clam fishery show an increasing trend between 1981 and 1985, indicative of the two large year classes supporting the fishery. The CPUE index for 1985 (160 bushels/hr fishing) exceeded that for 1974, and is likely to remain high during the next several years, owing to the continued high resource abundance and the restrictive management regime being used to husband the known resource.

3.2. Ocean Quahog

Whereas surf clam populations have exhibited pronounced fluctuations in both abundance and corresponding landings, ocean quahog stocks remained extremely stable during 1965–1986. Analyses of trends in research survey indexes indicated extremely poor recruitment throughout the period, but high and stable standing stocks of a very long-lived resource (Murawski and Serchuk 28). Research on the age, growth, mortality, and biomass of ocean quahog (Murawski and Serchuk 28, 29; Serchuk and Murawski 30; Murawski et al. 31) has established that ocean quahogs are among the slowest growing and longest lived of exploited animals on the continental shelf. Mid-Atlantic populations are dominated by animals from 40 to 80 yr of age, with a substantial proportion of individuals in excess of 100 yr. The age composition of the ocean quahog resource has received close scrutiny from scientists and fishermen alike owing to the implications of the extreme age and slow growth rate on fishery management policies. Thus a variety of techniques have been employed to validate aging studies (Murawski et al. 31; Ropes et al. 32), including mark–recapture, length–frequency analysis, and intra-annual variability in external banding patterns of small individuals.

Because of the relative stability in survey abundance indexes (in numbers) for ocean quahogs over time (Murawski and Serchuk 28), the survey data were grouped into four periods spanning 1965–1982. Estimates of ocean quahog standing stock in the Georges Bank–Cape Hatteras area for 1980–1982, the most recent period, are based on stratified mean weight per tow indexes expanded by the ratio of the area "swept" by a standard tow to the total area surveyed (Table 2). Area-swept population estimates (in meat weight) assume complete retention of animals in the path of the survey gear and thus are minimum calculations of the total stock available for harvest. Based on the pooled 1980–1982 data (three surveys), the total standing stock of quahogs in the Georges Bank–Cape Hatteras region was estimated to be 1.2 million tonnes of meats (Table 2). Of that total, approximately 55% occurred in southern New England–Georges Bank waters, 38% off Long Island and New Jersey, and 7% from Delmarva south to Cape Hatteras.

A series of yield-per-recruit (Y/R) analyses were conducted to determine optimum exploitation rates for maximizing the yield potential of ocean quahog cohorts. These calculations assumed very slow growth rates (Murawski et al. 31) and instantaneous natural mortality rates of 0.01–0.03, consistent with a population in which a substantial proportion of individuals survive to >100 years (Murawski and Serchuk 28). The Y/R analyses indicated that exploitation rates greater than 2–5%/yr (depending on the assumed age at first capture) would result in growth overfishing of the stock. These analyses assumed constant annual recruitment to the stock. Survey data indicate, however, that recruitment during the past two decades has been extremely poor. Thus optimal exploitation rates

TABLE 2. Calculation of Minimum Ocean Quahog Populations Based on NMFS Hydraulic Dredge Surveys, 1980–1982[a]

Assessment Subregions	Survey Strata Included[b]	Area of Region (km^2)	Number of Survey Tows[c]	Mean Weight per Tow (kg)	Population Estimate (tonnes meats)	Percentage of Total
Southern Virginia– North Carolina	80, 81, 1–3, 5, 6	10,654	74(3)	0.0533	1,548	0.13
Delmarva	82–86, 9–11, 13–16	20,326	250(4)	1.6233	89,981	7.43
New Jersey	87–90, 17–28	26,071	360(4)	3.2628	231,980	19.17
Long Island	91–93, 29–31, 33–35	15,360	150(4)	5.4111	226,652	18.72
Southern New England	94–95, 37–39, 41–43, 45–47	19,380	96(4)	5.9470	314,294	25.97
Georges Bank	54, 55, 57, 59–63, 65, 67–74	27,223	59(3)	4.6595	345,928	28.58
Totals		119,014	989	3.7293	1,210,383	100.00

[a]Average area swept by the dredge is calculated to be 367 m^2, based on a 1.52-m-wide dredge towed for an average of 241 m.

[b]Survey strata as illustrated in Fig. 8.

[c]Numbers of survey tows used to compute stratified mean weight per tow index. Number of surveys conducted is given in parentheses. For the southern Virginia–North Carolina area data are for the 1975–1979 period. For all other areas, data for the 1980–1982 period are used.

indicated by Y/R analyses would not be expected to result in stable ocean quahog stock sizes. It has been suggested (Murawski and Serchuk 28; Murawski et al. 31) that because the EEZ ocean quahog resource was essentially virgin until 1976, the extremely slow growth rate and poor recruitment reflect a resource at the carrying capacity of the ecosystem. If density-dependent population regulatory mechanisms are contributing to the low productivity of the ocean quahog stock, then areas subjected to intense fishing pressure should exhibit increased growth and recruitment rates as quahog density is reduced. Current surveying procedures for ocean quahogs are intended to sample heavily exploited areas intensively to determine the response of the populations to harvesting.

The current harvest quota for ocean quahogs (27,000 tonnes) is approximately 2% of the overall standing stock. This harvest rate is consistent with the lower values calculated from Y/R analyses. However, virtually all of the FCZ landings are being derived from the New Jersey and Delmarva subareas, where only 17% of the regionwide resource exists. If the areal pattern of landings persists for the next several years, the ocean quahog resource in these areas will decline substantially. As CPUE declines, the fishery is expected to expand to other more productive areas off Long Island and New England.

4. MANAGEMENT OF OCEAN CLAM RESOURCES IN THE EEZ

Management plans for the EEZ surf clam and ocean quahog fisheries were fully implemented beginning in late 1977 (Mid-Atlantic Fishery Management Council 33). Major elements of the management program during 1977–1985 are given in Figure 11. Explicit objectives of the initial fishery management plan (FMP) were to:

1. Rebuild declining surf clam populations to allow eventual harvesting to approach 23,000 tonnes, which was the average annual catch during 1960–1976.
2. Minimize short-term economic dislocations to the extent possible with (1), and promote economic efficiency.
3. Prevent the harvest of ocean quahogs from exceeding biologically sound levels, and direct the fishery toward maintaining optimum yield (Mid-Atlantic Fishery Management Council 33).

These objectives were adopted in response to several prevailing factors extant during the mid-1970s. The depletion of the surf clam resource off SVA–NC and the hypoxic water event off NNJ both led to a general consensus among industry and government representatives (as well as within the scientific community) that resource abundance had drastically declined. Although clam landings fell by 49% between 1974 and 1976, value of the catch *increased* 91% due to the higher

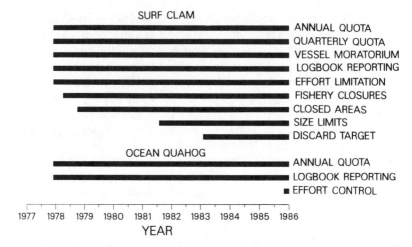

Figure 11. Time lines of the implementation of major fishery management measures for the Surf Clam–Ocean Quahog Fishery Management Plan, 1977–1986. Management authority for these resources in federal waters was initiated in 1977 under the extended jurisdiction laws of the United States.

dockside prices paid by processors for an increasingly scarce resource (Table 1). This prompted further increases in fleet size (Fig. 7), particularly of larger, more powerful vessels. The management objectives established by the Regional Management Council were, and continue to be, the basis of a very restrictive program aimed at constraining catch and fishing effort in order to rebuild the surf clam stocks, and to distribute economic benefits among the participants in the fishery fairly.

Initial management tactics included annual quotas on surf clams and ocean quahogs, with quarterly subquotas on surf clams to spread catch throughout the year. Spreading catch throughout the year was intended to avoid long fishery closures and the attendant unemployment of vessel crews and particularly of skilled processing plant workers. A moratorium on the entry of new vessels into the surf clam fishery was established, as well as mandatory logbook reporting requirements for vessels and processing plants (Fig. 11). A limitation on the amount of time each vessel could fish weekly was also implemented in order to spread catches throughout each quarter. Initially, an annual surf clam quota of 13,600 tonnes was established; well below the maximum catch levels exhibited during the mid-1970s. Nonetheless, the fishery was (and currently remains) significantly overcapitalized with respect to the harvesting capacity necessary to take the available quota. As a result, short-duration fishery closures became necessary when, despite the weekly effort limits imposed on each vessel, quarterly quotas were still being significantly exceeded (Fig. 11).

Surf clam quotas have increased since 1977 in response to improved resource conditions. The annual surf clam quota for the Mid-Atlantic management area for

1985–1987 was 20,400 tonnes. Separate quotas have also been established for the southern New England and Georges Bank resources. These regions have recently become more extensively utilized by industry and are now included in the research vessel survey program.

Additional management regulations for surf clams were implemented during 1978–1986 in response to changes in the resource itself. When the large 1976 and 1977 cohorts were first identified in the surveys, concern was expressed that significant quantities of very small clams would inevitably be caught. This led to areas of high densities of small clams being closed to fishing until the average shell length of clams in these areas exceeded explicit target sizes. A minimum size limit (initially, 14-cm shell length) was likewise established to restrict the catch of small individuals outside the closed areas. Nevertheless, fishery catch rates (CPUE) increased steadily during 1981–1985, as did the catch of undersized clams. Although some small clams were landed, large quantities were discarded at sea (Murawski and Serchuk 17). Dockside interview sampling for scientific purposes documented increasing numbers of sublegal-sized clams being landed, and vessel captain estimates of fraction discarded increased markedly during this period. Because discard mortality can be high (Haskin and Starypan 34; H. H. Haskin, personal communication), fishery managers became concerned that production was being wasted by excessive discarding to meet the minimum size requirement. Accordingly, in 1983, a "target" discard rate of not more than 30% of the landed portion of the catch was established. Subsequently, minimum size requirements have been adjusted downward to achieve this target (during late 1985 the minimum size was set at 12.7 cm). Because surf clam Y/R is maximized at shell sizes of 12–13 cm (with small declines in Y/R for a given fishing mortality at 11 or 14 cm), the effect of lowering the minimum size on surf clam Y/R has been negligible (Murawski and Serchuk 20). The original minimum size of 14 cm resulted in about a 5% loss in Y/R at F_{max} relative to size limits of 12–13 cm. This is because the 14-cm size was originally established to promote harvesting of large clams (> 14 cm), which economically are higher valued because of the large foot muscle that can be sliced for frying. Smaller clams are used primarily for lower-valued products, such as soups, which require only minced clam meat.

With the recovery of the Mid-Atlantic EEZ surf clam stocks, quotas have increased and CPUE has escalated (Fig. 6). Although quotas increased 50% between 1977 and 1985, CPUE values have risen by several hundred percent. As a result, fishing time per week has been shortened in order to prevent the quarterly quota allocations from being exceeded. For most of 1985 and all of 1986 each EEZ vessel was allowed 6 hr of fishing time every 2 weeks; this limited fishing time has been sufficient to catch the quotas.

Although quotas could clearly be increased in the short term without detrimental effects on the stock, the managers have elected to husband the known resources of 1976 and 1977 year classes because more recent cohorts are relatively weak. Given that known resources will have to support the fishery at least until the mid-1990s (clams are not recruited until 6–7 yr of age and the 1986 survey indicates poor prerecruit abundance), managers have sought to maintain moderate catch levels with the expectation that the 1976–1977 cohorts will necessarily be the major contributors to landings well into the next decade.

Compared to surf clams, ocean quahog fishing has proceeded relatively unimpeded. Calculations of MSY have been based on "area-swept" population estimates with biological reference points derived from Y/R studies. Recognizing the limited productivity of the resource, managers have chosen to maintain relatively low exploitation rates ($\sim 2\%$/yr) and closely monitor the effects of fishing. Effort regulations have only been required in late 1985; in prior years, annual quotas were generally not attained by the fishery.

5. BIOECONOMIC OVERVIEW OF BIVALVE FISHERY MANAGEMENT

What have we learned from the historical development of the surf clam and ocean quahog fisheries and from the intensive programs instituted to manage bioeconomic aspects of these and similar fisheries? Clearly, our knowledge of the population biology of these resources has expanded. Equally, new and more complete data have become available on these fisheries as a result of mandated reporting requirements. Because of the economic importance of both species, considerable scientific resources have been directed at answering key questions relating to population dynamics (i.e., growth, recruitment, and abundance). The management scheme initially implemented for these resources utilized MSY concepts derived from finfish management, tempered with the desire to (1) freeze the number of vessels in the surf clam fishery to limit further overcapitalization; (2) preserve anticipated long-term economic benefits gained from restrictive management of the surf clam fishery for those who would suffer short-term losses; and (3) *not* to repeat the boom and bust cycle of overfishing on the still undeveloped ocean quahog fishery. Unbeknownst to the fishery managers who were developing the initial FMP during 1977, very strong year classes would quickly rebuild the depleted surf clam resource. Thus the task of managers would shift from preserving a low-level fishery on small stocks to managing an increasing surf clam resource to preclude repeating the 1970s scenario of resource collapse and fishery overcapitalization. As can be seen from the timing of adoption of various management measures (Fig. 11), the regulation of the surf clam resource has adapted to changing resource conditions, particularly as related to recruitment of large year classes. Protection of small clams to maximize *economic* Y/R was initiated through closed areas and minimum size regulations. However, owing to increased utilization and discard of undersized clams and an apparent compensatory change in surf clam growth rate, the minimum size limit was reduced to achieve higher aggregate resource yields.

Both surf clams and ocean quahogs are relatively long-lived, have a low natural mortality rate for animals surviving their first year of life, and infrequently exhibit strong recruitment. For surf clams, instantaneous natural mortality rate (M) on clams $> \sim 20$ cm is ≤ 0.1; for ocean quahogs $M = 0.01$–0.03. Given these low M values, it is possible to "stockpile" biomass to sustain the fisheries on these bivalves through extended periods of poor recruitment. In both fisheries, large financial investments in harvesting and particularly processing facilities are required; equally, specialized equipment is necessary to catch and prepare clam products for market. Thus in the long term it may be economically more advantageous to forego *maximum* yield potential of particularly strong year classes if annual yields are likely to become highly oscillatory. Traditional biological reference points for fishery management (i.e., F_{max}, $F_{0.1}$) may not be appropriate for stocks such as these in which M is low, strong year classes are infrequent, recruitment strength seems independent of spawning stock size, and the long-term capital invested in the fishery is great. From an investment perspective, there is a financial advantage in knowing that catch rates and landings can be maintained at moderate levels for an extended time interval (several years). This is particularly important given a multi-year debt amortization schedule necessary for the substantial investment in processing/harvesting technology. Barring catastrophic natural mortality on surf clam and ocean quahog, harvestable biomass of both stocks is adequate to sustain current landings levels well into the 1990s.

The historical development of the surf clam fishery illustrates the fundamental vulnerability of fisheries on sessile organisms to extreme overcapitalization. This is due to the high unit value of shellfish, and the fact that dense concentrations, once located, can be systematically explored and fished utilizing modern harvesting and navigation equipment. The rapid technological development of such gear has resulted in a continued increase in the effective fishing power of the surf clam fleet despite a moratorium on new vessel entrants. Thus established annual quotas for surf clams are caught with a very short allocation of fishing time per vessel. In the absence of absolute individual property rights to a fixed portion of the resource, each vessel competes for a larger fraction of the available total quota. Many vessels are rigged with larger dredges, or tow two dredges simultaneously, to increase the *relative* catchability of their individual vessels. This naturally leads to lower available fishing time for the fleet as a whole if the quota is not to be exceeded. Thus even though the EEZ surf clam fleet was capped in 1977, effective effort continues to increase owing to vessel replacement and gear modifications to increase catchability. The fishery remains significantly overcapitalized with respect to the harvesting sector. Several alternative schemes to assign individual, transferable quota allocations to each vessel, based on prior productivity or other measures, have been explored and continue to be debated by fishery participants and managers. Such a scheme could allow for a more efficient use of participating vessels, with a lower aggregate overhead cost in vessels and harvest equipment.

Finally, given the high level and considerable expense of management control exerted in the EEZ surf clam–ocean quahog fisheries, one might ask, "Has the management program been a 'success'?" The initial objectives of the FMP were primarily related to the rapid decline in surf clam abundance and the ensuing overcapitalization of the fishery in the 1970s. Clearly, the surf clam resource has been rebuilt and the management program has allowed the economic benefits of the fishery to accrue to those initially impacted by low annual quotas. The aggregate value of the catch from these fisheries increased 134% between 1976 and

1986 (an average of 12%/yr, well above the average annual inflation rate for the period). Total landings of both species was a record 56,300 tonnes in 1986, 127% above the 1976 total. Employment in the surf clam harvesting sector has declined, however, with the advent of effort control strategies. For vertically integrated companies one vessel crew may be used for several different vessels, because current fishing time per boat is 6 hr every 2 weeks. Employment in the processing sector has increased in recent years, particularly with the large increases in ocean quahog landings. Opportunities for fishing for ocean quahogs have also increased, thereby replacing some jobs lost in the surf clam harvesting sector. Although the number of surf clam vessels remains well in excess of the numbers required to harvest the annual quota, total effort exerted in the fishery has been reduced dramatically since the mid-1970s, and industry and government continue to seek equitable methods to reduce fleet size. By these standards, the management program has successfully met its initial stated objectives. Continued ''success'' of this management program will require the establishment and attainment of long-term bioeconomic objectives consistent with the demands of the marketplace for ocean clam products, and the unique population dynamics of these bivalve resources.

REFERENCES

1. R. M. Yancey and W. R. Welch, The Atlantic coast surf clam—with a partial bibliography. *U.S., Fish Widl. Serv., Circ.* **288:**1–13 (1968).

2. P. S. Parker, History and development of surf clam harvesting gear. *NOAA Tech. Rep., NMFS Circ.* **364:**1–15 (1971).

3. J. R. Westman, *On the Origins, Development, and Status of the Surf Clam Industry, 1943–1945.* New York Dept. of Conservation, Bureau of Marine Fisheries, New York, 1946, 12 pp.

4. C. H. Lyles, Historical catch statistics (shellfish). *U.S., Fish Widl. Serv., Cur. Fish. Stat.* **5007:**1–116 (1969).

5. W. C. Neville, coordinator, *The Ocean Quahog Fishery of Rhode Island.* R. I. Dept. Agric. Conserv., Div. Fish and Game, Providence, RI, 1945, 31 pp.

6. M. Ruggiero, Equipment note No. 9—The surf clam fishery of New Jersey. *Commer. Fish. Rev.* **23(8):**11–13 (1961).

7. J. W. Ropes, The Atlantic coast surf clam fishery, 1965–1969. *Mar. Fish. Rev.* **34(7–8):**20–29 (1972).

8. J. W. Ropes, The Atlantic coast surf clam fishery, 1965–1974. *Mar. Fish. Rev.* **44(8):**1–14 (1982).

9. A. S. Merrill and J. W. Ropes, The general distribution of the surf clam and ocean quahog. *Proc. Natl. Shellfish Assoc.* **59:**40–45 (1969).

10. A. Bakal, W. F. Rathjen, and J. Mendelsohn, Ocean quahog takes supply spotlight as surf clam dwindles. *Food Prod. Dev.* **12(1):** 4 pp, February (1978).

11. A. Bakal, P. Garber, and J. Mendelsohn, Ocean quahog invades surf clam domain, wins quality objectives. *Food Prod. Dev.* **12(3):** 2 pp, April (1978).

12. J. W. Ropes, A. S. Merrill, S. A. Murawski, S. Chang, and C. L. MacKenzie, Jr.,

Chapter 11. Impacts on clams and scallops. Part 1. Field survey assessments. In R. L. Swanson and C. J. Sindermann, Eds., *Oxygen Depletion and Associated Benthic Mortalities in New York Bight, 1976.* Prof. Pap. 11. U.S. Dept. of Commerce, NOAA, Washington, DC, 1979, pp. 263–275.

13. S. A. Murawski and F. M. Serchuk, *An Assessment of the Georges Bank Surf Clam Resource—Summer 1984,* Woods Hole Lab. Ref. Doc. No. 84-28. U.S. Dept. of Commerce, Natl. Mar. Fish. Serv., Woods Hole, MA, 1984, 23 pp.

14. S. A. Murawski and F. M. Serchuk, *An Assessment of the Surf Clam Resource in FCZ Waters off Southern New England—Spring 1983*, Woods Hole Lab. Ref. Doc. No. 83-20, U.S. Dept. of Commerce, Natl. Mar. Fish. Serv., Woods Hole, MA, 1983, 21 pp.

15. A. S. Merrill and J. R. Webster, 1964. Progress in surf clam biological research. *U.S., Fish Wildl. Serv., Circ.* **200:**38–47 (1964).

16. F. M. Serchuk, S. A. Murawski, E. M. Henderson, and B. E. Brown, The population dynamics basis for management of offshore surf clam populations in the Middle Atlantic. In *Proceedings of the Northeast Clam Industries: Management for the Future*, MIT Sea Grant SP-112. Coop. Ext. Serv., University of Massachusetts, Cambridge, 1979, pp. 83–101.

17. S. A. Murawski and F. M. Serchuk, *Assessment Update for Middle Atlantic Offshore Surf Clam,* Spisula solidissima, *Populations—Autumn 1984.* Woods Hole Lab. Ref. Doc. No. 84–32, U.S. Dept. of Commerce, Natl. Mar. Fish. Serv., Woods Hole, MA, 1984, 40 pp.

18. J. W. Ropes and L. O'Brien, A unique method of aging surf clams. *Bull. Am. Malacol. Union* pp. 58–61 (1979).

19. F. M. Serchuk and S. A. Murawski, *Assessment and Status of Surf Clam* Spisula solidissima *(Dillwyn) Populations in Offshore Middle Atlantic Waters of the United States*, Woods Hole Lab. Ref. Doc. No. 80-33. U.S. Dept. of Commerce, Natl. Mar. Fish. Serv., Woods Hole, MA, 1980, 46 pp.

20. S. A. Murawski and F. M. Serchuk, *Assessment and Current Status of Offshore Surf Clam,* Spisula solidissima, *Populations off the Middle Atlantic Coast of the United States*, Woods Hole Lab. Ref. Doc. No. 81-33. U.S. Dept. of Commerce, Natl. Mar. Fish. Serv., Woods Hole, MA, 1981, 50 pp.

21. S. A. Murawski and M. J. Fogarty, A spatial yield model for bivalve populations accounting for density-dependent growth and mortality. *ICES C.M.* **1984/K:26** (1984) (mimeo).

22. M. J. Fogarty and S. A. Murawski, 1986. Population dynamics and assessment of exploited invertebrate stocks. In G. S. Jamieson and N. Bourne, Eds., *North Pacific Workshop on Stock Assessment and Management of Invertebrates*, Can. J. Fish. Aquat. Sci. Spec. Publ. 92. 1986, pp. 228–244.

23. P. S. Parker and L. A. Fahlen, Clam survey off Virginia (Cape Charles to False Cape). *Comm. Fish. Rev.* **39(1):**25–34 (1968).

24. S. A. Murawski, *Assessment Updates for Middle Atlantic, New England, and Georges Bank Offshore Surf Clam,* Spisula solidissima *Populations—Summer 1986,* Woods Hole Lab. Ref. Doc. No. 86-11. U.S. Dept. of Commerce, Natl. Mar. Fish. Serv., Woods Hole, MA, 1986.

25. R. J. Smolowitz and V. E. Nulk, The design of an electrohydraulic dredge for clam surveys. *Mar. Fish. Rev.* **44(4):**1–18 (1982).

26. J. W. Ropes, Reproductive cycle of the surf clam, *Spisula solidissima*, in offshore New Jersey. *Biol. Bull. (Woods Hole, Mass.)* **135:**349–365 (1968).

27. M. L. Botton and H. H. Haskin, Distribution and feeding of the horseshoe crab, *Limulus polyphemus*, on the continental shelf off New Jersey. *Fish. Bull.* **82(2):**383–389 (1984).

28. S. A. Murawski and F. M. Serchuk, *An Assessment of the Ocean Quahog,* Arctica islandica, *Resource and Fishery in FCZ Waters off the Northeastern USA—Autumn 1983,* Woods Hole Lab. Ref. Doc. No. 83-25. U.S. Dept. of Commerce, Natl. Mar. Fish. Serv., Woods Hole, MA, 1983, 32 pp.

29. S. A. Murawski and F. M. Serchuk, Distribution, size composition, and relative abundance of ocean quahog, *Arctica islandica*, populations off the Middle Atlantic coast of the United States. *ICES C.M.* **1979/K:26** (1979) (mimeo).

30. F. M. Serchuk and S. A. Murawski, *Evaluation and Status of Ocean Quahog,* Arctica islandica (Linneaus) *Populations off the Middle Atlantic Coast of the United States,* Woods Hole Lab. Ref. Doc. No. 80-32. U.S. Dept. of Commerce, Natl. Mar. Fish. Serv., Woods Hole, MA, 1980, 7 pp.

31. S. A. Murawski, J. W. Ropes, and F. M. Serchuk, Growth of the ocean quahog, *Arctica islandica*, in the Middle Atlantic Bight. *Fish. Bull.* **80(1):**21–34 (1982).

32. J. W. Ropes, D. S. Jones, S. A. Murawski, F. M. Serchuk, and A. Jearld, Jr., Documentation of annual growth lines in ocean quahogs, *Arctica islandica*, Linne. *Fish. Bull.* **82(1):**1–19 (1984).

33. Mid-Atlantic Fishery Management Council, *Fishery Management Plan for Surf Clam and Ocean Quahog Fisheries*. Mid-Atlantic Fishery Management Council, Dover, DE, 1977, 136 pp.

34. H. H. Haskin and G. Starypan, *Management Studies of Surf Clam Resources off New Jersey*, State/Federal Contract Rep. SC74-1-NJ-(2)-1. Rutgers University/National Marine Fisheries Service, New Brunswick, NJ, 1976, 42 pp.

22 CLAM FISHERIES WITH HYDRAULIC DREDGES IN THE ADRIATIC SEA

Carlo Froglia

Instituto Ricerche Pesca Marittima (CNR)
Ancona, Italy

I. INTRODUCTION

Clams of the family Veneridae inhabit particulate sediments, mostly in the infra-littoral and the circalittoral zones of temperate and tropical regions. Some of the economically most important clam species in the world belong to this family and their utilization dates from the early periods of human civilization. At present more than 400,000 tons of venerid clams are harvested yearly in the world (1). In the Mediterranean basin at least five species of Veneridae are commercially exploited, but the fishery with hydraulic clam dredges for *Chamelea gallina* (L.) along the western coast of the Adriatic Sea, with an average landing of 100,000 tons/yr of raw clams, is by far the most important and exceeds the sum of all other clams produced in the region.

Outside Italy, *C. gallina* is presently exploited along the Albanian coast (south-east Adriatic), in the Marmara Sea (Turkey), along the Spanish and Moroccan coasts of the Alboran Sea, and along the Spanish coast of Catalonia, mostly south of the Ebro delta. Very recently Albania and Turkey with the aim of developing their national fishery, imported from Italy some boats equipped with hydraulic clam dredges (reports of the Italian Port Offices). In Spain and Morocco hydraulic dredges are not yet used. Vives and Suau in 1955 (2) estimated the annual landings of *C. gallina* along the Catalonian coast to be some hundreds of tons.

In Italy "Vongola" is the official name for *C. gallina* (Ministerial Decree 15 July 1983, published in Italian Official Gazette No. 210/1983), but several other vernacular names are still in use locally. In English-speaking countries this species is known, together with other clams, under the more generic commercial name of "baby clam."

2. BIONOMICS

The western Adriatic Sea, characterized by shallow bottoms and soft sediment, is markedly influenced by inflow from rivers of the Alps and Apennines, which provides up to 34% of the total freshwater input in the Mediterranean basin (3), and, owing to these sources of nutrients and growth factors, its primary production is rather high (4). A "Venus community," already described by Vatova (5) in his classic work on Adriatic benthic communities, as "Zoocenosi a Chione gallina," is established in a belt of shallow (3–13-m-deep) sand bottom up to 2 miles wide that extends from the western part of the Gulf of Trieste to the southern part of the Gulf of Manfredonia.

In this environment, the bivalve *Chamelea gallina* (L.) is usually so abundant that it characterizes a facies that extends over 1600 km^2. From grab samples Vatova (5) estimated for *C. gallina* an average biomass of 239 and 294 g/m^2 off the Venetian lagoons and in the central Adriatic, respectively. Recent surveys in the central Adriatic made with an experimental hydraulic clam dredge (Froglia, unpublished data), estimated average densities of 20–100 g/m^2 for the commercial sizes of this species.

Immediately offshore of the *C. gallina* facies, where terrigenous muds replace sand, *C. gallina* is replaced by *Venerupis aurea* (Gmelin). Barilari et al. (6) showed that in the Gulf of Venice *C. gallina* does not extend into sediments with an electric potential below +50 mV and a percentage of sand less than 90%. Colantoni (7) recorded *C. gallina* in 84% of the 44 stations in front of the Po delta characterized by coastal sand, and in only 13% of the 137 stations characterized by silts.

Two other bivalves —*Ensis minor* (Chenu) and *Donax trunculus* (L.)— are common inshore and characterize the so-called "biocoenosis of fine sands in very shallow waters" (8). On some offshore grounds of the northern Adriatic and south of the Gargano peninsula two other venerids —*Venus verrucosa* (L.) and *Callista chione* (L.)— are common on coarse detritic sands. All the above-mentioned species live buried in sediments, and along the Italian coast are presently harvested with hydraulic clam dredges.

3. BIOLOGY

Chamelea gallina is a suspension feeder that profits from the high primary production of west Adriatic coastal waters. It can reach a maximum size of 5 cm and an age of more than 6 yr. Growth rate is rather high in the first years of life; nevertheless the minimal marketable size, fixed at 2.5 cm by Italian law since 1969, is reached only at the end of the second year (9, 10). Bottom temperatures recorded during investigations of Vongola biology in different Adriatic coastal districts show seasonal fluctuations typical of shallow temperate waters, with maxima in summer (24–25°C) and minima in winter (7–8°C) (9–11). In a growth study based on analysis of modal class progressions Froglia (10) revealed that the growth season for *C. gallina* is restricted to months with temperatures above 10°C.

The species is gonochoristic, and the sex ratio is close to 1. Sexual maturity is reached by a percentage of the population at the end of the first year and by all individuals within the second year of life; specimens as small as 10 mm with mature ovaries have been reported by Corni et al. (12). The spawning season is rather long, with a main peak in late spring (9, 11–14), and sometimes another one in later summer (10, 15).

Clam larvae are common in the neritic plankton of the Mediterranean in summer and early autumn (16), but no data are available for the Adriatic. Pelagic life of *C. gallina* probably does not extend beyond 20–30 days; from rearing experiments on the closely related species *Venus striatula* (Da Costa), Ansell (17) reports larval settlement at a size of 200 μm after 16 days' growth at 20°C.

Nothing is known about predation on larval stages of *C. gallina*, and until now no specific studies on its benthic predators have been carried out. From occasional observations it appears that *Astropecten bispinosus* and *A. jonstoni* can prey on spats of 2–5 mm, and remains of larger clams are observed occasionally in the stomach contents of *Gobius niger* and more frequently in *Lithognathus mormyrus* (18). During the 1984 and 1985 surveys of clam resources in the central Adriatic, in areas with high densities (10–20 specimens/100 m^2) of *Neverita josephina*, a remarkable number of dead clam shells were observed drilled by this predatory gastropod (C. Froglia, unpublished data). At present levels of resource exploitation, however, humans are probably the main predator of this species. Catastrophic mortalities of *C. gallina* and other benthic species have been reported in coincidence with anoxic conditions associated with summer phytoplankton blooms (19).

4. THE VONGOLA FISHERY BEFORE THE INTRODUCTION OF HYDRAULIC DREDGING

The fishery for *C. gallina* in the Adriatic Sea dates from long before the introduction of modern techniques of hydraulic clam dredging. Olivi (20) in 1792 mentions that *Venus gallina* was exported from the Venetian Republic to the neighboring Romagna (Papal States), where the species was more appreciated. In the catalog of fisheries products of the Venetian area prepared for the Berlin international fishery exhibition in 1880, Ninni (21) reports *V. gallina* as "very abundant along the shore, but eaten only by lower social classes." Unfortunately no information is available on landings in those earlier times.

Along the Adriatic coast Vongola was harvested until 16 yrs ago by hand rakes (1.5 m wide) on a long pole (up to 8 m long), operated from boats moved by warping on a big anchor (Fig. 1a). Fishermen used the long pole to set the rake and, by rocking it, to sift the sediment; at the end of each tow (of less than 100 m) the rake was hauled and the catch sorted through a hand-operated sieve. Each boat had a crew of two or three fishermen.

Fishing effort on the resource was rather low, and was a function of a fisherman's strength; moreover, clam populations in depths greater than 8 m were unexploited. Daily catch per boat was well below half a ton, and owing to the low price of Vongola, the income of fishermen was rather low. This could be the origin of the name Poverassa—that is, "poor thing"—under which *C. gallina* is still known in some northern Adriatic areas.

Clams were consumed fresh, mostly in the coastal regions, and until the end of World War II only one small cannery, which had been established at the beginning of the century operated along the central Adriatic coast.

Hydraulic dredging gear first appeared in the Adriatic clam fishery in the late 1960s. In the beginning a series of nozzles were added to the traditional rake, through which water under pressure was ejected both in front of the rake and inside its basket. This technological innovation greatly reduced human toil and increased gear efficiency, thus making it possible to exploit virgin areas at greater depths (8–10 m). Boats equipped with the new hydraulic harvesting gear were able to land several tons of Vongola per day. Mechanical sieving equipment was introduced to sort such large catches.

Like all technological novelties, this new fishing gear generated social conflicts and in some maritime districts local fisheries commissions tried to ban their use. However, this conservative effort ended in failure as adjacent districts authorized the new gear, which was also supported by fishery scientists. In the early 1970s, as a result of the large quantity of Vongola available on the market, the average price decreased, making harvesting with traditional rakes unprofitable.

Since the appearance of the first hydraulic harvesting gear several other technological innovations have been introduced. In 1974 38% of the 380 boats equipped with hydraulic harvesting gear had already adopted hydraulic dredges (see Fig. 1b) that closely resemble the "Nantucket dredges" used along the Atlantic coast of the United States (22), and only 60 were still fishing with hand-operated rakes (15).

At the same time, new trade possibilities were discovered. Several new canning plants were established along the Adriatic coast, and the processing capacity of the existing ones considerably increased. First-grade Vongola were, and still are, exported fresh to Spanish markets where the species, known there as "chirla," is greatly appreciated.

Figure 1. (a) Boat with hand rake used in the Vongola fishery in the first half of the century (see the hand-operated winch and the long pole of the rake) (courtesy FotoRiccione). (b) Boat equipped with hydraulic clam dredge as presently used in the Adriatic Vongola fishery.

5. FISHING GEAR AND FISHING TECHNIQUES

At present the boats involved in hydraulic clam dredging are 8–12 m long, with a GRT slightly below 10 tons, engines of 50–150 hp, and a crew of 2 or 3 fishermen. Introduction of hydraulic gear has greatly reduced the fisherman's toil, but the sequence of fishing operations remains unchanged and is summarized here from Froglia and Bolognini (23) (see also Fig. 2).

HYDRAULIC DREDGING OPERATION

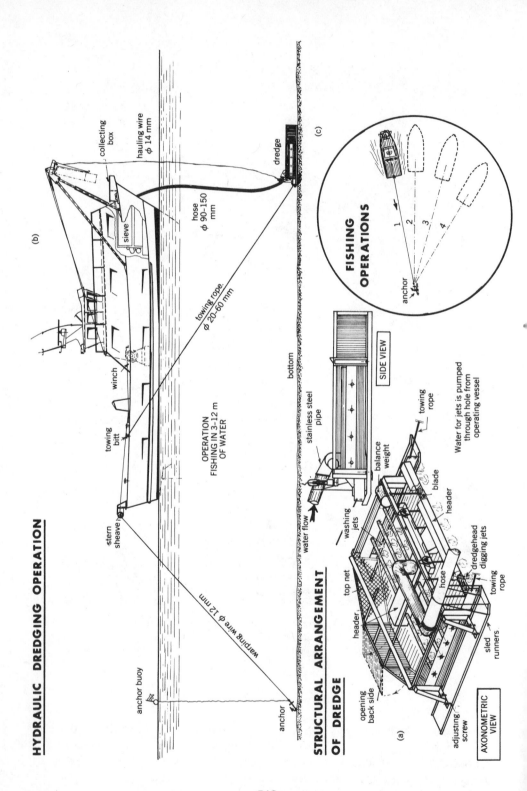

(b)

collecting box

hauling wire ⌀ 14 mm

sieve

hose ⌀ 90-150 mm

towing rope ⌀ 20-60 mm

winch

towing bitt

OPERATION
FISHING IN 3-12 m
OF WATER

dredge

bottom

stern sheave

warping wire ⌀ 12 mm

anchor buoy

anchor

(c)

FISHING OPERATIONS

1 2 3 4

anchor

STRUCTURAL ARRANGEMENT
OF DREDGE

SIDE VIEW

stainless steel pipe

balance weight

water flow

washing jets

blade

header

Water for jets is pumped through hole from operating vessel

towing rope

top net

header

hose

dredgehead digging jets

towing rope

sled runners

opening back side

adjusting screw

(a)

AXONOMETRIC VIEW

When the boat reaches a suitable fishing ground the stern anchor is cast, and 250–300 m of steel cable is paid out as the boat moves forward. The dredge is then lowered, and water under pressure is injected through a hose that connects the dredge to a centrifugal water pump on board. Recently submersible water pumps have been used successfully.

Two adjustable sled runners prevent the dredge from digging too deeply into the bottom. In the Vongola fishery the dredge is made to dig into the bottom to a depth of 4–6 cm; in the razor clam fishery it digs in 8–15 cm. The dredge is secured to the boat by two towing ropes, their length being about twice the fishing depth. The boat tows the dredge by warping on the anchor (i.e., moving backward).

At the end of the tow the dredge is hauled onto a frame placed on the bow, its rear surface unhinged, and the whole catch dumped into a collecting box. The catch is then conveyed to a sieve, usually consisting of two or three coaxial revolving gridded drums, that sorts clams by commercial size classes. Undersized clams and other small organisms are directly conveyed, by water jets, from the riddle back into the sea.

Each tow lasts 10–15 min at a towing speed of 0.6–1.0 knots. If the skipper judges the ground commercially profitable the anchor is not hauled and the next tow is made on a course divergent by a few degrees from the preceding one; thus at the end of the fishing operations at a given locality, the dredger has covered one or more circular sectors (see Fig. 2c). During a tow, the previous catch is sieved and packed in plastic net bags.

The main engine operates the two-drum winch that is used to warp on the anchor and haul the dredge, as well as providing power to the centrifugal water pump and rotating riddle.

From some preliminary observations made by scuba divers sampling with a suction sampler in the dredge track, the gear efficiency seems to be close to 100%. The towing technique (warping on the anchor) makes the dredge motion very regular on flat sandy bottom, and this probably contributes to the high gear efficiency.

6. HYDRAULIC DREDGING FOR VONGOLA AND ITS MANAGEMENT

In the early 1970s, hydraulic clam dredges rapidly became widespread all over the Adriatic in a local version of the gold rush. It soon became evident to most marine biologists and fishermen that management procedures able to prevent overexploitation of the Vongola stock were urgently needed.

Figure 2. (a) Details of dredge structure; (b) boat and fishing gear; (c) schema of the fishing operations. (From Froglia & Bolognini, G.F.C.M. Studies and Reviews, courtesy Food and Agriculture Organization of the United Nations.)

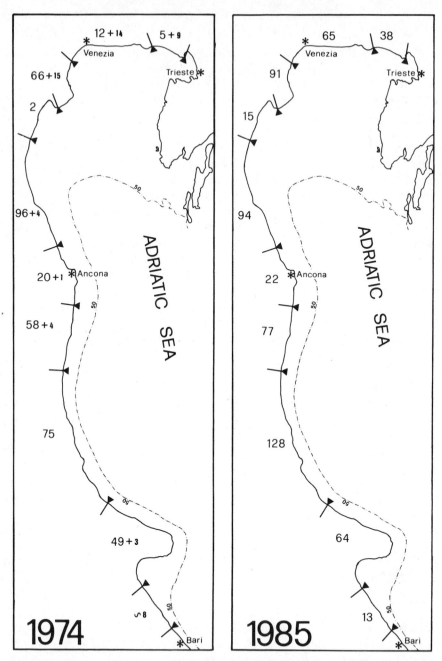

Figure 3. Evolution of the fishing effort in terms of the number of boats involved in the Vongola fishery in the Adriatic districts (Compartimenti). 1974: number of boats equipped with hydraulic gear or hand rakes (smaller number), as recorded in a census made in autumn 1974 all along the Adriatic coast. 1985: number of licensed boats with hydraulic gear, as obtained from the vessel registers kept in the district port offices.

In the first years of this new fishery, local port authorities, who in Italy are responsible for fishery controls, followed suggestions of the Institute for Fishery Technology in Ancona in issuing ordinances on dredge and sieve "meshes" and introducing a fishery closure in June. This was an attempt to prevent landing of undersized clams (smaller than 2.5 cm) and to reduce the possible extra mortality induced by dredging on newly settled juveniles. Unfortunately, provisions controlling gear selectivity were weakly enforced, because undersized clams, in the 2.0–2.4-cm range, owing to their lower price and slightly higher yield of meat were still in high demand by processing plants. In some districts (Compartimenti), difficulties were also encountered in persuading fishermen and merchants to accept the closed period.

An independent census of the existing fleet carried out in 1974 all along the Italian Adriatic coast (15) counted 380 boats equipped with hydraulic gear (Fig. 3), which were mostly irregular with reference to selectivity specifications. In the same survey (15) 13 processing plants were recorded (Fig. 4) where clam meat was steam extracted, then frozen or canned. The overall processing capacity of all plants was estimated at 180 tons/day of raw clams.

No statistics were available then for the Vongola fishery, because this species was not recorded separately in official statistics, and moreover only minor quantities were sold through fish markets. Data obtained in 1974 from export agencies, processing plants, fishermen's organizations, and wholesalers brought the combined estimate of yearly production to about 80,000 tons.

Until 1979, the fishery grew without any state regulation, although regulations were frequently requested by fishermen and merchants. When grounds close to the home port were depleted by too heavy exploitation or by annual failure in recruitment, fishermen from some districts were accustomed to move to other more productive grounds, with consequent friction between established fishermen in an area and the newcomers. Records of these fishermen conflicts can be easily traced in the Italian newspapers of those years.

Finally in 1979 a state decree was issued initiating management of the Vongola fishery, and for the first time in the history of Italian fishery legislation, a limited-entry policy was introduced to avoid excessive growth in fishing effort. Licenses were granted to all fishing units already established in the Vongola fishery, and no additional licenses were to be issued, except for boats already under construction as of August 1979.

C. gallina is a sedentary resource; thus in an attempt to tie fishermen to the fate of the local resource they are exploiting, and to involve them directly in a proper area management scheme, each licensed boat was given the right to fish only within its district (Compartimento) of registration. To avoid overexploitation, a daily quota of 2.5 tons per boat was also introduced, and to protect recruits in the early phase of setting, the Vongola fishery was closed in June all along the Italian Adriatic coast. At the same time a catch allowance of 10% undersized clams (smaller than 2.5 cm) was introduced. To overcome the lack of statistical data, as a condition of the fishing license, each dredger owner was required to supply monthly infor-

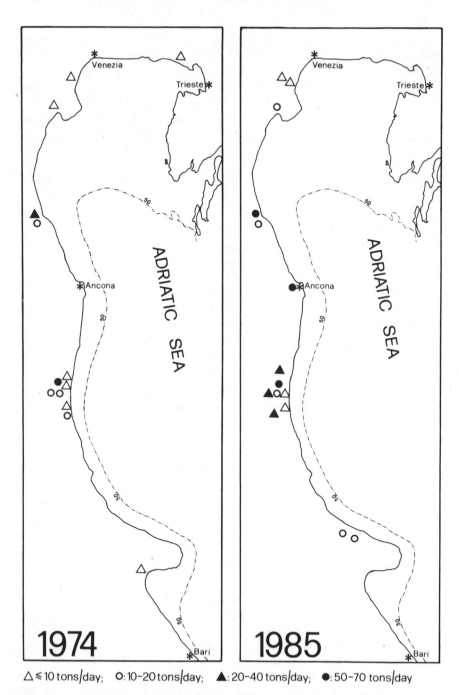

Figure 4. Location of processing plants with daily processing capacity (tons of raw clam per day) in 1974 and 1985.

mation on the quantity caught and the fishing locality. A central committee for the management of this rather special Adriatic resource was also appointed.

To halt landings of undersized clams, the following year the characteristics of dredge and sieve grids were fixed by law according to the results of selectivity investigations (24). Also, to avoid further increments in fishing capacity, maximum size (front width) of the dredges was fixed at 3 m; this was a compromise between the intention to control fishing effort and the need to accommodate most of the existing fishing gear. Fishing gear of all licensed boats was checked by local port officers within a given year; nevertheless, landing of undersized clams is still a problem because fishermen can very easily, and very quickly, reduce or restore sieve selectivity.

In 1981, the maximum size of new dredgers built to replace old ones was fixed at 10 tons GRT, and maximum engine power fixed at 150 hp. Obviously larger boats already active were allowed to maintain their licenses. As already recorded all over the world for other limited-entry fisheries (25, 26), these limitations alone are inadequate to freeze total fishing effort, because with time, total fishing power further increases as a result of the replacement of older boats by new ones that are almost always larger and have more powerful engines. An example of this increment in terms of engine power is given in Figure 5 for the three selected districts (Rimini, Ancona, and S. Benedetto del Tronto) that make up one-third of the whole Adriatic fleet. In 1979, most of the boats were equipped with engines of 60–80 hp; in the graph for 1985 two modal values are evident, one made up of newly built boats of around 150 hp and another at 60–80 hp, including older boats. Boat size had also increased proportionally. This increase in engine power means a reduction in travel time, increased fishing speed (~ 1.0 knot in 1985 versus ~ 0.6 knot in 1974), and a higher water delivery capacity of the pump, as well as the possibility of working in slightly rougher seas. In combination, these changes add up to an increase in total fishing effort. Since 1979 (and introduction of the limited-entry license scheme), the average engine power of dredgers has nearly doubled, and it is easy to forecast that this trend toward allowed maxima in engine power and boat size will continue in the future.

Unfortunately not only global fishing effort has shown this "expected" increase: although the criteria for establishing eligibility for licensing had been clearly specified in the law that introduced the limited-entry license procedure, the efforts of several port officers to enforce this measure rigorously were in vain against the strong political and economic pressures pushing to obtain additional licenses. The consequences of such pressures are shown in Figure 6, based on official data supplied by central fishery authorities to the Management Committee. By autumn 1985, the number of hydraulic dredges licensed to fish for *C. gallina* reached 607 units; in other words, in 5 yrs the number increased by 20%.

In the early 1980s the Vongola fishery was still the most profitable fishing enterprise in the Adriatic. The license, originally issued free of charge, then became very valuable in private transactions because it was transferable with the boat. The intrinsic value of a Vongola fishing license quickly rose to more than U.S. $10,000. The legality of this sale and purchase of a right to exploit a public resource became

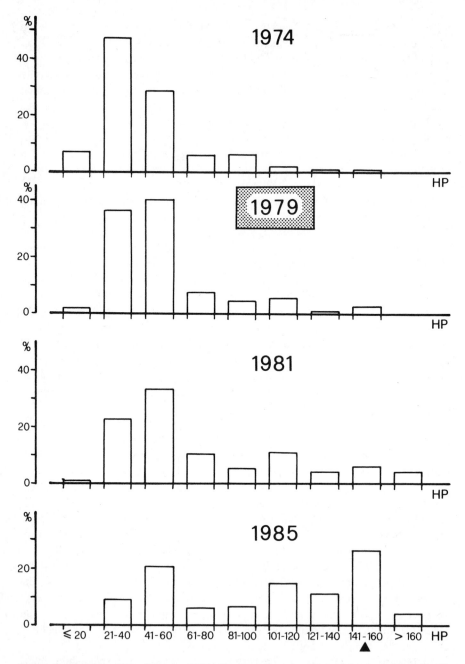

Figure 5. Evolution from 1974 to 1985 of engine power on board boats equipped with hydraulic gear fishing in the districts of Rimini, Ancona, and S. Benedetto (central Adriatic). A limited-entry scheme was introduced in 1979 and the upper limit of 150 hp in 1981.

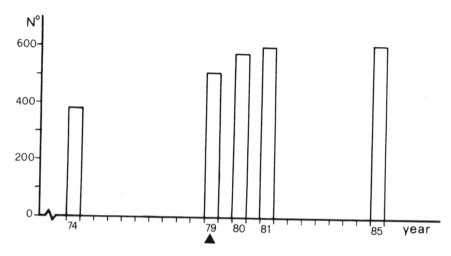

Figure 6. Evolution of the number of boats equipped with hydraulic gear in the Vongola fishery. (Arrow marks 1979, when the limited-entry scheme was introduced.)

a focus of debate for all those who opposed the limited-entry policy. The paper by Meany, "Should Licences in a Restricted Fishery be Saleable?" (27), although published 5 yrs earlier and dealing with Australian limited-entry fishery for rock lobsters, can be considered a complete summary of the long dispute on this subject that took place among representatives of fishermen's organizations in Italy.

In 1984 a new decree made it possible to transfer a license between boats with the same owner, but not to sell it with the boat. In this case the license had to be returned to the fishery authority. This decree was repealed the following year, but some restrictions were introduced to avoid the development of a company-owned fishing fleet, and to favor the owner–operator in clam dredging. This intense phase of legislation was not followed by a similar enforcement of its provisions. Fishermen's cooperatives, fighting one against the other, frequently claimed that a breakdown of provisions on minimum marketable size and on daily quotas had taken place. The Management Committee met only from 1980 to 1982, and never had enough data to evaluate the status of the resource. Statistical data were supplied by less than two-thirds of the license owners and it was nearly impossible to check their quality.

Despite these deficiencies in management, until 1984, hydraulic dredging for Vongola was a highly profitable fishery, as evidenced by the strong pressures to obtain new licenses and the high values of existing ones. Clam processing factories showed a parallel growth, only in part attributable to the increased living standards and consequent increase in the use of frozen and canned foods in Italy. The processing capacities of the 15 plants active along the Italian Adriatic coast in 1985 (Fig. 4) were estimated at more than 320 tons/day of raw clams (i.e., nearly double that for 1974).

TABLE 1   Estimates of Vongola landings (tons), Estimated from Export, Processing, and National Marketing Data

Year	Export Fresh to Spain	Processing Plants	Italian Market Fresh	Overall Landings
1974[a]	27,000	41,000	10,000–15,000	~ 80,000
1980	22,000[b]	58,000[c]	15,000–20,000	~ 95,000
1981	24,000[b]	48,000[c]	15,000–20,000	~ 90,000
1982	26,000[b]	50,000[c]	15,000–20,000	~ 95,000
1983	29,000[b]	55,000[c]	15,000–20,000	~ 100,000
1984	27,000[b] 31,000[c]	60,000[c]	15,000–20,000	~ 105,000
1985	25,000[c]	41,000[c]	10,000–15,000	~ 80,000

[a]Froglia (15).
[b]"Asociacion Nacional de Empresarios Exportadores e Importadores de Moluscos" Spain.
[c]"Associazione Nazionale Conservieri Ittici e delle Tonnare" Italy, Marricchi (28).

As shown in Table 1, where data from different sources are summarized, annual landings in the 1980–1984 period probably fluctuated around 100,000 tons—that is, they had increased by 20% since 1974—whereas fishing effort for the same period had more than doubled. At the same time the average size of clams had probably decreased, but there is inadequate experimental data to support this impression.

By 1984, the fleet had probably already expanded beyond the capacity needed to harvest the potential yield efficiently; a rough estimate based on the surface dredged daily per boat showed that the equivalent of the whole fishing ground was being swept once per year.

It was realized in 1984 that only experimental surveys over the whole area of the Venus community could give a realistic estimate of the status of the resource. A 3-yr program was launched involving five Adriatic research institutes, with a financial grant from the research fund set up by the 1981 fishery plan. Preliminary data from surveys showing rather low clam densities in some districts became available in the second half of 1985, and at the same time an irrefutable indicator of a serious shortfall in the resource became evident to everyone: Vongola prices skyrocketed; doubling in a few months (Fig. 7). National and export markets for fresh clams, which from the beginning paid prices nearly double those of the processing industry, took nearly the whole available production, and processing plants were faced by a shortage of supplies and an increase in price, so that their products were no longer competitive with similar products from outside the Mediterranean area.

Fortunately, an alternative species, *Venerupis aurea*, up until then almost unexploited, whose official name is also Vongola, was available. Despite its fragile shell, which makes it unsuitable for marketing fresh, it has proved very profitable for industrial processing. At the moment, probably half of the fishing effort of the hydraulic dredges has shifted to this latter resource and most of the processing

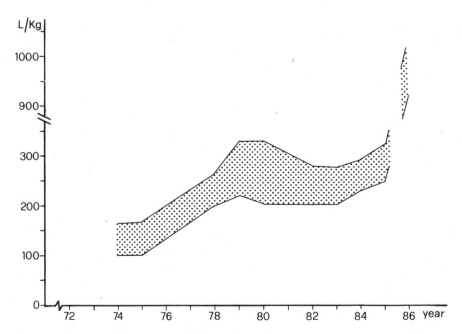

Figure 7. Evolution of prices (lira per kilogram) paid by processing plants from 1974 to the end of 1985. (Prices are converted to the 1985 value of italian lira using coefficients of inflation provided by the Italian Central Institute for Statistics); (1 U.S.$ = 1700 L in 1985.)

plants now work this species, which has a lower price and a meat yield 2% higher than *C. gallina*.

In October 1985, a new fishery act was issued that regulates all hydraulic clam dredging activities along the Italian coast (not only in the Adriatic). For the Adriatic Vongola fishery it maintains most of the provisions of the former decrees, but with the agreement of most of the fishermen's organizations, the daily quota per boat has been reduced by one-half, to 1.2 tons boat/day; a new Management Committee has been appointed; and the limited-entry policy has been extended to all hydraulic clam dredgers.

The first measure adopted by the new Management Committee has been to increase by 10% the number of licensed boats for the Adriatic Vongola fishery on the assumption that this increment in the total fishing effort is counterbalanced by the reduction of the daily quota per boat. When the difficulties experienced in past years in controlling landings and the apparently overharvested status of the resource are considered, this decision seems rather questionable. A precautionary freezing of the number of licensed boats would have been more appropriate from both biological and economic points of view, but other social considerations and the strong pressure from some fishermen's organizations led the Italian fishery authorities to take these management actions.

It may be that the new possibilities offered by the exploitation of *Venerupis aurea* will allow fishermen to divert effort from *C. gallina* beds whenever these are depleted, either by overexploitation or recruitment failure, or by catastrophic mortalities from algal blooms. Thus the economic rent from the whole fishery can be maintained at an acceptable level.

Probably this new equilibrium will be reached at higher costs, but with more employment opportunities generated. However, if a weakness in fisheries management and enforcement results in a continuous increase of total fishing effort with breakdown of daily quotas and so forth, we could again see over the next few years a shortage of available product for processing plants, with consequent unemployment problems in both processing and fishing sectors.

I believe that if hydraulic clam dredging is still to remain a profitable fishing activity, and one that operates without government subsidies, it will be largely a result of the efforts already made to bring the existing provisions into operation. In order to achieve this, it was necessary to develop a management philosophy among people who, until a few years ago, believed the resources of the sea were unlimited and uninfluenced by human activities.

ACKNOWLEDGMENTS

Through the years, many people from the Istituto di Ricerche sulla Pesca Marittima, Ancona, the Adriatic fishermen's organizations, the cannery industries, the port offices, and the Ministry of Merchant Marine, Fisheries Division, helped me in my work and in gathering the data presented herein. To all of them, my sincere thanks. The permission from Food and Agriculture Organization of the United Nations to reproduce Figure 2 from G.F.C.M. Studies and Reviews is also gratefully acknowledged.

REFERENCES

1. FAO, *Yearb. Fish. Stat. Catches Landings* **56**:1–394 (1984).

2. F. Vives, and P. Suau, Sobre la chirla (*Venus gallina* L.) de la desembocadura del rio Ebro. *Invest. Pesq.* **21**:145–163 (1962).

3. A. Artegiani, Seasonal flow rates of the Italian rivers having outlets in the northern and in the central Adriatic. *FAO Fish. Rep.* **290**:81–83 (1984).

4. M. Buljan, Osnovne karakteristike Jadrana kao produkcionog bazena. *Acta Adriatica* **16**:31–62 (1974).

5. A. Vatova, La fauna bentonica dell'Alto e Medio Adriatico. *Nova Thalassia* **1**:1–110 (1949).

6. A. Barillari, A. Boldrin, C. Mozzi, and S. Rabitti, Alcune relazioni tra natura dei sedimenti e presenza della vongola *Chamelea* (*Venus*) *gallina* (L.) nell'alto Adriatico presso Venezia. *Atti Ist. Veneto Sci. Lett. Arti Cl. Sci. Mat. Nat.* **137**:19-34 (1979).

7. P. Colantoni, Ricerche sui Molluschi dei fondali antistanti il delta del Po. *G. Geol. Ser. 2* **38**:513-528 (1972).

8. J. M. Peres, The Mediterranean benthos. *Oceanogr. Mar. Biol.* **5**:449–533 (1967).

9. L. Poggiani, C. Piccinetti, and G. Piccinetti Manfrin, Osservazioni sulla biologia dei molluschi bivalvi *Venus gallina* L. e *Tapes aureus* Gmelin nell'Alto Adriatico. *Note Lab. Biol. Mar. Pesca, Fano* **4**:191–209 (1973).

10. C. Froglia, Osservazioni sull'accrescimento di *Chamelea gallina* (L.) ed *Ensis minor* (Chenu) nel medio Adriatico. *Quad. Lab. Tecnol. Pesca, Ancona* **2**:37–48 (1975).

11. G. Marano, N. Casavola, C. Saracino, and E. Rizzi, Riproduzione e crescita di *Chamelea gallina* (L.) e *Venus verrucosa* (L.) (Bivalvia: Veneridae) nel Basso Adriatico. *Mem. Biol. Mar. Oceanogr.* [N.S.] **12**:97–114 (1982).

12. M. G. Corni, O. Cattani, L. Mancini, and G. Sansoni, Aspetti del ciclo biologico di *Venus gallina* L. in relazone alla tutela degli stocks esistenti. *Pubbl. Consorzio Cent. Univ. Studi Ric. Risorse Biol. Mar.: Cesenatico* 1–12 (1980).

13. G. Salvatorelli, Osservazioni sul ciclo riproduttivo annuo di *Venus gallina* (Molluschi Lamellibranchi). *Ann. Univ. Ferrara, Sez. 13* [N.S.] **2**:15–22 (1967).

14. G. Valli and G. Zecchini-Pinesich, Alcuni aspetti della riproduzione e della biometria in *Chamelea gallina* (L) (Mollusca Bivalvia) del Golfo di Trieste. *Atti Conv. Unità Operative Afferenti Sottoprogetti Risorse Biol. Inquinamento Marino, 1981* pp. 343–351 (1982).

15. C. Froglia, Aspetti biologici, tecnologici e statistici della pesca delle Vongole (*Venus gallina*). *Dagli Incontri Tecnici* **9**:7–22 (1975).

16. J. P. Guérin, Contribution à l'etude systématique, biologique et écologique des larves meroplanctoniques de Polychètes et de Mollusques du Golfe de Marseille. 2. Le cycle des larves de Lamellibranches. *Tethys* **5**:55–70 (1973).

17. A. D. Ansell, Reproduction, growth and mortality of *Venus striatula* (Da Costa) in Kames bay, Millport. *J. Mar. Biol. Assoc. U. K.* **41**:191–215 (1961).

18. C. Froglia, Feeding of *Lithognathus mormyrus* (L.) in central Adriatic sea (Pisces, Sparidae). *Rapp. Comm. Int. Mer Medit.* **24(5)**:95–97 (1977).

19. L. Mancini, La produzione dei molluschi bivalvi in Italia: Attualità e prospettive. *Rass. Econ. C.C.I.A.A., Forli* **12**:1–30 (1983).

20. G. Olivi, *Zoologia Adriatica, ossia Catalogo ragionato degli animali del golfo e delle lagune di Venezia; preceduto da una dissertazione sulla storia fisica e naturale del golfo; e accompagnato da memorie ed osservazioni di fisica, storia naturale ed economia.* Bassano, 1792, 334 pp.

21. A. P. Ninni, Saggio dei prodotti e della industria della pesca nelle Lagune e nel Mare di Venezia, inviato alla Esposizione internazionale della Pesca in Berlino. *Esposizione internazionale di Pesca in Berlino 1880, Sezione Italiana, Catalogo degli espositori e delle cose esposte.* 1880, pp. 169–187.

22. R. J. Stokes, E. A. Joyce, Jr., and R. M. Ingle, Initial observations on a new fishery for the sunray venus clam, *Macrocallista nimbosa* (Solander). *Tech. Ser.—Fla., Dep. Nat. Resour., Mar. Res. Lab.* **56**:1–27 (1968).

23. C. Froglia and S. Bolognini, Clam fishery with hydraulic dredges in the Adriatic sea. *G.F.C.M. Stud. Rev.* **62**:37–40 (1987).

24. C. Froglia and M. E. Gramitto, Considerazioni sulla selettività dei dispositivi di setacciatura utilizzati nella pesca delle vongole (Venus gallina). *Quad. Lab. Tecnol. Pesca, Ancona* **3**:37–46 (1981).

25. G. R. Morgan, Increases in fishing effort in a limited entry fishery the western rock lobster fishery 1963–1976. *J. Cons., Cons. Int. Explor. Mer* **39(1)**:82–87 (1980).

26. P. H. Pearse, Regulation of fishing effort. With special reference to Mediterranean trawl fisheries. *FAO Fish. Tech. Pap.* **197**:1–82 (1980).

27. F. Meany, Should licences in a restricted fishery be saleable? *Aust. Fish.* **37(8)**:16–21 (1978).

28. C. Marricchi, Andamento dell'industria conserviera ittica nel 1985. *Assoc. Naz. Conserv. Ittici Tonnare*, Roma: 1–21 (1986).

23 THE MUTSU BAY SCALLOP FISHERIES: SCALLOP CULTURE, STOCK ENHANCEMENT, AND RESOURCE MANAGEMENT

Sadao Aoyama
Aquaculture Center
Aomori Prefecture
Aomori-Ken, Japan

1. MAIN FEATURES OF THE HYDROGRAPHY OF MUTSU BAY

Mutsu Bay, the area of principal importance for scallop culture in the world, is a semiclosed body of water located in the northern part of Japan (Fig. 1). Its surface area of 1660 km^2 forms an almost self-contained environment, opening onto the

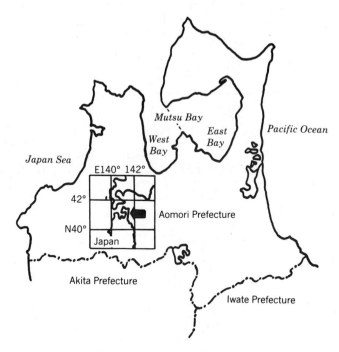

Figure 1. Location of Mutsu Bay.

Tsugaru Strait through a narrow (11-km-wide) passage. This bay has a fairly flat sea floor with mean depth of 38 m, covered mainly with sand and gravel along the shore line, turning into muddy sand toward the center. Mutsu Bay is relatively calm; that is, 82% and 92% of the wave heights (of an average of the three highest waves observed during a 20-min period in every 2-hr interval from 1973 to 1975), were less than 1.2 m in the East Bay and the West Bay, respectively (1).

The annual change in water temperature at a depth of 1 m is in the range of 3 to 25°C, and salinities range between 32 and 34%. There is a certain annual variation in the COD value and values averaging less than 0.5 mg/L have been recorded throughout the year (usually oxygen concentration ranges from 0.5 to 2 mg/L and rarely exceeds 3 mg/L). As a general phenomenon, the chlorophyll concentration in Mutsu Bay rises during winter and drops during summer.

2. CHARACTERISTICS OF THE SCALLOP FARMED IN MUTSU BAY

The scallop farmed in Mutsu Bay is *Patinopecten yessoensis* (Jay) and this is the only species dealt with in this chapter. On maturity, the gonad of the female scallop turns reddish orange in color and that of the male pale yellow. The diameter of a fully mature egg is about 80 μm, and fecundity varies with age (2) but is

TABLE 1 Fecundity of Scallops
(Yamamoto 2, 5)

Age (yr)	Shell Length (mm)	Live weight (g)	Weight of Gonad (g)	No. of Eggs ($\times 10^{-4}$)
2	122	211	39	11,440
2	108	162	29	8,448
2	111	191	37	11,000
3	137	294	43	12,584
3	127	282	40	11,704
3	131	299	45	13,200
3	126	288	44	11,088
4	148	367	53	15,488
4	156	352	59	17,248
4	138	330	52	15,224
4	142	341	61	17,864
5	148	359	60	17,600
5	149	381	62	18,128
6	151	392	59	17,248
6	152	420	64	18,744

usually in the range of 8 to 18×10^7 eggs/female (Table 1). Water temperature is a critical factor for gonad maturation, and it is reported (3) that gonadal development to spawning is mainly affected by changes in water temperature during September through December in the previous year (Fig. 2).

The spawning season lasts from March through April once the water temperature reaches 6–8°C. Fertilized eggs hatch and swim up as freely swimming trochophores in 3–4 days. Planktonic life lasts for about 40 days, and when larvae reach 280–300-μm shell size, they attach to seaweed or other substrates using a byssus excreted by the byssal gland. After attachment, formation of the spat shell is initiated, and development of the posterior adductor muscle become significant while the anterior adductor muscle degenerates. When spat reach 8–10-mm shell length (usually after 40–70 days), they proceed to a benthic life by severing the byssal thread and sinking to the sea floor; however, under natural conditions, only a small proportion of spat survive this stage in Mutsu Bay. By 30-mm shell length, juveniles acquire the ability to select their habitat (4), and once they have settled into a good habitat, they rarely move.

Scallops dig shallow depressions on the sandy bottom and recess themselves with the brown left shell facing up and white right shell down, feeding on phytoplankton and detritus collected by the ciliated gills. The level of physiological activity depends mostly on water temperature, and the optimum temperature for feeding judged from gill ciliary movement is in the 5–20°C range (5). If the water temperature deviates from this range, the growth of the scallop slows down or stops, and a resting mark will be formed on the shell. This situation usually occurs in Mutsu Bay during summer. Optimum temperature for their maximal growth is between 10 and 15°C. The life span of these scallops is 10–12 yr.

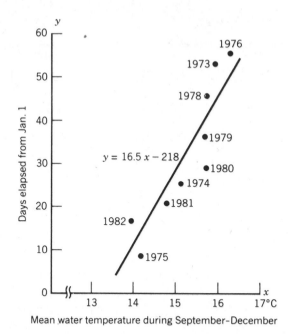

Figure 2. Relationship between water temperature in the preceding year and gonadal development [expressed as days required to reach the gonad index of 20% (Hirano 3)].

3. HISTORY OF SCALLOP FARMING

Because Mutsu Bay is the southernmost habitat of scallops as far as commercial-scale-fishing is concerned, its annual production formerly fluctuated greatly under natural conditions. As shown in Figure 3, scallop production largely depended on unusually high survival following a period of abnormal fecundity. These conditions have occurred every 10 or 20 yr up until the mid-1960s. Following such outstanding years, more than 10,000 tons of scallops were harvested for 2–5 yr, but then the production dropped back to the 200–300-ton level again.

In 1963–1964, an epoch-making method using a net onion bag for seed collection was developed in Mutsu Bay, which is the prototype of the natural seed collector presently in use. This collector consists of a cedar spray or branch as the attaching material, enclosed in a net onion bag. Larvae are able to penetrate through the net and attach to the material inside, but they are too large to pass back out through the mesh when they are ready to begin a benthic life, and consequently they are trapped in the bag. Use of this method made collection of a large quantity of spat possible. Owing to the spreading use of this collector, annual scallop yields from Mutsu Bay dramatically increased, and in 1974 reached the highet level in history, 47,000 tons. However, production showed a decline from the summer of 1975 owing to mass mortality and had dropped back to 16,000 tons by 1977.

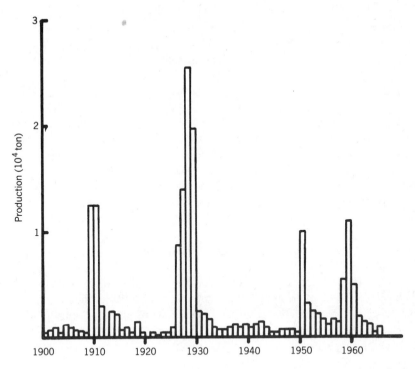

Figure 3. Historical production of scallops in Mutsu Bay prior to the culture method.

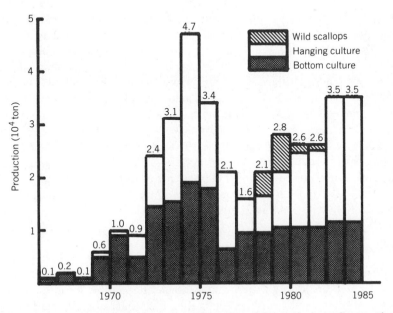

Figure 4. Annual production of scallops in Mutsu Bay after introducing culture method.

Various theories such as the occurrence of poor environmental conditions, overpopulation, disease, or genetic defects were proposed as causes of the mass mortality. In retrospect, through the series of field tests conducted by the Aquaculture Center, Aomori Prefecture (the Center), regulation of the total number of spat being introduced each year to a limit of 700 million for the whole area of Mutsu Bay was enforced. The application of a technique to increase scallop survival developed from field studies by the Center, improved the situation and had resulted in an increasing upward trend in production by 1978. In 1982, annual production reached 35,000 tons and the scallop industry had thus been revived (Fig. 4). Right now there are about 1900 firms farming scallops in Mutsu Bay. Volume and value of the farmed scallops account for 70 and 80%, respectively, of all fishery products for the bay, and thus scallop farming is the largest industry in this area.

4. TECHNIQUES OF SCALLOP FARMING

The life cycle of scallops during farming is shown in Table 2. Common techniques can be applied for both hanging culture and bottom culture during the period from seed collection to intermediate rearing. When scallops reach 3–5-cm shell length, some of them are thinned out and released for on-bottom sowing, but the majority are kept for hanging culture.

4.1. Collection of Natural Seed

After about 40 days of planktonic life, larvae attach to substrates by byssal threads and can be collected by hanging collectors in the sea during this period. Collection of natural spat for experimental purposes in Mutsu Bay was initiated by the Center in 1937 using sheets of hemp. Since then, the method of seed collection has been modified as shown in Figure 5. The collectors now most commonly used consist of worn out synthetic fishing net inside onion net bags. Under natural conditions, the secretory ability of the byssus deteriorates when the spat reaches a shell size of 8–10 mm, and the spat then falls off the collector and is reared in a pearl net (a net originally used for rearing pearl oysters). This is a wire frame net used for intermediate rearing of scallops. In the 1960s, a few hundred or, at most, a few thousand spat were collected on one collector. There is still a certain annual fluctuation; however, more than 10,000 spat per collector can now be consistently harvested owing to both an adequate size of the spawning stock and precise forecasts of the state and timing of spatfall (Table 3).

4.2. Forecast of the State of Spatfall

The prompt and pertinent forecast of spatfall, and associated detemination of stage and density of larvae, is a major technical factor responsible for the stabilization of a high level of natural seed collection. Surveys of planktonic larvae and spat in

TABLE 2 Life Cycle of the Scallop During Farming[a]

		Stage					
		Larval Shells					
	Fertilization	Trochophore	Attached Spat	Benthic Spat	Juvenile	Young	Adult
Month	March	April	May	July	Oct.–Dec.	May	Oct. —
Time after fertilization	0		40 days	4 months	8–10 months		18–20 months
Shell length	80 μm		300 μm	10 mm	30 mm	6 cm	10 cm
Farming method							
Hanging culture						Hanging culture →	
Bottom culture						Releasing bottom culture →	
		Seed collection (natural) →		Intermediate rearing →			
Farming area							
Hanging culture							
Bottom culture				Beneath the surface →			
				Sea floor			

[a]Based on the method used in Mutsu Bay. Artificial seed production in hatcheries is no longer practiced on a commercial scale.

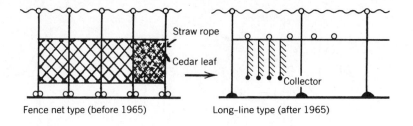

Fence net type (before 1965)   Long-line type (after 1965)

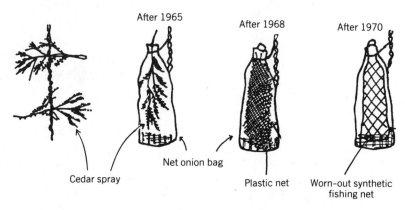

Cedar spray   Net onion bag   Plastic net   Worn-out synthetic fishing net

After 1965   After 1968   After 1970

Figure 5. Designs of natural seed collectors and their deployment.

TABLE 3 Progress in Spat Collection

Year	No. of Spat Harvested per Collector ($\times 10^{-3}$)	Year	No. of Spat Harvested per Collector ($\times 10^{-3}$)
1965	0.1	1975	69.2
1966	0.1	1976	3.5
1967	1.8	1977	15.8
1968	2.9	1978	38.8
1969	0.4	1979	34.6
1970	10.4	1980	30.6
1971	15.4	1981	59.2
1972	28.3	1982	1.6
1973	0.5	1983	35.1
1974	52.4		

Mutsu Bay were begun in 1941. Since 1950, the survey and research efforts have been continued with the objective of providing forecasts of available spatfall (6). At present the Center, together with a group studying scallop biology, is conducting a comprehensive survey of spawner maturity, larvae and spats, and environmental conditions, and a system providing precise forecasts has been established. These activities consist of a sequential survey of:

1) the gonad index for the spawning stock during December through May;
2) monitoring abundance of larvae during April through June;
3) spat during June through July; and
4) environmental conditions with automatic monitoring buoy robots throughout the year.

Maximal figures for larval density observed in Mutsu Bay are 6000 scallop larvae per ton of water. If more than 100 individuals per ton are observed, then in practice more than 400 individuals per collector are likely to be harvested. The collection of this information covers the entire area of Mutsu Bay, and data obtained are analyzed immediately and published promptly by radio, television, and laboratory leaflet.

4.3. Intermediate Rearing

Rearing of spat from collectors up to a shell length of 3–5 cm in pearl nets is called intermediate rearing. Intermediate rearing is usually carried out for two purposes: one is hanging culture (using cylindrical nets, pearl nets, pocket nets, or ear-hanging; that is, hanging from a hole drilled in the lateral ''ear'' of the shell) and the other is bottom culture (releasing scallops on the sea floor). Intermediate rearing starts from spat collection in July–August and lasts till December of the same year or until the following April, using facilities similar to the long line shown in Figure 6. The facilities used for hanging culture are described below. In both cases, scallops in hanging culture should be kept in conditions free from strong wave action.

4.4. Hanging Culture

Hanging culture is carried out over about 50,000 ha of the demarcated fishery zone (Fig. 7) according to the basic plan shown in Table 4. It takes about 1.5–2.5 yr to raise spat to marketable size. According to the standard growth curve shown in Figure 8, they usually attain 130-g meat weight in 2.5 yr. Culture methods most often employed are the use of a cylindrical net, a pearl net, and ear-hanging (Fig. 9). A comparison of these methods is summarized in Table 5.

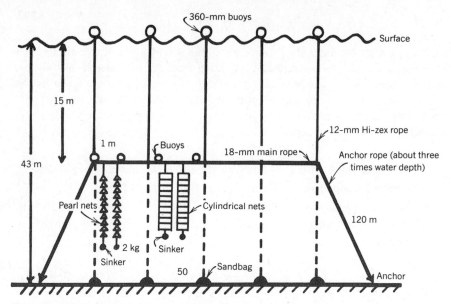

Figure 6. Intermediate rearing and hanging culture of scallops (mm = mm diameter). Hi-zex rope is a commercial name for a type of synthetic fiber. Distance between buoys is adjustable as the scallops grow.

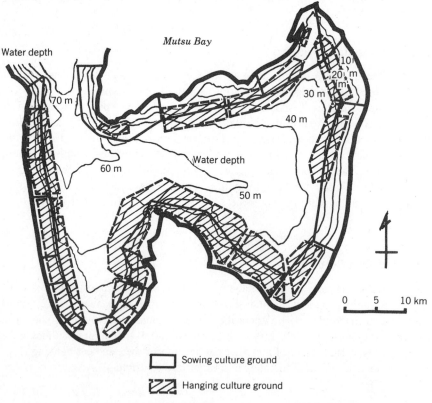

Figure 7. Hanging and sowing culture ground in Mutsu Bay.

TABLE 4   Basic Patterns of Scallop Culture in Mutsu Bay Established from the Feasibility Study

	July–Aug.	Sept.–Oct.	Feb.–April	Sept.–Oct.	Oct.–April
Ordinary case	Seed collection— 100 scallops per pearl net →	First thinning out to 20 scallops per pearl net →	Second thinning out to 100 scallops per cylindrical net or 100 scallops per ear-hanging set	⟶	Market
Modified case[a]				100 scallops per cylindrical net ⟶	Market

[a]Effective if stock not fully recovered from incidence of mass mortality.

4.5. Bottom Culture

There are about 23,000 ha of area suitable for bottom culture in Mutsu Bay, judging from various factors such as condition of the seabed, benthic fauna, and water movement. After intermediate rearing, spats are released in December of the same

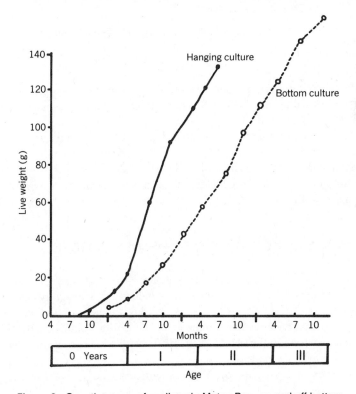

Figure 8. Growth curves of scallops in Mutsu Bay, on and off bottom.

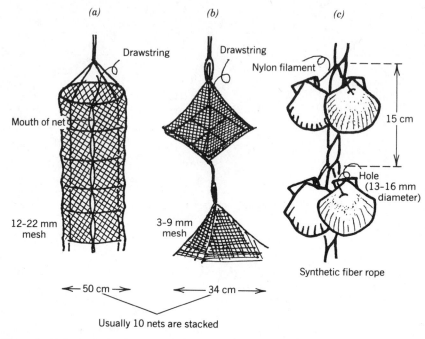

(a) *(b)* *(c)*

Drawstring

Drawstring

Nylon filament

15 cm

Mouth of net

Hole
(13-16 mm
diameter)

12-22 mm
mesh

3-9 mm
mesh

Synthetic fiber rope

←— 50 cm —→   ←——— 34 cm ——→

Usually 10 nets are stacked

Figure 9. Main methods of hanging culture used in Mutsu Bay. (*a*) Cylindrical net; (*b*) pearl net; (*c*) ear-hanging.

year or the following April as soon as they reach a shell length of 3–5 cm. They are harvested by a drag net 2.5–3.5 yr later. As is shown in the standard growth curve in Figure 8, they attain 150 g of live weight in 3 yr. The relationship between stocking density and production is shown in Figure 10. The maximum production per square meter in Mutsu Bay is 1200 g; therefore, less than 6 individuals/m^2 should be released to achieve normal growth until 200-g market size is reached (7). The proportion of abnormal individuals in net culture rises with stocking density (Fig. 11) (8).

5. PROPER MANAGEMENT OF THE RESOURCES

Fisheries cooperatives around Mutsu Bay possess the exclusive right to scallop farming and fishing operations, and the scallop industry in Mutsu Bay has achieved rapid progress since the mid-1960s, thanks to development of the technique of natural seed collection. However, tremendous damage was inflicted by the mass mortality that occurred in the period 1975–1977, but the industry has survived with the help of technical support.

Based on these experiences, our aim has shifted slightly toward looking for a

TABLE 5 Comparison of Culture Methods in Common Use in Mutsu Bay

Method	Materials	Size	Stocking Density	Cost of Materials	Durability	Cost of Materials/ Individual Scallop	Remarks
Net culture	Cylindrical net (10 nets are stacked)	200 × 50 cm	100 individuals/ net	1500 yen/net	5–6 yr	2.8 yen	Hanging of scallops is easy. High incidence of abnormal scallops if juveniles are in poor condition
	Pearl net (10 nets are stacked)	Pearl net: 34 × 34 cm Set length 2 m	6 individuals/ net 10 stacked/ set 60 individuals/ set	100 yen/net × 10 stacked = 1000 yen/set	5–6 yr	2.9 yen	Fairly free from trouble with attaching organisms. Provides more stable environment than a cylindrical net, but is easily clogged
Ear-hanging culture	Hanging rope or nylon filament	Set length 9 m	100 individuals/ set	Hanging rope: 90 yen/set Nylon filament: 20 yen/set	Hanging rope: 3–4 yr Nylon filament: 1 yr	0.4 yen	Material cost is low. Initial growth is good, but attaching organisms later on. Handling is time-consuming (2000 individuals/man-day)

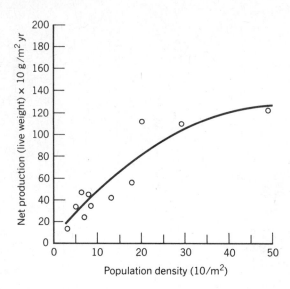

Figure 10. Relationship between population density of scallops and annual production (Yamamoto 7).

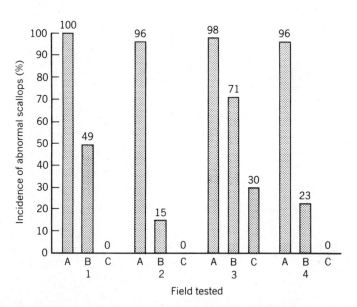

Figure 11. Relationship between stocking density and percentage of abnormal scallops. Data from Yokoyama (8). Stocking density (number of individuals per pearl net) after each thinning-out stage: (*A*) 500 juveniles at stocking (Aug.); 200 after first thinning out (Sept.); 100 after second thinning out (Oct.–Nov.); 50 after third thinning out (Dec.–Jan.); (*B*) 200 juveniles at stocking (Aug.); 20 after first thinning out (Oct.–Nov.); (*C*) 100 juveniles at stocking (Aug.); 20 after first thinning out (Sept.).

high but stable production. From past production records, from results of the feasibility studies conducted by the Center, and from data on the primary productivity
in Mutsu Bay, we concluded that the number of spat introduced annually at the
start of intermediate rearing should be limited to less than 700 million, in order to
avoid density-dependent growth, and that this limit should be strictly observed.

REFERENCES

1. Aomori Prefecture, *Final Survey Report on Environmental Conditions of the Fishing
   Ground and Basic Plan for the Fishery Development in Mutsu Bay* (in Japanese). 1976,
   pp. 45–46.
2. G. Yamamoto, *Scallop Farming in Mutsu Bay* (in Japanese). Fisheries Resource Conservation Association, Tokyo, 1964, pp. 6–19.
3. T. Hiranao, The state of scallop maturity (in Japanese). The news of *Aquacult. Cent.
   Aomori Pref.* **17**:3 (1982).
4. K. Takahashi, Activity of scallop (2) (in Japanese), *Tech. Rep. Aquacult. Cent. Aomori
   Pref.* **14**:296–306 (1985).
5. G. Yamamoto, *Scallop Farming in Mutsu Bay* (in Japanese). Japan Fisheries Resources
   Conservation Association, Tokyo, 1964, pp. 40–44.
6. S. Aoyama, Present state of scallop production in Mutsu Bay and related research topics,
   Bull. Jpn. Soc. Sci. Fish. Tohoku Branch **30**:7–63 (1982).
7. G. Yamamoto, *Marine Ecology.* Tokyo Univ. Press, Tokyo, 1973, pp. 127–174.
8. K. Yokoyama, Review of scallop culture technique study. *Aquacult. Cent. Amori Pref.*
   1982, pp. 65–68.

24 FISHERIES FOR GIANT CLAMS (TRIDACNIDAE: BIVALVIA) AND PROSPECTS FOR STOCK ENHANCEMENT

J. L. Munro
International Center for Living Aquatic Resources
Management, Coastal Aquaculture Centre
Honiara, Solomon Islands

1. Introduction
    1.1. Distribution
    1.2. Habitats
    1.3. Basic Biology
2. Fisheries
    2.1. Subsistence Use
    2.2. Commercial Exploitation
3. Cultivation
4. Potential Harvests
5. Reef Reseeding
6. Conservation Strategies
    References

1. INTRODUCTION

Giant clams (Tridacnidae; Bivalvia) have formed a significant part of the diets of the peoples of Oceania and Southeast Asia for thousands of years, although they have been known to scientists only for several centuries. Apart from the kidney, the entire clam is edible and the adductor muscle is a highly prized delicacy in several parts of Southeast Asia.

Much interest has been aroused in recent years in the development of mariculture techniques for giant clams (Heslinga and Watson 1), but there is very little information concerning the fisheries for these animals. In connection with maricultural prospects, Munro and Heslinga (2) suggested that

> the options range from small-scale hatcheries intended to supplement and conserve locally-consumed stocks, through large scale extensive maricultural enterprises employing many people on production and processing, to intensive high-technology cultivation systems operating on the near side of science fiction.

Within this range of options is the possibility of enhancing natural recruitment by supplementary stocking of reef areas with hatchery-reared juveniles. Such a system would depend greatly on mass production of extremely large numbers of spat that could be broadcast onto reefs over a wide area.

There are seven species of giant clams ranging in maximum size from less than 20 to 137-cm shell length (SL). By virtue of their size the species of principal commercial interest are *Tridacna gigas*, *T. derasa*, and, to a lesser degree, *Hippopus hippopus*. The smaller species, *T. maxima*, *T. squamosa*, *T. crocea*, and *H. porcellanus*, are utilized in subsistence fisheries and in the shell trade.

1.1. Distribution

Living species of tridacnids are confined to the Indo-Pacific region. Of the seven species, the geographical distributions of *T. maxima*, *T. squamosa*, and *T. crocea* apparently remain basically as reported by Rosewater (3) but the geographical ranges of the largest species have been dramatically attenuated, either by exploitation or by ecological factors.

In the case of *T. gigas*, the present centers of abundance are the waters of Palau, the coasts of New Guinea and adjacent islands, Solomon Islands, Marshall Islands, Kiribati, southern Burma, the west coast of Thailand, and the waters of northern and northeastern Australia. However, the species is now extinct around Java and eastern Sumatra and heavily exploited elsewhere in Indonesia. It is extremely rare in the Philippines and extinct in the Ryukyu Islands, Taiwan, the Marianas, Guam, New Caledonia, and Fiji. Its status in Vanuatu is unknown and only relict populations of a few individuals are known from Lamotrek and West Fayu Atolls in the Federated States of Micronesia.

T. derasa has a curious wedge-shaped geographical distribution stretching diagonally from the Philippines in the north to Tonga in the southeast and the Cocos-Keeling Islands in the southwest. It is now extremely rare in the Philippines, extinct in central Indonesian waters, and has confirmed centers of abundance only in Palau, northern New Guinea, Australia, Solomon Islands, New Caledonia, Fiji, and Tonga. There are no recent reports of living specimens from the Cocos-Keeling Atoll in the Indian Ocean or from Vanuatu in the Pacific. There is no evidence that this species has ever inhabited any part of Micronesia east of Palau.

According to Rosewater (3), *H. hippopus* originally had a distribution extending from the Andaman and Nicobar Islands in the west, through Southeast Asian waters to the Ryukyu Islands and Bonin Island in the northeast, to northern and north-eastern Australia and eastwards as far as Tonga, American Samoa, Tuvalu, Kiribati, and the Marshall Islands. It now appears to be extinct in Fiji, Tonga, Samoas, Taiwan, Gaum, and the Marianas. Ramadoss (4) found no evidence of its occurrence in the Andaman and Nicobar Islands. The abundance of this species in the Philippines has been greatly reduced by intensive exploitation—mostly for the shell trade.

H. porcellanus is found only in northeastern Indonesia, Palau, and the southern Philippines. In the latter area it has become very rare because its shell is more highly valued than that of *H. hippopus* (University of the Philippines 5).

Whether or not all of the extinctions have been caused by humans is unclear. The last *T. gigas* were harvested in Fiji about 15 yr ago and although it seems likely that their extinction can be largely attributed to overharvesting, the possibility that the contraction of the ranges of *T. gigas* and *H. hippopus* is a result of climatic or other ecological factors cannot be discounted. Subfossil shells of *H. hippopus* are found in the Samoas and in Tonga, but no living specimens have ever been collected and it is possible that some museum records are based on long-dead shells.

1.2. Habitats

The distribution of giant clams is largely contiguous with that of reef-building corals; they are normally found in close association with coral reefs. *Tridacna maxima*, *T. squamosa*, and *T. crocea* are all strongly attached to reefs thoughout their lives. However, *T. gigas*, *T. derasa*, *H. hippopus*, and *H. porcellanus* are all free-living as adults and it is clear that a hard substratum is needed for settlement and attachment only in the juvenile stages. For example, *T. derasa* is commonly found on sand at the base of large coral heads in Fijian waters and it appears that they settle on coral heads but drop off onto the sand at a certain size (A. D. Lewis and T. Adams, personal communication, 1985).

T. gigas occurs at all depths between the lower intertidal and around 15 m; there are unauthenticated reports of specimens being found in 30 m. *T. derasa* is found in shallow water, but apparently not intertidally, and extends to depths of 25 m in Fijian lagoons (Fisheries Department, Fiji 6).

H. hippopus is commonly found in intertidal areas, even on the windward sides of atolls, and also in lagoons and reefs to depths of 10 m (Munro 7). *T. squamosa* and *T. maxima* are normally associated with coral reefs, to which they are firmly attached by byssal threads throughout their adult lives. *T. crocea* erodes the coral heads to which it is attached and is normally totally embedded in the coral.

1.3. Basic Biology

The basic anatomical and biological features of tridacnids have been described in detail by Yonge (8,9). Their most important feature is their symbiotic association with a dinoflagellate algae, *Symbodinium microadriaticum*, which inhabits a thin layer of tissue in the exposed, fleshy mantle of the clam. Recent evidence (Munro and Gwyther 10; Fisher et al. 11) shows that the algae, called zooxanthellae, directly supply the clams with sufficient photosynthetic products to meet their basic nutritional needs and that external sources of food are not essential for survival and growth in shallow sunlit waters. The clams can therefore be described as phototrophic and essentially similar to plants in their basic requirements in that their needs from the external environment are for inorganic nutrient salts and sunlight. It is suggested that the main role of filter feeding in tridacnids is to provide additional nitrates and phosphates for use by the zooxanthellae. Their absolutely unique feature is that they are the only phototrophic, and thus self-feeding, potential farm animal known to humankind.

Giant clams are protandrous hermaphrodites and for most of their adult lives are normally capable of producing both eggs and sperm. Older clams might become solely female (Crawford et al. 12). Spawning appears to be year round in the equatorial regions but restricted to the summer months in higher latitudes (Crawford et al. 12). Under normal conditions a spawning event is one in which sperm is produced initially, followed by a quiescent period and then by egg production (Wada 13; Beckvar 14; Gwyther and Munro 15). Presumably self-fertilization is largely avoided by this process. Clams often produce sperm in response to stimulus but do not thereafter enter the egg production phase. The eggs have an associated chemical that is detectable in minute concentrations by other clams of the same species and that triggers the spawning cycle.

The spawning process thus facilitates fertilization in that eggs spawned by a clam will, when wafted across a reef, be fertilized if another clam of the same species is encountered and, during the spawning season, induce a series of spawning events across a reef. However, in common with other sedentary invertebrates, the same process makes clams highly vulnerable to overexploitation, because when populations are reduced below a certain level the proportion of eggs that are not fertilized increases dramatically and, despite the very high fecundity (Crawford et al. 12; Jameson 16), the numbers of larvae resulting might be curtailed and recruitment might in turn fall dramatically to negligible levels.

2. FISHERIES

Fisheries for giant clams in the Indo-Pacific region fall into two distinct categories:

(a) Subsistence gathering, which is usually done on an opportunistic basis by women and children over reef flats and shallows and by men in deeper water.

(b) Commercial harvesting, which has almost invariably been conducted by vessels from Taiwan fishing without permits within the territorial waters or exclusive economic zones of Pacific nations or, less often, on isolated oceanic reefs in international waters.

2.1. Subsistence Use

Tridacnid shells are major components of middens in coastal regions in many parts of the Indo-Pacific (Swadling 17) attesting to the fact that giant clams, particularly *H. hippopus*, *T. crocea*, and *T. maxima*, have been an important part of subsistence diets for many centuries.

Table 1 (18–26) (which is based upon responses to a questionnaire circulated by ICLARM throughout the region in 1983, personal contacts, published information, and direct observations) summarizes information on the status of giant clam stocks, local utilization, incidence of poaching by foreign vessels, and species recognized to be present in countries of the region. Indo-Pacific countries for which no information is available are omitted from Table 1. Species likely to occur in such countries can be determined by references to Figure 1. Apart from Singapore, Thailand, and Malaysia, where it is claimed that clams are harvested only for their shells, clams are a normal component of everyday diets, and in most areas are regularly offered for sale in marketplaces, usually alive but often either smoked or dried.

Access to giant clam stocks in the Indo-Pacific region is controlled in many areas by various reef tenure systems that have evolved over many years. The systems range from open access areas with no control over fishing activities or other extractive processes, through collective open access, in which the entire local community has access to all or most parts of the fishing grounds but outside groups are excluded. In some areas, particularly in Melanesia, personal or family ownership is claimed of specific reef and lagoon resources and outside access is totally excluded.

2.2. Commercial Exploitation

Starting around 1969, fisheries authorities in the Indo-Pacific region became aware of the presence of significant numbers of Taiwanese vessels fishing on reefs. Investigations soon showed that giant clams, particularly *T. gigas* and *T. derasa*, were the target species and that only the adductor muscles were being taken and the rest of the flesh discarded (Pearson 18).

The vessels engaged in the fishery function as mother ships to several dories manned by four or five divers, each equipped with basic skin-diving gear, wading shoes, a large chisel-like knife, and a string bag for the clam muscle (Pearson 18). Clams are harvested in shallow water by swimming or wading, and clams living in deeper water beyond easy skin-diving depths are usually not harvested.

It was reported that very high prices were received by the fishing vessels for the product, which was consumed in seafood restaurants in Taiwan. A part of the landings of the muscle was reported to be dried and sold to Chinese communities in Hong Kong, Singapore, and elsewhere at extremely high prices (Knox 27). However, the latter point has never been verified. Much of the dried product marketed in Hong Kong and elsewhere as ''yuen buz'', ''gong yiu chui'' or ''jan jui yuk'' is what the Japanese describe as ''kaibashira.'' This is the generic term

TABLE 1 Synopsis of Information on Status of Giant Clam Stocks, Local Utilization, and Species Recognized to Be Present[a]

Country	Species Present							Remarks on State of Stocks, Exploitation, and Utilization
	Tg	Td	Ts	Tm	Tc	Hh	Hp	
American Samoa			+	+				Ts very uncommon. Tm heavily exploited in all populated areas. Abundant protected stock of Tm at Rose Atoll. Used for subsistence. Eaten raw or sometimes boiled in coconut water
Australia	+	+	+	+	+	+		All species totally protected. Poaching has reduced stocks on Great Barrier Reef. Some use for subsistence by islanders (18)
Burma	+		+	+		+		Tg confined to southern waters. Marketed fresh for local consumption only
China (Taiwan)	?		?	?	?	?		Little suitable habitat
Cook Islands		+	+	+				Used for subsistence. Td recently introduced from Palau
Federated States of Micronesia	+	+	+	+		+		Relict populations of Tg reported only from Lamotrek Atoll and West Fayu. Td not reported in scientific literature. Now being introduced from Palau. Recent fossils of Tg abundant in Kosrae, Ponape, Truk, Yap. All species under heavy pressure from local exploitation. Eaten raw, marinated, sun-dried, or pickled in tuba vinegar. Remote reef areas may be fished by poachers (2)
Fiji		+	+	+				Tg and Hh are now extinct. Heavy artisanal and subsistence fishing in some areas close to major population centers. Other stocks relatively abundant. Much evidence of poachers recently reported. Clams are usually eaten fresh, marinated in lime juice and also coconut milk in some areas. Some is smoked but mostly marketed fresh
Guam				+				All other species thought to be extinct through local overexploitation (2)
India								
Andaman and Nicobar Is.			+	+	?			Exploited for subsistence only (4)
Laccadives			+	+				No information on utilization
Mainland			?	?				No information on utilization

TABLE 1 (Continued)

Country	Species Present							Remarks on State of Stocks, Exploitation, and Utilization
	Tg	Td	Ts	Tm	Tc	Hh	Hp	
Indonesia	+	+	+	+	+	+	+	All species are exploited and Tg and Td apparently extinct in Eastern Sumatra and Java. Undisturbed or moderately exploited stocks of all species in eastern waters. Flesh used mostly for subsistence and not extensively marketed. Shells used for making tiles (19,20)
Japan			+	+	+			Only confirmed extant in Ryukyu Islands where Tc is regularly marketed
Kiribati								All species regularly used for
Gilbert Is.	+		+	+		+		subsistence. Some artisanal fishing of
Phoenix Is.			?	+				Tm for sale in Tarawa for local
Line Is.				+				consumption only, dried, salted or
								fresh. Stocks of Tg very limited at Tarawa Atoll but moderate to good stocks elsewhere in Gilbert Islands group. No instances of poaching known (7,21)
Malaysia	+		+	+	+	+		Common in some parts of east coast of penninsular Malaysia and in Sabah. Little used as food—mostly dried. Shells marketed.
Marshall Is.	+	+	+	+		+		Numerous in some areas. Some poaching has occurred. Used for subsistence, either dried or cooked as a chowder with curry and coconut milk or cooked with coconut milk sauce. Td recently introduced from Palau
New Caledonia			+	+	+	+		Fossil Tg found but none alive. Stocks not overexploited. Local fishery sells fresh or iced clams. In 1982 a commercial company exported 30–40 tons of clam meat, but this fishery is now banned. No reports of poaching
Northern Marianas				+				Tg and other spp. extinct. Tm heavily exploited in past
Palau	+	+	+	+	+	+	+	All species used moderately for subsistence. Substantial stocks but poaching a continuing problem (22,23)
Papua New Guinea	+	+	+	+		+		All species used by subsistence fisheries. Td not found near to mainland. Tg not abundant on nearshore reefs or near main towns.

TABLE 1 *(Continued)*

Country	Species Present							Remarks on State of Stocks, Exploitation, and Utilization
	Tg	Td	Ts	Tm	Tc	Hh	Hp	
Papua New Guinea *(Continued)*								Some marketed smoked or fresh. Poaching a serious problem in remote areas and many foreign vessels arrested in recent years (24)
Philippines	+	+	+	+	+	+	+	Tg apparently almost extinct in most areas except in the southern and western periphery of the country. Td very rare. Hp confined to southern Philippines. All stocks heavily exploited for shells and meat. Mostly used for subsistence but some meat marketed
Polynesie Francaise				+				Used for subsistence, seldom marketed. Very abundant in lagoons, particularly if closed. Also scattered on outer slopes of high volcanic reef complexes
Saudi Arabia			?	+				Most stocks not utilized (25)
Singapore			+	+	+	+		All spp. very rare, on reef slopes of islands. Previously exploited by islanders but now only by recreational fishing. Not marketed
Solomon Islands	+	+	+	+	+	+		All stocks in good condition, especially in areas with religious constraints on eating "shellfish." Regularly consumed fresh for subsistence in many areas. Approximately 1 ton/yr of clam meat sold in Honiara at $0.50/kg. Shells used for craftwork. Clam poaching by foreign vessels is a problem
South Africa			+	+				Not utilized. Confined to extreme northeastern waters
Sri Lanka			+	+				Not consumed. Shells sold to tourists
Thailand	+		+	+		?		Stocks on west coast not heavily exploited. Shells are sold but meat not much used
Tokelau			?	+				Used for subsistence and some trade to Western Samoa. Heavily exploited
Tonga			+	+	+			Recent fossils of Hh can be found. Tm is most abundant species. Stocks heavily exploited for local consumption. Eaten raw or cooked. No poaching reported (26)

TABLE 1 *(Continued)*

Country	Species Present							Remarks on State of Stocks, Exploitation, and Utilization
	Tg	Td	Ts	Tm	Tc	Hh	Hp	
Tuvalu	+	?	+	+	?	+		No recent reports of Tg. Presence of Tc and Td not confirmed. Heavily exploited near villages for subsistence but healthy stocks elsewhere. No known poaching. Eaten raw
United States		+						Td recently introduced to Hawaii from Palau
Vanuatu	?	?	+	+	?	+		All species fished for subsistence and often sold in local markets. Sold fresh, soon after collection. No recent reports of live Tg or Td. No poaching reported
Western Samoa			+	+				Used for subsistence and some trade. Very heavily exploited. Hh believed to be extinct.

[a] Abbreviations are Tg = *T Tridacna gigas*, Td = *T. derasa*, Ts = *T. squamosa*, Tm = *T. maxima*, Tc = *T. crocea*, Hh = *Hippopus hippopus*, Hp = *H. porcellanus*. + = present confirmed, ? = presence unconfirmed, but within distributional range mapped by Rosewater (3). Published references noted in paretheses.

for dried bivalve adductor muscle (usually derived from the scallop, *Patinopecten yessoensis*). It is not known what fraction, if any, of the dried product was in the past composed of tridacnid muscle.

Most recently, it has been established that despite the decimation of stocks, the extension of exclusive economic zones, and improved surveillance, modest landings are still being made by Taiwanese vessels and that the price for frozen adductor muscle in Taiwan as of February 1986 was in the range of U.S. $20–30/kg (P. Philipson, Forum Fisheries Agency, personal communication).

In 1979 Taiwanese authorities listed 37 boats thought to be involved in clam fishing (Dawson 28). Within the next 6 months, 22 of these vessels were charged by the Taiwanese government with illegal clam fishing. Out of nine vessels for which the fishing area was known, six had been in Australian waters. Over the next 5 yr (1979–1984), nine of the 37 listed boats were apprehended in Australian waters plus another 10 that were not so listed. The foregoing suggests that in this period the Taiwanese clam boat fleet numbered around 80 vessels, for which Australia was the most favored destination, although acknowledged to afford a high risk of capture (Dawson 28). Boats apprehended on the Great Barrier Reef averaged more than 3 tonnes of clam muscle at capture (Pearson 18) and according to Dawson (28) sought loads of 10 tonnes before return to Taiwan. The number of trips made each year is unknown and no global estimate of clam harvests can be deduced other than to note that even if only a small fraction of these vessels

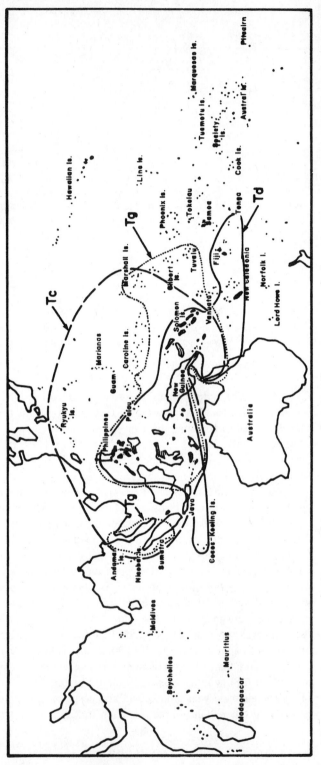

(a)

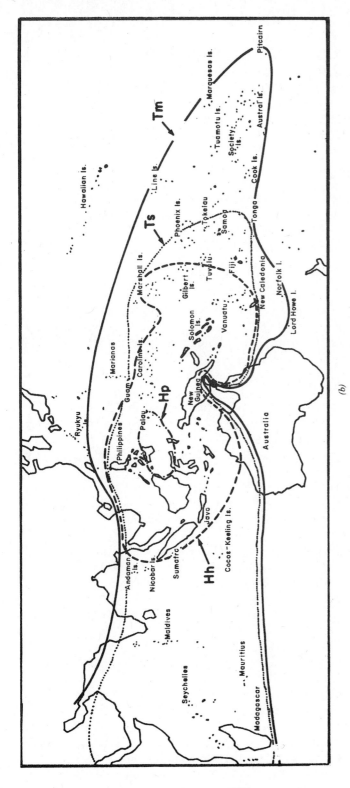

Figure 1. (a) Present geographical ranges of *Tridacna gigas* (Tg), *T. derasa* (Td), and *T. crocea* (Tc). (b) Present geographical ranges of *Tridacna maxima* (Tm), *T. squamosa* (Ts), *Hippopus hippopus* (Hh), and *H. porcellanus* (Hp).

managed to return to Taiwan with loads of 3–10 tonnes the overall harvest must have been several hundred tonnes of adductor muscle.

There is a substantial trade in ornamental giant clam shells in the Philippines (Wells 29) and shells are used in Indonesia for the production of high-quality tiles (Usher 19; Brown and Muskanofola 20).

3. CULTIVATION

On the basis of the work of Beckvar (14), Heslinga et al. (30), and Heslinga and Watson (1) it can unequivocally be stated that low-intensity hatchery techniques for *Tridacna derasa* have reached an advanced level. Recent work at James Cook University of North Queensland (James Cook University 31; Crawford et al. 12) suggests that high-intensity hatchery techniques for *T. gigas* and other species will soon become a reality. The low-intensity hatchery technique in its essentials (Heslinga and Watson, 1) requires nothing more than at least one tank with a minimum depth of 1.0 m and a volume of about 10 m^3, a few other tanks or raceways, plus a pump sufficient to produce a moderate flow of seawater for 8 hr/day. Such a system has the capacity to produce 22,000–55,000 12–17-mm *T. derasa* on a 5-month cycle (Heslinga and Watson 1).

High-intensity techniques require some basic laboratory facilities, including the provision of cultured microalgae to feed the larve, plus sufficient tanks in which recently metamorphosed larvae can attach to the substratum. The initial output from a high-intensity hatchery at James Cook University has been approximately 4 million newly metamorphosed juveniles/m^3 of hatchery tank, but the maximum density at which the spat can be induced to settle in shallow nursery tanks is not yet known.

The larval life is only 6–8 days (Heslinga and Watson 1; Crawford et al. 12) and juvenile *T. derasa* reach an average size of 5 mm in 4 months in Palau, at which time they can be transferred from settlement tanks into nursery tanks or raceways.

The ocean nursery stage in rearing of giant clams starts when clams are removed from land-based hatchery settlement tanks and are placed in semi-protected ocean situations. The size at which they can be moved offshore is likely to be around 0.5–1.5 cm and they can be expected to become relatively safe from most predators when they attain 10 cm, whereafter they can be removed from the ocean nurseries to relatively unprotected grow-out areas (Heslinga et al. 30). A primary option is to eliminate the ocean nursery phase and retain spat in tanks, ponds, or raceways until they attain about 10-cm length and then place them into unprotected ocean environments.

The ideal ocean nursery would appear to be a shallow (0.5–2.0 m), protected sandy sea floor with good water exchange but without significant wave action or environmental extremes. It should also be close to shore, thus rendering large, expensive boats unnecessary. However, areas such as this are few; in many areas less than optimal conditions will prevail but will have to suffice.

Ocean nurseries can be located on the sea floor, at mid-depths, or as floating structures. Sea-bed structures might be located subtidally or in the intertidal zone. The intertidal option is of much interest because of possible ease of access and maintenance but could subject the clams to environmental extremes (Munro 32). It would be possible to protect shallow ocean nurseries by simply enclosing the area in a net of sufficiently small mesh to exclude major predatory fishes. Tending a nursery of this sort would involve the care of the net enclosure and elimination of predators that have entered the system in their larval and juvenile stages.

The final grow-out stage is likely to be conducted on shallow sandy sea floors where the clams could be raised to a harvestable size.

4. POTENTIAL HARVESTS

Production in any fishery or aquaculture system is governed by recruitment or stocking rates and by subsequent growth and mortality. With shallow-water, sedentary organisms many of the usual concepts of fishing are inapplicable and the catch is merely gathered. If this is done systematically the total stock can be harvested from a given area, leaving no survivors. Growth rates of aquatic animals are usually described by means of the von Bertalanffy growth function (VBGF) and mortality rates are usually represented by the coefficient of the survival rate ($s = e^{-Z}$). Reliable estimates of these parameters for giant clams are few, and very little is known of the factors that govern growth rates or of the causes of mortality. Additionally, it appears likely that because of the special physiological features of giant clams caused by their symbiosis with dinoflagellate algae, the VBGF might not be entirely adequate to describe the detailed features of growth.

Table 2 shows trial computations of the biomasses that cohorts of 10,000 clams would attain under growth and mortality rates representing the "best" and "worst" survival rates under natural conditions and the "least-" and "most-favorable" likely growth parameters. VBGF parameters of $L_\infty = 63.0$ and $K = 0.20$ for *T. gigas* in Papua New Guinea have been estimated by Munro and Gwyther (33). Munro and Heslinga (2) summarized most available growth rate for giant clams and estimated the growth coefficient, $K = 0.136$, for *T. gigas* in Palau (based on limited data) when L_∞ was set at 100 cm. More recent work (Pearson and Munro 34) suggests parameters of $L_\infty = 78$ cm and $K = 0.11$ for the cooler waters of the Great Barrier Reef at latitude 17°S.

Munro and Heslinga (2) calculated a natural mortality coefficient of $M = 0.21$ for *T. maxima* on the Great Barrier Reef, the data being derived from the work of McMichael (35) on an unexploited population. Giant clams in a reef enclosure in Papua New Guinea had extremely low mortality rates; no *T. gigas* died over a 5-yr period and annual mortality rates in other species were only 3–6% (Munro and Heslinga 2).

As shown in Table 2, the biomasses vary greatly in response to the combinations of production parameters and, for example, after 5 yr an initial cohort of 10,000 10–12-cm clams might attain a biomass in the range from 40 to 142 tonnes.

TABLE 2 Biomasses of Cohorts of 10,000 *Tridacna gigas* Given Various Estimates of Growth and Mortality Parameters

	\multicolumn{8}{c}{Years after Stocking with 10,000 10–12-cm Clams}							
	1	3	5	7	9	11	13	15
Average individual length (cm)								
Case A: $K = 0.11$, $L_\infty = 78$, $t_0 = -0.02$	15.00	27.39	37.19	45.09	51.46	56.60	60.74	64.09
Case B:　$K = 0.20$, $L_\infty = 63$, $t_0 = 0.05$	23.00	33.87	43.22	49.53	53.78	56.65	58.58	59.89
Average individual weight (kg)								
Case A—poorest growth	0.65	4.32	11.32	20.78	31.51	42.52	53.12	62.88
Case B—best growth	2.49	8.44	18.18	27.93	36.20	42.64	47.39	50.80
Number of clams surviving								
Case C: $M = 0.21$	8106	5326	3499	2299	1511	993	652	429
Case D: $M = 0.05$	9512	8607	7788	7047	6376	5769	5220	4724
Biomass								
Rates A and C　(metric tons)	5.26	23.01	39.63	47.77	47.60	42.21	34.64	26.94
Rates A and D　(metric tons)	6.17	37.19	88.19	146.41	200.90	245.32	277.30	297.01
Rates B and C　(metric tons)	20.20	44.92	63.63	64.21	54.69	42.32	30.91	21.77
Rates B and D　(metric tons)	23.71	72.60	141.61	196.80	230.83	246.00	247.42	239.96

This highlights the need for more investigations to define these parameters accurately throughout the geographical ranges of the clams. The most important feature of Table 2 is that when natural mortality rates are low ($M = 0.05$) cohorts are still increasing in biomass after 15 yr. When mortality rates are relatively high ($M = 0.21$), maximum cohort biomasses are attained about 7 yr after stocking at 10-cm SL, irrespective of the growth parameters.

Current information suggests that there are no major constraints on stocking tridacnids in relatively high densities. For example, *T. maxima* has been recorded at densities of 50–70/m^2 at Reao Atoll in French Polynesia (Salvat 36). Heslinga et al. (30) have reared *T. derasa* of average size around 15-cm SL at densities of 16/m^2 with no apparent ill effects on growth and mortality.

Because giant clams are essentially phototrophic they would appear to be limited only by the physical space available for their shells and in which to spread their mantles. It is therefore likely that farming systems will be relatively compact and very large harvests per unit area appear to be attainable. For example, if it were decided to harvest clams 5 yr after stocking onto reefs the calculations in Table 2 show that between 3499 and 7788 clams of an initial cohort of 10,000 would survive that period. Because six clams of this size (37–43 cm SL) could be comfortably accommodated within each square meter, the initial cohort could be stocked onto an area of 583 m^2 (Case C) or 1298 m^2 (Case D), depending on the anticipated mortality rates. Because the flesh weight is 16.5% of the total weight, the average annual production would lie between 22 and 218 tonnes flesh/ha yr, depending on the growth rates.

5. REEF RESEEDING

The prime alternative to cultivating clams in clearly defined plots or leases controlled by individuals or organized groups is simply to reseed reef areas by dispersing small clams throughout reef systems. This might be the only course possible in areas where public opinion is strongly against the exclusive use of reef or lagoon areas by individuals or groups and open access fisheries must prevail. It is known that small (10–20 mm) clams placed in unprotected trays on reefs rapidly suffer total mortality as a result of predation (Heslinga et al. 30). Juvenile clams that are dispersed in a reef system might suffer a lower mortality rate as a result of being disaggregated, but this cannot be proved at this time.

Heslinga and Watson (1) have reported a monthly mortality rate of 5% in *T. derasa* reared in raceways with unfiltered seawater. Most of the mortality was attributed to muricid gastropods that entered the raceways as larvae. The clams were placed in the raceways at an age of 5 months (12–17 mm) and transferred to sea-floor cages at 8–9 months (30–40 mm). Twenty-five percent mortality in cages during the subsequent 15–16 months was reported by Heslinga and Watson (1). Finally, in the following 2–3 yr a mortality of 10% was reported.

These rates presumably can be taken to represent a lower limit to the mortality rates that would be experienced by clams stocked onto a reef. Clearly the upper bounds might be close to total mortality and the question is unresolvable without experimentation. The only conclusion that can be drawn at this point is that reseeding reefs with very small clams taken directly from shore-based hatcheries would probably be feasible only if hatchery costs were low and the hatchery output very great.

The alternative to stocking reefs with very small clams is first to rear the clams in ocean nurseries to a size at which the mortality caused by predation decreases markedly. This is at a length of 10–15-cm SL (Heslinga et al. 30). To do this would necessitate the development of economical ocean nursery systems.

6. CONSERVATION STRATEGIES

If reefs are restocked with clams by broadcasting seed or juvenile clams it would become necessary to institute some sort of minimum size limits to ensure optimum results. This could be very difficult to enforce in remote areas or where reefs are used intensively for subsistence fishing. However, if clams are to be harvested for commercial purposes, minimum size laws relating to shell length, weight of whole animal, or total flesh weight could be enforced in order to attempt to optimize harvests.

An alternative management measure where reefs are reseeded is to permit fishing only on an intermittent basis. This would appear to be possible only where the commercial purchase of clams is under a high degree of control to ensure that premature poaching of clam grounds did not occur. This would also have to be combined with size limits if reefs were reseeded on an annual basis.

Stocks of *T. gigas* and *T. derasa* are in danger of extinction in many parts of their range. In these areas it appears that the only conservation measure that is likely to be effective will be the aggregation of remaining individuals into strictly protected reserves where there is a chance of successful reproduction and dispersal of larvae over the surrounding areas. An extension of this concept is to use these aggregations as brood stock for a system of hatcheries for restocking of reefs. Genetic diversity has already been irretrievably reduced in many areas. Without immediate conservation measures these resources will be lost.

ACKNOWLEDGMENTS

The assistance of the following persons in providing information in response to inquiries about the status and utilization of giant clam stocks in various countries is gratefully acknowledged: T. Adams (Fiji), K. Alagarswami (India), Retno Andamari (Indonesia), P. T. Bacolod (Philippines), M. Batty (Tuvalu), W. Bour (New Caledonia), A. Cabanban (Philippines), Chai Hen Leong (Malaysia), M. Chapau (Papua New Guinea), Choo Poh Sze (Malaysia), Chou Loke Ming (Singapore), M. Coeroli (French Polynesia), C. Cross-land (Vanuatu), D. Evans (Solomon Islands), A. C. Gambang (Malaysia), M. Gawel (Federated States of Micronesia), R. Gillett (Federated States of Micronesia), E. Gomez

(Philippines), F. Herscheid (Tuvalu), G. A. Heslinga (Palau), R. Kaltongga (Vanuatu), S. Kawaguti (Japan), C. Kensler (Burma), T. F. Latu (Tonga), A. D. Lewis (Fiji), M. Lopez (Philippines), S. Muller (Marshall Islands), P. Nichols (Solomon Islands), Pensri Boonrung (Thailand), E. Pita (Tuvalu), R. Pullin (Philippines), R. Radtke (Samoa), U. Raj (Fiji), B. Sablan (Northern Marianas), Sann Aung (Burma), R. Stevens (Vanuatu), T. Tikai (Kiribati), N. Sims (Cook Islands), G. F. Usher (Indonesia), and R. Wass (American Samoa).

REFERENCES

1. G. A. Heslinga and T. C. Watson, Recent advances in giant clam mariculture. *Proc. 5th Int. Coral Reef Symp.* **5:**531–538 (1985).

2. J. L. Munro and G. A. Heslinga, Prospects for the commercial cultivation of giant clams (Bivalvia: Tridacnidae). *Proc. Annu. Gulf Carrib. Fish. Inst.* **35:**122–134 (1983).

3. J. Rosewater, The family Tridacnidae in the Indo-Pacific. *Indo-Pac. Mollusca* **1:**347–396 (1965).

4. K. Ramadoss, Giant clam resources. *CMFRI Bull.* **34:**1–108 (1983).

5. University of the Philippines, *Progress Report of the University of the Philippines. Annex M in the Culture of the Giant Clam* (Tridacna *sp.*) *for Food and Restocking of Tropical Reefs*, Prog. Rep. 2. ACIAR/JCUNQ, 1985.

6. Fisheries Division, Fiji, *Progress Report of Ministry of Primary Industries, Fisheries Division, Fiji. Annex K in the Culture of the Giant Clam* (Tridacna *sp.*) *for Food and Restocking of Tropical Reefs*, Prog. Rep. 2. ACIAR/JCUNQ, 1985.

7. J. L. Munro, *Status of Giant Clam Stocks and Prospects for Clam Mariculture in the Central Gilbert Islands Group, Republic of Kiribati*. ICLARM report to the Fisheries Division, Ministry of Natural Resources Development, Kiribati and the South Pacific Regional Fisheries Development Programme. UNDP, Suva, Fiji, 1986.

8. C. M. Yonge, Giant clams. *Sci. Am.* **232(u):**96–105 (1975).

9. C. M. Yonge, Functional morphology and evolution in the Tridacnidae (Mollusca: Bivalvia: Cardiacea). *Rec. Aust. Mus.* **33(17):**735–777 (1980).

10. J. L. Munro and J. Gwyther, Growth rates and maricultural potential of tridacnid clams. *Proc. 4th Int. Coral Reef Symp.* **2:**633–636 (1981).

11. C. R. Fisher, W. K. Fitt, and R. K. Trench, Photosynthesis and respiration in *Tridacna gigas* as a function of irradiance and size. *Biol. Bull. (Woods Hole, Mass.)* **169:**230–245 (1985).

12. C. Crawford, W. J. Nash, and J. S. Lucas, Spawning induction, and larval and juvenile rearing of the giant clam, *Tridacna gigas. Aquaculture* **58:**281–295 (1987).

13. S. K. Wada, Spawning in the tridacnid clams. *Jpn. J. Zool.* **11:**273–285 (1954).

14. N. Beckvar, Cultivation, spawning and growth of the giant clams *Tridacna gigas*, *T. derasa* and *T. squamosa* in Palau, Caroline Islands. *Aquaculture* **24:**21–30 (1981).

15. J. Gwyther and J. L. Munro, Spawning induction and rearing of larvae of tridacnid clams (Bivalvia: Tridacnidae). *Aquaculture* **24:**197–217 (1981).

16. S. C. Jameson, Early life history of the giant clams *Tridacna crocea*, *T. maxima* and *Hippopus hippopus. Pac. Sci.* **30:**219–233 (1976).

17. P. Swadling, Central Province shellfish resources and their utilization in the prehistoric past of Papua New Guinea. *Veliger* **19:**293–302 (1977).

18. R. G. Pearson, Impact of foreign vessels poaching giant clams. *Aust. Fish.* **36(7)**:8–11 (1977).

19. G. F. Usher, Coral reef invertebrates in Indonesia: Their exploitation and conservation needs. *IUCN/World Wildlife Fund Proj.* **1688, 2**:1–100 (1984).

20. J. H. Brown and M. R. Muskanofola, An investigation of stocks of giant clams (family Tridacnidae) in Java and of their utilization and potential. *Aquacult. Fish. Manage.* **1**:25–39 (1985).

21. A. H. Banner and J. E. Randall, Preliminary report on marine biology study of Onotoa Atoll, Gilbert Islands. Scientific investigations in Micronesia Report No. 13. *Pac. Sci. Board, Nat. Sci. Counc.* N7-onr-291:TO IV:1–42 (1949).

22. G. A. Heslinga, A checklist of giant clams. *Hawaii. Shell News* **27**:15 (1979).

23. J. T. Hardy and S. A. Hardy, Ecology of *Tridacna* in Palau. *Pac. Sci.* **23**:467–472 (1969).

24. J. W. J. Wankowski, *Report on a Preliminary Survey of Nuguria, Nukumanu and Takuu Atolls* (mimeo rep.). Research and Surveys Branch, DPI Fisheries, PNG, 1979, 27 pp.

25. A. Bodoy, An assessment of human impact on giant clam populations (*Tridacna maxima*) in the vicinity of Jeddah, Saudi Arabia. *Symp. Coral Reef Environ. Red Sea Jeddah* pp. 1–25 (1984).

26. J. L. McKoy, Biology, exploitation and management of giant clams (Tridacnidae) in the Kingdom of Tonga. *Fish. Bull. (Tonga)* **1**:1–61 (1980).

27. K. Knox, Giant clam in tinned crab—not likely. *Aust. Fish.* **36(10)**:17 (1977).

28. R. F. Dawson, *Taiwanese Clam Boat Fishing in Australian Waters.* Centre for Southeast Asian Studies, Griffith University, Brisbane, Australia. Research Paper 33, 1985, 46 pp.

29. S. M. Wells, Giant clams—a case for CITES listing. *TRAFFIC Bull.* **3**:60–64 (1981).

30. G. A. Heslinga, F. E. Perron, and O. Orak, Mass culture of giant clams (F. Tridacnidae) in Palau. *Aquaculture* **39**:197–215 (1984).

31. James Cook University, *Progress Report of James Cook University. Annex I in the Culture of the Giant Clam* (Tridacna *sp.*) *for Food and Restocking of Tropical Reefs,* Prog. Rep. 2. ACIAR/JCUNQ, 1985.

32. J. L. Munro, *Ocean Nurseries for Giant Clams: Options and Problems,* Int. Giant Clam Maricult. Proj., Discuss. Pap. 1 (mimeo. rep.) ICLARM, South Pacific Office, Townsville, Australia, 1985, 6 pp.

33. J. L. Munro and J. Gwyther, in preparation.

34. R. G. Pearson and J. L. Munro, in preparation.

35. D. F. McMichael, Growth rate, population size and mantle coloration in the small giant clam, *Tridacna maxima* (Roding), at One Tree Island, Capricorn Group, Queensland. *Proc. 2nd Int. Coral Reef Symp.* **1**:241–254 (1975).

36. B. Salvat, La fauna benthique du lagon de l'atoll de Reao (Tuamotu, Polynesie). *Cah. Pac.* **16**:283–321 (1972).

REVIEW

25 A PERSPECTIVE ON THE POPULATION DYNAMICS AND ASSESSMENT OF SCALLOP FISHERIES, WITH SPECIAL REFERENCE TO THE SEA SCALLOP, *Placopecten magellanicus* GMELIN

John F. Caddy
Food and Agriculture Organization
Rome, Italy

1. INTRODUCTION

The need to develop appropriate research and management options for scallop fisheries is well illustrated by experience in the several fisheries for the offshore or sea scallop, *Placopecten magellanicus*, in the northwest Atlantic, and in particular for the Georges Bank fishery, which is still the world's largest single natural scallop resource. Prior to the ratification of the Convention for the Law of the Sea, which led to extension of national jurisdiction to 200 miles offshore, this fishery came under the management authority of the former fisheries commission ICNAF (the International Commission for the Northwest Atlantic Fisheries), and its assessment by U.S. and Canadian scientists was subject to review and guidance by STACRES (the Standing Committee on Research and Statistics of ICNAF), made up of predominantly finfish assessment scientific experts from over a dozen countries. This accounts for the early application of stock assessment models to Georges Bank scallop stocks, when such models were not in vogue for management of molluscan shellfish resources. In particular, the yield-per-recruit model of Beverton and Holt (1) has been applied in assessing the likely effects of different size limits and fishing intensities on this fishery since the 1960s (2). At the same time, the special features of a fishery for a semisedentary, contagiously distributed resource, harvested by dredges, with their own special characteristics of a high incidental breakage rate and wide range of sizes at partial retention, must be taken into account. The fact that only the adductor muscle of the sea scallop is landed also poses a special problem that complicates application of conventional models. These features seem however, to offer several possibilities to the research worker in population assessment and management that have not as yet been fully exploited, such as the fact that scallops can be readily surveyed quantitatively by direct methods, and the fact that the growth and possibly also the past fishing pattern are recorded on the shell. As for other molluscan fisheries, modeling of scallop resource dynamics ideally should take these features of the biology and fishery into account.

2. POPULATION DISTRIBUTION AND STOCK UNITS

A recent review of the history of some 60 scallop beds in the North Atlantic (3) included the three economically most important species, *Placopecten magellan-*

icus, Pecten maximus, and *Chlamys opercularis,* and concluded for most stocks for which long-term landing data are available (20 to 100+ yr data series) that for all three species, these main fishable concentrations are persistent self-reproducing aggregations, each with its widely separated, hydrographically defined, geographic location. For each of them, the order of magnitude of average population size has remained constant, even though biomasses have shown wide fluctuations from year to year that are not all due to fishing. This suggests that scallop fisheries fall outside the stable category in the classification of fisheries time series proposed by Caddy and Gulland (4), into either cyclical, such as the Bay of Fundy fishery (5–7); or irregular, with peaks at apparently random intervals; or simply spasmodic, as suggested at the extreme northern (St. Pierre's Bank) (8) and southern limits of the range (off Virginia) (9).

In describing these apparently self-replenishing populations, Sinclair et al. (3) noted that commercially important beds are located in areas whose oceanographic characteristics permit retention of pelagic larvae in their vicinity for the approximately 2 months of larval life history. Such features include tidally induced movements of the water column, a degree of vertical stratification, and, in my opinion, for the larger stocks of *Placopecten,* the presence of a persistent tidal gyre (Fig. 1), as described for the Bay of Fundy stock (5) and even more evidently for Georges Bank (10). Such limited enzymatic work as has been reported on scallops does not support a clear genetic differentiation however, suggesting larval exchange between adjacent areas (11). For a semi-sedentary species such as the

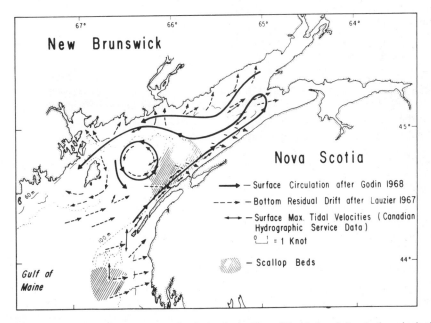

Figure 1. Surface and bottom water circulation in the Bay of Fundy in relation to the principal scallop grounds off Digby, Nova Scotia (from Caddy 5).

sea scallop, occupying a fairly broad ecological niche (12, 13) bathymetrically and by sediment type, the wide variability of morphometry cannot all be considered to reflect genetic characteristics, even though scallop shells from different beds can often be readily distinguished by eye.

The assumption, then, that the main scallop fisheries are self-sustaining megapopulations (14) is reasonable and conservative, and should be the basis for management measures for historically maintained centers of concentration. As noted by Caddy (15), because dynamic pool axioms (i.e., continuous mixing of the stock throughout its range as exploitation continues) do not apply to semi-sedentary shellfish, it makes sense to distinguish such self-replenishing populations in management and conservation from "fringe" populations that occur in a transitory fashion. This question is discussed further for molluscan shellfish in general (this volume) and later under yield models.

Within each megapopulation, the unit of stock that is important commercially is the scallop bed, an area with a degree of spatial continuity and similar internal densities, which are significantly higher than for surrounding areas. A scallop fishing ground may contain several discrete beds surrounded by low-density fringes, where populations may still be contagious, but overall densities are as low as 1/40 of that in the center of the bed (16). The scale of such aggregations is usually too small (17) to be taken into account in stratified sampling schemes such as are discussed later, and a decision has to be made early in a field investigation whether research is to be aimed at commercial concentrations or the whole population. The important point to be noted is that commercial fishing (by a boat or by a fleet) tends to concentrate on a given bed until the catch rate is depleted to "uncommercial" levels. Beds may occur (and recur?) in particularly favorable situations, hydrographically, bathymetrically, and by sediment type. Generally, beds are dominated by one or a few year classes that reflect past favorable reproductive and settlement episodes (see Chapters 21 and 22), although some overlapping of areas of recruitment in successive years occurs (18). Such beds, dominated by one or two year classes, were observed by me and surveyed by scuba and submersible (16, 19) in the same general area of Northumberland Strait in 1967 and 1968, and on the northern edge of Georges Bank by dredge-mounted camera in 1970 and 1971 (20, 21) (Fig. 2). Some statistics for these areas were as follows:

	Northumberland Strait	Georges Bank
Density inside patches	(1967) 4–6 (19)	(1970) 2–4 (15)
(per square meter)	(1968) 4–7 (16)	
Approx. dimensions (km)	(1968) 0.25 × 1+	(1970) 9 × 31
Modal size (mm)	(1967) 50–75 (19)	(1970) <72.5 (3)

What emerges from the literature on scallops of a variety of species, often as rather fragmentary observations, is that scallop beds are in general elongated, an observation (Fig. 3) made early during exploratory work in the Gulf of St. Lawrence and more recently (18) from contouring commercial catch records for

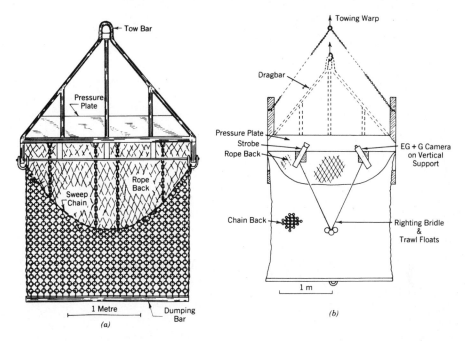

Figure 2. (a) Offshore scallop rake (dredge) used on Georges Bank seen from the belly. (b) Dredge modified for underwater photography in the path of the dredge.

Georges Bank. This observation also seems to apply for other scallop species, for example, calico scallops (22), and may reflect preference of scallops for specific sediments lying in partly sorted "streaks." In the Northumberland Strait, scallops occurred at lower densities on mud than on sand. An obvious sediment preference was not evident however, during underwater observations on predominantly coarser substrates in the Gulf of St. Lawrence and Bay of Fundy (12, 23). Two alternative and complementary hypotheses may help explain this phenomenon. The first is to suggest that as for oysters, spatfall occurs over a relatively limited period, in which a "swarm" of larvae ready to metamorphose are transported tidally close to the sea floor in a suitable settlement area, and "smeared" over the bottom, settling out in an oval or strip along the direction of current flow.

Alternatively, or in addition, secondary dispersal from some primary settlement site may occur, with active but random swimming of small scallops being preferential along tidal axes (24, 25), leading to an elliptical bed shape. Postsettlement dispersal, though limited even for active swimmers such as *Argopecten* (22), has been used by Japanese shellfish farmers in bottom culture of *Patinopecten yessoensis*. They place young scallops in the center of a lease, knowing that they will have dispersed to the margins by commercial harvest size (26) (Fig. 4). Circumstantial evidence that the swimming response may be important in density regulation owing to aggregation of predators on high scallop densities was provided

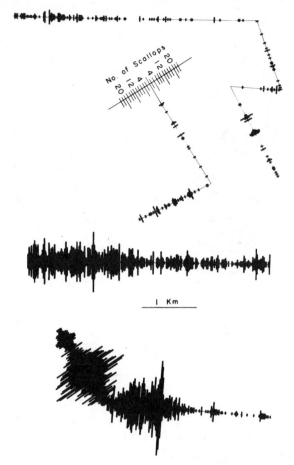

Figure 3. An example of the distribution pattern of scallops in areas of dense recruitment along dive tracks estimated by direct observation from a submersible in the Northumberland Strait (from Caddy 16). Scale shows density in numbers per square meter.

by an experiment in Passamoquoddy Bay in which large relatively inactive scallops placed at the center of a ''target'' marked on the bottom with lead rope dispersed over several days in directions that appeared related to the predominant tidal flows. This first experiment was of interest in that scallops that remained in the artificially high-density patch at the center of the target (Fig. 5) were attacked and consumed by predators that congregated, presumably in response to olfactory stimuli, within 48 hr of release; these predators were principally whelks (*Buccinum undatum*) and starfish (*Asterias* sp). This observation seems to explain why densities of *Placopecten* in nature may assume relatively low limiting values that decline with size and age, and suggests the existence of density-dependent mortality processes shown by other mollusks (Chapter 30).

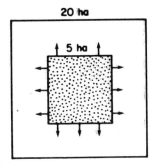 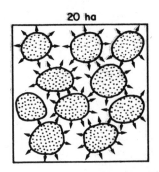

Figure 4. Two strategies used in sowing *Patinopecten yessoensis* seed (at around 3-cm shell size) in bottom culture in Japan, illustrating the initial distribution of scallops over the lease and their eventual distribution on harvesting 2 yr later. Redrawn from Quérellou (26).

Clearly, experimental work on the dynamics of formation and continuity of scallop beds will be difficult and costly, and will have to be carried out using some form of direct observation techniques, given that the dredge is only a semiquantitative tool and does not allow resolution of fine scale patterns. This type of work is of more than academic interest, however, and offers the main promise for improved understanding of scallop dynamics.

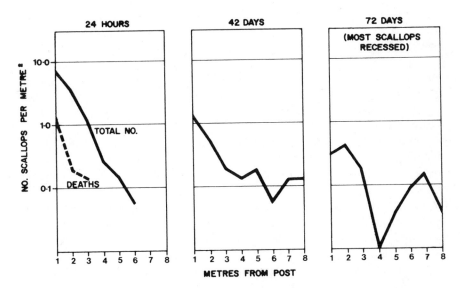

Figure 5. Dispersal of scallops from (and aggregation of predators, as shown by dead individuals) an artificially high density of adult scallops on gravel–mud bottom in Passamaquoddy Bay.

3. LIFE CYCLE, HABITAT, AND BEHAVIOR

3.1. Early Life-History Stages

The early preference of scallop postlarvae of a variety of species for filamentous or dissected settlement substrates has been well demonstrated; preferred substrates include eel grass (*Argopecten irradians*) (27), *Ulva* (*Chlamys tehuelcha*) (14), and a bryozoan (*Cellaria* for *Chlamys opercularis*) (28). The only natural substrates and locations where postlarval *Placopecten magellanicus* have been collected in abundance are also on a bryozoan (*Eucratea lorea*) in Penobscot Bay (29) and on red algae (30) in Newfoundland. It seems unlikely that there is an obligate settlement substrate for the species, for Naidu and Scaplen (31) harvested the spat on polyethylene film collectors in Newfoundland. The juveniles can crawl rapidly with an extended foot and, subsequent to becoming dislodged from the primary substrate, show cryptic behavior, attaching by a byssus in crevices, commonly inside dead paired shells of scallops or "cluckers." This behavior probably accounts for the lack of success with mechanical sorting of shell debris for young specimens. A great deal remains to be determined about the early postsettlement stages; it is assumed on little evidence that primary settlement is close to the adult stock, for there also seems to be little evidence for bathymetric displacements, especially for offshore populations of *Placopecten*. Over a shell height of 3 cm or so, young *Placopecten* may be found on substrates of muddy sand to gravel, where they show active and prolonged swimming responses to predators, divers, or scallop dredges (19).

It is interesting to contrast this situation with the European scallop, *Pecten maximus* L., which when adult is basically a species found in sand to muddy-sand bottoms, recessing itself until the flat (left) valve is level with the surface, and more likely to show locomotory movements if placed on hard bottom (32). By contrast, *Placopecten magellanicus* has a wide tolerance for different sediments, tending to be found at the highest densities on gravel and mixed gravel or hard (e.g., glacial till) bottoms, where it may attach byssally. It shows more tendency toward swimming behavior on sand–mud bottoms. Byssal attachment has been described only in the postmetamorphosis stage in *P. maximus*, where it serves, as for other pectinids, to fix the postlarvae and early juvenile stages to algae or colonial organisms. The use of a byssus by adults is still further developed in various species of *Chlamys*, which may occur in high densities attached with byssal threads to the bottom and each other (33). *Placopecten* can be seen as intermediate in its behavior in this respect; although it recesses on soft bottoms with its flat right valve downward, a byssal thread attached to even small pieces of shell at the bottom of the depression suggests that byssal attachment inhibits spontaneous swimming movements and possibly avoids undue dispersal from favorable environments. This hypothesis is indirectly confirmed by both active swimming behavior and byssal attachment dropping off rapidly with size, so that by 100–110-mm shell height, scallops in the Bay of Fundy rarely swim, and progressively fewer are attached by a byssus (29).

Although, as noted earlier, there is little information about early life histories, we know (34) that epidemic spawning may occur in *P. magellanicus* populations, and it seems inevitable from dispersion of gametes that the probability of fertilization (or of simultaneous triggering of spawning) between two or more spawning individuals of separate sex declines with distance apart and population density. This could imply (and needs further investigation for all "broadcast spawners") that the reproductive contribution of low-density fringes of population is of limited value. Such a conclusion, if true, carries its obvious management implications, to be discussed later.

Given the lack of a technique of spatfall prediction to date, recruitment indexes for incoming year class sizes prior to age 2–3 have not been developed, which is inconvenient given the dominant effect of recruitment on fishery performance.

3.2. Predators, Parasites, and Ecological Effects

The responses of scallops to predators have been well documented. Thus bay scallops, *Argopecten irradians* (35), are least sensitive to approaching predators on the dorsal side of shell periphery; predators are discriminated from other organisms chemically and mechanically. Their contact with the mantle leads to the scallop rotating to face away from the disturbance and swimming off. Similarly for *Placopecten*, a scallop approached from the front or ventral side detects the disturbance visually (a shadow response?) and shows a similar reaction (14). This leads to a proportion of small scallops being retained in fine-mesh covers placed over the rope back; small scallops have also on occasion been observed swimming upward as they entered the dredge.

Evidence for outbreaks of diseases in *P. megallanicus* are rarely unambiguous, in that although dead and dying scallops have frequently been reported in scallop dredges, especially at the limits of the species range, these in some cases can be primarily attributed to thermal shock when preferred temperatures have been exceeded. This may be due, for example, to thermocline overturns. Low oxygen levels, high sediment loadings following dredging on muddy bottoms, or incidental effects of fishing and discarding (36) may all lead to local mortalities. Pathogenic infestations have been reported (37, 38) that are associated with dark, watery, and stringy meats, however, and as for other causes of mass mortality, are detected by the dredging of abundant paired empty shells (cluckers or clappers). Dark stringy meats also occur, however, in old scallops and those from deep water.

In the Bay of Fundy, scallop digestive gland, gonad, and mantles retain high levels of paralytic shellfish poison throughout the year (39), even when the toxic dinoflagellate *Gonyaulax* is no longer in the plankton but is present as resting spores in bottom sediments. This supports the idea (14) that bottom-dwelling microflora and detritus are significant in scallop diets in addition to or instead of phytoplankton.

3.3. Swimming Behavior

R. H. Baird was the British fisheries worker whose practical concern with hydrodynamics led to development of the first scallop dredge built on hydrodynamic principles—the sledge or "sputnik" dredge (40), widely used in fisheries for many Pectinidae (but not commonly for *Placopecten*). He noted that *Pecten* is poorly designed hydrodynamically for swimming, and that "if it swam upside down" (i.e., with the flat valve downward) "and the other way round" (i.e., with the hinge forward instead of the gape), "it would be an efficient high lift hydrofoil." *Placopecten* is a far more efficient swimmer, and meets Baird's first specification with its flattened lower valve, but not the second, in that except for brief escape movements, it also shows its most spectacular swimming movements (19) when moving in a ventral direction propelled by exhalant streams of water expelled on each side of the hinge.

As for *Pecten*, *Placopecten* shows a preferred shell orientation to prevailing current direction, which changes with the tide. This regular rotation (around a fixed point, the byssus and foot) is probably responsible for the semicircular scratches often seen on the underside of specimens from hard substrates, and together with the mantle cleansing response (a sudden clapping of the shell valves), probably accounts for the recessing of adults, which on mud bottoms can be up to 10 cm below the sediment surface for large specimens. The elaborate recessing response shown by *Pecten* (32) does not seem to occur, but the same result is achieved: the upper shell valves of larger scallops on finer particulate bottom tend to be level with or below the sediment surface.

4. AGE COMPOSITION AND GROWTH RATES

Prominent shell rings on sea scallops are formed annually, as noted by Stephenson and Dickie (41), and have been used for aging various scallop species. Finer circular striae record more frequent events, and Silina (42) noted for *Patinopecten yessoensis* (Jay) that the small-scale sculptures on the upper shell each represent between 1 and 5–7 days' growth, depending on season. Annual rings occur on the lower valve, but mechanical damage tends to obscure these.

External rings of *P. magellanicus* become more difficult to read with distance southward; northern populations show very clear shell rings. For this reason, reading the rings on the resilium or shell hinge (43) is an alternative approach that avoids some of the confusion due to shock marks caused by dragging in heavily fished areas. Another problem is the identity of the first ring: at 6–8 mm, the size at first overwintering, it is often worn off, and U.S. practice (10) has been to refer to the second ring (at 18–25 mm) as ring 1. Other authors, for example, Roddick and Mohn (44), follow the convention in finfish research of assigning a birth date of January 1 in the same year as settlement (which typically occurs in early October on Georges Bank).

Perhaps a more serious source of bias in yield models for scallops is the pronounced Lee's phenomenon (44), which can result from several causes as discussed in Ricker (45), but here is surely due to size-selective fishing by a gear with partial escapement over most of the size range. The result is that survival is higher for individuals with lower growth rates, so that using all the shell ring measurements biases the estimate toward older shells, and underestimates the effective growth rate (K) for individuals at harvestable age. One solution is to use the second and the final rings only in fitting growth curves for shells of known age (44). This procedure yielded von Bertalanffy estimates of $K = 0.245$, $L_\infty = 139.31$, and $t_0 = 1.27$, which despite the high value of t_0, are quite similar to other estimates for Georges Bank scallops. The work of Jones (46) showed that in finfish, the degree of Lee's phenomenon might be an index of mortality. This may prove useful in estimating relative changes in Z at age.

The direct use of shell annuli in determining the age composition of commercial landings is necessarily limited by shucking at sea, and an age–length key derived from shucking individuals of a given age and some idea of the relationship of shell length to meat weight (e.g., Fig. 6) are both necessary. As illustrated in Figure 6,

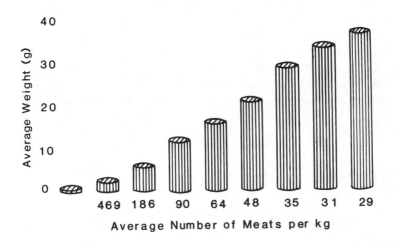

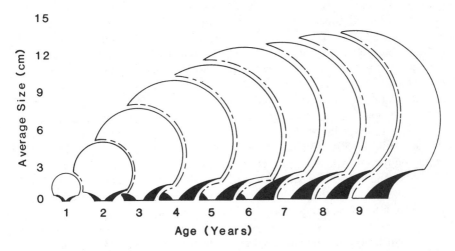

Figure 6. Graphic illustration of differential growth of shell size and meat (adductor) weight of Georges Bank scallops.

meat weight continues to show major gains with age even when shell growth has slowed down; an argument in favor of meat size as opposed to shell length regulations. A further potential problem is caused by the seasonality of (shell) growth in most scallop stocks, which has led some authors (47, 48) to fit seasonally adjusted von Bertalanffy curves incorporating a sine function to some species. This would also be valid for *Placopecten*, and could refine modeling of seasonal, intense fisheries.

There is considerable variability in growth form of *Placopecten*, particularly with depth (49) across different benthic communities (50), and this is most clearly shown in the Bay of Fundy. Apart from a significant decline in shell growth rate below 55 fathoms, shell curvature, shell thickness, and meat (adductor muscle) weight all fall off sharply with depth. Most of this differentiation presumably occurs during ontogeny in response to local conditions of which depth, feeding conditions, and density (14) seem likely to be the more important factors. As a result, meat yield in Fundy is low below 60 fathoms (49), and these specimens used to be exploited less intensely.

The large variations in meat weight for a given shell height (51) make it difficult to see how an absolute size limit could be related to some optimal age at first harvest for scallops in this area. Such a regulation, set for more productive components of the stock in moderate depths, would bar exploitation from deep water of older scallops having small meats.

5. FISHING GEAR AND GEAR SELECTIVITY

A good description of the type of gear and boats used in the offshore scallop fishery is given in Bourne (52) (Fig. 2). For the inshore fleet, the two principal ones are (1) the gangs of up to 13 digby or bucket dredges (2) over the side from a single bar (53). More recently, especially in the southern Gulf of St. Lawrence, lighter versions of the offshore dredge are fished from a stern gantry, similar to but less elaborate than that used with sputnik dredges in the Australian fishery. An entirely different approach to harvesting, namely, by scuba (54), is currently becoming widespread on many shallow-water scallop stocks, especially as unit prices rise; this approach avoids many of the incidental effects of dredging discussed later. Its application to deep-water stocks seems unlikely in the immediate future.

The efficiency of scallop dredges has been determined by direct and indirect observations, and is higher (55) for the offshore rake than for the bucket dredges (56) at around 15–20% and 5%, respectively, gear performance depending significantly on bottom type (57). The methodologies used to estimate dredge efficiencies have ranged from tag and recapture of scallops released in a known fishing area followed by intensive fishing (56), to comparing catches with densities photographed in front of the dredge (Fig. 2), and to the release of tagged scallops directly into the dredge. This may be conveniently associated with cover experiments and dredge selectivity experiments (55), or by diver observations in the track of the dredge compared with unfished areas (23, 57).

Overall dredge efficiency e is best defined as

$$e = \frac{\text{no. scallops caught}}{\text{no. in dredge path}}$$

In practice (55), it is useful to distinguish between two components of the overall dredge efficiency, efficiency of capture (E) and gear selectivity (s); $e = s \times E$, where

$$s = \frac{\text{no. scallops caught}}{\text{no. entering dredge}}; \quad E = \frac{\text{no. entering dredge}}{\text{no. in dredge path}}$$

Mesh selectivity (s) is directly measurable by catches in covers over the back of the dredge and [ideally in combination (55)] by release of marked scallops of known size into the dredge from fragile plastic bags attached inside the gear before shooting the dredge. This experiment showed that a significant proportion of gear selection occurs through the belly of the dredge. Capture efficiency can also have a size-related component. Large scallops are more deeply recessed; thus large *Pecten* are less likely to enter the dredge (57), and a similar phenomenon was found for *Placopecten* (55).

Similar gears have very different efficiencies on different grounds. Even with teeth, "bulldozing" of sediment in front of the dredge is a major factor (23, 57). For *Pecten* a 14.6% efficiency was found for standard dredges, but dredges with spring-loaded teeth, although similar in efficiency, pick up less trash and also allow some size selection between the teeth.

For a sedentary species it is especially important to distinguish gear selectivities as described above from the age-specific fishing mortality vector, which is also a function of fishing strategy (15, 17) and has been estimated by cohort analysis on meat size composition of commercial catches (17, 59).

The design of research vessel surveys (See Doubleday 58 for principles) also needs to take into account gear selectivity and contagious distribution of young scallops.

6. INCIDENTAL EFFECTS OF FISHING AND INDIRECT MORTALITY OR DAMAGE

The consistently highest densities of *P. magellanicus* on Georges Bank are associated with a "mixed" bottom of gravel and sand (17). In the Bay of Fundy, similar sediments thinly overlie glacial till with frequent ice-rafted boulders from glacial epochs. Such hard bottoms are associated with the high tidal energies favored by this species, and are one reason why fishing gear, especially offshore, is heavily built, unlike the sledge or sputnik dredges used in most fisheries for scallops associated with soft bottoms.

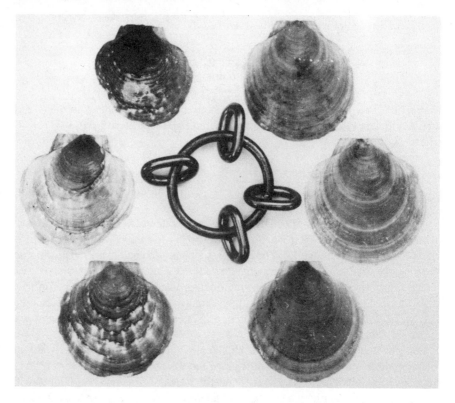

Figure 7. Shock marks on scallop shells from Georges Bank. The specimens on the upper left and lower left have experienced several episodes during the growth history of the shell. (Photograph reproduced courtesy of Bill McMullen, Fisheries and Oceans, Canada.)

Heavy gear inevitably imposes a high probability of incidental damage to scallops, whether retained, passing out through the ring or inter-ring spaces, or, if recessed; "run over" by the dredge. This type of damage is evident both from shell "shock marks" (Fig. 7) and from direct observations on performance of offshore and inshore dredges. Shell shock marks in the form of a prominent interruption of shell growth across the former shell periphery, with at least two separate loci or a quarter of the shell periphery chipped, were considered to be due to dredge damage (Fig. 8). This seems to be confirmed by the modal size and age corresponding closely to the point of 50% selection of the gear (when presumably damage would be greatest), pointing to incidental damage as the main causative agent. It is perhaps instructive to consider the question of incidental dredge damage in more detail, using the simple chain of probabilities shown in the schematic on the following page as a guide to future experimental work.

Apparently, the sum of probabilities for events marked (*) should be some function of F, and if discards are negligible, and undamaged shells passing through the dredge are a small proportion of the total (as for commercial sizes at above

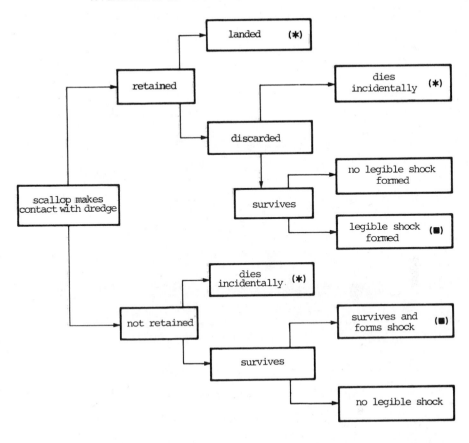

50% selection), then the ratio: shock marks/growth zones (events marked ■) is a direct index of fishing mortality. This seems to offer the possibility of parameter estimation.

In Chaleur Bay, Gulf of St. Lawrence, direct observations in 1971 on a hard gravel bottom with frequent cobbles and boulders similar to substrates photographed on the northern edge of Georges Bank (60), quantified incidental damage to unharvested animals in the dredge path (23). After allowing for efficiency of offshore dredge (15%) (19, 55), 13–17% of scallops in the path of the dredge were estimated to be lethally damaged but not caught. Predatory fish and crabs were attracted to the dredge track within 1 hr of fishing and fed on injured scallops (23, 61).

Four other effects of prolonged dragging over an area may be noted:

1. The accumulation of single shells from at-sea shucking. Under high-energy tidal regimes such as the Bay of Fundy, repeatedly discarded dead shells may accumulate upstream and downstream from the beds (49): (Fig. 8).

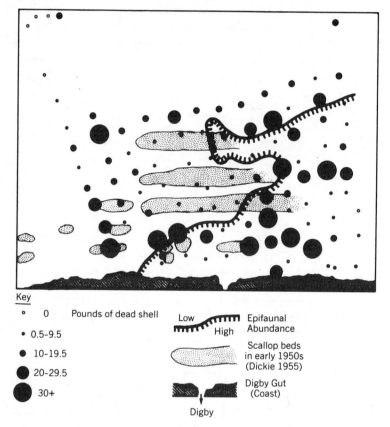

Key

∘	0	Pounds of dead shell	
•	0.5-9.5		
●	10-19.5		
●	20-29.5		
●	30+		

Low / High — Epifaunal Abundance

Scallop beds in early 1950s (Dickie 1955)

Digby Gut (Coast)

Digby

Figure 8. Distribution of dead shells in relation to the main beds over the Digby scallop grounds in the 1960s, and limits of distribution of epifauna (5, 61).

2. The sparsity of epifauna on scallop fishing areas, as revealed by submersible observation, contrasted with unfished areas further up the bay (12, 50). The long-term effects of epifaunal removal on subsequent recruitment is probably important but difficult to document.

3. Potentially important also for the long-term productivity of beds is the effect of repeated redisturbance on the grain size of sediments unprotected by epifauna from laminar water flow directly over the bottom.

4. One related feature of dredging on soft bottom was noted in the Northumberland Strait, from direct underwater observation in close vicinity of a scallop fishing fleet, namely, a dense layer of liquid silt drifting several centimeters above the bottom, and scallop mantle cavities containing large quantities of silt.

All these effects suggest the need for a careful reexamination of appropriate gear and fishing techniques in order to minimize unfavorable incidental effects.

7. ESTIMATES OF MORTALITIES

7.1. Total Mortality Rate, Z

Posgay (2) examined the age composition of regular samples of the commercial catch landed unshucked to estimate average mortality rates for the whole of Georges Bank in the 1960s. His estimate of $Z = 0.7$ for commercial sizes was obtained by catch curve analysis and (recognizing that aging of older scallops is uncertain) by comparing the catch rate of scallops of size corresponding to age n and older with those corresponding to age $n + 1$ and older the following year, giving Z as the log difference between the two estimates.

More direct estimates of total mortality rate were obtained for a local heavily fished population on the northern edge of Georges Bank. The area in question, a 274-km^2 patch of dense recruitment, was located on the northern edge in 1970 (20), and surveyed with drag and camera dredge. Mean densities of 0.98/m^2 of three-ring scallops with a modal size of 72.5 mm, and a smaller number of two-ring animals (modal size at or below 50 mm), supported a local standing stock in June 1970 of about 270 million scallops. It was fished heavily in 1970 and 1971 by the Canadian fleet, with a cull size of around 70 mm (62), and surveyed again 1 yr later using the same methods (21). By comparison of the two surveys, a high mortality rate ($Z = 1.2$ for three-ring scallops) was estimated, which also partly applied ($Z = 0.9$) to the 50–70-mm size range that were largely discarded by the fleet. Survivors of these smaller scallops were present in greatly reduced numbers at the new modal size the subsequent year, and showed no obvious sign of having dispersed to immediately adjacent areas. This isolated account illustrates the contagious nature of fishing mortality in scallops; presumably the average mortality rate for the whole stock was significantly lower. This experiment was also suggestive of a high direct and incidental fishing mortality, and deserves to be repeated; if confirmed, it has serious implications for management strategies on this resource, particularly suggesting the desirability of temporary closures of areas with large proportions of "undersized" animals.

7.2. Natural Mortality Rate, M

The approach to estimating natural mortality of sea scallops proposed by Dickie (56) and used by Merrill and Posgay (63), rests on the assumption that dead paired shells (cluckers) are caused by natural mortality only (shucked shells are separated before discarding). At equilibrium, it is postulated (56) that the number of paired shells being contributed to the population annually is balanced by the number separating because of degeneration of the ligament. This concept was used (63) to obtain values of $M = 0.1$ for Georges Bank scallops, observing the rate of detachment of paired shells from freshly killed scallops and following modal sizes of live scallops with cluckers from the same cohort. Given that there may be other components of fishing that result in paired dead shells (see earlier), if anything, this should provide an overestimate of M; on the other hand, some separation of cluckers by fishing may lead to the opposite bias.

Natural mortality may not always occur at a steady rate, especially in unfavorable conditions, and an index of mortality has been postulated (65) for recognition of unusually high natural mortality rates associated with mass mortalities: $I_m = C/(C + L)$, where C is the number of clappers and L the number of live shells in the dredge; values of I_m in the range 0.1–0.3 seemed to be normal for Narragansett Bay. Monitoring the live shell/clapper ratio is thus a way of estimating changes in the fishing mortality rate and/or the occurrence of past catastrophic mortalities. Dickie's method can also be simplified if a new year class enters the fishery, and its cluckers can be distinguished from those of earlier cohorts (from the lack of epifauna inside the clucker shell). Then

$$M_i \, Dt = \log_e (Cl_i + N_i) - \log_e (N_i)$$

where Dt is the time since formation of the last shell ring for a cohort with i rings, and Cl and N are the numbers of cluckers and live shell from the same cohort in the catch (assumed equally catchable). Merrill and Posgay (63) found some slight dependence of clucker/live shell ratios on density, but there is generally little evidence on Georges Bank for predation on (commercial sized) sea scallops. Given a fishing strategy that exerts most effort in high density areas, one estimate of M would presumably result from the mean age of scallops from low-density unfished areas, that is,

$$M = Z \simeq \frac{1}{\text{mean age} - \text{age at 50\% capture}}$$

or by catch curve analysis for areas that, from logbooks or port interviews, have received little or no fishing.

7.3. Fishing Mortality Rate, *F*

A direct estimate of fishing intensity is obtainable directly from logbook data, and may be used as the basis for a contouring program (66), to obtain fishing intensity (effort assumed proportional to F). Swept-area methods obviously have some potential for directly estimating stock density and an index of F can be devised for sedentary species. Thus in 1970, total days fished (U.S. + Canada) were 10,321, at 14.1 hr gear on bottom/day, for two dredges with combined width 25 ft, towed at 3.75 knots average speed (52). This is roughly equivalent to 2241 nm² swept. This is a small portion of the 10,800 nm² that Posgay gives as the area of the bank receiving some fishing in 1944–1961, and if scallops were dispersed evenly over the bank, this would not give rise to an $F = 0.7$; showing again that fishing effort is very contagious. This is confirmed from the predicted mean densities derived by this approach. A total catch of 12.2 million lb in 1970 is roughly equivalent to 0.00158 lb/m². With an average of 50 meats/lb, this is about 0.55 scallops/m², approaching average densities estimated by photography over productive scallop beds (see above).

Fishing mortality rate (F) can also be obtained directly by subtracting the natural from the total mortality rate, $F = Z - M$, or given M, by cohort analysis. In the procedure described by Mohn and Robert (67), the meat size composition determined from port sampling commercial catches, after adjusting up to the total catch, is converted to shell heights using a shell height–meat weight relationship constructed from current research cruise data, and then a von Bertalanffy growth curve to estimate age compositions. Obviously this last step is less accurate, but also much less costly in time and effort than producing an annual age–length key by reading shell rings, which as noted earlier leads to a pronounced Lee's phenomenon. To simplify procedures and avoid the last-mentioned bias, they used the first legible ring (the second) and the final ring to fit the growth curve, simply counting the number of rings between. Cohort analyses carried out with these two procedures (17, 67) suggest that the vector of fishing mortality of age for Georges Bank (Fig. 9) (68) rises to age 5 (the highest cohort biomass per unit area in most years consisting of ages 4–5), before dropping off with age for less dense survivors of previous year classes. This vector is similar to that predicted from simulation YRAREA (69) for moderately high levels of fishing intensity (Fig. 10). Care should be taken, however, in interpreting fishery-related changes by cohort analyses if a significant incidental fishing mortality is the case, for these deaths would not show

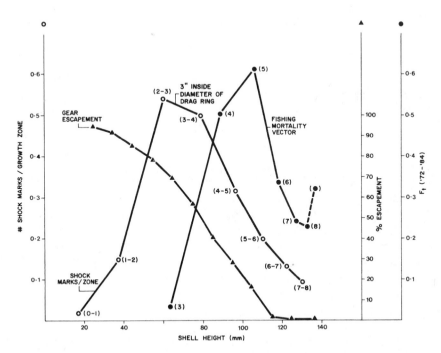

Figure 9. Vector of fishing mortality at size and age determined from cohort analysis (Mohn et al., 68); distribution of shock frequency marks with size and age (unpublished information); and selection curve of the dredge (from Caddy 21).

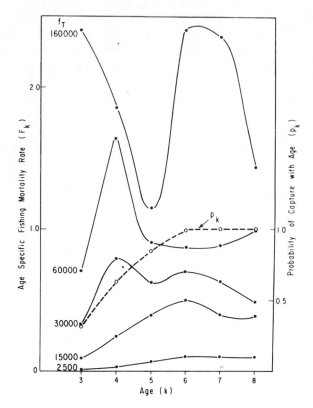

Figure 10. Vector of *F* values predicted for different intensities of fishing from spatial model YRAREA, assuming clumped recruitment, and effort directed at centers of highest density (from Caddy 15).

up in the observed catches. As noted above, there seems some direct evidence from underwater surveys, and at least strong circumstantial evidence from shock marks, that the *F* vector from cohort analysis (e.g., in Fig. 9) is underestimated for prerecruits. For commercial sizes, however, the high incidental breakage rates reported earlier may cause less bias in *F* as calculated from total removals (*C*) over mean stock numbers (*N*) in the interval, (that is, $F = C/N$), than for prerecruits, because of the strategy of fishing followed by commercial scallop fleets, which as described below, leads to a dome-shaped *F* vector.

7.4. Fishing Strategy of Scallop Fleets

Usually a boat (or a fleet) locates a promising bed and fishes it repeatedly over a period of days or weeks until catch rate drops below a commercially acceptable level. By this time, scouting by one or several boats may have located a new concentration. In practice, and this has been described for other species of scallops that recess in the bottom (54,70), the earlier tows made along a particular "line" are "developing" and even smoothing the bottom by dislodging scallops from their recesses into higher profile areas where they are more liable to capture. Those scallops damaged, even lethally, by fishing may still be captured in later tows, so that the proportion damaged that ultimately remain uncaptured may be significantly less over the whole (short) period of fishing than for a single tow. Despite this, there is likely to be a significant "invisible" mortality rate in dredge fisheries for scallops, and this must impose some limits on the accuracy with which original stock numbers can be back-calculated from catches.

Because processing scallops on board is a laborious manual procedure, the effective fishing power of a boat in conditions of high abundance is a function of crew size (71), with the consequence that catch per hour towed, but not catch per day is a true measure of local abundance (72).

8. YIELD MODELS

A variety of calculations of yield per recruit for sea scallop stocks have been made following the methodologies of Beverton and Holt (1) and Thompson and Bell (in Ricker 45), but most arrive at similar conclusions to those of Posgay (2), namely, that age at first capture for Georges Bank stocks should be raised from current levels of 3–4 yr at first capture to in excess of age 6, and that fishing mortality rates should be decreased, if yield per recruit is to be maximized. These measures would also increase spawning potential and are generally valid. However for scallop fisheries a number of other considerations apply that affect both the appropriate yield-per-recruit model to use and the desirable management criteria to be followed.

8.1. Temporal Fluctuations in Population Size

It is characteristic of most scallop fisheries throughout the world that recruitment shows significant and large fluctuations from year to year: in the extreme case (as in the recent Peruvian fishery for *Argopecten purpuratus*) (73), large populations have recently occurred, possibly in response to low development of the oceanic phenomenon of El Niño, where they had not been abundant previously. Similar fluctuations are shown by scallop fisheries in southern Australia. Intermittent recruitment is also typical of marginal areas such as St. Pierres Bank (74), Brown's Bank, and off Virginia (9) for sea scallops. To a less extreme extent, irregular recruitment seems also typical of Georges Bank, where for periods of 3–4 yr centered on 1961 and 1978 in the recent catch record, scallop landings were at least three times higher than in intervening and subsequent years (67). The influence of a relatively few excellent year classes played the major role here, and corresponding increases in catch rates probably preceded the effort pulses by 1–2 yr (15). A more regular series of cyclic fluctuations has been observed in the Bay of Fundy fishery (5), which it was suggested are tied to 9-yr long-term tidal periodicity. This phenomenon and its explanation appear to have been supported by more recent observations (6) pointing to long-term tidal phenomena. To date, no causal mechanisms have been suggested for recruitment fluctuations in the Georges Bank stock, but environmental, in particular hydrographic, factors affecting larval retention must also be prime suspects.

The importance of such marked temporal fluctuations in affecting the appropriate management strategy is that maximizing yield per recruit may be less important than conserving some spawning stock at high densities in order to take advantage of future favorable hydrographic conditions (see Chapter 30). From an economic perspective, spreading the effort exerted on unusually large cohorts (which must support the fishery for many years) over a longer period of time should minimize economic disruption caused by a "pulse" fishery by a specialized fleet with few alternative resources available.

8.2. Incidental Mortalities and Yields

The other problem in the direct application of straightforward knife-edged yield-per-recruit assessments is that if it is impossible to avoid a large by-catch of smaller scallops with unselective gear, and a significant fraction is lethally damaged on discard and on bottom, scallops that do not show up in the catch but are killed by fishing are presumed to do so if Y/R calculations are applied without correction. The impact of such incidental mortalities ideally must be taken into account (see Figure 9 and Chapter 30).

8.3. Spatial Considerations

All important processes in fisheries for sedentary shellfish have a spatial component, in particular, recruitment and fishing strategy, but so do presumably, density-dependent growth, mortality, and spawning success, for which little information is yet available for sea scallops. The potential impacts of geographic variations in recruitment and fishing strategy were addressed in the spatial model YRAREA proposed for Georges Bank scallops, in which patches of new recruitment were added at random to each of the 10' unit areas of the bank, and each unit area allowed to develop an age composition determined from its past history of recruitment and subsequent fishing pressure, with fishing effort directed preferentially at locally high densities of individuals. Under these circumstances, yield per recruit can be calculated incorporating the observed strategy of the fleet prior to introduction of meat size regulations, namely, concentrating effort on high densities of recruits in years of good recruitment, and dispersing to fish lower densities of older scallops in years of poor recruitment (72). This long-term strategy, which seems typical of West Atlantic sea scallop fisheries, still applies to the Bay of Fundy fishery (3), where the good recruitment of the early 1980s has been followed by heavy fishing, depletion, and dispersal of the fleet to less important beds toward the mouth of the Bay of Fundy (7); a pattern of displacement in time which has also occurred in the past (56).

Although published information (9) has suggested that for the U.S. fleet on

Georges Bank fishing has usually been selective for larger scallops, the strategy of some components of the Canadian Georges Bank fleet in the early 1970s was to concentrate on smaller scallops (62). This was somewhat modified, subsequent to the introduction of the average meat count regulation, to taking scallops from high-density patches of new recruits, but then searching peripheral areas of low density for larger scallops, to "blend" the catch so as to maintain the overall meat count of the landings below the statutory level. This procedure has been criticized; for taking small scallops undoubtedly reduces the yield per recruit from the population as a whole. However, by comparison with unrestricted fishing, the slow increase in mean size captured since introduction by Canada of legal minimum landed meat counts of 60/lb in 1973 (up to 40/lb in 1976), and of 40/lb by the United States in 1982 (up to 35/lb in 1983) has certainly increased the yield per recruit compared with earlier years. The legal minimum average count has had a number of other impacts on fishing strategy, however, that are not characteristic of a simple minimum size limit.

In fact, it may be questioned for a resource with a proven instability of recruitment, fished by a fleet of vessels not adapted to alternative resources or fishing methods, whether an absolute minimum size is a viable concept as a strategy. It would for example, effectively close exploitation of deeper-water components of the stock with small meats. Such a strategy associated with the elevated size limits predicted from Y/R considerations and high levels of exploitation, in absence of quotas, would also accentuate the inevitable fluctuations in supply, prices, employment, and market availability that such a type of fishery is liable to. To some extent, these effects are smoothed out under the current meat size regulation. Reducing effort drastically is apparently the strategy that would have the most effect in permitting a more stable fishery; it would also allow other alternative management strategies such as rotational harvesting to be considered. A minimum size limit without a viable system of protecting dense patches of new recruits by local area closures would also return the fishery to the destructive practice of fishing areas of predominantly small shell in order to cull out the few larger scallops, while discarding and damaging on deck and on bottom more scallops than are retained. Because some harvest of young scallops seems inevitable following periods of low density, this should be minimized and vessels encouraged to fish fringe areas for larger shell. To some extent, this is what occurs under the current type of regulation, which effectively "forces effort over more age classes" (67).

The impacts of all the above factors in practice lead to a yield per recruit under actual fishing conditions that is markedly below that predicted from steady-state, dynamic pool, knife-edge calculations (Fig. 11).

As have earlier workers, Mohn et al. (58) stressed that none of the current management measures is adaptive by itself to aid reconstruction of presently depleted stocks significantly. A further increase in size at first capture followed by reduction of effort continue to be recommended as priorities. A less emphasized high priority for scallop fisheries would be to explore alternative ways of making more efficient use of a valuable resource, involving alternative management strategies and developing better gear. In my opinion, the low efficiency, poor selec-

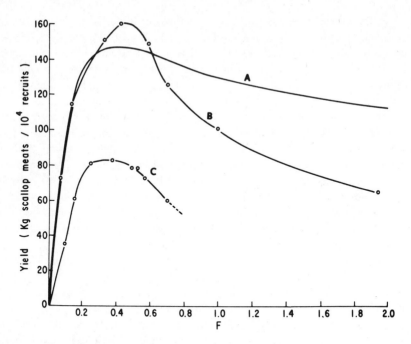

Figure 11. Yield-per-recruit calculation under dynamic pool assumptions (*A*) and that predicted from the same parameters (*B* + *C*) with different assumptions as to recruitment patchiness and effort aggregation (see Caddy 15 for details).

tivity characteristics, and high incidental damage caused by current gear designs make fishing gear development one of the two most potentially effective approaches to improving yield from this fishery. The other is to see if harvest strategy cannot be brought more into line with the contagious resource distribution. Such a program would necessitate in situ observations with various modifications to gear design, but is well within current technical capabilities for underwater research.

Despite the high individual fecundity of scallops, the suggestion of Sinclair et al. (3) that a viable spawning stock should be maintained is a reasonable one, especially if density-dependent spawning success is the case, and they demonstrate that recent fishing strategy on Georges Bank has resulted in the biomass of gonads dropping significantly, particularly with low legal meat sizes. These types of fecundity per recruit considerations at the population level are important, as shown by the increased regularity of spatfall for *Patinopecten* in Mutsu Bay, where massive scallop culture operations greatly increased spawning biomass. For a contagiously distributed, sedentary species, an alternative or supplementary hypothesis for the maintenance of adequate spawning levels is that the critical density of mature animals located in an area where hydrographic features allow a high probability of larval return should be maintained. This would support the maintenance of refugia (closed areas) in areas of high density where multiple age groups provide evidence for past successful recruitments (see Chapter 30 for suggested criteria for recognizing such areas).

9. MULTISPECIES CONSIDERATIONS

For the smaller inshore boats in the southern Gulf of St. Lawrence, the increase in prices for scallops in the late 1970s closely followed license limitation and quotas for other important inshore resources and effectively led to diversification of boats into fishing sea scallops full or part time. A somewhat similar phenomenon was reported later for the northern Gulf and Labrador for *Chlamys islandicus*, as a "pulse fishery" (75) occurring when higher relative prices and abundance for Iceland scallops and alternative resources made harvesting this species attractive. Something similar has occurred in relation to St. Pierre's Bank, where the Georges Bank fleet has on occasions harvested both species of scallops when abundance on Georges Bank was low.

10. SURVEY METHODOLOGY AND STATISTICS

Early studies of the feasibility of developing a stratified-random survey design for scallop stocks inside the 1200 square miles of Georges Bank investigated the various approaches to sampling, bearing in mind the very contagious type of distribution and the availability of detailed catch and effort data for each unit area. Caddy and Chandler (76) used data on effort and landings per unit area recorded the preceding year by U.S. and Canadian fleets, to select 10' unit areas of latitude–longitude, separated into shallow and deep areas and each of a divergent sediment type, in an attempt to investigate the best approach to stratification. Commercial catch followed by depth proved the two most useful parameters for weighting unit areas; mean commercial catch per day was a poorer index, because handling time is high under conditions of high abundance. A divergence in stratification methods has been followed in recent years by U.S. and Canadian scientists; the former stratifying by depth, the latter by commercial catch rate (kilograms per hour exerted). This can be contoured using the procedure of Jamieson and Chandler (39), with poststratification of the data being frequently carried out after the survey, which in the U.S. surveys, uses dredges with 3.8-cm mesh liners (77). Judging from the aforementioned paper and Mohn et al. (58), there appears little clear difference in performance between the two approaches to stratification. The U.S. approach probably has the advantage for estimating total biomass, and the Canadian for estimating that proportion of the biomass that is fishable.

As an alternative, systematic as opposed to random approaches to sampling have been widely used for scallop surveys (74), and with the addition of new statistical techniques for estimating abundance, these appear to be now coming back into favor.

11. SOCIOECONOMIC CONSIDERATIONS AND QUALITY CONTROL

As would be expected for a high-unit-value resource where price per pound has risen dramatically since 1970, the unsatisfied demand for scallops has had a significant impact on investment in fleet size and on the resultant fishing effort exerted by the fleet (78). The Georges Bank fishery, being the single largest source of supply of scallops in North America, has played a significant role in determining market prices for other smaller scallop fisheries, despite the increase in scallop imports from Alaska, South America, Europe, and other localities since the 1960s, and not just in North America. It is evident that there is a need for bioeconomic modeling to determine appropriate strategies, founded on a firm base of biological information, in evaluating, for example, strategies of limited entry.

12. APPROACHES FOR FUTURE DEVELOPMENT AND FISHERY ENHANCEMENT

The recent settlement at the International Court of the boundary dispute between Canada and the United States will have a major impact on the way that the Georges Bank fishery is carried out by the Nova Scotia- and New England-based fleets in future years. The settlement allocates the northeastern corner of the bank, containing one of the two main areas of high density, to Canada, and the larger area, containing the productive Cultivator Shoals, to the United States. At the time of writing, the major impact this is likely to have on management and enforcement of what may be two areas of exclusive national access is not clear. Some reduction of each fleet's options to expand into alternative areas in poor recruitment years will presumably be the case. The need to maintain meat counts at a low level, to address the question of a limitation on removals, and most difficult of all, to decide on an allocation of allowable harvest, will all become more pressing, as will the requirement of maximizing the yield per recruit from the existing population. This will eventually necessitate better gear and more precise fishing strategies if wastage due to indirect mortality is to be minimized. If at-sea enforcement is required along the international boundary, this may eventually lead to some scheme of rotational closure such as that envisaged by Gales and Caddy (69). In the more distant future, despite the considerable difficulties of working on Georges Bank, some form of individual user rights could lead to stock enhancement in order to make better use of the large area of apparently suitable bottom on the bank that rarely receives a viable spatfall.

　　The possibility of moving to some form of culture/enhancement operation is becoming more feasible as scallop prices increase, given that the technology for catching spat on polyethelene film and rearing spat in Japanese hanging bags using standard Japanese technology is apparently feasible (31). In all probability, these methods developed for *Patinopecten yessoensis* culture (Chapter 23), if only for logistic reasons, will first succeed for *Placopecten* in inshore areas (79), especially because meat yield is higher in shallow water.

REFERENCES

1. R. J. H. Beverton and S. J. Holt, On the dynamics of exploited fish populations. *U.K. Minist. Agric. Fish., Fish. Invest. (Ser. 2)* **19**:1–533 (1957).

2. J. A. Posgay, Maximum yield per recruit of sea scallops, *Placopecten magellanicus* (Gmelin). *Int. Commn. Northwest. Atl. Fish. Ser.* **1016**:Doc. 73 (1962).

3. M. Sinclair, R. K. Mohn, G. Robert, and D. L. Roddick, Considerations for the effective management of Atlantic scallops. *Can. J. Fish. Aquat. Sci., Tech. Rep.* **1382**:1–113 (1985).

4. J. F. Caddy and J. A. Gulland, Historical patterns of fish stocks. *Mar. Policy* October:267–278 (1983).

5. J. F. Caddy, Long-term trends and evidence for production cycles in the Bay of Fundy scallop fishery. *Rapp. P.-V. Reun., Cons. Int. Explor. Mer.* **175**:97–108 (1979).

6. M. J. Dadswell, R. A. Chandler, G. Robert, and R. Mohn, Southwest New Brunswick and Grand Manan scallop stock assessment, 1983. *CAFSAC* Res. Doc.* **84/28**:1–21 (1984).

7. G. Robert, M. J. Lundy, and M. A. E. Connolly, Recent events in the scallop fishery of the Bay of Fundy and Approaches. *CAFSAC Res. Doc.* **84/71**:1–41 (1984).

8. K. S. Naidu and F. M. Cahill, Status and assessment of St. Pierre's Bank scallop stocks 1982–83. *CAFSAC Res. Doc.* **84/69**:1–56 (1984).

9. F. M. Serchuk, P. W. Wood, J. A. Posgay, and B. E. Brown, Assessment and status of sea scallop (*Placopecten magellanicus*) populations off the Northeast coast of the United States. *Proc. Natl. Shellfish Assoc.* **69**:161–191 (1979).

10. J. A. Posgay, Population assessment of the Georges Bank sea scallop stocks. *Rapp. P.-V. Run., Cons. Int. Explor. Mer* **175**:109–113 (1979).

11. K. S. Naidu and J. T. Anderson, Aspects of scallop recruitment on St. Pierre Bank in relation to oceanography and implications for resource management. *CAFSAC Res. Doc.* **84/29**:1–15 (1984).

12. J. F. Caddy and J. A. Carter, Macro-epifauna of the lower Bay of Fundy—Observations from a submersible and analysis of faunal adjacencies. *Can. J. Fish. Aquat. Sci., Tech. Rep.* 1254, 1–35 (1984).

13. C. L. MacKenzie, Jr., Biological and Fisheries Data on Sea Scallop, *Placopecten magellanicus* (Gmelin), Tech. Ser. Rep. No. 19. Sandy Hook Laboratory, Northeast Fisheries Center NMFS, NOA, U.S. Dept. of Commerce, Washington, DC, 1979, 34 pp.

14. J. M. Orensanz, Size, environment and density: the regulation of a scallop stock and its management implications p 195–227 In: Jamieson G. S. and N. Bourne (ed.), 1986.

*CAFSAC = Canadian Atlantic Fisheries Scientific Advisory Committee: CAFSAC Secretariat, Bedford Institute of Oceanography, P.O. Box 1006, Dartmouth, Nova Scotia, Canada.

North Pacific Workshop on stock assessment and management of invertebrates. *Can. Spec. Publ. Fish. Aquat. Sci.* **92**:430p.

15. J. F. Caddy, Spatial model for an exploited shellfish population, and its application to the Georges Bank scallop fishery. *J. Fish. Res. Board. Can.* **32**:1305–1328 (1975).

16. J. F. Caddy, A method of surveying scallop populations from a submersible. *J. Fish. Res. Board Can.* **27**:535–549 (1970).

17. R. K. Mohn, G. Robert and D. L. Roddick, Research sampling and survey design for sea scallops (*Placopecten magellanicus*) on Georges Bank. *J. Northw. Atl. Fish. Sci.*, vol. 7:117–121 (1987).

18. G. Robert, G. S. Jamieson, and M. J. Lundy, Profile of the Canadian offshore scallop fishery on Georges Bank, 1978 to 1981. *CAFSAC Res. Doc.* **82/15**:1–33 (1982).

19. J. F. Caddy, Underwater observations on scallop (*Placopecten magellanicus*) behaviour and drag efficiency. *J. Fish. Res. Board Can.* **25**:2123–2141 (1968).

20. J. F. Caddy, Recent scallop recruitment and apparent reduction in cull size by the Canadian fleet on Georges Bank. *Int. Comm. Northwest. Atl. Fish. Redbook* Part III:147–155 (1971).

21. J. F. Caddy, 1972. Size selectivity of the Georges Bank offshore dredge and mortality estimate for scallops from the Northern Edge of Georges in the period June 1970 to 1971. *Int. Comm. Northwest. Atl. Fish. Redbook* Part III:79–85 (1972).

22. M. J. Broom, Synopsis of biological data on scallops (*Chlamys (Aequipecten) opercularis* Linnaeus), *Argopecten irradians* (Lamarck), *Argopecten gibbus* (Linnaeus). *FAO Fish. Synop.* 114 (Doc. FIRS/S114):1–44 (1976).

23. J. F. Caddy, Underwater observations on tracks of dredges and trawls and some effects of dredging on a scallop ground. *J. Fish. Res. Board Can.* **30**:173–180 (1973).

24. J. A. Posgay, Movement of tagged sea scallops on Georges Bank. *Mar. Fish. Rev.* **43(4)**:19–25 (1981).

25. G. D. Melvin, M. J. Dadswell, and R. A. Chandler, Movement of scallops *Placopecten magellanicus* (Gmelin 1791) (Mollusca: Pectinidae) on Georges Bank. *CAFSAC Res. Doc.* **85/30**:1–29 (1985).

26. J. Quérellou, Exploitation des coquilles Saint-Jacques, *Patinopecten yessoensis* Jay, au Japon. Milieux, méthodes, résultats, organisation de la production. *Publ. Assoc. Dev. Aquicult.* **75**:82 (1975).

27. J. S. Gutsell, Natural history of the Bay Scallop. *Bull. Fish.* **45**:569–632 (1930).

28. A. R. Brand, J. D. Paul, and J. N. Hoogesteger, Spat settlement of the scallops *Chlamys opercularis* (L.) and *Pecten maximus* (L.) on artificial collectors. *J. Mar. Biol. Assoc. U.K.* **60**:379–390 (1980).

29. J. F. Caddy, Progressive loss of byssus attachment with size in the sea scallop, *Placopecten magellanicus* (Gmelin). *J. Exp. Mar. Biol. Ecol.* **9**:179–190 (1972).

30. K. S. Naidu, Reproduction and breeding cycle of the giant scallop *Placopecten magellanicus* (Gmelin) in Port au Port Bay, Newfoundland. *Can. Soc. Zool.* **48**:1003–1012 (1970).

31. K. S. Naidu and R. Scaplen, Settlement and survival of the giant scallop, *Placopecten magellanicus*, larvae on enclosed polyethelene film collectors. *FAO Tech. Conf. Aquacult., 1976*, FIR Aqu. Conf. 76(**E7**):1–5 (1976).

32. R. H. Baird, On the swimming behaviour of escallops (*Pecten maximus* L.). *Proc. Malacol. Soc. London* **33(2)**:67–71 (1958).

33. O. Vahl and B. Clausen, 1981. Frequency of swimming and energy cost of byssus production in *Chlamys islandica* (O. F. Muller). *J. Cons., Cons. Int. Explor. Mer* **39(1):**101–103 (1981).

34. J. A. Posgay and K. D. Norman, 1958. An observation on the spawning of the sea scallop, *Placopecten magellanicus* (Gmelin), on Georges Bank. *Limnol. Oceanogr.* **3:**478 (1958).

35. C. J. Ordzie and G. C. Garofalo, Behavioural recognition of molluscan and echinoderm predators by the Bay Scallop, *Argopecten irradians* (Lamark) at two temperatures. *J. Exp. Mar. Biol. Ecol.* **43:**29–37 (1980).

36. J. C. Medcof and N. Bourne, Causes of mortality of the sea scallop, *Placopecten magellanicus. Proc. Natl. Shellfish. Assoc.* **53:**33–50 (1964).

37. J. C. Medcof, Dark-meat and the shell disease of scallops. *Fish. Res. Board Atl., Prog. Rep.* **45:**3–6 (1949).

38. G. Gulka, P. W. Chang, and K. A. Marti, Prokaryotic infection associated with a mass mortality of the sea scallop, *Placopecten magellanicus. J. Fish Dis.* **6:**355–364 (1983).

39. G. S. Jamieson and R. A. Chandler, The potential for research and fishery performance data isopleths in population assessment of offshore, sedentary, contagiously distributed species. *CAFSAC Res. Doc.* **80/77:**1–32 (1980).

40. R. H. Baird, Factors affecting the efficiency of dredges. In: Kristjonsson H., *Modern Fishing Gear of the World.* Fishing News Books Ltd., London, 1959, pp. 222–224.

41. J. A. Stevenson and L. M. Dickie, Annual growth rings and rate of growth of the giant scallop, *Placopecten magellanicus* (Gmelin) in the Digby area of the Bay of Fundy. *J. Fish. Res. Board Can.* **11(5):**660–671 (1954).

42. A. V. Silina, Determination of age and growth rate of Yezo Scallop by the sculpture of its shell surface. *Sov. J. Mar. Biol.* **4(5):**827–836 (1978).

43. A. S. Merrill, J. A. Posgay, and F. E. Nichy, Annual marks on shell and ligament of sea scallop (*Placopecten magellanicus*). *Fish. Bull.* **65:**299–311 (1966).

44. D. L. Roddick and R. K. Mohn, Use of age–length information in scallop assessments. *CAFSAC Res. Doc.* **85/37:**1–16 (1985).

45. W. E. Ricker, Computation and interpretation of biological statistics of fish populations. *Bull. Fish. Res. Board Can.* **191:**1–382 (1975).

46. R. Jones, Lee's phenomenon of "apparent change in growth rate" with particular reference to haddock and plaice. *Int. Comm. Northwest. Atl. Fish., Spec. Publ.* **1:**229–242 (1958).

47. L. Antoine, P. Arzel, A. Laurec, and E. Morize, La croissance de la coquille Saint-Jacques (*Pecten maximus* L.) dans les divers gisements francais. *Rapp. P.-V. Reun., Cons. Int. Explor. Mer* **175:**85–90 (1979).

48. D. I. Heald and N. Caputi, Some aspects of growth, recruitment and reproduction in the Southern Saucer Scallop, *Amusium balloti* (Bernardi 1861) in Shark Bay, Western Australia. *Fish. Res. Bull., Dep. Fish. Wildl.* (West. Aust.) **25:**1–33 (1981).

49. J. F. Caddy, R. A. Chandler, and E. I. Lord, Bay of Fundy scallop surveys 1966 and 1967 with observations on the commercial fishery. *Tech. Rep.—Fish. Res. Board Can.* **168:**1–9 (1970).

50. J. F. Caddy, Records of associated fauna in scallop dredge hauls from the Bay of Fundy. *Tech. Rep.—Fish. Res. Bd. Can.* **225:**1–11 (1970).

51. H. Hidu, M. S. Richmond, and A. H. Price, II, Morphological variability in sea

scallops, *Placopecten magellanicus* (Gmelin) related to meat yield. *Proc. Natl. Shellfish. Assoc. Med.*, **67,** 75–79 (1977).

52. N. Bourne, Scallops and the offshore fishery of the Maritimes. *Bull. Fish. Res. Board. Can.* **145:**1–60 (1964).

53. J. S. MacPhail, The inshore fishery of the Maritime Provinces. *Fish. Res. Board. Can., Atl. Biol. Stn. Circ., Gen. Ser.* **22:**1–4 (1954).

54. D. Hardy, *Scallops and the Diver Fisherman.* Fishing News Books Ltd., London, 1981, 134 pp.

55. J. F. Caddy, Efficiency and selectivity of the Canadian offshore scallop dredge. *ICES C.M.* **1971/K:25,** (1971) (mimeo).

56. L. M. Dickie, Fluctuations in the abundance of the giant scallop, *Placopecten magellanicus* (Gmelin) in the Digby area of the Bay of Fundy. *J. Fish. Res. Board Can.* **12(6):**797–857 (1955).

57. C. J. Chapman, J. Mason, and J. A. M. Kinnear, Diving observations on the efficiency of dredges used in the Scottish fishery for the scallop, *Pecten maximus* (L.). *Scott. Fish. Res. Rep.* **10:**1–16 (1977).

58. Doubleday W. G. (ed) Manual on groundfish surveys in the Northwest Atlantic NAFO *Sci. Coun. Studies*, **2:**7–55.

59. J. F. Caddy and G. S. Jamieson, Assessment of Georges Bank (ICNAF subdivision 5Ze) scallop stock 1972–76 incorporated. *CAFSAC Res. Doc.* **77/32:**1–21 (1977).

60. R. L. Wigley, Bottom sediments of Georges Bank. *J. Sediment. Petrol.* **31(2):**165–188 (1961).

61. R. W. Elner and G. S. Jamieson, 1979. Predation of Sea Scallops, *Placopecten magellanicus*, by the rock crab, *Cancer irroratus*, and the American Lobster, *Homarus americanus. J. Fish. Res. Board Can.* **36:**537–543 (1979).

62. J. F. Caddy and A. Sreedharan, The effect of recent recruitment to the Georges Bank scallop fishery on meat sizes landed by the offshore fleet in the summer of 1970. *Tech. Rep.—Fish. Res. Board Can.* **256:**1–10 (1971).

63. A. S. Merrill and J. A. Posgay, Estimating the natural mortality of the sea scallop (*Placopecten magellanicus*). *Int. Comm. Northwest Atl. Fish., Res. Bull.* **1:**88–106 (1964).

64. L. M. Dickie and J. C. Medcof, Causes of mass mortalities of scallops (*Placopecten magellanicus*) in the southwestern Gulf of St. Lawrence. *J. Fish. Res. Board Can.* **20:**451–482 (1963).

65. K. A. Marti, J. M. Hoenig, and S. M. Saila, A Catastrophic Decline in a Population of Sea Scallops (*Placopecten magellanicus*), Shellfish Comm. (mimeo) **1982/Doc K:20.** Int. Cons. Explor. Mer, Copenhagen, 1982, 4 pp.

66. G. S. Jamieson and R. A. Chandler, Paralytic shellfish poison in Sea Scallops (*Placopecten magellanicus*) in the West Atlantic. *Can. J. Fish. Aquat. Sci.* **40:**313–318 (1982).

67. R. K. Mohn and G. Robert, Comparison of two harvesting strategies for the Georges Bank scallop stock. *CAFSAC Res. Doc.* **84/10:**1–35 p (1984).

68. R. K. Mohn, G. Robert, and D. L. Roddick, Georges Bank scallop assessment—1983. *CAFSAC Res. Doc.* **84/12:**1–28 (1984).

69. L. E. Gales and J. F. Caddy, YRAREA: A program to demonstrate effects of exploi-

tation on a contagiously distributed shellfish population. *Tech. Rep.—Fish. Mar. Serv. (Can.)* **582**:1–79 (1975).

70. J. Mason, *Scallop and Queen Fishing in the British Isles*, A Buckland Found. Book. Fishing News Books Ltd., London, 1983, 144 pp.

71. J. F. Caddy, Some recommendations for conservation of Georges Bank scallop stocks. *Int. Comm. Northwest. Atl. Fish., Res. Doc.* **72/6**:1–8 (1972).

72. J. F. Caddy and E. I. Lord, Recent developments in the Georges Bank scallop fishery. *Int. Comm. Northwest. Atl. Fish. Redbook* Part III:£à-93 (1969).

73. M. Wolff. Population dynamics of the Peruvian Scallop *Argopecten purpuratus* during the El Niño phenomenon of 1983 Can. J. Fish. Aquat. Sci., vol. 44, 1684–1691 (1987).

74. K. S. Naidu and S. J. Smith, A two-dimensional systematic survey of the Iceland scallop, *Chlamys islandica* in the Strait of Belle Isle. *CAFSAC Res. Doc.* **82/4**:1–24 (1982).

75. S. K. Naidu, F. M. Cahill, and D. B. Lewis, 1982. Status and assessment of the Iceland scallop, *Chlamys islandica* in the northeastern Gulf of St. Lawrence. *CAFSAC Res. Doc.* **82/02**:1–66 (1982).

76. J. F. Caddy and R. A. Chandler, 1979. Georges Bank scallop survey, August 1966: A preliminary study of the relationship between research vessel catch, depth and commercial effort. *Fish. Mar. Serv. Manuscr. Rep. (Can.)* **1054**:1–13 (1979).

77. F. M. Serchuck and S. E. Wigley, Evaluation of USA and Canadian Research Vessel Surveys for Sea Scallops (*Placopecten magellanicus*) on Georges Bank. *J. Northw. Atl. Fish. Sci.* vol 7, 1–13 (1987).

78. M. A. Altobello, D. A. Storey, and J. M. Conrad, The Atlantic sea scallop fishery: A descriptive and econometric analysis. *Mass., Agric. Exp. Stn., Res. Bull.* **643**:1–80 (1977).

79. K. S. Naidu and F. M. Cahill, Culturing giant scallops in Newfoundland waters. *Can. Manuscr. Rep. Fish. Aquat. Sci.* pp. 1–23 (1986).

CASE STUDIES

C / CEPHALOPODS

26 MANAGING AN INTERNATIONAL MULTISPECIES FISHERY: THE SAHARAN TRAWL FISHERY FOR CEPHALOPODS

J. Bravo de Laguna
Instituto Español de Oceanografia Madrid, Spain

1. INTRODUCTION

The area off the northwest coast of Africa is one of the most important and productive upwelling systems of the world. Especially along the coastline of the western Saharan desert, and on the continental shelf between parallels 27°N and 20°N, several artisanal and industrial fisheries for demersal and pelagic species have a common area of action, and in 1976 produced up to 2.2 million tonnes of fish and invertebrates.

Economically the most important of the industrial fisheries is the trawl fishery for cephalopods. It has developed over the past two decades as a result of a change in dominance on the fishing grounds, from previously dominant finfish species to cephalopods. The high commercial value of the cephalopods and the high yields obtained in this area have attracted, among others, long-distance fleets coming from Europe and the Far East, making this fishery economically the most important industrial fishery of the eastern central Atlantic.

In spite of this importance, studies on the marine biology and assessment of fish stock in this area are quite recent. It was clear from the ICES/FAO Symposium on the Living Resources of the African Continental Shelf between the Strait of Gibraltar and Cape Verde (La Laguna, Tenerife, March 25–29, 1968) that most of the faunal studies done until then were descriptive and taxonomic, and that considerable efforts should be made to obtain a better understanding of the dynamics especially of these cephalopod populations (Bravo de Laguna 1,2). The International Program of Cooperative Investigations on the North East Central Atlantic (CINECA) contributed to an increase in our knowledge of the ecological features of this area, but its contribution to our understanding of the exploited marine populations was limited.

The establishment of the FAO Committee for the Eastern Central Atlantic Fisheries (CECAF) in 1969 and its scientific advisory body, the Working Party on Resources Evaluation, substantially improved the situation. Scientific institutes and fisheries administrations of the coastal and distant-water fishing countries concerned found a common framework for a better understanding and management of the resources. In the case of the cephalopods, the establishment of an ad hoc Working Group on the Assessment of Cephalopod Stocks in 1977 and the work accomplished during its four meetings (1; FAO 3–6) made available the biological and fisheries information required to establish the basis for an appropriate management scheme for this fishery.

2. THE ENVIRONMENT

The main fishing activities in this area take place on the continental shelf, which is quite extensive and varies from 20 nautical miles wide off Cape Bojador to 100 off Punta Leven (Bravo de Laguna 7). Between Cape Juby and Cape Blanc its area is 22,200 square nautical miles, of which 94% is suitable for trawling operations.

The high productivity of this area has its basis in some critical meteorological and hydrographic features (Belveze and Bravo de Laguna 8). The meteorology of the region has three basic aspects. The average annual temperature is 2°C below that to be expected at this geographical lattitude, the barometric pressure is intermediate between the Azores anticyclonic and equatorial cyclonic situations, and the region is affected by the North East trade winds.

From a hydrographic perspective, the region is under the influence of the Canary Current, which transports cold water at a temperature below that corresponding to this latitude (Le Floch 9). In combination, these effects, especially the trade winds, produce upwellings in the region that are especially strong during the summer. The surface temperatures may then descend below 16°C, and the salinity reaches values of 35.7‰. The winds are constant throughout the whole year, but decrease in intensity from October to January. During these months there is a displacement of the upwelling towards the south.

3. THE FISHERY

The cephalopod fisheries of northwest Africa take place on the continental shelf approximately between parallels 24°N and 9°N, along the coasts of the Western Sahara, Mauritania, Senegal, the Gambia and Guinea Bissau, and Guinea (see Fig. 1). The oldest and most productive fishery has been that located along the Saharan coast between Cape Bojador (26°N) and Cape Timiris (19°N).

The origin and evolution of this fishery have been described at intervals during

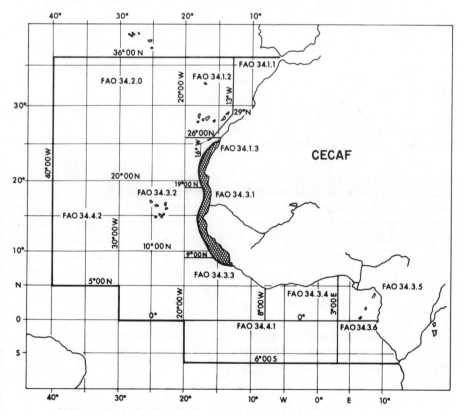

Figure 1. Area classification for cephalopod fisheries off northwest Africa.

the past 25 years of its history (Bravo de Laguna 1; Garcia Cabrera 10–12; Bas et al. 13; Pereiro and Bravo de Laguna 14; Caddy 15). The fishing grounds located on the continental shelf off the northwest African coast between parallels 24°N and 25°N have historically been very productive for sea bream (Sparidae). That fishery started in the fifteenth century, when the first Spanish and Portuguese settlements were established along the coast. Even today, the Spanish fishermen refer to the different fishing grounds by the names of different kinds of sea bream.

During World War II several important areas of the northeast Atlantic were closed to fishing operations. As a result, several trawlers moved to the Saharan continental shelf in order to find species to substitute for those traditionally marketed. They succeeded to the extent that shortly afterward it was possible to find "African cod," "African hake," and so forth on the Spanish market. Navarro et al. (16) presented the results of one of these expeditions.

After the war these trawl fisheries not only continued, but were augmented by new trawlers coming from eastern Europe. In a very few years a decrease in yields was noted, at the same time that an unwanted and troublesome by-catch became more and more abundant in the trawl operations, that is, large numbers of octopuses and cuttlefish. Although appreciated on Spanish markets, the prices were then very low for these species, and they did not compensate for the costs of freezing and transport to markets.

In the early 1960s a long-distance Japanese fleet joined this fishery, with a trawl fishery aimed mainly at sea bream and cephalopods. Prices increased, and in 1963 a Japanese factory vessel started operating, the F/S *Awazu Maru*, which worked in cooperation with fresh fish trawlers and pair trawlers, mainly under the Spanish flag. In 1967 two factory vessels were operating, F/S *Awazu Maru* and the F/S *Galicia*, and the number of trawlers fishing for cephalopods was around 70, mainly Spanish and Japanese. The latter took their peak catches in this fishery in 1967 and 1968, with 51,200 and 72,460 tonnes, respectively.

In the early 1970s, a significant number of Spanish freezer trawlers joined the fishery and slowly replaced the fresh fish trawlers, and consequently Spanish production increased from 30,700 tonnes in 1969 to 67,300 tonnes in 1970. In this decade the fishing grounds were expanded south of parallel 23°N, reaching Cape Timiris and Nouakchott in Mauritania. At the same time, cephalopods replaced sea bream as the dominant catch component. From the negligible catches achieved during exploratory fishing by the Spanish trawlers M/V *Abrego* and M/V *Cierzo* in the 1940s (Navarro *et al.* 16), and 3% octopuses in the total cephalopod catch recorded in the trawl survey by the French vessel *Thalassa* in 1962 (Maurin and Bonnet 17), the contribution by cephalopods to the total commercial catch by Spanish trawlers had increased to around 90% by 1976, of which 49% were octopuses (Bravo de Laguna et al. 18). This pronounced change was also noticeable in the Japanese catches (El Gharbi and Idelhadj 19). In 1959, 93% of the catch were sea bream and 1% cephalopods; these were cuttlefish only. In 1967, sea bream made up only 16% of the catches, whereas cephalopods had increased to 65%, of which 33% were cuttlefish, 12% squid, and 55% octopuses. In 1975, 51% of the Japanese catch were cephalopods, and sea bream made up only 13%.

There are three main fishing grounds between Cape Bojador (26°N) and Nouak-chott (18°N)(FAO 3, 5), which all extend down to 250 m: Cape Garnett (24°30′) to Cape Barbas (22°N); off Cape Blanc (21°N) to Cape Timiris (19°30′N); and off Nouakchott (18°N) (Fig. 2). The first has been the oldest and most productive fishing area, and has traditionally supported the highest fishing effort.

The cephalopods presently most sought in the Saharan trawl fishery are octopus (*Octopus vulgaris* Cuvier, 1797), four cuttlefish species (*Sepia officinalis officinalis* Linnaeus, 1758; *S. officinalis hierredda* Rang, 1837; *S. bertheloti* Orbigny 1838; *S. elegans* Blainville, 1827; *S. orbignyana* Ferrusac, 1826), and two squid species (*Loligo vulgaris* Lamark, 1798 and *L. forbesi* Steensrup, 1856). According

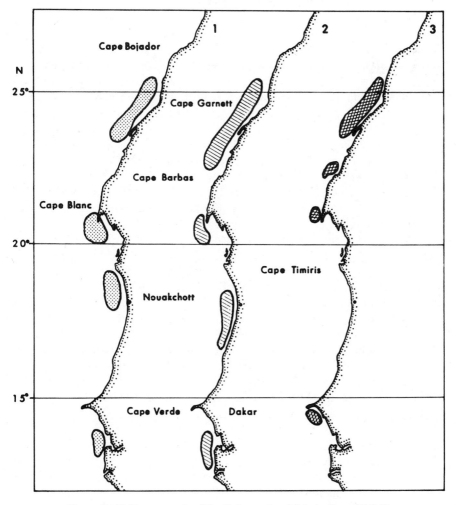

Figure 2. Fishing grounds of the Saharan trawl fishery (from FAO 5).

to Ariz (20), the proportions of the key species in the commercial catch of Spanish freezer trawlers during the 1975–1982 period were 59% octopuses, 19.5% cuttlefish, and 8.9% squid, with the rest made up of several fish species. The composition of landings by the international fleet registered in the 1985 Las Palmas survey (Roest and Frielink 21) was 29.6% octopuses, 27.3% cuttlefish, 21.4% sea bream, 17% other fish, 2.5% squid, 1.43% croaker, and 0.79% pelagics.

The main countries fishing in the Saharan trawl fishery historically have been Japan, Spain, Morocco, Korea, and Mauritania. A certain number of other countries such as Libya, Italy, and Soviet Union, Greece, Panama, and Taiwan have also participated in recent years (FAO 6).

Japanese trawlers began operations in this area in the early 1960s, and in 1959 two vessles started fishing off the northwest coast of Africa between Cape Bojador and Cape Roxo (Hatanaka 27). In 1965 there were already 41 vessels in operation, and they reached their peak number of 83 Japanese trawlers fishing cephalopods in 1972. Operations started in the so-called Villa Cisneros area between Cape Garnett (24°30′) and Cape Barbas (22°N). Here, the highest yields were obtained during the 1966–1968 season with 101,000 tonnes. As yields decreased, they expanded activities toward the south, starting to exploit the fishing grounds off Cape Blanc in 1965. Since then they have concentrated their activities there, alternating with fishing grounds off Nouakchott until they left the fishery in 1983 (FAO 6). Their catches were around 30,000 tonnes annually, reaching peak values in 1974–1976 with 366,000 tonnes, of which 58,600 tonnes was octopus.

The Spanish fishery specializing in cephalopods started in 1963 with the operation of pair trawlers supplying factory vessels. In the late 1960s freezer trawlers joined the fishery, and 39 of them were in operation in 1969. This number increased to 297 in 1980, although no more than 126 vessels were operating at the same time; all were in the area of Villa Cisneros (FAO 6). Before 1979, but only for short periods subsequently, the Spanish fleet also operated off Cape Blanc. Catches have oscillated in recent years between 67,000 tonnes in 1983 and 86,000 tonnes in 1981, with a peak value for the Spanish catch in this fishery of 112,200 tonnes in 1972. Recently, the Spanish fleet has been declining in size from 48,590 GRT in 1983 to 29,063 GRT in 1986.

The Moroccan fishery started in 1977–1978 with 43 freezer trawlers, although only 11 have been actively in operation (El Gharbi and Idelhadj 19). Since then, national capacity has continuously increased to the present level of 164 freezer trawlers, most of them larger than 250 GRT. Their area of operation is the fishing grounds located north of Cape Blanc, and their cephalopod catch has ranged between 736 tonnes in 1978 and a peak value of 50,700 tonnes in 1983.

The Mauritanian fishery for cephalopods began in 1980 as a joint venture operation (FAO 23), with only 15 freezer trawlers in operation initially, but the fleet has since increased to 52 in 1983 and 60 in 1984. Up to 26 fresh fish trawlers have also been supplying cephalopods to freezing plants in Nouadhibou. Their area of operation has been the grounds located between Cape Blanc and Cape Timiris and off Nouakchott, including the Bay of Levrier.

The Korean fishery began in 1976 and replaced the Japanese trawlers in the fishery (FAO 6), In 1979, 87 fishing vessels were operating. Since then their numbers have continuously decreased from 80 vessels in 1980 to 41 in 1985, as they have been transferred to joint venture operations. Their contribution to the total cephaloped catch in this fishery in 1985 was 20,000 tonnes.

The total catches of octopus in the northern subregion of responsibility of CECAF, the Committee for Eastern Central Atlantic Fisheries (Fig. 1), have remained around 100,000 tonnes despite changes in fishing effort. The most productive statistical division has been 34.1.3 (Fig. 1); it includes the most important fishing grounds described above and contributed more than 90% to the total catch in 1983 (Table 1).

This statistical division is also the most productive for cuttlefish (see Table 2). Their total catches in the subregion have been of the order of 40,000–50,000 tonnes, contributing from 65 to 94% of the total catch in the Sahara coastal division. The catches of division 34.3.1 have fluctuated widely (FAO 6).

During the last few years, squid catches, which used to be a target species of the Spanish pair trawlers before the cephalopod fishery started, have decreased drastically (Table 3). Their fishery, off Capes Bojador, Garnett, Barbas, and Blanc, was seasonal; it took place in the deeper waters of CECAF statistical division 34.1.3 and produced between 82 and 97% of the total catch. This used to be

TABLE 1 Catches ($t \times 10^{-3}$) of Octopus in the Saharan Trawl Fishery

Year	Division				Not Identified	Total
	34.1.1	34.1.2	34.1.3	34.3.1		
1966	0.4		67.2	1.2	3.0	71.8
1967	0.5		94.6	2.4	2.7	100.2
1968			115.5	1.8	3.6	120.9
1969			83.7	3.6	3.9	91.2
1970			69.2	2.1	2.8	74.1
1971			104.4	2.8	2.9	110.1
1972			119.2	4.1	1.9	125.2
1973	2.9	4.9	66.8	2.7	4.6	81.9
1974	2.0	0.5	90.0	4.0	7.3	103.3
1975	2.8		86.3	3.0	10.7	102.8
1976			106.2	1.5	19.9	127.6
1977			97.5	2.6	0.1	100.2
1978			87.3	4.1	0.1	91.5
1979	0.3	0.4	60.6	5.2	0.2	66.7
1980	0.4	0.2	67.7	2.2		70.5
1981	1.0		113.8	6.2		121.0
1982			91.5	3.6		95.1
1983			101.2	1.4		102.6
1984			100.5	4.1		104.6

Source: FAO 6.

TABLE 2   **Catches ($t \times 10^{-3}$) of Cuttlefish in the Saharan Trawl Fishery**

Year	34.1.1	34.1.2	34.1.3	34.3.1	Not Identified	Total
	Division					
1966	1.3		28.9	2.1	2.3	34.6
1967	1.8		38.5	1.7	2.8	44.8
1968	2.7		36.6	2.4	2.5	44.2
1969	1.3		30.0	4.3	3.9	39.5
1970			17.9	2.7	2.9	23.5
1971			24.3	6.7	5.1	36.1
1972			27.0	8.0	4.9	39.9
1973	2.6	14.1	16.1	6.4	9.5	48.7
1974			35.8	6.0	5.1	46.9
1975	1.8	0.3	18.8	7.2	5.2	33.3
1976			17.0	2.9	6.4	26.3
1977		0.2	24.0	4.4	2.3	30.9
1978	0.5	0.3	41.6	9.0	1.5	52.9
1979	0.6	0.5	42.1	10.1	2.1	55.4
1980	5.6		38.0	5.4		49.0
1981	1.0		51.3	2.0		54.3
1982			40.0	7.9		47.9
1983			40.7	3.9		44.6
1984			28.7	14.4		43.1

Source: FAO 6.

between 20,000 and 25,000 tonnes, but in recent years has declined to 6000–7000 tonnes.

4. HABITAT, BIOLOGY, AND BEHAVIOR

4.1. Octopus

The common octopus (*Octopus vulgaris* Cuvier, 1797) is the main resource sought in this fishery. It is a benthic species with worldwide distribution in temperate and tropical waters (Roper et al. 24). The species is distributed across the continental shelf from the coast down to 200 m; in the eastern central Atlantic it is found on rocky and sandy bottoms and in grass beds, and is common down to 110-m depths. Tagging experiments showed no extensive migrations in this area.

Two spawning seasons have been detected annually in this area, the first, in May–June, and the second in September. Sizes at first maturity are 260 g and about 8-cm mantle length for males, and 1100 g and 12–13-cm mantle length for females (FAO 3). Spawning takes place all over the continental shelf, the females depositing 130,000–250,000 benthic eggs (exceptionally 400,000) grouped in clusters on the bottom; subsequently they show brooding behavior to protect the eggs. This

TABLE 3 Catches ($t \times 10^{-3}$) of Squids Trawl Fishery

Year	Division 34.1.1	34.1.2	34.1.3	34.3.1	Not Identified	Total
1966	0.3		12.2	0.7	1.5	14.7
1967	1.3		18.6	0.8	0.8	21.5
1968	0.6		13.9	0.5	1.3	16.3
1969	0.7		13.7	1.1	2.7	18.2
1970			12.6	0.8	5.0	18.4
1971			18.4	1.2	4.7	24.3
1972			21.2	1.5	2.7	25.4
1973	1.5	0.2	19.0	1.4	6.9	29.0
1974	1.4	1.1	29.1	1.2	6.4	39.2
1975	2.8	0.2	17.6	0.5	5.7	26.8
1976	1.6		9.5	0.1	5.9	17.1
1977	1.5	0.2	19.0	2.4	0.1	23.2
1978	2.5	0.3	21.1	2.6	0.1	26.6
1979	0.8	0.2	16.7	2.9	0.1	20.7
1980	3.6	0.3	10.8	1.4		16.1
1981	0.8		15.9	2.7		19.4
1982			10.0	0.8		10.8
1983			5.6	0.2		5.8
1984			6.9	0.2		7.1

Source: FAO 6.

parental behavior probably compensates for the low fecundity per unit of biomass in comparison to most teleosts. The brooding period lasts from 20 to 25 days at 25°C and to 125 days at 13°C (Pereiro and Bravo de Laguna 14). The larvae have a pelagic phase and become benthic after about 40 days at a size of about 12 mm.

A high seasonal, geographic, and bathymetric variability in the octopus sex ratio has been observed on the Saharan shelf. The most recent and complete information indicates that below 20-cm mantle length, the sex ratio is about 1 : 1. Subsequently, the proportion of females decreases until at above 30-cm mantle length, females almost disappear from the fishery (FAO 5).

Octopuses feed on bivalves, small fish, crustaceans, and cephalopods. Hatanaka (22), who studied the stomach contents of octopuses caught off Cape Blanc, noted that shellfish account for 45–60% in the diet, finfish 19–34%, crustaceans 7–16%, and cephalopods 4–13%. Two main feedings seasons were noted, summer and winter (Hatanaka 22), which are associated with fishing seasons in the region. During the period of spawning and brooding, yields decrease drastically off Cape Blanc.

The main predators and competitors in this area during the pelagic phase are squid, jellyfish, and small pelagic fish. During the adult phase, the main predators are sea bream, conger eels, and grouper.

No clear partition into separate octopus stocks has been justified on biological grounds. Nonetheless, for fish stock assessment purposes, octopuses on the different fishing grounds have been separated into two stocks, the Dakhla and the Cape Blanc stock (FAO 5). Those octopuses caught in the fishing grounds between Cape Bojador and Cape Blanc are considered part of the Dakhla stock. The others taken from Cape Blanc to Nouakchott are considered as the Cape Blanc stock.

4.2. Cuttlefish

Several species of cuttlefish are commonly caught in the Saharan trawl fishery. The most important are the common cuttlefish (*Sepia officinalis* Linnaeus, 1758, varieties *officinalis* and *hierredda*), and the African cuttlefish (*S. betheloti* Orbigny 1839). Ikeda (25) indicates that north of Cape Blanc, *S. officinalis officinalis* is most common, whereas further south, *S. officinalis hierredda* dominates in shallow waters. Other cuttlefish currently caught but in much smaller quantities are the pink cuttlefish (*S. orbignyana* Ferrusac, 1826) and the elegant cuttlefish (*S. elegans*, Blainville, 1827).

The common cuttlefish is a demersal species distributed in the eastern Atlantic from 60°N to South Africa, including the Baltic Sea and the Mediterranean Sea; it is found on sandy bottoms from the coast down to 200-m depth (Roper et al. 24). On the Saharan continental shelf it is mainly found from the coastline down to 150-m depth, where larger individuals occur.

Cuttlefish migrate to spawn in shallow waters; Ikeda et al. (26) identified three main spawning grounds along the Saharan continental shelf by counting the frequency of egg clusters during commercial trawling operations. The higher concentrations occurred from Cape Garnet to Dakhla May–June and August–September at between 30- and 40-m depth, and off Cape Blanc in February and from June to August from 10- to 30-m depth. Further studies have shown that spawning takes place year round, being more intensive from May to September (Hatanaka 22, 27).

The size at first maturity for *S. officinalis* is 12-cm mantle length off Dakhla and 14-cm off Cape Blanc and Nouakchott for males, and 14-cm mantle length for females in all areas (FAO 3). Depending on size, females produce 250–1400 eggs of 8–10-mm diameter (see Bakhayokho 28). After 30–90 days a larva of 7–8 mm hatches. Postspawning mortality for females is high. As for other cephalopods in the area, *S. officinalis* shows a drastic decrease in percentage of females with mantle length greater than 25–29 cm (FAO 5).

The African cuttlefish is much less abundant in this fishery because this is the northern part of its geographic range between 27°N and 14°S; it occurs on the continental shelf between 50- and 150-m depth. The spawning season last from early summer to the end of autumn (Roper et al. 24). According to size, females deposit up to 100 eggs. The life-span is 1–2 yr.

The food of both common and African cuttlefish consists mainly of mollusks, crustaceans, and small demersal fishes.

4.3. Squid

Two species of squid are caught in this area, the European squid (*Loligo vulgaris* Lamarck, 1798) and the veined squid (*L. forbesi* Steenstrup, 1856). The European squid is a neritic and semipelagic species that is distributed in the eastern Atlantic from 55°N to 20°S, including the North Sea and the Mediterranean Sea. It occurs from the surface down to 500 m, being most abundant between 20 and 250 m depth (Roper et al. 24).

Along the northwest African coast, squid concentrate near capes and on the deeper part of the continental shelf from the surface down to 400-m depth, but predominantly about 100-m depth. Size generally increases with depth, with 43–45-cm specimens usually below 100 m, and those of around 5-cm mantle length in very shallow waters (Bravo de Laguna 1).

Vertical and horizontal migrations occur, according to environmental conditions and season. Off the Saharan coasts squid migrate offshore in winter, and approach the coasts for spawning in spring and in October–November.

Mantle length at first maturity is around 13 cm in males and 16 cm in females, which may produce up to 20,000 eggs with a diameter of 2 mm. Larvae hatch after 25 and 45 days and recruitment takes place in February–March and July–September. As with other cephalopods in the area, there is a marked decrease of females with mantle length greater than 26 cm (FAO 5).

The veneid squid is much less abundant on the Saharan continental shelf, being a species of subtropical and temperate waters distributed between parallels 20°N and 60°N in the Mediterranean and Red Seas and the eastern Atlantic, at depths between 100 and 400 m (Roper et al. 24). In the Saharan trawl fishery this species is mainly caught in the deeper trawl operations.

This squid shows seasonal and diurnal migrations, although little is known of its movements off the northwest coast of Africa. The size as first maturity is about 25–30 cm mantle length, which is reached in less than 1 yr. Postspawning mortality is very high, which means that the life-span is very short.

Both squid species feed mainly on other squid, small pelagic fish, and crustaceans.

5. SAMPLING, SURVEYING, AND MONITORING

The design of an information-gathering system for these fisheries and species must take into account their characteristics. On the one hand, it must be remembered that the fishery is carried out by freezer trawlers, highly sophisticated technically, that immediately freeze their catches when they arrive on deck, after sorting by cephalopod species and after a rapid gutting operation in the case of octopus. When the vessels arrive in port, the landing operation is very rapid, with a freeze box spending less than 30 min between the transport vessel and the cold store. Given this situation, the possibilities for biological sampling are very limited. Other possibilities are buying a part of the catch, which is a very expensive procedure, or sampling at sea, which is not always possible owing to limitations of space and the long duration of the trips. This would mean the allocation of several full-time technicians to work on samples for each vessel. This will have to be done of course, in order to gather data on sex, maturity, and other relevant biological information.

Taking into account this situation and the existence of quality control in Las Palmas (Canary Islands, Spain) from where most of the cephalopod fleet has been operating, a sampling and monitoring system was developed by Spanish scientists that allows information gathering on size composition of the catches and recruitment (Ariz and Balguerias 29; Ariz et al. 30,31). This is based on the accuracy of the sorting operations on board, which follow the classification in Table 4.

The catch by commercial category for each boat is available at customs and/or the shipping agents. In this way it is possible to know with high precision the catch of each commercial category by month, quarter, year, and so on. A recuitment index for octopus on the Saharan fishing grounds has been followed for many years by monitoring the catch rate (CPUE) of category 8, that is, those less than 0.3 g (FAO 6).

The next step is to determine the size composition of the total catch. As previously noted, quality-control agents take samples of the various commercial categories, and these samples are carefully analyzed after defrosting the boxes; afterward, among other parameters, the individual sizes of specimens in the sample are measured. This information allows "weighting up" to determine the size composition within each commercial category and hence in the total catch.

Trawl surveys on cephalopod stocks have provided useful information about the relative biomass and indexes of abundance for these stocks. Surveys have been carried out periodically by both Spanish and Moroccan research vessels (Guerra 32, Idelhadj 33; Ariz et al. 34–36). The strategy applied consists basically of monitoring catches over a grid of predetermined stations with time. When applied correctly, this procedure has been especially useful in determining areas of concentration, spawning grounds, and so forth, and has led to the establishment of more

TABLE 4   Classification by Size of Cephalopod Catches

Category	Octopus (kg)	Cuttlefish (kg)	Squid (cm)
0	—	—	>31
1	>4	>2	26–30
2	3–4	1–2	21–25
3	2–3	0.7–1	16–20
4	1.5–2	0.5–0.7	11–15
5	1–1.5	0.3–0.5	>10
6	0.5–1	0.2–0.3	—
7	0.3–0.5	0.1–0.2	—
8	<0.3	<0.1	—

appropriate sampling schemes, that is, a stratified-random sampling scheme, which allows better estimates of biologically relevant parameters.

6. GROWTH, MORTALITY, AND SELECTIVITY

Studies in aquariums show that octopus become benthic 30–40 days after hatching at 24.7°C (Itami et al. 37), at an individual weight of 0.1–0.25 g. After 90 days they average 39 g, and 135 days later, they may have reached 5-cm mantle length. At an age of 12–15 months, the males may weigh 2.5 kg and the females 2 kg.

Growth studies of octopus in this area have been based on size distributions from commercial operations and bottom trawl surveys. Ariz (38) analyzed the data from five bottom trawl surveys between February 1980 and March 1982, and followed catches of the smallest commercial category of octopus (less than 300 g) by the Spanish freezer trawler fleet between 1976 and 1982. Growth parameters were estimated by the Beverton and Holt (39) method.

Values obtained for the von Bertalanffy's growth (Ariz 38) which are currently used in assessment are

$$l_t = 34.5\{1 - \exp[-0.37(t + 0.095)]\}$$

This equation predicts mantle lengths of 13, 19, 23, and 26 cm, respectively, for specimens of 1, 2, 3, and 4 yr of age.

Natural mortality estimates have been derived by several authors (FAO 5, 6; Pereiro and Bravo de Laguna 14; El Gharbi and Idelhadj 19; Hatanaka 27); the values most accepted fall between 0.5 (FAO 6) and 0.8 (El Gharbi and Idelhadj 19) on an annual basis. From a biological perspective, the intermediate position of octopus in the trophic chain makes the existence of these relatively low natural mortality rates quite understandable.

In a situation of overfishing and uncertainty, recruitment is a key parameter to monitor. This has been attempted with different degrees of success for the Dakhla octopus stock by means of trawling surveys, catch rates of commercial fisheries, and cohort analysis. So far, monitoring of octopus CPUE for the smallest size category, those of less than 300 g, has been an invaluable tool for this purpose (FAO 5,6), and this recruitment index has been monitored since 1977 (Fig. 3). There have been oscillations in annual recruitment, with minimum values in 1979–1981 and peak values after 1982. This can be considered a good index as long as juvenile catchability remains constant.

So far, selectivity studies for octopus in the Saharan fishery have been conducted with Spanish bottom trawls designed for capturing cephalopods (Guerra 32; Ariz and Fernandez 40; Idelhadj 41). Selectivity at size is not constant and fluctuates seasonally and geographically. According to the last information provided by Idelhadj (41), selection factors for males oscillate between 0.73 and 1.06, and size at first capture from 4.4- to 6.4-cm mantle length. In females, the selection factors are between 0.7 and 1.08, and size at first capture from 4.2- to 6.5-cm mantle

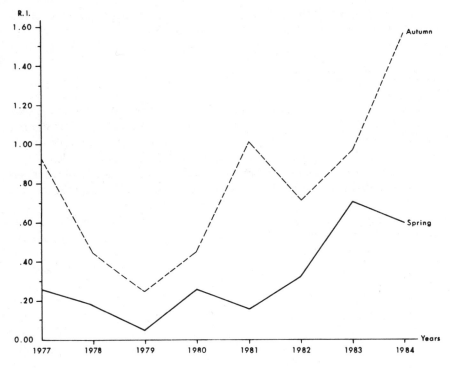

Figure 3. Recruitment indexes for octopus in the Saharan trawl fishery (from FAO 6).

length. Average values obtained with a 60-mm mesh size gear were 0.79 for the selection factor and 4.2–5.6 for the size at first capture.

Little is known about growth and mortality of cuttlefish and squid in this area, except for some information on *S. officinalis hierredda* off Senegal, where Bakhayokho (28) obtained some preliminary information from the size distribution of the catches. These data indicate that this species has a high growth rate and a short life-span. Females reach their maturity around 5–6 months old. Using Petersen's method Bakhayokho estimated a growth rate of 2.6 cm/month up to 10-cm mantle length, and 2.1 cm/month from 10- to 25-cm mantle length.

7. YIELD MODELS

Production models have been traditionally used in the assessment of CECAF cephalopod stocks (see Bravo de Laguna 2; FAO 3, 4, 6, 23; El Gharbi and Idelhadj 19).

The last updated information on octopus, cuttlefish, and squid was prepared during the third meeting of the ad hoc Working Group on Cephalopod Stocks, using the generalized production model (Pella and Tomlinson 43). Effort data were

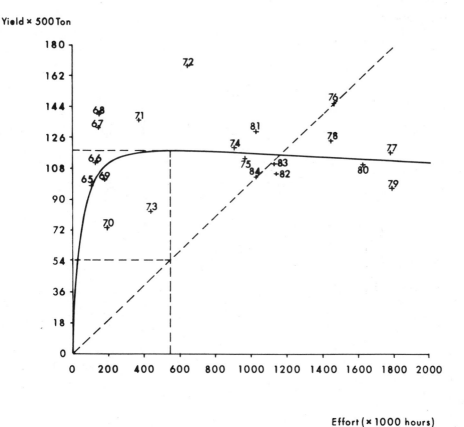

Yield × 500 Ton

Effort (× 1000 hours)

Figure 4. Generalized production model for octopus of Dakhla stock (from FAO 6).

expressed in Spanish standard trawling hours. Total effort was estimated from the standard Spanish CPUE series.

The best adjustment for octopuses of the Dakhla stock was obtained with the shape parameter of the generalized production model set at $m = 0.1$ (Fig. 4). This corresponds to an MSY of 59,200 tonnes, equivalent to 545,000 trawling hours annually. This is about half of the fishing effort applied in the period 1983–1984; that is, $f_{83-84}/f_{MSY} = 1.98$. This confirms the heavily overexploited situation of this stock, which has been evident for many years (FAO 5), even though there has been a certain reduction in fishing effort and management measures closing areas within 12 nautical miles of the coast to fishing have been adopted.

The information available for cuttlefish does not allow a separate production model analysis for each fishing ground, which has therefore been conducted for the whole of CECAF statistical division 34.1.3 (Fig. 1) for the period 1969–1984. The best adjustment was obtained for $m = 0.3$ (Fig. 5); it corresponds to an MSY

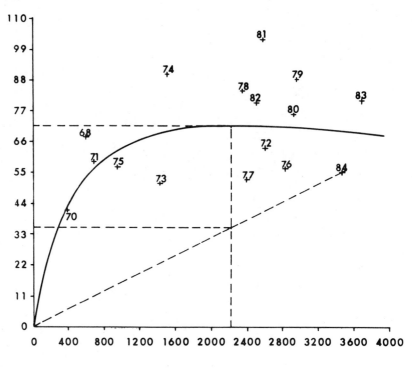

Figure 5. Generalized production model for cuttlefish in CECAF statistical division 34.1.3 (from FAO 6).

of 35,900 tonnes and is equivalent to an optimum effort of 2,217,000 Spanish standard trawling hours. The ratio of fishing effort applied in 1983–1984 and optimum effort is 1.64, which also indicates an overfishing situation.

The application of the same production model to the squid catch and effort series of CECAF statistical division 34.1.3 (Fig. 6) showed this stock to be in poor condition, with serious danger of depletion. The best adjustment in this case was obtained for $m = 1$, and MSY was calculated at 25,600 tonnes for an optimal effort of 1,214,000 Spanish standard trawling hours, far less than that applied in the period 1983–1984 ($f_{83-84}/f_{MSY} = 2.54$).

Assessments of Cape Blanc octopus stocks by production modeling have encountered several difficulties owing to apparent anomalies in the catch data for the period 1979–1984 (6,23). Until 1980, this stock was in a very healthy situation (5), but since then, the new data points included in the model fitting yield a completely different diagnosis.

According to FAO (23) the best fit of a generalized production model is obtained

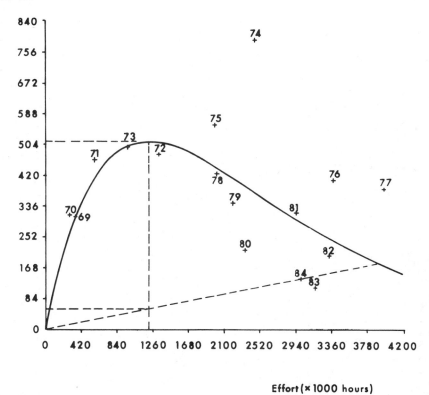

Figure 6. Generalized production model for *Loligo* squid in CECAF statistical division 34.1.3 (from FAO 6).

for $m = 0.5$, giving an MSY of 43,100 tonnes and an optimal effort of 281,000 standard Japanese trawling hours (Fig. 7). If this is accepted, the stock is an underfished situation ($f_{83-84}/f_{MSY} = 0.43$).

Other analyses done for the same period (FAO 6) show apparent anomalies in the data for the four years (1981–1984). In this case the best adjustment would be with $m = 2$ (the Schaefer model), corresponding to an MSY of 35,600 tonnes and an optimal effort of 232,000 Mauritanian standard trawling hours.

These apparent discrepancies are difficult to clarify at this stage. There may be several biases involved originating either in the transformation of the Japanese effort units used until 1980 into the Mauritanian effort units now in current use. Poor quality of fishing effort data, underestimates of catches over the last few years, and changes in recruitment independent of fishing activity are all possible causes.

With the one possible anomaly mentioned, all models show that the cephalopod stocks in CECAF statistical division 34.1.3 have been in an overfished state, and

Yield × 500 Ton

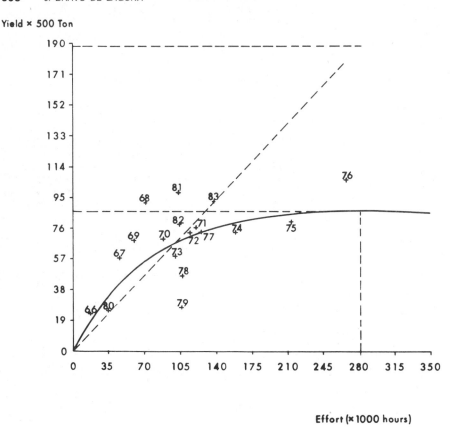

Effort (× 1000 hours)

Figure 7. Generalized production model for octopus of Cape Blanc stock (from FAO 6).

that a substantial reduction in fishing effort will be necessary in order to obtain higher yields and improve economic performance of the fleet.

8. MANAGEMENT

Management of a cephalopod fishery faces basically the same difficulties as those for finfish stocks. Notwithstanding that there are certain biological characteristics of cephalopods that make assessment difficult and hence complicate advice to fishery administrators (Caddy 44), basically the main problems arise from our poor knowledge of cephalopod biology and from their short life-spans. The latter feature means that any management measure will have rapid effects and that there are few possibilities of changing the fishing strategy for any cohort once it enters the fishery.

Other biological characteristics that should be taken into consideration in any analytical asessment, for example, are the occurrence of a postspawning mortality, the almost complete absence of competition between successive cohorts, the possible relationship between spawning stock size and that for the next year class, changes in catchability because of spawning behavior, and the effects of competition with fish species.

The management advice of the last CECAF ad hoc Working Group on Cephalopod Stocks (FAO 6) was to advocate a reduction in fishing effort. This advice has been based on the results of production modeling and, although largely indicative, strongly suggests reduction in the fishing effort applied in 1984 by 49% for octopus in the Dakhla stock, by 38% for cuttlefish of CECAF statistical division 34.1.3, and by 60% for squid. Only in the case of octopus in the Cape Blanc stock could fishing effort apparently be increased by up to 232%, but other considerations expressed above advice caution in placing too much credibility in this last figure, which depends heavily on the data points of the past few years.

An overall reduction in the 1984 level of fishing effort on cephalopod stocks would have rapid effects on increasing yields and spawning stock sizes and should maintain catches near present levels. Added to a real enforcement of the 60-mm mesh size recommended by CECAF and the existence of closed zones of 12 nautical miles from the coast, this measure would probably allow the optimal yields from this fishery to be approached.

REFERENCES

1. J. Bravo de Laguna, Distribution and abundance of demersal resources of the CINECA region. *Rapp. P.-V. Reun., Cons. Int. Explor. Mer* **180**:432–446 (1982).

2. J. Bravo de Laguna, Los recursos pesqueros del area de afloramiento del NW africano. *Simp. Int. Alf. O. Afr., Inst. Invest. Pesq.*, *1985* **2**:761–798 (1985).

3. FAO, Fishery Committee for the Eastern Central Atlantic. Report of the *ad hoc* Working Group on the Assessment of Cephalopods Stocks. *FAO, Rome, CECAF/ECAF Ser.* **78/11**:1–149 (1979).

4. FAO, Comité des Pêches pour l'Atlantique Centre-Est. Conclusions du sous-groupe de travail sur les cephalopodes. *FAO Fish. Rep.* **244**:35–67 (1981).

5. FAO, Fishery Committee for the Eastern Central Atlantic. Report of the Special Working Group on Cephalopods Stocks in the Northern Region of CECAF. *FAO, Rome, CECAF/ECAF Ser.* **82/24**:1–178 (1982).

6. FAO, Rapport du Troisième Groupe de travail *ad hoc* sur l'évaluation des stocks de cephalopodes dans la région nord du COPACE. *FAO, Rome, CECAF/ECAF Ser.* **86/41**. 101 p.

7. J. Bravo de Laguna, La pesca en el Banco Sahariano. *El Campo* **99**:69–74 (1985).

8. H. Belveze and J. Bravo de Laguna, Les resources halieutiques de l'Atlantique centre-est. Deuxième partie: Les ressources de la côte ouest-africaine entre 24°N et le Detroit de Gibraltar. *FAO Fish. Rep.* **186.2**:1–64 (1980).

9. J. Le Floch, Quelques aspects de la dynamique et de l'hydrologie des couches superficielles dans l'ouest marocain. Compagnes CINECA-CHARCOT I et III. *Thethys* 6(1-2):53–68 (1974); through *FAO Fish. Rep.* **186.2**:3 (1980).

10. R. C. Garcia Cabrera, Biologia y pesca del pulpo (*Octopus vulgaris*) y choco (*Sepia officinalis*) en aguas del Sahara español. *Publ. Tec. Dir. Gen. Pesca Marit.* **7**:141–198 (1968).

11. R. C. Garcia Cabrera, Pulpos y calamares en aguas del Sahara español. *Publ. Tec. Dir. Gen. Pesca Marit.* **8**:75–103 (1969).

12. R. C. Garcia Cabrera, Espèces du genre *Sepia* du Sahara espagnol. *Rapp. P.-V. Reun., Cons. Int. Explor. Mer* **159**:132–139 (1970).

13. C. Bas, B. Morales, and J. M. San Feliu, Pesquerias de cefalopodos en el Banco Sahariano. *Publ. Tec. Dir. Gen. Pesca Marit.* **9**:129–151 (1971).

14. J. A. Pereiro and J. Bravo de Laguna, Dinamica de la poblacion y evaluacion de los recursos de pulpo del Atlantico Centro Oriental. *FAO, Rome, CECAF/ECAF Ser.* **80/18**:1–53 (1980).

15. J. F. Caddy, Some factors relevant to management of cephalopod resources off West Africa. *Dakar, CECAF Proj.* CECAF/TECH/81/37:1–46 (1981).

16. F. de P. Navarro et al., La pesca de arrastre en los fondos de Cabo Blanco y Banco de Arguin. (Africa Sahariana). *Trab. Inst. Esp. Oceanogr.* **18**:1–225 (1943).

17. C. Maurin and M. Bonnet, Le chalutage au large des côtes nord-ouest africaines. Résultats des campagnes de la THALASSA. *Sci. Pech.* **177**:1–17 (1969).

18. J. Bravo de Laguna, M. A. R. Fernandez, and J. C. Santana, Actividad de la flota pesquera española en aguas del Banco Sahariano durante 1975. Estado actual de las poblaciones de animales de interes comercial. *Inf. Pesq. Inst. Esp. Oceanogr.* **6**:1–34 (1976).

19. R. El Gharbi and A. Idelhadj, Caracteristiques bio-économiques et évolution recente de la pêcherie marocaine aux cephalopodes. *Dakar, CECAF Proj.* CECAF/TECH/86/73:1–63 (1986).

20. J. Ariz, Descripcion de la actividad de la flota española que explota la pesqueria de cefalopodos de Africa NO. *Simp. Int. Afl. O Afr., Inst. Invest. Pesq., 1985* **2**:889–904 (1985).

21. F. C. Roest and A. B. Frielink, Las Palmas survey 1985. Landings of cephalopods and fish from West African waters. *Dakar, CECAF Proj.* CECAF/TECH/86/60:1–33 (1986).

22. H. Hatanaka, Spawning season of the cuttlefish *Sepia officinalis officinalis* Line off the Nw coast of Africa. *FAO, Rome, CECAF/ECAF Ser.*, **87/11**:63–74 (1979).

23. FAO, Comité des Pêches pour l'Atlantique Centre-Est. Description et évaluation des ressources halieutiques de l ZEE Mauritanienne. Rapport du Groupe de travail CNROP/FAO/ORSTOM Naouadhobou, Mauritanie, 16–27 septembre 1985. *FAO, Rome, CECAF/ECAF Ser.* **86/37**:1–310 (1986).

24. C. F. E. Roper, M. J. Sweeney, and C. E. Nauen, FAO species catalogue. Vol. 3. Cephalopods of the world. An annotated and illustrated catalogue of species of interest to fisheries. *FAO Fish. Synop.* **3(125)**:1–277 (1984).

25. I. Ikeda, Rapport sur les recherches japonaises (1972). *FAO Fish. Rep.* **158**:41–45 (1973).

26. I. Ikeda, H. Hatanaka, S. Kawahara, and T. Inada, Observations on egg clusters of *Sepia* spp. caught by the commercial trawlers in the CECAF waters. *FAO Fish. Rep.* **183**:65–71 (1976).

27. H. Hatanaka, Studies on the fisheries biology of the common Octopus off the NW coast of Africa. *Bull. Far Seas Fish. Res. Lab.* **17**:13–124 (1979).

28. M. Bakhayokho, Life history of *Sepia officinalis* off the Senegalese coast. *FAO Fish. Tech. Rep.* **231**:204–263 (1983).

29. J. Ariz and E. Balguerias, Composicion de las capturas de *Octopus vulgaris* realizadas por la flota española en la division 34.1.3. de CECAF en el periodo 1976 a 1980. *FAO, Rome, CECAF/ECAF Ser.* **82/24**:45–64 (1982).

30. J. Ariz, E. Balguerias, and J. C. Santana, Composicion de las capturas de *Loligo vulgaris* realizadas por la flota española en la division 34.1.3. de CECAF en el periodo 1976 a 1980. *FAO, Rome, CECAF/ECAF Ser.* **82/24**:65–82 (1982).

31. J. Ariz, E. Balguerias, and J. C. Santana, Composicion de las capturas de *Sepia spp.* realizadas por la flota española en la division 34.1.3 de CECAF en el periodo 1976 a 1980. *FAO, Rome, CECAF/ECAF Ser.* **82/24**:83–102 (1982).

32. A. Guerra, Selectividad de *Octopus vulgaris* del Atlantico Centro-Oriental. *FAO, Rome, CECAF/ECAF Ser.* **78/11**:75–81 (1979).

33. A. Idelhadj, Analyse de la pêche des cephalopodes de la zone de Dakhla (26°N–22°N) et résultats des études biologiques effectuées lors des campagnes du navire de recherche Ibn Sina de 1980 a 1983. *Trav. ORSTOM Doc.* 42 (1984); through R. El Gharbi and A. Idelhadj, Caractéristiques bio-économiques et évolution recente de la pêcherie marocaine aux cephalopodes. *Dakar, CECAF Proj.* CECAF/TECH/86/73:1–63 (1986).

34. J. Ariz, E. Balguerias, and J. C. Santana, Distribucion de los rendimientos de *Octopus vulgaris* obtenidos en las campañas de colaboracion cientifica hispano-marroqui efectuadas en la division 34.1.3. de CECAF por el B/O IBN SINA. *FAO, Rome, CECAF/ECAF Ser.* **82/24**:103–111 (1982).

35. J. Ariz, E. Balguerias, and J. C. Santana, Distribucion de los rendimientos de *Sepia officinalis* obtenidos en las campañas de colaboracion cientifica hispano-marroqui efectuadas en la division 34.1.3. de CECAF por el B/O IBN SINA. *FAO, Rome, CECAF/ECAF Ser.* **82/24**:112–120 (1982).

36. J. Ariz, E. Balguerias, and J. C. Santana, Distribucion de los rendimientos de *Loligo vulgaris* obtenidos en las campañas de colaboracion cientifica hispano-marroqui efectuadas en la division 34.1.3. de CECAF por el B/O IBN SINA. *FAO, Rome, CECAF/ECAF Ser.* **82/24**:121–129 (1982).

37. H. Itami et al., Notes on the laboratory culture of the octopus larvae. *Jpn. Soc. Sci. Fish.* **29**:514–520 (1963); through J. A. Pereiro and J. Bravo de Laguna, *FAO, Rome, CECAF/ECAF Ser.* **80/18**:1 (1980).

38. J. Ariz, Nota sobre la edad y crecimiento del pulpo (*Octopus vulgaris* Cuvier, 1797) del Atlantico Centro Oriental (25°N–22°N). *Simp. Int. Afl. O. Afr., Inst. Invest. Pesq., 1985* **2**:969–976 (1985).

39. R. J. H. Beverton and S. J. Holt, On the Dynamics of exploited fish populations. *Fish. Invest. (Ser. 2)* **19**:1–533 (1957).

40. J. Ariz and M. A. R. Fernandez, Selection on octopus (*Octopus vulgaris*) and seabreams of the Spanish cephalopods bottom trawl fishery off Northwest Africa. *ICES C.M.* 1980/K:35 (1980).

41. A. Idelhadj, Etude de la selectivité du chalut de fond espagnol utilisé dans la pêcherie de cephalopodes du Sahara marocain. *FAO, Rome, CECAF/ECAF Ser.* **82/24**:140–148 (1982).

42. FAO, Evaluation of the fishery resources of the Eastern Central Atlantic. Report of the 3rd session of the Working Party on resource evaluation of CECAF. Rome, 4–13 February 1976. *FAO Fish. Rep.* **183**:1–135 (1976).

43. J. J. Pella and P. K. Tomlinson, A generalized stock production model. *Bull. Int.-Am. Trop. Tuna Comm.* **13**:419–496 (1969).

44. J. F. Caddy, The cephalopods: Factors relevant to their population dynamics and to the assessment and management of stocks. In: J. F. Caddy (ed). Advances in assessment of world cephalopod resources, *FAO Fish. Tech. Pap.* **231**:416–452 (1983).

27 POPULATION ASSESSMENT, MANAGEMENT AND FISHERY FORECASTING FOR THE JAPANESE COMMON SQUID, *Todarodes pacificus*

Mamoru Murata

Hokkaido Regional Fisheries Research Laboratory
Kushiro, Hokkaido, Japan

1. INTRODUCTION

In Japan, the cephalopods (squid, cuttlefish, and octopus) have been important fishery resources since ancient times. In recent years (1971–1983), the catches of cephalopods have ranged from 547,000 to 676,000 tons (metric ton) and make-up a significant portion (47.3% on average) of the world's catch of these organisms.

Of all the cephalopods taken in Japanese coastal waters, the Japanese common squid, *Todarodes pacificus* Steenstrup, is the most important in terms of the volume of catch. This species accounted for 70–80% of the total annual domestic catch of cephalopods prior to the 1970s. However, annual catches have been declining since then.

The Japanese common squid is caught virtually throughout the coastal waters of Japan, more than 90% of the total catch being taken by the squid jigging fishery. The catches vary considerably, however, from year to year as well as seasonally, as do the locations of the main fishing grounds. Consequently, studies were initiated in the late 1950s to forecast the time and amount of recruitment likely to take place on the various fishing grounds.

Before the 1960s, when the Japanese common squid was very abundant, it was generally believed that the annual variations in catch were mainly due to natural causes. However, the sharp decline in catches since the 1970s could no longer be explained solely on the basis of natural causes. Thus the relationship between stock abundance and fishing intensity, as well as the problem of overfishing, came under intense scrutiny at about this time. On the other hand, there was a great increase in fishing intensity as squid jigging vessels increased in size and as more sophisticated navigational and fishing equipment was developed. It soon became possible for larger vessels to locate and pursue the seasonally migrating schools of squid over considerable distances and for vessels to select those areas of densest squid concentration to carry out their fishing operations. As a result, the excessive fishing effort deployed in certain limited areas and the resulting decrease in productivity have recently become serious problems.

This chapter begins with a discussion of the history of Japan's squid jigging fishery and the biological characteristics of Japanese common squid. The chapter also touches on the fishery forecasting, assessment, and management of the Japanese common squid resources.

2. HISTORY OF THE JAPANESE SQUID JIGGING FISHERY

The localities of fishing grounds and the fishing season in recent years are shown in Figure 1. There seems to be a general trend for squid catches to be made earliest on the southernmost fishing grounds but to end first on the northernmost fishing grounds. The main fishing season runs from late spring (May) through winter (February, March) in the waters south of the Noto and Boso Peninsulas, and from summer (June, July) through late autumn (October–December) in the waters north of 40°N. In the more southerly fishing ground, where squid catches often show a temporary decline between August and September, the fishing season is usually divided into two periods of summer and autumn–winter seasons.

According to data on the number of days fished by squid jigging vessels by year and by vessel size class (Fig. 2), the motorized vessels of all size classes increased their fishing effort (number of days fished) between the 1950s and the early 1970s. This increasing trend in fishing effort was especially noticeable in the case of vessels

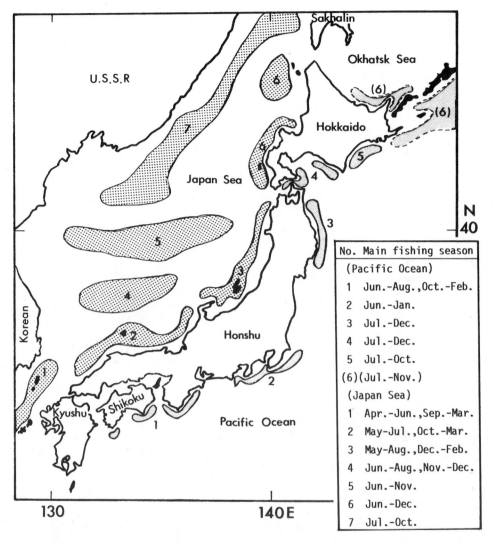

No.	Main fishing season
(Pacific Ocean)	
1	Jun.-Aug.,Oct.-Feb.
2	Jun.-Jan.
3	Jul.-Dec.
4	Jul.-Dec.
5	Jul.-Oct.
(6)	(Jul.-Nov.)
(Japan Sea)	
1	Apr.-Jun.,Sep.-Mar.
2	May-Jul.,Oct.-Mar.
3	May-Aug.,Dec.-Feb.
4	Jun.-Aug.,Nov.-Dec.
5	Jun.-Nov.
6	Jun.-Dec.
7	Jul.-Oct.

Figure 1. Location of jigging fishing grounds and main fishing seasons for *T. pacificus* in recent years.

larger than 20 GRT. On the other hand, the fishing effort deployed by nonmotorized vessels was very high during the 1950s, but has declined sharply since the 1960s.

Because fishing efficiency naturally differs between the various size classes of vessels, the average catch per fishing day (1953–1974) was first calculated for the vessels in each size class, and the annual variation in total fishing days then standardized. The following indexes were obtained, and are illustrated in Figure

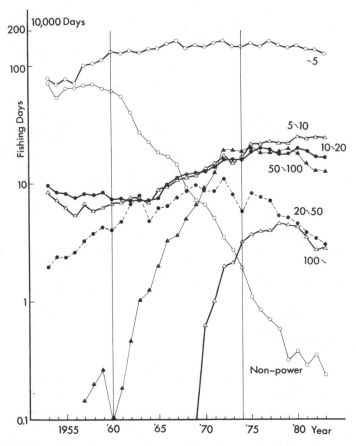

Figure 2. Annual changes in fishing days by size classes of vessel (GRT) for Japanese jigging fisheries around Japan in 1953–1983.

3. The average catches obtained in this manner were as follows, assuming an index of 1.0 for vessels of the 5–10 GRT class: nonmotorized vessels, 0.03; under 5 GRT, 0.17; 10–50 GRT, 2.82; and over 50 GRT, 5.74 (Murata and Araya 1). As illustrated by Figure 3, fishing effort increased very sharply from the late 1950s to the early 1970s (mean effort for 1972–1974/mean effort for 1953–1956 = 4.63), and then leveled off or even decreased slightly at 20×10^5 to 30×10^5 days/yr.

The fishing effort (number of days fished) by marine area is not clearly known because of the lack of detailed spatial data. However, judging from the proportion of catches made on the Pacific and Japan Sea fishing grounds, it was estimated that before 1970 about 80% of the total fishing effort had been expended in the Pacific fishing grounds, and in the 1970s, 60–90% of the total effort had been on the Japan Sea fishing grounds. In recent years the squid jigging vessels have been

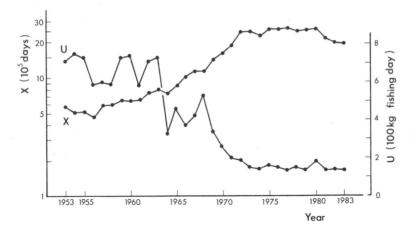

Figure 3. Annual changes in fishing day (*X*) and catch per fishing day (*U*) for Japanese jigging fisheries. *X*: Standardized by indexes obtained from catch per fishing day by size classes of vessel.

spending some time on both the Pacific and the Japan Sea fishing grounds in scouting for squid schools; after 1974, some fishing days in the Pacific have also been devoted to scouting for schools of flying squid, *Ommastrephes bartrami*.

As discussed above, it seems that the total fishing effort for Japanese common squid increased greatly in the 1950s and 1960s in the Pacific (centered on the Sanriku and Hokkaido fishing grounds) and in the early 1970s on the Japan Sea fishing grounds (mainly the offshore fishing grounds). However, the fishing effort in the Pacific has declined very sharply since the early 1970s. In the Japan Sea the fishing effort has remained rather steady since the latter half of the 1970s, but began decreasing in the 1980s.

Annual variations in total catch of Japanese common squid during the 84-yr period (1900–1983) are shown in Figure 4. The catches prior to 1951 also included species other than Japanese common squid (less than 10–20%). Before 1946, catches fluctuated widely between 20,000 and 200,000 tons, but generally remained at a low level, averaging around 80,000 tons; whereas in 1947 and afterward, catches increased sharply and reached approximately 600,000 tons in 1951–1952 and 1963, and 670,000 tons in 1968. Until the 1960s, the catches fluctuated around a relatively high level, averaging 427,000 tons, but began dropping rapidly in the 1970s to a lower level of 180,000–240,000 tons in recent years (1977–1983, except for 1980). There have also been increased catches of other species of squid in recent years, such as the flying squid in the northern part of the Pacific as well as various other species from distant-water fishing grounds.

The Pacific catches of Japanese common squid from the Sanriku and Hokkaido fishing grounds (Fig. 1, areas 3–6) made up 70–90% of the total catch during 1952–1969. However, the Pacific catches gradually declined in the 1970s and

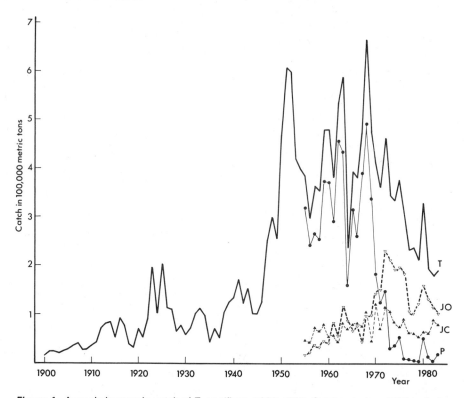

Figure 4. Annual changes in catch of *T. pacificus*, 1900–1983. Catches before 1951 include other squid and cuttlefish. *T*, total by all kinds of Japanese fisheries; *P*, in the Pacific by jigging fisheries; *JC*, in the coastal waters in the Japan Sea by jigging fisheries; *JO*, in the offshore waters in the Japan Sea by Japanese and Korean jigging fisheries.

averaged only around 21,000 tons (1973–1984); constituting a mere 6% of the 1960–1969 average catch of 352,000 tons.

The catches of Japanese common squid in the Japan Sea (including catches made by Korea) increased substantially from a low level of 66,000 tons in 1955 to some 200,000 tons in 1963 and again in 1968, then jumped further to establish a new record of 350,000 tons in 1972. After that, the catches settled at around 280,000 tons between 1974 and 1976, then decreased to 170,000–230,000 tons in and after 1977. An examination of the catches in the various coastal and offshore fishing grounds (Fig. 1, areas 1–3, 6 and areas 4, 5, 7) shows that the catches in the coastal fishing grounds have ranged from 50,000 tons to 120,000 tons (average 76,000 tons in 1955–1983) and have not shown particularly wide fluctuations. On the other hand, the catches on the offshore grounds have nearly doubled each year after 1966 and by 1972 reached 200,000 tons. After that, however, the catches began to decline gradually and fell to 60,000–120,000 tons/yr in 1977–1983.

3. BIOLOGICAL CHARACTERISTICS OF THE JAPANESE COMMON SQUID

On the basis of the spawning seasons, growth types, and migratory patterns, three groups (subpopulations) of Japanese common squid are believed to exist in Japanese waters. These are the winter-spawning, autumn-spawning, and summer-spawning groups (Araya 2; Okutani 3; Osako and Murata 4). The migratory patterns of the former two groups are shown in Figure 5.

The winter-spawning group is distributed most widely, and constitutes the main portion of the catch of this species in Japanese coastal waters. Its stock size was the largest among the three groups and recruitment to the coastal waters of Sanriku and Hokkaido was especially high until around 1970. The spawning grounds are situated in the waters off southwestern Japan, centered around the central and northern East China Sea, and the main spawning season is in the period January–March.

The autumn-spawning group is mainly distributed in the offshore waters of the Japan Sea. This group has been the principal component of the catch of Japanese

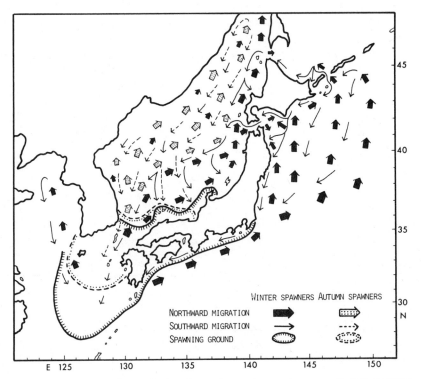

Figure 5. Schematic diagram of migration routes and spawning grounds of *T. pacificus*.

common squid since the 1970s when it replaced the winter-spawning group. The mantle length of the mature individuals is the largest among the three groups. The spawning ground extends from the northern East China Sea to the southwestern Japan Sea, and the peak spawning season is believed to be from September through November.

The summer-spawning group should probably be referred to as the so-called "local" group. It is distributed in the coastal waters of Japan and its stock size is extremely small compared to the other two groups. The mantle length of the mature individuals is the smallest of all three groups. The spawning grounds are located in the coastal waters of southwestern Japan Sea and in the Pacific coastal waters of central Honshu. The peak spawning season is believed to be from May through August.

The growth in mantle length of the winter-spawning group may be represented by a sigmoid curve fitted to the average monthly mantle lengths of squid caught off southern Hokkaido (May–November) and near Tsushima Islands (January), as seen by the equation (Araya and Ishii 5)

$$L = 24.5/(1 + \exp(0.3282 - 0.7171t))$$

where L is the mantle length (cm) and t is an integer between -4 and 7 (*n*th calendar month $- 5$).

From this and the fact that the mean mantle lengths of sexually mature individuals of the winter-, autumn-, and summer-spawning groups were approximately 24, 27, and 23 cm, respectively, the growth patterns for the three groups are as shown in Figure 6.

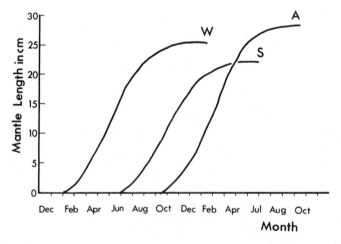

Figure 6. Growth patterns for the three groups of *T. pacificus*. *W*, winter-spawning group; *S*, summer-spawning group; *A*, autumn-spawning group.

The following equation shows the relationship between mantle length and body weight for catches off southern Hokkaido (Murata 6):

$$W = 0.091L^{3.2472}$$

where W is the body weight (g) and L is the mantle length (cm).

The males mature about 2–3 months earlier than the females and enter into copulatory behavior with the immature females. Spawning is believed to take place around 2 months after copulation. Following copulation and/or spawning, both males and females die of physical exhaustion (Hamabe 7; Hayashi 8). Judging from the above information, as well as results of large-scale tagging experiments, it appears that the life-span of the Japanese common squid is just about 1 yr (Araya 2; Okutani 3).

4. FORECASTING FISHING CONDITIONS

4.1. Methodology

Fishing condition forecasts are being carried out principally for the north Pacific fishing grounds (generally north of 38°N), including the Okhotsk Sea, and for the Japan Sea fishing grounds. The forecasts are made twice each year in June–July and in September. The first is for the summer fishing season (June–September) and the second for the autumn–winter season (October–January). The fishing condition forecasts are made separately for the Pacific and the Japan Sea fishing grounds by fisheries scientists of various official research organizations under the leadership of the Hokkaido Regional Fisheries Research Laboratory and the Japan Sea Regional Fisheries Research Laboratory, respectively. However, forecasts made beyond a year in advance, as well as the very short-term forecasts made only 5–10 days in advance, are not included here because they are still experimental in nature.

The total amount of recruitment is predicted through the synthesis of the various informations listed below in paragraphs 1–6. Based on these data, together with information from paragraph 7, the start and end of the fishing season, location of the main fishing ground, catch per unit effort (CPUE), and total catch are predicted for the various fishing grounds. There are three fishing grounds in the Pacific: Sanriku, south of Hokkaido, and east of Hokkaido. There are four fishing grounds in the Japan Sea: the respective coastal waters of Kyushu, Honshu, and Hokkaido, and the offshore waters (Fig. 1). The main types of information needed for the forcasts are as follows.

1. Biological Characteristics. Among the aforementioned biological information, knowledge of the distribution of the various population groups, their migratory pattern (Fig. 5), and the ecology and growth at the various developmental stages of the squid are quite important.

2. Changes in Stock Size and in the Abundance of Adult Squid. Trends in stock size may be examined separately by sea area and population group using the method described in the next section. Because the data that ideally would be needed to estimate the abundance of adult squid are quite sparse, the catches made in a specified period and sea area or the corresponding CPUEs are currently used as indexes of abundance. In particular, abundance of the adults of the winter-spawning group is indicated by the amount of winter catch in the coastal waters of southern Japan, or by the catch per fishing per vessel (U) in the coastal waters of northern Japan. U value in a given year has been reported to show a close relationship to the larval index (the index of larval density, L_p) in January–March of the following year for the Pacific coastal waters of southern Japan, which suggests that these are adults and larvae from the same population group. A positive relationship was found between U of one year and L_p of the following year (coefficient of correlation $R = 0.63$; 5% level of significance) (Okutani and Watanabe 9).

L_p was calculated from the results of plankton surveys (net with 60-cm mouth diameter), involving vertical tows from 150 m to the surface, using the equation

$$L_p = \Sigma \ (N_i \times A_i)/\Sigma \ A_i$$

where N_i is the average number of larvae collected per tow in the ith 1° square of latitude and longitude, A_i is the area of the ith square block, and $\Sigma \ A_i$ is the total area of the waters surveyed (east of 130°E).

There was also a very close relationship between the total catch (C) of adult squid in the offshore waters of the Japan Sea in a given year, and the larval index (L_j) during the autumn season of the same year in the northern East China Sea and the southwestern Japan Sea, which indicates that these were adults and larvae from the autumn-spawning group. There was a very strong positive correlation (Fig. 7) (Japan Sea Regional Fisheries Research Laboratory 10). L_j was calculated from the results of larval collections, oblique tows by plankton net (with 80-cm mouth diameter) from 75 m to the surface, by means of the equation

$$L_j = \Sigma \ (L_i \times B_i)/\Sigma \ B_i$$

where L_i is the number of larvae per 1000 m^3 of water and B_i is the number of 10′ × 10′ squares of latitude and longitude within each contoured level of L_i.

3. Abundance of Larvae. The values of L_p and U in the same year were reported to correspond to larval abundance and total recruitment of the winter-spawning group to the Pacific fishing grounds. There was a positive relationship ($R = 0.68$; significant at the 5% level) between the two variables (Okutani and Watanabe 9). On the other hand, the abundance of larvae and the total recruitment of the autumn-spawning group in the Japan Sea reportedly corresponded to L_j of one year and C of the following year. A positive relationship was found between the two variables (Fig. 7).

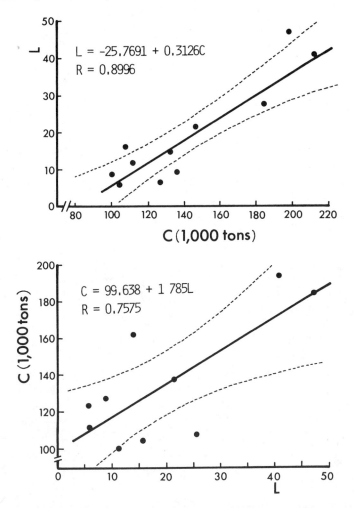

Figure 7. Relationship between total catch (C) of *T. pacificus* in the offshore waters in the Japan Sea and larval index (L) for the waters from the northern East China Sea to the southwestern Japan Sea in autumn (Japan Sea Regional Fisheries Research Laboratory 10). Upper figure: relationship between C and L in the same year, 1973, 1975–1984. Lower figure: relationship between L and C in the next year, 1973, 1975–1983. R = correlation coefficient.

4. Information Obtained from the Squid Jigging Fishery. Information such as the total catch and the CPUE on the various fishing grounds was compiled, and the relationship between the fishing grounds examined.

5. Experimental Fishing by Research Vessels. Experimental jig fishing has been carried out in the Japan Sea and in the Pacific since 1972 by 5–10 research vessels in June and in August–September. The vessels surveyed over a grid of stations. The survey results were useful in determining the distributional pattern of the squid, as well as in estimating the prerecruitment standing stock in the offshore areas. The stock size index (P) and the stock density index (D) were calculated as follows:

$$P = \Sigma \, (U_i \times B_i) \qquad D = P/\Sigma \, B_i$$

where U_i is the catch in numbers per automatic jigging machine per hour of fishing at the various survey stations, and B_i is the number of 1° squares of latitude and longitude within each contoured level of U_i.

The values of both variables for June as well as September in the Japan Sea were very closely related to the catch per fishing day per vessel in the summer and autumn fishing seasons (Fig. 8). The annual variation in the values for August and September in the Pacific generally corresponded to the annual variations in the total Pacific catches (Murata 11; Araya 12).

6. Biometric Information. By means of biometric data obtained from the catches, it is possible to examine the population structure, growth, and migrations of the recruiting cohort.

7. Oceanographic Conditions. The start and end of the fishing season and the position of the main fishing ground can be examined by means of information on oceanographic conditions. In other words, the recruitment of the Japanese common squid onto the Sanriku and Hokkaido fishing grounds in the Pacific during summer and in autumn–winter is considered to be closely related to the movements of the Kuroshio warm water branch current, and the southerly moving branch of the Oyashio (Araya 13). The position and movements of the Polar Front and of the warm and cold water masses over the Japan Sea fishing grounds are considered very important oceanographic factors (Kasahara 14).

4.2. The Present Situation

Table 1 shows an extract from the 1983 forecasts. Resources trends and the basic information used in making the forecasts are reported at the same time but are not included here. Furthermore, only the forecasts of fishing conditions for the offshore area of the Japan Sea are presented here.

The forecast of total recruitment is shown as a relative value and is believed to be quite accurate. For example, in reference to the accuracy of the 1972–1983 total recruitment forecasts for the Pacific fishing grounds, Murata and Shingu (15) divided the results of these forecasts into three categories as follows: "on target," "off target," and "results uncertain," and found that in the first series of forecasts, 58% were "on target," whereas in the second series the score had improved to 100% "on target." Similar results were obtained in the Japan Sea forecasts, with the first series scoring lower in accuracy than the second.

The forecasts of recruitment to various fishing grounds are shown as the numer-

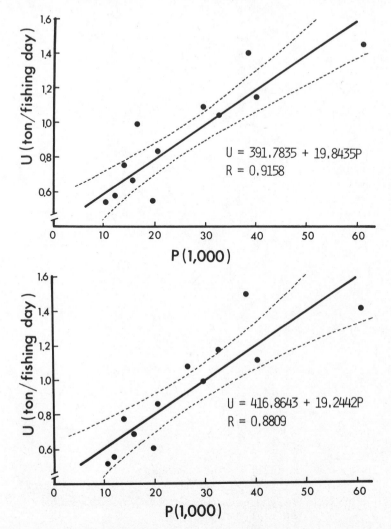

Figure 8. Relationships between stock size index (*P*) calculated from the results of jigging grid survey and catch per fishing day (*U*) by medium-sized vessel for *T. pacificus* in the Japan Sea, in 1972–1984 except 1977 (Japan Sea Regional Fisheries Research Laboratory 10). Upper figure: relationship between *P* in June and *U* in May–August. Lower figure: relationship between *P* in June and *U* in May–December.

ical values of catches and CPUE by fishing ground. However, because catches and CPUE are both affected by oceanographic conditions as well as by fishing effort, the accuracy of these forecasts is often lower than when predicting total recruitment. Furthermore, the accuracy is lower in the *first* forecast than in the *second*, and there is also the tendency for the accuracy to be lowest for the fishing grounds near Hokkaido, which lie at the extremity of the squid distribution.

TABLE 1 Portion of Fisheries Forecasts in 1983

Forecast	Pacific Fishing Grounds off the Sanriku–Hokkaido	Offshore Fishing Grounds in the Japan Sea
First	(Published on July 6) 1. Total recruitment: almost equal to those (very low level) in 1976–1982 except 1980 2. Total catch in July–September: 2000–12,000 tons	(Published on June 24) 1. Total recruitment: same level as those in 1981 and 1982, or a little lower 2. Catch in June–September: Around 34,000 tons, or a little lower
Second	(Published on September 21) 1. Total recruitment: same as that in first forecast 2. Total catch in October–December: 2000–7000 tons	(Published on September 28) 1. Total recruitment: same level as that in 1982, or a little higher 2. Catch and catch per fishing day in October–December: around 27,000 tons and 890–1100 kg/vessel · day, or a little higher

5. STOCK ASSESSMENT AND RESOURCE MANAGEMENT

5.1. Annual Variations in Stock Density

Annual variations in stock density can be examined from research and commercial statistics and from long-term fishery trends, leading to the estimation of stock size in the next section.

1. Analysis of Catch Statistics. Judging from the distribution and migration of Japanese common squid, from population analysis, and from the history of the development of the squid jigging fishery, it appears that annual catches in the Pacific since 1960 and on the offshore fishing grounds of the Japan Sea since 1973 are representative of the stock sizes of the winter- and autumn-spawning groups, respectively. The annual variations in total catch were discussed above.

The average value of catch per fishing day per vessel (U) during a fishing season is another indicator of stock size and is analyzed separately by vessel size class and by fishing ground. Figure 3 shows the annual variation in U values standardized for all national fishing grounds from 1953 to 1983. According to this figure, after 1964, U values declined and fell to an extremely low level in the 1970s. The relationship between the total annual number of days fished and the total catch on the Pacific fishing grounds off northern Japan shows clearly that U values in 1979–1984 were very much lower than those during 1953–1963 (Fig. 9). The annual

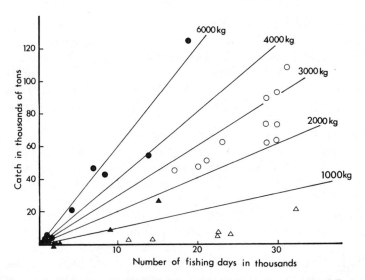

Figure 9. Relationships between fishing days by jigging boats and catch of *T. pacificus* in the Pacific coast of Japan by years and areas. The lines in the figure show catch per fishing day in kilograms. (1) In the period 1953–1963 (20–50 GRT jigging boats; Shingu et al. 18): ●, off eastern Hokkaido; ○, in the waters from southern Hokkaido to northeastern Honshu. (2) In the period 1979–1984 (5–99 GRT jigging vessels): ▲, off eastern Hokkaido; △, in the waters from southern Hokkaido to northeastern Honshu.

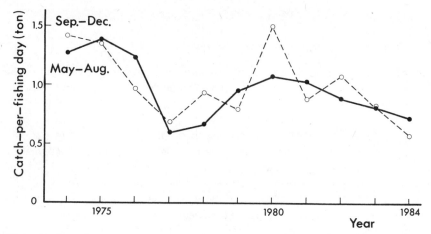

Figure 10. Annual changes in catch per fishing day by 99-GRT jigging boat for *T. pacificus* offshore in the Japan Sea, 1974–1984.

variations in *U* values shown in Figure 3 and 9 are believed to represent variations in stock density of the winter-spawning group. On the other hand, *U*-values of offshore fishing grounds in the Japan Sea are considered to reflect the stock density of the autumn-spawning group. According to Figure 10, *U* values of the 99 GRT vessels for the summer- and autumn-fishing seasons have both declined since 1976–1977.

U-values of medium-sized fishing vessels, tabulated by 10-day periods in 30′ × 30′ squares of latitude and longitude, were totaled for the entire fishing ground and fishing season to calculate the stock size index (*P′*). *P′* represents one of the indexes of total recruitment into the fishing ground, and *P′* values for the entire Japan Sea fishing grounds in 1978–1984 show a decline after 1981. The average *P′* value for 1982–1984 was only around 60% of that for 1978–1980.

2. Analysis of the Research Vessel Survey Results. Judging from the results of larval collections by NORPAC net and 80-cm plankton net in the northern portion of the East China Sea and the southwestern Japan Sea during the autumn, the main distribution of larvae prior to around 1973 was in the waters west of the Tsushima Straits. However, subsequent to 1977, the larval density in this area declined to virtually nothing (except for 1980) (Fig. 11), and the larval density index (*L_j*) of 16–47 in 1973–1976 declined to 6–16 in 1978–1984 (except 1980).

From the grid fishing survey in the Japan Sea, the stock size index (*P*) was calculated as shown in Table 2. Although there were some differences in the survey area from year to year, *P* values generally declined each year between 1971 and 1976, increased slightly in 1979 and 1980, and again declined to a low level in and after 1981. The above results are believed to represent the stock density of the autumn-spawning group in the Japan Sea.

In the Pacific, the larval density index (*L_p*) in the Kyushu–Shikoku offshore

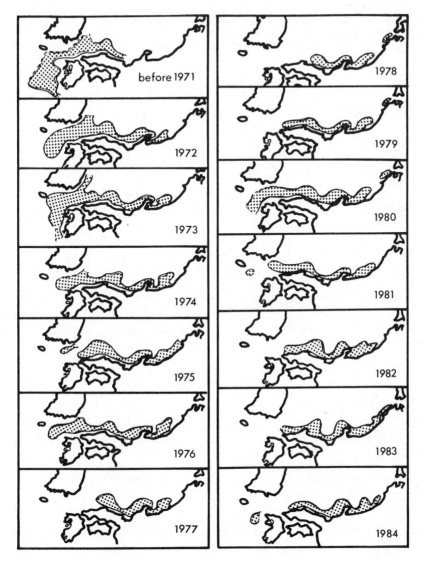

Figure 11. Annual changes in the range of distribution of *T. pacificus* larvae bred in autumn (Japan Sea Regional Fisheries Research Laboratory 10).

areas in January–March as well as the stock size index (P) and stock density index (D) based on the grid survey results in the Sanriku–Hokkaido areas all represent the stock density of the winter-spawning group. However, because variations in L_p and in P and D values for June are considerable, these are not always the most suitable data for analysis, and improvements in the survey methods are evidently needed. Despite differences in survey areas among the years, however, the values

TABLE 2 Annual Changes of Stock Size Index (P) of T. pacificus Based on Grid Survey Results in the Japan Sea in 1971–1984

Year	P × 1000 June[a]	July[b]	September[b]	Year	P × 1000 June[a]	July[b]	September[b]
1971	100.9	—	67.3	1978	19.9	27.4	—
1972	61.0	—	66.6	1979	20.8	34.5	36.0
1973	38.5	—	71.9	1980	26.5	39.3	44.2
1974	32.7	64.8	57.3	1981	14.0	23.5	42.3
1975	40.3	30.2	49.3	1982	15.8	27.9	26.0
1976	29.6	43.5	29.3	1983	12.2	32.5	38.5
1977	—	—	—	1984	10.7	22.7	27.9

[a]Survey area: East of 130°30'E, 36°–42°30'N, exclusive of the 200-mile limits of the Soviet Union and North Korea since 1978.

[b]Survey area: East of 130°30'E, 36°–45°30'N, exclusive of the 200-mile limits of North Korea since 1978.

TABLE 3 Annual Changes of Stock Size Index (P) and Stock Density Index (D)[a]

Year	A	P × 100	D	Year	A	P × 100	D
1968	275	104.5	38.06	1978	830	0.3	0.04
1969	287	80.5	28.06	1979	830	0.5	0.06
1973	1000	7.8	0.72	1980	830	4.6	0.55
1974	832	5.1	0.61	1981	830	3.3	0.39
1975	690	21.1	3.06	1982	830	0.9	0.11
1976	590	0.6	0.10	1983	830	3.6	0.44
1977	830	1.6	0.19	1984	830	6.0	0.72

[a]Based on the grid survey results in the Sanriku–Hokkaido area in August–September, 1968–1984. A = total number of 10' × 10' squares of latitude and longitude in survey area; survey area includes the 200-mile limits of the Soviet Union in 1968–1975, but excludes them since 1976; based on the results of only one research vessel.

of P and D for August and September in 1968 and 1969 were considerably higher than those for the years after 1976 (Table 3).

3. Resource Trends. From the various kinds of information discussed above, it appears that the stock density on the Pacific fishing grounds was at a relatively high level from the 1950s to the 1960s. However, it had declined in the 1970s, falling to very low levels after 1973. This kind of variation seems to reflect the stock size of the winter-spawning group. On the other hand, the stock density in the Japan Sea declined after 1977, but recovered temporarily in 1980, only to drop to former low levels thereafter. This variation seems to follow the stock size of the autumn-spawning group.

5.2. Estimation of Stock Parameters and Causes of Stock Variations

1. Estimation of Stock Parameters. The main stock parameters of Japanese common squid include stock size N and biomass C, exploitation ratio (E), fishing mortality coefficient (F), and natural mortality coefficient (M). These parameters were estimated as follows.

Stock size biomass was estimated for the Japan Sea fishing grounds at the end of June (1972–1974) by the De Lury method (Kasahara 16). The estimated N values ranged from 1496×10^6 to 1920×10^6, and the corresponding stock biomasses ranged from 313×10^3 to 421×10^3 tons. The stock size at the start of the fishing season (May) in the offshore fishing grounds of the Japan Sea has been estimated by means of cohort analysis utilizing the monthly catches in number (assuming $M = 0.05$/month, $F = 0.05$/month) (Fishery Agency of Japan 17). The estimated values of N ranged from 1298×10^6 to 1363×10^6 for 1978–1980, 1014×10^6 to 1076×10^6 for 1981 and 1983, and 1636×10^6 for 1982.

On the other hand, the stock size in weight (C) was estimated for the Sanriku and Hokkaido fishing grounds in the Pacific for the start of the fishing season (July) from changes in the monthly accumulated catches in number (Shingu et al. 18). The estimated values ranged from 800×10^3 to 1100×10^3 tons for 1964–1966, 700×10^3 to 1000×10^3 tons for 1967–1969, 200×10^3 tons for 1970–1975, and 50×10^3 to 150×10^3 tons for 1976–1981.

The monthly F and M as well as the value of E for the fishing season were estimated from the relationship between the annual total number of days fished by the jigging vessels and the annual average catch per fishing day per vessel. Araya (19) reported an increase in E (for an approximately 8-month period) of the winter-spawning group from 0.38–0.45 in 1952–1959 to 0.40–0.50 in 1965–1971. From the relationship between recruitment (in numbers) at the start of the fishing season as derived by the aforementioned De Lury method, and the total catch in numbers in July–November, the value for the exploitation ratio E in the Japan Sea (mainly the autumn-spawning group) was estimated at 0.36–0.45 for 1972–1974. On the basis of the biomass analysis method as well as by assuming a decrease in the stock size of winter-spawning group to one-tenth of the original level between 1968 and 1978, Doi and Kawakami (20) estimated $F = 0.025$, $M = 0.431$, and $E = 0.054$ (June to February) for the latter half of the 1970s. From the results of tagging experiments carried out in the Japan Sea in 1979, Machinaka et al. (21) estimated $Z = 0.13$–0.14, $F = 0.0063$–0.0069, $M = 0.13$, and $E = 0.047$–0.049 (over a 5-month period). However, the authors pointed out that their estimated values of F are much too small, and their estimate of M seems too much large. Shingu et al. (18) estimated the values of Z, F, and E, based on changes in the monthly accumulated catches in number and the value of $M = 0.25$ for the northern part of the Pacific. They reported that the Z, F, and E (6-month period) had increased from 0.3, 0.05, and 0.14 for 1964–1966 to 0.6, 0.35, and 0.57 for 1973–1981, respectively.

2. Causes of Stock Variations. In the annual variations in catch from 1910 to 1970, periodic cycles of approximately 9 and 27 yr have been surmised (Murata and Araya 1; Kawasaki 22; Yasui 23). Furthermore, the 9-yr cycle has been used as a basis for forecasting total recruitment of the winter-spawning group (Araya 12; Yasui 23). However, as mentioned above, the stock density of the winter-spawning group declined sharply after 1970, as did the stock density of the autumn-spawning group after 1977. On the other hand, it seems that the total fishing effort in the squid jigging fishery increased steeply during the 1960s and had reached a very high level in the 1970s. The above seems to suggest that natural factors and increased fishing intensity both have been closely tied to the variations in the stock abundance of this species.

Available information on natural mortality is very poor. There are only a few reports on the predation on larval and juvenile squid by large predators such as bluefin tuna (Matsuda et al. 24), or on the relationship between long-term changes in oceanographic conditions and natural mortality (Yasui 23; Kitano 25). There is a positive relationship between stock density in a given year and larval abundance, as well as the amount of recruitment in the following year, for both the winter-spawning and autumn-spawning groups. However, the correlation coefficient between stock density and larval abundance was reported to be higher than that between larval abundance and subsequent total recruitment (Okutani and Watanabe 9; Japan Sea Regional Fisheries Research Laboratory 10). Judging from the above, and considering the ecology of the Japanese common squid, it seems that the natural mortality of larval squid prior to recruitment may be very high in certain years. However, it is estimated that the natural mortality of juvenile and adult squid is in general very small as compared with the recent level of fishing mortality.

Since the introduction of efficient fishing methods in the latter half of the 1960s both the start of the fishing season and its peak have tended to arrive earlier in the year. For example, the April–June catch in 1952–1961 made up only 27% of the total catch from the coastal waters of the Japan Sea, but this proportion increased to 49% in 1971–1975, indicating an increase in early-season catch. Furthermore, the peak monthly catch in the vicinity of Hokkaido occurred in October–November during 1952–1966, but had shifted to October in 1967–1974 (Murata and Araya 1). Similarly, the mode in the monthly catches made in offshore waters of the Japan Sea occurred in August in 1971 but shifted to June in 1975 (Kasahara 16). The autumn–winter catches in the coastal waters of western Japan and of Korea, which are mainly composed of squid migrating southward from waters off northern Japan, have fallen off very sharply in recent years.

From the various types of information discussed above and from past knowledge of the biological characteristics of the squid, I have made the following assumptions. Stock variations in Japanese common squid were largely caused by natural factors prior to the 1960s. However, the 1970s, which happened to correspond to a naturally "low period" for the resources, also saw a rapid increase in fishing intensity that caused the winter-spawning group to decrease very sharply. This was later followed by a similar decrease in the stock of the autumn-spawning group in the Japan Sea. In summer–autumn, the winter-spawning group occurred mainly in a relatively narrow area in the coastal waters of northern Japan at depths of around 200 m. By contrast, the autumn-spawning group occurred in a very broad area in the offshore areas of the Japan Sea, and thus was not as greatly affected by the increased fishing intensity as was the winter-spawning group.

5.3. Resource Management

1. Historical and Present Situations. Japan's squid jigging fishery began as a "free fishery" in which anyone could participate without restrictions. However, with new developments in the method of tracking migrating squid in the late 1960s, the vessels began to converge in large numbers on productive fishing grounds. This resulted in increased competition among squid fishing vessels. In order to resolve this, the Ministry of Agriculture, Forestry and Fisheries established a "ministry approval system" regulating the operations of large (larger than 100 GRT) and medium-sized (30–99 GRT) squid jigging vessels. The regulations required (a) ministry approval for the operations of large vessels after 1969 and medium-sized vessels after 1972, (b) specified approved fishing grounds, and (c) established periods (March and April) when no fishing would be permitted.

However, these ministry regulations were intended primarily for preventing disputes among vessel operators and for improving the management of fishery operations. The protection of the resources and the effective utilization of fishery resources through regulation of fishing effort or catches were not the main consideration at that time, and neither have they been to this day. The same thing can be said for most of the other coastal fisheries in Japan. Rather than resource management, the priority has been placed in the efficient management of fishery operations. This has traditionally been the basic fishery policy in Japan.

2. Future Directions in Resource Management. Doi and Kawakami (20) examined the relationship between fishing mortality coefficient and reproduction potential and postulated that the drastic decline in the winter-spawning stock between the late 1960s and the 1970s was due to an increase in fishing effort of about 2.5 times during that period.

From the relationship between total catch and fishing effort and their annual variations in the northern Japan Sea, Araya (12) reported that an upper limit to fishing intensity was attained late in the 1960s.

Shingu et al. (18) examined the relationship between yield per recruit and the fishing mortality coefficient and recommended that the beginning of the fishing season for the winter-spawning group should be moved to August–September (Fig. 12).

Hasegawa (26) calculated the economic break-even point occurred when the supply of Japanese common squid was around 600,000 tons and reported that the corresponding fishing effort necessary to yield the maximum sustained economic yield (MEY) occurred at around the 1965 level of effort, or at roughly half the fishing effort exerted in 1979 and 1980.

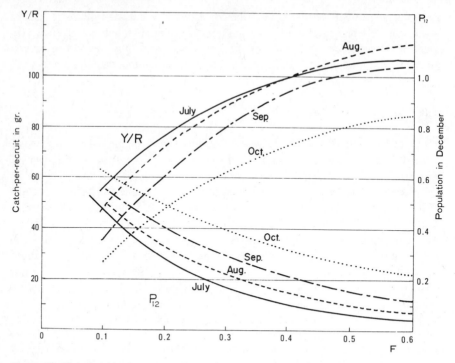

Figure 12. Relationships between fishing mortality (*F*) and yield per recruit (Y/R, upper lines) and population in December (P_{12}, lower lines) expressed as ratio to population in July. Months represent the beginnings of first capture (Shingu et al. 18).

From the above, I believe that the following kinds of countermeasures are necessary in order to bring about a recovery of the resources of the Japanese common squid and to improve the management of the fisheries:

(a) The reduction of total fishing effort by squid jigging vessels to about one-half of the present level.

(b) The prohibition of fishing in the Pacific fishing grounds for 1 or 2 yr.

(c) The extension of the prohibited period (period of no fishing) to December–June.

Because the life-span of this squid is about 1 year, the measures listed above should begin showing results in 1 or 2 yr. Therefore, future resource management should be responsive to these results, as well as to future resource trends, and fishing effort and catch should be "adjusted" by trial and error in order to achieve the optimal level. The vital part played by forecasts of fishing conditions in the management of squid fisheries is also emphasized from experience with this fishery.

ACKNOWLEDGMENTS

The author wishes to thank Dr. T. Otsu of Hawaii and Dr. T. Sato, Head of Resources Division of Hokkaido Regional Fisheries Research Laboratory, Fisheries Agency, for reading the manuscript with suggestions for its improvement.

REFERENCES

(JJ), title and text in Japanese; (Je), in Japanese with English title; (JE), in Japanese with English title and summary.

1. M. Murata and H. Araya, Some problems on winter population stock of common squid, *Todarodes pacificus*. Report of 1976 annual meeting on resources and fisheries of squids. *Jpn. Sea Reg. Fish. Res. Lab.* 1–13 (1977) (JJ).

2. H. Araya, *Resources of Common Squid*, Todarodes pacificus *Steenstrup in the Japanese Waters*, Fish. Res. Ser., Vol. 16. Jpn. Assoc. Mar. Resour. Conserv., Tokyo, 1967, p. 60 (Je).

3. T. Okutani, *Cephalopod Life Cycles*, Vol. 1. Academic Press, London, 1983, pp. 201–214.

4. M. Osako and M. Murata, *Advance in assessment of the world cephalopod resources. FAO Fish. Tech. Pap.* **231**:55–144 (1983).

5. H. Araya and M. Ishii, Population structure of common squid in the waters around Hokkaido. *Res. Rep., Tech. Couns. Agric. For. Fish.* **57**:192–205 (1972) (JJ).

6. M. Murata, The relation between mantle length and body weight of the squid, *Todarodes pacificus* Steenstrup. *Bull. Hokkaido Reg. Fish. Res. Lab.* **43**:33–51 (1978) (JE).

7. M. Hamabe, Exhaustion process of the genital organs of common squid, *Ommastrephes sloani pacificus*. *Bull. Jpn. Sea Reg. Fish. Res. Lab.* **11**:53–64 (1963) (JE).

8. Y. Hayashi, Studies on the maturity condition of the common squid. III. Ponderal index and weight indices of internal organs during maturations and exhaustion. *Bull. Jpn. Soc. Sci. Fish.* **37**:960–963 (1971) (JE).

9. T. Okutani and T. Watanabe, Stock assessment by larval surveys of the winter population of *Todarodes pacificus* Steenstrup (Cephalopoda: Ommastrephidae), with a review of early works. *Biol. Oceanogr.* **2**:401–431 (1983).

10. Japan Sea Regional Fisheries Research Laboratory, Data on the fisheries and oceanic conditions of common squid in the Japan Sea in 1985. Research data on *Jpn. Sea Reg. Res. Lab.* **85-02**:1–40 (1985) (JJ).

11. M. Murata, Quantitative assessment of oceanic squid by means of jigging surveys. *Biol. Oceanogr.* **2**:433–456 (1983).

12. H. Araya, History of squid jigging fishery and changes in stock size of *Todarodes pacificus*. Report of 1984 annual meeting on resources and fisheries of squids. *Hokkaido Reg. Fish. Res. Lab.*   :1–6 (1985) (JJ).

13. H. Araya, Migration and fishing ground of winter sub-population of the squid, *Todarodes pacificus* Steenstrup, in the northern waters of Japan. *Bull. Hokkaido Reg. Fish. Res. Lab.* **41**:119–129 (1976).

14. S. Kasahara, Descriptions of offshore squid angling in the Sea of Japan, with special reference to the distribution of common squid (*Todarodes pacificus* Steenstrup); and on the techniques for forecasting fishing conditions. *Bull. Jpn. Sea Fish. Res. Lab.* **29**:179–199 (1978).

15. M. Murata and C. Shingu, Present status and future aspects of fishery forecasting for Japanese common squid, *Todarodes pacificus* in the Pacific. Report of pelagic fish section ("Ukiuo-Bukaiho") of *Fish. Resour. Invest. Sci. Fish. Agency Jpn. Gov.* **17**: (1986) (JJ).

16. S. Kasahara, Some problems on autumn population stocks of common squid, *Todarodes pacificus* in the Japan Sea. Report of 1976 annual meeting on resources and fisheries of squids. *Jpn. Sea Reg. Fish. Res. Lab.*   :25–37 (1977) (JJ).

17. Fishery Agency of Japan, *Stocks of Common Squid in the Japan Sea*. Report of investigations on major fishing resources in Japanese coastal waters in 1983. Fishery Agency, Research Division, 1984, pp. 175–84 (JJ).

18. C. Shingu, M. Murata, and M. Ishii, Changes in catch of common squid, *Todarodes pacificus* Steenstrup, in the Pacific coasts of Japan. *Bull. Hokkaido Reg. Fish. Res. Lab.* **48**:21–36 (1983) (JE).

19. H. Araya, Some problems of the conservation for the jigging fishery resource of a common squid, *Todarodes pacificus* Steenstrup in Japan. *Rep. Fish. Resour. Invest. Sci. Fish. Agency Jpn. Gov.* **16**:71–78 (1974) (JE).

20. T. Doi and T. Kawakami, Biomass of Japanese common squid *Todarodes pacificus* Steenstrup and the management of its fishery. *Bull. Tokai Reg. Fish. Res. Lab.* **99**:65–83 (1979) (JE).

21. S. Machinaka, M. Miyashita, H. Miyajima, and S. Kasahara, Tagging experiments for the common squid (*Todarodes pacificus*) in the offshore areas of the Japan Sea, with estimation of parameters in dynamics of squid population. *Bull. Ishikawa Prefect. Fish. Exp. Stn.* **3**:37–52 (1980) (JE).

22. T. Kawasaki, Periodicity in the fluctuation of the fisheries resources. *Mar. Sci., Kaivo Shuppan Co.* **10**:50–54 (1973) (JE).

23. T. Yasui, Fluctuations in abundance of the winter sub-population of *Todarodes pacificus*. Expert consultation on fishing for squid and other cephalopods. *FAO Fish Rep.* **170**(Suppl. 1):24–29 (1976).

24. S. Matsuda, F. Hanaoka, T. Kato, and M. Hamabe, Recruitment and underlying mechanism of common squid stock in the southwestern waters of Japan. *Res. Rep., Tech. Couns. Agric. For. Fish.* **57**:10–30 (1972) (JJ).

25. K. Kitano, Note on the fluctuation tendency of the total catch of the common squid *Todarodes pacificus* Steenstrup in the light of the unusual oceanic conditions. *Bull. Hokkaido Reg. Fish. Res. Lab.* **44**:73–76 (1979).

26. A. Hasegawa, *Economic Structure of Squid Jigging Fishery*. Jpn. Fish. Assoc., Tokyo, 1982, pp. 353–390 (JJ).

CASE STUDIES

D / OTHER INVERTEBRATES

28 PRECIOUS CORAL FISHERIES OF THE PACIFIC AND MEDITERRANEAN

Richard W. Grigg
Hawaii Institute of Marine Biology
University of Hawaii at Manoa
Coconut Island, Kaneohe, Hawaii

1. Introduction
2. Natural History of Precious Corals
3. Methods of Fishing
4. Catch Statistics
5. Conservation and Management
6. Future Outlook of the Industry
   References

1. INTRODUCTION

The precious coral fishery is one of the oldest fisheries known to man. Records of human use of precious coral date back to Paleolithic times, about 25,000 years ago (Tescione 1). Presently, the fishery has two major centers, the Mediterranean Sea and the far western Pacific Ocean. The Mediterranean fishery has been in existence for almost 5000 years, although at a very low level until about A.D. 1000 when a dredging device called the engegno was invented by the Arabs. Precious coral was not discovered in the Pacific until the early nineteenth century and did not flourish until after the Meiji Reform in Japan in 1868, which permitted fishermen to engage in private enterprise. A more detailed account of the history of the fishery can be found in Liverino (2) and Grigg (3).

Annual global production of precious coral is somewhat variable depending on supply and demand, but in recent years it has ranged approximately between 100

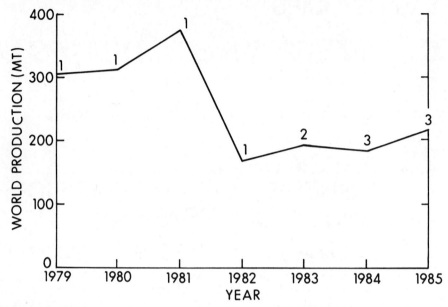

Figure 1. Approximate world annual production of precious coral (all species of *Corallium*) for the 1979–1985 period. Numbers refer to data sources: 1, Grigg 3; FAO 4. 2, Grigg 10; Masso 11. 3, In literature of C. Chu, Liven Coral Co, Taiwan; and H. Yokoyama, All Nippon Coral Fishing Association, Tokyo and Kochi, Japan.

and 400 tonnes (See Fig. 1). Most of the production consists of species of *Corallium*. The production of all other species (Table 1) combined consists of less than 60 tonnes. For about the last 20 years, more than 70% of the world's harvest has come from the Pacific Ocean, largely from the Emperor Seamounts north of Midway Island, in the Hawaiian–Emperor chain.

The variable nature of annual production illustrates the boom and bust nature of the fishery and emphasizes the need for conservation and management of the resource. Variable production is caused primarily by an imbalance between rates of harvest and rates of growth and recovery of the resource. Like most fishery resources, successful management requires a full understanding of the life history of the resource as well as the socioeconomics of the fishery. This chapter is organized to consider initially the natural history of various species, followed by a description of the fishery, recent trends in yield, health of the stocks, attempts to manage the resource, socioeconomics of the fishery, and the future of the industry including recommendations for conservation.

2. NATURAL HISTORY OF PRECIOUS CORALS

Precious corals consist of a variety of species of red or pink coral belonging to the genus *Corallium* (Fig. 2) as well as several other groups including black (*Antipathes* spp.), gold [*Gerardia* (=*Parazoanthus*) spp.], and bamboo corals (*Lepidisis* and

TABLE 1 Major Species of Precious Coral in the World

Species Name	Common Name(s)	Geographic and Depth Ranges
Corallium rubrum	Red coral, noble coral	Mediterranean Sea and eastern Atlantic between Portugal and Senegal; 5–300 m
C. japonicum	Aka-sango, moro	Japan, Okinawa, Bonin Islands, south to 26°N; 100–300 m
C. elatius[a]	Momo, momoiro-sango, cerasualo, scotch, boke magai, misu	Northern Philippines to Japan, 26–36°N; 100–300 m
C. konojoi	Shiro-sango	Northern Philippines to Japan, 26–36°N; 50–150 m
C. secundum	Angel skin, pelle d'ange	Hawaiian archipelago from Hawaii (20°N) to the Milwaukee Banks (36°N); 350–475 m
Corallium sp. nov.	Midway deep-sea coral	Midway and Emperor Seamounts, 28–36°N; 900–1500 m
C. regale	Garnet	Hawaii to north of Midway, 20–32°N; 400–700 m
Antipathes dichotoma	Black coral	Hawaii and Indo-West Pacific; 30–100 m
A. grandis	Black coral, pine, umimatsu	Hawaii, 45–110 m
Antipathes spp.	Black coral	Caribbean Sea; 40–80 m
Cirrhipathes spp.	Wire coral, black coral	Philippines and Indo-West Pacific to Hawaii; 30–50 m
Primnoa spp.	Gold coral	Alaska; 50–800 m
Gerardia (= *Parazoanthus*) sp.	Gold coral	Hawaiian archipelago and Indo-West Pacific; 300–400 m
Lepidisis olapa	Bamboo coral, (unbranched)	Hawaiian archipelago; 375–400 m
Acanella spp.	Bamboo coral, (branched)	Hawaiian archipelago to Indo-West Pacific, 375–450 m

[a]Probably represents at least five subspecies of taxonomic uncertainty.

Figure 2. *Corallium rubrum*, the red coral of commerce, with polyps fully extended. Depth = 40 m, Marseille, France. Photograph by J. G. Harmelin.

Acanella spp.) (Table 1). The most valuable species all belong to the genus *Corallium*, the so-called red coral of commerce (Fig. 2). The precious coral fishery worldwide is in fact sometimes thought of as only species of *Corallium*. The major precious corals of the world including species in other groups are listed in Table 1.

Most species of precious coral have life histories characterized by low rates of growth, fecundity, recruitment, and mortality and are consequently relatively long lived. In the jargon of *r–k* selection, precious corals can generally be characterized as *k*-selected species. One exception to this pattern is *Corallium rubrum*, which has a moderately high rate of recruitment. It is a slow-growing but fairly short-lived species (30 yr) and reaches reproductive maturity at an age of about 4–5 yr (FAO 4). Other more long-lived corals with life-spans on the order of 75 yr do not reach reproductive maturity until approximately 10–12 yr (on the order of $\frac{1}{6}$ their longevity, Grigg 3). Most species of precious coral are dioecious; that is, sexes are separate. Fertilization is generally external and the larval stage is on the order of days to weeks. Species with low rates of growth, recruitment, and mortality produce populations that turn over very slowly and therefore are quite vulnerable to overexploitation. Management implications of this are considered in a later section.

Habitat requirements of precious corals are varied but always include a firm (rocky) substrate and moderately strong current. All species are ahermatypic; that is, they lack symbiotic zooxanthellae. Many species therefore are found in deep water below the euphotic zone. Most commercial species of black corals are found below 25 m in the Pacific. Most species of *Corallium* are found between 100 and 400 m. The three species that depart from this behavior are *C. rubrum* in the Mediterranean (5–300 m), *C. regale* in Hawaii (400–700 m), and Midway deep-sea coral (900–1500m). Some species of gold coral in the genus *Primnoa* also are found near 1500 m in Alaska (Cimberg et al. 5).

Little is known about the feeding habits of precious corals, although those species that have been studied are herbivores and detritivores (Grigg 6; Lewis 7; Warner 8). The most common predators are boring sponges of the genera *Cliona* and *Haliclona* (FAO 4) and the mollusk *Neosimnia* (Liverino 2); both increase mortality owing to weakening or encrustation of the skeleton. In many habitats, mortality of precious corals is caused by smothering due to sedimentation. Strong bottom current is important in sweeping substrates free of sediment as well as providing food to sessile colonies.

3. METHODS OF FISHING

Most species of precious coral are harvested with dredging devices. The most famous of these, the engegno, also known as Saint Andrew's Cross, was invented by the Arabs in the tenth centry A.D. to harvest red coral in the Mediterranean. It consists of a wooden or metal cross about 4–5 m wide with nets about 8 m long attached to its extremities. The cross is dragged across the bottom, breaking and entangling coral. Japanese fishermen use a similar but less cumbersome device called a tangle mop, which consists of stones or other heavy weights with attached netting. Dredging with tangle devices is both destructive and inefficient. Small and immature colonies are destroyed and only about 40% of the coral broken from the bottom is entangled (Grigg et al. 9).

Some precious corals have been harvested selectively with the use of scuba (Fig. 3) (black corals and *C. rubrum*) or a submersible (various species of *Corallium*), although without management guidelines and enforcement, overharvest of the resource can also easily occur.

Submersibles have also been used extensively in the Mediterranean, Hawaii, Japan, and Taiwan for research puposes (see Liverino 2). In general, the operational costs of harvesting precious corals with a submersible exceed the value of the catch.

4. CATCH STATISTICS

The precious coral fishery has been described as a boom or bust industry (Grigg 3) and is characterized by large fluctuations in local supplies. Cycles in production

Figure 3. The black coral *Antipathes dichotoma* is harvested in Hawaii by divers from depths of 30–80 m. State regulations prohibit the taking of colonies less than 48 in. (1.22 m) high. Photograph by R. Grigg.

are caused by the discovery of new grounds, which are invariably overharvested and rapidly depleted (Grigg 3). Rates of growth and recruitment are insufficient to sustain production. Geographic nonconcurrence of "good" and "bad" years tends to damp annual fluctuations on a global scale (Fig. 1) (Grigg 10; Masso 11). Efforts to manage the resource have largely failed because of difficulties in enforcement. Most coral beds in the Pacific exist within international waters where no regulations apply, and in areas within territorial waters that are difficult to patrol.

The variable nature of annual production is well illustrated by the past 25 yr, during which five major new grounds were discovered (Grigg 3). The last of these was Midway deep-sea coral at depths between 900 and 1500 m on the Emperor Seamounts, which led to a worldwide glut of precious coral. Prices of raw material fell drastically in 1982 and many Taiwanese and Japanese vessels did not fish the Midway grounds in that year. Consequently, production fell. Supply would have further dropped in 1982 and 1983 were it not for a large discovery in the Mediterranean in the Alboran Sea. By 1984–1985, the marketplace had started to absorb the glut caused by deep-sea coral, and supply in the Mediterranean began to decline, leading to firmer prices. Hence in 1984–1985 effort and catch again started increasing in the Midway area. In 1985, a new productive area consisting of seamounts north of Gardner Pinnacles was found in the Hawaiian archipelago. Hence the cycle of discovery, overexploitation, and depletion continues to plague the industry. The only solution to this problem would seem to be management of the fishery or control of global supply in the marketplace by cartels. In the meantime, it is unclear as to how much longer the discovery of new grounds will continue to sustain the industry.

5. CONSERVATION AND MANAGEMENT

Various efforts to manage the harvest of precious coral have been largely unsuccessful. For example, during the tenth century the Arabs practiced rotation of fishing grounds requiring 9 yr between permit years. A variation of this approach is now practiced in Sardegna and Spain, where areas are closed for 5 and 25 yr, respectively (FAO 4). Closures of 25 yr obviously allow for greater recovery but poaching in such areas has been a serious problem. Another approach using limited entry was attempted in Okinawa in 1963 but lack of enforcement undermined its success. A number of other approaches including absolute moratoria on the taking of corals, natural reserves, benign neglect, and size and weight quotas, have been described by Grigg (3); however, all depend on adequate enforcement, which seems to be a difficult problem universally.

In Hawaii a yield–recruit model adapted from the model of Beverton and Holt which requires knowledge of growth, recruitment, and mortality rates for particular species has been developed and applied to the black and pink coral fisheries with reasonable success (Fig. 4). In this case, the method used to control fishing effort is a size limit that allows enforcement to be applied at the level of the buyer; for example, it is illegal to possess corals under a certain size. This method has the advantage of not requiring field surveillance.

Another method that has been suggested by the Japan Fisheries Association is limited entry, to be controlled at the international level under a multilateral treaty arrangement and enforced by each country, making it illegal to sell precious corals without a permit. Given the difficulties inherent in enforcement, this approach appears to have the most promise. Another means of monitoring production would be to place all species of precious coral in Appendix II of CITIES (Convention on International Trade in Endangered Species of Wild Flora and Fauna, United Nations). CITIES records of exports and imports could serve as a measure of annual trade of both raw and finished products. Presently only black corals are listed on Appendix II.

Recommendations from several independent bodies have been made to IUCN (International Union for the Conservation of Nature) to place *Corallium* spp., *Gerardia* sp., *Lepidisis olapa*, and *Acanella* spp. in Appendix II as well.

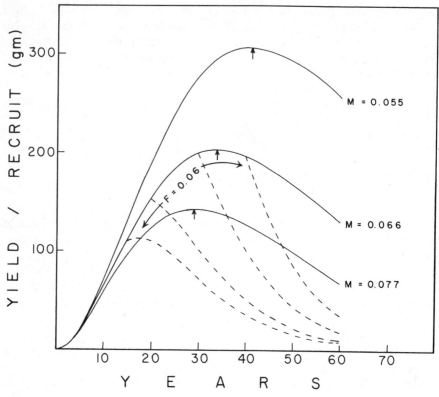

Figure 4. Yield-per-recruit curves of precious coral (*Corallium secundum*) in Hawaii at different rates of natural mortality (solid lines). *M* = levels of natural mortality, *F* = level of fishing mortality, dashed lines = yield of a cohort resulting from *F* applied at different ages of first capture for *M* = 0.07. Maximum yield (arrows) at average *M* is about 200 g/recruit and occurs at an age of 34 yr.

6. FUTURE OUTLOOK OF THE INDUSTRY

In spite of a history of overproduction and resouce depletion, the precious coral industry has retained a degree of stability owing to the control of auction sales of raw material by cartels. As to the future of the resource itself, it is fortunate that economic extinction occurs well before biological extinction; however, this may not necessarily assure an undiminished supply in the future. Because production has historically depended on the discovery of new grounds as opposed to the replenishment of old, future supply should eventually fall below demand, leading to inevitable price increases. The ability of the major producing countries (Taiwan, Japan, Italy, Spain, and France) to develop a multilateral treaty perhaps holds the best promise for future management. Lacking that, the future of the industry will probably depend on forces of supply and demand dictated by the marketplace.

REFERENCES

1. G. Tescione, *Il corallo nella storia e nell'arte*. Montanino Editore, Napoli, Italy, 1965, 405 pp.

2. B. Liverino, *Il corallo*. L. Causi Editore, Banca di credito populare, Torre Del Greco, Italy, 1983, 229 pp.

3. R. W. Grigg, Resource management of precious corals: A review and application to shallow water reef building corals. *Mar. Ecol.: Prog. Ser.* **5:**57–74 (1984).

4. FAO, Technical consultation on red coral resources of the Western Mediterranean. *FAO Fish. Rep.* **306:**1–142 (1984).

5. R. L. Cimberg, T. Gerrodette, and K. Muzik, *Habitat Requirements and Expected Distribution of Alaska Coral*. Final report to NOAA Contract 27–80. NOAA, Boulder, CO, 1981.

6. R. W. Grigg, Ecological studies of black coral in Hawaii. *Pac. Sci.* **19(2):**244–260 (1965).

7. J. B. Lewis, Feeding mechanisms in black corals (Antipatharia). *J. Zool.* **186:**393–396 (1978).

8. G. F. Warner, Species descriptions and ecological observations of black corals (Antipatharia) from Trinidad. *Bull. Mar. Sci.* **31:**147–163 (1981).

9. R. W. Grigg, B. Bartko, and C. Brancart, University of Hawaii Sea Grant Program. A new system for the commercial harvest of precious corals. UNIHI-SEAGRANT-AR-73-01:1–6 (1973).

10. R. W. Grigg, Precious corals: An important seamount fisheries resource. *NOAA Tech. Rep.*, *NMFS* 43:43–44 (1986).

11. C. Masso, *Coral-84*. Inst. Espanol de Oceanografia, Ministerio de Agricultura, Pesca y Alimentacion, Alcala 27, Madrid-14, Spain 1984, 67 pp.

29 WORLD FISHERIES FOR ECHINODERMS

C. Conand
Université de Bretagne Occidentale
Brest, France

and

N. A. Sloan
Pacific Biological Station
Nanaimo, British Columbia, Canada

1. INTRODUCTION

Echinoderms, particularly sea urchins and sea cucumbers, have been harvested for thousands of years from many nearshore temperate and tropical areas. Information on these small-scale fisheries is generally scattered and landings estimates are often inaccurate, classified under miscellaneous categories, or unreported. There are reviews of echinoderm fisheries worldwide (Sloan 1), according to region or species of sea cucumbers (2–7) and of sea urchins (8–14). Annual world echinoderm landings are now approximately 80,000 t (Table 1). Sea star fisheries are very with mainly Danish landings being reported for reduction into domestic animal feed stocks (Sloan 1). Pest control of sea stars occurs in some areas to decrease predation on commercial shellfish beds (*Asterias* spp.) or on coral reefs

TABLE 1   Estimated World Landings of Echinoderms (tonnes, live weight) According to Class and Main Areas or Nations

Landings by Class and Area or Nation	Year					
	1978	1979	1980	1981	1982	1983
Echinoidea						
Sea urchin total (t)	41,206	50,333	48,566	53,181	47,372	48,825
% landed by						
Japan	63	53	50	46	55	55
Chile	17	26	28	30	26	24
Northeast Pacific[a]	18	19	21	22	17	18
Holothurioidea						
Sea cucumber total (t)	23,514	23,246	27,480	24,960	25,768	27,125
% landed by	43	40	33	32	33	32
Japan	11	14	11	14	14	15
South Korea	29	32	40	42	42	43
West Central Pacific[b,c] West Indian Ocean[c,d]	17	13	15	11	11	10
Asteroidea						
Sea star total (t)	3,977	1,726	333	184[e]	1117[f]	149
% landed by Denmark	90[g]	99	90	89	99	92
Estimated World Total (t)[h]	71,666	78,150	79,914	82,614	79,961	81,773

[a] Canada/United States/Mexico.

[b] Indonesia/Malaysia/Philippines/South Pacific Islands.

[c] FAO data increased by an order of magnitude because it is probable that most statistics of tropical sea cucumbers are recorded from dried product (bêche-de-mer) rather than live weight.

[d] Kenya/Madagascar/Mozambique/Tanzania/India/Sri Lanka.

[e] Signified incomplete estimate by FAO.

[f] Danish landing of 1100 t (V. H. Jacobson, personal communication) not recorded by FAO.

[g] 10% of landings were reported from Federal Republic of Germany and the Netherlands.

[h] World total includes some landings unspecified as to class by FAO.

Source: Data are from various management agencies and references (1, 7, 13, 15).

(*Acanthaster planci*). There are also curio fisheries for dried sea stars, which are very poorly documented (Sloan 1).

Despite the numerous studies on the biology of some commercial species, there is a general lack of fisheries-related research and yield models are not currently in use for any species. This chapter provides brief histories of sea cucumber and sea urchin fisheries and information on their fisheries biology, and suggests approaches for resource assessment and future management.

2. HOLOTHUROIDS (SEA CUCUMBERS)

2.1. Introduction

The world's sea cucumber harvest is dominated by two ancient fisheries, the multispecies tropical fishery throughout the Pacific and Indian Oceans, and the temperate northwest Pacific (Japan, the Koreas, China, Soviet Union) fishery for *Stichopus japonicus* (Table 2) (1-7,16,17). In Japan and Korea the gutted body walls of *S. japonicus* are consumed raw or pickled, and a range of specialized products are processed from their viscera (Choe 2; Mottet 4). The Chinese dry the body walls for a product called hoi-som. Approximately 10 tropical species, whose body walls are generally massive, are gutted, boiled, and sun-dried or smoked to produce bêche-de-mer or trepang (Fig. 1) in a wide range of grades according to species, size, and appearance (4,7,18,19). The main producing countries are in the west central Pacific, the Philippines, Indonesia, and Malaysia. Minor production areas are in the Indian and south Pacific Oceans. The principal species are the "sandfish" *Holothuria scabra* and the "teatfish" *H. nobilis* and *H. fuscogilva*

TABLE 2   The Main Commercial Sea Cucumbers and Characteristics of Their Fisheries

SPECIES	FISHING AREA	TYPE	VALUE[a]
Temperate			
Stichopus japonicus	Northwest Pacific	Unprocessed or processed	1
Parastichopus californicus–P. parvimensis	Northeast Pacific	Processed	2
Tropical			
Holothuria scabra–H. s. var. *versicolor*	West Central Pacific and West Indian	Processed	1
H. nobilis–H. fuscogilva		Processed	1
Actinopyga echinites–A. miliaris		Processed	2
Thelenota ananas		Processed	2

[a] 1 = high commercial value, 2 = medium value.

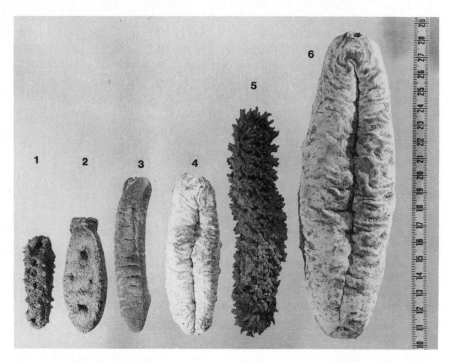

Figure 1. Bêche-de-mer market samples. 1, *Stichopus japonicus*; 2, redfish *Actinopyga echinites*; 3, sandfish *Holothuria scabra*; 4, 6, teatfish *H. nobilis*; 5, prickly redfish *Thelenota ananas*. (Photographs by C. Conand.)

(Conand 7). A minor fishery for *Parastichopus* spp. started recently in the northeast Pacific (Sloan 13).

2.2. History of the Fisheries, Harvesting, and Trade

For more than a thousand years, the Chinese explored the seas bordering India, Indonesia, and the Philippines for sea cucumbers. They taught the natives their processing methods but the trade remained in their hands. In the eighteenth century Japanese exported bêche-de-mer to China, the Malaysians fished along the north coast of Australia, and harvesting began in East Africa. At the turn of the century, European, Australian, and American merchants explored for bêche-de-mer in the Pacific Islands to trade for exotic products for domestic markets. In the nineteenth century sea cucumber fishing expanded greatly and the first commercial statistics indicate the predominance of China as importer, with 1000 t annually. Further studies on the fisheries and trade during the first half of the twentieth century show that 15 countries produced about 3,000 t of bêche-de-mer a year (Conand 7). After the closure of the Chinese market, Hong Kong and Singapore became the main markets. Recently interest has grown in assessing and managing this resource (Sachithananthan 3; Conand 7; South Pacific Commission 18).

Sea cucumbers are harvested worldwide from shallow waters by wading, free diving, spearing, and dredging from small boats. In the Soviet Union and North America, scuba diving is used (Sloan 1,13), and the species are therefore vulnerable to overharvesting.

Processing is vital to a successful fishery; the price the fishermen get depends on how careful the processing has been. Traditional methods are still in use (Conand 7; South Pacific Commission 18). During the processing, length is reduced roughly by half and weight by 90–95% (Conand 19).

The market is complicated because some countries are exporting whereas others are exporting and also consuming, or only importing (Sachithananthan 3). In addition, a large proportion of the imports into the main trading centers is reexported to other countries. Nevertheless, trade statistics from Hong Kong and Singapore are useful to help follow production and market trends (Sachithananthan 3; Conand 7; Sella and Sella 16). Hong Kong is the main market and during the last 10 yr imports have increased from about 500 to 2,900 t in 1984. Now more than 80% of the product comes from Indonesia and the Philippines and the reexports have been destined for China, the Republic of Korea, Taiwan, and Singapore. Singapore, the second market, has shown a slower increase; during the last decade the imports passed from 500 to 600 t coming from the Philippines, East Africa, India, Sri Lanka, and the South Pacific Islands. Most of the reexports in 1984 have been to Malaysia, Hong Kong, and Taiwan. Prices vary greatly with the species, country of origin, and grade so that in November 1985 bêche-de-mer was sold in Singapore for U.S.$1–15/kg.

2.3. Life History of Commercial Species

As with other fisheries resources, knowledge of the population parameters of sea cucumbers is necessary to establish yield models, but owing to the technical difficulties of size measurements of sea cucumbers and the low economic value of the fisheries, there is still a lack of extensive biological studies. The main references on morphometrics, growth, mortality, reproduction, and recruitment are presented in Table 3. The life history of *S. japonicus* has been recorded in some detail (2,6,20,21). Its annual cycle is divided into four phases: after a period of activity and growth, a pre-aestivation period and an aestivation period with cessation in feeding, are followed by a recovery period. Some progress has recently been achieved in clarifying the biology of the North American *Parastichopus* spp. (13,22–24). The systematics and biogeography of the tropical species are well documented (Clark and Rowe 25; Cherbonnier and Feral 26). Recent studies from the South Pacific region, Australia (Harriot 27,28), Papua New Guinea (Shelley 29,30), and New Caledonia (7,19,31–34) have increased our knowledge of morphometrics, reproductive cycles, size at first sexual maturity, and fecundity for the common species. Growth rate still remains undefined apart from *Actinopyga echinites* and *H. scabra*, whose growth parameters were estimated from monthly size-frequency distributions (Shelley 29,30). Modal analysis could not be applied successfully to other species as a modal progression was not evident in the size

TABLE 3   Key References on Life Histories of Commercial Sea Cucumbers

| Species | Morpho-metrics | Growth | Reproduction | | | Recruit-ment |
			Cycle	First Maturity	Fecundity	
S. japonicus	2, 6	2, 20	2, 21	2	2	2
P. californicus	22		22, 13, 4	22		
P. parvimensis	23, 24	24	24			24
H. scabra	7, 29, 27	30	35, 28, 20, 7	7	7	
H. s. var. versicolor	7		7	7	7	
H. nobilis	31		31	31	7	
H. fuscogilva	31, 36		31	31	7	36
A. echinites	32, 29	30, 33	29, 32	32	7	
A. miliaris	7					
T. ananas	31		31	31	7	

frequencies. The data from tagging experiments in New Caledonia (Conand 7,33) will provide estimates of growth of some species in spite of a high rate of tag loss. More populations studies are needed because the paucity of juveniles and the apparently unimodal length distributions widely observed raise the problem of recruitment, which has yet been little studied (Ebert 37). Natural mortality rate is unknown, but the main predators of adults, apart from man, are probably asteroids and gastropod mollusks (Bakus 38).

2.4. Resource Assessment

Commercial sea cucumbers are sedentary megabenthic organisms living in varied shallow-water habitats. Because fishing and processing are generally small-scale localized enterprises, statistics are scarce and little resource assessment has been undertaken (Sloan 1; Conand 7). Four methods are useful for evaluating standing stocks in relation to their habitats:

1. Quantitative sampling with various gears (dredge, trawl, grab) has been used, for example, in Japan with dredge to estimate the stocks of *S. japonicus* in an area, by applying a corrective factor for dredge efficiency (Choe 2).

2. Studies of catch per unit effort (CPUE) based on fishery statistics or on scientific evaluations can be defined as the number (or fresh weight) of sea cucumbers collected per diver hour (or day) or per tow (trawling). Some anecdotal data are available from India (James 5) from the South Pacific Islands (Conand 7; Shelley 29) and from the logbooks used to record catches in the northest Pacific fishery (Sloan 13). They provide some indications of stock size, but standardized measures should be collected regularly to allow reliable comparisons and stock assessments.

3. Direct visual evaluation of numbers in quadrats used for the tropical multispecies fisheries provide more precise evaluation of density in relation to

habitat (Harriot 28; Conand and Chardy 34). In New Caledonia, sea cucumbers show increased densities and biomasses from the Barrier Reef to the coast and from the slopes to the reef flats (Fig. 2). Three major groups of species were determined by inertia analysis (Conand and Chardy 34) corresponding to three habitats: the first is found in coral zones, outer and inner slopes, and passages, without terrestrial influence. The populations of the characteristics species *H. nobilis*, *H. fuscogilva*, and *Thelenota ananas* are small, but the mean size of individuals is large. The second inhabit inner reef flats, and are particularly developed on islets and fringing reefs coming under terrestrial influence. The populations of *H. scabra*, *A. echinites*, and *A. miliaris* can be very dense and in spite of their small individual size, the biomass can be very high. The third group correspond to inner lagoons, bays, and estuaries. In these biotopes, the only species of commercial value, *H. scabra* var. *versicolor*, shows intermediate densities and biomasses. Other habitats support species of little commercial value.

4. Indirect visual evaluation of densities from underwater photographs has not yet been attempted for littoral species.

2.5. Fisheries Regulations and Socioeconomic Considerations

The general lack of fisheries management is reflected by the paucity of documented regulations. In Japan, management of the *S. japonicus* fishery is decentralized and regulations vary according to the concerns of local cooperatives within each prefecture (Sloan 1; Mottet 4). For example, seasonal closures to protect breeding stock, area closures, catch quotas, and gear restrictions are decided within cooperatives for their designated coastal areas. In China, the seasonal closure corresponds to the spawning and to the aestivation seasons. The tropical fisheries for bêche-de-mer are generally localized, unregulated enterprises, under little centralized control. In some countries such as Madagascar, Sri Lanka, and India, there are regulations on the size of the processed products for export but they are variable within the region for the same species. Because the small-sized product is classified as lower grade with little commercial value, a combination of economic factors and population dynamics ideally leads to management measures based on size limits. Other restrictive options such as catch quotas, seasonal or area closures, and licenses are probably unrealistic for those fisheries taking place in developing countries with diverse social organizations and different coastal-area tenure systems.

2.6. Future of Sea Cucumber Management

The assessment and management of a marine resource requires fisheries statistics (catch, fishing effort) for the use of surplus–yield models and for evaluation of the population parameters for analytical models. The lack of standardized data has hindered the establishment of such models for sea cucumber fisheries. In temperate fisheries only one species is generally exploited, whose annual cycle including a breeding season followed by an aestivation period is now documented. This has

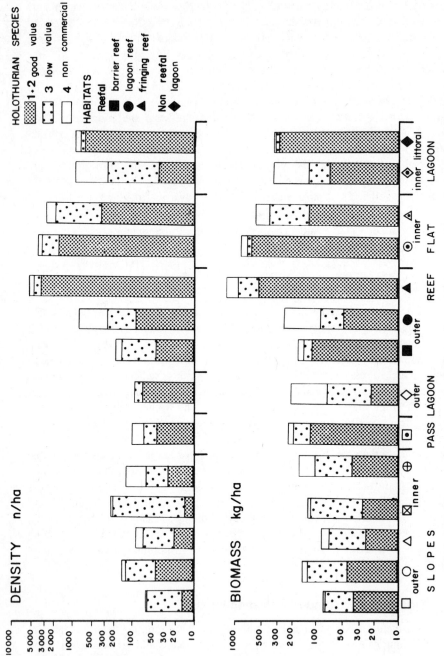

Figure 2. Densities and biomasses of sea cucumbers in relation to different habitats (New Caledonia).

led to some regulations such as the closed fishing seasons in Japan. Various methods have long been used there to increase the stocks or give shelter to juveniles (Sloan 1; Mottet 4). Recent mariculture research in Japan (Anonymous 39) and in China (Sui et al. 40) looks promising.

The more complicated tropical fisheries are multispecies artisanal enterprises for export. At present, there are still major gaps in our knowledge of population parameters and yield cannot be predicted without estimates of growth, recruitment, and mortality. There is also a need to record catch statistics and exports. In these cases, sea cucumber management should be considered as a part of overall coral reef management programs.

3. ECHINOIDS (SEA URCHINS)

3.1. Introduction

Ancient fisheries for sea urchin roe (Fig. 3) occur in western Europe, Mediterranean countries, and the north Pacific, and more recent ones in Chile and Pacific North America (1,8–10,13). Many small-scale artisanal fisheries occur throughout the tropics as well. Urchins are harvested intertidally by hand, or from the shallow subtidal by divers or by use of rakes, spears, or dredges.

Figure 3. Roe exposed in the open test of a ripe *Strongylocentrotus franciscanus*. (Photograph by R. Harbo.)

Landings are reported mostly from temperate areas where urchins are harvested seasonally. Some roe is locally consumed, although most of it is exported to Japan as fresh, raw product. A small proportion of the catch is processed by canning or smoking. Eighteen genera and many species are fished, particularly members of the families Echinidae and Strongylocentrotidae (Mottet 10). The most important species are *Strongylocentrotus intermedius* (north Japan), *S. franciscanus* (Pacific North America), and *Loxechinus albus* (Chile). Approximately 50,000 t of urchins is landed annually worldwide (Table 1), and Japan dominates in both domestic landings and imports (Sloan 1).

There is a voluminous literature on urchin biology and the ecological importance of urchin herbivory. With the exception of literature from Japan, however, very little is available on urchin fisheries biology and management. Examples from the scattered literature, recently summarized (Sloan 1,13), are used here to generalize on the scientific basis of urchin fisheries. There is a worldwide curio fishery for the tests of both regular and irregular (biscuit-like sand dollar) urchins, which can be damaging to urchin populations. Available data are so limited, however, that these are not mentioned further.

3.2. Urchin Life-History Characteristics Relevant to Fisheries

The life-history characteristics relevant to urchin fisheries are known from extensive studies on the heavily exploited *S. intermedius* from northern Japan (Table 4) and the evaluation by Dix (41,42) of unexploited *Heliocidaris erythrogramma* around Tasmania. Overharvesting and stock declines of *S. intermedius* (Fuji 43) have stimulated much research as summarized in Table 4. Having acquired much information on the nutritional physiology, life history, and ecology of *S. intermedius*, the Japanese can now estimate the potential urchin productivity of a site and are concentrating on stock enhancement through mariculture.

Dix (41,42) evaluated virgin Tasmanian *H. erythrogramma* populations strictly from the viewpoint of diving fishermen and processors. Firstly, urchin distribution and abundance (catch per diver hour) were estimated all around Tasmania. Sites were graded for exposure, bottom type, and bottom profile according to depth and urchin densities. Site evaluation did not include assessment of urchin diets or algae food availability. Second, roe yield was estimated from sites supporting commercial quantities of urchins. The reproductive characteristics of *H. erythrogramma* were thoroughly examined, including seasonal gonad cycle, test diameter at sexual maturity, gonad size related to test size, gonad index of large adults, urchin size and gonad index related to depth, and gonad color.

As agents of biological disturbance through intense grazing, urchins can profoundly influence kelp ecosystems and their associated fisheries. Moreover, mass mortalities of urchins through diseases or storms can initiate revegetation and marked increase in the productivity of nearshore food chains (44). These nearshore multispecies fisheries concerns are a topic of considerable debate and have been recently reviewed (1,13,44).

TABLE 4 Urchin Life-history Criteria Important to Their Fisheries: Research on *Strongylocentrotus intermedius* from Hokkaido, Northern Japan

Reference	Main Themes	Specific Life-history Topics
43	• History of Japanese urchin fisheries • Urchin metabolism, growth, and reproduction related to food intake • Assessment of urchin productivity via energetics "bioeconomics" • Annual adult urchin cycle in terms of feeding, somatic growth, and gonad growth • Need for urchin–algae ecological studies • Need for harvest regulations and mariculture to offset chronic overfishing	• Gonad cycle; gonad index; test size to gonad weight relationship; size at sexual maturity; age determination using test size frequencies; feeding rates preference and activities; metabolism of digestion and assimilation; maintenance of food levels; energetics of somatic and gonad growth
45	• Population model including growth, recruitment, immigration, emigration, and natural mortality • Habitat requirements and seasonal variations in distribution of different urchin life stages • Energy (algal food) needs of an entire urchin population	• "Balance of (urchin) population metabolism"; estimated algae and urchin production of a site on a g/m², yr basis.
9	• Review of previous work, emphasizing urchin life cycle, distribution, and abundance • Options of fisheries management and/or mariculture to enhance recruitment and algal food production of harvest sites • Use of five habitat types by different urchin life stages • Different urchin stock estimation techniques	• Seasonality of larval abundance; duration of planktonic phase and metamorphosis; aging improved using rings in genital plates; diets of juvenile urchins; mortality of different life stages; between-site and depth-related differences in spawning, recruitment, growth, overfishing, and CPUE
48	• Mariculture technique for urchin early life stages in aid of releasing juveniles for local stock enhancement	• Design and deployment of larval collectors according to area and season; on-growing of settled juveniles; release of juveniles

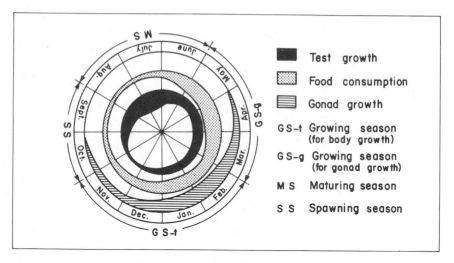

Figure 4. Schematic diagram of the relationship between gonad development, food consumption, and test growth in adult *Strongylocentrotus* from northern Japan (from Fuji 43).

3.3. Extrinsic Biotic and Abiotic Factors Important to Fisheries

The key extrinsic biotic factor to roe production is algal food availability. Fuji (43) attempted to define the life cycle of *S. intermedius* in terms of metabolic processes and illustrated interactions among food consumption, gonad growth, and somatic growth of adults on an annual scale (Fig. 4). The energetics approach to urchin–algal ecosystems (45,46) may therefore become a key urchin management tool in predicting roe production.

Important abiotic factors affecting roe production are first, site-specific criteria of exposure, substrate, and depth for algal growth, and second, the influence of season on algal growth and urchin gonad cycles.

3.4. Surveillance of Urchin Fisheries

Regular monitoring of urchin resources is rudimentary in application with the exception of regional Japanese management. Generally, urchin landings and stock assessments are not well recorded. The *S. franciscanus* fishery is indirectly monitored through a mandatory logbook program (Sloan 13). This provides data on landings and fishing effort according to region. Regional abundances and test size frequencies from grounds or market samples are gathered by shellfish management biologists (Kato and Schroeter 14). Data on algal food abundance, urchin abundance, test size frequencies, and gonad index should be gathered regularly on a site-specific basis to establish the well-being of local stocks, predict roe yield, and estimate optimal timing of harvesting.

3.5. Criteria for Achieving Optimal Yield

Certain aspects of urchin life history are amenable to modeling in aid of fisheries management. One is the description of roe yield according to urchin size (age). This requires a knowledge of gonad weight relative to test dimensions or whole dry weight, using an index that must allow for different gonad growth rates during different life stages (Greenwood 46). Gonad size increases rapidly relative to test size from early life to some point soon after sexual maturity. After this, gonad index does not increase or may slightly decrease with increasing urchin size or age (Gonor 47). Indeed, "reproductive senility" occurs in old individuals of some species, and lower roe quality in large, old urchins can decrease their market value. Obtaining the best roe yield on a sustainable basis requires harvesting before the protracted period of relatively poor gonad growth in later life but not before the animals have reproduced at least once. Moreover, after sexual maturity, habitat so strongly influences urchin size that age–size relationships become unreliable (Gonor 47).

Local habitat-related variations in urchin populations are important to potential roe yield. Between-site differences in spawning times, recruitment, food supply, and somatic and gonad growth (46–48) will lead to differing harvest potentials. Gonor (47) cautions that in order to assess roe production between sites, similar size ranges of urchins must be sampled.

Harvest techniques are highly selective because urchins are basically gathered by hand. Intertidal populations can be completely denuded (Southward and Southward 49). Subtidal populations are precisely fished by divers or by hand-operated equipment deployed from small vessels. Unselective technique such as dredging are rarely used.

The largest urchin fisheries occur in temperate climates and are seasonal according to urchin gonad cycles. Spawning urchins yield low-quality roe and harvesting is best for prime-quality roe during the recovery period of high roe growth, as, in autumn through winter in the northeast Pacific for *S. franciscanus* (Sloan 13).

Dix (41) evaluated sites, with sufficient densities of commercial-sized urchins, using catch rates (kilogram per diver hour) and a gonad index for a commercial rating in kilogram of roe, or dollars, per diver hour. He also compared incomes of diving scallop and abalone harvesters with potential urchin-derived incomes to assess the viability of an urchin industry. Economic evaluations of this sort along with biological characteristics of local urchin populations should be encouraged.

3.6. Regulations and Major Markets

Despite widespread overharvesting of urchins (1,8,9,13,43), fisheries regulations are often minimal and generally not well documented. This is especially true for the many artisanal tropical fisheries. The two basic management approaches are as follows.

1. *Common Property Resources*. Stocks are equally accessible to all harvesters licensed by a centralized government authority such as the northeast Pacific fishery for *S. franciscanus* (regulations reviewed in Sloan 13). Because of the common property nature of local urchin stocks, fishermen have little incentive to implement stock enhancement. There can be great variability within this management approach; for example, the southern California fishery is virtually unregulated, with the market controlling effort. Recent landings declines have stimulated industry requests for regulations. The British Columbia fishery, on the other hand, is managed in a preemptive fashion to control harvesting through a suite of area closures, seasonal closures, area landings quotas, and a size limit. Surveillance and enforcement or regulations are difficult because harvesters are highly mobile over large coastal areas.

2. *Exclusively Owned Resources*: Stocks are managed by nongovernmental, regional cooperatives with access only to members, such as in the Japanese urchin fisheries (Sloan 1; Mottet 10). Local cooperatives have a vested interest in their local urchin stocks and set their own regulations such as gear restrictions, area or seasonal closures, size limits, and stock enhancement measures. The Japanese urchin management is a patchwork of local regulation regimes whose surveillance and enforcement are likely to be tightly controlled because of the clearly defined management area and restricted mobility of harvesters.

The economics of urchin fishing are not well known. Japan absorbs all its domestic landings and must import vigorously from many countries such as Chile to satisfy internal markets. The Pacific North American fishery exists solely to satisfy the Japanese market for fresh roe in the sushi trade. Within Europe, France dominates the urchin trade and must import *Paracentrotus lividus* (usually whole, live product from Ireland) to augment domestic landings, which are now mostly from the Mediterranean rather than Brittany (1,8,49). Mariculture of *Psammechinus miliaris* is being investigated (P. Le Gall, personal communication).

3.7. Future of Urchin Management: Wild Stock Enhancement and Mariculture

Widespread overfishing and stock declines suggest that urchins are a vulnerable resource. The Japanese in particular have recognized this, using various stock enhancement techniques such as habitat improvement, for more than 75 yrs (1,43,48). There have been numerous enhancement procedures: (1) reef building for algae food settlement and urchin shelter habitat; (2) increased urchin food production using edible kelp culture techniques over urchin grounds; (3) deployment of larval urchin collectors, culturing young stages, and releasing juveniles; (4) transplanting adult urchins to good feeding sites for increased gonad growth; and (5) culling large, relatively low roe-yield individuals from local populations.

Roe production in urchin populations is strongly site-specific and future urchin management should take this into account. Assessment of urchin feeding, growth, gonadal cycle, and recruitment according to region will be required. Improved assessment of urchin–algae relations will eventually lead to prediction of regional roe productivity, thus helping managers to assign appropriate harvest levels.

ACKNOWLEDGMENTS

We thank Drs. P. A. Breen and F. Conand for reviewing early drafts.

REFERENCES

1. N. A. Sloan, Echinoderm fisheries of the world: A review. In B. F. Keegan and B. D. S. O'Connor, Eds., *Proceedings of the Fifth International Echinoderm Conference.* A. Balkema, Rotterdam; 1985, pp. 109–124.

2. S. Choe, Biology of the Japanese Common Sea Cucumber, *Stichopus japonicus* Selenka. Kaibondo, Tokyo, 1963, 226 pp. (in Japanese).

3. K. Sachithananthan, South Pacific Islands bêche-de-mer fishery. FAO FI:DP/RAS/69/102/11, 1–32 (1972).

4. M. G. Mottet, The fishery biology and market preparation of sea cucumbers. *Tech. Rep.—Wash., Dep. Fish.* **22**:1–57 (1976).

5. D. B. James, The Bêche-de-mer resources of India, *Proc. Symp. Living Resources from the Seas around India*, pp. 706–711 (1973).

6. V. S. Levin, Japanese Sea Cucumber. USSR Acad. Sci., Vladivostock, 1982, 191 pp. (in Russian).

7. C. Conand, Les ressources halieutiques des pays insulaires du Pacifique. Deuxième partie: Les Holothuries. *FAO Doc. Tech. Peches* **272.2** (1986), 108 pp.

8. J. Y. Allain, La pêche aux oursins dans le monde. *Peches Marit.* 1133:625-630 (1972); *Can. Fish. Mar. Serv. Transl. Ser., Ottawa* 4259 (1978).

9. K. Kawamura, Fishery biological studies on a sea urchin, *Stronglylocentrotus intermedius* (A. Agassiz). *Sci. Rep. Hokkaido Fish. Exp. Stn.* **16**:1-54 (1973).

10. M. G. Mottet, The fishery biology of sea urchins in the Family Strongylocentrotidae. *Tech. Rep.—Wash, Dep. Fish.* **20**:1–66 (1976).

11. R. Deppe and C. A. Viviani, La Pesqueria artesanal del erizo comestible *Loxechinus albus* (Molina) (Echinodermata, Echinoidea, en la region de Iquique. *Biol. Pesq. Chile* **9**:23-41 (1977).

12. D. Nichols, The Cornish sea-urchin fishery. *Corn. Stud.* **9**:5-18 (1981).

13. N. A. Sloan, World jellyfish and tunicate fisheries and the northeast Pacific echinoderm fishery. *Can. Spec. Publ. Fish. Aquat. Sci.* **92**:23–33 (1986).

14. S. Kato and S. C. Schroeter, Biology of the red sea urchin and its fishery in California. *Mar. Fish. Rev.* **47(3)**:1–20 (1985).

15. FAO, *Yearbooks of Fisheries Statistics*, Vols. 52, 54, and 56 resp. FAO, Rome, 1982, 1983, 1984.

16. A. Sella and M. Sella, L'industria del Trepang. *Thalassia (Venice)* **4(5)**:1–116 (1940).

17. A. Panning, Die Trepangfischerei. *Mitt. Hamb. Zool. Mus. Inst.* **49**:2–76 (1944).

18. South Pacific Commission, Bêche-de-mer of the tropical Pacific., Handbook No. 18. SPC, Noumea, New-Caledonia., 1979, 29 pp.

19. C. Conand, Bêche-de-mer in New-Caledonia: Weight loss and shrinkage during processing of three species of holothurians. *Fish. Newsl. S. Pac. Comm.* **19**:14–18 (1979).

20. K. Mitsukuri, Note on the habits and life-history of *Stichopus japonicus*, Selenka. *Annot. Zool. Jpn.* **5**:1–21 (1903).

21. Y. Tanaka, Seasonal changes occurring in the gonad of *Stichopus japonicus*. *Bull. Fac. Fish.*, *Hokkaido Univ.* **9**:29–36 (1958).

22. J. L. Cameron and P. V. Fankboner, Reproduction biology of the commercial California sea cucumber *Parastichopus californicus* (Stimpson) (Echinodermata: Holothuroidea). I. Reproductive periodicity and spawning behaviour. *Can. J. Zool.* **64**:168–175 (1986).

23. J. Y. Yingst, Factors influencing rates of sediment ingestion by *Parastichopus parvimensis* (Clark), an epibenthic deposit-feeding holothurian. *Estuarine Coastal Shelf Sci.* **14**:119–134 (1982).

24. A. M. Muscat, Population dynamics and the effect of the infauna of the deposit-feeding holothurian *Parastichopus parvimensis* (Clark). Ph.D. Thesis, University of Southern California, Los Angeles, 1983, 328 pp.

25. A. M. Clark and F. W. E. Rowe, Monograph on the shallow-water IndoWest Pacific Echinoderms. British Museum (Nat. Hist.), London, 1971, 238 pp.

26. G. Cherbonnier and J. P. Feral, Les holothuries de Nouvelle-Calédonie. Deuxième contribution. *Bull. Mus. Natl. Hist. Nat.*, *Ser. 4*, **6(3)**:659–700 (1984).

27. V. J. Harriott, The ecology of holothurian fauna of Heron Reef and Moreton Bay. M.Sc. Thesis, University of Queensland, Australia, 1980, 153 pp.

28. V. J. Harriott, The potential for a bêche-de-mer fishery. *Aust. Fish.* **44(6)**:18–21 (1985).

29. C. Shelley, Aspects of the distribution, reproduction, growth and fishery potential of holothurians (bêche-de-mer) in the papuan coastal lagoon. M.S. Thesis, University of Papua, New Guinea. 1981, 165 pp.

30. C. Shelley, Growth of *Actinopyga echinites* and *Holothuria scabra* (Holothuroidea: Echinodermata) and their fisheries potential (as bêche-de-mer) in Papua-New-Guinea. *Proc. 5th Int. Coral Reef Congr.* **5**:297–302 (1985).

31. C. Conand, Sexual cycle of three commercially important holothurian species (Echinodermata) from the lagoon of New Caledonia. *Bull. Mar. Sci.* **31**:523–543 (1981).

32. C. Conand, Reproductive cycle and biometric relations in a population of *Actinopyga echinites* (Echinodermata: Holothuroidea) from the lagoon of New-Caledonia. *Proc. Int. Echinoderms Conf.* pp. 437–442 (1982).

33. C. Conand, Methods for studying growth in holothurians (Bêche-de-mer), and preliminary results from a Bêche-de-mer tagging experiment in New Caledonia. *Fish. Newsl. S. Pac. Comm.* **15**:36–48 (1983).

34. C. Conand and P. Chardy, Les holothuries aspidochirotes du lagon de Nouvelle-Calédonie sont-elles de bons indicateurs des structures récifales? *Proc. 5th Int. Coral Reef Congr.*, **5**:291–296 (1985).

35. S. Krishnaswamy and S. Krishnan, A report on the reproductive cycle of the holothurian *Holothuria scabra* Jaeger. *Curr. Sci.* **36**:155–156 (1967).

36. M. Gentle, The fisheries biology of Bêche-de-mer. *S. Pac. Comm. Bull.* **29:**25–27 (1979).

37. T. A. Ebert, Recruitment in echinoderms. In M. Jangoux and J. M. Lawrence, Eds., *Echinoderm Studies. I.* A. A. Balkema, Rotterdam, 1983, pp. 169–203.

38. G. J. Bakus, The biology and ecology of tropical holothurians. In O. A. Jones and R. Endean, Eds. *Biology and Geology of Coral Reefs*, Vol. 2. Academic Press, New York, 1973, pp. 325–367.

39. Anonymous, 1983. Takarao morise kawamura namako (Artificial Production of Holothurian Larvae), Rep. No. 12 (1983) 200–204 (in Japanese).

40. Sui, X.-L. et al., Preliminary report on artificial ripening of parental sea-cucumber. *Fish. Sci.* **4(3):**28–32 (1985) (in Chinese).

41. T. G. Dix, Survey of Tasmanian sea urchin resources. *Tasmanian Fish. Res.* **21:**1–14 (1977).

42. T. G. Dix, Reproduction in Tasmanian population of *Heliocidaris erythrogramma* (Echinodermata: Echinometridae). *Aust. J. Mar. Freshwater Res.* **28:**509–520 (1977).

43. A. Fuji, 1967. Ecological studies on the growth and food consumption of Japanese common littoral sea urchin, *Strongylocentrotus intermedius* (A. Agassiz). *Mem. Fac. Fish, Hokkaido Univ.* **15:**83–160 (1967).

44. R. J. Miller, Seaweeds, sea urchins, and lobsters: A reappraisal. *Can. J. Fish. Aquat. Sci.* **42:**2061–2072 (1985).

45. A. Fuji and K. Kawamura, Studies on the biology of the sea urchin VII. Bio-economics of the population of *Strongylocentrotus intermedius* on a rocky shore of Southern Hokkaido. *Bull. Jpn. Soc. Sci. Fish.* **36:**763–775. (1970).

46. P. J. Greenwood, Growth, respiration and tentative energy budgets for two populations of the sea urchin *Parechinus angulosus* (Leske). *Estuarine Coastal Mar. Sci.* **10:**347–367 (1980).

47. J. J. Gonor, Gonad growth in the sea urchin, *Strongylocentrotus purpuratus* (Stimpson) and the assumptions of gonad index methods. *J. Exp. Mar. Biol.* **10:**89–193 (1972).

48. Anonymous, Ezobafun-uni no tennen saibyo, chukan ikusei, shubyo horyu ni tsuite. Hokusuishi Geppo (On the natural seed collection, intermediate culture and release of the sea urchin, *Strongylocentrotus intermedius*). *J. Hokkaido Fish. Exp. Stn.* **41:**270–315 (1984); *Can. Transl. Fish. Aquat. Sci., Ottawa* 5200 (1985).

49. A. Southward and E. Southward, Endangered urchins. *New Sci.* **66:**70–721 (1975).

30 RECENT DEVELOPMENTS IN RESEARCH AND MANAGEMENT FOR WILD STOCKS OF BIVALVES AND GASTROPODS

John F. Caddy
Food and Agriculture Organization
Rome, Italy

1. INTRODUCTION

1.1. Categories of Molluscan Stocks from the Perspective of Population Dynamics

The wide range of form, mode of life, and habitat of the Mollusca have been commented on in the extensive literature on molluscan biology (for example, Wilbur 1), and should have some parallels in any attempt to summarize those factors of importance in assessing and managing commercial species. This short review concentrates on the population dynamics of those predominantly sedentary and semisedentary mollusks that are members of the classes Gastropoda and Bivalvia. Some of the approaches outlined for Crustacea (Chapter 5) also have some application here, but are touched on only briefly in this chapter in order to avoid repetition.

An earlier attempt (2) to review the dynamics of that other commercially important group of Mollusca, the cephalopods, showed that these have more functional similarities to finfish than to other mollusks from the point of view of stock assessment, despite a number of striking differences whose practical significance is not yet fully clear, and they are touched on only briefly here. However, Chapters 26 and 27 deal with two of the best studied and most important fisheries for cephalopods, and Chapter 25 is a more detailed review of the population dynamics of those other actively swimming mollusks, the Pectinacea. As indicated by the title, therefore, we concentrate here principally on the more sedentary Bivalvia, and secondarily, on the rather less well studied Gastropoda. This still encompasses a wide range of life forms as well as fisheries, from sedentary (clams, oysters) to actively crawling organisms (e.g., whelks, conch); from hand gathering in the intertidal to harvesting by scuba in the sublittoral; and through a wide variety of fisheries using traps, dredges, and other specialized gear; to industrial-scale operations offshore with dredges and hydraulic harvesters.

1.2. Stock Units and the Dynamic Pool Assumption

The concept of the unit stock as that subset of the population of a commercially exploitable species that for management purposes can be considered as distinct and capable of being treated independently from the rest of the species is a necessary "working abstraction" that allows the methodologies of population dynamics to be applied in assessing the state of a resource. In particular, it allows the dynamic pool axiom to be invoked. This latter concept greatly simplifies stock assessment calculations, because it assumes (a) that survivors in an exploited population are being continually mixed by movement, and/or (b) that activities of harvesting units are random, so that a properly designed sampling scheme can lead to a single estimate of current population characteristics, such as catch rate (mean abundance), growth rate, and mortality rates, that is characteristic of the whole. This basic axiom is routinely applied to both main approaches to stock assessment now in vogue, namely, surplus production models and analytical models (for reviews, see Ricker 3; Gulland 4).

The concept of the dynamic pool, as used for example, in finfish stock assessment, means that during each harvesting period, the catch rate, age composition, and other characteristics of the population sample, should continue to reflect the overall characteristics of the stock during this period. As pointed out by various authors (e.g., Hancock 5; Caddy 6) there are going to be a number of difficulties in applying conventional stock assessment methodologies to sedentary or semi-sedentary species if a clear idea of the subunits of population being considered is not kept in mind.

1.3. Stock Units versus Genetically Self-Sustaining Populations?

Recent usage in fish population dynamics has generally been to leave the distinction between a unit stock and a self-reproducing population rather vague, in part because the methods of identifying the limits of genetically distinct self-reproducing units are poorly developed, and partly because a significant degree of genetic mixing appears to be the case between adjacent "stocks" of most molluscan species; for shellfish this is particularly true because the structure of megapopulations or unit stocks has been little discussed (see Chapters 18 and 24). Particularly for estuarine mollusks, a degree of genetic isolation is apparent, coupled perhaps for coastal species with the expectation of some genetic drift along an extended coastline. The findings of early workers such as Jorgensen (7) for those mollusks that have high fecundity and longer planktonic lives seem at first to imply that because the larval stage is the only dispersive agent for sedentary species, attempting to distinguish separate, self-generating populations is not a worthwhile exercise. More recent reviews, however (e.g., Sinclair et al. 8), and the growing application of methods of genetic discrimination (Chapters 15 and 18) are beginning to suggest that although stock units are primarily determined by a locality's hydrographic characteristics, local, discrete populations may be maintained over long periods of time, and could perhaps have a degree of genetic separation. Such an assumption of course underlies the life table, stock–recruit, and Leslie matrix approaches touched on below, because quantitative estimates at each life-history stage would be meaningless unless survival during the life cycle can be followed quantitatively as a "closed loop." A conservative approach to management is valid, therefore, and obviously applies with particular emphasis to those species with lower fecundity and short or no planktonic stages, for example, the whelk *Buccinum undatum*.

This whole area requires a great deal more work, and with the above proviso, the more restrictive definition of a unit stock as a geographically distinct group of adults that is not necessarily genetically isolated is perhaps still the most practical approach, as long as parental populations are not seriously depleted. This is especially true given that not the least of the difficulties of establishing an overall stock–recruit relationship for a population, as we have seen, is identifying the population itself! The above definition allows yield-per-recruit analysis to remain useful in resolving the best strategy for harvesting a given number of local recruits (assumed constant, or at least largely independent of parent population size) that enter the stock in a year, and calculations of fecundity per recruit may also provide warnings of excessive depletion of spawning stocks. Such analyses may be carried out separately by unit areas and combined subsequently according to one or more hypotheses. Stock units for sedentary mollusks may generally be seen then as having a geographic identity that can correspond to as large an area as a sea or gulf, major offshore bank, or estuary, to as small an area as a single shellfish bed for one species. Thus the effectiveness of the dynamic pool assumption as a reasonable approximation to reality in yield calculations and consequent harvesting strategies will be improved if we deal with relatively small subunits of stocks (even perhaps individual shellfish beds in the extreme case) for sedentary species.

As a generalization that seems to apply to most sedentary species within a fairly wide range of parental population sizes, recruitment fluctuations from year to year are most affected by environmental and interspecific biotic factors. Parental stock size for fecund species is usually considered less important than available space and suitable conditions for settlement (e.g., Hancock 9). It is surprising, however, given the growing importance of mariculture of marine mollusks, that the impact of these operations on natural recruitment in the vicinity has not been more systematically studied. As a broad generalization, the changes in yield per recruit caused by changes in size or age at harvest, although having an economic impact (given the variability of unit price with size for many shellfish), are less significant than the often larger year-to-year fluctuations in recruitment.

In some respects, highly fecund bivalve populations seem to perform as *r*-selected (opportunistic) organisms (MacArthur and Wilson 10), characterized by high fecundity but very variable recruitment success, which is largely determined by environment and the availability of suitable settlement surfaces. As discussed later, this can lead to apparent spawner–recruit relationships (Hancock 9), especially when effort intensifies, and yield overfishing is replaced by recruitment overfishing.

1.4. Sampling Strategy and Distribution Pattern

As a general conclusion that depends, however, on the size of the sampling unit (see Pielou 11), most sedentary invertebrate populations show a degree of population contagion or clumping over scales of tens or thousands of meters; the scale of aggregations varies with species, and is measurable by special sampling techniques (see Gardefors and Orrhage 12, and chapter 21). In consequence, their sampling distribution is often best described (Fig. 1) by a negative binomial distribution (Pielou 11; Elliott 13), as opposed to the Poisson distribution characteristic of randomly distributed populations. This must be taken into account in developing a proper statistical design for sampling (see Saila and Gaucher 14). In fact, the existence of high-density clumps is the feature that makes harvesting sedentary mollusks economically feasible in most cases, and has led to the realization that a significant proportion of the area occupied by the population (and a much smaller proportion of the stock) is not harvestable, economically speaking, at any given time. The significance of this for shellfish assessments and shellfish biology has been little investigated to date, but is illustrated in Figure 1.

Beginning at the largest spatial scale (the stock or even the genetically isolated population) we have, in decreasing order of dimension, the fishing ground, the shellfish bed, and those local patches or mosaics of higher than background density that occur both within and outside shellfish beds. Mapping the main fishing areas is often most economically done in an exploited population from the data recorded in logbooks or by interviews with harvesters. Bearing in mind the needs for greater accuracy for some purposes (and the fact that fishermen for obvious reasons do not always wish to disclose specific fishing locations!), systematic surveys may be necessary however. These may use transects if the extent of grounds is unknown, or random or stratified-random sampling, if a map already exists showing locations of promising and less promising depths, tidal levels, or sediment types. A method of position fixing is of course implied here, ranging from standard surveying approaches intertidally to one of the various electronic navigation systems now available for subtidal resources. Such a map can form the basis for a stratification scheme, leading to estimates of stock sizes or biomasses and a calculation of the variance of sampling estimates. Fogarty (15) noted that separating zero records from positive occurrences of a species in a survey haul reduces greatly the skewness and variance of the density estimate and allows sediment preferences to be more

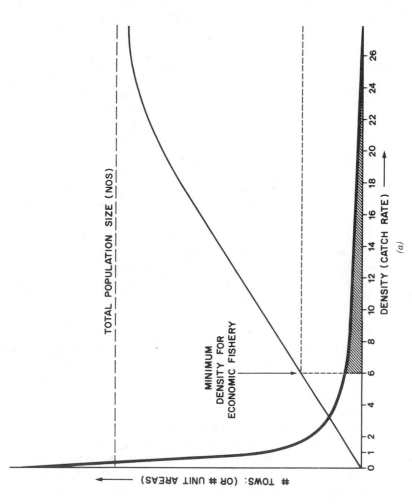

Figure 1. (a) Distribution of densities and cumulative numbers characteristic of many sedentary molluscan species, and the concept of threshold density for harvesting, which can divide a population into economically fishable (shaded) and effectively unfished (unshaded) components.

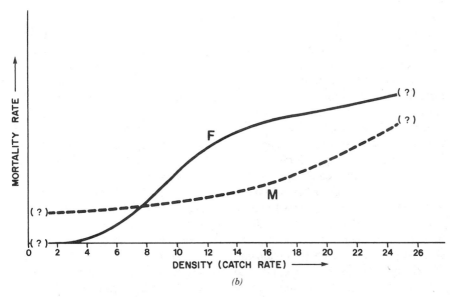

Figure 1. (*Continued*) (b) Both natural (*M*) and fishing (*F*) mortality components may be expected to show some increase with density.

closely determined. Nickerson (16) described a survey procedure for population assessment of intertidal clams by a hand-held hydraulic clam digger, following stratification of the beach by tidal level and sediment type based on a knowledge of habitat preferences. More recently, Conan (17) has pointed to geostatistical techniques developed for mapping mineral deposits (where samples are taken at points across a surface presumed to support a population concentration), as a viable alternative sampling design to random stations for mapping shellfish beds; especially subtidally. Approaches to stock size and abundance estimation using similar strategies to those employed by predators may be an appropriate alternative where populations consist of beds or discrete aggregations (Mangel and Beder 18).

With respect to temporal changes in population structure, repeated sampling at fished and unfished locations on a seasonal basis should allow estimates of Z and M, respectively, to be obtained (Hancock and Urquart 19), as well as estimates of recruitment.

2. AGE DETERMINATION AND GROWTH

2.1. Age Determination in Mollusks

A variety of approaches have been refined in recent years that depend on analysis of time sequences of marks on molluscan shells, and others that do not. In the first category is the use of shell sections (e.g., MacDonald and Thomas 20 for *Mya arenaria*) and acetate peels where growth rings are close together, as in the giant burrowing bivalve, *Panope generosa* (which may reach more than 100 yr of age with adult rates of natural mortality of the order of $M < 0.02$; Harbo et al. 21). More detailed verification of the seasonability of rings in shell sections by means of isotope analysis is possible (Jones et al. 22), and although such technology-intensive methods are unlikely to be necessary for standard age reading, in general there is a need for cross comparison of different methods of growth estimation in early stages of investigation. These methods may include seasonal sampling to follow annulus formation, perhaps with tagging, shell notching, or the use of Alizarin red dye (Nickerson 16), accompanied by size–frequency analysis to follow shell modes (see, e.g., Murawski et al. 23). Shell or internal banding of the operculum may be used for aging of some gastropods (Miranda 24).

Considerable progress has been made in recent years for both fish and mollusks in refining and extending age-reading techniques to time intervals shorter than 1 yr, particularly reading daily annuli, visible only with high-powered magnification. Thus, age reading can (with significant effort!) be extended to juveniles and short-lived species, both for organisms with an exoskeleton (e.g., Bivalvia) and for reading sections through squid statoliths (Dawe et al. 25). There are complications, however; in the common European cockle, Richardson et al. (26) found that internal shell banding may occur semidiurnally when suspended below rafts, but stronger, overriding bands may be formed on each tidal emersion of intertidal populations, thus illustrating the need for careful validation when working with short-term events recorded in the shell structure.

For species where direct age-reading is not practical, or simply too time consuming to carry out, a variety of new approaches to growth and mortality estimation have been developed that seem to have application to molluscan populations. These include cohort analysis with size frequencies (Jones 27, catch curve analysis Pauly 28), and other approaches that assume that size frequencies represent an equilibrium between growth and mortality rates, for example, the approaches of Powell (29) and Fournier and Breen (30) for species such as abalone where age is not readily determined.

2.2. Density-dependent and Environmentally Dependent Growth

One of the characteristics of sedentary or semisedentary mollusks is their plasticity in growth, even within the same species and within the same stock, as a response to the range of environmental conditions in which individuals may find themselves. Such lability can produce wide variations in growth rate or meat yield for the same population as a function of sediment type as well as tidal height and latitude (Appeldoorn 31), depth (see Chapter 25), water flow rate (Hadley and Manzi 32; Wildish and Kristmanson 33), and sediment type (Newell 34), or as a result of variations in food availability in the environment, which may be density-dependent or position-related. Density-dependent growth is well established for most bivalves (e.g., Walker 35 and Fig. 2). "Nonlethal predation" in the form of siphon cropping by demersal fish also can reduce growth rate (Peterson and Quammen 36). Analysis

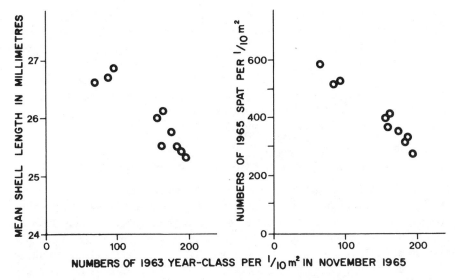

Figure 2. Growth in shell length of 2-yr cockles (left) and density of spat (right) as functions of density of 2-yr cockles, *C. edule* (from Hancock 9).

of the relative impact of the many factors that contribute to high growth variability in sedentary organisms susceptible to location-specific effects can be complex; principal component analysis is a promising tool for sorting these out (Appeldoorn 31).

The conventional von Bertalanffy growth model has been found to be a good description of molluscan growth (e.g., Sasaki 37), but one of the consequences of a lack of motility in an environment subject to wide seasonal variability (e.g., the intertidal zone) is that growth can be a seasonally fluctuating process (e.g., Bodoy 38; Farrow 39). This is true with respect to growth in shell size, but it also affects factors related to spawning and feeding conditions, which for many mollusks such as oysters affects their marketability at least as much as does size. Various approaches to modeling seasonal growth in length have been proposed, some of them involving incorporating a sine function in the von Bertalanffy growth formulation.

3. MORTALITY ESTIMATION

3.1. Total Mortality Rate (Z)

In discussing mortality estimation in mollusks, it is convenient to distinguish those mollusks with external, especially annual, growth rings on the shell, resilium, or operculum, for which the conventional age-reading procedures developed for marine fish apply, from those species where mortality (and growth rates) have to be deduced from size–frequency distributions and their changes with time. In the former category, mollusks have the added advantage of usually possessing externally legible annuli, so that age and current and former size at age can be read directly from living specimens, without some of the problems of back-calculation from otolith rings to total fish length that one faced in developing age–length keys for finfish species (Ricker 40). Despite this, an extended period or size range of partial size selectivity by the fishing gear can still lead to a pronounced Lee's phenomenon (see Ricker 40; Hancock 41, and chapter 25).

3.2. Population Estimates or Estimates for the Fishable Population?

In order to estimate overall mortality rate for the whole population, it follows that we must first define the limits of the population, and then (Yap 42), estimate an overall mortality rate Z as the log difference of the numbers of individuals N_i' remaining after some time interval Δt of those initially present N_i in each unit area i of the population, that is, mean mortality rate for the whole population would be

$$Z\Delta t = \log_e \left(\sum_i N_i \right) - \log_e \left(\sum_i N_i' \right)$$

From the point of view of practical stock assessment, however, it may be of more interest to consider what is the mortality rate for the fishable concentrations only (this will certainly be higher), and to carry out yield calculations for each statistical unit, density stratum, or fishing area within the population separately; combining them later as necessary.

3.3. Natural Mortality Rate (*M*)

One of the most widely used approaches to mortality estimation in bivalves has been the procedure of Taylor (43), who provides an expression for M as a function of 95% of the asymptotic length, $A_{0.95}$, as

$$M = 2.996/A_{0.95}$$

This may be a useful rule of thumb, but is unlikely to be valid for all latitudes, and is not a substitute for a direct estimate of M, which seems quite feasible for sedentary species. Especially for intertidal species, an experimental approach using cages or enclosures, with direct monitoring of year class abundance inside and outside areas closed to fishing and to predators, can offer a more direct approach to estimating mortality rates, an approach that is not available to finfish biologists (Hancock and Urquart 19).

The impact of predators, including birds (e.g., oyster catchers *Haematopus ostralegus* on cockles *Cerastoderma edule*; Sutherland 44), fish (e.g., rays on estuarine shellfish beds; Merriner and Smith 45), and especially crabs (Yap 42;

Whetstone and Eversole 46; Kraeuter and Castagna 47; Boulding and Hay 48) and predatory gastropods (Carriker 49), has long been a major source of concern to shellfish farmers, and there is a significant literature on methods of predator control (Loosanoff 50; MacKenzie 51). There is certainly good evidence that predators control the structure of bottom communities (Edwards et al. 52). At least in theory, then, natural mortality in shellfish stocks is a parameter that is at least potentially controllable, although usually at a significant cost; this is not true for most marine fish stocks.

From detailed observations on the feeding activity of one predator, the oyster catcher, Sutherland (44) was able to establish that feeding rate of wading birds is aimed at maximizing biomass intake rate per bird; the highest feeding rate (and hence, presumably, the highest local value for natural mortality) occurs on the highest density of large cockles, with negligible mortality on small individuals and those occupying low-density areas. A similar density-dependent predation by crabs on littleneck clams was observed by Boulding and Hay (48), which suggests that this may be a general phenomenon with respect to sedentary invertebrates. Predators also have an overriding influence on size composition of unexploited mollusks (Chilton and Bull 53), and the concept of a size refuge for larger individuals if they attain a certain size has also been demonstrated (Commito 54).

The relevance of the numerous observations of density-dependent natural mortality is in part that it suggests that thinning of very dense patches of juveniles may be a way of reducing predator impact, and in part that it parallels closely the harvesting strategy of hand gathering by shellfish harvesters (see below). Switching between alternative prey as a function of predator preference and relative abundance also occurs for gastropod predators (Katz 55). It also suggests that estimates of M based on age composition of what are largely unexploited stocks might be obtained by estimating the slope of the right-hand limbs of catch curves (Ricker 3) from samples taken in low-density "fringe" areas outside the main fishing beds. These, and estimates based on longevity (see Hoenig 56) might be useful even if the values obtained would likely be on the low side for calculations on the harvestable stock.

3.4. Juvenile Mortality Rates

Although a secondary postsettlement stage of dispersal to the eventual adult habitat is one of the features of some sedentary or semisedentary species such as mussels (Bayne 57), cockles (Baggerman 58), and other infaunal bivalves, one of the inevitable corrolaries of the sedentary nature of many molluscan shellfish is that the act of settlement in many species definitively chooses the eventual adult habitat. It is clear that favorable hydrographic regimes may maintain planktonic larval stages within range of suitable adult habitats, and gregarious larval settlement behavior in fixed bivalves such as oysters (Hidu et al. 59) may also result in a degree of selection of appropriate habitats. The relevance of this observation in this context is that mortality rates of spat based on log-density ratios may be biased by transport processes.

3.5. Seasonal and Catastrophic Mortalities

The timing of settlement may itself take advantage of seasonal low points in predator feeding activity on newly settled spat (Jorgensen 7), which is generally higher for younger specimens. Mortality rates, although generally high for small spat, tend to drop at a close to exponential rate with size and age (Fig. 3), as well as with density. In addition to biotic factors, high mortality of spat owing to an unfavorable physical environment in the first few months of benthic life may result from a number of abiotic factors and lead to mass mortalities. These include anoxic bottom water conditions (Haskin et al. 60), unfavorable temperatures, and low salinities (Yap 42); all of which can result in catastrophic mortality events that appear characteristic of some species and environments and greatly complicate assessment of stocks. Such effects have been revealed in analyses of long-term shellfishery data (Ulanowicz et al. 61). There is also some evidence that intraspecific competition between adults and spat may lead to mortality of early postsettlement stages (Hancock and Urquart 19). Prior depletion of older year classes by mass mortality (Haskin et al. 60) in habitat-limited situations may in fact be a precondition for unusually good recruitment and good year classes; thus mass mortality may reduce early mortality of the next year class. Presumably fishing could have a similar effect in freeing up suitable settlement surfaces. It has also been recognized that natural mortality rate varies seasonally, and that fall transplantation of young clams could avoid higher predation rates at warmer times of the year and allow smaller clams to reach less vulnerable sizes by the next summer (presupposing of course, that catastrophic deaths due to winter storms or spring runoff of fresh water are not major problems). These well-documented size-, density-, and season-specific variations in natural mortality rate should ideally be taken into account in population modeling, but have not often been to date.

The close relationship between age, size, and mortality rate due to predation in sedentary bivalves, as we have seen, reflects both their higher density and a higher specific attraction to predators; with ambient temperature acting as a key modulating factor, presumably on the metabolic rate. If this last supposition is correct, the relationship given by Bayne and Newell (62), metabolic rate at size, $U(w) = aW^{-b}$, might as a first approximation give an order of magnitude for the change in activity of (poikilothermic) predators and prey and hence the response of natural mortality with size also; this idea seems worth experimental investigation, noting that values of b generally lie in the range -0.33 to -0.25.

We might expect then that the "attractiveness" of a given area of clam flats to predators might be a function of density and the rate of production of metabolites; leading to an expression easily derived from the above equation.

3.6. Experimental Estimation of *M*

The possibility of directly estimating that most difficult to measure of population parameters in marine fish, namely, the natural mortality rate, must be one of the attractions of working on the dynamics of sedentary species such as infaunal bivalves, and the more precise experimental designs used in agricultural science

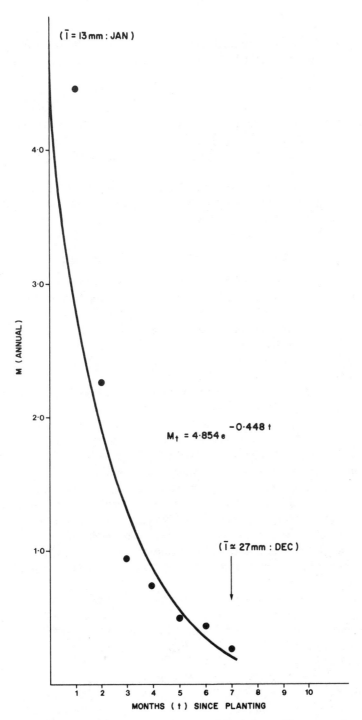

Figure 3. Decline in natural mortality rate for spat of hard clams, *M. mercenaria*, with size and age, subsequent to planting in trays maintained at constant density (recalculated from Whetstone and Eversole 46).

such as ANOVA or Latin square designs (e.g., Kline 63 for *Saxidomus giganteus*) probably have not yet been taken full advantage of for this purpose. Hancock and Urquart (19) and Hancock (64) carried out large-scale covered mesh experiments to distinguish natural mortality components due to different predators, and cage experiments have confirmed the key role of predators in early benthic survival. Other methods such as tagging (e.g., Boulding and Hay 48) or examination of the age structures of dead shells with predator perforations (e.g., Franz 65) have also been used to establish size selection of predators. Quantitative sampling of predators over shellfish beds and subsequent stomach content analysis is another indirect approach that led to, for example, the conclusion that the small estuarine fish, *Fundulus heteroclitus*, may be the most serious predator on young *Mya arenaria* spat in the Gulf of Maine (Kelso 66).

3.7. Analysis of Dead Shells for Mortality Parameters

Chapter 25 expands on an approach to estimation of natural mortality rates based on the number of paired dead shells of scallops. This approach, pioneered by Dickie (67) is capable of further generalization. Yap (42) found a significant correlation between size-specific mortality rates calculated for living clams and the size frequency of dead shells in the same bed, which gives a direct estimate of probability of death at size (Fig. 3).

Bivalves, some gastropods, and cephalopods are provided with hard parts that remain after death (shells in the first two cases and in the Sepioidea, and beaks for all cephalopods), which provide information on the size and possibly also the age at death, and for squid beaks in fish stomachs or bivalve shells with bored holes or characteristic breakage marks, information also on the nature of the mortality component as well as (in the case of bored shells) the size of the predator (Wiltse 68). Analysis of this material would appear to provide a fertile field for research into natural mortality rates that has hardly been touched on, although material, (for example, dead shells taken incidentally in the dredge or onshore shell middens from past human harvesting), is often readily available (see Fig. 4 for some preliminary approaches). As for live mollusks, caution obviously must be exercised in sampling dead shells, in that the assumptions of constant availability at size and constant rates of dispersal or decomposition with size following death may not be correct in all cases.

In the following brief summary, three situations are considered:

1). Dead shells due to natural causes only (unfished populations): number of survivors at age t:

$$N_{t+1} = N_t \exp - M_t$$

Number of deaths (i.e., number of dead shells produced annually):

$$D_t = N_t[1 - \exp(-M_t)]$$

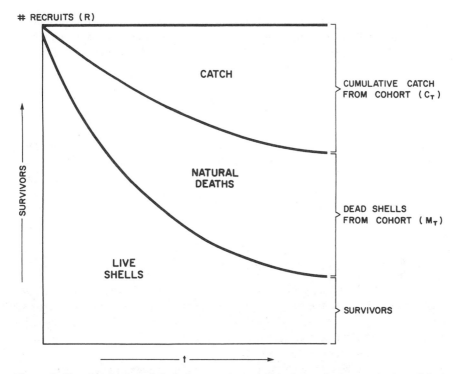

Figure 4. Showing the fate of shells from a cohort of R shelled mollusks recruited to a fished stock, where dead shells from the cohort are recognizable and remain intact as an index of the natural rate of death.

and the ratio of deaths in successive ages (if equally available), is then given by

$$\frac{D_t}{D_{t+1}} = \frac{N_t[1 - \exp(-M_t)]}{N_t \exp\left(-M_t\right)\left[1 - \exp\left(-M_{t+1}\right)\right]}$$

This simplifies to

$$\frac{D_t}{D_{t+1}} = \frac{\exp M_t - 1}{1 - \exp(-M_{t+1})}$$

If the natural death rate of one age group is known, or the mean natural death rate, the relative abundance of empty paired, left, or right valves (for bivalves) can in theory be solved for the death rate at age.

2. In the case of a fished population, where shucking at sea is practiced, dead shells reflect both fishing and natural deaths, and total mortality rate at age (Z) can be substituted for M in the above equation.

3. If catches for a series of years are mixed in a shell midden, the only information that can be derived is on the vector of fishing mortality at age using cohort analysis by sizes on the combined size frequency (see Jones 27), assuming estimates of M/K and L_∞ for the stock are available.

3.8. Prerecruit Mortality

Estimating prerecruit mortality is a matter of key concern to shellfish workers, and poses some obvious problems in estimation so that in practice it has usually been approached indirectly (see Section 5.1 for details on methods). Thorson (69) estimated for *Mercenaria mercenaria* that roughly 98.6% mortality occurs in the postsettlement period, and Muus (70) found minimum mortalities over the same period of 67% for Oresund bivalves; as noted by Brousseau et al. (71), these estimates do not include losses of eggs at fertilization and of larvae in the plankton.

3.9. Mortality Caused by Fishing, F

In general, fishing mortality rates (F) can be determined directly or indirectly for sedentary mollusks. In the first category are direct estimates of density before and after harvesting, and area-swept methods: both rely on the relative facility with which gear performance can be estimated for most sedentary species. Following Baranov (72) and later developments (Caddy 6; Alverson and Pereyra 73), fishing mortality rate can be estimated directly based on knowledge of gear selection (see Chapter 25). The formulation $F_{ij} = f_i, p_j \, ac_j/A$ is the one generally used, where i refers to year and j to age group. In this formulation, f_i is the annual total of effort units exerted, each of which sweeps an area a of the total surface A, with a (measurable) probability of capture p_j, and a probability of retention c_j at age. (This equation can of course be further subscripted for a grid of areas, each receiving its own level of effort.) One of the problems with this approach is that it does not take into account the less than random allocation of effort inside area A, although here the choice of unit areas is critical. Also, when obtaining indirect estimates of fishing mortality such as by cohort analysis or by subtraction of existing estimates of total and natural mortality rates, $F_t = Z_t - M_t$, where these are available from existing studies, attention should be paid to assuring that mortality estimates refer to the same density, area of exploitation, and age group (see Chapter 25). Consideration should also be given to the possibility of indirect fishing mortality due to damage by gear or, in discard, biasing mortality estimates. If the latter component appears to be important, it may be possible to determine it experimentally as a proportion a_t of the fishing mortality at age, so that the catch equation predicting yield in weight Y at age t of N_t individuals of a given mean weight w_t would then have to be modified to

$$Y_t = \frac{w_t N_t F_t}{F_t'' + M_t} \{1 - \exp\left[-(F_t'' + M_t)\right]\}$$

where $F_t'' = F_t(1 + a_t)$.

A final category of potential approaches that should provide a direct index of fishing mortality rate that may be worth exploring is the relative frequency of shell shock marks at age (see Chapter 25) if shock marks can be correctly attributed to fishing: (it should be feasible to test this hypothesis, given separate estimates of fishing intensity and shock mark frequency over a series of areas).

4. FISHING EFFORT, FISHING INTENSITY, AND GEAR SELECTION

4.1. Units of Fishing Effort and Estimation of Fishing Power and Catchability

A variety of considerations apply to the definition of fishing effort units for the range of gear types used in invertebrate fisheries, and a number of these are summarized by Caddy (74). In general, the approach is to identify those effort units (f) that are most clearly proportional to fishing mortality, and multiply them by the appropriate catchability coefficient (q) to give an index that is closely proportional to F:

$$qf \rightarrow F'$$

However, one problem in this simple formulation occurs as we noted when discussing predator-induced mortalities, namely, that the pattern of fishery-induced fishing intensity and mortality across a sedentary population of commercial shellfish is non uniform, corresponding to the strategy of fishing of harvesters, which depends on their accuracy of location of high-density patches of commercial-sized animals. This, incidentally, is one of the sources of bias when using Leslie–DeLury methods (Rhodes et al. 75), which should be carried out separately for individual areas with more or less uniform density, before combining individual estimates of stock size. One mathematical description of effort allocation that seems to broadly describe that for a scallop fishery is referred to as proportional allocation (Caddy 6). Here the fraction of the total effort f_i expended in each of the N unit areas of a population is allocated proportionally to the local potential catch rate at the start of the time interval; that is

$$f_i = f_T \mathrm{pr}_i \bigg/ \sum_i^N (\mathrm{pr}_i)$$

where $\mathrm{pr}_i = \Sigma_k (\mathrm{CPUE}_k)$. The second summation (by age k) is over the (partially) recruited age classes. In fact, for some fisheries this strategy probably greatly underestimates the fisherman's ability to locate productive concentrations and avoid wasting time in unproductive areas except for occasional scouting trips, and it has been suggested that for such fisheries for sedentary species, a "minimum economic catch rate" might be used to partition the population into two segments (Fig. 1).

An approach that specifically recognizes the occurence of pulse fishing as a typical harvesting strategy for shellfish is described by Sluczanowski (76).

If a system of logbooks or fishermen's interviews is in place, an index of effort concentration such as that of Rothschild and Robson (77) can be used to estimate the degree of concentration of effort on the stock (Caddy 6).

An approach that could be carried out on management subunits having the same basic productivity per unit area with constant fishing is that referred to as the composite production model (Section 6.3; Caddy and Garcia 78; Caddy 79).

4.2. Selectivity of Fishing Gear

A wide variety of types of fishing gear are used in the capture of commercial molluscan species, but particular emphasis is placed in the following brief summary on hydraulic dredges, traps, and hand gathering, both intertidally and by diving. A brief review of the different selectivity characteristics of invertebrate fishing gear follows.

4.2.1. Dredges

Dredges of a variety of types are the principal method of harvesting most subtidal bivalves and they share a number of common characteristics with trawls, both being swept-area gears, although dredges are more rapidly "saturated" (selective processes are rapidly reduced to a minimum on bottoms with abundant epifauna that clogs the dredge apertures). As a result it is fairly characteristic to find a wide range of sizes over which partial selection is the case. For those dredges made up of circular rings joined by one or more links, there are two effective routes for escaping individuals: the inter- and intra-ring spacing; with the latter being the smaller. Further details of experimental determination of selectivity of scallop dredges are given in Chapter 25.

4.2.2. Hydraulic Harvesters

Hydraulic harvesters are used to fish a variety of infaunal bivalves, using high-pressure jets of water (from either the dredge mouth or a propeller deflector in shallow water) to take advantage of the thixotropic characteristics of muddy sand. In the resulting semiliquid state, sand is readily sieved or carried on board by a wide-diameter pipe mechanically assisted by air or water lift for subsequent sieving or size grading of clams. Experimental observations show that there is a possibility with poorly adjusted depths of cut, of a significant degree of incidental damage in harvesting, and this is certainly true for the older style of rocker dredge (Medcof and Caddy 80), which was observed to cut cleanly through *Arctica islandica* if set too shallowly.

A variety of simple methods are used to harvest (at high tide) intertidal and shallow subtidal bivalves, which are essentially hydraulic methods. One is the practice of driving an anchored motorboat in decreasing circles on sand flats at high tide; the resulting centripetal processes dislodging bivalves from the substrate and concentrating them for collection at low tide or by dredge. Some concern has been expressed relating to the potentially high incidental mortalities resulting from this practice, and from hydraulic harvesting in general (Adkins et al. 81), as well as its impact in destabilizing beach surfaces; but hand gathering can cause similarly high incidental death rates and smothering of seed, and hand-pushed hydraulic harvesters for *Mya arenaria* are claimed to reduce this type of incidental damage, as well as permitting active reburying of juveniles. Diver-operated hydraulic harvesters have been used for harvesting and quantitative sampling of subtidal clam populations; they appear to be highly efficient.

4.2.3. Traps

Harvesting various types of gastropods (e.g., whelks) by pots is a common practice in various parts of the world, as it is for *Octopus* in various artisanal fisheries (although here it seems to be a sheltering as opposed to a feeding response as in predatory gastropods). Predatory snails can detect bait at considerable distance, and as for other trap fisheries, the problem of estimating the fishing power of individual traps—especially when, as often, they are set in a string on a ground line—adds to the already considerable difficulties of estimating range of action and effectiveness of a gear that depends on physical responsiveness of the organism being fished. This in turn depends on physiological responses, which change with state of tide, time of day, water temperature, and metabolic state.

In general, the experience gained from crustacean and tropical finfish fisheries using traps seems applicable here, and leads to several conclusions:

1. Trap catches tend to approach an asymptote C' with time (Munro 82), in which cumulative catch C_t on day t is given by

$$C_t = C'[1 - \exp(-Rt)]$$

2. Harvesting and investigating trap effectiveness are best carried out at multiples of time of the prevailing tidal cycle (activity of many foraging organisms seems tied to the tidal period).

3. The "random walk," or nondirectional movement response of an organism in response to a baited trap, converts to a directional response when the odor threshold of the organism is crossed and the bait is detected upstream. For crustaceans, (Miller 83) calibrated the effective area fished within the odor threshold of the trap in order to estimate prey density. Brêthes (84) found that biological factors play an important role in trap performance, and that distance to the trap can sometimes be much higher than expected. These results probably apply also to predatory mollusks.

4. As noted for groundfish long lines, the effectiveness of an individual trap depends on the spacing between traps on a line: too close a spacing results in a confusion between odor trails; too far apart loses the "synergistic" action of adjacent traps. This has been used as a basis for estimating catchability in longline fisheries (Eggers et al. 85), and the method would presumably also apply to gastropod pots set in a string.

4.2.4. Hand Gathering

Handgathering is widely used for molluscan harvesting, and a number of experiments have been reported, especially for abalone, on the calibration of capture efficiency for divers, using recapture of tagged shellfish to detect diver acuity, as well as estimate natural mortality and/or migration (Beinssen and Powell 86). Selectivity of hand-gathering equipment (e.g., rakes and sieves) is discussed by Hancock (87).

5. SPAWNING AND RECRUITMENT

In his extensive review of the literature on invertebrate stock and recruitment, Hancock (9) concluded that there is little specific evidence for a direct relationship between spawning stock size and recruitment, although other intraspecific mechanisms may be important that can lead to statistical relationships similar to those described by Ricker (3). Apparently, heavy spatfall may occur in a wide range of circumstances, including those when adult stock is high or low, when predation has been reduced, or when conditions for larval survival and settlement are particularly good, or a combination of all three. After Thorson (69), Hancock distinguished between epifaunal species (e.g., oysters), where larvae are attracted to and settle in the presence of adults; and infaunal bivalves, where aggregation of juveniles in areas of high adult density, when it occurs, can be catastrophic, either through their being inhaled in the feeding current or by being less competitive than adults in the subsequent struggle for food and space. In Hancock's extensive work on *Cardium* (*Cerastoderma*) *edule* (see, especially, (19); among the most exhaustive reports on a molluscan shellfish population), he notes that the mechanism of population control appears to act through the competition for existing space between "adults" (especially the previous year's recruits) and the newly settling year class (Figs. 2; 5). This shows up as an inverse correlation between abundance of successive year classes on the same grounds. A high density of cockles also results in slower growth, stunting, and higher predation rates. These relationships, which both have spatial components, may explain the long period that often occurs between good year classes in many clam populations, and the good year classes often noted after extensive winter kills of intertidal shellfish. An upper ceiling for population density for infaunal and some epifaunal species also seems to apply, and has been used (Caddy 6) in modeling spatial characteristics of shellfish populations.

5.1. Population Fecundity and the Effects of Environment on Recruitment

Brousseau et al. (71) used an indirect approach for estimating settlement rate for *Mya*, based on age-specific estimates of fecundity and survivorship, and for estuarine shellfish (Ayers 88); stable conditions for stock replacement depend on survival to maturity and spawning success, but also on environmental conditions

OLDER THAN SECOND YEAR

SECOND YEAR

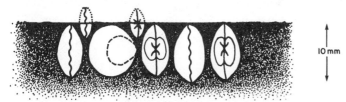

10 mm

Figure 5. Higher probability of competitive eclusion of spat from the substrate at high densities when ground shared with 2-yr cockles than with older animals (after Hancock 9).

and on the configuration of the estuary (Andrews 89). Notable here for *Mya arenaria* is the flushing rate (f) characteristic of a given estuary, which determines the probability of return of locally produced larvae (89). For *Mya arenaria*, the life table given by Brousseau et al. shows the following:

15×10^6 eggs and larvae produced per spawning pair over 5-yr average period of maturity, leading to
- 200 spat that settle and survive (i.e., 40 spat/yr), and
- 1 second-generation pair that survives to maturity and spawns.

The combination of flushing (f) and larval mortality rates (M) that allow a settlement success of at least 40 spat/yr per breeding pair can be considered as limiting for successful population maintenance, and can be calculated from

$$N_t = N_0(1 - f)^t \exp(-Mt)$$

where for *M. arenaria*, $t = 28$ tidal cycles ($= 2$ weeks larval life). This leads to a definition of those limiting conditions under which recruitment either is successful in maintaining a stable or increasing population, or leads to population decline. Treatment of equilibrium recruitment rates using a modified Leslie matrix is also possible (Brousseau et al. 90).

In theory at least, considerations similar to these apply to open-sea populations, although measurement of dilution rates (or of more relevance, probabilities of return of water masses or larvae) is less simply accomplished. Iles and Sinclair (91) have shown that many stocks of pelagic fish are associated with hydrographic conditions that lead to retention of water masses and hence larvae. For sedentary species such as shellfish, the size of the population is determined of course by the area of suitable adult habitat, but presumably also by the size of the gyre or other oceanographic system resulting in water mass retention. For sedentary organisms also, the age structure of the population at a given point (in an unharvested or lightly harvested population) is an index of the probability of successful recolonization of that area by a larva released from some point on the grounds. In the case of a population showing wide fluctuations in abundance, areas where all year classes are represented in all years are probably loci with a high probability of recolonization because of hydrographic as well as other conditions favorable to settlement (Fig. 6). We might then postulate an "index of recolonization" as simply a function of

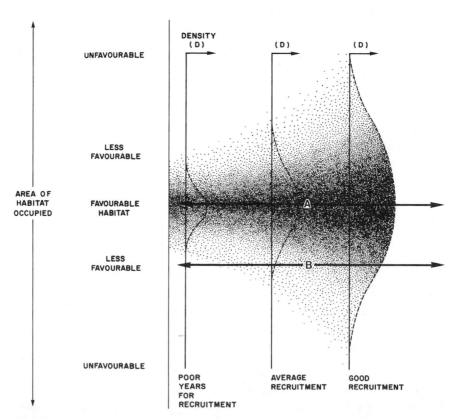

Figure 6. Diagrammatic representation of variations in the spatial extent of spatfall in good and bad years of recruitment over favorable and unfavorable loci.

the number of year classes (n) present at each locality compared with the local life-span N.

$$\text{i.e.} \qquad I_c = (N - n + 1)^{-1}$$

Of course, other indexes weighted by year class abundance could be considered which would give a more precise measure of the regularity of recruitment in each locality. The utility of such indexes in a population that shows wide fluctuations in recruitment (e.g., Fig. 6) might be to distinguish those areas that consistently produce recruits, even in poor years, from those that receive recruitment only occasionally. Such areas might be worth special conservation measures, such as local fishery closures as special brood stock areas, or might be useful localities for stock enhancement exercises.

5.2. Indexes of Recruitment and Annual Fluctuations in Recruitment

There is good evidence for several mollusk resources for which long series of catch statistics are available, that recruitment shows regular long-term fluctuations superimposed on the short-term variability discussed earlier. A variety of hypotheses have been put forward for different species; thus for *Mya arenaria*, Tinsman (92) found predation to be the causative factor. For scallops in the Bay of Fundy, a long-term cyclic recruitment with a periodicity of 9 yr was estimated by Caddy (93), with long-term tidal periodicities as the putative control factor. Diseases (Sindermann and Rosenfield 94) and catastrophic mortalities can also play a major role in population fluctuations, especially at high densities. Various statistical procedures for identifying climatic factors correlated with shellfish recruitment have been proposed (e.g., Ulanowicz et al. 61).

6. YIELD MODELS AND HARVESTING STRATEGY

6.1. Yield-per-Recruit Models

Yield-per-recruit models have been applied to assessment of a variety of bivalve populations, based on a knowledge of the growth and mortality characteristics of the population. Two main approaches to yield-per-recruit calculation seem possible in a managed shellfish fishery for defining optimal harvesting strategy:

1. The approach commonly used in finfish yield-per-recruit analysis, which applies when age or size classes are widely mixed, is to define the optimal rate of harvesting above a given minimum size, if harvesting is to occur continuously at a constant rate (i.e., at a constant level of fishing mortality, F). Such an approach has been adopted for a number of molluscan stocks (see Fig. 7; Wood and Olsen 95).

2. The formulations of Beverton and Holt (96) are most simply employed using either the yield tables (Beverton and Holt 97) or various computer programs written

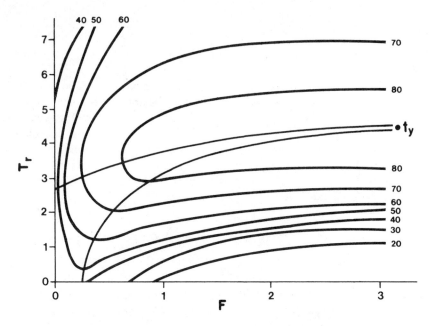

Figure 7. Yield-per-recruit isopleths for the queen conch *Strombus gigas*, redrawn from Wood and Olsen (95) to show the point t_y.

for 'his purpose (e.g., Sims 98). If time intervals are irregular, and rates differ throughout life, Ricker's approach (Paulick and Bayliff 99) is perhaps more suitable. However, the straightforward approach associated with Thompson and Bell and described in Ricker (3) appears to be one of the most flexible, and is illustrated in Table 1. This calculation is of course being repeated for different vectors of F; or, in the case of an increase in mesh size, with different values of

TABLE 1 Simple Yield-per-Recruit Calculation[a]

Age (i)	M_i	p_i	$\overline{w}$ (g)	F_i	Popula-tion	Deaths	Catch Nos.	Yield (kg)
					(R =			
1	0.6	0.10	12.2	0.08	13,000)	5922	78	0.95
2	0.3	0.50	27.0	0.40	7078	2785	1114	30.07
3	0.2	0.80	32.3	1.30	4293	3051	2559	82.65
4	0.2	0.95	44.5	1.86	1242	1068	959	42.67
>4	—	—	—	—	(174)	—	—	—

Total yield   156.34 kg

Average yield per recruit   12 g

[a]For a shellfish population having four exploited year classes, an average recruitment of 13,000 clams, and age-specific natural and fishing mortality rates, where $D_i = N_i\{1 - \exp[-(F_i + M_i)]\}$, $C_i = D_i F_i/(F_i + M_i)$, $F_i = p_i F$, and $Y_i = C_i \overline{w}_i$.

p_i. The potential change in yield per recruit can then be calculated for each change in harvest strategy.

6.2. Yield-per-Recruit Calculations and Parameter Variability

There has been little consideration of variability of growth and natural mortality of sedentary shellfish species in fisheries assessment and management, and the importance of carrying out calculations over the likely range of parameter values, even if these are not precisely known, seems worth emphasizing. For populations showing a geographic range of growth and mortality characteristics, calculations should ideally be described separately for each spatially distinct stratum, and the predictions (for example, the calculated yield from a given strategy) integrated afterward. Thus if the population is made up of j smaller units, each subject to a different harvesting intensity and size at first capture, we have

$$Y_T = \sum_j (Y/R)_j (R)_j$$

This may have particular relevance to assessment of some patchily distributed or widely scattered shellfish resources (e.g., clam beds in a series of estuaries along a coastline); and an approach to yield-per-recruit modeling that takes into account spatial factors has been proposed by Caddy (6).

These procedures used for wild finfish stocks may be simplified for some commercial shellfish if a single year class predominates in a local population, owing either to annual recruitment fluctuations or to geographic segregation of year classes. This situation can also occur as a result of past catastrophic mortalities of earlier year classes, can be naturally caused, or can be due to a (rotational) havesting regime, as in an artificially relayed population. In these circumstances it may then be desirable to harvest a bed completely over a short period of time when a dominant year class reaches an optimal size. In terms of the physical yield resulting, this strategy is better than the "continuous harvesting regime" characteristic of finfish management. This provides a yield approximating to the potential yield with infinite fishing effort (Kutty and Qasim 100), which corresponds to the right-hand margin of a yield isopleth diagram (see point t_y on Fig. 7). The size at optimal harvesting has been defined by the last-mentioned authors and further discussed by Alverson and Carney (101). It can be simply calculated by looking for the maximum value of $B_t = N_t w_t$ with age, given the natural mortality and growth rates of the population, where

$$N_t = R \exp [-M(t - t_0)] \quad \text{and} \quad w_t = aL_t^b$$

Several simple algorithms for finding the critical size and age, t_{opt}, are suggested in Kutty and Qasim (100) and in Alverson and Carney (101); thus for species showing allometric growth, t_{opt} can be found by solving

$$\exp\left(K(t_{\text{opt}}) - t_0\right) = (bK + M)/M$$

where t_0 and K are parameters of the von Bertalanffy equation.

Other criteria than simply the age at maximum physical yield may be used in deciding on an optimal harvesting strategy, however, and more complex two-stage harvesting schemes seem likely to be more productive if density-dependent growth and mortality are pronounced (e.g., Hannesson 102); for example, in the early culling of small cherrystone clams before main harvesting later at a larger size. The effects of various harvest strategies on the physical and economic return can be determined given a knowledge of population parameters, using a simulation approach.

6.3. Simple Production Modeling Theory and Its Application to Mollusks

In the absence of age–composition data, and given information on annual catch and effort (or catch rate and total landings), the production modeling approach (for a brief review, see Gulland 4), provides a cost-effective way of assessing large homogeneous populations, with the added advantage that extensions of this type of model to take into account economic theory are well developed (see, e.g., Clark 103). Other approaches that replace fishing effort by mortality rates have been proposed that seem potentially useful for molluscan assessments (Csirke and Caddy 104) and are discussed by Caddy (79), but a major drawback to all of these dynamic pool approaches is, as noted earlier, the systematic changes in harvesting intensity that may occur over a patchy, immotile population. The composite production model approach suggested by Munro (105) and Caddy and Garcia (106) seems worth considering for application to molluscan shellfish fisheries. Here data are collected in the same or successive years from a series of grounds or beds with similar basic productivities but different harvesting intensities, in order to define optimal harvesting intensities and production per unit area.

The model is fitted by regression of CPUE on intensity of fishing, in which different estimates for area (i) of extent A_i in the same or different year (j) are combined, and the model fitted is either

(a) Schaefer: $Y_{ij}/f_{ij} = A - B(f_{ij}/A_i)$
(b) logarithmic: $\log_e (Y_{ij}/f_{ij}) = A - B(f_{ij}/A_i)$

Both of these approaches necessarily suppose a good map of the grounds being harvested, and some prior ideas of productivity and standing stock for the resource in each subarea.

6.4. Closed Seasons, Closed Areas, and Rotational Harvesting

Shellfish closures (or inversely, open seasons) may be instituted for a number of reasons, as summarized for fisheries in general by Caddy (107). Of particular interest for shellfish are:

Closure for economic reasons of poor landed price.

Closure because of poor meat condition.

Closure for reasons of shellfish toxicity; e.g., high coliform counts or paralytic shellfish poison (PSP) contamination.

Closure to conserve populations and promote stock buildup.

It may be noted that although any motive for closure indirectly promotes the objectives of the last mentioned, this measure, as for daily bag limits used in some fisheries, is not particularly effective in the absence of some overall limits to access.

We may wish to single out the second motive for closure as being of particular interest for shellfish resources where meat condition, and especially stage of gonad development, often affects marketability. Ideally, a closed season should be related to the normal cycle of gonad condition, in order to improve quality and probably overall revenue from the fishery.

The possibility of rotational harvesting as a yield strategy is an attractive one for sedentary mollusks, and McHugh (108) has suggested that dividing the harvest area into three zones for rotational cropping is a real possibility; an age-structured computer model that permits exploration of these strategies is given by Gales and Caddy (109). The importance of adequate enforcement of the necessary regulations, even with the exclusive user rights implied by this strategy, imposes a real cost, however.

In the case of rotational harvesting, as indicated above, it is at least theoretically possible to realize close to instantaneous harvesting of a cohort when its population biomass is maximal, and this age at maximum biomass has been calculated (e.g., Hughes and Bourne (110) at between 9 and 13 yr for *Spisula polynyma* in the southeast Bering Sea).

6.5. Optimal Harvesting Levels

The selection of an optimal level of fishing mortality and a size at first entry to the fishery which balance the rate of harvesting against the rates of growth and natural mortality have always been the two main objectives of yield-per-recruit analysis. In more recent years, concern has switched from the impacts of this classic "growth overfishing," in which the yield-per-recruit analysis for a given number of recruits falls below some preset criterion, to a concern with "recruitment overfishing;" signifying that level of fishing where the average number of new recruits produced by the stock under the current harvesting regime has fallen well below previous levels. Just as for production modeling estimates where the maximum sustainable yield (MSY) is now recognized to occur at harvesting intensities above those providing the maximum economic yield (MEY), so for yield-per-recruit analysis, criteria leading to a lower level of harvesting with a higher catch rate of larger animals than at the maximum yield per recruit (Y_{max}) are now widely applied. One of the most popular alternative criteria in current use, although arbitrary, seems to come close to MEY: the fishing mortality at the $F_{0.1}$ level (Gulland and Boerema 111). This is the fishing rate for a given size selection pattern that results in the yield per recruit rising at one-tenth of the rate prevailing at the start of fishing. Another alternative approach in production modeling is to try to define the point of maximum biological production (MBP) of the population (again falling to the left of MSY), as the point of maximal robustness of the population to the effects of harvesting (Caddy and Csirke 112).

From practical experience, both MSY and Y_{max} seem to lead to a reduced or more irregular level of recruitment than the lower harvesting rates suggested above as alternatives. Another source of concern is the fall of size at first capture below the size at 50% sexual maturity at high fishing intensities: under these conditions, it may be more relevant to calculate the fecundity per recruit (number of eggs produced by those recruits that survive to maturity at each harvesting level), to ensure that this has not fallen too drastically.

7. STOCK ENHANCEMENT AND SOCIOECONOMICS

7.1. Planting of Seed

The factors that must be taken into account in successful replanting of seed bivalves have been the subject of an extensive literature. These range from obvious considerations such as site and substrate selection, the size of seed and season of planting, and, to a significant extent, the abundance of predators (e.g., Flagg and Malouf 113).

In a number of cases, this attractive approach to population enhancement has been suggested while ignoring the basic density-dependent mechanisms in play in shellfish populations whereby, as we have seen, natural mortality increases and growth declines, for example, as a function of high density. The basic economics of labor-intensive cultivation procedures also has to be taken into account. Other rationales for transplanting, for example, of hard clams from cool northern waters to the warmer-water New York fishery, with the intention of prolonging the presence of spawning in the population (and hence the possibility of successful recruitment), seem to have been less than successful (e.g., Kassner and Malouf 114), both because of the variability of spawning time of native stock and presumably because of rapid acclimatization of transplants.

7.2. Control of Access and Harvesting Intensity

A wide variety of approaches to management of shellfish resources are now practiced (Table 2), ranging from various definitions of TURF (Territorial User Rights for Fishermen), which can range from the exercise of traditional rights of access by individuals or communities, to various forms of limited licensing, or to formal leasing arrangements. An alternative approach is to define an overall quota, usually in conjunction with a license scheme, in order to attempt to control the level of removals; this usually involves some system of catch allocation by company, gear and vessel type, and so on, which at the final level of elaboration

TABLE 2 Brief Summary of Strategies Applied for Management of Shellfish Fisheries

Management Methodology	Nature of Access Right	Responsibility for	
		Control of Harvest Intensity	Control of Access
Shellfish lease	Exclusive to lessee	Lessee	Lessee
Traditional user right	More or less exclusive	Management authority or user?	
Limited license	Exclusive to licensee	Management authority	
Catch quota	All or licensee	Management authority	Annually or quarterly limit?
Quota allocation	Licensee	Management authority	Annually or quarterly limit?
Bag limit	All or licensee	Management authority	Trip limit

can even be refined to annual boat or fishermen allocations. This final group of methodologies has rarely been used to date in molluscan shellfisheries, but has some analogies in the daily bag or creel limit used in a number of local clam and cockle fisheries. All these methodologies have as their stated or unspoken objectives the promotion of conservation of the resource and/or the assurance that opportunity of access for all participants conforms to some preset arrangement. The procedure chosen depends to a large extent on the political structures in place, as well as the attitudes of fishermen and those in authority. In terms of the need to control the level of removals, bag limits alone will be successful only if some control of access is also exerted. In practice, this also seems to be the case for quotas, although in recent years a large but diffuse literature on this subject representing diverse points of view has sprung up; the reader is referred to this literature for further details (see, e.g., Grover 115). Clearly, the success of all these measures (even leasing, where the management authority needs at least to maintain lease records, keep up bench marks for lease boundaries, and collect rent), depend for their success to a varying degree on the level of expenditures by the management authority, and this in turn has to take into account the economic return to the authority from the fishery as a whole.

For many intertidal and nearshore shellfish resources, stock management areas are small, and may fall under the jurisdiction of regional or local government or municipalities (Townsend 116), whereas shelf or high-seas resources such as squid or offshore scallops, with generally larger unit stocks, may come under management of national or international bodies. This situation poses certain advantages for determining the optimal management of nearshore resources, in that statistical replication will, at least in theory, allow a comparison of the effectiveness of different alternative management schemes if the necessary data can be collected. As indicated earlier, composite production models can be used in this way to compare yield per hectare against fishing effort per hectare of a productive shellfish bed, in order to define an empirical optimum.

7.3. Predator Control

In practice the methodologies developed over the past century or so range from mechanical elimination, for instance, of starfish from oyster beds by towing rope mops, to the chemical elimination of sea urchins and predatory gastropods. Although quite effective, particularly on a small scale, few accounts of the economic performance of these methods seem to be available, although some perspectives on the techniques are given by Hancock (64), MacKenzie (51), and Loosanoff (50). Much attention, however, has been directed at the energetics of predator–prey relationships in a number of invertebrate systems, and it seems clear that the presence of alternative prey for predators, but more especially the density of commercial prey species, may be the key factors; the latter is probably the main limiting factor for intensive bottom culture and one of the major incentives for raft or hanging cultivation of mollusks. Predation is one of the factors that has contributed to failure of artificial transplantation of clam seed, although the view has been expressed (e.g., McHugh 108) that the scale of introductions when trying to bring a new area into production is often too small, and should be large enough to smother existing predators. This general area requires a knowledge of predator/parasite life histories (e.g., Hancock 87), and falls within the field of multispecies interactions which have been discussed more extensively in other chapters (e.g., Chapters 17 and 26).

REFERENCES

1. K. M. Wilbur, ed., *The Mollusca*, Vol. 1–6. Academic Press, New York, 1983.
2. J. F. Caddy, (Ed.) Advances in assessment of world cephalopod resources. *FAO Fish. Tech. Pap.* **231**:1–452 (1983).
3. W. E. Ricker, Computation and interpretation of biological statistics of fish populations. *Bull. Fish. Res. Board Can.* **191**:1–382 (1975).
4. J. A. Gulland, *Fish Stock Assessment: A Manual of Basic Methods*, Vol. 1. Wiley, New York, 1983, 233 pp.
5. D. A. Hancock, Population dynamics and management of shellfish stocks. *Rapp. P.-V. Run., Cons. Int. Explor. Mer* **175**:8–19 (1979).
6. J. F. Caddy, Spatial model for an exploited shellfish population, and its application to the Georges Bank scallop fishery. *J. Fish. Res. Board Can.* **32(8)**:1305–1328 (1975).
7. C. B. Jorgensen, 1946. Reproduction and larval development of Danish marine bottom invertebrates. 9. Lamellibranchia. *Medd. Komm. Dan. Fisk.- Havunders., Ser. Plankton* 4(1) p 277–311 (1946).

8. M. Sinclair, R. K. Mohn, G. Robert, and D. L. Roddick, Considerations for the effective management of Atlantic scallops. *Can. J. Fish. Aquat. Sci. Tech. Rep.* **1382**:1–113 (1985).

9. D. A. Hancock, The relationship between stock and recruitment in exploited invertebrates. *Rapp. P.-V. Reun., Cons. Perm. Int. Explor. Mer* **164**:113–131 (1973).

10. R. H. MacArthur and O. E. Wilson, *The Theory of Island Biogeography*. Princeton Univ. Press, Princeton, NJ, 1967, 25 pp.

11. E. C. Pielou, *An Introduction to Mathematical Ecology*. Wiley (Interscience), New York, 1969, 286 pp.

12. D. Gardefors and L. Orrhage, Patchiness of some marine bottom animals: A methodological study, *Oikos* **19**:311–321 (1968).

13. J. M. Elliott, Some methods for the statistical analysis of samples of benthic invertebrates. *Sci. Publ.—Freshwater Biol. Assoc.* **25**:1–144 (1971).

14. S. B. Salia and J. A. Gaucher, Estimation of the sampling distribution and numerical abundance of some mollusks in a Rhode Island salt pond. *Proc. Natl. Shellfish Assoc.* **56**:73–80 (1966).

15. M. J. Fogarty, Distribution and relative abundance of the Ocean Quahog *Arctica islandica* in Rhode Island Sound and off Martha's Vineyard, Massachusetts. *J. Shellfish Res.* **1(1)**:33–39 (1981).

16. R. B. Nickerson, A study of the littleneck clam (*Protothaca staminea* Conrad) and the butter clam (*Saxidomus giganteus*, Deshayes) in a habitat permitting coexistence, Prince William Sound, Alaska. *Proc. Nat. Shellfish Assoc.* **67**:85–102 (1977).

17. G. Y. Conan, Assessment of shellfish stocks by geostatistical techniques. *ICES Meet. Doc., Shellfish Comm.* C.M. 1985/**K:30**:1–18 (1985).

18. M. Mangel and J. H. Beder, Search and stock depletion: Theory and applications. *Can. J. Fish Aquat. Sci.* **42(1)**:150–163 (1985).

19. D. A. Hancock and A. E. Urquart, The determination of natural mortality and its causes in an exploited population of cockles (*Cardium edule* L.). *Fish. Invest. (London)* (2) **24(2)**:1–40 (1965).

20. B. A. MacDonald and M. L. H. Thomas, Age determination of the soft-shell clam *Mya arenaria* using shell internal growth lines. *Mar. Biol. (Berlin)* **58(2)**:105–109 (1980).

21. R. M. Harbo, B. E. Adkins, P. A. Breen, and K. L. Hobbs, Age and size in market samples of geoduc clams (*Panope generosa*). *Can. Man. Rep. Fish. Aquat. Sci.* **1714**:1–81 (1983).

22. D. S. Jones, D. F. Williams, and M. A. Arthur, Growth history and ecology of the Atlantic surf clam, *Spisula solidissima* (Dillwyn), as revealed by stable isotopes and annual shell increments. *J. Exp. Mar. Biol. Ecol.* **73(3)**:225–242 (1983).

23. S. A. Murawski, J. W. Ropes, and F. M. Serchuk, Growth of the ocean quahog, *Arctica islandica*, in the middle Atlantic Bight. *Fish Bull.* **80(1)**:21–34 (1982).

24. B. O. Miranda, Crecimiento y estructura poblacional de *Thias* (*Stromanita*) *chocolata* (Duclos 1823), en la Bahia de Mejillones del Sur, Chile. *Rev. Biol. Mar.* **15(3)**:263–286 (1975).

25. E. G. Dawe, R. K. O'Dor, P. H. Odense, and G. V. Hurley, Validation and application of an ageing technique for short-finned squid (*Illex illecebrosus*). *J. Northwest. Atl. Fish. Sci.* **6**:107–116 (1985).

26. C. A. Richardson, D. J. Crisp, and N. W. Runham, An endogeneous rhythm in shell deposition in *Cerastoderma edule*. *J. Mar. Biol. Assoc. U.K.* **60(4):**991–1004 (1980).

27. R. Jones, Assessing the effects of changes in exploitation pattern using length composition. *FAO Fish Tech. Pap.* **256:**1–118 (1984).

28. D. Pauly, Some simple methods for the assessment of tropical fish stocks. *FAO Fish. Tech. Pap.* **234:**1–52 (1983).

29. D. G. Powell, Estimation of mortality and growth parameters from the length frequency of a catch. *Rapp. P.-V. Reun., Cons. Int. Explor. Mer* **175:**167–169 (1979).

30. D. A. Fournier and P. A. Breen, Estimation of abalone mortality rates with growth analysis. *Trans. Am. Fish Soc.* **112:**403–411 (1983).

31. R. S. Appeldoorn, Variation in the growth rate of *Mya arenaria* and its relationship to the environment as analysed through principal components analysis and the omega parameter of the von Bertalanffy equation. *Fish Bull.* **81(1):**75–84 (1983).

32. N. H. Hadley and J. J. Manzi, Some relationships affecting growth of seed of the hard clam *Mercenaria mercenaria* in raceways. *J. Shellfish Res.* **3(1):**92 (1983) (abstr. only).

33. D. J. Wildish and D. D. Kristmanson, Control of suspension feeding bivalve production by current speed. *Helgol. Meeresunters.* **39:**237–243 (1985).

34. C. R. Newell, The effects of sediment type on growth rate and shell allometry in the soft-shell clam *Mya arenaria* Linne. *J. Shellfish Res.*, **3(1):**98 (1983).

35. R. L. Walker, Effects of density and sampling time on the growth of the hard clam, *Mercenaria mercenaria*, planted in predator-free cages in coastal Georgia. *Nautilus* **98(1):**114–119 (1984).

36. C. H. Peterson and M. L. Quammen, Siphon nipping: Its importance to small fishes and its impact on the growth of the bivalve *Protothaca staminea* (Conrad). *J. Exp. Mar. Biol. Ecol.* **63(3):**249–268 (1982).

37. K. Sasaki, Growth of the Sakhalin surf clam, *Spisula sachalinensis* (Schrenk), in Sendai Bay. *Tohoku J. Agric. Res.* **32(4):**168–180 (1981).

38. A. Bodoy, Croissance saisonniere du bivalve *Donax trunculus* (L.) en Mediterranée nord-occidentale (France). *Malacologia* **22(1-2):**353–358 (1980).

39. G. E. Farrow, Periodicity structures in the bivalve shell: Experiments to establish growth controls in *Cerastoderma edule* from the Thames Estuary. *Palaeontology* **14(4):**571–588 (1971).

40. W. E. Ricker, Effects of size-selective mortality and sampling bias on estimates of growth, mortality, production and yield. *J. Fish. Res. Board Can.* **26:**479–541 (1969).

41. D. A. Hancock, Growth and mesh selection in the edible cockle (*Cardium edule* L.). *J. Appl. Ecol.* **4:**137–157 (1967).

42. W. G. Yap, Population biology of the Japanese little-neck clam, *Tapes philippinarum*, in Kaneohe Bay, Oahu, Hawaiian Islands. *Pac. Sci.* **31(3):**223–244 (1977).

43. C. C. Taylor, Temperature, growth and mortality—The Pacific cockle. *J. Cons., Cons. Int. Explor. Mer* **26:**117–124 (1960).

44. W. J. Sutherland, Spatial variation in the predation of cockles by oystercatchers at Traeth Melynog, Anglesey. II. The pattern of mortality. *J. Anim. Ecol.* **51(2):**491–500 (1982).

45. J. V. Merriner and J. W. Smith, *A Report to the Oyster Industry of Virginia on the*

Biology and Management of the Cownose Ray (Rhinoptera bonasus, Mitchill) in Lower Chesapeake Bay, Spec. Rep. Appl. Sci. Ocean Eng., VA, 1979.

46. J. M. Whetstone and A. G. Eversole, Predation on hard clams, *Mercenaria mercenaria* by mud crabs, *Panopeus herbstii*. *Proc. Natl. Shellfish Assoc.* **68:**42–48 (1978).

47. J. N. Kraeuter and M. Castagna, Effects of large predators on the field culture of the hard clam, *Mercenaria mercenaria*. *Fish. Bull.* **78(2):**538–541 (1980).

48. E. G. Boulding and T. K. Hay, Crab response to prey density can result in density-dependent mortality of clams. *Can. J. Fish. Aquat. Sci.* **41(3):**521–525 (1984).

49. M. R. Carriker, Critical review of biology and control of oyster drills, *Urosalpinx* and *Eupleura*. *U.S., Fish Wildl. Serv., Spec. Sci. Rep.—Fish.* **148:**1–150 (1955).

50. V. L. Loosanoff, Recent advances in the control of shellfish predators and competitors. *Proc. Gulf. Caribb. Fish. Inst.* **13:**113–127 (1960).

51. C. L. MacKenzie, Jr., Use of quicklime to increase oyster seed production. *Aquaculture* **10:**45–51 (1977).

52. D. C. Edwards, D. O. Conover, and F. Suter, III, Mobile predators and the structure of marine intertidal communities. *Ecology* **63(4):**1175–1180 (1982).

53. N. B. Chilton and C. M. Bull, Influence of predation by a crab on the distribution of the size-groups of three intertidal gastropods in South Australia. *Mar. Biol. (Berlin)* **83(2):**163–170 (1984).

54. J. A. Commito, Naticid snail predation in New England: The effects of *Lunatia heros* on the population dynamics of *Mya arenaria* and *Macoma balthica*. *J. Shellfish Res.* **3(1):**87 (1983) (summary only).

55. C. H. Katz, A nonequilibrium marine predator-prey interaction. *Ecology* **66(5):**1426–1438 (1985).

56. J. M. Hoenig, Empirical use of longevity data to estimate mortality rates. *Fish Bull.* **82(1):**898–903 (1984).

57. B. L. Bayne, Primary and secondry settlement in *Mytilus edulis* L. (Mollusca). *J. Anim. Ecol.* **33:**513–523 (1964).

58. B. Baggerman, Spatfall and transport of *Cardium edule* L. *Arch. Neerl. Zool.* **31:**1–108 (1953).

59. H. Hidu, W. G. Valleau, and F. P. Veitch, Gregarious setting in European and American oysters—Response to surface chemistry vs. waterborne pheromones. *Proc. Natl. Shellfish Assoc.* **68:**11–16 (1978).

60. H. H. Haskin, E. S. Wagner, and M. L. Tarnowski, The surf clam along the New Jersey coast: Population size, recruitment, growth rates. *J. Shellfish Res.* **3(1):**93 (1983).

61. R. E. Ulanowicz, M. L. Ali, A. Vivian, D. R. Heinle, W. A. Richkus, and J. K. Summers, Identifying climatic factors influencing commercial fish and shellfish landings in Maryland. *Fish Bull.* **80(3):**611–619 (1982).

62. B. L. Bayne and R. C. Newell, Physiological energetics of marine molluscus. In A. S. M. Saleuddin and K. M. Wilbur, Eds., *The Mollusca*, Vol. 4, Part I. Academic Press, New York, 1983, pp. 410–516.

63. T. C. Kline, The effect of population density on the growth of the butter clam *Saxidomus giganteus*. *J. Shellfish. Res.* **3(1):**112 (1983).

64. D. A. Hancock, The role of predators and parasites in a fishery for the mollusc

Cardium edule. Proc. Adv. Study Inst. Dyn. Numbers Popul., 1970, pp. 419–439 (1970).

65. D. R. Franz, Size and age-specific predation by *Lunatia heros* (Say, 1822) on the surf clam *Spisula solidissima* (Dillwyn, 1817) off western Long Island, New York. *Veliger* **20(2):**144–150 (1977).

66. W. E. Kelso, Predation on soft-shell clams, *Mya arenaria*, by common mummichog, *Fundulus heteroclitus. Estuaries* **2(4):**249–254 (1979).

67. L. M. Dickie, Fluctuations in abundance of the giant scallop, *Placopecten magellanicus* (Gmelin) in the Digby area of the Bay of Fundy. *J. Fish Res. Board Can.* **2(6):**797–857 (1955).

68. W. I. Wiltse, Predation by juvenile *Polinices duplicatus* (Say) on *Gemma gemma* (Totten). *J. Exp. Mar. Biol. Ecol.* **42(2):**187–199 (1980).

69. G. Thorson, Bottom communities (sublittoral or shallow shelf). Treatise on Marine Ecology and Palaeoecology. Vol. 1. *Mem.—Geol. Soc. Am.* **67:**461–534 (1957).

70. K. Muus, Settling, growth and mortality of young bivalves in the Oresund. *Ophelia* **12:**79–116 (1973).

71. D. J. Brousseau, J. A. Baglivo, and G. E. Land, Jr., Estimation of equilibrium settlement rates for benthic marine invertebrates: Its application to *Mya arenaria* (Mollusca: Pelecypoda). *Fish. Bull.* **80(3):**642–644 (1982).

72. T. I. Baranov, On the question of the biological basis of fisheries. *Proc. Inst. Icthiol. Invest.* **1:**81–128 (1918).

73. D. L. Alverson and W. T. Pereyra, Demersal fish explorations in the northeastern Pacific Ocean—An evaluation of exploratory fishing methods and analytical approaches to stock size and yield forecasts. *J. Fish. Res. Board Can.* **26:**1985–2001 (1969).

74. J. F. Caddy, Some considerations underlying definitions of catchability and fishing effort in shellfish fisheries, and their relevance for assessment purposes. *Fish. Mar. Serv., Manuscr. Rep.* **1489:**1–18 (1979).

75. R. J. Rhodes, W. J. Keith, P. J. Eldridge, and V. G. Burrell, Jr., An empirical evaluation of the Leslie-Delury method applied to estimating hard clam, *Mercenaria mercenaria*, abundance in the Santee River estuary, South Carolina. *Proc. Natl. Shellfish Assoc.* **67:**44–52 (1977).

76. P. R. Sluczanowski, A management oriented model of an abalone fishery whose substocks are subject to pulse fishing. *Can. J. Fish. Aquat. Sci.* **41:**1008–1014 (1983).

77. B. J. Rothschild and D. S. Robson, The use of concentration indices in fisheries. *Fish. Bull.* **70(2):**511–514 (1972).

78. J. F. Caddy and S. Garcia, Production modelling without long data series. National reports and selected papers presented at the third session of the working party on assessment of marine fishery resources of the Western Central Atlantic Fishery Commission (WECAFC). *FAO Fish. Rep.* **278(Suppl.):**309–313 (1983).

79. J. F. Caddy, Stock assessment in data-limited situations—the experience in tropical fisheries and its possible relevance to evaluation of invertebrate resources. *Can. J. Fish Aquat. Sci., Spec. Publ.* **92:**379–392 (1986).

80. J. C. Medcof and J. F. Caddy, Underwater observations on performance of clam dredges of three types. *ICES Meet. Doc. C.M.* **1971/B:**10:1–5 (1971).

81. B. E. Adkins, R. M. Harbo, and N. Bourne, An evaluation and management consid-

erations of the use of a hydraulic clam harvester on intertidal clam populations in British Columbia. Can. *Manuscr. Rep. Fish. Aquat. Sci. (Can.)* **1716**:1–38 (1983).

82. J. L. Munro, The mode of operation of Antillean fish traps and the relationships between ingress, escapement, catch and soak. *J. Cons., Cons. Int. Explor. Mer.* **35(3)**:337–350 (1974).

83. R. J. Miller, Density of the commercial spider crab, *Chionoecetes opilio*, and calibration of the effective area fished per trap using bottom photography. *J. Fish. Res. Board Can.* **32**:761–768 (1975).

84. J.-C. Brêthes, R. Bouchard, and G. Desrosiers, Determination of the area prospected by a baited trap from a tagging and recapture experiment with snow crabs (*Chionoectes opilio*). *J. Northwest. Atl. Fish. Sci.* **6**:37–42 (1985).

85. D. M. Eggers, N. A. Rickard, D. G. Chapman, and R. R. Whitney, A methodology for estimating area fished for baited hooks and traps along a ground line. *Can. J. Fish. Aquat. Sci.* **39**:448–453 (1982).

86. K. Beinssen and D. Powell, Measurement of natural mortality in a population of Blacklip Abalone. *Notohaliotis ruber. Rapp. P.-V. Run., Cons. Int. Explor. Mer* **175**:23–26 (1979).

87. D. A. Hancock, The ecology of the molluscan enemies of the edible mollusc. *Proc. Malacol. Soc. London.* **34(3)**:123–143 (1960).

88. J. C. Ayers, Population dynamics of the marine clam, *Mya arenaria. Limnol. Oceanogr.* **1(1)**:26–34 (1956).

89. J. D. Andrews, Transport of bivalve larvae in James River, Virginia. J. Shellfish Res. **3(1)**:29–40 (1983).

90. D. J. Brousseau, J. A. Baglivo, and G. E. Land, Jr., Estimation of equilibrium settlement rates for benthic marine invertebrates: Its application to *Mya arenaria* (Mollusca: Pelecypoda). *Fish Bull.* **80(3)**:642–644 (1982).

91. T. D. Iles and M. Sinclair, Atlantic herring: Stock discreteness and abundance. *Science* **215**:627–633 (1982).

92. J. C. Tinsman, Hard clam cycles. *Del. Fish. Bull.* **2(2)**:13 (1981) (abstr.).

93. J. F. Caddy, Long-term trends and evidence for production cycles in the Bay of Fundy scallop fishery. *Rapp. P.-V. Run., Cons. Int. Explor. Mer* **175**:97–108 (1979).

94. C. J. Sindermann and A. Rosenfield, Principal diseases of commercially important marine bivalve mollusca and crustacea. *Fish. Bull.* **66(2)**:335–385 (1967).

95. R. S. Wood and D. A. Olsen, Application of biological knowledge to the management of the Virgin Islands conch fishery. *Proc. Annu. Gulf. Carribb. Fish. Inst.* **35**:112–121 (1982).

96. R. J. H. Beverton and S. J. Holt, On the dynamics of exploited fish populations. *Fish Invest. Minist. Agric. Fish. Food, Ser. 2* **19**:1–533 (1957).

97. R. J. H. Beverton and S. J. Holt, Manual of methods for stock assessment. Part II. Table of yield functions. *FAO Fish. Tech. Pap.* **38(Rev. 1)**:1–10 (1966).

98. S. E. Sims, Ed., Selected computer programs in FORTRAN for fish stock assessment. *FAO Fish. Tech. Pap.* **259**:183 (1985).

99. G. J. Paulik and W. F. Bayliff, A generalised computer program for the Ricker model of equilibrium yield per recruitment. *J. Fish. Res. Board. Can.* **24**:249–252 (1967).

100. M. K. Kutty and S. Z. Qasim, The estimation of optimum age of exploitation and

potential yield in fish populations. *J. Cons. Cons. Int. Explor. Mer.* **32(2):**249–255 (1968).

101. D. L. Alverson and M. J. Carney, A graphic review of the growth and decay of population cohorts. *J. Cons. Cons. Int. Explor. Mer.* **36(2):**133–143 (1975).

102. R. Hannesson, Optimal thinning of a year class with density dependent growth. *Can. J. Fish Aquat. Sci.* **43:**889–892 (1986).

103. C. W. Clark, *Mathematical Bioeconomics: The Optimal Management of Renewable Resources.* Wiley (Interscience), New York, 1976, 352 pp.

104. J. Csirke and J. F. Caddy, Production modelling using mortality estimates. *Can. J. Fish Aquat. Sci.* **40:**43–51 (1983).

105. J. L. Munro, Stock assessment models: Applicability and utility in small-scale fisheries. In Stock assessments for tropical small-scale fisheries. S. B. Saila and P. M. Roedel, Eds., *Proceedings of the International Workshop, University of Rhode Island.* Univ. of Rhode Island Press, Kingston, 1979, pp. 35–47.

106. J. F. Caddy and S. Garcia, Production modelling without long data series. *FAO Fish. Rep.* **278(Suppl.):**309–313 (1983).

107. J. F. Caddy, Indirect approaches to regulation of fishing effort. *FOA Fish. Rep.* **289(Suppl. 2):**63–75 (1984).

108. J. L. McHugh, Recent advances in hard clam mariculture. *J. Shellfish Res.* **1(1):**51–55 (1981).

109. L. E. Gales and J. F. Caddy, YRAREA: A program to demonstrate effects of exploitation on a contagiously distributed shellfish population. *FRB Tech. Rep.* **582:**1–72 (1975).

110. S. E. Hughes and N. Bourne, Stock assessment and life history of a newly-discovered Alaska surf clam, (*Spisula polynyma*) resource in the Southeastern Bering Sea. *Can. J. Fish Aquat. Sci.* **38(10):**1173–1181 (1981).

111. J. A. Gulland and L. K. Boerema, Scientific advice on catch levels. *Fish. Bull.* **71(2):**32–35 (1973).

112 J. F. Caddy and J. Csirke, Approximations to sustainable yield for exploited and unexploited stocks. *Oceanogr. Trop.* **18(1):**3–15 (1983).

113. P. J. Flagg and R. E. Malouf, Experimental plantings of juveniles of the hard clam, *Mercenaria mercenaria* (Linne) in the waters of Long Island, New York. *J. Shellfish Res.* **3(1):**19–27 (1983).

114. J. Kassner and R. E. Malouf, An evaluation of "spawner transplants" as a management tool in Long Island's hard clam fishery. *J. Shellfish Res.* **2(2):**165–172 (1982).

115. J. H. Grover, Allocation of fishery resources. *Proceedings of a Technical Consultation Sponsored by EIFAC of FAO of the UN; Department of Fisheries and Oceans, Canada; U.S. Department of the Interior, Fish and Wildlife Service and the U.S. Department of Commerce NOAA/NMFS.* Published by Auburn University, Montgomery, AL, by arrangement with FAO, Rome and the American Fisheries Society, 1980, 623 pp.

116. R. E. Townsend, An economic evaluation of restricted entry in Maine's soft shell clam industry North American *J. Fish. Manag.* **5:**57–64 (1985).

31 FORECASTING YIELD AND ABUNDANCE OF EXPLOITED INVERTEBRATES

Michael J. Fogarty
National Marine Fisheries Service
Woods Hole, Massachusetts

1. INTRODUCTION

Predicting yield or abundance of an exploited species may serve a number of objectives. For example, projected estimates of abundance are often required to set annual catch quotas. Forecasts with a broader time horizon may be used to guide investment decisions in a particular fishery. Comparison of yield or abundance predicted by a conceptual or mathematical model with observed levels serves as a test of the model. Each of these objectives may require a different forecasting strategy. Ultimately, our goal is to predict based on an understanding of the dynamics of the system. The complexity of natural systems, however, often precludes a complete specification of this sort and alternative approaches are necessary. Here I make a distinction between structural and heuristic models. Structural models represent the system in some simplified way that nevertheless captures its essential dynamics. In contrast, heuristic models do not imply causality but rather are based on empirically derived relationships or recognition of recurrent patterns in a time series. Structural models of fishery systems are discussed elsewhere in this volume; accordingly, I focus primarily on empirical models. Structural and heuristic models play complementary roles in the ecological sciences (Rigler 1; Pielou 2). It is often possible to provide more accurate point predictions with the latter, whereas the former are indispensable in abstracting general principles.

For the purposes of this discussion, the terms prediction and forecasting are used interchangeably, although some authors use these terms in different contexts (e.g., Brown 3). A number of excellent texts describe empirical forecasting and modeling strategies (Brown 3; Whittle 4; Box and Jenkins 5; Chatfield 6). The interested reader should consult these references for a fuller description of the technical aspects of model construction. Some of the terminology used may be unfamiliar; accordingly, a brief glossary is provided at the end of this chapter.

2. HEURISTIC MODELS

An experimental approach is the preferred method of studying the dynamics of natural systems. Unfortunately, it is often impossible to manipulate systems on a broad spatial (and/or temporal) scale in a strictly controlled fashion. It is, however, these larger scales that are relevant in most of fishery research. In many instances, the only possible approach is to examine relationships among variables for which there is reason to suspect a causal association. Considerable care is necessary, however, in evaluating empirical relationships of this type. Seemingly related variables may follow similar trends by chance. Alternatively, some unknown factor may underlie common trends in the variables. Nevertheless, heuristic models do play an important role in defining possible relationships among components of a system and serve as the foundation for further research.

A number of heuristic methods have been used to examine possible relationships between catches of commercially important invertebrates and environmental variables. Qualitative comparisons have long been used in this way (Templeman and Fleming 7; Taylor et al. 8) and in some instances, very detailed comparisons have been made (Kitano 9). Other approaches range from correlation and simple or multiple linear regression to more complex methods such as regression on principal components, distributed delay models, path analysis, and time series models (see examples in Table 1) (10–44). Clearly, considerable effort has been devoted to developing predictive models for invertebrate fisheries.

It is important to distinguish between changes in catch levels and changes in population size. Often, catch is used inappropriately as a proxy for abundance (Sissenwine 45). Changes in population levels clearly influence catch rates; however, many other factors also play important roles (e.g., changes in fishing

TABLE 1 Examples of Heuristic Methods Used to Relate Landings of Selected Invertebrates to Explanatory Variables[a,b]

Species	Area	Period	Method	Explanatory Variable(s)	Source
Mollusca					
Crassostrea virginica	Maine	1951–1967	Correlation	Temp	Dow (10)
	Maryland	1938–1976	Multiple regression	Temp Precip Sal	Ulanowicz et al. (11)
Mercenaria mercenaria	Maine	1932–1970	Correlation	Temp	Dow (12)
	Maine	1939–1962	Correlation	Temp	Dow (10)
	New England	1928–1971	Correlation	Temp RD	Sutcliffe et al. (13)
Mya arenaria	Maine	1887–1970	Correlation	Temp	Dow (12)
	Maine	1940–1967	Correlation	Temp	Dow (10)
	New England	1928–1971	Correlation	Temp	Sutcliffe et al. (13)
	Maryland	1952–1977	Regression	Temp Sal	Ulanowicz et al. (11)
Placopecten magellanicus	Maine	1939–1967	Correlation	Temp	Dow (10)
	Bay of Fundy	1936–1953	Correlation	Temp	Dickie (14)
	Bay of Fundy	1922–1971	Correlation	Temp	Caddy (15)
	New England	1928–1971	Correlation	Temp RD	Sutcliffe et al. (13)
Mytilus edulis	Maine	1939–1967	Correlation	Temp	Dow (10)
Littorina littorea	Maine	1940–1967	Correlation	Temp	Dow (10)
Loligo pealii	Maine	1939–1967	Correlation	Temp	Dow (10)
Illex illecebrosus	Nova Scotia	1979–1981	Multiple regression Path analysis	Temp Time	Coelho and Rosenberg (16)

TABLE 1 (Continued)

Species	Area	Period	Method	Explanatory Variable(s)	Source
Crustacea					
Homarus americanus	Maine	1939–1949	Correlation	Temp	Taylor et al. (8)
	Maine	1939–1967	Correlation	Temp	Dow (17)
	Maine	1939–1970	Regression	Temp	Flowers and Saila (18)
	Maine	1937–1975	Dist/delay	Temp Effort	Orach-Meza and Saila (19)
	Maine	1928–1984	Transfer function	Temp	Fogarty (20)
	Gulf of St. Lawrence	1912–1972	ARIMA model	None	Boudreault et al. (21)
H. gammarus	Europe (Total)	1950–1972	Correlation	Temp	Dow (17)
	Europe (Total)	1955–1975	Regression	Temp	Dow (17)
Nephrops norvegicus	Europe (Total)	1950–1973	Correlation	Temp	Dow (22)
	Europe (Total)	1950–1972	Correlation	Temp	Dow (22)
	Europe (Total)	1955–1975	Correlation Regression	Temp	Dow (17)
	Scotland	1959–1982	Correlation Regression	Temp	Chapman (23)
Panulirus cygnus	W. Australia	1971–1984	Regression	Recruit	Caputi and Brown (24)
	W. Australia	1969–1980	Regression	Recruit	Phillips (25)
Jasus edwardsii	New Zealand	1963–1974	Harmon regr ARIMA model	None	Saila et al (26)
Cancer magister	N. California/Washington	1949–1972	Correlation	Upwell	Peterson (27)
	N. California/Washington	1949–1972	Correlation	Upwell	Botsford and Wickham (28)
	California	1940–1976	Correlation	Pred	Botsford et al. (29)
	N. California/Washington	1951–1980	Correlation	Wind	Johnson et al. (30)

Species	Location	Years	Method	Variables	Reference
	N. California/Washington	1935–1975	Correlation	Temp Pressure Upwell	Wild et al. (31)
C. irroratus	Maine	1939–1949	Correlation	Temp	Dow (10)
Callinectes sapidus	Maryland	1960–1979	Multiple regression	Irrad RD Temp Sal Spawners	Tang (32)
Pandalus borealis	Maine	1939–1967	Correlation	Temp	Dow (10)
Crangon crangon	Netherlands	1960–1968	Correlation	Temp	Boddeke (33)
	United Kingdom	1946–1974	Multiple regression	Temp Precip Pred	Driver (34)
Pandalus montagui	United Kingdom	1948–1962	Correlation	Temp	Warren (35)
Penaeus esculentus	W. Australia	1970–1983	Multiple regression	Precip Spawners	Penn and Caputi (36)
P. setiferus	Texas	1937–1951	Correlation	Precip	Hildebrand and Gunter (37)
	Texas	1927–1964	Correlation	Precip	Gunter and Edwards (38)
	Louisiana	1927–1964	Correlation	RD	Gunter and Edwards (38)
	Texas/Louisiana	1961–1979	GMDH	Wind RD Temp Sal	Prager and Saila (39)
	Texas/Louisiana	1960–1976	Regr prin comp	Wind RD Temp Sal	Walker and Saila (40)

TABLE 1 (Continued)

Species	Area	Period	Method	Explanatory Variable(s)	Source
Crustacea					
P. aztecus	Texas	1948–1964	Correlation	Precip	Gunter and Edwards (38)
	Louisiana	1956–1964	Correlation	RD	Gunter and Edwards (38)
	Gulf of Mexico	1960–1966	Regression	Recruit	Berry and Baxter (41)
	North Carolina	1978–1981	Multiple regression	Temp	Matylewich and Mundy (42)
				Sal	
				Precip	
				RD	
				Time	
P. merguiensis	N. Australia	1970–1983	Regression	Precip	Staples (43)
	N. Australia	1970–1979	Multiple regression	Precip	Vance et al. (44)
				RD	
				Temp	
				Wind	
				Spawners	

[a]When several time periods were used as an analysis, the largest span is indicated.

[b]Abbreviations used for variables were temp (water temperature), precip (precipitation), sal (salinity), RD (river discharge), recruit (recruitment index), upwell (upwelling), irrad (solar irradiation), pred (predation index), wind (wind stress), and time (time index—month, etc). Abbreviations used for methods were dist delay (polynomial distributed delay model), ARIMA (autoregressive integrated moving average), harmon regr (harmonic regression), GMDH (group method of data handling), regr prin comp (regression on principal components).

intensity, availability, catchability). Catch cannot be taken as an index of abundance if fishing effort or availability is changing. Catch per unit effort (CPUE) is traditionally used as an index of population size and is preferable to catch as a measure of abundance.

An example will illustrate some potential problems. Dow (22) reported an inverse relationship between seawater temperature and landings of Norway lobster (*Nephrops norvegicus*), lagged by 6–7 yr. However, Chapman (23) demonstrated that increases in fishing effort were in fact responsible for increased *Nephrops* landings; during this period, water temperature declined. The relationship between CPUE and water temperature was shown to be positive at lags of 3–6 yr (Chapman 23). Approximately 3–6 yr are required for *Nephrops* to grow to harvestable size, indicating that temperature may affect survival rates during the early life-history stages. This example highlights some of the difficulties in establishing meaningful relationships between environmental variables and measures of fishery performance without additional biological and fishery-related information.

Time series present special difficulties because the data are often autocorrelated (i.e., a data point is not independent of one or more previous values). It is well known that apparent relationships between two autocorrelated series can be spurious (Quenouille 46; Durbin and Watson 47; Draper and Smith 48). In this chapter I deal principally with a class of heuristic models that were developed for modeling discrete time series. Box and Jenkins (5) devised an approach for developing predictive models using a class of linear stochastic difference equations. The Box–Jenkins approach obviates many of the problems associated with other heuristic methods when applied to time series data.

Box–Jenkins models incorporate autoregressive and/or moving average terms; the former represent *memory* in the system (the effect of past values) and the latter reflect the influence of random *shocks* or past perturbations. Below, I provide a brief nontechnical description of the method and illustrate the approach for several invertebrate fisheries. I then demonstrate multivariate time series (transfer function) and intervention models. The latter are useful for modeling the effects of a discrete impact such as a change in fishery regulations. For a detailed description of these methods, see Box and Jenkins 5).

2.1 Autoregressive Integrated Moving Average Models

Suppose we have measurements of a variable (e.g., catch or abundance) taken at equally spaced points in time. It is probable that there will be some similarity in the values taken by this variable in adjacent years. For example, in multiage class stocks, a cohort will contribute to the fishery over several years. If mortality rates are relatively constant, we would expect that catches in one year would be related to those in the next. In addition, adjacent year classes may be similar in size if environmental or biotic factors (e.g., spawning stock or predator levels) affecting recruitment are autocorrelated. We might regress catches in one year on those of one or more previous years to develop a predictive index. We would then have an autoregressive model (the variable has been regressed on itself, hence the term autoregressive). Catches in previous years will not provide a completely accurate predictor of catches in the following year; there will remain some unexplained variation. For instance, random changes in catchability from year to year will clearly have an effect on catch levels. We might consider this source of variability to be a random "shock." Random variability in environmental factors may affect survival of early life-history stages—this too might be considered a random shock. Catches or abundance levels might therefore reflect the impact of random perturbations in previous years. This is the "moving-average" component. A model with both autoregressive and moving-average terms can be expressed

$$
\begin{aligned}
y_t = \phi_1 y_{t-1} + \phi_2 y_{t-2} + \cdots + \phi_p y_{t-p} + a_t \\
- \theta_1 a_{t-1} - \theta_2 a_{t-2} - \cdots - \theta_q a_{t-q}
\end{aligned}
\tag{1}
$$

where y_t is the value taken by variable y in year t, ϕ and θ are autoregressive and moving average parameters, respectively, and a_t is a random error term that is normally distributed with zero mean and constant variance σ^2.

If we have specified the model correctly, the a_t will represent a white noise process; that is, there will be no autocorrelation in the error terms. The process must also be stationary (i.e., there must be no trend in level or variance). For a full discussion on this point, see Box and Jenkins (5). Because many catch and abundance series do appear to exhibit trends, this may seem to be quite restrictive. Fortunately, it is often possible to transform the data to meet these assumptions. Logarithmic or other transformations can be used to stabilize the variance. To remove trends in the series, we can "difference" the observations (i.e., form the new series $z = y_t - y_{t-r}$, where r is the period of differencing). For notational convenience, we specify the backshift operator B, and represent the differenced series as $z = (1 - B^r) y_t$. This convention is used throughout the chapter. The differenced rather than the original series is then modeled. If the series is reduced to white noise after differencing, we have an "integrated" process. A model of a differenced series that contains autoregressive and moving average terms is an autoregressive integrated moving average (ARIMA) model.

The Box–Jenkins approach to modeling time series comprises three distinct phases: (1) model identification, (2) parameter estimation, and (3) diagnostic checking. In the first phase, a tentative model structure is identified based on certain characteristics of the autocorrelation and partial autocorrelation functions (after differencing and/or transformation, as required). An autocorrelation function that decays very slowly with successive lags indicates a nonstationary series. A first-order autoregressive process is characterized by an autocorrelation function that decays exponentially from a lag of one and a significant partial autocorrelation only at the first lag. A first-order moving average process would be characterized by the reverse (a decaying exponential partial autocorrelation function and significant autocorrelation value at a lag of one). We might also have a mixed autoregressive moving average process with an exponential decay in both the autocorrelation and partial autocorrelation functions. Box and Jenkins (5) illustrate autocorrelation and partial autocorrelation functions for more complicated models. Series with seasonal or periodic patterns can also be identified and modeled based on their distinctive autocorrelation and partial autocorrelation functions.

Following tentative model identification, parameters are estimated by either conditional least-squares or maximum likelihood methods. The tentative model is checked by examining the autocorrelation and partial autocorrelation functions of the residuals (a_i). If the model is adequate, there will be no significant values of either function at any lag and we conclude that the residuals are white noise. Conversely, significant autocorrelations or partial autocorrelations indicate that the model is not yet complete. Additional terms are then added to the model to correct this deficiency and the estimation and checking phases are repeated until an adequate model is found. Examples of this approach are provided below to illustrate these concepts.

2.2. Applications

Landings of Dungeness crab, *Cancer magister*, in northern California have fluctuated dramatically over the past 40 years. A number of mechanisms have been proposed to explain the observed periodic oscillations. Physical factors (Peterson 27; Johnson et al. 30; Botsford 49) and biological mechanisms (Botsford 49; Botsford and Wickham 50) have been advanced as possible causes; effects of the former have been examined primarily using heuristic methods (correlation and regression) whereas the latter have been inferred from structural models. Botsford (49) provides an overview and critique of these studies and a synthesis of the two approaches. Mechanisms underlying the apparent cycles must be understood to guide management decisions. Currently, there is a systematic attempt underway to unravel the complex interaction of environmental and population-level factors affecting Dungeness crab stocks. Here, I illustrate the use of the Box–Jenkins approach to model northern California Dungeness crab landings. This analysis is not intended as a substitute for mechanistic studies, but rather as an adjunct. This method may provide more precise short-term forecasts than a structural model as well as insights into underlying population processes.

We begin by examining the autocorrelation and partial autocorrelation functions of the 1943–1983 landings series (Fig. 1). Although a slight upward trend in landings was evident, it was not necessary to difference the series. Note that the autocorrelation function is sinusoidal, reflecting the periodicity in the series. Peaks in the autocorrelation function occur at lags of 1 and 9 yr (although only the former is significant at the 0.05 level). The partial autocorrelation function is characterized by significant values at lags of 1 and 9 yr (Fig. 1). We therefore tentatively identify an autoregressive model with coefficients reflecting the previous year's catch and the catch 9 yr earlier. The apparent 9-yr cycle has been previously described (e.g., Botsford and Wickham 50). Estimating the parameters and checking the residuals of the model, we find that the autoregressive model is adequate (that is, the residuals are not significantly different from white noise).

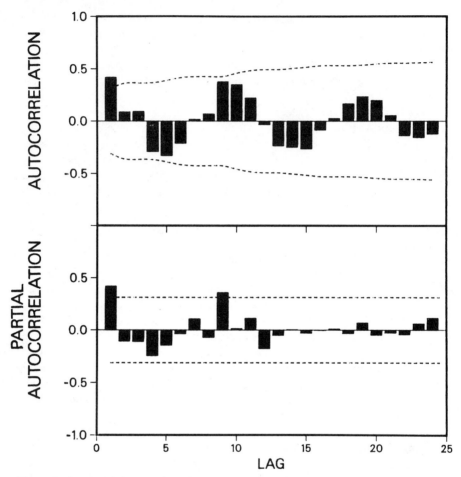

Figure 1. Autocorrelations and partial autocorrelations for Dungeness crab catch (1943–1983). Values extending beyond the dashed lines are significant at the $p = .05$ level.

The final model is

$$(1 - 0.373B^1 - 0.784B^9)C_t' = a_t \qquad (2)$$

where C_t' is the "centered" catch series ($C_t' = C_t - \mu$), where μ is the mean of the series and B is the backshift operator.

A comparison of the observed and fitted series is provided in Figure 2. It is clear that the simple autoregressive model is capable of describing the dynamic behavior of the series. Interpretation of the 9-yr cycle has been extensively debated. For example, an overcompensatory form of population regulation can cause population cycles with a period of approximately twice the mean generation time.

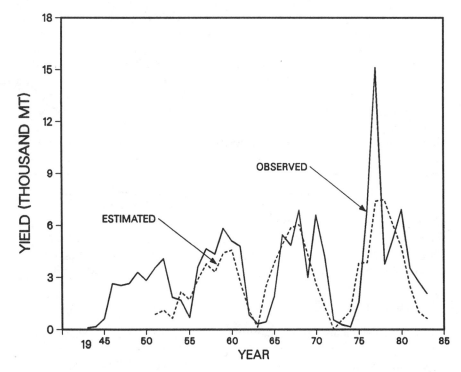

Figure 2. Comparison of observed (——) and estimated (– – –) catches and Dungeness crab (1943–1983).

Alternatively, a periodic environmental forcing factor with the appropriate wavelength might underly the cycles (see Johnson et al. 30; Botsford 49).

2.3. Transfer Function Models

The ARIMA models described above are based entirely on identifiable regularities in a single time series. The effects of other variables are not explicitly considered. Frequently, we wish to relate yield or abundance to one or more variables (e.g., environmental factors that may affect recruitment). The predictive power of the model is enhanced if such a relationship can be established. To construct a time series model with multiple inputs, we develop a "transfer function" model. A transfer model with a single input would be expressed:

$$
\begin{aligned}
y_t - \delta_1 y_{t-1} - \delta_2 y_{t-2} &- \cdots - \delta_m y_{t-m} \\
&= \omega_0 x_{t-b} - \omega_1 x_{t-b-1} - \cdots - \omega_n x_{t-b-n} + N_t
\end{aligned}
\tag{3}
$$

where y_t is the "output" variable (e.g., catch), x_t is an input variable, δ and ω are model parameters, and N_t is an error term that may be modeled as an ARIMA process. The general form of the transfer function is determined by examining the correlations between the variables at various lags. These cross-correlations are used to identify the model structure in much the same way that the autocorrelation and partial autocorrelation functions are used to identify the structure of univariate models. Box and Jenkins recommend that the series be "prewhitened." To do this, a univariate time series model is first fit to the input variable. If an adequate model has been specified, the model residuals are independent. We next apply this same model to the output variable and cross-correlate the residuals rather than the original series. Prewhitening removes the deterministic signal from the input series by reducing it to white noise. If the input and output series are causally related, applying the univariate ARIMA model developed for the input series to the output series should remove the deterministic component of the output variable as well. After the parameters of the tentative model are identified and estimated based on the cross-correlation function, the residuals of the model are checked to see if they are uncorrelated. If significant autocorrelations or partial autocorrelations are indicated, additional autoregressive and/or moving average terms are added to the model. This process is repeated until the residuals are reduced to white noise.

2.4. Applications

An example of this approach is provided for American lobster (*Homarus americanus*) landings in Maine (Fogarty 20). Annual Maine landings from 1945–1983 were related to mean annual water temperature at Boothbay Harbor. Relationships between Maine lobster landings and water temperature have been extensively studied using a variety of heuristic methods (see Table 1). If we consider the cross-correlation between the original catch and temperature series, we note apparently significant relationships between landings and temperature at lags of 3–10 yr (Fig. 3) in addition to several apparently significant correlations at higher lags. We might infer that temperature has a delayed effect on landings (i.e., yield is related to temperature several years previously). If we prewhiten the series, a markedly different picture emerges.

The temperature series is well described by a second-order autoregressive model. Applying this model to both series and cross-correlating the residuals, we find that significant correlations are found at lags of 0 and 6 yr. That is, there appears to be a positive relationship between catch and temperature in the same year and at a lag of 6 yr. Recall that the cross-correlation between the original series and temperature showed no significant correlation at a lag of zero. After initial model identification, estimation and checking, the final transfer function model was

$$C_t' = (524.58 + 461.72B^6)T_t + a_t/(1 - 0.65B) \tag{4}$$

where C_t' is the centered catch series and T_t is the temperature series, B is the backshift operator, and a_t is a white noise random variable.

A significant correlation between landings and temperature in the same year is biologically reasonable. Changes in temperature levels are known to affect the

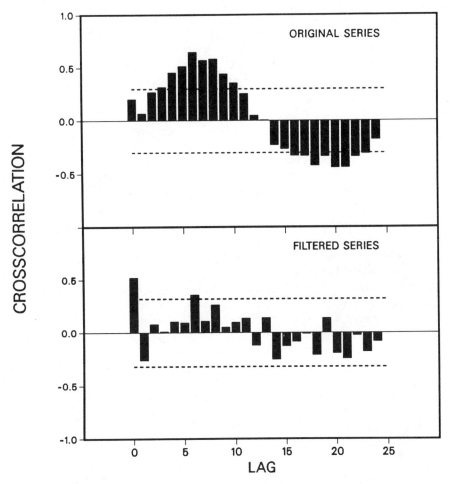

Figure 3. Cross-correlation between annual Maine lobster catch (1945–1983) and mean annual water temperature using the original (upper panel) and prewhitened or filtered (lower panel) series.

activity and catchability of the American lobster (McLeese and Wilder 51). In high temperature years, lobsters are more active and therefore have higher metabolic needs. Increased levels of foraging behavior might be expected, resulting in an increased probability of capture in baited traps. It is also known that increased temperature results in an increased probability of molting. The lobster fishery is characterized by high exploitation rates. Accordingly, the fishery depends on recruitment, and increased molting rates from the prerecruit to the fully recruited size ranges will increase the short-term supply of lobster.

The marginally significant correlation at a lag of 6 yr may reflect an increase in survival in the early life-history stages. The expected time for a lobster to reach legal size is 5–7 yr (Wilder 52) and therefore the observed delay is appropriate if temperature effects in the larval or early juvenile stages are important. Duration of the larval phase is inversely related to temperature. Assuming the larval stages are highly vulnerable to predation and other sources of mortality, a reduction in duration of this phase would be beneficial, possibly explaining the positive temperature effect with a delay of 6 yr. A comparison of the observed and fitted series is provided in Figure 4.

2.5. Intervention Analysis

Catch or abundance levels may be subject to marked changes over relatively short time intervals. For example, the imposition of new fishery regulations might have an abrupt effect on yield by reducing the total allowable catch. The occurrence of a dominant year class could result in a sharp increase in population size and catch rates. Intervention analysis is a useful approach for modeling such impacts. The intervention may take the form of a single event which has either a temporary or a lasting impact on the series. Alternatively, the intervention may have a gradual effect. The transfer function models described previously relate two or more continuous variables. In contrast, we might think of the intervention as an on–off switch taking on values of 0 or 1. Before the event, the intervention term is 0, whereas afterward we assign it a value of 1. An impact that has an immediate full effect is modeled using a zero-order intervention model

$$y_t = \omega_0 I_t + N_t \tag{5}$$

where ω is the intervention term, I is the step function (0 or 1), and N_t is the noise component, which can be modeled as an ARIMA process. A gradual effect is modeled using the first-order intervention model

$$y_t = (\omega_0/(1 - \delta_1 B)) I_t + N_t \tag{6}$$

where the term δ is restricted to the interval -1 to $+1$; this coefficient represents the graduated effect of the intervention. As δ approaches -1 or $+1$, the effect becomes more gradual. For an example of intervention analysis applied to a Dungeness crab population, see Noakes (53).

2.6. Applications

The intervention model approach will be illustrated using a CPUE series for pandalid shrimp in the Gulf of Maine. The northern shrimp, *Pandalus borealis*, supports worldwide fisheries in boreal waters. A small but commercially important fishery for northern shrimp occurs in the Gulf of Maine. Landings in this fishery varied dramatically during the past four decades. Catches declined to extremely low levels during 1950–1965, apparently in relation to high water temperature (Dow 10). The exact mechanism underlying the apparent response to water temperature is not currently known. Catches subsequently increased during the

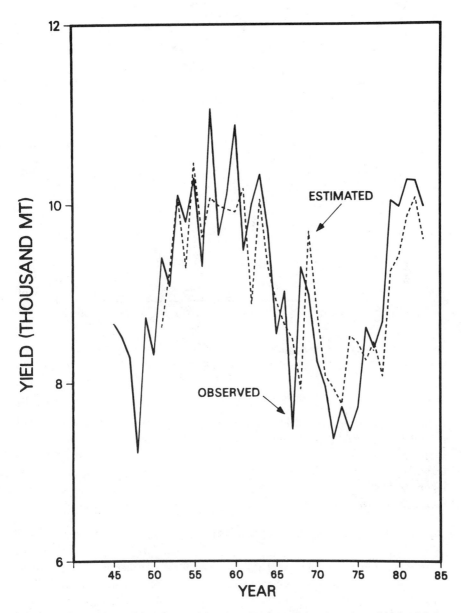

Figure 4. Comparison of the observed (——) and estimated (– – –) catches of Maine lobster (1945–1983).

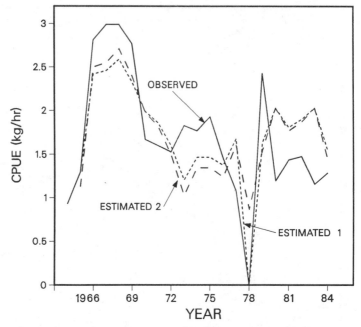

Figure 5. Comparison of the observed (——) CPUE and estimated CPUE using an intervention model (— — —, 1) and a transfer function model (---, 2) for Gulf of Maine northern shrimp.

later 1960s. Concerns over declining catches, however, prompted a closure of the fishery during 1978. This closure represented an intervention and may be modeled accordingly. Here, we are concerned with an intervention that had an immediate temporary (1 yr) effect. The tentative model is therefore a zero-order intervention model. A temperature term has also been included in the model. We then have a mixed transfer-intervention model. The final model was

$$\text{CPUE}_t = 4.834 - 1.081 I_t - 0.431 T_t + a_t \tag{7}$$

where I is the intervention term (1 for 1978 and 0 for all other years). The coefficient for the intervention component represents an estimate of the decline in CPUE in 1978 resulting from the closure. In this case, of course, CPUE was reduced to 0 during the intervention. For comparison, predicted CPUE using a transfer function model with a temperature term is also depicted in Figure 5.

3. STRUCTURAL MODELS

Structural models such as surplus production and yield-per-recruit (Y/R) models can be used to predict changes in yield or Y/R with variations in fishing intensity.

Stock–recruitment models may be used to predict recruitment levels for a given spawning stock size. Matrix projection models are often used to predict changes in population size based on specific assumptions about the survival rates and reproductive output. In their usual form, however, these models are deterministic. That is, we assume that there is no variation in the parameters. The success of the models as predictors depends on the extent to which their assumptions are met and whether the parameters are in fact constant. Stochastic analogues have been developed for many types of structural population models (e.g., Poole 54; Nisbet and Gurney 55), but these are mathematically complex.

It is possible to develop "hybrid" models with a structural biological basis but also incorporating empirically determined factors. For example, suppose we wish to predict recruitment levels for a given species. It may be known or hypothesized that recruitment depends on spawning stock size and certain environmental factors. For example, Penn and Caputi (36) described a Ricker recruitment model with explicit representation of environmental variables:

$$R = aS \exp\left(-bS + \Sigma c_i V_i\right) \tag{8}$$

where R is recruitment, S is spawning stock, the V_i are environmental variables, a is a density-independent parameter, b is stock-dependent mortality, and the c_i are environmental parameters.

Given time series of recruitment, spawning stock, and appropriate environmental variables, we may estimate the parameters of the model by nonlinear regression or by linearizing equation 8. Note that the structure of the model implies that the environmental factors (V_i) have a multiplicative, density-independent effect on recruitment. An alternative approach might be to express the density-independent parameter (a) as a function of environmental factors (see Tang 32). It is somewhat difficult to interpret equation 8 as a compensatory recruitment model if the expression in parentheses is positive (i.e., $\Sigma c_i V_i > bS$). As before, it will be important to examine the residuals of the fitted model for evidence of nonindependence and trends.

To illustrate these points, I considered a series of recruitment and spawning stock estimates for tiger prawn, *Penaeus esculentus*, and a set of environmental factors (data from Penn and Caputi 36). Environmental variables that might affect recruitment include rainfall during January and February (an index of cyclone activity) following hatching. Penn and Caputi (36) provide an interesting analysis of these data and a full discussion of the biological rationale for the model. The recruitment model with environmental terms fitted by nonlinear squares was

$$R = 3.732P \exp(-0.054P + 0.0027V_1 - 0.0030V_2) + e_t$$

where V_1 and V_2 represent January and February precipitation, e_t is a normally distributed random error term with mean 0 and constant variance, and all other terms are defined as before. These estimates are very similar to those provided by Penn and Caputi (36) based on the linearized model. The principal advantage of the nonlinear regression approach used here is that it is easier to obtain unbiased predictions for recruitment.

A comparison of the observed and estimated recruitment levels is given in Figure 6. Model residuals were tested for independence by computing autocorrelations at lags up to 10 yr and testing the null hypothesis that the correlation was 0 using $1/n^{1/2}$ as an approximate standard error. No significant autocorrelation was detected in the residuals, nor was there any evidence of trends (Fig. 6). To test the residuals for normality, I grouped them into five equal-sized classes and used a Kolmogorov–Smirnov one-way test. The hypothesis of normality could not be rejected ($d = 0.180$; $p > .01$). Note that if the residuals were not normally distributed, a

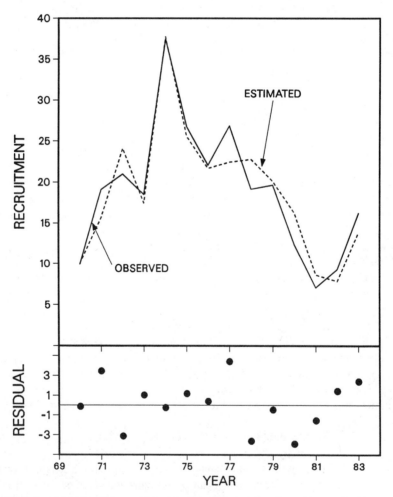

Figure 6. Comparison of observed (———) and estimated (– – –) recruitment for tiger prawn (upper panel). Lower panel shows residuals from fitted model.

different error structure for the model would have to be considered. For example, an alternative specification might be

$$R_t = aS \exp(-bS + \Sigma c_i V_i + e_t)$$

where now we have a log-normally distributed error structure. The linearized model has an implicit error structure of this general form. It is useful to emphasize that these models are in fact stochastic models—we have specified a random error component and we can provide not only a point prediction but a full probability distribution for the prediction. The principal distinction between the regression model described above and the time series models discussed previously is that (1) it is nonlinear and (2) it is based on a priori assumptions about the form of the relationships among the control variables.

4. CONCLUSIONS

Prediction is an important objective of ecological research (Poole 54, 56; Peters 57; Pielou 2). It may be possible to obtain more accurate forecasts using an heuristic model than a structural model. Difficulty in modeling all relevant components of the system and estimating the parameters often precludes development of a structural model capable of accurate prediction. Development of heuristic models, however, requires particular care. The frequently cited failure of correlations to stand over time undoubtedly reflects, in part, a failure to consider fully the assumptions underlying the analysis (for further discussion, see Sissenwine 45; Shepherd et al. 58). Many fishery-related and environmental observations are autocorrelated. Correlative studies should be undertaken only after evaluation of autocorrelation in each series. Difficulties are particularly likely to occur if both series exhibit unidirectional trends. The series can be "detrended" by fitting a model to the data with time as the independent variable; the residuals rather than the original series are then used in subsequent analyses. The results may be less satisfactory, however, than those obtained by differencing the series. For regression models, the residuals must be checked for autocorrelation, trends, and normality to test the adequacy of the model.

A common approach in exploratory data analyses relating two or more variables is to examine relationships among the variables at different lags and/or with combinations of variables. It is important to note that if a large number of such combinations are considered, some will be "significant" because of chance alone. We would expect one such apparent correlation at the 95% significance level for every 20 combinations examined. Each of these potential pitfalls should be carefully considered before accepting an heuristic model of any type.

The Box–Jenkins approach provides a formal structure for developing predictive models in an extremely adaptable framework. It has been shown that these models are often the best possible predictors [for discussion on this point in an ecological context see Poole (54, 56), Saila et al. (26), and Mendelsson (59)]. A strategy of model building using both structural and time series models may provide the best possible approach to forecasting yield and abundance of exploited invertebrates.

ACKNOWLEDGMENTS

I would like to express my appreciation to R. Methot and S. Clark for graciously providing data for the Dungeness crab and northern shrimp fisheries. I am also grateful to J. S. Cobb and J. F. Caddy for their many helpful comments on the manuscript.

GLOSSARY

Autocorrelation. The autocorrelation *coefficient* is a measure of the correlation between successive observations of a single variable. For example, the first-order autocorrelation coefficient is the correlation between observations one time unit apart; the fifth-order autocorrelation coefficient is the correlation between observations five time periods apart.

Autoregressive Process. In an autoregressive process, a time series is represented as a linear function of past values of the series. For example, in the first-order autoregressive model, the variable at time t is expressed as a function of its value at time $t - 1$.

Backward Shift Operator. The backward shift operator is defined for notational convenience. By definition, $y_{t-r} = B^r y_t$, where B is the backshift operator and the superscript represents the period of the shift (r units). To represent the differenced series, $z = y_t - y_{t-2}$, we may write $z = (1 - B^2) y_t$.

Cross-Correlation. The cross-correlation between two variables is the correlation at successive lags between the variables.

Differencing. It is often possible to remove trends effectively in a time series by forming the new series $z = y_t - y_{t-r}$, where r is the period of differencing. Then the relative *change* in the series is analyzed rather than the series itself. A nonstationary series may be made stationary by appropriate differencing.

Heuristic Method. An heuristic method is one that provides direction in the solution of a problem but is not necessarily based on causality. For the purposes of this discussion, an heuristic model and an empirical model are equivalent.

Independence. If two or more observations or variables are unrelated (uncorrelated), they are independent. Strictly speaking, we require the covariance between the variables to be zero.

Moving Average Process. In a moving average model, a time series is modeled as a linear combination of past forecast errors. The errors are attributable to past random shocks or perturbations.

Prewhitening. An important component of the identification process in developing a transfer function model is evaluation of the cross-correlation between the dependent and independent variable(s). Box and Jenkins (5) recommend that the deterministic component of the input (independent) series first be removed by fitting an ARIMA model to the series. The same model is applied to the output (dependent) series and the *residuals* from both series are then cross-correlated. Reducing both series to residuals from the fitted model is "pre-whitening" the series.

Residual. A residual is the difference between the observed and a predicted value (i.e., $y_t - \hat{y}_t = a_t$).

Structural Model. A structural model is a model based on known or inferred causal relationships between one or more variables.

Stationarity. A stationary time series is one that exhibits no trends in level or variance. That is, the series fluctuates about a constant mean and the variance does not change over time.

REFERENCES

1. F. H. Rigler, Recognition of the possible: An advantage of empiricism in ecology. *Can. J. Fish. Aquat. Sci.* **39**:1323–1331 (1982).

2. E. C. Pielou, The usefulness of ecological models: A stock-taking. *Q. Rev. Biol.* **56**:17–31 (1981).

3. R. G. Brown, *Smoothing, Forecasting and Prediction of Discrete Time Series*. Prentice-Hall, Englewood Cliffs, NJ, 1963, 468 pp.

4. P. Whittle, *Prediction and Regulation by Linear Least-Square Methods*. Van Nostrand, New York, 1963.

5. G. E. P. Box and J. Jenkins, *Time Series Analysis: Forecasting and Control*, 2nd ed. Holden-Day, San Francisco, CA, 1976, 575 pp.

6. C. Chatfield, *The Analysis of Time Series*, 3rd ed. Chapman & Hall, Londong, 1984, 286 pp.

7. W. Templeman and A. M. Fleming, Long term changes in hydrographic conditions and corresponding changes in the abundance of marine animals. *Int. Comm. Northwest. Atl. Fish. Annu. Proc.* **3**:78–86 (1953).

8. C. C. Taylor, H. B. Bigelow, and H. W. Graham, Climatic trends and the distribution of marine animals in New England. *Fish. Bull.* **57**:293–345 (1957).

9. K. Kitano, Note on the fluctuation tendency of the total catch of the common squid *Todarodes pacificus* (Steenstrup) in the light of the unusual ocean conditions. *Bull. Hokkaido Reg. Fish. Res. Lab.* **44**:73–76 (1979).

10. R. L. Dow, Effects of climatic cycles on the relative abundance and availability of commercial marine and estuarine species. *J. Cons., Cons. Int. Explor. Mer.* **37**:274–380 (1976).

11. R. E. Ulanowicz, M. Liaquat Ali, A. Vivian, D. R. Heinle, W. A. Richkus, and J. K. Summars, Identifying climatic factors influencing commercial fish and shellfish landings in Maryland. *Fish. Bull.* **80**:611–619 (1982).

12. R. L. Dow, Fluctuations in Gulf of Maine sea temperature and specific molluscan abundance. *J. Cons., Cons. Int. Explor. Mer.* **34**:532–534 (1972).

13. W. H. Sutcliffe, Jr., K. Drinkwater, and B. S. Muir, Correlations of fish catch and environmental factors in the Gulf of Maine. *J. Fish. Res. Board Can.* **34**:19–30 (1977).

14. L. M. Dickie, Fluctuations in abundance of the giant scallop, *Placopecten magellanicus* (Gmelin) in the Digby area of the Bay of Fundy. *J. Fish. Res. Board Can.* **12**:797–857 (1955).

15. J. F. Caddy, Long term trends and evidence for production cycles in the Bay of Fundy scallop fishery. *Rapp. P.-V. Reun., Cons. Int. Explor. Mer* **175**:97–108 (1979).

16. M. L. Coehlo and A. A. Rosenberg, Causal analysis of some biological data for *Illex illecebrosus* from the Scotian Shelf. *NAFO Sci. Counc. Stud.* **7**:61–66 (1984).

17. R. L. Dow, Relationship of sea surface temperature to American and European lobster landings. *J. Cons., Cons. Int. Explor. Mer* **37**:189–191 (1977).

18. J. M. Flowers and S. B. Saila, An analysis of temperature effects on the inshore lobster fishery. *J. Fish. Res. Board Can.* **29**:1221–1225 (1972).

19. F. L. Orach-Meza and S. B. Saila, Application of a polynomial distributed lag model for the Maine lobster fishery. *Trans. Am Fish. Soc.* **197**:402–411 (1978).

20. M. J. Fogarty, Population dynamics of the American lobster (*Homarus americanus*). Ph.D. Dissertation, University of Rhode Island, Kingston, 1986, 277 pp.

21. F. R. Boudreault, J. N. Dupont, and C. Sylvain, Modéles linéaires de prédiction des débarquements de homard aux Iles-de-la-Madeleine (Golfe du Saint-Laurent). *J. Fish. Res. Board Can.* **34**:379–383 (1977).

22. R. L. Dow, Effects of sea-surface temperature cycles on landings of American, European and Norway lobster. *J. Cons., Cons. Int. Explor. Mer.* **38**:271–272 (1978).

23. D. Chapman, Relationship between temperature and Scottish *Nephrops* landings. *Int. Counc. Explor. Mer., C.M.* 1984/K:**34**:1–4 (1984).

24. N. Caputi and R. S. Brown, Relationship between indices of juvenile abundance and recruitment in the western rock lobster (*Panulirus cygnus*) fishery. *Can. J. Fish. Aquat. Sci.* **43**:2131–2139 (1986).

25. B. F. Phillips, Prediction of commercial catches of the western rock lobster. *Can. J. Fish. Aquat. Sci.* **43**:2126–2130 (1986).

26. S. B. Saila, M. Wigbout, and R. J. Lermit, Comparison of some time series models for the analysis of fisheries data. *J. Cons., Cons. Int. Explor. Mer* **39**:44–52 (1980).

27. W. T. Peterson, Upwelling indices and annual catches of Dungeness crab (*Cancer magister*) along the west coast of the United States. *Fish. Bull.* **71**:902–910 (1973).

28. L. W. Botsford and D. E. Wickham, Correlation of upwelling index and Dungeness crab catch. *Fish. Bull.* **73**:901–907 (1975).

29. L. W. Botsford, R. D. Methot, Jr., and J. E. Wilen, Cycle covariation in the California king salmon, *Oncorhyncus tshawytscha*, silver salmon, *O. kisutch*, and Dungeness crab, *Cancer magister*, fisheries. *Fish. Bull.* **80**:791–802 (1982).

30. D. F. Johnson, L. W. Botsford, R. D. Methot, Jr., and T. C. Wainwright, Wind stress and cycle in Dungeness crab (*Cancer magister*) off California, Oregon, and Washington. *Can. J. Fish. Aquat. Sci.* **43**:838–845 (1986).

31. P. W. Wild, P. M. W. Law, and D. R. McLain, Variations in ocean climate and the Dungeness crab fishery in Northern and Central California. *Calif. Fish Game Fish. Bull.* **172**:175–180 (1983).

32. Q. Tang, Modification of the Ricker stock recruitment model to account for environ-

mentally induces variation in recruitment with particular reference to the blue crab fishery in Chesapeake Bay. *Fish. Res.* **3**:13–21 (1985).

33. R. Boddeke, Forecasting the landings of brown shrimp (*Crangon crangon*) in the Netherlands. *Int. Counc. Explor. Mer. C.M.* 1968 K:8 (1968).

34. P. A. Driver, Prediction of fluctuations in the landings of brown shrimp (*Crangon crangon*) in the Lancashire and western sea fisheries district. *Estuarine Coastal Mar. Sci.* **4**:567–573 (1976).

35. P. J. Warren, *The Fishery for the Pink Shrimp* (Pandalus borealis) in the Wash. Lab. Leaf. (New Ser.) 28. Minist. Agric. Fish. Food, Lowestoft, U.K., 1973.

36. J. W. Penn and N. Caputi, Stock recruitment relationships for the tiger prawn, *Penaeus esculentus*, fishery in Exmouth Gulf, Western Australia, and their implications for management. In P. C. Rothlisberg, B. J. Hill, and D. J. Staples, Eds., *Proceedings of the Second Australian National Prawn Seminar*. Simpson, Halligan, Brisbane, Australia, 1985, pp. 165–173.

37. H. H. Hildebrand and G. Gunter, Correlation of rainfall with the Texas catch of white shrimp, (*Penaeus setiferus* (Linnaeus). *Trans. Am. Fish. Soc.* **82**:151–155 (1953).

38. G. Gunter and J. C. Edwards, The relation of rainfall and fresh-water drainage to the production of the penaeid shrimps (*Penaeus fluviatilis* Say and *Penaeus aztecus* Ives) in Texas and Louisiana waters. *FAO Fish. Rep.* **57**:875–892 (1969).

39. M. Prager and S. B. Saila, Predictive GMDH models of shrimp catches: Some practical considerations. In S. J. Farlow, Ed., *Self Organizing Methods in Modelling. GMDH Type Algorithms*. Dekker, New York, 1984, pp. 179–197.

40. H. A. Walker and S. B. Saila, Incorporating climatic and hydrographic information into shrimp yield forecasts using seasonal climatic component models. *Tex. A&M Univ. Sea Grant Program [Rep.] TAMU-SG* **TAMU-SG-86-110**:56–99 (1986).

41. R. J. Berry and K. N. Baxter, Predicting brown shrimp abundance in the northwestern Gulf of Mexico. *FAO Fish. Rep.* **57**:775–798 (1969).

42. M. A. Matylewich and R. R. Mundy, Evaluation of the relevance of some environmental factors to the estimation of migratory timing and yield for the brown shrimp of Pamlico Sound, North Carolina. *North Am. J. Fish. Manage.* **5**:197–209 (1985).

43. D. J. Staples, Modelling the recruitment processes of the banana prawn, *Penaeus merguiensis*, in the southeastern Gulf of Carpentaria, Australia. In P. C. Rothlisberg, B. J. Hill, and D. J. Staples, Eds., *Proceedings of the Second Australian National Prawn Seminar*. Simpson, Halligan, Brisbane, Australia, 1985, pp. 175–184.

44. D. J. Vance, D. J. Staples, and J. D. Kerr, Factors affecting year-to-year variation in the catch of banana prawns (*Penaeus merguiensis*) in the Gulf of Carpentaria, Australia. *J. Cons., Cons. Int. Explor. Mer.* **42**:83–97 (1985).

45. M. P. Sissenwine, Why do fish populations vary. In R. May, Ed., *Exploitation of Marine Communities*. Springer-Verlag, Berlin, 1984, pp. 59–109.

46. M. H. Quenouille, *Associated Measurements*. Butterworth, London, 1952, 242 pp.

47. J. Durbin and G. S. Watson, Testing for serial correlation in least squares regression. II. *Biometrika* **38**:159–178 (1951).

48. N. R. Draper and H. Smith, *Applied Regression Analysis*, 2nd ed. Wiley, New York, 1981, 709 pp.

49. L. W. Botsford, Effects of environmental forcing on age-structured populations:

Northern California Dungeness crab (*Cancer magister*) as an example. *Can. J. Fish. Aquat. Sci.* **43**:2345–2352 (1986).

50. L. W. Botsford and D. W. Wickham, Behavior of age-specific density-dependent models and the northern California Dungeness crab (*Cancer magister*) fishery. *J. Fish. Res. Board Can.* **35**:833–843 (1978).

51. D. McLeese and D. G. Wilder, The activity and catchability of the lobster (*Homarus americanus*) in relation to temperature. *J. Fish. Res. Board Can.* **15**:1345–1354 (1958).

52. D. G. Wilder, The growth rate of the American lobster (*Homarus americanus*). *J. Fish. Res. Board Can.* **10**:371–412 (1953).

53. D. Noakes, Quantifying changes in British Columbia Dungeness crab (*Cancer magister*) landings using intervention analysis. *Can. J. Fish. Aquat. Sci* **43**:634–639 (1986).

54. R. W. Poole, *An Introduction to Quantitative Ecology*. McGraw-Hill, New York, 1976, 531 pp.

55. R. M. Nisbet and W. S. C. Gurney, *Modelling Fluctuating Populations*. Wiley, New York, 1982, 379 pp.

56. R. W. Poole, The statistical prediction of population fluctuations. *Annu. Rev. Ecol. Syst.* **9**:427–448 (1978).

57. R. H. Peters, From natural history to ecology. *Perspect. Biol. Med.* Winter, 1980. 191–203 (1980).

58. J. G. Shepherd, J. G. Pope, and R. D. Cousens, Variations in fish stocks and hypotheses concerning their links with climate. *Rapp. P.-V. Reun., Cons. Int. Explor. Mer.* **185**:255–267 (1984).

59. R. Mendelsson, Using Box-Jenkins models to forecast fishery dynamics: Identification, estimation, and checking. *Fish. Bull.* **78**:887–896 (1981).

INDEX